Quality Management Handbook

QUALITY AND RELIABILITY

A Series Edited by

1. Designing for Minimal Maintenance Expense: The Practical Application of Reliability and Maintainability, *Marvin A. Moss*
2. Quality Control for Profit: Second Edition, Revised and Expanded, *Ronald H. Lester, Norbert L. Enrick, and Harry E. Mottley, Jr.*
3. QCPAC: Statistical Quality Control on the IBM PC, *Steven M. Zimmerman and Leo M. Conrad*
4. Quality by Experimental Design, *Thomas B. Barker*
5. Applications of Quality Control in the Service Industry, *A. C. Rosander*
6. Integrated Product Testing and Evaluating: A Systems Approach to Improve Reliability and Quality, Revised Edition, *Harold L. Gilmore and Herbert C. Schwartz*
7. Quality Management Handbook, *edited by Loren Walsh, Ralph Wurster, and Raymond J. Kimber*
8. Statistical Process Control: A Guide for Implementation, *Roger W. Berger and Thomas Hart*
9. Quality Circles: Selected Readings, *edited by Roger W. Berger and David L. Shores*
10. Quality and Productivity for Bankers and Financial Managers, *William J. Latzko*
11. Poor-Quality Cost, *H. James Harrington*
12. Human Resources Management, *edited by Jill P. Kern, John J. Riley, and Louis N. Jones*
13. The Good and the Bad News About Quality, *Edward M. Schrock and Henry L. Lefevre*
14. Engineering Design for Producibility and Reliability, *John W. Priest*
15. Statistical Process Control in Automated Manufacturing, *J. Bert Keats and Norma Faris Hubele*
16. Automated Inspection and Quality Assurance, *Stanley L. Robinson and Richard K. Miller*

17. Defect Prevention: Use of Simple Statistical Tools, *Victor E. Kane*
18. Defect Prevention: Use of Simple Statistical Tools, Solutions Manual, *Victor E. Kane*
19. Purchasing and Quality, *Max McRobb*
20. Specification Writing and Management, *Max McRobb*
21. Quality Function Deployment: A Practitioner's Approach, *James L. Bossert*
22. The Quality Promise, *Lester Jay Wollschlaeger*
23. Statistical Process Control in Manufacturing, *edited by J. Bert Keats and Douglas C. Montgomery*
24. Total Manufacturing Assurance, *Douglas C. Brauer and John Cesarone*
25. Deming's 14 Points Applied to Services, *A. C. Rosander*
26. Evaluation and Control of Measurements, *John Mandel*
27. Achieving Excellence in Business: A Practical Guide to the Total Quality Transformation Process, *Kenneth E. Ebel*
28. Statistical Methods for the Process Industries, *William H. McNeese and Robert A. Klein*
29. Quality Engineering Handbook, *edited by Thomas Pyzdek and Roger W. Berger*
30. Managing for World-Class Quality: A Primer for Executives and Managers, *Edwin S. Shecter*
31. A Leader's Journey to Quality, *Dana M. Cound*
32. ISO 9000: Preparing for Registration, *James L. Lamprecht*
33. Statistical Problem Solving, *Wendell E. Carr*
34. Quality Control for Profit: Gaining the Competitive Edge. Third Edition, Revised and Expanded, *Ronald H. Lester, Norbert L. Enrick, and Harry E. Mottley, Jr.*
35. Probability and Its Applications for Engineers, *David H. Evans*
36. An Introduction to Quality Control for the Apparel Industry, *Pradip V. Mehta*
37. Total Engineering Quality Management, *Ronald J. Cottman*
38. Ensuring Software Reliability, *Ann Marie Neufelder*
39. Guidelines for Laboratory Quality Auditing, *Donald C. Singer and Ronald P. Upton*
40. Implementing the ISO 9000 Series, *James L. Lamprecht*
41. Reliability Improvement with Design of Experiments, *Lloyd W. Condra*
42. The Next Phase of Total Quality Management: TQM II and the Focus on Profitability, *Robert E. Stein*
43. Quality by Experimental Design: Second Edition, Revised and Expanded, *Thomas B. Barker*
44. Quality Planning, Control, and Improvement in Research and Development, *edited by George W. Roberts*
45. Understanding ISO 9000 and Implementing the Basics to Quality, *D. H. Stamatis*
46. Applying TQM to Product Design and Development, *Marvin A. Moss*
47. Statistical Applications in Process Control, *edited by J. Bert Keats and Douglas C. Montgomery*
48. How to Achieve ISO 9000 Registration Economically and Efficiently, *Gurmeet Naroola and Robert Mac Connell*
49. QS-9000 Implementation and Registration, *Gurmeet Naroola*
50. The Theory of Constraints: Applications in Quality and Manufacturing: Second Edition, Revised and Expanded, *Robert E. Stein*
51. Guide to Preparing the Corporate Quality Manual, *Bernard Froman*
52. TQM Engineering Handbook, *D. H. Stamatis*
53. Quality Management Handbook: Second Edition, Revised and Expanded, *edited by Raymond J. Kimber, Robert W. Grenier, and John Jourdan Heldt*

ADDITIONAL VOLUMES IN PREPARATION

Quality Management Handbook

Second Edition,

Revised and Expanded

edited by

Raymond J. Kimber
Ohmite Manufacturing Co.
Skokie, Illinois

Robert W. Grenier
Consultant
Springfield, Illinois

John Jourdan Heldt
Free Lance Reliability Service
San Jose, California

MARCEL DEKKER

MARCEL DEKKER, INC. NEW YORK • BASEL • HONG KONG

Library of Congress Cataloging-in-Publication Data

Quality management handbook / edited by Raymond J. Kimber, Robert W. Grenier, John Jourdan Heldt. — 2nd ed., rev. and expanded
p. cm. — (Quality and reliability ; 53)
Includes bibiographies and index.
ISBN 0-8247-9356-0 (hardcover : alk. paper)
1. Quality control—Handbooks, manuals, etc. I. Kimber, Raymond J. II. Grenier, Robert W. III. Heldt, John J. IV. Series.
TS156.Q363 1997
658.5'62—dc21

97-22485
CIP

The publisher offers discounts on this book when ordered in bulk quantities. For more information, write to Special Sales/Professional Marketing at the address below.

This book is printed on acid-free paper.

MARCEL DEKKER, INC.
270 Madison Avenue, New York, New York 10016
http://www.dekker.com

Current printing (last digit):
10 9 8 7 6 5 4 3 2 1

PRINTED IN THE UNITED STATES OF AMERICA

To the memory of W. Edwards Deming.
His voice is no longer with us, but his ideals are.

Series Introduction

The genesis of modern methods of quality and reliability will be found in a sample memo dated May 16, 1924, in which Walter A. Shewhart proposed the control chart for the analysis of inspection data. This led to a broadening of the concept of inspection from emphasis on detection and correction of defective material to control of quality through analysis and prevention of quality problems. Subsequent concern for product performance in the hands of the user stimulated development of the systems and techniques of reliability. Emphasis on the consumer as the ultimate judge of quality serves as the catalyst to bring about the integration of the methodology of quality with that of reliability. Thus, the innovations that came out of the control chart spawned a philosophy of control of quality and reliability that has come to include not only the methodology of the statistical sciences and engineering, but also the use of appropriate management methods together with various motivational procedures in a concerted effort dedicated to quality improvement.

This series is intended to provide a vehicle to foster interaction of the elements of the modern approach to quality, including statistical applications, quality and reliability engineering, management, and motivational aspects. It is a forum in which the subject matter of these various areas can be brought together to allow for effective integration of appropriate techniques. This will promote the true benefit of each, which can be achieved only through their interaction. In this sense, the whole of quality and reliability is greater than the sum of its parts, as each element augments the others.

The contributors to this series have been encouraged to discuss fundamental concepts as well as methodology, technology, and procedures at the leading edge of the discipline. Thus, new concepts are placed in proper perspective in these evolving disciplines. The series is intended for those in manufacturing, engineering, and marketing and management, as well as the consuming public, all of whom have an interest and stake in the products and services that are the lifeblood of the economic system.

The modern approach to quality and reliability concerns excellence: excellence when the product is designed, excellence when the product is made, excellence as the product is used, and excellence throughout its lifetime. But excellence does not result without effort, and products and services of superior quality and reliability require an appropriate combination of statistical, engineering, management, and motivational effort. This effort can be directed for maximum benefit only in light of timely knowledge of approaches and methods that have been developed and are available in these areas of expertise. Within the volumes of this series, the reader will find the means to create, control, correct, and improve quality and reliability in ways that are cost effective, that enhance productivity, and that create a motivational atmosphere that is harmonious and constructive. It is dedicated to that end and to the readers whose study of quality and reliability will lead to greater understanding of their products, their processes, their workplaces, and themselves.

Edward G. Schilling

Preface

When the editors of the first edition of the *Quality Management Handbook* were beginning to assemble manuscripts, the so-called quality revolution appeared to be fully underway. All signs seemed to be pointing toward a new and complete understanding of what quality was all about. In the intervening years, however, there has arisen a question as to just how well U.S. companies have joined in the quality revolution.

To be sure, we now have the Malcolm Baldrige Award and the ISO 9000 series. But there are more than a few practitioners troubled by the "acronym of the week" during the 1980s who now wonder if some of the current hot topics are not just a continuation of the glitz.

Each of the current editors is a long-time practicing quality assurance professional. Together, the three of us probably represent 100 years of quality assurance experience. And we are concerned. No, we are worried. We have gotten dirt under our fingernails, and we still work with our sleeves rolled up. We have been on the shop floors of a lot of companies. And we are not sure that the quality revolution express has really stopped at more than a relatively few well-publicized stations. There are still a great many U.S. companies whose quality philosophies and practices are right up there with those of the 1950s.

Why should this be so? With all the mathematical concepts, high-technology techniques, and computer-driven programs, why does it seem so hard for some companies to develop world-class quality? Why, in fact, are so many companies struggling to comply with the requirements of Handbook 51 and Military Standard 45208? We think it is because too many companies have failed to adopt the basic quality framework. What is the basic quality framework? It is an understanding that the first step is to find out—and document—what the customers want. Then it is a commitment by the entire company—from the most senior executive on down—that the company will provide what it has promised to its customers. The very sophisticated techniques can bridge the gap between 1000 defective parts per million and 50. But they are of precious little help to a company whose defect rate is 2%, or 5%, or 10%. As long-time quality practitioner David Crosby put it, "The concepts that a lot of companies need to learn are those that have been in place for 30 years."

And so we have the *Quality Management Handbook*—a basic encyclopedic reference. This second edition contains 21 completely new chapters and the remainder of the material has been revised, rewritten, and updated. Like the first edition, this is primarily a book for managers who are not quality managers. Quality managers, nevertheless, will find this volume equally helpful.

Raymond J. Kimber
Robert W. Grenier
John Jourdan Heldt

Contents

Section III. Tools

Contributors

John M. Aaron President, Milestone Planning and Research, Inc., Hickory Hills, Illinois

Dan D. Ashcraft Oklahoma State University, Stillwater, Oklahoma

Charles A. Cianfrani Principal, Casmar Consulting Group, Green Lane, Pennsylvania

Joseph J. Cieplak Acco Industries, Inc., Bridgeport, Connecticut

Philip B. Crosby Chairman, Career IV, Inc., Winter Park, Florida

David C. Crosby President, The Crosby Company, Glen Ellyn, Illinois

Anthony DeBellis Acco Industries, Inc., Bridgeport, Connecticut

Eugene W. Ellis* Consultant, Sanford, North Carolina

Lowell W. Foster President and Director, Lowell W. Foster Associates, Inc., Minneapolis, Minnesota

Robert W. Grenier Consultant, Springfield, Illinois

John Jourdan Heldt Medalist, Free Lance Reliability Service, San Jose, California

---

* Retired.

Harry S. Hertz National Institute of Standards and Technology, Gaithersburg, Maryland

Raymond J. Kimber Director, Quality Assurance, Ohmite Manufacturing Company, Skokie, Illinois

George F. Kleitz Principal, George Kleitz and Associates, Inc., Wheaton, Illinois

Ray A. Klotz President, World Class Quality Consulting Company, Escondido, California

William E. Land Principal Consulting Engineer, William Land Associates, Ione, California

Ronald A. Lavoie Consultant, Providence, Rhode Island

Jay W. Leek* President, PCA, Lecanto, Florida

C. S. McCamy Consultant in Color Science, Wappinger Falls, New York

Paul J. Mikelonis Consultant, New Berlin, Wisconsin

Richard Y. Moss Hardware Reliability Manager, Corporate Quality Department, Hewlett-Packard Company, Palo Alto, California

Ben A. Murray Consultant, Englewood, Florida

Srihari Nandyal Product Engineer, Superabrasives, CBN Market Development, Norton Company, Worcester, Massachusetts

James A. Nelson Manager, Educational Services, Buehler Ltd., Lake Bluff, Illinois

George O. Rice* Rockwell International/Boeing North America, Anaheim, California

Curt W. Reimann National Institute of Standards and Technology, Gaithersburg, Maryland

James Richard Santos Senior Quality Engineer, Department of Regulatory Affairs/Quality Assurance, Fidus Medical Technology Corporation, Fremont, California

Norman G. Siefert* Consultant, St. Louis, Missouri

Martin R. Smith Consultant, Management Sciences USA, Lawrenceville, Georgia

---

* Retired.

K. Subramanian Director, World Grinding Technology Center, Norton Company, Worcester, Massachusetts

Mae-Goodwin Tarver* Quest Associates Ltd., Forest Park, Illinois

Robert W. Traver Principal, Traver Associates, Averill Park, New York

Joseph Tsiakals Vice President of Quality, Fenwal Division, Baxter Heathcare, Round Lake, Illinois

Jon Turino Mixed Signal Test Station Marketing Manager, Integrated Measurement Systems, Beaverton, Oregon

Robert W. Vincent* Vice Chairman, Philip Crosby Associates, Inc., Winter Park, Florida

---

* Retired.

1
Quality Management

PHILIP B. CROSBY
Career IV, Inc.
Winter Park, Florida

I. INTRODUCTION

Anyone who has played golf knows that there are thousands of ways to do the thing wrong and only a few to do it properly. In golf, the results speak for themselves and do so immediately. Those who strike a ball with a club have a result in mind. That result is part of their overall plan for the complete match. A poor result of a single shot can spoil the entire day, or at least send it downhill.

In the hopes of improving their possibility of success, golf devotees seek help from equipment and techniques. They are forever reading about "tips" from more successful golfers, they purchase clubs of every type of description, and they have no loyalty to balls by brand name, switching at the first sign that one might fly truer or further than another. Devoted golfers will change courses and playing companions in the hopes of finding some arrangement that helps them. Money is no object, family obligations come in a limpid second. They want to be successful in their chosen sport.

Golfers are willing to do anything to make that happen. Anything, that is, except study the concepts of the game and work hard at learning how to do it. The thoughts of hitting several thousand balls a week, and understanding the mechanics of the swing to the extent that they can see what happens when the club hits the ball—well those thoughts are not welcome. They want something that does it all by itself; sweat and thought are not part of that.

There is a distinct parallel between quality management and golf in the way it is practiced. Those responsible for causing quality usually love techniques and activities, they blossom when exposed to regulations, and they abhor comprehension. The result today may be incomplete, but they know that tomorrow will bring some other device, usually with three initials, and everything will turn out all right. Like golfers, they regale each other with anecdotes about their good shots and smile away those that did not turn out so well. Golfers blame the producers of golf courses, they blame the wind, they blame the "rub of the green," they blame those who do not fix ball marks on the green, they curse those who leave spike marks in their line, and they yearn for the chance to play more. But playing does not improve the game. Understanding and practicing that understanding improves everything. Quality implementers blame management.

The conventional wisdom of doing quality management today reminds me of Prince Potemkin of Russia. He was assigned by Catherine the Great to create comfortable cities and lives for the populace. He spent a lot of money and time doing this. When the Empress wanted to see the result, he took her on a train ride. She saw villages with homes, schools, office buildings, factories, and in front of these structures stood happy peasants waving as she rode by on the train. Prince Potemkin was rewarded with many honors. However, the villages were only facades, one board thick. There was no substance behind them; the villagers were waving happily because they knew that was what he wanted.

The Potemkin villages of quality management are produced from techniques and activities such as empowerment, team building, benchmarking, ISO 9000, and lots of programs. Management rides by on the train of "show and tell" meetings and is impressed with all the activity. Managers do not look at the real results; what they are fed is anecdotal. Actually doing quality management correctly is much more rewarding and quite a bit easier than the conventional methods. The purpose of this chapter is to lay out the concepts and efforts that cause that to happen.

Those who are involved in quality management should take time to understand how important this assignment is to their personal future. Quality is the only operational function that touches every part of the organization. Having the opportunity to do it provides a tremendous advantage, and exposure, for those involved. They can help the company achieve its goals, they can make management look good, and they can cause the customers to admire and respect the organization. But the secret of doing this lies in personally comprehending the concepts of quality management and applying the proper direction of energy.

Business consists of transactions and relationships. When managed properly, the business routinely completes all transactions correctly; and relationships with employees, suppliers, and customers are successful. It is not reasonable to think that a packaged system, such as ISO 9000, could hope to accomplish this. Business is chaotic, not systematic. Many things can remain the same each day, but the world changes continually. Situations happen and have to be dealt with. We cannot run to look at a set of procedures each time we need to take action.

Let's define quality management:

Quality management is deliberately creating an organizational culture within which all transactions are completed correctly and all relationships are successful.

Quality management involves concepts and action.

We need to review the concepts by discussing the absolutes of quality management. Then we will list some of the actions that must be taken if quality is to become routine. When the absolutes are understood, it is possible to know how to deal with any situation that arises.

There are four absolutes of quality:

1. Absolute #1—Definition: Quality means conformance to requirements, not goodness.
2. Absolute #2—System: Quality is produced through prevention actions, not detection.
3. Absolute #3—Performance Standard: Quality comes from zero defects, not from acceptable quality levels.
4. Absolute #4—Measurements: Quality is measured by the price of nonconformance and the complete transactions rating (CTR), not statistical computations.

II. DEFINITION

People rattle on about quality without ever agreeing on exactly what it is. Lately we hear "delighting the customer" and other fine-sounding but unexplainable meanings. Quality is doing what we said we would do, which means conducting transactions in accordance with the agreed requirements. These requirements define the product or service we are offering. In order to lay out requirements, it is necessary to know what the customer wants, even if the customer may not know that. Customer needs change rapidly, and in order to comply, we have to create an organization that knows how to meet the requirement, since they will be changing regularly also.

We have to insist that our employees and suppliers understand the requirements clearly as they apply to them. They have to understand how to accomplish our requirements, and they have to know how important it is that we can count on them doing just that. "Conformance to agreed requirements" they can understand.

III. PREVENTION

People are vaccinated in order to prevent them from acquiring infectious diseases. Companies can also be vaccinated in order to avoid problems. This is accomplished through causing operations to communicate with each other as products and services are developed; through training and education so everyone understands the transactions they must accomplish; through corrective action that eliminates the root cause of problems; and through being close to the customer. When we look at the problems we have had, we find that most of them occur over and over. Some things come about anyway and have to be dealt with, but the vast majority of problems can be prevented once the cause is known.

IV. PERFORMANCE STANDARD

"Zero defects" is a symbolic way of saying "do it right the first time." There are those who think it pragmatic to allow for a certain amount of error, such as "six sigma." This standard permits 3.4 per million defects. That sounds like a pretty rigid goal, but when millions of components are placed on a chip, for instance, all of them become nonconforming. It is hard to explain to employees and suppliers why you do not expect everything to be exactly as ordered and agreed. Those who advocate statistical goals are just plain wrong. The result of such a policy is to produce the waste of money, time, and customers.

V. MEASUREMENT

Management talks money. Everything they do revolves around it, because it is their measuring stick. If quality wants to receive the proper attention, then it too must relate to money. Calculating the price of nonconformance (PONC) is the way to do this. The accounting department can determine how much is spent doing things over, servicing customers, and other results of not doing things right the first time. All they have to do is look at the account numbers and see what would not have to be spent if things were all done right. Most companies spend 25% or more of their revenue on the price of nonconformance. A sensible quality management process can reduce this by half in less than two years. Financial results are viewed continually by management, PONC puts the quality intensity right up front. This keeps the whole effort focused on real results, not stories about good happenings.

Complete, incomplete, and wrong transactions can be measured quickly. This lets us know which suppliers are taking our work seriously; which employees are producing conforming work. This can be done at the lowest levels of the organization once people are shown the way. When supplier results are calculated like this, it can be used to select those we wish to cultivate and grow (Fig. 1).

VI. THE TOUGH COMPANY

In *Quality Is Still Free*, the revision of *Quality Is Free*, I wrote about the Tough Company. That is the idea we need to keep in our head as we perform the quality manage-

	THIRD REICH	BANANA REPUBLIC	CONSTITUTIONAL MONARCHY	AMERICAN REPUBLIC	21ST CENTURY COMPLETENESS
Organizational Policy	The boss makes it up each day	Might makes right	Rule by the elite	The balance of power	Consent of the governed
Requirements Definition	The boss announces it each day	No one knows for sure	Governed by agreements of leadership	Described in depth	A clear description
Education	Teaching people to serve the organization	Not much of a concern	Available by class	Available to all as capable	Everyone keeps learning
Performance Measurement	What makes the boss happy	Stay useful and alive	Those who serve well	Meet requirements	A climate of consideration
Purpose of Organization	To glorify the leader	To make the junta rich	To have an orderly life	To keep people free	To make citizens successful

Figure 1 Completeness grid. (Reprinted with permission from *Completeness: Quality for the 21st Century*. © Philip B. Crosby.)

ment function. The millions of transactions that occur each day cannot possibly be checked or examined for correctness each time. We have to build integrity in, not lay it on top. In that same regard, we have to create an organization that can continue to grow and prosper whether we are in a recession, an earthquake, a boom, or whatever. Quality management is the key to the reality that makes this happen.

Here are the characteristics of a Tough Company:

1. Management is focused on results. This requires separating reality from fantasy.
2. Management knows the customer. Requirements originate with the customers, but they do not send you a neat list. We have to dig them out ourselves.
3. Management creates veteran employees. People with a high complete transaction rating (CTR) are selected and then treated well so they will stay. This results in a high revenue per employee base.
4. Management avoids distractions. They stick with a business they understand, they do not hop on each new management book that emerges. They learn continually.
5. Management has a common agenda. They speak the same language, understand goals the same way, and have agreed objectives. They may fuss at each other as things get accomplished, but that is purposeful.
6. Relationships are successful. This is discussed in detail in the book *Completeness*: *Quality in the 21st Century*.

When I entered the quality business in 1952, the purpose of having a quality function was to sort the good from the bad and then find a way to use the bad. Today, quality management should be based on never having any bad and not needing any sorting. We have to think of "wellness," not curing sickness (Fig. 2).

	COMATOSE	INTENSIVE CARE	PROGRESSIVE CARE	HEALING	WELLNESS
QUALITY	Nobody does anything right around here. *Price of Nonconformance =33%*	We finally have a list of customer complaints. *Price of Nonconformance = 28%*	We are beginning a formal Quality Improvement Process. *Price of Nonconformance = 20%*	Customer complaints are practically gone. *Price of Nonconformance = 13%*	People do things right the first time routinely. *Price of Nonconformance = 3%*
GROWTH	Nothing ever changes. *Return after tax = nil*	We bought a turkey. *Return after tax = nil*	The new product isn't too bad. *Return after tax = 3%*	The new group is growing well. *Return after tax = 7%*	Growth is profitable and steady. *Return after tax = 12%*
CUSTOMERS	Nobody ever orders twice. *Customer complaints on orders = 63%*	Customers don't know what they want. *Customer complaints on orders = 54%*	We are working with customers. *Customer complaints on orders = 26%*	We are making many defect-free deliveries. *Customer complaints on orders = 9%*	Customer needs are anticipated. *Customer complaints on orders = 0%*
CHANGE	Nothing ever changes. *Changes controlled by Systems Integrity = 0%*	Nobody tells anyone anything. *Changes controlled by Systems Integrity = 2%*	We need to know what is happening. *Changes controlled by Systems Integrity = 55%*	There is no reason for anyone to be surprised. *Changes controlled by Systems Integrity = 85%*	Change is planned and managed. *Changes controlled by Systems Integrity = 100%*
EMPLOYEES	This place is a little better than not working. *Employee turnover = 65%*	Human Resources has been told to help employees. *Employee turnover = 45%*	Error Cause Removal programs have been started. *Employee turnover = 40%*	Career path evaluations are implemented now. *Employee turnover = 7%*	People are proud to work here. *Employee turnover = 2%*

Figure 2 The Eternally Successful Organization Grid. (Reprinted with permission from *The Eternally Successful Organization*. © Philip B. Crosby.)

All of this comes about readily when a company has:

1. A quality policy: "We will deliver defect-free products and services to our customers and co-workers, on time."
2. Quality education: The "absolutes" and the individual's role in causing quality to be routine.
3. Management example: Showing integrity all day long under any circumstance.

Quality management causes companies to be successful.

FURTHER READING

Crosby, P. *The Absolutes of Leadership*. Pfeiffer, 1996.
Crosby, P. *The Art of Getting Your Own Sweet Way*, 2nd ed. New York: McGraw-Hill, 1981.
Crosby, P. *Completeness: Quality for the 21st Century*. Dutton, 1992.
Crosby, P. *Cutting the Cost of Quality*. Industrial Education Institute, 1967.
Crosby, P. *Cutting the Cost of Quality*, Anniversary Edition. The Quality College Bookstore, 1990.
Crosby, P. *The Eternally Successful Organization*. New York: McGraw-Hill, 1988.
Crosby, P. *Leading: The Art of Becoming an Executive*. New York: McGraw-Hill, 1990.
Crosby, P. *Let's Talk Quality*. New York: McGraw-Hill, 1989.
Crosby, P. *Quality Is Free: The Art of Making Quality Certain*, New York: McGraw-Hill, 1979.
Crosby, P. *Quality is Still Free*. New York: McGraw-Hill, 1995.
Crosby, P. *Quality Without Tears: The Art of Hassle-Free Management*. New York: McGraw-Hill, 1984.
Crosby, P. *Reflections on Quality*. New York: McGraw-Hill, 1995.
Crosby, P. *Running Things: The Art of Making Things Happen*. New York: McGraw-Hill, 1986.

2
Organization and Planning

RAYMOND J. KIMBER
Ohmite Manufacturing Company
Skokie, Illinois

I. INTRODUCTION

To at least some degree, the material in this book is directed to managers other than the quality manager. This chapter, however, is primarily directed to the concepts of organization and planning of what we would call "the traditional quality department." As such, the material is primarily directed to the current or prospective quality manager. Managers who interface with the quality department will, of course, also benefit from the material presented here.

II. THE RECENT HISTORY OF INSPECTION

Prior to and during much of World War II, quality assurance (QA) functions were largely those of inspection. So traditional had inspection become, in fact, that both buyers and sellers accepted one inspector for every ten production people virtually as a requirement. Even today, quite a lot of companies ask how many inspectors a supplier has. And, the head of the "quality" department is still called the chief inspector in many companies.

The introduction of what was to become "SPC" and "SQC" had already begun, but only a few practitioners and probably fewer companies were really interested. Inspection worked. Perhaps products were relatively simpler than they are today.

With the so-called "quality revolution" of the 1970s and 1980s, classic inspection appeared to be headed toward obsolescence. According to the most visible quality gurus, inspection was inefficient, ineffective, and unnecessary. Even the American Society for Quality Control (ASQC) considered dropping inspection as an operating segment.

Quality engineers armed with the latest techniques, it was touted, could reduce defects to zero by the simple expedient of eliminating all variation within the process. That this transformation failed to take hold everywhere is reflected in the myriad of U.S. manufacturing companies that are still struggling to master the requirements of as basic a program as MIL-I-45208.[1]

III. ORGANIZATION

An attempt to present and discuss all of the many kinds and styles of quality departments, their organization, and their techniques of planning would be impossible. So to try to stay within reasonable bounds, this chapter concentrates on one fairly narrow operation; the "traditional" quality department in a medium-sized (say $15–150 million annual sales) manufacturing company in the United States. Let us also assume that, as in most mid-size companies today, an inspection function is—or should be—the major part of the department's operations. Those of you who direct the quality activities of IBM, GM and GE or who are approved first-level suppliers to Motorola or AT&T won't need this material. Those of you in very small companies have more immediate needs.

To further direct our discussion, let's also assume that you have just been hired as Manager of Quality Assurance[2] for one of these mid-size companies.

IV. SOME OF THE RULES

At the onset, forget everything that you have heard from gurus, consultants, trade show conferences, and other professionals about how you are not responsible for quality in your company. As far as the outside world—i.e., the customer—is concerned, when it's time to fish or cut bait, the responsible person is not the head of operations, the manager of engineering, the vice president of marketing, or even the CEO. It's you.

When a major customer sends a letter stating that the last shipment was rejected and a formal corrective action statement is mandatory within 5 working days, the person on the hot spot is not the vice-president of human resources, it's the quality manager.

A. Irrefutable Rule #1: Walk the Walk; Talk the Talk

You accepted the hand you were dealt when you took the job. Play 'em like you got 'em. As quality manager, you are not likely to make many friends outside of

---

[1] As this book was being readied for publication, the U.S. Department of Defense announced that it would no longer support MIL-I-45208 and a number of other associated documents. Nevertheless, "MIL-I" continues to be a very useful tool for many companies.

[2] The title "quality manager" will be used in this chapter to identify the head of the quality function or operation. This function may be extremely sophisticated or primarily an outgoing inspection operation. The head of the operation might have the title director of quality, chief inspector, vice president of quality assurance, or any of a number of other titles.

your group, especially if you actually want to help your company improve quality. Tough and fair and thick skinned is the only way to go. Of course, it helps if you are an expert in quality assurance.

You are not likely to have the big office, and you probably will not have a secretary (they are getting to be a scarce commodity, anyway). You are not likely to be the "senior among equals" when you try to get the cooperation of the purchasing manager, the head of manufacturing, and your other peers. But you sure as heck can act like you've got all the aces.

Do not permit, in your presence, statements such as, "It's hung up in inspection," "QA is on a binge, again," and the like. If the statements are true, make changes. Quickly. If they are not true, make the person who dares utter them pay a price. Diplomatically. With a smile. Of course.

B. Irrefutable Rule #2: No One Gets Hurt

(Actually this should be irrefutable rule #1, but we needed to set some ground work.) No matter what else, prevent products liability exposure at virtually any cost. The definition of products liability exposure refers to shipping product that contains one or more critical defects.

The definition of a critical defect—the *only* definition of a critical defect—is a defect that could result in death or injury to one or more persons and/or serious damage to property. Other things could result in products liability litigation, including fraud, but injuries and property damage that could result from failure of your company's product are the big ones for you to watch out for. You may never be in a position where the product you ship could hurt people, but if you are, you have to play the game in a different way.

C. Irrefutable Rule #3: Tell Them What You Do

Immediately define your department's charter or mission and its goal; or someone else will. A really good mission statement is something like the following:

Mission Statement

> *It is the mission of the Quality Assurance Department to apply efficient and effective techniques of product and process measurement and evaluation; to generate statistical or other summaries that consolidate these measurements and evaluations; and to provide these summaries in addition to recommendations and counsel to cognizant personnel so that prompt and effective management decisions can be made regarding product disposition and/or process conformance to specifications.*

Translation: We evaluate and report.

Notice that no where in this mission statement does it say, "We are prepared to throw ourselves under the wheels of the truck. . . . ," or, "We are prepared to fight to the death. . . ." Too many QA/QC/inspection groups have allowed a situation to develop in which there is literally constant warfare between them and production—and sometimes engineering as well. Some bosses even encourage such warfare. If you can't prevent the warfare, talk to the boss.[3] If he/she can't or won't end the warfare, look for another opportunity. Somewhere else.

Departmental Goal

As part of the mission statement, a statement of departmental goal is often useful. A good goal statement is something like the following:

> *It is the goal of the Quality Assurance Department to perform its mission so as to aid the company to improve operational efficiency and to meet or exceed product and service requirements expected by customers.*

Translation: We evaluate and report—to the boss.

A common misconception in industry is that inspection departments, or the inspection groups of quality assurance departments, exist to reject products and otherwise make life difficult for production departments. Although there may be somewhere in the United States an inspection group whose charter is to hold back as much product as possible, it is very unlikely that any QA or inspection group anywhere is paid a bonus or commission for rejecting products.

The question, then, is why should there be an inspection function in the first place?

The answer to that question lies with what the real charter of a quality assurance department is, since inspection is a major part of quality assurance. Again, there is much misconception. Some people believe that quality assurance groups are responsible for adding quality to the product in the same sense that finishing departments are responsible for putting paint on the product. With the possible exception of improving the quality of product that isn't made to specifications through rejection followed by effective production sorting or rework, quality assurance departments cannot make good products out of products with problems. *Translation*: *You probably can't make a silk purse out of a . . . .* well, you know the rest.

If quality assurance departments don't make good quality, then what do they do? They do almost exactly what the finance departments of most companies do. The finance department doesn't make profits for the company. Marketing and sales knowledge of the customer and efficient production techniques make profits. The finance department evaluates and reports on the company's financial conditions—including profits—so managers can determine if and how changes are to be made.

The quality assurance department of modern companies evaluates and reports on the status of production efficiency as a function of how well products and processes meet the requirements that are specified for them. Inspection is an evaluation technique. Nonconforming material reports[4] are a reporting technique. Skilled inspectors aren't happy about finding problems in the company's products.

---

[3] In this chapter, the term *boss* refers to the highest level manager at the location. Titles might be general manager, chief executive officer, president, and the like.

[4] These notifications of product or process problems are often called discrepant material reports. Avoid the use of the term *rejection report*. By doing so, you will greatly minimize confrontation. You are not rejecting the product—or even worse, the operator—but simply reporting on compliance to specifications.

Finance departments aren't happy about reporting poor profit margins. But when quality problems or low profits happen, reporting them to managers is part of the job.

D. Irrefutable Rule #4: Put It in Writing

Learn how to write a good memo, fast. Depending on the circumstances, you may have to write four or five letters or memos a day. Maybe more. You are in the reporting business, remember? You can't do this if it takes you an hour to compose a memo and an hour to get a secretary (yours or someone else's) to type it. Put your departmental computer in your office and load a word processing package into it. Take a college level course in effective writing. Write, write, write.

You will not be popular. Behind your back you may be referred to as the memo fairy. Or worse. But if your company has the kinds of problems that by far most mid-size companies have, you may well be the way of staying in business. Figure 1 illustrates a typical product problem report memo. It's a real memo, only the names have been changed.

To: W. A. Hyuhp, Operations Manager
From: Al Atsee, Quality Manager
Subject: Circuit Connections

CiCan Units

Attached is a copy of an RMA from AceAcme Co. regarding a malfunctioning circuit assembly. Upon opening the can, QA found that the weld from the jumper wire to the center lead had failed.

Recently, during an evaluation of units for AcmeAce prior to installing the cans, QA found that the welds were also not satisfactory. Rewelding apparently did not solve the problem because during an evaluation of the units for high contact resistance we found another failed weld. In fact, failed or very weak welds on these units are relatively common.

Part, but not all, of the problem is due to the use of standard terminals. The holes used to externally wire the terminals of non-sealed units make good welds on sealed units very difficult.

It is recommended that all sealed units be made only using nonperforated terminals. In addition we will need to more closely monitor the terminal welding operations.

Figure 1 Problem reporting memo.

E. Irrefutable Rule #5: SPC Is Not the Sword Excalibur

(And you already know this isn't Camelot.)

Since the late 1970s, many quality practitioners—and several of the most visible gurus—have stated that SPC is the way to solve every problem. To "old-timers" in the quality profession, there is only one definition for the acronym. SPC—typically undefined, but by original definition the Shewhart X-bar and R control chart—is one of the most powerful process evaluation tools we have available. But only one.

Where it can be used to evaluate process stability and capability, there are few tools that are as good. The X-bar and R control chart is based on real mathematics, not a consultant's try at stardom. The statistical techniques associated with the X-bar and R chart can pinpoint a process that is being affected by assignable causes of variation, where the process is centered, and, by extension, whether or not the process is capable. But it can't do everything. And in some cases, it is the wrong tool altogether.

Although some practitioners have tried to adapt the X-bar and R chart to short runs, the basic premise that two or more short runs of different products (for example) can be combined to make one sufficiently long process run to permit the use of X-bar and R chart has been rejected by many practitioners.

Even more to the point, the company's structure must be such that something (internal corrective action) will be done to the process if the evaluation indicates that the process is not stable (in statistical control) and capable (able to produce product to the customer's requirements).

F. Irrefutable Rule #6: It's Wednesday; Time for the New Program

Steadfastly ignore the Acronym of the Week. In the United States, there must currently be 50 (150?) or more quality programs, each of which is guaranteed to eliminate scrap, end quality problems, satisfy customers, and increase profits. Among these programs are: TQM—total quality management; QFD—quality function deployment; TCWQA—total companywide quality assurance; TQC—total quality control; TCS—total customer satisfaction; and on and on. There are even some people who think ISO 9000 (series) should be added to this list.

But before anyone thinks that this kind of thing is only a problem of industrial quality assurance, it should be pointed out that there are probably far more fad diets than there are fad quality programs; for example, the high-protein diet, the grapefruit diet, the Jenny Craig diet, the Slimfast diet, the Ultra Slimfast diet, the Dexatrim diet, and on and on. It appears, too, that the same situation exists with stop-smoking programs.

Now there is no question that thousands of people have successfully lost weight using one or the other of the "fad" diets. And it is probably true that lots of people have stopped smoking after being hypnotized or after using nicotine patches. And it is also true that more than a few companies in the United States have turned poor performance and unhappy customers into great profit margins and happy customers with one or the other of the "fad" quality programs. But there are two secrets. The first secret is: the cold remedy Dristan works for some people and not others; nicotine patches work for some people who want to quit smoking and not others; Jenny Craig helps some overweight people lose weight, and not others. And quality function deployment works for some companies and not others. Programs become fads when they are touted as the one and only road to salvation.

If it has been determined that there is a problem, either a runny nose, too large a waistline, a cigarette cough, or too many unhappy customers, there is probably a program that can help. The second secret is: don't buy into just any program; especially not on the spur of the moment. Not all "programs" are "fads" in the generally accepted sense of the word; but like a good suit or dress, a program has to fit properly and be appropriate to the situation.

As quality manager, you should have a wide variety of tools at your disposal. Pick the tools, and the program, that fits your company's personality and market situation.

G. Irrefutable Rule #7: I Do Too Know What Quality Is; I Can Measure It

For more than a few years *quality* has been a term used more and more often. Slogans such as "Quality is Job 1" (Ford) and "The Quality Goes in Before the Name Goes On" (Zenith) have become as well known as the companies that use them. But what do we really mean by the word *quality*?

Recently, quality professionals—who, themselves, often strongly disagree about almost everything—have begun to agree that there are two definitions for quality. The first definition applies to the supplying company as a whole. That definition is: quality means customer satisfaction. This definition is the definition of company quality. The definition says that there is more to quality in the eyes of the customer than just the product or service they receive. Of course, the customer sees the quality of the product or service. But the customer also sees delivery dates made or missed; pricing that is or is not equitable; and how the supplier's people act on the phone and by mail.

To be a definition, customer satisfaction must be measurable. Sales volume, market share, market research studies, and, perhaps, benchmarking can quickly measure how satisfied customers are with a supplier. These are measurements that marketing or sales would make.

But how about the second definition? Since production can't control all of the things that go into customer satisfaction, a different measure is required for them. That measure, that definition, is: quality is conformance to specifications. That is the definition of product quality and is a much better definition for production. On the surface, it should be very easy. We have a blueprint or drawing (the specification), we make the part the way the blueprint or drawing says to make it (conformance to the specification), and presto, we've got product quality. Well, maybe and maybe not.

H. Irrefutable Rule #8: We Don't Make the Rules

(Well, almost never.)

One of the criticisms often heard about quality control or inspection groups is that they are too "nit picky." In other words, inspection groups spend too much time finding little things wrong that everyone knows really aren't a problem. Another criticism is that inspection groups "expect perfection," whereas we live in an imperfect world. These criticisms and others like them can be heard in many, perhaps most, U.S. companies even today.

If the criticisms are correct, how can the inspectors be allowed to waste time and hold up product that customers are waiting for? How can the inspectors be allowed to decide what is and is not acceptable? Who makes the rules?

In many companies, all the rules about what the product (or service) should and should not look like are set by the customer. In these companies, a market survey and/or a benchmarking study are completed before the product is introduced to the marketplace. Or, sometimes, marketing or sales simply asks the customer what they want. The results of these studies and surveys help the company determine what it is that customers want in the product or service.

Perhaps in as many other companies, little or no outside advice is sought. In these companies, engineering departments and/or management set the rules for what the product should look like. And of course in more than a few companies, the production line is turned on and whatever comes out the end is boxed and shipped. (Motto: If it fits in the box, it's fit to ship.)

So if the customer or engineering or the boss makes the rules, when does QA or inspection make the rules? The answer? They don't. With one exception. That exception is when a decision needs to be made about the conformance of a feature of the product or service and there are no established rules to guide the decision. Then, skilled quality assurance personnel will use their experience and judgment to make decisions about the acceptability. But only with reluctance. Their job is to compare the product (or, often, the processes) with the requirement; that is, to evaluate, not define, the requirement. Actually, quality assurance groups only expect perfection when perfection is called out in the rules.

I. Irrefutable Rule #9: Stand by your Man—and Women

Protect your people (yes, we know they're not "your people," but it has a nice ring to it), no matter what. Such statements as "It's hung up in inspection", "How come inspection didn't catch it?", "What's inspection rejecting it for now?", and "Give it to inspection and have them sort it right away" cannot be tolerated. And, needless to say, you never blame one of your staff for a problem. To once again paraphrase poor Pogo, We have met the leader and he is us. Since you are the head of the QA department, you are responsible for what happens—and what goes wrong—in the department. The rule is easy: They get the credit, you get the blame.

J. Irrefutable Rule #10: Never Stop Training

Your people (yes, yes, we already said they're not "your" people) must continue to grow. If your company is involved with threaded fasteners, conduct training in pitch diameters, tensile strength, and plating requirements. If you are involved with electronics, make it Ohm's law, resistors, and capacitors. Certainly, training in the use of measuring equipment, common terminology in the business, and X-bar/R and P charts—if they will be of value in your company—are mandatory. If you absolutely can't do the training yourself, hire a teacher from a local school.

Take attendance; pass out study material; hold exams. At the end hold a final exam. Those who pass get a formal certificate. Not typed. Calligraphed. In a formal presentation. It's a big deal, it should be handled that way. Those who don't pass the first time take the course over. Training results should be kept in each person's personnel file. If you are into ISO 9000, this is required anyway. If you're not into ISO 9000 yet, you'll have a leg up on it.

Everyone in the company—everyone—should be in formal training either in plant or in school. At least keep your people involved. Oh, yes, "everyone in the company" includes you.

K. Irrefutable Rule #11: Corrective Action Never Means Disciplining the Inspector

Unless your company is quite unusual, you will have the opportunity to respond to requests from customers for corrective action as a result of their having received defective parts, late parts, the wrong part number, or something similar. Your customer will be looking for (1) a piece of paper with your company's letterhead, preferably with some writing on it; (2) a letter containing a statement more or less to the effect that you are personally embarrassed and chagrined by the whole episode and you will personally see to it that the careless inspector is sent home for at least three days without pay; or (3) a technically sound description of the root cause of the occurrence and the steps that your company will take to eliminate the root cause and prevent further occurrences.

Unless your company has an exceptionally enlightened marketing/sales/customer-service group, you will somehow have to figure out how to handle it. No you can't delegate the responsibility to your newest staff member.

Requests or demands for corrective action have always been annoying, especially when the customer finds one bad part or returns a shipment with a statement such as: "don't work," "don't fit," "wrong color," and the like. Or when they buy stock units from the catalog and find they don't have a 20,000 hour MTBF at 350°C. And then you receive the letter stating, "Respond with corrective action within five days." But, as the experts have often said, rejections, returns, customer complaints, and the like provide actual failure criteria that you can use to improve the quality function in your company.

If they are not already in place:

- Build a returned goods evaluation function in your department.
- Personally open a review file for every customer complaint and request for engineering evaluation of product failures or malfunctions (or situations of unacceptable service).
- Personally supervise the evaluation of the complaint or request and personally close the file only when the customer has been completely satisfied and the necessary internal communications have been distributed and acknowledged.
- Write a professional quality letter to the customer summarizing what has been done and, if at all possible, specifying the corrective action taken. An example of a real customer response letter is shown in Figure 2.
- Continually work toward a full technical corrective action program in which operations and engineering develop the corrective action under quality

Mr. Joel Grey
ACE/ACME Corp.
7 Worchestershire Place
Berwyn, IL

Dear Mr. Grey:

Quality Assurance has evaluated the returned parts and the request for corrective action we received on RMA ZZ-30042. The complaints refer to mixed parts (44T units mixed with the correct 42T units), stained parts, and parts with crooked or bent terminals.

All of the problems referred to above occurred during a process conversion from manual to automatic operation of an insertion machine. During the conversion program a number of checks and procedures were inadvertently bypassed. As a result, some parts of wrong value were added to this shipment you received, as were a number of parts that would normally have been discarded.

Operations management has reestablished the temporarily bypassed checks and procedures and we are confident that future shipments will comply with your requirements. Please accept our thanks for the opportunity to review and correct this occurrence.

Yours truly,

Wiliford P. Smith
Director, Quality Assurance

Figure 2 Customer response letter.

assurance coordination. In many companies, this is a difficult and often unrewarding task. It has to be pursued, however, because operations (production) and engineering (design and/or manufacturing) are the only ones who can implement successful corrective action.

V. STAFFING

No matter how sophisticated the company, one way or another everything that gets done gets done by people. The work of the quality department is no exception. Unless you have been exceptionally lucky and your predecessor had it all put together, your department staffing will have to be reviewed. Quickly determine what needs to be done, and whether you have enough trainable people to accomplish the work. If you don't have the staff, get them. But remember, you are not building an empire. Keep the staff small but sharp. Everyone should have just slightly more work to do than they can leisurely handle.

One of your main goals is to build a tightly knit group of cross-trained people who understand that teamwork comes before personalities. You want people who are inquisitive and energetic and who will continually look for things that are not being done right; especially in the QA department. All the rest is easy.

VI. WHAT TO DO

If your company does not have an established quality program, get copies of MIL-I-45208, *Inspection System Requirements*, and MIL Handbook H-51, *Evaluation of a Contractor's Inspection Systems*. And since you can't make good measurements with a bad measurement system, get a copy of MIL-STD-45662, *Calibration System Requirements*.[5] (A good gage calibration and control software package will make life easier.) Build your system to the standards. Most of your customers who evaluate supplier quality programs will accept full compliance to these protocols as acceptable basic supplier quality standards.

If "MIL-I" is already in place, work toward MIL-Q-9858. It's also on the U.S. Department of Defense's hit list—but that doesn't affect its value as a quality program guide.

It starts getting complex after "MIL-Q". There's TQM and ISO 9000 (series) and others like them.

VII. A LAST WORD

The final decision about the kind of quality program that goes into your company is a business decision, not a quality management decision. You can't implement ISO 9000 unless the boss wants ISO 9000. You probably won't even be able to put in MIL-I-45208 unless the boss wants it. That's the way it is.

Your job is to determine the kind of quality system that you think is best for the company—then go to work to convince the boss to let you do it. If you really know quality assurance, there's a lot you can do no matter which business decision is made.

---

[5] MIL-STD-45662 has to some degree been superseded by ANSI/NCSL Z540-1-1994, *Calibration Laboratories and Measuring and Test Equipment General Requirements*. This standard is available from the National Conference of Standards Laboratories, 1800 30th St., Suite 3053, Boulder, CO 80301.

3
Asset Management

JAY W. LEEK*
PCA
Lecanto, Florida

I. INTRODUCTION

Within the structures of business management, there are many systems and procedures developed to perform line functions. These line or direct functions range from marketing to design and production. Additionally, there are staff or support functions that assist the line functions. One of these support systems is asset management. Support systems for the management of assets are complex, because they are designed to interface with many other facets of business management. The assets discussed in this chapter are typically referred to as capital equipment and for the purposes of this chapter refer primarily to measuring equipment and/or instrumentation used by engineering, manufacturing, and quality organizations to measure, assess, and evaluate processes and products. The purchasing process related to these assets requires careful planning. Asset management may be defined as the planning, acquisition, and control of capital equipment.

II. ORGANIZATION

Organization for the management of assets can be accomplished by a centralized or decentralized system. The elements affecting the determination of which system is used should be product lines, company size, product sophistication, process complexity, and company policy regarding rental agreements versus purchases.

---

* Retired.

Decentralized asset management typically lends itself to smaller companies with less complex processes and to those that prefer rental agreements to purchases. (The company policy in this regard is usually based on tax advantages.) In a smaller company, the volume of assets is much less and control can be handled by individual departments; so planning, acquisition, and control of these assets are less formal, with little documentation required.

In the centralized form of asset management, each company organization submits its requirements through one group. This group has the responsibility for administering the asset management system. It may also have the responsibility for implementing the planning function of capital assets and for controlling these assets after purchase.

As a service organization within the company, the quality assurance group works closely with the operational and other service organizations and therefore should be a prime contender for handling assets management. Whichever group is assigned responsibility for assets management must stay abreast of technological changes and be able to assess requests for new or additional equipment. Experience shows that assets management is best operated in a centralized control form for inventory control and consolidation of annual capital planning.

The remainder of this chapter describes and discusses the planning, acquisition, and control of capital equipment. Covered are capital equipment/assets relative to the long- and short-range plan, justification, acquisition, utilization, and how to plan equipment and when and why to plan for it. Also covered are the acquisition process; the benefits received from proper justification and processing requests for approval to purchase capital assets; the proper control of assets; and the advantages received from proper and well-coordinated controls. A note of caution here. The system described is not the only system of asset management; many systems will work in the right environment with the right people to maintain the proper disciplines. The system described here is an operating system that is offered as an effective method of asset management.

III. PLANNING

Planning is perhaps the most important problem faced in asset management. Sound, logical, well-thought-out planning can be a direct road to successful operation; on the other hand, poor planning invariably results in disorganization and failure. There is no simple formula to establish criteria for good planning. It is a matter of optimizing the critical path to accomplish the task, then considering alternative methods to accomplish the same end. When establishing the critical path, set milestones and alternative paths for each. Then if the optimum path encounters an obstacle, it is only necessary to return to the last major milestone and select an alternative. The return on the investment of time spent in good planning is realized in effective and efficient achievement of the intended goals.

In asset management, there are essentially two types of planning necessary: long range, sometimes referred to as *strategic*; and short range, sometimes called *tactical*. Strategic planning generally covers 5 years. It should be understood, however, that the longer the time frame, the less detailed and accurate planning becomes. The plan is updated each year, bringing more reality to the current year.

In long-range planning, adjustments can be made to the plan with ample time to return to the original path or to set up a different task to arrive at the desired point. The original goal is never lost, yet can be examined from several angles before execution. On the other hand, short-range or tactical planning is sensitive and responsive to changes, but it has limited options. The objectives and milestones of the short-range plan typically cover the next 12 months, and because of reaction time afford less opportunity to effect corrective actions on the path to overall achievement of the objectives.

A. Long-Range Planning

Generally, the purpose of a long-range plan (LRP) is to provide a strategic route and timetable for the company to acquire new business and revenue. The LRP varies from company to company but typically covers 5 years, with the first year representing a relatively firm picture and each succeeding year of the plan being less precise. In the LRP, the company's executives develop and communicate a consolidated program for business growth or for sustaining existing levels. Through this planning, all of the aspects of the business are examined in detail. Just as the number of years the plan covers vary, so can the contents vary.

An element critical to the plan's effectiveness is capital expenditures. Decisions on capital planning are complex and are best discussed with the various operating organizations. When the LRP is established, it is essential that each organization be given the opportunity to project and communicate their capital requirements. At the beginning of a financial year, the organization that has the responsibility for capital plan administration provides each operating organization with a detailed review of the business philosophy, product mix, and baseline for the years covered by the LRP. The various years' planning is reviewed individually and collectively to assure continuity of requirements and business interfaces among contributing organizations. The first year of the plan should be well detailed and specific. Each following year will be less certain and thus less specific. As each year is completed, each succeeding year will, in turn, become more certain.

The Plan

Figures 1 and 2 show typical forms used in an operating system to document the asset management planning operation. In using these forms, each organization provides consistent information to communicate the objectives of the planned acquisition. Figure 1 is the worksheet and Fig. 2 is a summary of the information provided on the worksheets. The instructions to complete Figure 2 are provided in Figure 3.

The business profile, outlining objectives and milestones, is provided to the operating organizations. Each organization then establishes supporting objectives within its own area of responsibility. These supporting objectives are documented in the justification block of the planning worksheet (see Fig. 1) and become the basis for the selection of capital equipment. When that has been done, the equipment description is provided in the description block. Typically, planning for the later years is difficult with regard to the details of the equipment needed. In this case, a logical generic description is provided; then in the following years, additional

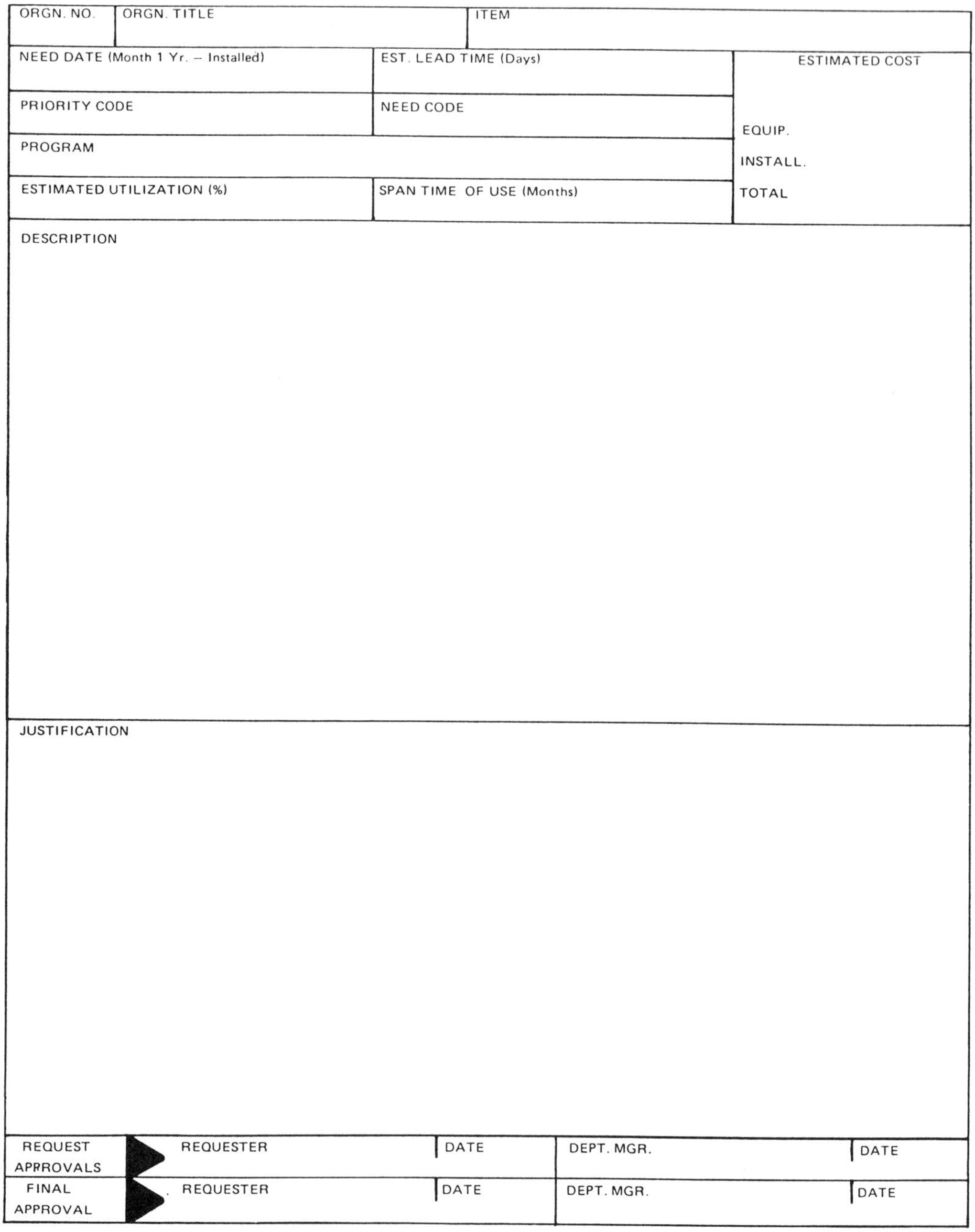

ORGN. NO.	ORGN. TITLE	ITEM	
NEED DATE (Month 1 Yr. – Installed)	EST. LEAD TIME (Days)	ESTIMATED COST	
PRIORITY CODE	NEED CODE		
PROGRAM		EQUIP.	
		INSTALL.	
ESTIMATED UTILIZATION (%)	SPAN TIME OF USE (Months)	TOTAL	
DESCRIPTION			
JUSTIFICATION			
REQUEST APPROVALS	REQUESTER	DATE	DEPT. MGR. / DATE
FINAL APPROVAL	REQUESTER	DATE	DEPT. MGR. / DATE

Figure 1 Capital assets planning worksheet.

information is provided as details become available. The same situation holds for cost identification. If the specific equipment is known, obtaining a quotation well in advance is often of little value when it is time to purchase. In estimating costs, however, projections should be as accurate as possible.

ORGANIZATION NO.			ORGANIZATION NAME						SHEET OF		
BUDGET LINE CODE	NEED CODE	PRODUCT LINE CODE	ITEM DESCRIPTION	YEAR	PRIOR. CODE	EQUIPMENT COSTS	NEW INST. COSTS	TOTAL COSTS	AUTHORITY FOR EXPENDITURE		
									NUMBER	AMOUNT	DATE

Figure 2 Capital assets planning form.

BLOCK NO.	DESCRIPTION
1	Consecutive numbers for each item (1, 2, 3, etc.).
2	Capital Asset Need Code: 1. Code 1 - Firm Business 2. Code 2 - Planned Business 3. Code 3 - Potential Business
3	Product Line Code (Examples): 1. Product Line A 2. Product Line B 3. Etcetera
4	Item Description (Make, Model, etc.).
5	Calendar year for which request is made (use last 2 digits, e.g., 81).
6	Capital Asset Priority (see Definitions).
7	Costs in thousands (e.g., 1918.4.).
8	To be completed during AFE* processing.

* Authority for Expenditure.

Figure 3 Plan item descriptions.

Two remaining items significant in long-range planning are *priority* and *need* codes. These items signify to the reviewer the application planned. There may be certain information known only to senior company executives, such as program risks or the probability of success for a new product line. In such cases, the decision to consider acquisition relative to particular programs may be deferred.

The information on the worksheet (see Fig. 1) is summarized on the capital asset planning form (see Fig. 2), thus yielding a planning sheet for each organization. At this point, finalization of the LRP may differ from one company to the next. For a company that is structured vertically, it may be advantageous to summarize the plans on a single capital asset planning sheet at predetermined levels to afford a logical review. For horizontally structured companies, each operating organization may submit plans independent of the others. In a horizontally structured organization, the individual plans can be summarized on a single planning sheet. Once completed, the capital plan can be included in the other portions of the company's long-range plan. Each year, as the LRP is updated, the capital plan can also be reviewed and updated. The plan is a tool to aid management in executing the rigorous task of operating a business.

Standardization

In developing the asset management system, it is important that the goals of the LRP be kept in mind. For example, equipment should not be chosen arbitrarily without careful thought being given to how, where, how frequent, and by whom the equipment is to be used. Considerations related to equipment utilization and standardization should also become a part of the LRP. Standardization of equipment is influenced by many facets of business performance. If any of these are overlooked, a "white elephant" may be purchased, resulting in an additional, unpredictable expenditure in trying to make the equipment usable. Elements requiring standardization include:

- Facilities
- User experience level
- Maintenance capability
- Calibration sophistication
- Spare-parts inventory

Facilities

When considering standardization relative to facilities, consideration should be given to the physical constraints and utilities. When considering the acquisition of large equipment, space becomes a major factor. If two organizations are planning major installations simultaneously, they should either be closely coordinated or prioritized to accommodate one or the other. If there is a section in the LRP identifying floor space, appropriate mention should be made to alert others to the need for expanded or new facilities. The utilities requirements for a planned acquisition may not present a problem unless there is excessive demand (i.e., electric, gas, water, air) to the extent that a major modification is required. In this event, additional planning should be accomplished at the earliest date.

User Experience Level

The second element of standardization is the users' experience level. With the installation of new equipment, it is frequently necessary to consult the manufacturer and obtain recommendations relative to required operator experience level as well as available training. Typically, large equipment manufacturers support training. In any event, it may be necessary to make additional investments to assure adequate operator training.

Maintenance Capability

Considerations similar to those for operator training should be given to maintenance training. An equipment manufacturer may provide maintenance training for several years, after which it is necessary to be self-sustaining or to purchase a service contract. These decisions have an impact on the life-cycle cost of the acquisition and if not considered may cause extensive recurring costs. It is good asset management to acquire as few unique or special devices as possible.

Calibration Sophistication

If the manufacturing department plans the acquisition of automated process equipment requiring calibration accuracy not presently available within the facility, additional planning is necessary by the support organization responsible for calibration. Major process equipment typically does not lend itself to calibration outside the facility; therefore, the capability should be available within the company. In meeting these requirements for calibration, additional acquisitions by the support organizations may be necessary.

The purpose of standardization planning is to facilitate growth in capability and technology. Communicate the needs through the capital plan portion of the LRP, in such a way that support functions can do the same level of planning. This results in a coordinated, goal-oriented approach to solving company growth problems. Standardization is attained through a concentrated effort of communicating intent in the LRP. When an item first appears on the LRP, it may simply state that "to meet process rate requirements it is necessary to automate subassembly and final assembly test." It might be estimated that existing utilities and calibration capabilities will suffice, but that additional floor space is required. This type of entry can be updated each year as the plan matures until the equipment is placed in order.

B. Short-Range Planning

Updating the LRP

As pointed out in Sec. III.A, the LRP is a means of communicating the intent of implementing the requirements of the company. When it comes to planning the following year's activities, it is no longer a matter of communicating intent, but a matter of identifying and implementing reality.

Figures 1 and 2 are essentially an extension of broader material found in the LRP. The forms are used in identifying capital equipment requirements for the 5-year plan and the short-range plan (SRP) provides the detail necessary to support the more immediate acquisitions. After having an objective materialize over a period of years, it should not be difficult to identify the current requirements and then execute its implementation.

Utilization

As standardization was a major consideration for long-range planning of capital equipment, utilization becomes critical when finalizing the SRP or tactical plan. Planning for utilization becomes critical in the short-range plan because this information provides the final justification for the purchase and demonstrates how the new equipment is to work for the company: how the acquisition will create a return on investment (ROI). Utilization is assuring that the equipment will be in service and in use a maximum amount of time to obtain the greatest return on investment. In some instances, a decision to lease or rent the equipment might prove to be the best economical choice. These options should always be considered prior to acquisition by purchase if utilization proves uneconomical.

Utilization of a standard test instrument need not be 100%, nor need it be utilized through its full service life. Standard electrical/electronic instrumentation typically has a useful service life of 5–7 years, whereas mechanical devices have a considerably longer service life. If a user requests a device and identified only a 2- or 3-year need, upon completion of this period the device will be returned to inventory and utilized elsewhere later. Another type of limited need is a requirement of only 8–10 hours of use per week; it does not mean that the acquisition would not be economical; instead, it could be made available the remainder of the time for others to use unless it is dedicated to a special station, in which case economics would prevail.

C. Scheduling

Scheduling the acquisition of assets is another function of the asset management planning process. The heading *Need Date* of Figures 1 and 2 is the date that must be met for the commitments of the program to be satisfied considering the equipment manufacturer's lead time, installation, and start-up problems. It may be necessary to order instrumentation 12–18 months in advance to allow for delays. Predictable delays can be exemplified when considering the acquisition of a computer-controlled test station and the time required for writing and verifying programs. Buying this type of equipment at the time it is needed is shortsighted and should be avoided. Since predictable delays must be considered when establishing need dates, it is also necessary to examine other possible contingencies. In doing this, the data identified on the acquisition form presents as small a risk as possible. There should be coordination with the equipment suppliers to determine the best delivery information.

IV. ACQUISITION

The acquisition process contains three elements; justification, document processing, and receipt. Justification is a part of the documentation prepared for acquisition. Since convincing management of the need is the most important phase of the acquisition process, considerable attention has been given to this subject.

A. Justification

Before discussing the details of justifying capital equipment, there is a general consideration worthy of mention. When a person determines that a new piece of equipment is needed, there usually is good justification. But there is often difficulty in communicating this justification to those responsible for approval. In many cases, the equipment is so directly related to the person's duties that justification for the equipment becomes a defense of the job. This sometimes breeds resentment and a resistance to preparing effective justification because of pride. Too often it is said: "Isn't it enough that I say I need the equipment—doesn't my judgment mean anything?" Of course it does, and astute management takes the requestor's experience and judgment into consideration. No requestor should fear embarrassment or castigation for making a sincere request. Management does not challenge the need. Rather, they weigh the degree of need for one request with all others to arrive at a total equipment requirement consistent with available funds and company objectives. The advice to requesting and using organizations in this regard is to be objective and impersonal. Another person or department which was objective and impersonal may get the dollars for an item of lesser need simply because management was given more meaningful facts for review.

Justification for capital equipments usually falls into the following categories: (1) must have, (2) needed, and (3) desired. The amount of effort and detail put into justification varies for each of these categories. However, it is wise to cover all the points listed below which are applicable to the case. It should not be assumed that management is aware of the facts relative to an acquisition. Present them all. Many justifications for urgent items encounter difficulties with management if the "must" characteristics of the equipment is not clearly and decisively established. The points to be considered in a complete and sound justification are listed below and then discussed in detail.

- Description versus justification
- Listing of specific facts
- Effect on work if not obtained
- Amount of usage
- If unique, who designated or prepared specifications? Who will operate?
- Growth potential
- Analysis of total cost
- Intangible benefits

Description versus Justification

Give a complete, detailed description of the item desired and what it does. Catalog or vendors' numbers (e.g., C-210 and B-300) may be clear to the requestor but may not be to management. A short, clear statement describing the function is appreciated by a busy executive. The requestor may know what an "exciter" does, but if mention is not made of vibration, for example, the general manager or finance manager may not know what the item is for or, even worse, assume that it is for a different function.

Also, a description should not be used as the justification: This is the most common failing of equipment justifications. Much detail is often given regarding the type of test to be performed or measurement to be taken in a justification statement. For a complete justification, the reason the test is to be performed or measurement taken should be provided. An example of this is a detailed description of the capability of a vibration system to perform up to 3 kHz, whereas the present equipment provides vibration to only 2 kHz. So far, so good, but the most important information is the explanation of why the 3 kHz is needed—new specification, contract change, design trend, and so on. As noted, this is the most common failing in justification preparation.

Listing of Specific Facts

The word *more* should be avoided in a justification. Examples of this are *more* accurate, *more* complete, *more* sensitive, *more* range, *more* economical. This is practically meaningless to top management and may offend them by giving the impression that vague facts are being forced on them. If the equipment wanted is *more* accurate, give the details. For example, old equipment may be accurate to only 1% and the equipment desired is 0.1% accurate. Then follow up with a reason for greater accuracy. The same applies to comparable characteristics, such as sensitivity, range, completeness, and economy. Economical parameters should be substantial by providing actual and estimated costs and anticipated savings. Get help on this score from financial specialists if it is felt that one person cannot do a complete job.

Effect on Work If Not Obtained

If the effect on work is not treated in the justification, an astute general manager will ask about it during the review. Justification must be available or the case will probably die on the spot. If there is no significant effect on the work, the only result may be embarrassment at taking up time in preparing, presenting, and reviewing, a justification for an unneeded acquisition. The effects of not obtaining the desired equipment have been assessed and should be covered in detail. All the positive reasons given may not be as strong as the negative effects on nonacquisition, such as nonperformance on a specification requirement, requiring customer waiver or outside subcontract, product liability exposure, inefficient or uneconomical operation involving schedule slippage or cost overruns, loss of competitive position, and acquiring the reputation of being behind the state of the art. The decision to acquire new equipment should be based not only on what the equipment will do but also on the penalties that may result if the equipment is not acquired.

Amount of Usage

Be factual when identifying the usage of the proposed equipment. This is similar in financial analysis to return on investment. Indicate the planned work that is firm and that which is anticipated. Utilization (as discussed in Sec. III.B.2) for most equipment is the *total* time it is "tied up" on a job, including setup, calibration, running time, holding time for quick-look review of data, and teardown time. Management may think only in terms of running time; 10% running time for a vibration system may represent 100% utilization, but if this is not thoroughly explained, management may interpret this to mean that the equipment will be idle 90% of the time.

If Unique, Who Designed or Prepared Specifications?

If the equipment being requested is pushing the state of the art, management should be convinced that the technical homework has been done. There is a natural reluctance on the part of management to invest large funds in an untried item. Be sure, therefore, that a coordinated input is received from specialists within the organization or from outside consultants if necessary. Many justifications fall into stormy waters when management determines that the experts available within the organization have not been consulted on a developmental acquisition. Encourage competent people to review the design, specifications, or requirements. Identify the areas of technical risk. For unique facilities, identify the operational features. Are trained personnel available to operate the equipment? If so, they should be identified. If not, discuss the detailed plans that have been developed to achieve this capability: for example, have the supplier train the personnel; send personnel to appropriate training classes; or hire a consultant to operate and train people on the job. Such training requires funds and time. Management should be made aware of this and their approval requested.

Growth Potential

Before management gives its approval to spend a large amount of money, they generally want some assurance that there is little likelihood for a near-term repeat performance. Consult marketing or advanced systems organizations. Determine the trends that might affect this acquisition. Are the items being tested getting bigger, specifications getting tighter, acceptance quality level increasing, frequencies getting higher? If advice is unavailable within the company, do some independent research or seek help outside the company. Acquiring growth potential in a facility may increase its current cost, but it may be good business, since costs are continually on the rise. Buying tomorrow's capability today may save money and obtain a favorable competitive position. It is easier to win business with capability than with statements of intent to acquire it.

Analysis of Total Cost

One of the pitfalls many equipment justifications encounter is an incomplete listing of an item's cost. The purchase price is only a part. What about the maintenance and operating cost? Make sure that equipment listed is acceptable to the standardization committee within the company (see Sec. III.A), and whenever possible, assure minimizing life-cycle costs. If these costs are anticipated to be lower than those of similar existing equipment, describe the cost saving. It should help to sell the case. If maintenance and operating costs are going to be higher, describe, explain, and justify these higher costs. It is better to do it now. Otherwise, a critical general manager may bring this out during a review of your justification, and your integrity may be challenged. Even if the equipment is acquired, the higher maintenance and operating costs will not go away. If the costs are high, trouble may result even though the high cost may be justified, since the initiator will be explaining things after the fact. Also, management may feel that there has been a deliberate withholding of information which they should have had to make a proper and complete assessment of the request.

Intangible Benefits

Intangible benefits can be the icing on the cake. Include such things as enhancement of company image, attractiveness to technical applicants, meeting customer's unusual requirements, beneficial effects on employee morale, and stress unusual or unique characteristics. Know management and use what appeals to them. If the requestor has done a thorough and complete job on items 1 through 7 of Figure 1, no "selling" is required. However, remember that top management looks at a broad base, and these types of intangibles enhance a broadened overview.

The essentials of a good justification, when reviewed, seem simple and straightforward. If there is a magic formula, it is simply "give management *all* the facts" and success will be more frequent.

B. Documentation Process

The documentation process should be properly coordinated and closely monitored. The process described in this chapter, although complex, has been effectively implemented. The system described may need tailoring depending on a company's organization. Achieving the goals of asset management and providing control is of significance; the system ultimately implemented is not.

To begin with, the short-range plan, once approved, becomes the source of the majority of information necessary to complete the form and initiate the request for acquisition approval. Considering the forms discussed previously (see Figs. 1 and 2), there are two ways of initiating a request. The first is controlled by the administrator. The administrator disassembles the capital plan (previously assembled by organization/department) and reassembles it according to need dates (as explained earlier, this considers manufacturers' lead times). Thereafter, at the beginning of each month, the administrator initiates the acquisition request form and returns it to the user for signature and further processing. Processing in this fashion, however, does not allow for program slips or changing requirements; it assumes that the plan was accurate. The second system is to allow the initiator to execute the plan and order what is needed when it is needed as long as it stays within the initial plan.

Figure 4 shows one type of form used to document the "authority for expenditure." The form serves to collect the information leading to the final decision to make the acquisition. The reviewer is provided with the rationale necessary to support the acquisition. The second page of Figure 4 provides additional information:

- Disposition of equipment being replaced (if applicable) (item 16)
- Alternative action (item 18)
- Commentary (additional, intangible benefits received as a result of this acquisition) (item 19)
- Approval signatures (item 20)

Regardless of the system used, once the requirement is verified, the request-for-acquisition form is signed by the initiator. The missing element then is control, which is discussed in Sec. V. Having obtained the initiator's signature, the document is next submitted to appropriate levels of management for review and approval. It should be noted that this is the last time that these managers will have an opportunity to review the document for administrative and technical details before the final review by senior approval authority reviews or challenges it. After the department manager has completed his or her review, the request is returned to the administrator for approval and entry into budget commitments. Recording the commitment assures a continuous status of capital asset commitments for reporting to upper management.

The document is then forwarded to the financial organization and, depending on the company's structure, perhaps to the chief executive officer for final review and approval. The flow described above is shown in Figure 5. Typically, the form is multicopy, and after the order is placed, the form is distributed to persons in the approval loop.

C. Receipt/Inspection

Once the equipment has been received, it is necessary to assure its satisfactory operating condition. The equipment may be delivered to the intended user, who in conjunction with the group charged with maintenance and calibration, performs

PAGE 1 OF 2

1. BUDGET LINE ITEM	2. DATE PREPARED	3. PURCHASE ORDER NO	4. SHIP TO PLANT	5. AFE NO

6. REQUESTED BY	7. PHONE EXT.	8. ORGANIZATION TITLE	9. ORGN NO.	10. DATE REQUIRED

11. VENDOR	ADDRESS	CITY / STATE	12. NO. BIDS REC'D

ITEM NO.	QTY	PART/MODEL NO.	DESCRIPTION	UNIT COST	TOTAL COST
			13. INCLUDE ALL SUPPLIES, SERVICES, REQUIREMENTS AND INSTALLATION COSTS		

14. REASON FOR EXPENDITURE

15. SAVINGS/PAYBACK ACCOMPLISHED BY REASON OF EXPENDITURE

DISTRIBUTION: WHITE–ACCOUNTS PAYABLE, GREEN–PURCHASING,
YELLOW & PINK–PROPERTY ADMINISTRATION,
GOLD–REQUESTING ORGANIZATION

Figure 4 Authority for expenditure.

functional testing, exercising all parameters of the equipment. When exotic or state of the art equipment is involved, the equipment should be exercised to the manufacturer's specifications to preclude problems after the warranty expires. Many equipment manufacturers will install and test newly delivered systems at no additional cost. The user organization and those with maintenance responsibility should avail themselves of this service when available from the manufacturer.

The organization having responsibility for maintenance and calibration normally is equipped to exercise equipment of a routine nature to the fullest extent of the manufacturer's specification. As part of the asset management systems, newly purchased equipment should be calibrated and certified prior to use. This receiving

PAGE 2 OF 2	17. AFE NO

16. EXISTING ASSETS REPLACED AS A RESULT OF EXPENDITURE (INDICATE BY NAME AND ASSET NUMBER)

18. ALTERNATE ACTION

19. COMMENTARY

20. APPROVALS	DATE
a. ORIGINATOR	
b. SECTION MANAGER	
c. DEPARTMENT MANAGER	
d. LABORATORY SERVICES	
e. FACILITIES	
f. PROPERTY ADMINISTRATION	
g. FINANCIAL MANAGEMENT	
h. VICE PRESIDENT & GENERAL MANAGER	
CORPORATE APPROVALS	DATE
DIRECTOR, FACILITIES DEPARTMENT	
PROPERTY ADMINISTRATION	

21. ACCOUNT NO. ORG. NO.
a. ☐ LEASE
b. ☐ EXPENSE

22. ACCOUNT NO. ORG. NO.
☐ CAPITAL ASSET ....
a. ☐ NEW ☐ USED USEFUL LIFE YEARS
b. COMSUMPTION PATTERN: ☐ STRAIGHT LINE ☐ DECLINING BALANCE
c. ESTIMATED SALVAGE (RESIDUAL) VALUE AT END OF USEFUL LIFE ____________ % OF ORIGINAL COST

23. ☐ ASSET ☐ EXPENSE

	TOTAL DSD	ORG
a. AMOUNT IN 19_____BUDGET	$	$
b. TOTAL COMMITTED/INCLUDING THIS REQUEST	$	$
c. BUDGETED BALANCE AVAILABLE	$	$

DISTRIBUTION: WHITE–ACCOUNTS PAYABLE; GREEN–PURCHASING
YELLOW & PINK–PROPERTY ADMINISTRATION
GOLD–REQUESTING ORGANIZATION

Figure 4 Continued

inspection/certification serves a dual purpose; it assures that equipment is not damaged in transit or contains latent defects at the time of receipt, and calibration assures accuracy of measurements.

V. CONTROL

Once the equipment has been received, inspected, and calibrated, it is necessary to assure that proper systems exist to control it through its life cycle. The management control system to be discussed includes:

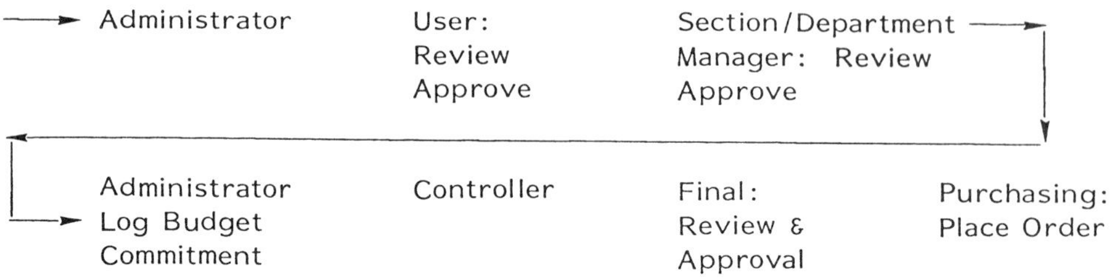

Figure 5 Approval flow chart.

- Custodial control
- Maintenance and calibration
- Inventory
- Training
- Spare-parts control
- Budgetary control
- Equipment disposition.

To facilitate a management control system, a series of asset identification numbers is generated and issued for each piece of equipment and a log maintained by a central property administrator. The asset numbering system need only provide that each piece of equipment have its own identifier. Frequently, there is more than one piece of a given type of equipment within a facility, and the equipment serial numbers are generally small and often internal, requiring disassembly to locate. Using a unique numbering system facilitates affixing the number externally on the equipment and results in a uniform, easily monitored, identification system.

A. Custodial Control

Once the system of identification is established, someone must be charged with the responsibility for "care and feeding" after the equipment is put into use. Within the plant or facility, one person from each operating department should be assigned responsibility as equipment custodian. Departments can be combined if small. The equipment custodian is responsible for the continuous status of the equipment, including where it is being used, assuring proper operating condition and calibration requirements. (The calibration cycle is discussed below.)

When equipment is assigned to area custodians, they sign the "asset transfer" (Fig. 6). This acknowledgment signifies that the custodian accepts responsibility for the equipment, is aware of its intended use, and is aware of the calibration schedule that has been established. Figure 6 shows a sample form with asset number, issuing activity, recipient, equipment description (serial number is optional), and calibration cycle. The content or information on the form should be provided to meet individual system requirements. The information contained on this receipt document becomes a part of the property administration log, where the central control documents are maintained for the asset inventory system.

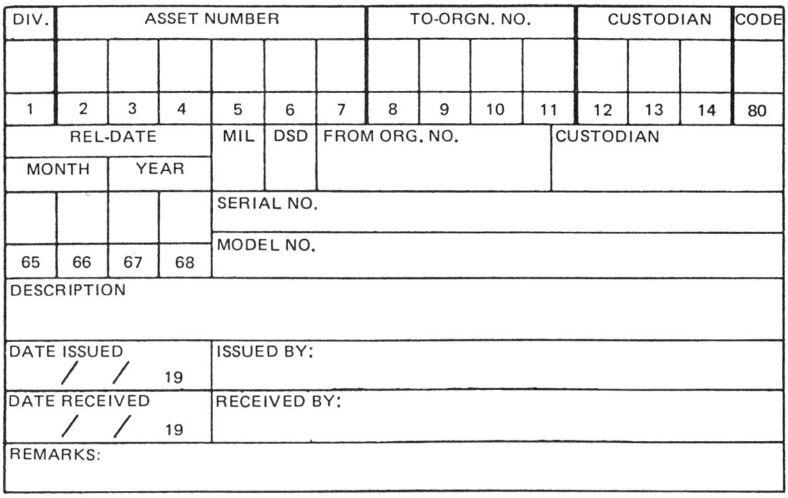
DIV. | ASSET NUMBER | TO-ORGN. NO. | CUSTODIAN | CODE
1 | 2 | 3 | 4 | 5 | 6 | 7 | 8 | 9 | 10 | 11 | 12 | 13 | 14 | 80
REL-DATE | MIL | DSD | FROM ORG. NO. | CUSTODIAN
MONTH | YEAR
SERIAL NO.
MODEL NO.
65 | 66 | 67 | 68
DESCRIPTION
DATE ISSUED / / 19 | ISSUED BY:
DATE RECEIVED / / 19 | RECEIVED BY:
REMARKS:

Figure 6 Asset transfer ticket.

B. Maintenance and Calibration

The next element of control is maintenance and calibration. Systems and organizational responsibilities established to perform this function range from very basic to very elaborate. This control is necessary to cycle the equipment periodically for calibration and maintenance. This periodic cycling is documented through updating of an "equipment history record" (EHR) (Fig. 7).

This EHR becomes the document to maintain a history, or fingerprint, describing both routine and significant events, including required maintenance throughout its service life. The calibration results may be shown in terms of attribute data; however, maintaining variable data allows statistical analysis to predict maintenance requirements, wear-out, calibration frequencies, and so on. More specifically, if the measurement is correct and within its tolerance limitations, it suffices to say "No cal. required"; however, if the measurement is beyond its tolerance limits or variable data are maintained, the specific measurement and action taken to correct it should be entered on the EHR. Again, the variable data taken are necessary for proper assessment of equipment stability and therefore become an important consideration in determining succeeding calibration cycles. For additional information regarding calibration or metrology systems, refer to Chapter 35.

C. Inventory

One of the most effective tools in asset management is "inventory control." This refers not only to equipment in storage but also to equipment in use within the facility. An effective inventory is usually necessary and should be dynamic in nature. When a department makes a request for new equipment, it would be simple enough to buy the new equipment at the least price with the best delivery; conversely, effective asset management comes from being able to research inventory and find an instrument with the same or similar capability. Should such equipment be avail-

TYPE MODEL CAT. NO.	MFG. S/N	P/O	COST	DATE RECEIVED	ASSET NUMBER
MFGR:					
FUNCTIONAL DESCRIPTION:					CALIBRATION PROCEDURE

CODE	CALIBRA-TION INTERVAL	DATE CHECKED	DATE DUE	LOCATION OR CUSTODIAN	INSTRU-MENT CONDITION	COMMENTS	CERTIFIED BY

Figure 7 Equipment history record.

able from inventory, arrangements can be made either to loan it to accomplish the task or to issue it permanently. If the equipment is available under the custodial control of another department and utilization there is low, a sharing arrangement or a transfer may be arranged. Moving equipment between departments is generally not an easy task. However, once managers become familiar with the practice of a dynamic inventory, they usually cooperate; the important point here is that a dynamic inventory is a two-way street. This is not to say that once the initial equipment has been purchased further acquisitions are not necessary—quite the opposite. Equipment lives out its useful life, gets damaged beyond economical repair, does not meet new requirements, or there is not enough to go around. Any of these reasons constitutes the need for acquisition. The important point to remember is to check inventory before making a financial commitment; the cost savings can be put to use in other areas.

D. Training

The next element of control is training for both equipment operation and maintenance. Training can be of generic type, such as electronics or mechanics; of a specific type, such as meters, coordinate measurement systems, and differential gage systems; or operator training, as described by the manufacturer. Manufacturer's training can be performed either in-house or at the manufacturer's facilities and in-house combined. Again, it is very important to consider training when computing the cost of an acquisition; if not, the group may end up with equipment it can neither operate nor maintain.

E. Spare-Parts Control

The next points concern the control of spare-parts inventory. Deciding on quantity and type of spares begins with establishing a "spares policy." That is, spares must coincide with the maintenance policy; maintaining the equipment, purchasing a service contract, or somewhere in between. Once the policy is established, it should remain constant across all equipment. By remaining consistent, other departments will learn what to expect and will have a better understanding of what is necessary to keep their equipment in top operating condition.

F. Budgetary Control

Budgetary controls are usually established and tailored to a specific operating practice of a company and therefore are too detailed and varied to be discussed here. The point to be made, however, is that once the annual "capital budget" has been established, it is essential that it be controlled in a manner similar to the assets themselves. The group with the responsibility for equipment inventory control, standardization, maintenance and calibration, and spares is best equipped to oversee the budgetary control of capital equipment. Information relating to basic price, option cost, deliveries, and rental costs makes this organization a natural focal point which can provide timely, accurate data as to status. Again, the critical point is "consistency." If each department is handling its own controls, the entire system can be unwieldly. On the other hand, if a central department has this responsibility, the total system becomes much more effective.

G. Equipment Disposition

The final element of control is that of equipment disposition after it has completed its useful life or when it is damaged beyond repair. The first decision to be made is whether to replace the equipment, and, if so, whether it be the same type or an improved model. Essentially, the planning phase of a new acquisition process begins at this point. However, in a working system of asset management, the planning would have been completed and, in fact, would have predicted the end-of-life condition. If the asset is damaged beyond repair, the equipment should be scrapped and, where possible, component parts that may be usable as spares should be salvaged. If, the equipment is still operational but no longer possesses the accuracy or stability necessary to support its intended use, the asset should be sold or otherwise disposed of; for example, by being given to a school. The point here is that once the device has completed its useful life, it should be removed from inventory. Carrying it on the inventory is of no value and adds unnecessarily to the controls required, and may actually be a tax liability.

4
Management Communication

MARTIN R. SMITH
Management Sciences USA
Lawrenceville, Georgia

I. INTRODUCTION

One of the major problems confronting quality professionals today is that of opening up and maintaining effective communication links with various levels of management. Too often, for example, the same quality performance report is issued to everybody, from the first-line supervisor through the president. Results are generally predictable. The president needs summary information, contrasting goals, and actual results. The supervisor needs details relating to performance by operator, shift, and machine. Issuing the same report to both persons is self-defeating; only one of these levels will be fully able to use the information in a given format. The astute quality professional will therefore provide each level of management with the information best suited to convey results, highlighting areas in need of attention so that each level will be able to act and to take corrective action.

The entire structure of communications between quality assurance and management is vital to the success of the quality program. The best quality program available can only be as potent as the ability of company management to have the information it needs to make things happen. Without this vital communication network, quality will falter.

There are certain ways to establish effective communication links within the company that will enhance the quality information flow. The first of these is concerned with the relationship of quality assurance to top management.

II. QUALITY ASSURANCE AND TOP MANAGEMENT

"Top management must commit itself to achieving quality or it will never happen." These are good words but they are meaningless without tangible methods for top management to demonstrate that support. It is simply not enough to have the president tell the staff that he or she wants quality. The mechanism must be established which sets quality goals and then measures progress against those goals.

Nothing is as hard to achieve as quality without the active involvement of top management. Quality must be planned, measured, and controlled in similar fashion to sales, costs, productivity, and other company needs. Quality just will not happen by itself; it needs the full thrust of management commitment, not just well-intentioned statements.

There are four very practical tools that can make that top management commitment a reality:

1. A quality policy
2. A company quality plan
3. A quality board
4. A quality assurance reporting relationship

A. Quality Policy

The first step is to think through exactly what top management expects from the quality function. This should include the expected level of involvement in different areas of the company. The quality policy answers such questions as:

1. "Should our product be the best in the industry, or should it merely be competitive? Where, in fact, do we want it to be?" (Each of these positions demands a certain price which individual companies must recognize.)
2. "Are quality practices and techniques to extend to design and service, or do we want to confine quality assurance to manufacturing?"
3. "To which management level should quality assurance report?"
4. "What level of costs are acceptable in achieving the quality program?"
5. "How will customer quality needs be satisfied in after-sales service? How flexible should the company be in settling customer claims for substandard quality?"
6. "What degree of control should be exercised over vendors?"

Stating the Policy

An effective company quality policy can generally be expressed in just a few pages. The policy is not designed to include quality methodologies and techniques—it simply sets the stage for achievement of the quality program. A typical quality policy for a capital goods manufacturer is shown in Figure 1.

Notice that in the first paragraph the company specifically states its desire to be "among the leaders in the industry." This clearly indicates the company's desire to have its products above the average competitive level, but does not lock it in the beat *all* its competitors. That allows the company to produce a first-class product, but it does not require the company to incur the extra costs necessary to move ahead of all competitors.

Notice also that the criteria for competitive quality rest in the customer's eyes through the words "judged by our customers." Nothing could be clearer. The success of the company's quality efforts has been given a definitive measurement that can be checked by customer surveys.

QUALITY GOODS, INC.

QUALITY POLICY

Policy statement

1. It is the policy of Quality Goods, Inc. to deliver products whose performance quality is judged by our customers to be among the leaders in the industry.

2. Our goal is tc provide the proper environment in which product quality meets or exceeds contractual obligations with our customers, and which gives the necessary service level to customers to fulfill the intended function of our products for a period of time to satisfy customer requirements. Customer claims for defective workmanship will be evaluated and settled within the time limits expressed in individual contracts.

3. It will be the responsibility of company general managers, in cooperation with their respective quality assurance managers, to establish an effective quality function which achieves the company's quality program.

4. To assure effectiveness of the quality program in the most objective manner possible, each quality assurance manager will report directly to the general manager of the operating unit and be placed at the same organizational level of manufacturing, engineering, materials and marketing.

5. Final product acceptance decisions, based on company quality requirements and specifications, are the sole responsibility of each unit's general manager.

6. Quality programs are intended for application to all functional areas (marketing, engineering, materials, and manufacturing) involved with shipping and servicing specified quality products for customers.

7. Quality programs will be achieved within specified quality cost budgets established by unit general managers which have been approved by the company president.

Reference Procedures

QA 202	Quality Board Responsibilities
QA 203	The Company Quality Plan
QA 300	Cost of Quality Reporting
QA 315	Vendor Quality
QA 402	Design Quality and Reliability
QA 403	Pre-Production Quality
QA 500	Product and Process Quality
QA 600	Quality Information Systems
QA 703	Field Service Quality
QA 805	Quality Measurement Systems
QA 830	Quality Surveys
QA 900	Quality Training

Figure 1 Quality policy statement.

The second paragraph describes the level of service to the customer without leaving room for doubt. There are no ambiguities. Customer claims are to be settled within time spans indicated in customer contracts.

The third and fourth paragraphs establish the organizational relationships that quality assurance has with other company components and places control of internal quality in the hands of the general manager.

The fifth paragraph places responsibility for shipped quality on the shoulders of the general manager—where it must be to assure achievement of quality policy.

The sixth paragraph assures that quality will be applied to such functional company components as manufacturing, purchasing, warehousing, design, and service.

The last paragraph establishes the necessity for achieving the quality program goals within financial boundaries consistent with company cost goals. In practice, cost-of-quality goals are established based on those tasks necessary for meeting company quality policy.

Finally, references are made to specific procedures designed to assure compliance to the policy.

B. The Company Quality Plan

Although quality *policy* is the formalized commitment of management to quality, the company quality *plan* is a detailed expression of how that policy will be achieved. The quality plan is normally the creation of the quality assurance manager with the blessing of the general manager. It describes quantifiable quality goals (cost of quality, warranty, etc.) and supports those goals with specific plans of action.

The quality plan is similar to a general's tactical plan in a combat zone. The tactical plan (quality plan) is an extension of military strategy (quality policy) and is an expression of how that strategy will be achieved. As such, the quality plan should encompass all aspects of the quality program from design, through manufacturing, and into service.

The time span of the quality plan is flexible and should reflect the period of time it takes to accomplish major tasks. Three years is typical; 5 years is not unusual.

Figure 2 is a page from the quality plan of the same capital-goods manufacturer discussed earlier. Notice that very definitive problems (sometimes called "opportunities") are described together with a plan of action to respond to the problem/opportunity. Specific people are assigned to assure completion of the tasks by stated completion dates. Similarly, component plans of action are written for manufacturing, materials, and marketing. A list of plans of action are, of course, dependent on those problems unique to the company and areas of improvement which will result in lowered scrap, rework, and warranty charges.

Each contributing manager should approve his or her section of the plan, and the final quality pin should then be approved by the unit general manager. This will assure that the plan is workable and gains management support. Progress must be reviewed periodically against scheduled completion dates by the quality assurance manager. The quality assurance manager and general manager must insist that "behind-schedule" items be brought up to date or the goals will not be achieved.

QUALITY GOODS, INC.

QUALITY PLAN

1993-1995

Section #6: Engineering Quality Systems

Problem/Opportunity	Plan of Action	Person(s) Assigned	Time Frame
There is no pre-production quality planning resulting in excess scrap/rework and warranty problems.	Develop a procedure which includes design review, reliability predictions, life-cycle testing, etc.	Director of Engineering, Manager of Quality Engineering	10/93
Quality characteristics are not specified on engineering drawings which creates confusion in manufacturing as to which characteristics are most important.	Identify quality characteristics by critical - major - minor categories on engineering drawings.	Manager Design Engineering, Quality Manufacturing	1/94
There are constant errors made in manufacturing due to specification of wrong parts.	Define bill-of-materials system.	Director of Engineering, Manager Manufacturing Engineering, Manager Quality	3/94

Figure 2 Portion of a quality plan.

C. The Quality Board

Once a quality policy has been adopted and the quality plan approved by top management, execution of the plan starts. Unfortunately, it is during the management, execution of the plan starts. Unfortunately, it is during the crucial state of execution that many quality programs fail to deliver.

The fault for that can be laid at many doorsteps, but frequently the problem arises because top management thinks the quality program is now on "automatic." They often believe that a quality policy and quality plan are all that are needed to make things happen.

But the successful execution of quality plans is much the same as the successful execution of plans aimed at productivity, costs, and other critical company needs. Plans are simply not enough to get the job done. Controls and follow-up are the remaining prime ingredients.

The quality board is a proven way to keep top management in constant touch with progress—or lack of progress—of the quality plan. It opens doors of communication with companywide management and provides an opportunity for the quality assurance manager to keep programs on track.

The quality board should have the responsibility to support the implementation of the quality plan and to provide continuing direction for the company's quality programs. Other subjects for discussion can range from policy matters to the handling of major company problems and any other matters essential to successful accomplishment of the quality policy.

Typically, the board would be composed of the directors of quality assurance, manufacturing, engineering, and marketing, with the general manager as chairman. By placing chairmanship of the committee in the hands of the general manager, the quality board is assured of a "balanced" look at the issues presented. Since the general manager is the person vested with primary responsibility for achievement of quality goals, it makes good sense to have him or her run the top quality committee of the company.

D. Quality Assurance Reporting Relationships

The final major ingredient of effective communications with top management involves organization. If quality assurance reports to anybody other than the general manager, communications will be hampered. There is no contesting the fact that the quality message will not come across as clearly; it will be filtered through the eyes and ears of another manager (typically of manufacturing or engineering), who has several other balls to juggle. The message the quality assurance manager wants to get across is bound to be difused and weakened.

Another strong argument can be made for a direct reporting relationship of quality assurance to the general manager. It is simply human nature to expect the general manager to place more emphasis on those functions reporting directly to him or her. Should quality assurance be relocated to a lower management level, the quality program will just not receive the time and attention it demands to assure its contribution to profitability. If the general manager is going to be held responsible for attainment of quality goals, he or she must have direct access to the quality function, and it must be placed on a level that assures its objectivity. That objectivity can be misplaced if quality assurance reports to an organizational component (such as engineering or manufacturing) which has other primary goals.

III. REPORTING TECHNIQUES

While one of the historic problems of quality assurance has been communications with top management, another concern has been reporting techniques. What information needs to be reported to control quality, how often, and how should it be presented? Getting the right answers to those questions has caused many a headache for quality professionals.

Too often, either too much or too little information has been supplied by quality assurance. Failure to utilize some basic principles of reporting control has, in essence, shut a communications door—a door that quality assurance—needs open.

Improvement in reporting techniques can be obtained through a study of:

- The selection of information needed to control operations
- How to present that information
- Reporting that information to different organization levels

A. Selecting the Information to Report

Other than special reports, there are three basic and essential performance reports: (1) cost of quality, (2) lot acceptance rates, and (3) process average defective.

Cost of Quality

Cost of quality (COQ) is a valuable tool in the quality practitioner's kit. It allows him or her to communicate with all company levels and functions regarding just how well the quality program is progressing. It is expressed in costs, and costs are universally understood. If a company does not use COQ, its quality manager is at a severe disadvantage. His or her claims to accomplishments will probably be subjective, and subjective claims are usually challenged. COQ—a quantifiable measure—removes all the subjectivity.

When starting a COQ program, the quality assurance manager may need to gather the numbers personally to show people how it is done. When the COQ report gets off the ground, however, it is best to have it published by the finance department. That will lend the numbers respectability; they will remain unquestioned. A quality assurance manager who publishes his or her own numbers is always suspect in others' eyes.

Lot Acceptance Rate

The lot acceptance rate is a measure of importance in any business or industry where parts can be measured in discrete numbers. It is a number of significance to the production facility (and that could be a clerical station as well as a manufacturing operation). The lot acceptance rate tells production people how much work must be sorted to cull defective parts. It is, therefore, a rough measure of their labor costs for defective work, as well as an indication of the scrap and rework problems being experienced. If, for example, a machine shop processes 1000 lots during the week, and if the lot rejection rate is 10%, manufacturing then knows that it will incur excess labor costs for 100 lots to sort out defective work. The lot acceptance rate is a prime method of communication with manufacturing management, because they will understand the penalty incurred for poor quality. The manufacturing manager will probably know within 5% the costs that are being absorbed, because the manager's staff did not pay close attention to the quality aspects of their jobs. Sorting should be both handled by and charged to manufacturing so that they are penalized for poor-quality work. If sorting is performed by quality assurance, manufacturing will not pay close attention to quality, and the lost acceptance rate criterion will lose much of its meaning.

Process Average Defective

The process average defective is a measurement of how well a given process is turning out acceptable quality products. Thus it is a measurement of great significance to the quality practitioner as well as the manufacturing engineer and product engineer. It is an indication of where they must concentrate their energies.

The process average defective is based on the number of samples measured by inspectors (or operators). It is explained as follows:

Week: April 10th
Department: Drilling
Samples inspected: 500
Samples rejected: 25
Process average defective: 5%

In this example, inspectors in the drilling department of a machinery manufacturer inspected 500 parts during the week and found 25 pieces defective. Dividing the 25 defective pieces by the 500 sampled pieces and multiplying $\times$ 100 results in a process average defective of 5%. The process is therefore estimated to be producing 95% acceptable parts (assuming, of course, use of valid sampling plans and random selection procedures).

This figure is useful to the manufacturing engineer, because it relates the adequacy of tools, gages, and fixtures, and demonstrates over a period of time the condition of the equipment. The product engineer is similarly interested, because it demonstrates the effectiveness of the design and the adequacy of the tolerances. The quality engineer will also watch the process average defective so as to reduce scrap and rework on those processes exhibiting high defect rates.

B. How to Present the Information

One basic rule should govern the presentation of quality information aimed at getting corrective action: Keep it simple. Reports should be easily understood by all levels of the organization, and information presented should not be cluttered with peripheral information. It is helpful to present information in graphical form so that trends can be quickly detected.

Figure 3 is a monthly lot rejection report for Quality Goods' machine shop. The top left-hand side of the report displays the salient numbers: current monthly performance, year-to-date performance, and the previous year's average. What could be simpler? At a glance the reader can compare current performance with last year's rate to see if any improvement has been made.

The next section displays the graph. It is apparent that lot rejections have stabilized somewhat during the year and that every month was below the previous year's average. Again, the simplicity of the salient numbers and the graph provide for good communications. It would be difficult to misinterpret their meaning.

The final section of the lot rejection report for the machine shop shows the top defects and top causes of defects for the month. Notice that the top two defects

MACHINE SHOP

Month Ending 11/30/92 Plant: Houston

Current Month: 4.8%
Year-to-Date: 4.1%
Previous Year Avg.: 6.0%

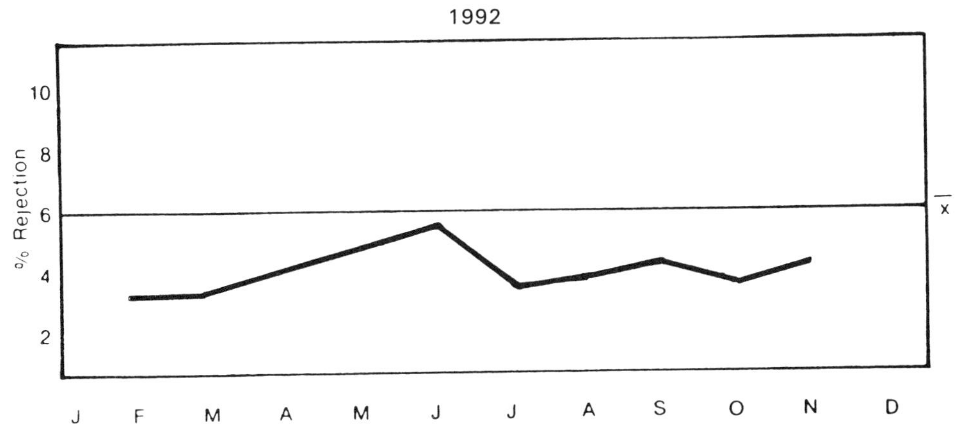

November Top Defects	
Drilling - Incorrect Hole Location	46%
Dimensional Errors - Millers	29%
Missed Operations	9%
Other	16%
	100%

November Top Causes	
Operator Error	60%
Tooling Problem	25%
Materials Problem	10%
Other	5%
	100%

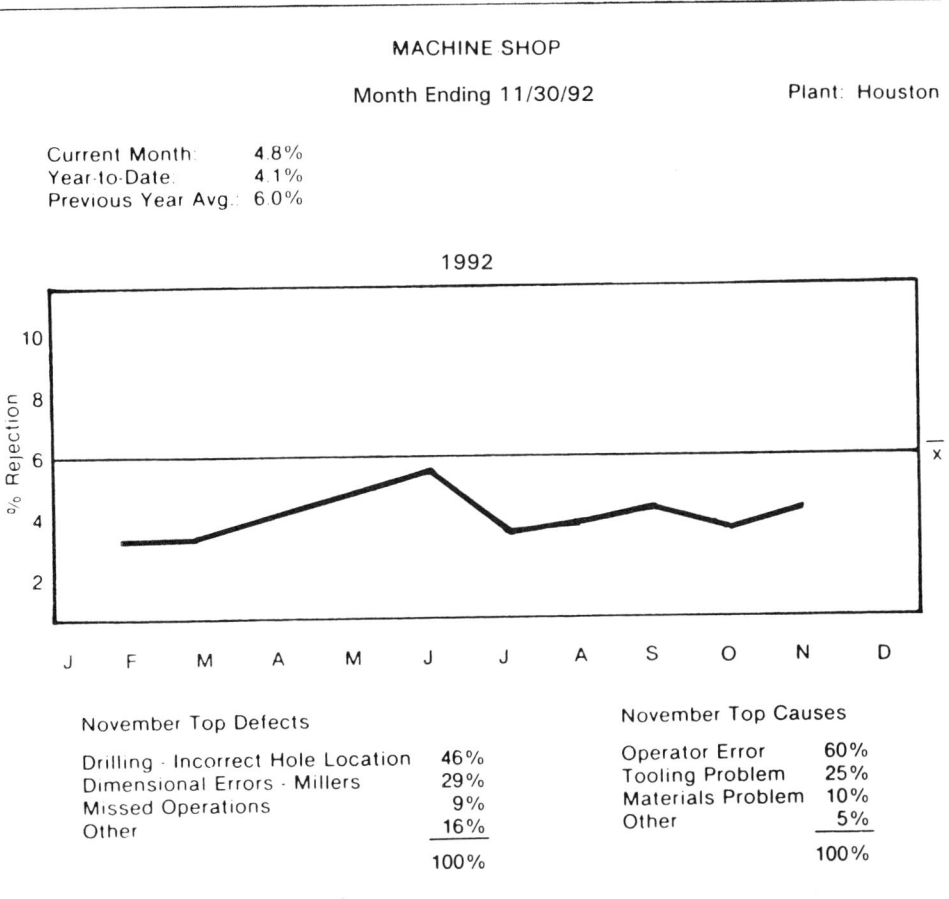

Figure 3 Quality Goods, Inc., lot rejection rates.

for November accounted for 75% of the rejections and that the top two causes constituted 85% of the causes for rejection.

What could be simpler? The entire report can *easily* be interpreted by anybody from machine operator to company president, yet all the information needed to generate action is present. From the report, you can see where you have been, where you are now, and where you are heading. You can also see the top problems and their causes. Communication is instant and clear.

C. Management Reports

The larger a company becomes, or the more complex its operations are, the greater the need to present succeeding levels of the company organization with differing amounts of reporting detail. Since the purpose of reporting is for control and improvement, the president will need summary information of all operations and the production supervisor will need detailed information of his or her operation alone. Figure 3 is a good example of the kind of information the production manager needs to control the shop's quality performance. The machine shop supervisor, on the other hand, will need the same information as shown, but applied to his or her individual sections. The machine shop supervisor will also need specific listings of quality performance by operator and by machine. That level of detail allows zeroing-in on problem people and equipment. Pareto information by part number will also help focus on major quality problems.

Normally, summary reports for general managers will take the form shown in Figure 4. Section A of the report summarizes quality performance for all of the individual plants of Quality Goods, Inc., and shows the total plant average. Although this report is issued monthly, many companies prefer to have weekly reports. The shorter the time interval of control reports, the faster the response in correcting problems. A good example is December's 12.5% lot rejection rate for the Pittsburgh plant. If this report had been issued weekly, it would have received prompt manage-

QUALITY GOODS, INC.

QUALITY PERFORMANCE

Month Ending 11/30/95

A. Summary of Month & YTD

	Lot Rejection %			Process Avg. Defective %		
Plant Location	Nov.	YTD	95	Nov.	YTD	95
Houston	7.4	8.2	7.7	3.4	2.9	3.9
Pittsburgh	12.5	6.5	7.3	5.7	2.8	3.6
Los Angeles	4.5	7.1	8.4	4.2	3.3	3.7
All Plant Avg.	7.8	7.3	7.8	4.4	3.0	3.7

B. Outstanding Problems, Solutions, Opportunities

1. Heavy rejection rate at Pittsburgh - assembly and automatic screen machines. New station fixtures now being installed at assembly and screw machines now in planned maintenance overhaul.

2. The new preventive action program aimed at reducing scrap and refinish has just completed its third month at Los Angeles, and the results have been encouraging. Lot rejections were down to 4.5%, the lowest rejection rate ever.

Figure 4 General manager's summary report.

ment attention and quite possibly the rate would have been lower by the time the month ended.

Section B of the report explains the problems, solutions, and opportunities in quality performance. Note the explanation for the problems found at the Pittsburgh plant, as well as an explanation of the low 4.5% November lot rejection rate at the Los Angeles plant.

A refinement to this report could include lot rejection and process-average-defective goals by plant. Some general managers prefer to issue specific goals, whereas others use the previous year's average as a base and ask for a lower rate. It all depends on the extent of the quality problems as well as the "gold mines" available.

IV. COMMUNICATION FOR PREVENTIVE ACTION

Quality assurance is bombarded with multitudes of problems but should not surrender to the impulse to spend its time dashing around madly putting out all the fires. Quality assurance *must* always take the lead, steering other departments along the path of problem resolution and problem prevention. Nobody else is going to do it, at least not to the extent that quality assurance will. Putting out fires is not enough. The fire will come back.

Preventive action always starts with identification of areas of opportunity (the major, high-cost quality problems). The action must be organized to attack the problems successfully. This organization usually involves a focusing of efforts on (1) customer quality problems; and (2) the costs of scrap, rework, and sorting of products in-house.

A. Customer Quality Problems

These problems involve complaints, warranty, retrofits, upgrades, and any other mechanism by which customers express their dissatisfaction with quality. It should be noted that many times customers and the company may look upon quality standards differently. Shoddy quality can mean different things to customers than to a company. If, for example, a trend in customer complaints is noted for paint problems (sags, runs, thickness, peeling), a peek at company specifications might show some latitude for quality acceptance—an obvious dichotomy with customer expectations. Yet these types of differences tend to occur frequently, if only because specifications are dynamic. Specifications can change for many reasons; notably for cost reduction, but specification changes may overlook potential customer complaints.

Some method, therefore, needs to be devised to look at outgoing products with "customer eyes." The product audit, which provides management with an indication of the outgoing quality level, is probably one of the best methods. It is composed of quality characteristics obtained from marketing's field service and quality assurance (Fig. 5). Those characteristics are a reflection of what must be controlled to assure customer satisfaction. As such, they are not derived from product specifications. Where the two do not mesh, quality assurance, marketing,

QUALITY GOODS, INC.

PRODUCT AUDIT

Plant: L.A. Date: 10-15 Auditor: L. Jones

Machine Audited: 4021-R Possible Points: 470

Customer: RLF Actual Points: 435 Rating: 93%

Quality Characteristic	Possible Points	Actual Points
Control Unit Tension Device	100	100
Control Unit Actuator	50	40
Control Unit Gear Train	100	100
Automatic Lubricator	50	50
Cosmetics - Paint	50	25
Cosmetics - Sheet Metal Finish	10	10
Cosmetics - Welds	50	50
Support Block Positioner	10	10
Turnaround Device	50	50
Total Points:		435

Critical (100 points) - Could cause personal injury, result in machine downtime of 48 hours or more, or result in warranty claims over $500.

Major (50 points) - Could cause machine downtime under 48 hours, result in warranty claims under $500, or create customer dissatisfaction with cosmetics.

Minor (10 points) - Minor defects in cosmetics or workmanship not resuling in warranty claims or machine downtime.

Figure 5 Quality characteristic ratings.

and engineering jointly decide what specification changes are necessary to agree with customer expectations for the product.

Product audits are taken randomly by quality assurance. As shown in Figure 5, each quality characteristic is rated as critical, major, or other and assigned a possible point value. Those point values are explained on the lower third of the form.

The product auditor evaluates the product selected and assigns earned points based on what type of job was done. The auditor can assign portions of the possible point total if, in his or her judgment, some of the job was done correctly. The actual rating is made characteristic by characteristic on the middle section of the form.

A final point total is tabulated and shown opposite "actual points" on the top of the form. This total is divided by the "possible points" total and multiplied by 100 to arrive at the rating. In this example, the rating was 93%.

The product audit is a workable device to communicate outgoing quality levels with management people and operators alike. It alerts the organization to the needs and desires of customers. The product audit is an effective tool for corrective action.

Since the product audit is an in-house tool, quality problems found by customers must be handled differently. There are numerous methods for investigation, compilation, and reporting of customer quality problems. One of the major roadblocks to resolving these problems and invoking preventive action occurs because of this sheer mass and a lack of organization in attacking the problems.

There are two basic methods of handling customer quality problems. One involves use of a small group of marketing, manufacturing, product engineering, and quality assurance representative. Their purpose is not to solve the problems, but to assign priorities for corrective action to the functional departments bearing responsibility. This method assures not only good communications regarding customer complaints, but also a total business approach to the problem resolution. It will, in other words, allow the company to deal with the most pressing problems first, and handle the balance in a descending order of importance.

The other method assigns quality engineers to product lines, allowing them to assign priorities for corrective action, investigate and analyze customer complaints, and work with the functional departments for corrective action.

B. Internal Failures

There are also several different professional approaches from which to select when combating scrap, rework, and sorting costs incurred in manufacturing. Either the team approach or the quality engineer approach mentioned earlier is feasible. In companies with extensive failure costs, both approaches are utilized successfully. The team approach should focus on specific manufacturing engineering, quality assurance, and materials. Either method will generate results and open up communication channels for preventive action techniques.

5
Managing Quality Costs

JOHN JOURDAN HELDT
Free Lance Reliability Service
San Jose, California

I. INTRODUCTION

Good quality costs less. No matter what type of business you are in,[1] it will always be more efficient and more cost-effective when it isn't impeded by the wastes caused by such problems as scrap and rework. Businesses that are already profitable might realize profits multiplied by four without increasing revenue. Some firms might be able to reap a 30% or more increase in the services they are able to provide or in the use of their facilities.

An added benefit of making "quality pay" is better product and service quality. It's obvious that products are always better when they haven't been reworked, and services are always more satisfying when performance is first rate the first time.

It is difficult to establish an effective quality cost system. It is down right aggravating trying to tweak that system for the best use of money and "man" power. But the "gold"[2] derived from such a system is a bonanza that more than repays the effort. Figure 1 describes the general guidelines typically used in a quality cost program.

[1] Optimizing quality costs enriches service organizations as well as commercial companies. Nonprofit groups as well as commercial ventures can benefit by eliminating the waste of scrap and rework.

[2] J. M. Juran first described the Avoidable Costs of Quality as GOLD in the MINE in Ref. 1.

1. GRAPH QUALITY COSTS
   Collect data
   - Present cost data as a ratio to revenue, manufacturing costs, or sales
   - Or select an index for your activity

2. ANALYZE DATA FOR RESULTS AND TRENDS
   Report trends to management
   Interpret graphs for affected groups
   Develop historical data to be used as standards

3. SELECT FERTILE AREAS FOR SAVINGS
   Stack cost reduction targets on a Pareto Chart
   Review inspection tasks for potent savings

4. DRIVE THE SAVINGS PROGRAM
   Oversee the Cost Savings Teams
   Reward work well done
   Update management of all achievements

5. MAINTAIN AND REFINE THE WINNING FORMAT

Figure 1 Guidelines.

II. KINDS OF QUALITY COSTS

Generally, quality cost categories and cost elements are defined in this chapter in the same context as defined by Feigenbaum (2):

A. Costs of Controlling Quality

These costs are generally broken into two categories:

1. Appraisal costs. Appraisal costs are the costs of inspection and test. Note that inspection and test are appraisal costs only when they are *first time* inspection or test. Inspection or test becomes a failure cost when it is reinspection or retest (i.e., the inspection of rejected and reworked parts is a failure cost).
2. Prevention costs. Prevention costs are generally considered to be the costs of quality management (but not necessarily the quality *manager*): that is, those things that are done to keep defects from happening. Prevention costs include quality planning, test engineering, reliability engineering, data analysis, and—especially—training.

B. Costs of Failure

These costs also are generally broken into two categories:

1. Internal failure. Internal failure costs are the costs of product defects and failures that are detected before the product or service is delivered to the customer. Typical internal failure costs include the cost of scrap, the cost of repair or rework, retest and reinspection; material review board costs, and the like.
2. External failure. external failure costs are incurred when products or services of less than acceptable quality are delivered to the customer. Typical external failure costs include warranty cost, field engineering expenses, field failure expenses, returned goods costs, complaint adjustment costs, corrective action costs, allowances, and other charge backs.

There is nothing sacred about using these four listings. [Philip Crosby, for example, uses two categories he calls the price of conformance (POC) and the price of nonconformance (PONC)]. The four categories used here, however, as outlined by Feigenbaum and modeled by Frank Gryna (see Figs. 4.1 and 4.2 in Ref. 3), have become almost universal in their use. This universality means that data from one company can be directly compared to that of another company—at least when similar products are involved. The ratio comparisons, however, have little meaning when different quality cost categories are used.

For example, in recent years, there has been a trend to add a category that has to do with white collar quality costs. Sometimes the added category is simply unique to that industry. When "nonstandard" quality cost systems are evaluated, the ratios between categories are used for historical data comparison. That is, past data are used only for comparison with current data rather than as measures in themselves.

III. GATHERING QUALITY COST INFORMATION

Most planned quality cost analysis programs die a bornin', because there are no quality cost data base to analyze. Normally, the controller or chief accountant is expected to gather and chart quality cost data. Then, the quality manager can make the analysis and set up the system to begin to reduce the causes of such problem areas as scrap and rework.

Controllers and accountants tend to assign low priorities to quality cost data tabulation, because they are not aware of how much profit can be derived by a potent system for waste reduction. The finance and accounting people, however, are more likely to get involved when the value of collecting and tabulating quality costs data is revealed to them. Remember, there is no way to divine how much gold is in the mine with no data to analyze.

One way to start a quality cost program in the face of apathy is to dig the facts out of the existing records. For areas without records, estimates can be based on an audit of the category. Cost of quality estimates that are at least 85 or 90% accurate are sufficient to show that reductions in the costs of scrap and rework will certainly result in higher profits.

These estimates may be used to convince the CEO (as well as controllers, accountants, and other managers) that hard, smart work applied to the reduction of wasted dollars can make a big difference in profits is as little as six months.

IV. COST OF QUALITY ILLUSTRATIVE EXAMPLE

First Estimate: January

Figure 2 is a cost of quality worksheet for a hypothetical company and is used to illustrate how quality cost data can be assembled. Most of these data can be considered to have been extracted from reports published by the accounting department. In the report shown in Figure 2, the data that were taken directly from accounting reports is shown in boldface type.

Under *prevention* costs, the costs associated with planning quality tasks (i.e., *QE planning*, *test engineering*, and *reliability engineering*)[3] are considered to have been taken directly from budget and payroll data. These costs would be very close to 100% accurate.

Data system costs, the fourth item under prevention costs, are estimated. These costs are considered to have come from a data processing department report that lists all data processing costs but does not break out the costs of individual tasks. The portion of the costs allocated to the quality data system was assessed at $560. This figure is judged to be within ±5% of the actual cost.

Similarly, all of the data under *appraisal* cost, except that for *calibration* services, was considered to have been taken from payroll wage reports. Calibration work is contracted to an outside supplier. The $3019 cost figure for calibration work came from purchasing department data. Appraisal category data thus are actual and not estimated. "Actual" data values are shown in boldface type.

Manufacturing rework and *inspection of reworked product* are internal failure costs, because they are costs of repairing and retesting failed product. Since these elements were not recorded separately in the accounting report, their costs are estimated. The estimated costs were judged to be at least 85% accurate. *Engineering change rework and repair*, *scrap*, and *MRB* (material review board) are considered to be accurate costs, since they would have been taken from the existing accounting

---

[3] When the actual expense item is a wage figure, there is usually a *burden* factor connected to it. This burden factor means that a quality engineering (QE) planner's cost to our company is much more than the amount that he or she is paid. The burden figure represents the cost of things like health insurance, vacation pay or holidays, sick time, stock benefits, and management and administrative salaries. In this case, the burden for a QE planner is 1.5 times his or her hourly wage. Thus, the burden factor for the QE planner is 2.5 (i.e., the wage plus 1.5 times the wage or 2.5 times his wage). The burden factor is usually different for differing wage categories. For instance, the manufacturing test burden factor is 4, probably because the manufacturing test wages (under appraisal costs) are lower than other kinds of wages, which makes the burden a higher percentage of the manufacturing wage plus burden.

Month's Revenue:	$3,541,115		Month's Profit:	$141,701	
			Profit as percent of Revenue:		4.00%
	Cost of Quality Illustrative Analysis				
JANUARY		Actual	Burden	Total	Percent
First Estimate		Expense	Rate	Expense	of
		(When Applicable)			Revenue
Prevention Costs:					
	Q E Planning	**$4,600**	**2.5**	**$11,500**	
	Test Engineering	**$6,400**	**2.1**	**$13,440**	
	Reliability Engineering	**$900**	**2.5**	**$2,250**	
	Data System Costs	$560	n/a	$560	
	Total Prev Cost:			$27,750	0.78%
Appraisal Costs:					
	Inspection	**$13,822**	**2.3**	**$31,791**	
	Manufacturing Test	**$4,162**	**4**	**$16,648**	
	Calibration Services	**$3,019**	**n/a**	**$3,019**	
	Quality Audit	**$3,271**	**2.5**	**$8,178**	
	Total Appraisal Cost:			$59,635	1.68%
Failure Costs:					
Internal Failure Cost:					
	Manufacturing Rework	$12,376	3.5	$43,316	
	Eng Change Rework & Repair	**$3,611**	**3.5**	**$12,639**	
	Inspection of Reworked Product	$1,196	2.3	$2,751	
	Cost of Scrap Material	**$5,642**	**n/a**	**$5,642**	
	Material Review Board Costs	**$2,555**	**n/a**	**$2,555**	
	Total Internal Failure Cost:			$66,902	1.89%
External Failure Cost:					
	Product returns	**$334,711**	**n/a**	**$334,711**	
	Field Engineering Costs	$52,760	3.5	$184,660	
	2% fund (Warranty Reserve)	**$70,822**	**n/a**	**$70,822**	
	Allowances (i. e. Discounts)	**$15,892**	**n/a**	**$15,892**	
	Total External Failure Cost:			$606,085	17.12%
TOTAL QUALITY COSTS:				$760,373	21.47%
	Total dollar amount of COQ Estimated:			$231,287	
	Percent of Cost of Quality Estimated:			30.42%	

Figure 2 Cost of quality worksheet (January, first estimate).

reports. *Field engineering* costs have been estimated. Half of these costs are charged to marketing and the remainder to external failure.

First Approximation—Rule of Thumb: To make a quick estimate of an organization's wasted money, audit the value of all returned goods and multiply that amount by two. In the discussion above, the value of returned goods is $334,711 times 2 = $669,422. This estimate is relatively close to the calculated value of $760,373.

V. OPTIMUM QUALITY COST RATIOS[4]

The generally accepted optimum ratio of quality costs is given by:

Prevention + Appraisal = Internal Failure + External Failure
or a relationship of 1 : 1.

The calculated relationship from the data of Figure 2 is:

$$27750 + 59635 \stackrel{?}{=} 66902 + 606085$$

By constructing the ratio of 27750 + 59635/66902 + 606085 = ?, the result is 87385/672987, or a relationship of about 7.7 to 1.0. This ratio is way out of line. A nearly 8 to 1 ratio of failure costs to appraisal and prevention costs clearly indicates that a little extra money spent on prevention and/or appraisal in the short run will pay off as a fairly large reduction in the total cost of quality in the long run.

The immediate action should be a proposal to top management outlining detailed plans of the changes that can be made in the prevention and appraisal tasks coupled with detailed improvements expected in the internal and external cost areas. When the "pitch" is made, however, remember that managers and certainly accountants pay more attention to the language of money than to any other words.

Show management the expected return on investment that can be realized by putting your plan into action, and they will back your action. Also show them how much money is being thrown away on poor quality. Outline a sound strategy for starting an effective tracking system. Then get everyone involved in closing up the ratholes where the wasted money has been going. No one says that it will be easy, but substantially improving the bottom line will make all of the tough going more than worthwhile in the end.

VI. COST OF QUALITY ILLUSTRATIVE EXAMPLES

A. April

Figure 3 shows the April cost of quality worksheet for the improvement program begun in January. The April analysis was chosen because the improvements due to increased prevention measures are beginning to show. The data have been *normalized*[5] to hold revenue constant. Figure 4 shows an example of data normalizing. Data are normalized by adjusting the various entries to reflect what would be expected to occur if the base against which the data is compared did not, in fact, change from period to period. Normalization is done by multiplying all of the entries by a normalization factor. The normalization factor is the ratio of the base value, measured in the base period, to the comparable value in the evaluation period. For example, the base value used in this chapter is revenue.

---

[4] Based on Figures 4.1a and 4.2 in Ref. 3.

[5] Normalization of quality cost data makes comparison easy. To make it easy to understand this powerful tool, the normalization procedure is illustrated in Figure 4.

Month's Revenue: $3,541,115		Month's Profit:	$284,091	
		Profit as percent of Revenue:		8.02%
Cost of Quality Illustrative Analysis				
APRIL	Actual Expense	Burden Rate	Total Expense	Percent of Revenue
	(When Applicable)			
Prevention Costs:				
Q E Planning	**$5,400**	**2.5**	**$13,500**	
Test Engineering	**$7,040**	**2.1**	**$14,784**	
Reliability Engineering	**$900**	**2.5**	**$2,250**	
Data System Costs	$616	n/a	$616	
Total Prev Cost:			$31,150	0.88%
Appraisal Costs:				
Inspection	**$13,822**	**2.3**	**$31,791**	
Manufacturing Test	**$4,162**	**4**	**$16,648**	
Calibration Services	**$3,019**	**n/a**	**$3,019**	
Quality Audit	**$3,271**	**2.5**	**$8,178**	
Total Appraisal Cost:			$59,635	1.68%
Failure Costs:				
Internal Failure Cost:				
Manufacturing Rework	$9,282	3.5	$32,487	
Eng Change Rework & Repair	**$2,890**	**3.5**	**$10,115**	
Inspection of Reworked Product	$957	2.3	$2,201	
Cost of Scrap Material	**$4,520**	**n/a**	**$4,520**	
Material Review Board Costs	**$2,555**	**n/a**	**$2,555**	
Total Internal Failure Cost:			$51,878	1.47%
External Failure Cost:				
Product returns	**$207,769**	**n/a**	**$207,769**	
Field Engineering Costs	$52,760	3.5	$184,660	
2% fund (Warranty Reserve)	**$70,822**	**n/a**	**$70,822**	
Allowances (i. e. Discounts)	**$11,428**	**n/a**	**$11,428**	
Total External Failure Cost:			$474,679	13.40%
TOTAL QUALITY COSTS:			$617,343	17.43%
Total dollar amount of COQ Estimated:			$219,964	
Percent of Cost of Quality Estimated:			35.63%	

Figure 3 Cost of quality worksheet (April).

January's revenue is $3,541,115. The value to be compared is April's revenue, which was $3,895,227. Dividing January's revenue by April's revenue results in a normalization factor of 0.91. To normalize April's data, all of April's quality costs are multiplied by 0.91. The result is quality costs that would have been expected to occur if April's revenue had been the same as January's. Normalization

Month's Revenue: $3,541,115 (1)		Month's Profit:	$141,701	
		Profit as percent of Revenue:		4.00%
Cost of Quality Illustrative Analysis				
JANUARY First Estimate	Actual Expense	Burden Rate (When Applicable)	Total Expense	Percent of Revenue
Prevention Costs:				
Q E Planning	$4,600	2.5	$11,500	
Test Engineering	$6,400	2.1	$13,440	
Reliability Engineering	$900	2.5	$2,250	
Data System Costs	$560	n/a	$560	
Total Prev Cost:			$27,750	0.78%
Appraisal Costs:				
Inspection	$13,822	2.3	$31,791	
Manufacturing Test	$4,162	4	$16,648	
Calibration Services	$3,019	n/a	$3,019	
Quality Audit	$3,271	2.5	$8,178	
Total Appraisal Cost:			$59,635	1.68%
Failure Costs:				
Internal Failure Cost:				
Manufacturing Rework	$12,376	3.5	$43,316	
Eng Change Rework & Repair	$3,611	3.5	$12,639	
Inspection of Reworked Product	$1,196	2.3	$2,751	
Cost of Scrap Material	$5,642	n/a	$5,642	
Material Review Board Costs	$2,555	n/a	$2,555	
Total Internal Failure Cost:			$66,902	1.89%
External Failure Cost:				
Product returns	$334,711	n/a	$334,711	
Field Engineering Costs	$52,760	3.5	$184,660	
2% fund (Warranty Reserve)	$70,822	n/a	$70,822	
Allowances (i. e. Discounts)	$15,892	n/a	$15,892	
Total External Failure Cost:			$606,085	17.12%
TOTAL QUALITY COSTS:		(3)	$760,373	21.47%
Total dollar amount of COQ Estimated:			$231,287	
Percent of Cost of Quality Estimated:			30.42%	

(a)

Figure 4 Data normalization. (a) Actual January data (normalization base). (b) Actual April quality cost data. (c) Normalized April quality cost data.

is often necessary to clarify increases or decreases in quality cost categories from period to period.

Figure 4a shows the actual January quality cost data, which are used as the base on which the April data are normalized. Figures 4b and c show the actual April data and the normalized April quality cost data.

Month's Revenue:	$3,895,227 (1)	Month's Profit:	$313,204		
		Profit as percent of Revenue:			8.04%
Cost of Quality Illustrative Analysis					
APRIL Profit Doubled		Actual Expense	Burden Rate	Total Expense	Percent of Revenue
		(When Applicable)			
Prevention Costs:					
Q E Planning		$5,940	2.5	$14,850	
Test Engineering	(2)	$7,744	2.1	$16,262	
Reliability Engineering		$990	2.5	$2,475	
Data System Costs		$678	n/a	$678	
Total Prev Cost:				$34,265	0.88%
Appraisal Costs:					
Inspection				[illegible]4 970	
[illegible]				[illegible]	
2% fund (Warranty Reserve)		$77,905	n/a	$77,905	
Allowances (i. e. Discounts)		$12,571	n/a	$12,571	
Total External Failure Cost:				$522,147	13.40%
TOTAL QUALITY COSTS:			(3)	$679,077	17.43%
Total dollar amount of COQ Estimated:				$241,961	
Percent of Cost of Quality Estimated:				35.63%	

(b)

Month's Revenue:	$3,541,115	Month's Profit:	$284,731 (4)		
		Profit as percent of Revenue:			8.04%
Cost of Quality Illustrative Analysis					
APRIL Profit Doubled		Actual Expense	Burden Rate	Total Expense	Percent of Revenue
		(When Applicable)			
Prevention Costs:					
Q E Planning	(2)	$5,400	2.5	$13,500	
Test Engineering		$7,040	2.1	$14,784	
Reliability Engineering		$900	2.5	$2,250	
Data System Costs		$616	n/a	$616	
Total Prev Cost:				$31,150	0.88%
Appraisal Costs:					
Inspection				[illegible]31,791	
[illegible]ield Engineer[illegible]				[illegible]	
2% fund (Warranty Reserve)		[illegible],822	n/a	$70,822	
Allowances (i. e. Discounts)		$11,428	n/a	$11,428	
Total External Failure Cost:				$474,679	13.40%
TOTAL QUALITY COSTS:			(3)	$617,343	17.43%
Total dollar amount of COQ Estimated:				$219,964	
Percent of Cost of Quality Estimated:				35.63%	

(c)

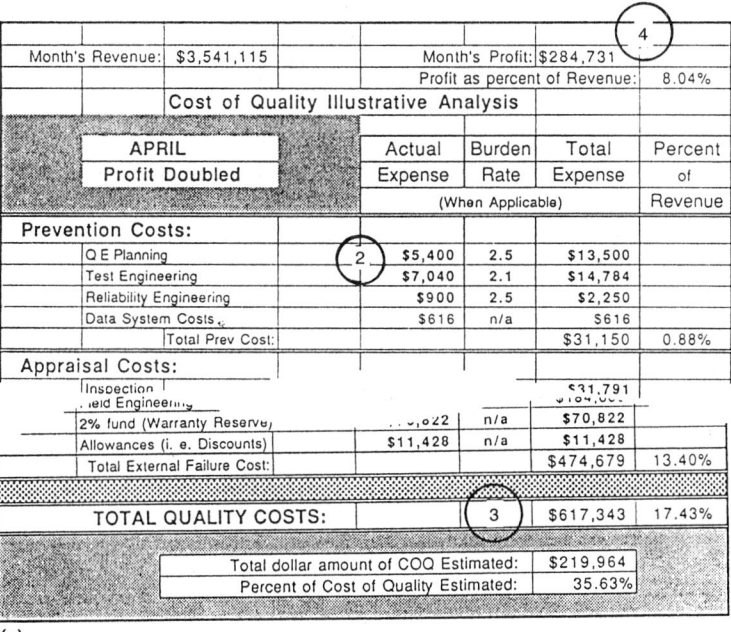

Figure 4 Continued

Summary of Normalization Steps (Circled numbers shown here refer to circled number on Figures 4a–c):

1. Divide January's revenue ① by April's actual revenue ① to determine the April normalization factor.

   $$\frac{3541115}{3895227} = 0.9090908$$

2. Multiply April's actual revenue by the normalization factor to get April's normalized revenue.

   ① (3895227)(0.9090908) = 3,541,115

3. Multiply each April's actual quality cost value ② to get each April's normalized quality cost value ②.
4. Subtract April's normalized total quality cost from January's total quality cost.

   760373 − 617343 ③ = 143,030

5. Add this difference (143,030) to January's Profit (141,645)

   143,030 + 141,701 = 284,731 ④

Since these normalized April values show the same revenue, the April profit figure reflects what would have happened if all sales conditions had been held steady. Actual profits for April are a little higher than this, but normalizing the data gives a little better picture of improvements due to success in reducing quality costs.

The April spread sheet shows a modest increase in prevention costs (about a 10% increase in QE planning and test engineering). The same number of people are doing inspection and testing—i.e., there is no increase in appraisal costs—but somebody must be doing it better and smarter, because failure costs, both internal and external, have been reduced. Total failure costs are down about 20%.

The difference between January's and April's total quality costs ($761,013 − $617,927 = $143,086) has been added to the April profit total. Net waste reduction goes directly to the bottom line—there is no place else for it to go. When this increase is shown as a percentage of revenue, the effect is dramatic. In this example, with no increase in sales revenue or production, the company's profit in April is more than 8% compared to 4% in January: This is an increase of 2 to 1 in the bottom line over a 4-month period.

B. December

Continued improvements bless the quality cost program. The December analysis is shown in Figure 5 to illustrate the ongoing trends. As in the April analysis, the data have been normalized in order to hold revenue constant.

It appears that failure costs have been lowered, probably because of increased appraisal activity. Relatively modest cost increases in inspection, manufacturing,

Month's Revenue: $3,541,115		Month's Profit:	$587,658	
		Profit as percent of Revenue:		16.60%
Cost of Quality Illustrative Analysis				
DECEMBER	Actual Expense	Burden Rate	Total Expense	Percent of Revenue
	(When Applicable)			
Prevention Costs:				
Q E Planning	$5,400	2.5	$13,500	
Test Engineering	$7,040	2.1	$14,784	
Reliability Engineering	$900	2.5	$2,250	
Data System Costs	$616	n/a	$616	
Total Prev Cost:			$31,150	0.88%
Appraisal Costs:				
Inspection	$15,316	2.3	$35,227	
Manufacturing Test	$4,523	4	$18,092	
Calibration Services	$3,019	n/a	$3,019	
Quality Audit	$3,564	2.5	$8,910	
Total Appraisal Cost:			$65,248	1.84%
Failure Costs:				
Internal Failure Cost:				
Manufacturing Rework	$3,692	3.5	$12,922	
Eng Change Rework & Repair	$1,723	3.5	$6,031	
Inspection of Reworked Product	$433	2.3	$996	
Cost of Scrap Material	$1,277	n/a	$1,277	
Material Review Board Costs	$2,555	n/a	$2,555	
Total Internal Failure Cost:			$23,780	0.67%
External Failure Cost:				
Product returns	$43,115	n/a	$43,115	
Field Engineering Costs	$22,330	3.5	$78,155	
2% fund (Warranty Reserve)	$70,822	n/a	$70,822	
Allowances (i. e. Discounts)	$921	n/a	$921	
Total External Failure Cost:			$193,013	5.45%
TOTAL QUALITY COSTS:			$313,192	8.84%
Total dollar amount of COQ Estimated:			$92,689	
Percent of Cost of Quality Estimated:			29.59%	

Figure 5 Cost of quality worksheet (December).

test, and auditing have helped to remove half of the waste from the internal and external failure costs. Again, the reduction in total quality costs has been added to the profit line to show that the percentage of profit has been doubled in the interim between April and December. This is remarkable when we remember that the profit had already been doubled between January and April. The gist of this illustrative example is that an intelligently applied quality cost program, starting from scratch, can often double the profit line and then double it again in a very short period of time.

One thing should be noted about this scenario: These analyses are merely a way of tracking improvements. The reasoning that was used to explain the technique might make it appear that miracles can happen with just a little more effort in the right place.[6] Generally speaking, nothing could be further from the truth.

A number of steps were required to achieve the gains described above:

- Improvements to the prevention system must include a solid method for listing and analyzing the kinds of defects and the modes of failure.
- The appraisal system must provide a classification of defects and modes of failure system to show which areas are the ripest for cost savings; i.e., use a Pareto Chart to identify the "vital few."
- A product improvement program must be started and pursued to eliminate the scrap and rework costs, thus transferring most of these costs directly into the profit structure. This task requires everyone's commitment to the effort. Top management's whole-hearted support is especially needed. Quality's role in this effort is more or less that of facilitator. In other words, Quality's job is more like aiming the effort rather than leading it; acting as a moderator instead of director.

VII. OVERVIEW: WHITE COLLAR QUALITY COSTS

"White collar" employees also make mistakes.[7] Mistakes and any lack of supervision and good procedures are directly related to the quality cost system. This discussion is included to provide a head start for those organizations in which white collar quality costs require differentiation.

Most of the material presented so far has dealt with the "conventional" quality cost program. Conventional quality cost programs deal almost exclusively with products and their manufacture. In recent years, however, emphasis has increasingly been placed on what are often referred to as "white collar quality costs." In fact, some agencies have added a fifth quality cost category for white collar quality costs.

White collar quality costs are generally categorized as those costs that occur when errors are made by the "office staff."

If a member of the sales department misreads a memo and grants a distributor an extra 15% discount rather than the 5% discount that was negotiated, the cost to the company is dollars that are just as real as those wasted in scrap. If a customer service representative enters part number 12J39 instead of 21J39, the parts are as useless to the customer as if they had arrived broken. Errors made by an engineer that result in prototype failures and a canceled contract are every bit as severe a

---

[6] That is, spending a little more money on prevention and appraisal will make a great improvement in the cost of rework and failures.

[7] Look no further than the spell checker on your word processor. (Most software suppliers even include a grammar checker program as a bonus!)

product failure as are weak welds made in the shop. Although they continue to be ignored by many companies, white collar quality costs are just as real as those created by manufacturing problems.[8]

REFERENCES

1. J. M. Juran. *Quality Control Handbook*, 1st ed. New York: McGraw-Hill, 1951.
2. A. V. Feigenbaum. *Total Quality Control*, 4th ed. New York: McGraw-Hill, 1983.
3. J. M. Juran. *Quality Control Handbook*, 4th ed. New York: McGraw-Hill, 1951.
4. H. J. Harrington, *Poor Quality Cost*. New York: Marcel Dekker, 1987.

FURTHER READING

D. J. Costa and J. J. Heldt. *Quality Pays*. Carol Stream, IL: Hitchcock, 1988.

J. J. Heldt, *Controlling Quality Costs*. White Plains, NY: The MGI Management Institute, 1986.

J. J. Heldt. *More Than Ever, Quality Pays*. Carol Stream, IL: Quality, 1994, p. 31.

---

[8] H. J. Harrington's *Poor Quality Cost* (4) outlines details for controlling white collar quality costs. This book is required reading for everyone who wishes to pursue this avenue for added cost reduction.

6
Supplier Cost Recovery

ROBERT W. GRENIER
Consultant
Springfield, Illinois

I. INTRODUCTION

Questions often asked by manufacturing managers include the following: Can we recover labor and burden dollars that are lost because of defective material that we received from a supplier, in addition to the basic cost of the material or parts? Can we also recover any other costs that are incurred while adding value to product that eventually proves to be unusable because of a supplier's defective material.

The answer to these questions is, yes. A well-managed purchasing department, assisted by the quality assurance department, can and should recover the majority of these costs. There well may be times when these costs have to be negotiated, but recovering supplier-defective material costs is definitely a purchasing responsibility.

Purchasing departments typically provide a number of reasons such as economical conditions or a single-source supplier for not recovering losses due to supplier-defective product. They should not, however, be influenced by outside sources to evade the responsibility for recoverying those costs. As for difficulties in dealing with suppliers, good, cooperative, suppliers will react cooperatively to requests for negotiation efforts as they have responded cooperatively in the past.

II. PURCHASING DEPARTMENT RESPONSIBILITY

The first premise that has to be made is that the purchasing department has the functional responsibility for procurement quality. Any supplier-defective product costs, therefore, should be part of the purchasing department's expense budget. This budget becomes one of the main measures of the department's annual activities. It is erroneous for general management to believe that major improvements can be made in reducing supplier-defective costs if these dollars are charged to the quality department's budget. The quality department does not make the choice of supplier, nor does it negotiate the recovery of supplier-defective dollars.

What is included in the term *gross supplier-defective dollars*? The category consists of the base costs of purchased materials and any associated labor and burden costs expended upon receipt; plus any additional lost labor, with its associated burden, if purchased parts or materials are found to be defective after the purchaser has performed work on them. This area would include removal and replacement of a defective item in an assembly area or at a customer site.

III. QUALITY ASSURANCE RESPONSIBILITY

What role does the purchaser's supplier quality assurance (SQA) group have in these negotiations if it does not have the main responsibility for the quality of purchased items? SQA has a major responsibility in assisting the purchasing department in reducing supplier defects and in recovering supplier-defect costs when these occur. Let's examine how a supplier-defect cost-recovery system would work.

Suppose that you accepted a position in the CJD Co. as quality manager. Let's also assume that CJD had no previous quality systems in place other than a basic inspection and sorting program.

In the initial systems review, one of the first things you should do is review the departmental expense budget for the last 12 months. If all the defective-purchased material costs of the company are listed in your budget, you must be ready to sell your boss on the idea that those costs are the responsibility of the departments that are responsible for taking action to reduce or eliminate the costs. As far as make versus buy decisions are concerned, the quality department does not design the parts; does not buy the parts; does not make the parts; does not assemble the parts; and does not service the product at the customer site.

What then, does the quality department do? The main activity of the quality department is to aid the other company departments in implementing prevention activities and in day-to-day routine quality responsibilities. With luck, your boss will understand and if supplier-defect costs are currently assigned to the quality department, they will be eventually allocated to the proper departments.

IV. ANALYSIS OF DEFECT COSTS

All supplier-defective costs should be analyzed to determine the various product quality levels. Special attention should be given to the gross supplier-defective costs. Divide these costs by total sales over the same time period to obtain a supplier quality-level percentage. Alternatively, the total dollar amount of purchased goods can be used to calculate the number rather than sales. Using purchase dollars rather than sales dollars eliminates nonappropriate costs such as marketing and engineering design that are not included in purchase dollars.

Once supplier-defect costs are established, the next step is to review your in-house supplier control system. It is important to determine how rejected incoming material is handled.

Many companies reject defective material at the location at which it is first discovered. The defective material is then dispositioned at this site by representatives of quality and other management functions—within 24 hours whenever possible. Too often defective supplier material becomes intermixed with internal discrepant material. The resulting pile is susceptible to gathering dust if there is not a proceduralized and strictly enforced defective-material system. Perhaps even worse, the scattered and intermixed supplier-defective material and in-plant defective material is sent to the scrap bins together. The result is that crucial information that should be fed back to the supplier for correction and retrieval of recovery costs is lost.

V. CONTROL SYSTEM RECOMMENDATIONS

Once supplier material has been found to be defective, it must be properly identified and immediately forwarded to a central area for disposition. This area should be under lock and key to prevent the defective material from entering the production stream.

On occasion, supplier material will be found to be defective at a customer site, usually when the customer has had problems and has requested failure analysis and corrective action. If at all possible, supplier representatives should be requested to review and analyze the material at the customer's site. This review can enhance cost-recovery opportunities. Then this material should be removed from the customer's site and returned to the manufacturer's central location for further processing.

If the agreed-on disposition calls for scrapping the parts of materials or returning them to the supplier, the purchasing department should process the paperwork as quickly as possible. The defective parts or materials should be on their way back to the supplier a maximum of 15 days from the date of the rejection. Any longer will impede immediate corrective action on future deliveries.

When there is an urgent need to meet production schedules, the parts or materials may need to be reworked in-house. The amount of rework time and other related costs must be estimated.

Armed with this information, the purchasing department, can then contact the supplier and notify them of the rejection and of the estimated cost of repairing or reworking the rejected materials. If these costs are reasonable, many suppliers will agree to the in-house rework or repair. It is often cheaper to pay the customer to do the repairs than to pay the costs of freight, paperwork processing, and the costs of repair/rework at the supplier's facility.

When the disposition is "use as is," a design engineer should sign off, denoting acceptance of the deviation. The supplier should be notified under any circumstances for future corrective action procedures.

During the continuing initial review of the defective materials–control system, the reject form should be analyzed to make certain that it is not simply a common scrap ticket. The form should provide a clear description of the discrepancy and subsequent disposition of "use as is," "rework," "repair," "scrap," and "return to vendor." If the form is large enough, it can be used as the method for describing the defect and as a return notice for corrective action on subsequent material. The complete history is now on one sheet. If the form is simply a scrap ticket, defective material designed "use-as-is," "rework," or "repair" is never processed for corrective action by the supplier. There is no impact on correcting future shipments.

The rejection declaration must be precise and clear, avoiding such statements as "bad," "porous," or "doesn't fit." Say what the problem is and be specific about dimensional or functional parameters.

Make sure that the responsible personnel are properly trained in the internal material controls system for rejected materials. This includes documentation procedures implemented by the material control department (or its equivalent) as well as the system for processing forms for supplier-defective material that would be issues by the purchasing department.

The next step is to analyze rejections. During the initial review of the system, one will find that the Pareto law of "significant few and trivial many" has definite application. The analysis will show that a few of your suppliers are generating the majority of your rejections. List those rejections along with the causes, the supplier's responsible, and the associated costs of labor, material, and burden. Once this is done, forward the data to the purchasing manager for his or her use in contacting the suppliers to improve their level of quality.

Even with a limited number of personnel, improvements can be made quickly by communicating solely with the major-problem suppliers.

Do not try to shot-gun activities with all supplies, as this will provide a limited payback and the improvement of incoming quality will be slow.

On a continuing monthly basis, the supplier quality assurance engineers, with their purchasing department counterparts, should review the progress of the problem suppliers. As a matter of good business, in reviewing the corrective action timetable, always try to use the supplier's plan. Do not dictate an unreasonable time frame, because it may be completely unachievable and destroy the desired corrective action efforts.

Another important step related to purchased material is the implementation of a procedure for dealing with new suppliers. A quality system survey should be conducted by representatives from the purchasing department and supplier quality assurance. After the initial contact by the purchasing department has been made and an agreement to the survey has been reached, perform the survey as quickly as possible. The following is a representative survey checklist:

Instructions

Rank each question 0–10 on observation of activity and stage of implementation as listed below. Summarize survey on summary sheet.

- 0 No; or the procedure or document does not exist or is not being performed.
- 1–2 Yes; but the procedure or document is outdated or is poorly or seldom performed.
- 3–4 Yes; but the procedure or document is not completely up-to-date or is not always performed.
- 5–7 Yes; the procedure or document is essentially up-to-date and is generally performed well.
- 8–9 Yes; the procedure or document is excellent, up-to-date, and virtually always performed very well.

10 Yes; the procedure of document is exceptional and is always performed extremely well.

(N/A = not applicable).

Section 1: Quality System

1. Is the quality function a separate and distinct part of the suppliers organization?
2. The quality organization is comprised of:
   _____ Inspectors _____ Technicians _____ Engineers
3. Are adequate written procedures established to define quality control operations and responsibilities?
4. Is there sufficient data available to confirm the quality control procedures are operational?
5. Does the supplier have an adequate drawing and change control program in effect?
6. Are adequate drawings and specifications available to manufacturing and inspection?
7. Are customer drawings and specifications utilized? Or does the supplier utilize own drawings and specifications which have been approved by the customer?
8. Is quality cost information reported and utilized by management?
9. Is classification of characteristics utilized and/or understood by supplier personnel.
10. Are quality personnel trained or qualified for specific assignment?

Section 2: Calibration Problem

1. Is a tool calibration program in operation?
2. Is all information equipment identified and traceable to records.
3. Are records of calibration maintained and up-to-date?
4. Are master standards traceable to National Bureau of Standards?
5. Are personal tools calibrated the same as company tools?
6. Are tools, gages, and test equipment adequately handled and protected to prevent damage?
7. Are metrology requirements periodically evaluated and new equipment added to upgrade capabilities?

Section 3: Metallurgical Controls

1. Are adequate metallurgical laboratory facilities available and utilized? Source is _______________
2. Are laboratory facilities and personnel adequate for the type of product produced?
3. Are laboratory procedures and instructions documented?
4. Are records of laboratory tests maintained and traceable to the products?

5. Laboratory capabilities include:

physical test	microstructure
chemistry	magnetic particle
hardness	Non Destructive Testing (NDT) (list)

Section 4: Incoming Material Control

1. Are quality requirements communicated to suppliers and subcontractors?
2. Are all incoming shipments submitted to inspection prior to storage or use?
3. Is each new job, issue change, or initial shipment from a new source submitted to FACI?
4. Is raw material tested to determine conformance to specifications?
5. Are incoming inspection instructions documented?
6. Are incoming inspection records maintained and up to date?
7. Is the validity of certifications verified periodically?
8. Are adequate gages and test equipment available to incoming inspection and is their calibration timely?
9. Is the acceptance status of each incoming shipment identified? How: Rough _____ Semifinished _____ Finished
10. Is supplier quality performance evaluated and information fed back to suppliers?

Section 5: In Process Controls

1. Are work instructions documented and available on the job?
2. Is the required inspection tooling specified for each job and is it calibrated and available to the operator?
3. Is production tooling and gaging qualified prior to or at the time of first use?
4. Are inspection instructions documented and do they include specific characteristics and checking frequencies?
5. Are applicable drawings and specifications available at each work station where required?
6. Is each setup qualified prior to running production? Arranged by: Operator _____ Supervisor _____ Inspector _____
7. Do operators maintain a record of inspection and test that they perform?
8. Is acceptance criteria established for all products and processes?
9. Are special process controls utilized where applicable?
10. Are all units reviewed by inspection prior to storage or use?
11. Is material identity and acceptance status maintained?
12. Do inspection frequencies follow acceptable practices?
13. Have process capability studies been performed and is there evidence that capabilities are known?

Section 6: Acceptance Inspection

1. Is all outgoing product submitted for final acceptance inspection?
2. Are inspection instructions documented and in use?

3. Are records of inspection and test maintained?
4. Is acceptance status clearly identified for all material accepted after inspection?
5. Are test procedures and acceptance criteria documented where applicable?
6. Are inspection and test records traceable to the product?
7. Are special identification, protection, and packing requirements documented and available to shipping?
8. Do final inspections and tests correlate with use or function of product?

Section 7: Information Feedback

1. Is nonconforming material adequately identified to prevent use or shipment?
2. Is rejected material promptly reported to management and action taken?
3. Do inspection and test records indicate the nature and number of observations as well as number of defects?
4. Are customer complaints recorded and is an appropriate corrective action initiated?
5. Is customer returned material evaluated to determine cause of rejection and plan corrective action?
6. Is customer advised of failure analysis findings and corrective action?
7. Are deviation requests and applicable authorizations documented?
8. Are records of rejections analyzed to segregate repetitive problems and plan corrective action?
9. Are records of corrective action maintained?

Section 8: Miscellaneous/Other

1. Housekeeping.
2. Lighting in inspection.
3. Lighting in manufacturing.
4. Controlled operations.
5. Condition of information equipment.
6. Condition of manufacturing facilities.
7. Materials handling and storage.

Survey Summary Sheet

Selection	Potential	Score	%
1. Quality system	X10		
2. Calibration program	X10		
3. Metallurgical controls	X10		
4. Incoming material control	X10		
5. In-process controls	X10		
6. Acceptance inspection	X10		
7. Information feedback	X10		
8. Miscellaneous/Other	X10		
Overall Rating	—		

Upon completion of a supplier-system survey, the auditors should state whether or not the potential supplier is acceptable, conditional (i.e., certain system deficiencies must be corrected), or not acceptable. If a supplier is considered to be not acceptable, purchasing must send out inquiries to other reliable sources. A "good" purchasing department will buy only from reliable sources regardless of the economy. Reliable suppliers will also acknowledge their deficiencies, will try to correct the problems and will accept the cost incurred.

When the quality-system survey has concluded that a given supplier has the potential to furnish acceptable materials, a preaward conference would be recommended. Why a preaward conference? The majority of errors that a reliable supplier makes are traceable to lack of communication. This communication failure takes many forms including incomplete drawings or blueprints, supplier misunderstanding about the function of the part or material in the final assembly, poor packing and packaging, and the like.

Purchasing, quality, design engineering, manufacturing, and the supplier's sales, quality, and perhaps representatives of their manufacturing group should meet prior to the issuance of all but the smallest purchase orders. The meeting agenda must include a detailed review of all proposed requirements and a schedule for the discussion of recovery costs. Clarify any questions that the supplier might have and establish ground rules for the definitions of quality acceptance criteria, manufacturing methods and processes, required packing and packaging protection, and the protocol for handling defective material. At this time, recovery costs should be discussed and the details agreed to by both parties. Once this total agreement has been reached, the minutes of the meeting should be signed by representatives of your purchasing department and the supplier's sales department as binding clauses to your future working relationships.

As part of the continuing efforts to increase supplier recovery dollars, the purchasing manager should provide the quality department with copies of the major purchase orders with present suppliers and any original quotes sent to potential suppliers prior to issuance of the purchase orders or acceptance of the quotes. The quality manager should review the documents to assure that proper and complete quality information is forwarded to the supplier.

The next step in the implementation of the supplier quality system must deal with new product acceptance. Upon receipt of the first lot of a new part or material, a complete first-article-complete-inspection (FACI) should be performed. The FACI will provide an evaluation of the dimensional, chemical, metallurgical and physical conformance of the parts or materials. Under no circumstances should a new part or material be evaluated unless accompanied by the supplier's complete inspection results that indicate conformance to your requirements. In essence, this is the supplier saying, "I say these are OK; do you agree Mr./Ms. Buyer?"

Once the interaction systems are functioning, begin tracking "gross supplier-defective dollars" on a chart. The recommended base is the dollar amount of purchased goods received and the data should be evaluated for two-year periods. A typical graph and the associated data are shown in Figure 1. The lefthand data column (Dollar Receipts) in Figure 1a is the dollar value of received product by month. The second column data lists the gross supplier-defective dollars by month. The third column percentage of received material that was found to be defective by month. Year-to-date summary data are shown in the remainder of the data

	MONTHLY			YEAR TO DATE		
YEAR X	DOLLAR RECEIPTS	SUPPLIER GROSS DEFECT COSTS	PERCENT RECEIPTS	DOLLAR RECEIPTS	SUPPLIER GROSS DEFECT COST	PERCENT RECEIPTS
JAN	4,440,000	79,763	1.80	4,440,000	79,763	1.80
FEB	5,100,000	65,750	1.29	9,540,000	145,513	1.53
MAR	6,838,000	127,464	1.86	16,378,000	272,977	1.67
APR	5,072,000	107,733	2.12	21,450,000	380,710	1.77
MAY	5,770,000	82,630	1.43	27,220,000	463,340	1.70
JUN	6,532,000	118,874	1.82	33,752,000	582,214	1.73
JUL	3,857,000	82,846	2.15	37,609,000	665,060	1.77
AUG	4,455,000	91,913	2.06	42,064,000	756,973	1.80
SEP	7,900,000	135,858	1.72	49,964,000	892,821	1.79
OCT	6,714,000	144,960	2.16	56,678,000	1,037,781	1.83
NOV	6,471,000	158,870	2.45	63,149,000	1,196,660	1.89
DEC	7,600,000	150,143	1.98	70,749,000	1,350,514	1.90
YEAR Y						
JAN	4,300,000	29,984	0.70	75,049,000	1,380,497	1.83
FEB						
MAR						
APR						
MAY						
JUN						
JUL						
AUG						
SEP						
OCT						
NOV						
DEC						

(a)

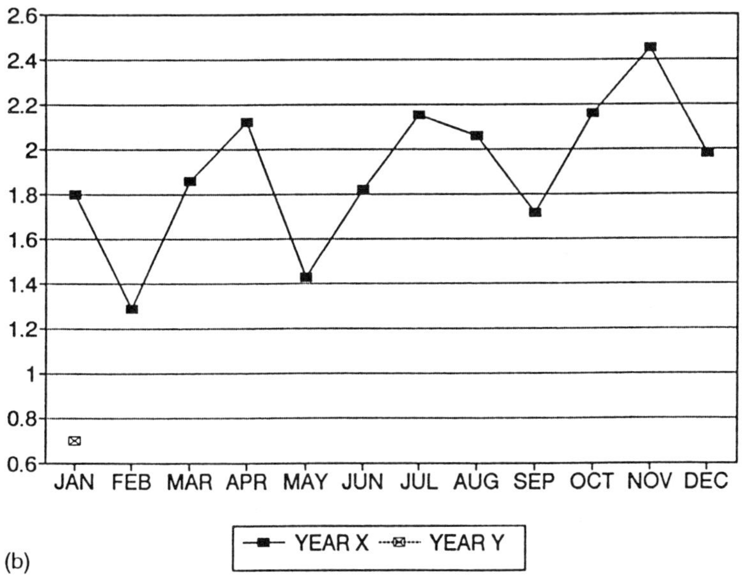

(b)

Figure 1 Tracking supplier defect dollars: (a) data; (b) chart.

	MONTHLY			YEAR TO DATE		
YEAR X	DEFECT DOLLARS	RECOVERY DOLLARS	PERCENT OF DEFECTS	DEFECT DOLLARS	RECOVERY DOLLARS	PERCENT OF DEFECTS
JAN	79,763	49,016	61.5	79,653	49,016	61.5
FEB	65,750	55,725	81.7	145,513	102,741	70.5
MAR	127,464	77,141	60.7	272,977	179,882	66.0
APR	107,733	76,713	71.2	380,710	256,595	67.4
MAY	82,630	74,492	90.2	463,340	331,087	71.5
JUN	118,874	91,786	77.3	582,214	422,873	72.6
JUL	82,846	62,277	75.2	665,060	485,150	72.9
AUG	91,913	58,413	63.6	756,973	543,563	71.8
SEP	135,848	92,494	68.1	892,821	636,057	71.2
OCT	144,960	972,968	67.6	1,037,781	733,755	70.7
NOV	158,879	126,135	79.4	1,196,660	859,890	71.9
DEC	150,143	157,679	105.0	1,350,514	1,017,569	75.3
YEAR Y						
JAN	29,983	22,942	76.5	1,380,497	1,040,511	75.4
FEB						
MAR						
APR						
MAY						
JUN						
JUL						
AUG						
SEP						
OCT						
NOV						
DEC						

(a)

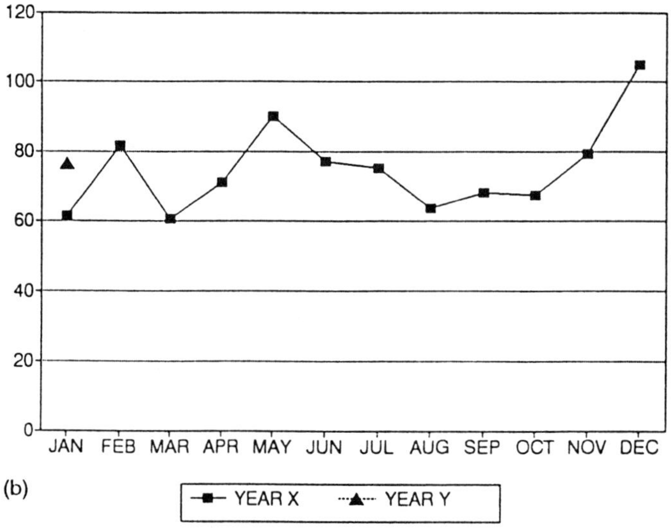

(b)

Figure 2 Two-month rolling average of recovered supplier-defective costs percentage: (a) data; (b) chart.

table. The two-month moving average graph of the data shown in the third column is plotted in Figure 1b).

Figure 2 shows the recovered supplier-defective dollars data. The first data column of Figure 2a is the gross supplier-defective costs; the second column shows the percentage relationship between supplier-defective costs and recovered costs. The remainder of Figure 2a is year-to-date summary data. Figure 2b is the graph of the recovered cost percentage data.

Using data such as that shown in Figures 1 and 2, an aggressive purchasing department can typically recover 75–80% of supplier-defect costs.

VI. SUMMARY

The following things can be done to help increase the amount of recovered supplier-defective costs:

1. A good purchasing department buys from reliable sources. Regardless of the economy, reliable sources will acknowledge their mistakes and generally accept the costs associated with their defective product.
2. Whenever possible, return defective material to the supplier. If you must repair a defective product submitted by a supplier, document the repair instructions and negotiate repair/rework cost with the supplier ahead of time. You will find that if you tell them how much it will cost to correct the situation in question, the supplier will often agree, especially when they tally the costs of transportation, processing, delays, and the like.
3. When you do, return defective materials to the supplier in the shortest time possible. Set your goal as a maximum of 2 weeks from the date of rejection.
4. Set up a separate staging area for supplier-defective material. Promptly move all "return to vendor" material to this central area for control. Do not leave material scattered all over the place. Make certain that a procedure is written on handling defective material so that everyone knows what is expected of them.
5. Be very definitive in listing the cause for rejection. Do not simply write "Defective" on your report. If possible, identify the defective area on a sample to be returned with the rejection report.

Experience has shown that it is possible to recover 75–80% of gross supplier-defect dollars. It is justifiable to charge lost labor and burden costs to the supplier. Recovering these costs is a legitimate and important part of an effective purchasing operation.

7

Managing ISO 9000

NORMAN G. SIEFERT*
Consultant
St. Louis, Missouri

I. INTRODUCTION

Throughout most of the post–World War II history of quality in the United States, company managements have routinely asked their quality professional, "Don't you have a recipe that lists the ingredients for a real quality program? If you can find one, our quality will get better and we'll get those unhappy customers off our backs." Experienced quality professionals quickly learned that quality recipes often didn't work.

In recent years, however, through the efforts of an international committee of quality professionals and others with a variety of business and manufacturing expertise, there has developed a recipe for quality that more and more managers believe may finally work. That recipe is a set of documents called the ISO 9000 series.[1] The people responsible for creating these documents were completely business wise and profit motivated.

Properly referred to as the ISO 9000 *Series of Quality Standards*, the series contains five parts; ISO 9000[2] through ISO 9004, each in a separate booklet. In a

---

* Retired.

[1] *Editor's note*: The letters *ISO* in ISO 9000 are not an acronym for the International Organization for Standardization even though that body is the sponsor of the standards. Rather, ISO, here, is a Greek prefix meaning "equal" or "the same." Whether intentional or accidental, the ISO in ISO 9000 can be considered to stand for the well-known "level playing field." Technically, the complete title of the series discussed in this chapter, the first version released, is ISO 9000-1987. Chapter 8 discusses the latest revision, which is titled ISO 9000-1994.

[2] Only ISO 9000, the first standard in the series, would be referred to as ISO 9000 in the singular.

relatively few pages, these documents present the basic ingredients of a standardized quality system. The extent, depth, and effectiveness of each ingredient used will depend on the user's degree of seriousness about quality and customer satisfaction. The degree of seriousness will be influenced by the company's then current situation and chosen direction.

The adaptability, flexibility, and international acceptance is the uniqueness of the standards, and this is what makes them so useful in comparison to other quality systems in current use. However, versatile and accepted ISO 9000 series is today, the standards are not without weakness, nor are they free from criticism.

II. ISO 9000—A SUMMARY

The ISO 9000 series of quality standards is a set of descriptions of the basic elements of a modern quality system.[3] The series had been in use—more or less in its current form—in the United Kingdom for some years under the designation BS (British Standard) 5750.[4]

ISO 9000 (series) provides a guideline as to *what* a basic quality system must contain. The operative documents—ISO 9001, 9002, and 9003—differ in the substance of the required program as a function of what *elements* are involved in the company's operations. As shown in Table 1, ISO 9001, the most comprehensive document, references 20 elements and would be used by companies that perform sales, design, manufacturing, and provide after-sales service. The standards do not describe *how* to treat each element but only *what* elements need to be treated.

Whichever of the three standards is chosen can be used as part of a more complete program that might include, for example, application for the Malcolm Baldrige National Quality Award. This composite program is a powerful means of identifying system mechanics and may well be the best guide for an overall quality system available today.

III. MARKETPLACE DRIVEN

The standards, as they are currently being driven by the marketplace, provide a common language for a quality system as agreed upon by two parties—typically a customer and a supplier.

It is entirely possible that some companies have embraced the series as only one ingredient of an overall quality improvement plan, and thus were not targeting third-party registration as their strategic objective.

---

[3] *Editor's note*: In addition to the quality standards, the ISO 9000 *family* includes ISO 9004-1 through 9004-8, *Quality System Elements*; ISO 9000-1, *Quality Concepts*; and ISO 9000-2 and -4. In addition are a number of documents numbered 10005 to 10016. ISO 9000-3 is the *Application of ISO 9001 in the Software Industry*.

[4] *Editor's note*: Some practitioners have indicated that BS 5750 was itself derived from the U.S. standard MIL-Q-9858, *Quality System Standards*.

Table 1 ISO 9000

1.	Clear management responsibility
2.	Quality system
3.	Contract review
4.	Design control
5.	Documentation control
6.	Purchasing specification control
7.	Supplier (purchased goods and services) Conformance
8.	Product identification and traceability
9.	Process control
10.	Inspection and testing
11.	Measurement and test equipment Control
12.	Inspection and test status
13.	Control of nonconforming product
14.	Corrective actions
15.	Handling, storage, packaging, and delivery
16.	Quality records
17.	Internal quality audits
18.	Training
19.	Servicing
20.	Statistical techniques

All 20 of these elements appear only in ISO 9001, *Quality Systems—Model for Quality Assurance in Design/Development, Production, Installation and Servicing.*

Those companies that have incorporated the standards because it was a good idea for quality have elected to do so for the noblest of reasons. Their rationale is that incorporating the ISO 9000 concept because of internal motivation—rather than external pressure—is a way of doing business that will provide quicker results, be more entrenched in daily habits, and be longer lasting. This, however, should in no way be construed as a criticism of those companies that yielded to the external pressure exerted by their customers or a third-party registrar.[5]

The standards form a set of guidelines based on a common set of elements. We can communicate with one another and measure the application of those 20 elements irrespective of the product, industry, or culture. Appropriately applied third-party certification assures a level of confidence in the quality systems. A benefit of third-party certification is that independent and continuous audit (usually twice a year) assures that the system is maintained. A further benefit of third-party certification is that of providing continuity for the company's quality system. Managements and priorities often change over time. Successful maintenance of the

[5] The concept of the outside auditor or registrar as the "third party" derives from the concept of the buyer and seller as the first and second parties.

certified quality system is something that wise managers would not likely give up during their tenure.

IV. COMPARISON TO TOTAL QUALITY MANAGEMENT

ISO 9000 (series) does not include all of the elements typically found in total quality management (TQM) programs. ISO 9000 is intended primarily as a means of creating consistency in a quality system. Total quality management programs generally require additional "ingredients." However, the additions required for a TQM program were left out of ISO 9000 by intent, and this is not a weakness. Nor is the series static, and over time the responsible committees are likely to incorporate additional quality system elements. (See also Chapter 8.)

V. AN ISO 9000 OVERVIEW

ISO 9000 (series) is a set of five quality system standards—ISO 9000, ISO 9001, ISO 9002, ISO 9003, and ISO 9004—first published by the International Organization for Standardization in 1987.[6] The series of documents was subsequently adopted by most countries and is generally known in the U.S. as the ANSI/ASQC Q90 series. With the exception that the Q90 series uses "American English" and calls ISO 9001, Q91; ISO 9002, Q92; and so forth, the two series are the same.

A. ISO 9000 (ANSI/ASQC Q90)

Quality Management and Quality Assurance Standards—Guidelines for Selection and Use. This document provides definitions and explanations of basic quality system concepts along with guidelines for the use of 9001 (Q91) through 9004 (Q94). The document contains a cross reference of common paragraphs between the operating ISO 9000 series documents.

B. ISO 9001 (Q91)

Quality Systems—Model for Quality Assurance in Design/Development, Production, Installation, and Servicing. 9001 is the most comprehensive document in the series. It covers all 20 of the elements contained within the standard.

C. ISO 9002 (Q92)

Quality Systems—Model for Quality Assurance in Production and Installation. 9002 covers 18 of the quality elements—design and after sales service are not covered.

---

[6] As described in Chapter 8, the series was extensively revised in 1994.

D. ISO 9003 (Q93)

Quality Systems—Model for Quality Assurance in Final Inspection and Test. 9003 covers only 12 of the quality elements and is often considered to be a minor contribution to the series. It is generally accepted that 9003 will have minimal use and there are some registrars that will not certify to it as a standard.

E. ISO 9004 (Q94)

Quality Management and Quality System Elements—Guidelines. 9004 describes a basic group of system elements that are useful for establishing quality management within an organization. The document is expected to be used in conjunction with 9001, 9002, 9003 or 3 as an expansion of concepts and application of the system elements in those standards. It further describes quality system aspects that are in addition to 9001, 9002, and 9003.

ISO 9004 is often referred to as an *internal* quality systems guide. It is written to provide a guide for establishing a focused quality system that will assure that the product or service meets requirements. The document includes a discussion of the cost of quality, a topic that is not discussed elsewhere in the series.

ISO 9001, 9002, and 9003 are referred to as *external* or *contractual* quality system standards. They define the provisions of a quality system that must be effectively treated for third-party certification.

VI. DEFINITION OF TERMS

A number of terms are defined for use in ISO 9000 series[7]

Supplier: As used in ISO 9001, 9002, and 9003, this word means your own organization. As used in 9004, the word means a person or firm that provides products or services to your organization.

Purchaser: As used in ISO 9001, 9002, and 9003, a purchaser is your customer.

Elements: Elements are the requirements, clauses, or paragraphs used to identify items requiring attention under the ISO program (see Fig. 1).

VII. HOW TO UTILIZE ISO 9000 (Series)

The two fundamental questions that are routinely raised in organizations are: Should we accept ISO 9000 as our quality program? and Should we seek third party certification? For purposes of this chapter, we will assume that the answer to both questions is "yes." It will be assumed that your company has decided to undertake ISO 9000 (series) certification because there is a business reason and a market value for doing so.

The implementation of ISO 9000 is somewhat dependent on the extent of quality system integration and documentation that already exist. If the workforce

---

[7] A significant number of terms have been redefined in ISO 9000-1994.

is well trained, if the systems are well defined with documented procedures, and if there is a clearly demonstrated top-to-bottom people involvement in the quality systems, there should be relatively little additional action required.

For the purposes of this chapter, however, the implementation and planning discussed are based on the supposition that there is an operating business, but there is minimal documentation of the system; the continuity of events is dependent on the time of the job; and righting wrongs—i.e., "firefighting"—is the routine.

A common outline of steps used to achieve registration (or, as it is often called, certification) would be as shown in Table 2. For ease of discussion, the program is shown divided into two segments: *Develop the Team* and *Implement the Plan*.

A. Develop the Team

Assign the Project Leader

The project leader is crucial to the success of the mission. The appointment should originate at the highest management level and the person chosen given the freedom and authority to do what is necessary to make the program a success. Conceivably, this appointment could be made after the management overview training session has occurred so that a more thorough understanding of what needs to be accomplished is achieved. Since a quality system extends far beyond the quality department, it is

Table 2 Implementation Plan Guideline Steps

Develop the Team
Assign a project leader
Have management orientation (an overview)
Assure complete management involvement
Assemble a team of people trained in depth
Implement the Plan
Develop a budget
Establish a steering committee to include management sponsors
Communicate about ISO 9000 throughout the organization
Interview and select a registrar
Begin self-assessment to determine status
Use self-assessment results to determine actions
Determine documentation format if not established
Train in writing procedures to your needs and format
Train internal system auditors
Compile a manual of procedures
Train people throughout in your procedures and how ISO 9000 applies to them
Begin internal system auditing
Have a preassessment from the registrar (optional) and react to results
Continue internal system auditing
Provide the manual to registrar and react to critique
Have site assessment by registrar
React to any findings
Celebrate receipt of registration certificate

often a good idea not to assign the quality manager as project leader. Assigning the position to the quality manager might send a signal that this is ". . . just another quality department program." The project leader may or may not go on to become the management representative—a position required by ISO 9000.

Management Orientation

It is generally beneficial to have an in-house executive management briefing before the program gets underway. The briefing may be handled by a consultant hired to assist in the implementation process. A division of a large or diversified corporation may be able to draw on the expertise of a division that has already been registered. The briefing should be short and designed to introduce the essential topics.

All 20 elements of ISO 9000 should be discussed. In addition, customer expectations, quality improvement opportunities, organizational involvement requirements, potential registration costs, ongoing registrar costs, and internal auditing certainly are necessary topics. What is happening in Europe regarding the standard, registrar reciprocity agreements, and the like may be quite unnecessary for a management overview.

Assure Management Support and Involvement

By means of the management overview and other means that may be appropriate to your organization, determine that there is, in fact, top down support. One way of determining this is to verify that implementation has or will become part of the operating business plan. Without this total participative support, the project has minimal opportunity for success.

Assign a Team

It is essential that more than one person be knowledgeable about all the requirements of the standard. A one-person interpretation is avoided, there is backup, and there is the opportunity for ownership in other areas affected by the project.

B. Implement the Plan

Preparation of a rather specific step-by-step plan is crucial to achieving a successful registration.

The front-end training and education of a team of people to ISO 9000 is the stage for developing the implementation plan. A plan's success will be predicated on cross-functional participation. One person may closet themselves and write a plan more rapidly than a team. The plan, however, must then still be explained and organizational "buy-in" achieved. ISO 9000, by its design, cuts across many departmental lines. Therefore, from the beginning, it is best to have the plan evolve from a trained cross-functional team. As in any good plan, the ISO 9000 plan should be specific about the actions, resource assignments, areas affected, and project dates.

Prepare a Budget

Without question, as complete a forecast of expenditures as is possible is a primary requirement for nearly any project, especially one as far reaching as ISO 9000.

Formal budgeting steps are appropriate, keeping in mind that in addition to the costs of the time of internal people (i.e., project leaders, team members, etc.), fees for an outside consultant and outside writers/editors (to write, compile, and organize the procedures, flowcharts, and other documentation) are likely to be needed as well as the fees of the auditing firm.

Establish a Steering Committee

The structure and membership of the steering committee is another of the more critical steps to success, as is the selection of the project leader. Proper selection of members and the organizational structure of the group will either accelerate or retard implementation. Proper selection of these key people will result in implementation either becoming part of managing the business or a superficial system with minimal inherent benefit.

ISO 9000 quality system requirements cross most functional departmental territorial boundaries within a company. Therefore, a steering group should consist of, at least, one member of each of the affected departments. An effective model is one in which an executive is the sponsor of one or more teams addressing sections of the ISO program.

The executive sponsor is the cheer leader and the driving authority who unblocks obstacles when necessary. Each sponsor should have accountability for one or more chairpersons leading subteams that are assigned various elements of the program (see Fig. 1).

The subsystem structure shown in Figure 1, is primarily oriented to a larger organization where it is essential that teams be cross functional; that is, having members who are all affected by those sections of ISO 9000 that will be assigned. Smaller organizations, in which people often have multiple responsibilities, nevertheless should follow the concept as closely as possible.

Key functions of the subteams are:

- To determine what current procedures (written or unwritten already exist within the organization.
- Write new procedures, if necessary, in a uniform, agreed upon format. The team should keep in mind that the business is, in fact, operating. Therefore a procedure is in effect. If a procedure is to be written, it may simply be written around the existing system. It is important that the team ultimately assure that where more than one person performs the same function, they all do it according to the agreed upon format.

Communicate About ISO 9000 Throughout the Organization

Explaining the content and concept of ISO 9000 throughout the organization is fundamental to the quality system's implementation. Further, a well-presented orientation session will enhance the opportunity for a successful execution of the plan. Generally, this in-house training is done by the project leader and other key people who are adequately trained.

The syllabus of the in-house general training consists of an overview of the standard supplemented with more detail about the specific ISO 9000 paragraphs that will affect the particular group being trained. Additionally, discussion of why

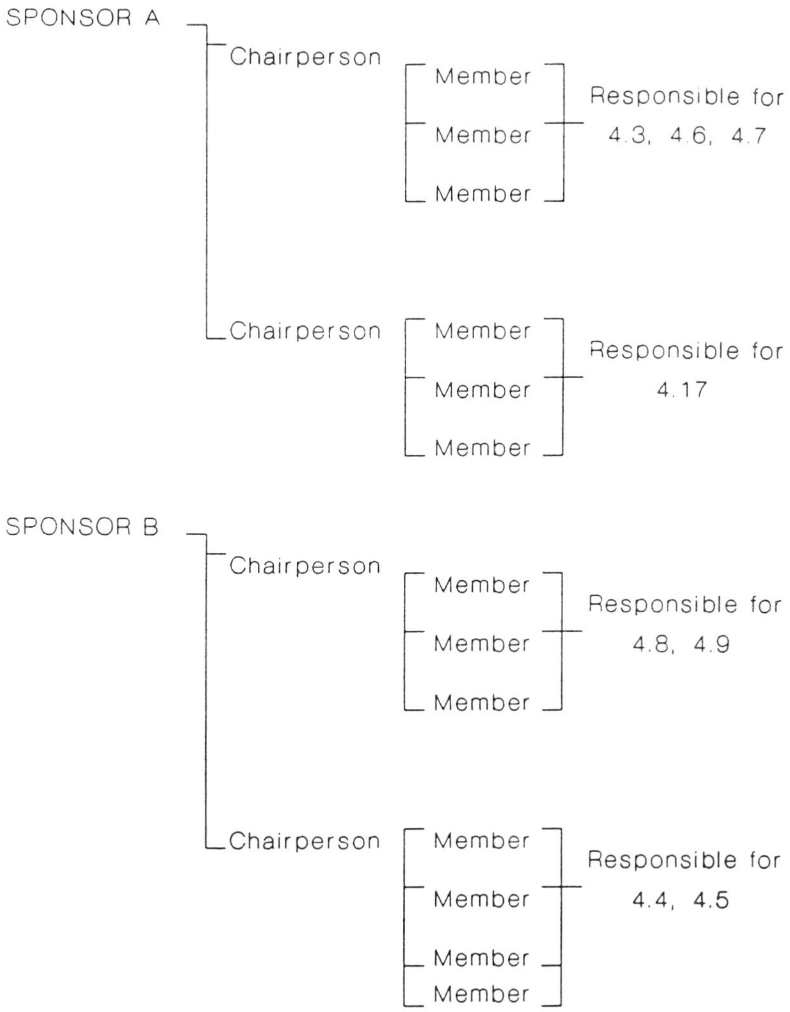

Figure 1 A subteam structure.

ISO 9000 is being adopted, what the expectations are, the registration process, and how the quality system will affect each individual should occur.

Interview and Select a Registrar

Many agencies and groups are establishing themselves as registrars. Like many other supplier/customer relationships, the selection of a registrar may be predicated on a combination of business and personality issues. If your products require certification or label from a particular agency, there is logic to use that agency if it is also an ISO 9000 registrar (third-party auditor).

When there is no preexisting alignment with an agency that is also a registrar, there are many alternatives. You have the responsibility to select the one that best suits your business needs and with whom you are comfortable. Since there are many registrars, you should interview a representative group of them and obtain competitive quotes. Be sure you recognize that your auditor must not be involved in consulting with you on how to implement the program.

Some questions to consider during the process of registrar selection:

- What other firms have they registered?
- What accreditation marks does the registrar have?
- What is the scope of the registrar's accreditation?
- What is the lead time for a registration assessment visit?
- Will your customers accept the registrar's accreditation?
- What are the registrar's requirements for surveillance audits and periodic full reassessment visits?
- What is the expiration period for the registration certificate and what happens at expiration?
- What are the initial assessment and surveillance fees?
- What certifications do the auditors have?
- Are the auditors employees of the registrar, or are they subcontracted as needed?
- What is the registrar's approach to auditing and what is their application of the sections of ISO 9000?

Begin Self-Assessment

After ISO 9000 education is complete, the period of self-evaluation (i.e., the assessment period) begins. A good place to start is in determining what documentation currently exists that could be applied to the requirements of the quality standard being targeted.

A documentation self-assessment work sheet (Fig. 2) serves a dual purpose. As it is completed, it presents an overview of where you are in relation to the

DOCUMENTATION SELF ASSESSMENT				
ISO 9001 CLAUSE	OUR APPLICABLE DOCUMENTS	TO BE WRITTEN		
		YES	NO	ASSIGNED TO
4.1 Mgmt. Responsibility				
4.2 Quality System				
4.3 Contract Review				

Figure 2 Example of a procedure self-assessment worksheet.

completeness of the documentation for each element. Then, for each element that requires work, it becomes a guide to outline the work that needs to be done.

The type and extent of documentation that must exist to support a controlled system for each element is one of the differences that may be encountered among auditors and registrars. One registrar may expect to see substantial evidence of detailed written procedures covering the application of each element. Another registrar will be satisfied with something less detailed in writing, but put more dependence on practical demonstration that a controlled system is in place.

In spite of the differences, your main considerations must be: What is best for my business? What do I need to do to best ensure consistency of doing the right thing. What do I need to do to best ensure that people are properly trained? The correct answer to all of these questions is to provide a good set of procedures that adequately describe what is expected and the best way to doing it. This approach is what fosters consistency, effectiveness, and continuity.

In addition to the documentation analysis, a detailed "micro" self-assessment of each element is valuable. The micro self-assessment is best accomplished using a detail checklist guide, which is typically supplied by the registrar. An example of one section of a micro checklist is shown in Figure 3. Additionally, flowcharting is a useful tool that can be used at this stage of the project.

Use Self-Assessment to Determine Actions

Now that the self-assessment is complete, it should be possible to determine the status of compliance to ISO 9000. An important step is an appraisal that answers such questions as:

- What procedures are in place?
- What documents need to be written?
- Do people who do the same thing do it the same way?
- To what extent is training or retraining required?
- What changes and additions to procedures are required?
- What is the timetable for implementation?
- What records need to be created, maintained, and/or updated?

Determine the Documentation Format

Documentation formatting can be an asset or a hindrance to the success of implementation. As used in this chapter, the word *documentation* will mean procedure writing and the creation or revision of a quality systems manual.

The following are some fundamental considerations in developing your documentation:

Policy manual. The policy manual is normally the upper tier document addressing each of the ISO 9001 elements as a matter of policy for your company. It is seldom changed and is general in its content.

Procedures. Procedures are the documents in which you address how you carry out the policy with respect to the various aspects of each ISO 9000 clause and your total system. Try to keep them simple and written around what you already do; after all, you have been in business for a long time. Flowcharting is an effective way to document or describe a system. At

ISO 9001/2	OUR REFERENCE DOCUMENTS	ASSESSMENT RESULTS
4.3 Contract Review - Is there an established procedure for reviewing contracts i.e. contracts and purchase orders/requirements from customers? - Does the procedure assure that we review the contract to ensure requirements are adequately defined and documented? - Do we assure that differing contract requirements are resolved.... How do we do it? - Assuring that we have the capability to meet the contract requirements. - That we have records of contract reviews.		

Figure 3 Example of a self-assessment checklist.

the time of self-assessment, flowcharting is an effective analytical method to determine the existing system. It will be of great assistance in preparing for this step and eventual implementation.

Numbering. If you have the option and are not locked into an existing number scheme, consider following the number pattern of ISO 9001/2. For example, any policy or procedure written to support paragraph 4.3, Contract Review, should be a 430 or 4300 or 300 document. This allows an obvious trail of policy and procedure back to the ISO 9000 clause. Following this technique is of immeasurable value later on to yourself and the third-party auditor. This simple number trail helps keep the manual user friendly.

Structure. Again, if not locked into an existing system, there is value in maintaining the quality systems manual separate from a general operations manual. The value of this separation is simplicity, user friendliness, and auditing ease (remember you pay for the time of third-party auditors), and it allows better control.

Train in Writing Procedures

For consistency, establish a format for documents to be used throughout the organization. Writing procedures is not always an easy task. Some people like to seize upon the opportunity to display their latent literary desires. Regrettably, these documents usually mean little to the user.

Training will be required in your writing format and how-to flowchart. Flowcharting a procedure provides simplicity, minimizes words, and is usually a better communication vehicle than a multitude of paragraphs. Flowchart where possible and then add descriptive sentences as needed.

Procedure development should be a cross-functional event. Teams of people should construct the procedure. The team should consist of people from departments affected by that element of the system.

Train Internal System Auditors

A rule of thumb is that it is most desirable to have your system operational and implemented approximately six months prior to calling in the registrar. During that interval of implementation action, at least one cycle of internal auditing should be completed. This cycle would include the resulting corrective action and follow-up verification. During this implementation interval prior to a registration visit, a management review must also occur and be documented.

Internal auditing is regarded as one of the major requirements of ISO 9000. Effective internal auditing has the opportunity to be one of the greatest value-added contributions of your quality system. Consequently, exercise discretion when assigning responsibility for internal auditing. In most companies, internal auditing will be an added task to people.

Whichever set of circumstances prevail in your organization, several people should be thoroughly trained as internal auditors. There are many public offerings of internal auditor training which may be used. Some of the course providers will do in-house training if that alternative best fits your needs.

Internal auditing should be operational prior to a registrar assessment. Internal auditing is your method of determining routine compliance with the procedures (system) you have established. This aspect of auditing does not perform product audits as a requirement to the standard.

How you enhance the internal auditing function to serve broader management goals is a quality business decision. At a minimum, the requirement is that audits be planned, documented (supported by a well-written procedure), scheduled, followed-up, and reported to management.

Compile a Procedures Manual

During the first several years after the release of the first issue of ISO 9000, there were discussions on whether or not an actual quality systems manual was a requirement. That argument/discussion is academic, because operational quality business logic allows one to conclude there needs to be a nesting place for the various written procedures of how you want to run the quality system—the system requirements. Therefore, a systems manual is a reasonable place to compile the procedure documents.

The procedures manual becomes the basic audit reference source for the internal auditing and for the registrar. One should always keep in mind that the auditing is measuring your degree of compliance to your own procedure. Another way of stating it is: Are you doing what you said you would do and did you say what you are doing in respect to the various elements of *your* quality system.

Train People Throughout in Your Procedures and How ISO 9000 Applies to Them

It is not a safe assumption to believe all people by now know what part they play in the quality system or that ISO 9000 is part of everyone's understood vocabulary.

Once procedures are documented and agreed upon, all employees should be trained in the content of the systems procedure they affect. This is part of the fine-tuning process to assure compatibility of procedure requirements to actual practice.

During the same training session, an overview of ISO 9000 should be presented simultaneously with a specific explanation of the sections applying to the group being trained. There is little value in discussing details of contract review with an audience oriented to the sections on process control or product identification and traceability.

The entire process of ISO 9000 implementation, registration, and reasons for use of the standard should be briefly reviewed for all employees during these training/communication sessions.

Begin Internal Systems Auditing

Generally, it is best to begin formal internal auditing after a manual is issued and people trained in the content and expectations of the quality system.

Internal auditing, properly executed, is also part of the fine-tuning process. Auditors are not to find fault with people. They are trained to find system faults by evaluating compliance to system requirements. This auditing definition is elementary. Auditing value goes beyond this elementary stage.

Effective auditing provides routine feedback not only on degree of compliance, but it is also used to (1) to evaluate problem areas, (2) to provide information on the effectiveness of system changes and product relocations, (3) as a problem solving tool, and (4) to detect problems before they become a crisis. How, or if, you enhance internal auditing is an issue to be resolved within your organization. Many excellent books and courses are available on auditing. Auditing should be one of your most valuable tools emerging from the quality system.

Have a Preassessment from the Registrar

A formal preassessment of your system by the designated registrar is optional. This is a topic to be thoroughly discussed with the registrar to review the pros and cons applicable to your particular circumstances and the registrars pricing of preassessment. Some registrars build in a preaudit visit with the client to establish rapport and obtain an overview of the quality system. This communication visit may be adequate for some organizations, but it is not a full preassessment visit.

More often than not a preassessment visit is worth the additional cost. It does provide a standard compliance measurement prior to an actual assessment audit. If there are major deficiencies detected during the actual assessment visit, a follow-up audit by the registrar will occur at additional cost. A preassessment visit and resulting corrective action can avoid an assessment failure and the need for a subsequent follow-up audit. Generally, preassessment is a good idea and worth the investment.

Continue Internal Auditing

If the option of a formal preassessment by the registrar was pursued, much was probably learned. Not only about your system but additional techniques of auditing. Internal auditing should be used to follow-up on the corrective action implemented as a result of the preassessment visit.

Provide the Manual to the Registrar

When a formal application is completed and submitted to the registrar, a copy of your quality manual will usually be requested. The registrar will often require that the manual be sent to their office for review prior to the assessment visit. The specific documents the registrar expects for this document review (often referred to as a desk audit) should be clearly established at the time of your initial interview.

The registrar will furnish a critique of the manual which may require corrective action prior to the assessment visit. Suitability and verification of the required corrective action should be done by your implementation task group prior to the registrar's assessment visit. Depending upon the severity of the registrar's manual review findings, a resubmittal of the manual prior to a visit may be requested.

Have Site Assessment by the Registrar

Site assessment is the event you have been preparing for and anticipating for perhaps 12–18 months. Final preparation activities should not now be underestimated or seem anticlimatic.

Escorts will accompany each of the registrar's audit teams to facilitate finding the right people and areas the auditor wants to visit. Since the sequence and time of each departmental visitation is not always firmly predictable, all levels of site management should plan to be readily available during the days of the audit. Escorts should be knowledgeable of the entire organization and have some preparation training. They should understand their role is to escort and not interfere or editorialize throughout the audit process. Their role is to observe and even take notes but only offer comments if solicited. If difficult circumstances arise, they should obviously facilitate the situation. Further, the escorts should not be defensive if any nonconformances surface during the audit.

Do everything reasonable to help the auditors perform their job rapidly and effectively. Assuring they have immediate access to all documents and a private room to caucus is essential. The entire organization should be aware of the scheduled audit days.

React to any Findings

The registrar holds a closing meeting prior to departing and presents a review of findings (nonconformances or noncompliances). This oral and written report is graciously accepted and a timetable for corrective action is agreed upon.

Since your preparation was probably effectively accomplished, only minor findings surfaced and no follow-up visit by the registrar is required prior to a favorable vote by the registrar's Certification Committee.

Celebrate Receipt of the Registration Certificate

Have a sitewide celebration and recognition of a successful registrations—why not?

VIII. RECERTIFICATION

How recertification is practiced should be an issue you discuss with your registrar. Registrars have different approaches. A registrar's accreditation mark may have some influence on how the issue is treated, but there is, as of this writing, no international standard. You can expect to find that one registrar may require a complete recertification every 3–4 years, whereas another depends entirely upon surveillance visits.

Things can change, and probably will.

FURTHER READING

Lamprecht, J. L., *ISO 9000; Preparing for Registration*. New York: Marcel Dekker, 1992.
Lamprecht, J. L., *Implementing the ISO 9000 Series*. New York: Marcel Dekker, 1993.
Todorov, B., *ISO 9000 Required*. Portland: Productivity Press, 1996.

8
ISO 9000—The Update

CHARLES A. CIANFRANI
Casmar Consulting Group
Green Lane, Pennsylvania

JOSEPH TSIAKALS
Baxter Healthcare
Round Lake, Illinois

I. INTRODUCTION

The first revision to the ISO[1] 9000 series of quality assurance standards was published in 1994. The revision was prepared over a period of several years and was approved during the summer of 1993 by the member countries of the International Organization for Standardization. The acceptance of the revisions to the ISO 9000 series by the member countries was practically universal, and this reflects the worldwide interest in the standards as well as the need to improve them continuously to make them even more useful.

II. UNIVERSAL ACCEPTANCE

The standards that comprise the ISO 9000 series are ISO 9000, 9001, 9002, 9003, and 9004. These standards are the most popular standards ever published in the international marketplace. They have been adopted as national standards by more than 80 countries, and they have become the common language for specifying quality requirements around the world.

Perhaps the most surprising aspect of the ISO 9000 series of quality standards is the degree to which organizations have voluntarily decided to demonstrate compli-

[1] The *ISO* in "ISO standards" is not an acronym for the International Organization for Standardization. It is a Greek prefix that means "equal" and is intended to convey the concept that one purpose of international standards is to foster an equal opportunity for all to participate in world trade.

ance to independent third-party auditors. By December 1996 there were over 200,000 organizations worldwide that had been formally registered by third-party auditors, and this number is growing daily. Such widespread use of the standards over the past several years across many industries as well as by public sector organizations also highlighted numerous areas where the contents of the standards could be improved.

The initial publication of the standards in 1987 was the result of work by technical experts with primary experience in the hardware sector. The consequence of the process that produced the initial version of the standards was that it was not "user friendly" to "products" other than hardware. Also, small businesses indicated that the documentation required by the standards was excessive. Further, the standards, i.e. ISO 9001, 9002, 9003, and 9004 did not have a common structure, thus creating difficulties for organizations attempting transition from one standard to another and to use the quality management guidance standard (ISO 9004) with the quality assurance requirements standards (ISO 9001, 9002, 9003).

There was also confusion about the use of various terms such as *supplier*, *purchaser*, and *customer*. And certain requirements for written descriptions of activities were not clear and were subject to widely varying interpretations by users, consultants, and third-party auditors. For these and other reasons, it became obvious that an update to the standards was necessary.

III. THE 1994 UPDATE

ISO directives require a review of standards every five years and publication of revisions to standards as appropriate. The process for reviewing and revising ISO standards is complicated, since consensus among ISO members is required. To generate a new or to review an existing standard typically takes above five years. In the case of the ISO 9000 series of standards, it was obvious soon after they were issued in 1987 that an early revision would be desirable. Therefore, ISO Technical Committee 176 started the review and revision process in 1990. In July 1994, the revisions were published, culminating four years of work by three different working groups in Technical Committee 176.

A strategic decision was made to review and revise the ISO 9000 series in two phases. This approach was chosen because there was a strongly perceived need to minimize changes, since the standard was still new to much of the world. Technical Committee 176 believed that it was desirable to have more experience and a better understanding of the standards through widespread use before considering any substantive changes. Accordingly, the first revision of the standards had a primary objective of retaining consistency with the original publication. The so-called Phase I revision was intended only to include changes that clarified the original standards.

The more comprehensive changes to the standards would be reserved for Phase II, which is targeted for completion in the late 1990s. In the Phase II revisions, it was anticipated that the technical committees and working groups would address not only further clarifications, but also more difficult questions such as how to integrate current best practices, including quality improvement and total quality management into the standards.

IV. GENERAL MODIFICATIONS TO THE STANDARDS

In the original (1987) version of the standards, there was considerable confusion over the use of terms such as *purchaser*, *supplier*, and *customer*. These terms were not used consistently. In the 1994 revision, the term *purchaser* is no longer used: there are either customers or suppliers. For example, in the 1987 ISO 9001, 9002 and 9003, there is a clause for "purchaser supplied product." The clause becomes "customer supplied product" in the 1994 update.

The 1987 version refers to "procedures", to "written procedures", and to "documented procedures." This created confusion with users of the standards. The original concept was that a systematic approach required written procedures. Many users of the 1987 standards created written procedures only when the words *documented* or *written* were included in the standard. In the 1994 standards, the word *procedures* only appears in the two-word phrase *documented procedures*. This is not meant to be an increased requirement but merely a clarification of the original intent of the standards.

Another clarification common to the use of the standards is the use of a new definition of the term *product*. The new definition establishes four product categories under the umbrella of "product." These categories are hardware, software, processed materials, and services. The standards are intended to apply just as readily to any of the categories or combinations of the categories. For example, a product such as a pH meter may contain hardware in the form of a printed circuit board, software in the form of code in a microprocessor, processed material in the form of a chemical solution used as a reference, and start-up service as a part of the original purchase. Thus the standards apply to combinations of categories just as they do to any single category.

The phrase *product/service* was eliminated from the standards, since the new definition of the word *product* includes service as well as hardware, software, and processed material. In addition, it was decided to have a common clause numbering system for ISO 9001, 9002, 9003, and to align the clause numbers with those of ISO 9004 as closely as possible to provide easy cross referencing. Also, the standards now state that they are suitable for assessment by an external party (i.e., a third-party auditor).

And, finally, the scope statement in ISO 9001, 9002, and 9003 explicitly states that the requirements specified are aimed primarily at achieving customer satisfaction by preventing nonconformity—an introduction of quality management in a subtle manner.

V. ISO 9000-1

ISO 9000 has been renumbered as ISO 9000-1. This renumbering is an element of an overall scheme for numbering all documents in the ISO 9000 series in a consistent and logical manner. The new numbering scheme, which is detailed in a document called *Vision 2000*,[2] requires that standards supporting quality management or quality assurance be designated as part numbers of either ISO 9000 or ISO 9004. Thus, the first revision to ISO 9000-1987 has become ISO 9000-1.

ISO 9000-1 is the guide to the ISO 9000 series. It contains an extensive description of each of the individual standards of the series; i.e., ISO 9000-1, 9001, 9002, 9003, and 9004-1. ISO 9000-1 expands the original 1987 standard and contains clear definitions of new terms such as *industry/economic sector*, *stakeholder*, *ISO 9000 family*, and *product*; i.e., the four generic categories discussed earlier.

There is also a much more detailed description of the various situations where quality system standards can be used, more information on the application of each standard in the ISO 9000 series, and three new annexes with information and definitions related to product and process factors. Guidance is also provided on the prevention of proliferation of standards.

ISO 9000-1 now addresses key quality concepts such as:

- Five key objectives and responsibilities for quality
- Four facets that are key contributors to quality
- The process model
- The network of processes in an organization
- The relation of quality systems to the network of processes
- The concept of stakeholders
- The difference between quality system requirements and product technical requirements
- Evaluation of quality systems

In addition to the principal concepts indicated above, ISO 9000-1 provides guidance regarding the importance and use of documentation, since documentation has proven to be one of the major areas of concern regarding achieving compliance with the requirements of the standards. In particular, the importance of documentation as a value-adding activity is emphasized. Guidance is also included to indicate how the documentation of procedures can be used to support quality improvement and to contribute to the consistency of implementation of procedures throughout the organization. ISO 9000-1 also stresses the importance of documented procedures and objective evidence so that external auditors can evaluate the adequacy and effectiveness of a quality system. A key concept regarding documentation that is presented in ISO 9000-1 is that documentation must add value. Quality system documentation should not add cost without adding value.

Annex E of ISO 9000-1 contains a very useful reference to all standards in the ISO 9000 family, which consists of the five standards that constitute the ISO 9000 series discussed earlier plus additional supporting standards. Fifteen standards and an ISO Handbook are listed in Annex E (see Table 1). Guidance is provided regarding the content and appropriate use of each of the standards.

---

[2] The *Vision 2000* document was prepared by a strategic planning group of the International Organization for Standardization. The purpose of the *Vision 2000* document is to provide a roadmap and framework to guide the evolution of the standards in the ISO 9000 family.

Table 1 Adapted from Annex E of ISO 9000-1-1994 and ANSI/ASQC Q9000-1-1994

ANSI/ASQC Q9001-1994, *Quality Systems—Model for Quality Assurance in Design, Development, Production, Installation, and Servicing.*
ANSI/ASQC Q9002-1994, *Quality Systems—Model for Quality Assurance in Production, Installation, and Servicing.*
ANSI/ASQC Q9003-1994, *Quality Systems—Model for Quality Assurance in Final Inspection and Test.*
ANSI/ASQC Q9004-1-1994, *Quality Management and Quality Systems—Guidelines.*
ANSI/ASQC Q10011-1-1994, *Guidelines for Auditing Quality Systems—Auditing.*
ANSI/ASQC Q10011-2-1994, *Guidelines for Auditing Quality Systems—Qualification Criteria for Quality Systems Auditors.*
ANSI/ASQC Q10011-3-1994, *Guidelines for Auditing Quality Systems—Management of Audit Programs.*
ISO 9000-2:1993, *Quality management and quality assurance standards—Part 2: Generic guidelines for the application of ISO 9001, ISO 9002 and ISO 9003.*
ISO 9000-3:1991, *Quality management and quality assurance standards—Part 3: Guidelines for the application of ISO 9001 to the development, supply and maintenance of software.*
ISO 9000-4:1993, *Quality management and quality assurance standards—Part 4: Guide to dependability programme management.*
ISO 9004-2:1991, *Quality management and quality system elements—Part 2: Guidelines for services.*
ISO 9004-3:1993, *Quality management and quality system elements—Part 3: Guidelines for processed materials.*
ISO 9004-4:1993, *Quality management and quality system elements—Part 4: Guidelines for quality improvement.*
ISO 10012-1:1992, *Quality assurance requirements for measuring equipment—Part 1: Metrological confirmation system for measuring equipment.*
ISO 10013:–*Guidelines for developing Quality Manuals (request publication date information from ISO headquarters or from ASQC).*
ISO Handbook 3: 1989 *Statistical Methods.*

There are four other Annexes in ISO 9000-1 (Annex A–Annex D). Annex A is a collection of terms and definitions that have been excerpted from ISO 8402, the ISO quality vocabulary standard.

The other annexes address product and process factors (B), proliferation of standards (C), and cross referencing of clause numbers (D).

Appendix D may be of particular interest to new users of the ISO 9000 series, since it contains a cross-reference list for ISO 9001, 9002, 9003, 9000-2, 9004-1, and 9000-1 to facilitate the use of the guideline standards in conjunction with the requirements standards.

VI. THE REQUIREMENTS STANDARDS

The quality assurance requirements standards—ISO 9001, ISO 9002, and ISO 9003—are the standards that contain quality system requirements. These standards can be used as the basis for establishing minimum quality assurance requirements for an organization, for conducting internal audits, and for demonstrating the organization's quality system capabilities to either customers or outside auditors; i.e., second-party audits or third-party audits. The formal titles of these standards are:

- ISO 9001-1994: Quality Systems—Model for Quality Assurance in Design, Development, Production Installation, and Servicing
- ISO 9002-1994: Quality Systems—Model for Quality Assurance in Production, Installation, and Servicing
- ISO 9003-1994: Quality Systems—Model for Quality Assurance in Final Inspection and Test

Choosing the appropriate standard for a manufacturing organization is simple—the scope of operation of the organization usually clearly defines which quality system model is most appropriate. If guidance is needed, see Section 7 of ISO 9000-1. In particular, it is inappropriate to exclude product design from the quality system if product design is within the scope of operation of an organization.

It is significant to note that the requirements contained in each of the standards are minimum requirements for an organization of any size and in any economic sector. Also, the standards describe minimum requirements for an organization without indicating or prescribing how to implement the quality system elements.

The clause structure of ISO 9001-1994 remains unchanged from the 1987 version, although some changes have been made to the subclause structure. Several clause and subclause titles have been modified to more accurately reflect the content of the quality system elements. Numerous changes have been made to the text to clarify meaning. Examples of specific changes include:

- The statement that quality systems are suitable for two-party contractual purposes has been expanded to also include assessment by external parties.
- The quality policy requirements have been strengthened to include reference to customer expectations and needs.
- The management representative must now be a member of the supplier's management team.
- The management review process of an organization must now take place at defined intervals.
- A quality manual is now required.
- A subclause on quality system procedures has been added to clarify the degree of documentation required for the quality system. The extent of documented procedures required for activities depends on the skills needed, the methods used, and the training acquired by personnel in performing the activity.
- Quality planning is required. Product quality plans are not necessarily required, however. The supplier can decide if quality plans are required by giving appropriate consideration to the needs of the organization.
- Contract review requirements are expanded and clarified.
- Design control has been expanded specifically to include validation as well as design review and design verification.
- Corrective and preventive actions have been separated with considerable new text added to describe the requirements for each type of action. Corrective action is focused on eliminating the causes of actual nonconformities while preventive action is directed at eliminating potential nonconformities.

- Preservation has been added as a new element of the subclause that deals with handling, storage, packaging, and delivery.
- The requirements for statistical techniques have been strengthened. As a minimum, the supplier must now evaluate the need for statistical techniques.

The structure of ISO 9002 has been altered to match identically the structure of ISO 9001; i.e., ISO 9002 now has 20 clauses for quality systems. All changes made to ISO 9001 have been made, word for word, to the corresponding clauses and subclauses of ISO 9002. As was true in the 1987 standards, the design control element does not apply to the ISO 9002 model. A place keeper has been inserted for the design control clause with a simple sentence indicating that design control is not included in the ISO 9002 standard. Servicing, which was not an element in the 1987 version of ISO 9002, is now included with identical wording to ISO 9001.

ISO 9003 now has the same clause structure as ISO 9001. ISO 9003 now contains 20 quality elements, including eight new clauses, four of which are used only as place keepers, and four of which contain new requirements. The new requirements are those that relate to contract review, control of customer supplied product, corrective action, and internal quality audits. Numerous other changes have been made to ISO 9003 that are similar to the requirements contained in ISO 9001. ISO 9003 applies only to organizations utilizing final inspection and test. Since ISO 9003 is essentially not used in the United States, the changes will not be discussed here.

VII. ISO 9004-1

ISO 9004 has been renumbered. It is now ISO 9004-1. This document provides guidance on designing and implementing a quality system to meet marketplace needs and to guide internal quality improvement. Since ISO 9004-1 is a guidance standard, its contents can address quality management as well as quality assurance, and consequently it is much more comprehensive in scope. It provides considerable "advice" on topics and issues that organizations should consider when developing and implementing a comprehensive quality management process in an organization. Primary changes to ISO 9004 include:

- Adding a new subclause on quality improvement to coordinate with the newly developed standard ISO 9004-4 entitled *Quality Management and Quality System Elements: Part 4—Guidelines for Quality Improvement.*
- Retitling the clause relating to the economics of quality to financial consideration of quality systems; this clause links to the contents of a standard being developed by ISO to address this issue.
- Revising the contents relating to configuration management to be consistent with a standard that is being developed in this area.
- Including several references to environmental considerations.
- Emphasizing marketplace input and feedback as a means of improving the quality system.
- Aligning the structure of the clauses relating to handling and storage to be more compatible with ISO 9001, 9002, and 9003.

As is true of the other standards in the family, extensive minor changes have been made in ISO 9004-1 to improve clarity and/or consistency. Some examples include the use of the term *defined and documented* in place of the term *develop and state*. And the term *quality management systems* now replaces the term *quality systems*. Overall, ISO 9004-1 provides an even more valuable reference source for organizations who are developing or expanding their quality assurance and quality management systems.

VIII. THE FUTURE

The ISO Technical Committee for the ISO 9000 series of standards completed the final draft of the 1994 edition in November of 1992. The process of international consensus consumed about eighteen months, thus resulting in the publication of the revised standard in July 1994. Although it may seem that this is a long time, consensus is a vital element of the standards generation process and consensus takes time.

Since the work on the 1994 revision was completed, ISO Technical Committee 176 has been developing the next revision of the standards, which has been titled Phase II. The same working groups that prepared the 1994 Phase I revisions are responsible for the Phase II revisions.

There are two current issues that must be resolved to achieve the initial objectives of the Phase II working groups. The first issue relates to finding a structure for the ISO 9000 series of standards that meets the needs of small, medium, and large businesses as well as the needs of organizations in the four product categories.

The original ISO 9000 series was based on a product life-cycle model. A life-cycle model is the basis of most traditional quality systems, including those utilized by the U.S. Food and Drug Administration (FDA), various military organizations, and, indeed, most hardware manufacturing organizations.

To accommodate the needs of all potential users of the standards, a process model appears to be more appropriate. Such a model is based on the concept that all work is a process. A process model applies to all work in all organizations and to all products. A process model has been applied effectively and successfully in such areas as the Malcolm Baldrige National Quality Award and the European Quality Award, as well as numerous other national and international quality awards.

The current thinking is that a process model will provide the best framework for the next—i.e., the Phase II—revisions of the standards. It is anticipated that between four and seven major processes would be necessary to accommodate all organizations and all product categories.

The use of the process model does lead to a further issue. The process model is strongly associated with quality improvement and total quality management concepts. The concern relating to quality assurance standards is that requirements will no longer be based on minimum requirements, reflected in a quality assurance standard, but rather on quality improvement and/or quality management requirements expressed in a quality management standard. Such a set of requirements would be difficult, if not impossible, to audit, and would be subject to wide variations in interpretation. Furthermore, no organization ever fully achieves its total quality management objectives, since the improvement process never ends.

In the near term, at least for the rest of the 1990s, there is no intention by ISO Technical Committee 176 to broaden quality assurance standards to include requirements for quality management. The U.S. delegates to Technical Committee 176 have assiduously labored to maintain a sharp distinction between the quality assurance standard, i.e., ISO 9001, and the quality management standard, i.e., ISO 9004.

IX. SUMMARY

The Phase I revisions to the ISO 9000 series of quality standards reflect years of work by some of the best quality thinkers in the industrial world. There has been a very high level of involvement and dialogue in proposing, discussing, and modifying ideas to enhance the usefulness of the quality standards that provide quality assurance requirements and quality management guidance to the world. Not only has there been a strong positive reaction to the modifications to the 1987 version of the standards by the user community, but there also has been a continuing growth in the number of registrations. This underscores both the growing support for quality systems around the world and the success of ISO Technical Committee 176 in identifying the needs of the users of the standards and in implementing the most easily addressed of these needs in the Phase I revisions. Users of the standards can anticipate that the Phase II revisions, targeted for completion by the year 2000, will result in even more powerful tools that all organizations can use to achieve desired levels of quality assurance and quality management.

9
Project Management

JOHN M. AARON
Milestone Planning and Research, Inc.
Hickory Hills, Illinois

I. INTRODUCTION

A project is a coordinated set of interdependent activities aimed at achieving a specific goal. A project always begins and ends, and it usually consumes both human effort and capital resources. Project management embodies the special set of management tools, concepts, and techniques that enable a project team to achieve their goal while simultaneously satisfying commitments to a schedule, a budget, and an appropriate level of quality.

The management of business projects is important. Projects are vital to the long-term success of organizations. They are the vehicles used by organizations to realize new opportunities, to improve operations, to cut costs, and to make strategic adjustments. Without projects, organizations would find it difficult to grow or advance. This importance of business projects heightens the need for all managers and leaders to understand the best practices of project management.

In practice, projects often turn out to be the school of hard knocks. It is not uncommon to see an ordinarily effective manager of a repetitive, routine process become ineffective in managing a project. The reason for this paradox lies in the special nature of projects. First, projects are by definition unique. People working on a project team often lack project-specific experience and are more likely to make mistakes and miscommunicate. This phenomenon known as the "experience curve" tends to work against project effectiveness.

Second, project activities are difficult to manage because they often cross departmental boundaries. In many organizations, the parochialism of the functional organization interferes with collegial teamwork. Marketing, for instance, may have a difficult time working with engineering; engineering has friction with manufacturing; and so on. Such rivalries and interpersonal conflicts that exist between departments can be difficult inhibitors to overcome.

Third, projects sometimes require individuals to perform activities that extend beyond their "normal" job descriptions. As a result, people often perceive that projects are intrusions into their real jobs. For example, we often hear members of quality improvement teams decry, "We don't have time to spend on quality; we have to get our real work done first." Many a project leader[1] has encountered a similar situation by finding something less than enthusiastic cooperation and support for his/her project.

II. ROLE OF THE PROJECT MANAGER

In order to be successful, a project must be cared for by a single person who is willing to take charge and be accountable for the project's success. Regardless of what we call this caretaker—be it project manager, project leader, project coordinator, or team leader—the need exists to have one person function as the focal point for information, coordination, control, and responsibility throughout the duration of the project.

The primary role of the project manager is to ensure that team members carry out their responsibilities in a timely, cost-effective and high-quality manner. Typically, project activities are performed by team members other than the project manager. However, on small projects, the project leader may be a "doer" as well. Team members perform their regular work through functional organizations and report to a department head. Accordingly, the project manager, who may not be a department head, likely does not have "line authority" over members and must use influence to get the job done. The successful project leader is the person that can steer a project team through unchartered waters. Clearly, having solid interpersonal, communication, and leadership skills is a valuable asset to anyone in charge of a project. However, the most important personal asset is having the skills to establish an effective disciplined project management process. The purpose of this chapter is to provide the project manager with the tools to establish that process.

III. DEFINING PROJECT SUCCESS

Success on a project is defined as accomplishing the project's mission within the sponsor's required time frame, with sufficient quality, and within a specified budget. This "triple constraint" of schedule, quality, and budget drives all projects.

Experienced project managers—especially those working on large, high-risk projects—develop elaborate, sophisticated techniques that help them plan, track,

---

[1] In this chapter the terms *project leader*, *project manager*, and *project coordinator* are synonymous and refer to the person to whom has been delegated the responsibility for successfully accomplishing the project. The terms *project owner* and *project sponsor* are also synonymous and refer to the person to whom the project manager would be subordinate with regard to the project. The relationship is analogous to that of the owner of a business and the person that the owner employs to manage the business's affairs.

and control all three elements of the triple constraint. Quite often, the sophisticated cost-management techniques, which may be expensive to implement, may be inappropriate to use on smaller projects such as quality improvement activities, special team assignments, and task forces. Fortunately, most small projects are primarily driven by quality and schedule. Typically, if a project achieves its mission with the right level of performance and within the allowable time frame, it is considered successful by its sponsor.

Because a strong cause-effect relationship exists between schedule and cost on most business projects, the manager can adequately control the triple constraint by focusing upon the schedule and quality dimensions. This emphasis on schedule rather than cost can make the job of managing a project much more cost effective and efficient.

IV. FUNDAMENTAL CONCEPTS OF PROJECT MANAGEMENT

In its most fundamental form, effective project management consists of two basic steps:

1. Make Realistic Commitments: One should begin the project management process by making only realistic commitments.
2. Keep the commitment: After making a realistic commitment to meet a specific scope of work, the project leader should regularly assess the performance of the project and inform all stakeholders[2] of the project's accomplishments versus plan. If either the rate of accomplishment or the sufficiency of accomplishment is below plan, the manager should identify the problem, isolate the root cause of the problem, and then work with the team members who need to take corrective action.

In cases when the project's performance falls below plan, the manager has three options: (1) Expedite the remaining work and make up for lost time, (2) reduce the scope of the work along with the corresponding expectations of what can actually be accomplished in the scheduled time frame; or (3) replan by extending the schedule and/or increasing the budget.

Figure 1 illustrates the two-step approach to project management. It shows that the role of the project manager is first to make realistic commitments through planning and then to ensure that those commitments are met through regular performance assessment and corrective action. If it ever becomes apparent that the originally planned commitments cannot be met, then the project leader should recommit to a revised plan.

V. COMMITMENT MAKING—THE FIRST KEY TO SUCCESS

One fundamental driver of project success relates to the project manager's skill at making realistic schedule commitments. On every project, a leader sooner or later

---

[2] Stakeholders are those who have a specific interest in the project. They are often called project "customers."

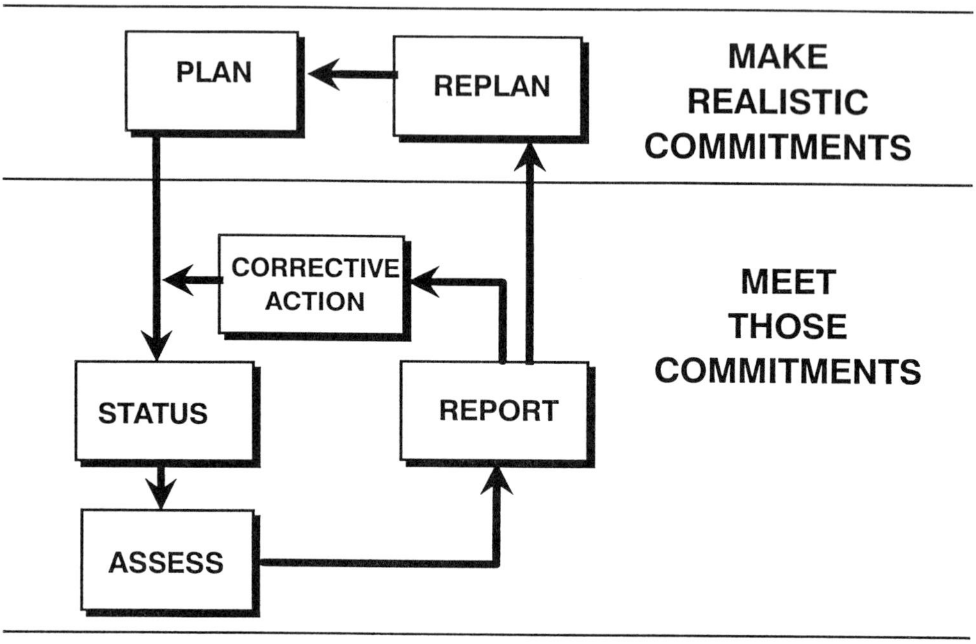

Figure 1 The project management cycle.

agrees to achieve a result through the execution of work performed by others. The role of project leadership, then, is to ensure that the commitments he/she makes are realistic and then, once made, the commitment is met. To a great extent, managing expectations by first ensuring that the expectations are realistic is the first rule to managing effectively.

Realistic expectations are developed when schedule commitments are made after planning by the project team. As part of the commitment-making process, the project manager is responsible for:

1. Ensuring that initial project requirements are documented, communicated, and understood as completely as possible.
2. Developing a project team that is appropriately staffed with individuals who are available, committed, and willing to be held accountable.
3. Developing, in conjunction with the project team, an appropriate project schedule.
4. Negotiating the schedule and contingency commitment with the project sponsor after the plan is developed, not before.

If the project sponsor and project leader cannot come to an agreement whereby the schedule produced by the project team does meet the sponsor's time requirements, the written plan can be used as an objective negotiating tool to trade-off time, scope of work, and staffing needs. Project leaders should never "cave in" to a project sponsor's demands unless the plan shows its feasibility. Only by first developing a plan, can a team negotiate schedules intelligently.

All of this sounds straightforward, but just stop and think for a moment how many times all of us have witnessed unrealistic schedules "set in stone."

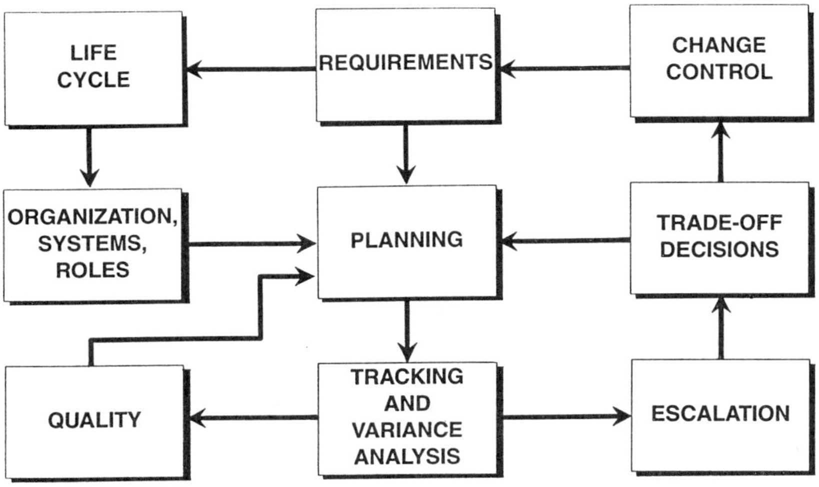

Figure 2 Elements of a project plan.

More likely than not in these situations, project due dates were established by wishful thinking and commitments were made by telling the boss what he/she wanted to hear. Beginning a project with an unrealistic plan is no way to run a successful project. Figure 2 shows the elements of a comprehensive project plan.

VI. CONTROL—THE SECOND KEY TO SUCCESS

After a project leader makes a rational commitment, the next requirement is adequately to control the project from the beginning to the end. The term *control* sometimes carries a negative connotation that suggests manipulation or coercion. The concept of project control refers to steering the execution of the project by evaluation of its performance, corrective action, and replanning. Specifically, the role of project team leader is to:

1. Regularly monitor the status of the project and report on its performance.
2. Instill a sense of discipline into the management of the project by identifying problems and conflicts and by ensuring steps toward corrective action are taken when actual performance falls behind planned performance.
3. Ensure that the project sponsor is regularly informed of actual project performance versus plan; similarly to provide the sponsor with adequate options to trade-off scope for time, time for cost, and so forth, when necessary.
4. Ensure that the sponsor understands the cost and schedule consequences when making changes to requirements or scope as the project processes through the life cycle.

It is important to note that a tendency exists within all projects for discipline to erode. Three forces operate that cause this erosion. First, team members tend to be optimistically biased in reporting actual work performance. Second, new information causes project sponsors to change their minds in mid-stream about requirements. Third, team members or sponsors may try to expand the scope of work without understanding the consequences to the schedule and cost.

The project leader can counter these tendencies through regular assessment and control. The project leader should establish a feedback loop in the process to spot drifting and to encourage corrective action. This is accomplished by effective planning, information gathering, status reporting, and status meetings.

VII. AN EXAMPLE OF A PROJECT OUT OF CONTROL

The first commandment for successful project management is to establish rationality when making schedule commitments. It sounds easy, but it really is difficult for people in the workplace to do this. Consider the all too typical exchange between a supervisor, Jane, and her subordinate, Jim, and later between Jim and his subordinate, Hank.

Jane: "Jim, I want your group to perform this market research study. It must be completed in six weeks. I know I can count on you."

Jim: "Sure, Jane. No problem."

Jim leaves Jane's office and walks over to his work area where he runs into his market research analyst, Hank.

Hank: "How'd the meeting go with Jane? Anything new?"

Jim: "Yeah. We've got another research project on our plate. It has to be finished in six weeks."

Hank: "Six weeks!! You've got to be kidding! We're already overloaded with five other assignments that are due that same time."

Jim: "Yeah, but you know Jane. She just won't take no for an answer. Whenever I object about an assignment, she thinks I am being uncooperative."

Hank: "But how are we going to finish the assignment in six weeks? I don't see any way of making that schedule without letting something else slip."

Jim: "Don't worry, the due date is six weeks away. By that time, Jane will have either forgotten the date I promised or she will change her mind about what she wants, which lets us off the hook."

Jim's secretary interrupts the conversation and tells Jim that Jane is holding on the phone for him. Jim takes the call and goes through the following dialogue.

Jim: "Hello Jane. How can I help you?"

Jane: "Oh Jim. I forgot to mention that the research study must include focus groups as well as consumer surveys. I know that your group usually relies upon questionnaires of individual consumers, but Jack (the president) is intrigued by focus groups. He wants them."

Jim: "Sure, Jane. No problem."

Jane: "Thanks, Jim. Bye."

The scenario shown above exemplifies a management process that is out of control. It begins with the desire of a sponsor ("Jane") to have assignment completed by a specific date. The sponsor communicates this desire to a project leader, Jim. Wanting to please the sponsor, the project leader immediately commits to finishing the assignment on the date requested. All too often, however, the commitment is made too hastily. When the promised time comes around, the job is often nowhere near complete. Subtly, the organization begins to adopt an implicit understanding that commitments are almost never met . . . "it's just the way things are around here." Accommodation has set in as part of the culture. It is an insidious merry-go-round that never stops.

A cycle seems to exist in Jane's case as shown in Figure 3. Jane has set a requirement date for assignment completion at six weeks. Jim agreed to the date hastily without verifying how feasible it really was. In all likelihood, as the promised date would come and go without completing the assignment, Jim would expect an apathetic response from Jane followed by her acceptance of his rationalizations as to why the date was missed. Worse yet, when Jane sets up her next assignment, she would once again go through a process of wishful thinking expecting that "this time" the commitment made by Jim would be met. In this scenario, Jane is the loser because the job does not finish on time.

An alternative cycle could exist where Jim (not Jane) pays the heaviest price. This is shown in Figure 4. As in Figure 3, Jim hastily agrees to meet Jane's requirement date and, as before, the promised date is missed. This time, however, Jane responds with anger. She blames Jim for the negative consequences and embarrassment that she faces due to the missed commitment. Jane remembers this incident and the next time she gives Jim an assignment, she will press him for an even tighter schedule. And so it goes, on and on. Continually, an inadequate amount of time will be budgeted to perform the job properly, and schedules are never believed.

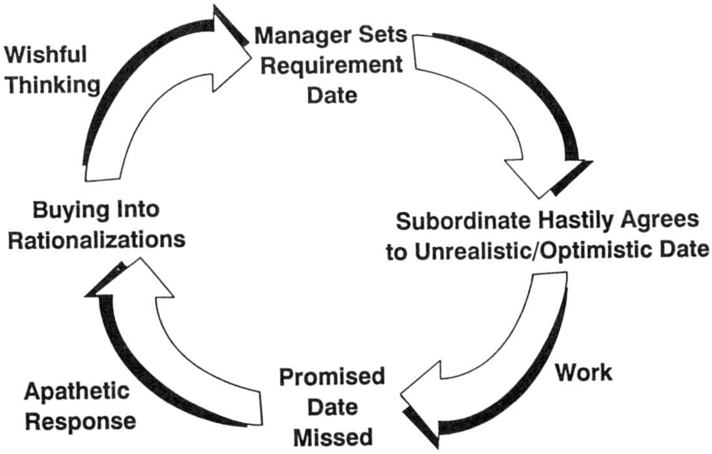

Figure 3 The vicious cycle, case 1.

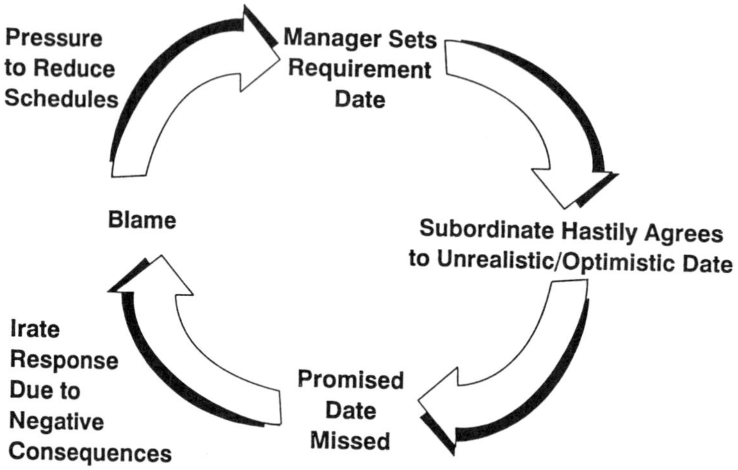

Figure 4 The vicious cycle, case 2.

VIII. GETTING THE PROCESS UNDER CONTROL BY USING PLANNING TOOLS

As a first step, Jim must break the vicious cycle that leads to consistently making unrealistic commitments. To do this, he must establish process discipline. He must ensure that:

1. No commitments to meet a specific deadline will be made until some form of project plan is first developed by the team.
2. After the project plan is developed, it will then be objectively reviewed and fairly negotiated with the project sponsor/owner.
3. Once agreement to the plan is reached, the deadline is treated as a serious commitment. The commitment is elevated in importance, and progress versus plan is reviewed regularly and communicated upward. A sense of time urgency is placed on the project. Schedule slips are identified and corrective action is expected to put the project back on track. If corrective action is not possible, the situation is rigorously assessed, appropriate trade-offs are made, and replanning is performed.

Let's apply our simple commitment rule to the case of Jane and Jim.

If Jim is on his game, he will resist making time commitments until he has planned the project and examined what impact Jane's project might have on other projects. So, the following dialogue might be appropriate from the beginning.

Jane: "Jim, I want your group to perform this market research study. It must be completed in six weeks. I know I can count on you."

Jim: "It sounds like this project is important to us, Jane. I want to take your six-week deadline as a serious requirement. Therefore, I need to determine exactly what is required and what needs to be done to satisfy those requirements. Also, I must examine who is available in my department to work on the project. To meet the six-week deadline, we may have to reprioritize some other projects."

Jane: "That's a good idea, develop a plan."

Jim: "Why don't I include in the plan a discussion of my understanding of your requirements. I'll also show in the plan how my department intends to perform the work. If the project conflicts with other projects already underway, we can determine priorities at that time."

Jane: "Great. I'll be out of town until next Monday. Let's meet on Tuesday at 2:00 p.m. to discuss your plan."

Jim: "Fine. I'm available at that time, too. I'll see you on Tuesday with my plan ready for review. Have a good trip."

Jane: "Thanks, Jim. Bye."

In this scenario, Jim has accomplished several things. First, he has escalated the seriousness of meeting the sponsor's needs, and accordingly has dampened the tendency to offer a knee-jerk commitment. He is protecting his department from the organizational forces that tug and pull at every project manager. Second, he is engaging in a rational process of setting expectations. By requiring the development of a preliminary plan before making a commitment to a specific deadline, Jim has established control over the process. Also, Jim is protecting both himself and Jane by checking to make sure that his commitment will be realistic and not just wishful thinking. The plan itself can become an excellent communications vehicle linking Jim with Jane as well as Jim with his project staff.

If performed correctly, Jim's planning process will immediately signal any potential resource conflicts due to other scheduled projects. Similarly, the plan will add focus to the work and will facilitate Jim and Jane to achieve a common understanding of the scope of work and the required tactics.

Having set the stage for rational commitment making, Jim must now utilize effective planning tools.

A basic project plan should utilize the following tools:

1. A requirements document.
2. A work breakdown structure that decomposes the work into manageable activities.
3. A realistic schedule in the form of either a Gantt chart or a listing of completion dates for key milestones (milestones signal the sufficient accomplishment of important project activities). In addition, the schedule should be supported by an assignment of accountabilities of team members who will execute the work activities and be responsible for schedule, cost, and quality targets.
4. A dollar budget in aggregate and an hourly budget in detail (often omitted on small projects).
5. A staffing plan (if additional human resources must be hired) and/or a responsibility assignment matrix (if team members are already known).

Planning adds focus, efficiency, and control to a project. It represents the difference between a "rifle" approach and a "shotgun" approach to the execution of work. Planning also forms the basis of communication and collegiality between the project leader, the project team, and the project sponsor. The plan forms a communication link to ensure that a mutual understanding exists between the project leader and sponsor regarding what is to be done and how it is to be done. Also, the process of planning facilitates assessment and control of project work.

IX. PROJECT MANAGEMENT TOOLS

A. Tool #1—Determining Requirements and Scope of Work

The role of the project manager is to answer the basic "what," "when," and "how much" questions. The project team determines "who" will do the work and "how" the work will be performed.

Let's examine how Jim and Hank develop the "what" elements of the project plan. They start by asking the basic "what" question.

Jim: "Hey, Hank. We have to develop a plan for another market research project that Jane has given to us."

Hank: "What do you mean by a plan? It sounds like we're supposed to put some sort of schedule together."

Jim: "A schedule is part of it, but first we have to make sure we understand Jane's objectives; in other words, the goals of the project. We have to know what her requirements are. We have to make sure we understand *what* she really wants and needs."

Hank: "Okay. *What* does she want?"

Jim: "She wants us to perform market research on a new product idea."

Hank: "Okay. But what are her specific objectives for the study, and what are the time constraints?"

Jim: "I have them written down on this requirements document."

This dialogue between Hank and Jim illustrates a value-added step to the project. Too often people begin work on projects before they fully understand the sponsor's requirements. By ensuring that they fully understand what Jane wants, Jim and Hank have nipped this potential problem in the bud. They recognize that they need more information describing what Jane really needs before they can finalize a realistic plan. But they have sufficient detail to get started with planning. For many projects, the completion of the requirements document (or one like it) shown in Figure 5 will force additional clarity to the requirements identification process.

In general, if the project leader knows the answers to the basic what and when questions, the leader understands the sponsor's requirements adequately to begin planning.

It is important to stress the necessity that the project manager commit project requirements to writing. First, by putting the requirements in writing, the team and the sponsor have the ability to communicate and be sure that they share an accurate, common understanding of what is really required. Second, the project manager must have a way accurately to communicate these requirements to all team members. Sharing a written document is an essential way to help ensure that the requirements have been communicated. Third, as a project progresses over time, team members and sponsors tend to forget what the original requirements really were.

PROJECT SPONSOR: ____________ **PROJECT LEADER:** ________________

PROJECT REQUIREMENTS DOCUMENT

What is to be done?	________________
What are the goals and benefits?	________________
What are the deliverables?	________________
What are the increments of deliverables?	________________
What are the standards of performance?	________________
What are the constraints impacting performance, time, cost?	________________
What are the risks to be aware of?	________________

Signatures of Acceptance

Project Leader: ________________ Project Owner: ________________

Date: ________________ Date: ________________

Figure 5 Blank project requirements document.

This process of forgetting tends to allow expectations to drift upward. The requirements document can be used as a tool to bring everyone's perspectives down to earth.

Determining requirements on a project is an iterative process and can be determined by interviewing various stakeholders of the project. The project sponsor (the owner or person requesting the assignment) should always be one of the first people interviewed. To uncover requirements, the leader should ask the sponsor to comment on the following seven questions:

- What exactly needs to be done?
- When is it due?
- What time and cost constraints exist?
- What are the goals and benefits to be realized by successful execution of the project?
- What deliverables (tangible outputs) are to be produced by the project?
- What are the standards of performance by which the quality constraint will be judged?
- What risks should we be aware of?

These questions should also be asked of all project stakeholders; that is, all persons who have a direct interest in the project.

After the project team completes the interviews of stakeholders, a requirements document should be developed that summarizes the answers to the key "what" and "when" questions. The sample document shown in Figure 6 is a useful tool to meet this documentation need. The sample document indicates what information Jim might receive if he interviewed project stakeholders. These individuals represent the project's "customers."

The requirements document may evolve over a period of time. Once a level of agreement is reached, the project leader is ready to move to identify the scope of work to be performed. A useful tool called the work breakdown structure (WBS) links the project's "what" with the "how."

B. Tool #2—The Work Breakdown Structure

The first step in developing a solid project plan is to decompose the work into manageable parts. The tool used to accomplish this is the work breakdown structure. A work breakdown structure is a hierarchical chart that shows what the planner believes to be the major components of work to be performed. It also shows how the work elements roll down into progressively greater levels of detail. Most projects are managed between the second and fourth levels of detail on the WBS.

Generally speaking, a project schedule is best planned at the second or third level of detail on the WBS chart. This level typically puts the project leader at the appropriate level of control between those performing the work and those sponsoring the work. It is usually a waste of time to go further down than five levels of detail. Project budgets are typically less detailed than schedules. Thus, a WBS level 1 or level 2 might be most appropriate when cost control is needed.

Figure 7 illustrates a work breakdown structure using Jim's market research project as an example. Note that the WBS includes all project activities that constitute the scope of work as the team understand it.

Development of the WBS is an excellent planning exercise, and it yields a document that is a convenient communications vehicle. The WBS should be developed by project team members jointly. Going back to our example, if Jim, the project leader, were to show this WBS to a project sponsor, an immediate dialogue could take place to share thoughts, to communicate expectations, and to find out

PROJECT SPONSOR: Jane **PROJECT LEADER:** ____________

PROJECT REQUIREMENTS DOCUMENT

What is to be done	A market research study that accurately shows consumer the responses to the new product xxx. The study must clearly indicate whether or not the company should proceed with product launch, and if so, what percent of the market could be achieved.
What are the goals and benefits?	To reduce the risk of product development. If consumers do not see the product as being of benefit over existing products, development will be stopped or product specs will be modified.
What are the deliverables?	The research results of a valid sample of potential customers that accurately reflects their probability of purchase of product xxx if it is commercialized.
What are the increments of deliverables?	A) General attitudes based upon responses to in-store questionnaires. B) Specific preferences in simulated buying situations in focus groups.
What are the standards of performance?	- Validity of results - Ensuring that the study answers the key questions required by marketing and development.
What are the constraints impacting performance, time, cost?	The research study must be completed within six weeks. Budget is not an issue unless costs exceed our normal costs for this type of research.
What are the risks to be aware of?	Lack of available staff time. We will probably use an outside market research firm. These outside costs must be determined and approved by Jane before the project is approved.

Signatures of Acceptance

Project Leader: ____________ Project Owner: ____________

Date: ____________ Date: ____________

Figure 6 Example project requirements document.

if both people view the scope of work in the same way. The WBS also communicates to the sponsor how the project leader plans to execute the work. In this way, it links the what and the how.

The WBS can also form the basis of cost, resource, and time estimates for project team members. A specific dollar amount or level of effort in hours can be

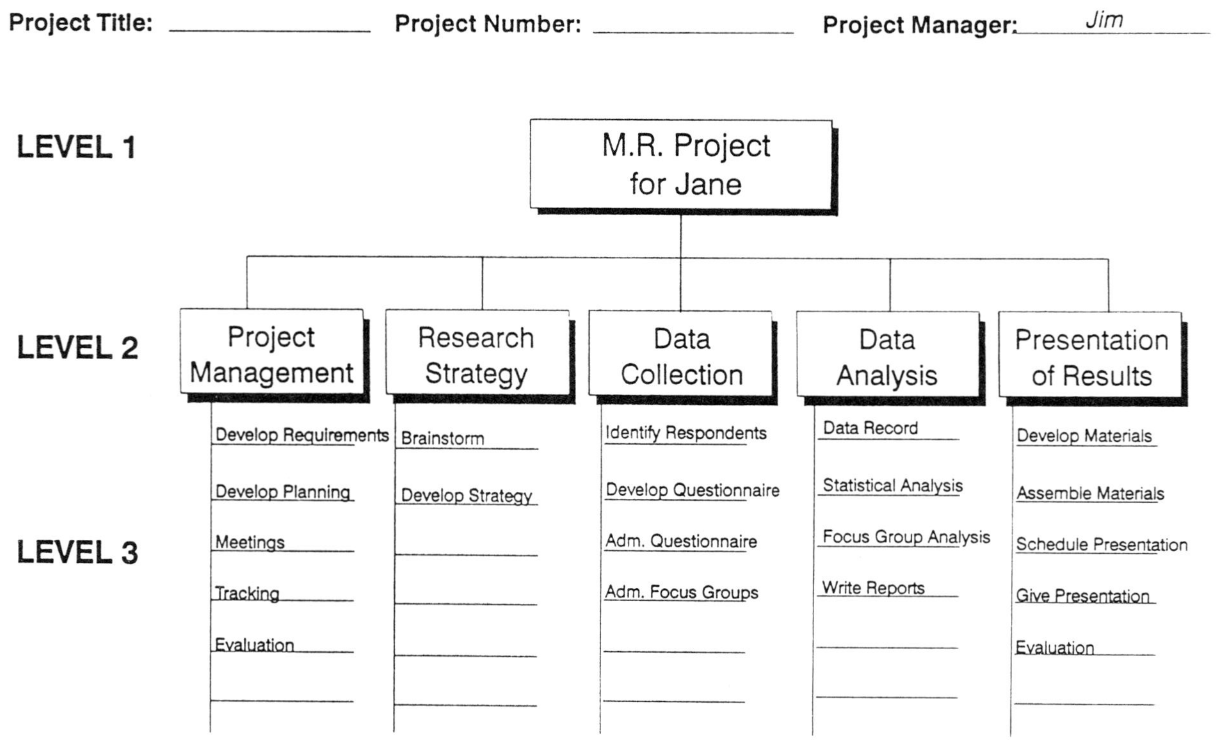

Figure 7 Project work breakdown.

assigned to each of the level 3 elements. These elements are useful for scheduling, resource planning and budgeting.

By developing a WBS, the project leader can smoothly transition from the project requirements to the project schedule. Now, let's pause briefly and review how Jim and Hank might use the requirements document with the work breakdown structure.

Hank: "Well, what do we do now?"

Jim: "We develop a plan that clarifies how the work that will lead to meeting Jane's requirements is going to be performed."

Hank: "I see. The next document describes *what* work is going to be executed and how it will be performed."

Jim: "Exactly, and it also links the what with *who* will perform the work and *how* it is going to be completed."

Hank: "I think we know *who* and *how* the work will be performed, but I'm worried about *when* the work will be performed. Doesn't it all depend upon our staff's availability?"

Jim: "Yes, it does, Jim. I suggest that we also develop a master schedule of all of our jobs and then determine if we will have adequate available staff to perform Jane's project."

Hank: "I have been meaning to develop a master schedule anyway. This is a good opportunity to get it done."

Jim: "But first things must come first. Our next step must be to develop a work breakdown structure that helps us develop a work plan."

Hank: "A work breakdown what?"

Jim: "A work breakdown structure, or WBS. It is a hierarchical chart that partitions the planned work into its basic elements. The WBS links Jane's requirements to our strategy to meet those requirements. The WBS also allows us to estimate how much of our manpower is going to be needed for Jane's project."

C. Tool #3—Project Scheduling

With a work breakdown structure developed, the project team is now ready to assemble a tentative project schedule. A schedule is typically developed by taking the lowest level tasks from the WBS and then building a network diagram from those tasks. It allows us to schedule by determining the precedence of tasks. The network diagram is a simple flowchart that shows:

1. What tasks must happen before other tasks
2. What tasks must happen after other tasks
3. What tasks can happen during the execution of other tasks

A network diagram for Jim's project is shown in Figure 8. After the network diagram is drawn, it is necessary to estimate the duration of each task and then to calculate the total project duration. The tasks that sum together to form the longest path through the network, start to finish, constitute the *critical path.* When added together, task durations on the critical path reflect the minimum project duration. This duration would be the earliest possible date for project completion. This

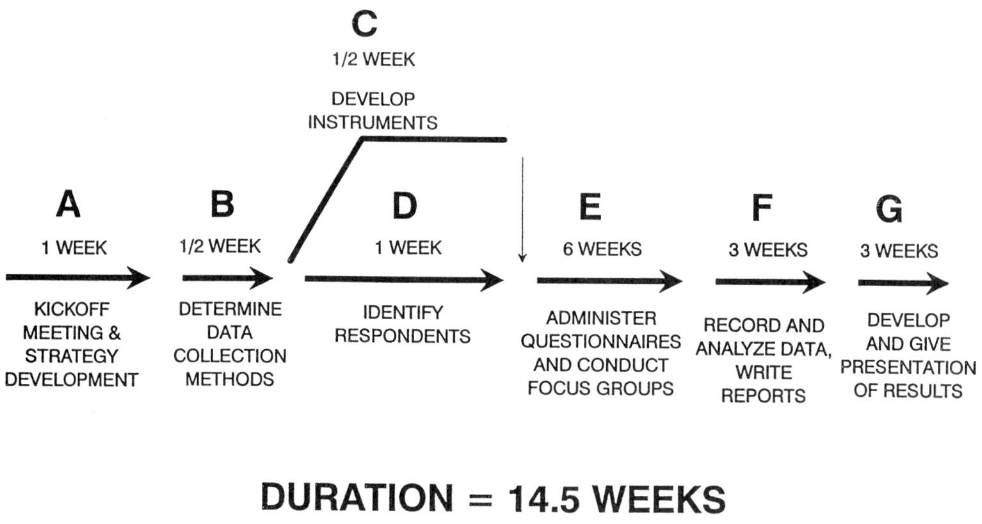

Figure 8 Network diagram for Jim's project.

completion date would form the basis for making a commitment. It is also wise to add time contingency to the schedule to cushion against unforeseen time slippage.

In the case of our manager, Jim, we see that his network diagram suggests that the project will take 14-1/2 weeks to complete. For the most part, all tasks follow a sequence one after the other. The tasks "Develop Instruments" and "Identify Respondents" can occur simultaneously, however.

It is also evident that the tasks A, B, D, E, F, and G form the critical path. Any extension of these tasks' duration would cause the project to slip beyond the estimated 14-1/2 week duration. Task C could finish 1/2 week later than planned without impacting the total project duration. For tasks such as task C, we say that it possesses *float* or *slack* that can absorb a delay.

A more preferred way of displaying a schedule is by developing a Gantt chart (bar chart). The Gantt chart enables us to put time estimates on an easy-to-read time line. Figure 9 shows a Gantt representation of Jim's proposed schedule. Gantt charts are excellent visual tools that facilitate the communication and negotiation of a project schedule. The Gantt charts also includes milestones (marked as Δ) that indicate major accomplishments along the way. Milestones help us track the project. Many project managers use scheduling computer software to plan and update schedules.

One planning caveat to consider concerns resource utilization. For example, before Jim commits to the 14-1/2 week schedule, he should first make sure that people are available to execute the work. Similarly, he should verify that his time estimates are valid. There is nothing more demoralizing for people than to be expected to execute someone else's plan without having input to it. Time estimates are no better than the availability and commitment of the people who are expected to do the work.

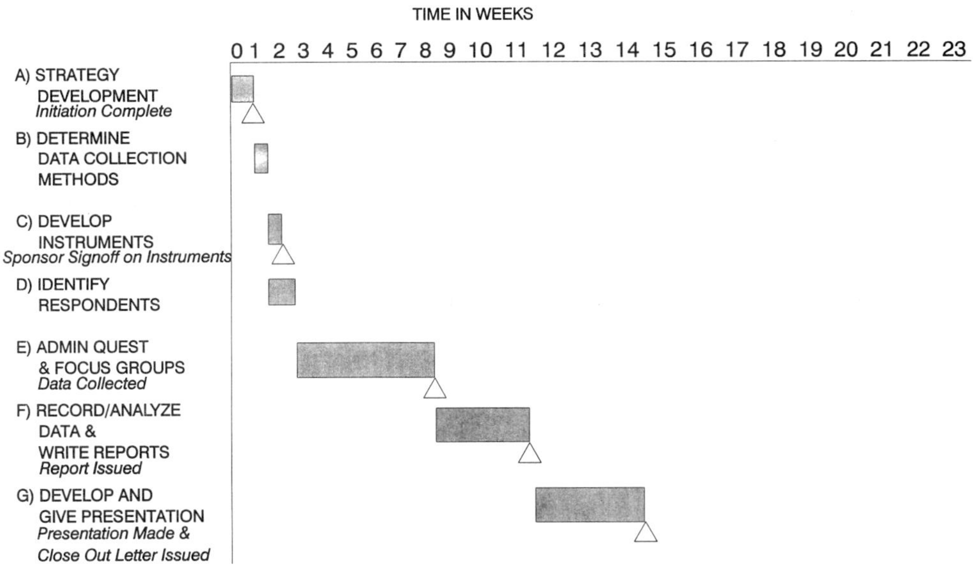

Figure 9 Determining milestones.

An effective device to control resource utilization is the master schedule. The master schedule is a display of all projects and the attendant summation of staff assigned to the projects. A well-constructed master schedule will signal an over commitment of staff.

In order for Jim to assess the impact of Jane's project on the master schedule, he needs to first determine his staffing needs for the duration of the project. Jim needs to estimate both the level of effort in hours to complete each task and the expected duration of each task. Then staffing needs are determined by dividing duration into effort. These estimates are shown in Table 1.

Table 1 Project Staffing Needs

Task	Estimated effort (hours)	Estimated duration (weeks)	Needed staff*
Kick-off meeting and strategy development	120	1	3.0
Determine data collection methods	16	1/2	0.8
Develop instruments	80	1/2	4.0
Identify respondents	40	1	1.0
Administer questionnaires and conduct focus groups	1000	6	4.2
Record data, analysis data, write reports	1200	3	10.0
Develop and give presentation	100	3	0.8

* One staff person can provide up to 40 hours of work per week maximum.

Jim should ask three critical questions at this point in the scheduling process:

1. Are enough people available to perform the job as scheduled?
2. Are the right people available to perform the job as scheduled?
3. What other jobs may have to slip in order to complete the project on time?

By loading Jane's project on the master schedule, Jim and Hank can answer these questions. They may have to adjust their schedules to accommodate staffing availability or adjust the assignment of priorities to release people from other projects. It could very well be that by elevating the priority of Jane's project, Jim will be required to delay someone else's project. This potential conflict must be worked out before Jim makes a commitment. Project leaders must often function as diplomats working between project sponsors each wanting high priority for his/her respective project.

Figure 10 shows a simple, one-project resource utilization chart. In this case, staff available exceeds staff requirements throughout the project's life. Figure 11 shows a multiple-project resource utilization estimate in which staff needs and availability are equal throughout the lives of all three projects. However, any increase in work or decrease in staff will require additional time to complete the projects. In Figure 12, project Y starts after project X and is followed by project Z. Apparently there is more than enough staff. Optimism in the initial planning, however, may result in insufficient resources later.

Optimism is a fact of life in project scheduling. In fact, optimism is built right into some project scheduling techniques. A wise planner takes this into considerations. One way to isolate planning optimism is by examining the assumptions made in the plan. For instance, in Jim's plan, a total of 10 people are required to record data, analyze data, and write a report. Jim should ask: Is that assumption realistic?

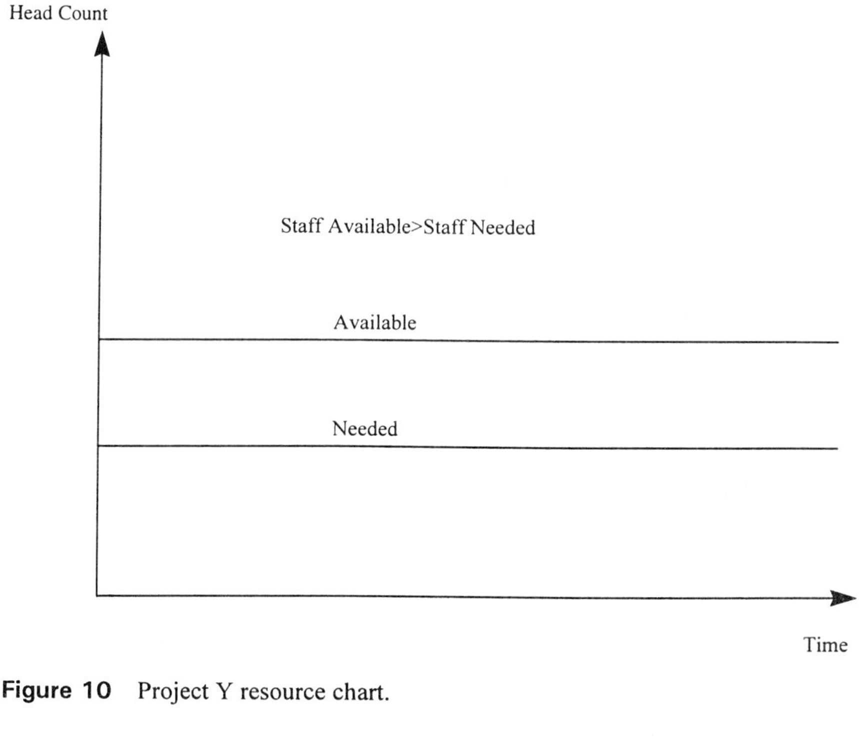

Figure 10 Project Y resource chart.

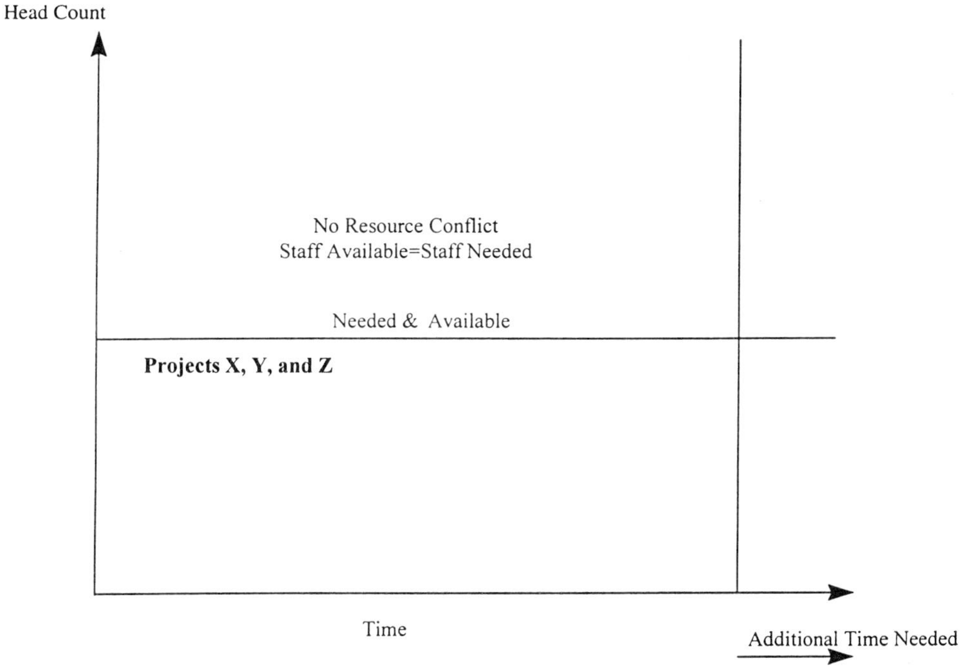

Figure 11 Example X, Y, Z resource chart.

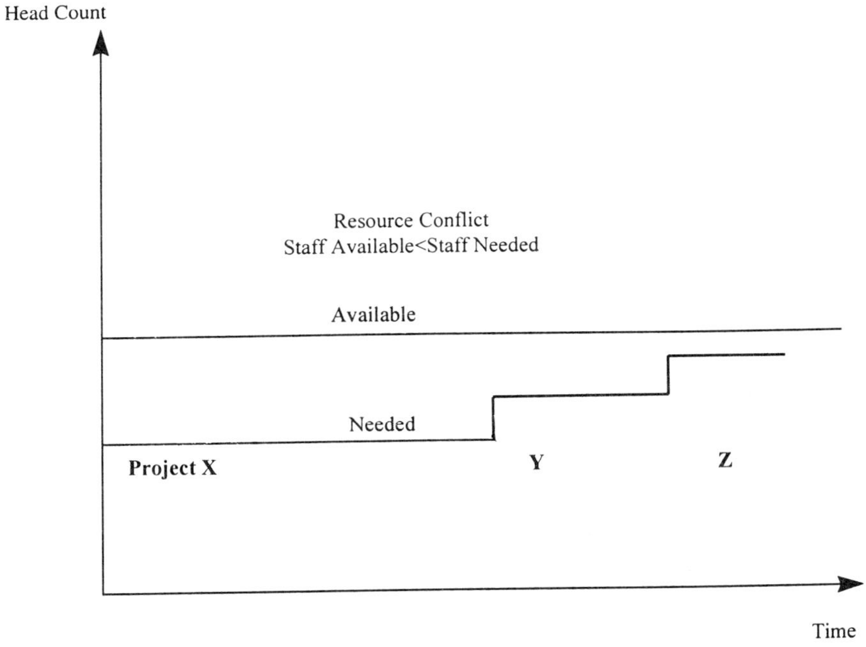

Figure 12 Multiproject example X, Y, Z resource chart.

If 10 people are not available, he should extend the estimated duration or plan to procure outside assistance. Similarly, it is unusual to see people actually spending 40 hours per week working on tasks. A work week of 28 hours per staff person is more realistic.

One way to compensate for optimism in planning is to add contingency to a schedule. Adding an additional 10–20% time on project schedule may be necessary to secure realism.

Finally, before submitting a schedule as a commitment, Jim should obtain a sign-off from each of the team members who will be performing the work. This sign-off is a contract of sorts that shows agreement to the schedule by those who must perform the work. A responsibility assignment matrix is sometimes used as the instrument for sign-off. A commitment does not lead to accountability until it is shown publicly. Sign-offs are useful for this purpose. A responsibility assignment matrix is shown in Figure 13.

Now with a schedule in hand, Jim is ready to negotiate with Jane and make a commitment to meet a specific deadline. If Jane requires a shorter schedule, she must show Jim how and where he can obtain the needed resources or agree to let other projects slip. Similarly, Jim should resist pressure to reduce project dura-

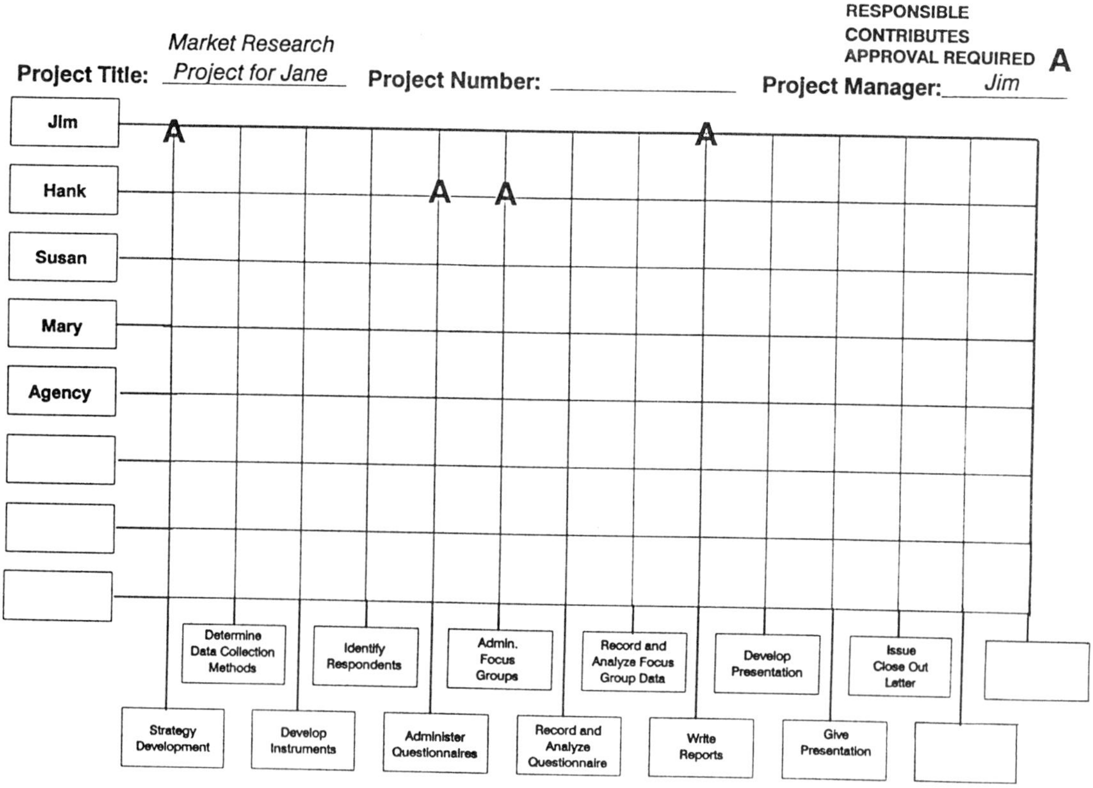

Figure 13 Project responsibility matrix.

tion unless a reduction of scope or an increased level of human resources go along with it.

X. CONTROLLING A PROJECT WITH INTERMEDIATE MILESTONES

Project control begins with the project plan. The project schedule (Gantt chart) should contain a listing of project milestones along with an association reference to the individual responsible and accountable for its successful completion. The successful completion of project milestones indicates that progress toward the project's final goal is actually being made.

It is important that the milestones selected reflect true achievement. Milestones should be anchored in sufficiency criteria that are concrete and tangible. This is necessary to prevent the creeping optimism that team members adopt once project work begins. Notice that the last four of Jim's five milestones are anchored to a tangible deliverable. The first milestone, "Initiation Complete", is not, however. The question arises: How do we know when initiation is actually complete? Without a tangible deliverable and documented sufficiency criteria that signals completion, there is too much room for differing interpretations as to whether the milestone is, in fact, complete. It would be better if, for example, "Initiation Complete" was marked by the issuance of a report or by the sign-off of a project sponsor, or the passing of a test. Assigning such sufficiency criteria to milestones helps remove ambiguity in knowing whether a milestone is achieved or not. Figure 14 describes milestones and sufficiency criteria for a typical quality improvement program.

Referring again to Jim's example, we find that his Gantt chart (see Fig. 9) contains five milestones. These milestones, whose completions are noted by triangles, are:

- Initiation complete
- Sponsor sign-off on instruments
- Data collected
- Analytical report issued
- Presentation made and close-out letter issued

The milestone dates become the basis of control for the project. Effective control over projects necessitates the performance of regular status reviews and team meetings. The use of milestones puts substance into project review meetings. Consider the sequence in which three project reviews are conducted at week 0, week 2, and week 3. Figure 15 illustrates the baseline schedule for the five project milestones just prior to the beginning of work at week 0.

At the first review period (week 2 of execution), the project leader meets with the project performers to check on progress. Figure 16 indicates that the initiation still has not been completed. The report further indicates that the responsibility party reports that initiation will be complete by the end of week 3 and that sponsor sign-off has been delayed 2 weeks. The revised dates also suggest, however, that recovery will take place to put the project back on track by the time the third milestone is due to be complete in week 9. The leader should view this assumption of recovery with suspicion and ask for a detailed explanation of how recovery will actually be accomplished. People are often quite unrealistic in their own assessment of chances for the future recovery of lost time.

	Milestone	Sufficiency Criteria
M1	Process management responsibilities are established	The process owner accepts responsibility as owner Process management team (PMT) is formed and holds a kick-off meeting PMT members accept responsibility *Deliverable*: Produce document that defines team mission, ownership, and team membership responsibilities. All process members sign the document.
M2	Define process and identify customer requirements	The process is defined in terms of its boundaries and interfaces High-level process work flows are defined on a block diagram (map) Customer requirements are defined and quantified Supplier requirements are defined and quantified Customer and supplier requirements are validated. *Deliverable*: Documentation of the above
M3	Process measures taken	Customer requirements are reviewed Effective and appropriate measures are determined Measures are validated Measures are installed as appropriate A measurement reporting system is established for the process A satisfaction feedback system is established with customers and suppliers *Deliverable*: Documentation of the above
M4	Conformance to customer requirements is assessed	Measurement data on process outputs are collected reviewed and evaluated A comparison is made between process performance and process requirements Gaps are identified *Deliverable*: Documentation of the above
M5	Process is investigated to identify improvement opportunities	Critical problems with process outputs are identified Internal process problems are identified Measurable data on problems are collected and Pareto analyzed Each process activity is examined and scrutinized

Figure 14 Sufficiency criteria for milestone completion.

Milestone	Sufficiency Criteria
	Deliverable: (1) A written description is prepared for each identified key process problem, and (2) documentation of the above steps
M6 Improvement opportunities are ranked and priorities are set	Criteria are established for ranking improvement opportunities For each high priority opportunity: Formulate objectives with reasonable targets and time frames Negotiate objectives with stakeholders Select the most immediate opportunities based upon ranking and objectives Document agreements *Deliverable*: Documentation of the above
M7 Execute activities to improve process quality	Develop required plans for quality improvement Identify root causes of problems (cause-effect diagrams) Begin implementation of solution plans Document improvements Hold gains as appropriate *Deliverable*: Documentation of the above
M8 Customer satisfaction achieved	Process demonstrates that it *consistently* meets customer requirements over an appropriate time period Document improvements Hold the gains *Deliverable*: Documentation of the above

Figure 14 Continued

At the next review period, week 4 as indicated in Figure 17, we see a clear pattern that something may be wrong with the project. The first milestone has still not been passed and once again the responsible person believes that recovery will ultimately occur. This may be just wishful thinking. Jim, the project leader, should be discussing the situation with those responsible for completion. The project leader and the team members should explore ways to correct the schedule slippage. If necessary, the project leader may need to raise a red flag to people at higher levels in the organization to resolve issues that impede progress on the project. If the schedule slippage cannot be made up, then the project schedule should be revised.

This process of regularly reviewing and taking prompt, corrective action on variances forms the basis of project control. It should occur throughout the execution phases of the project.

Reporting Period End: WEEK 0

MILESTONE	←EXPECTED COMPLETION DATES→ CURRENT	PREVIOUS	ORIGINAL	STATUS
Initiation Complete			WEEK 1	I
Sponsor Sign-off on Instruments			WEEK 2	I
Data Collected			WEEK 9	I
Data Analyzed and Report Issued			WEEK 12	I
Presentation Made and Close Out Letter Issued			WEEK 15	I

Figure 15 Project control baseline. (I = incomplete.)

XI. SUMMARY

Effective project management requires accountability, planning, and control. The tools of project management are available to assist the leader to achieve the mission while satisfying commitments for quality, schedule, and budget.

Reporting Period End: WEEK 2

MILESTONE	←EXPECTED COMPLETION DATES→ CURRENT	PREVIOUS	ORIGINAL	STATUS
Initiation Complete	WEEK 3		WEEK 1	I
Sponsor Sign-off on Instruments	WEEK 4		WEEK 2	I
Data Collected	WEEK 9		WEEK 9	I
Data Analyzed and Report Issued	WEEK 12		WEEK 12	I
Presentation Made and Close Out Letter Issued	WEEK 15		WEEK 15	I

Figure 16 Project control: first review. (I = incomplete.)

Reporting Period End: WEEK 4

MILESTONE	CURRENT	PREVIOUS	ORIGINAL	STATUS
	← EXPECTED	COMPLETION	DATES →	
Initiation Complete	WEEK 5	WEEK 3	WEEK 1	I
Sponsor Sign-off on Instruments	WEEK 6	WEEK 4	WEEK 2	I
Data Collected	WEEK 9	WEEK 9	WEEK 9	I
Data Analyzed and Report Issued	WEEK 12	WEEK 12	WEEK 12	I
Presentation Made and Close Out Letter Issued	WEEK 15	WEEK 15	WEEK 15	I

Figure 17 Project control: second review. (I = incomplete.)

FURTHER READING

Archibald, R. D. *Managing High-Technology Programs and Projects*. New York: Wiley, 1992.

Boehm, B. W. *Software Engineering Economics*. Englewood Cliffs, NJ: Prentice-Hall, 1981.

Covey, S. R. *The 7 Habits of Highly Effective People, Powerful Lessons in Personal Change*. New York: Fireside, Simon & Schuster, 1990.

Ellul, J. *The Technological Society*. New York: Random House, 1967.

Fisher, R., W. Ury. *Getting to Yes, Negotiating Agreement Without Giving In*. Boston: Houghton Mifflin, 1981.

Fleming, Q. W. *Cost/Schedule Control Systems Criteria, The Management Guide to C/SCSC*. Chicago: Probus, 1992.

Hersey, P., and K. H. Blanchard. *Management of Organizational Behavior, Utilizing Human Resources*. Englewood Cliffs, NJ: Prentice-Hall, 1988.

Humphrey, W. S. *Managing the Software Process*. New York: Addison-Wesley, 1990.

Janis, I. L., and L. Mann. *Decision Making, A Psychological Analysis of Conflict, Choice and Commitment*. New York: Free Press, 1977.

Juran, J. M., and F. M. Gryna, Jr. *Quality Planning and Analysis*. New York: McGraw-Hill, 1980.

Krone, R. M. *Essays For Systems Managers, Leadership Guidelines*. Bend, OR: Daniel Spencer, 1991.

Meredith, J. R., and S. J. Mantel, Jr. *Project Management, A Managerial Approach*. New York: Wiley, 1989.

Nicholas, J. M. *Managing Business & Engineering Projects, Concepts & Implementation*. Englewood Cliffs, NJ: Prentice-Hall, 1990.

Obradovitch, M. M., and S. E. Stephanou. *Project Management, Risks & Productivity*. Bend, OR: Daniel Spencer, 1990.

Pappas, J. L., and M. Hirschey. *Fundamentals of Managerial Economics*. Chicago: Dryden, 1989.

Pusateri, C. J. *A History of American Business*. Arlington Heights, IL: Harlan Davidson, 1988.

Randolph, W. A., and B. Z. Posner. *Effective Project Planning & Management, Getting the Job Done*. Englewood Cliffs, NJ: Prentice-Hall, 1988.

Scherkenbach, W. W. *The Deming Route to Quality and Productivity, Road Maps and Roadblocks*. Milwaukee: CEE Press, 1990.

Smith, P. G., and D. G. Reinersten. *Developing Products in Half the Time*. New York: Van Nostrand Reinhold, 1991.

Stephanou, S. E. *The Systems Approach to Societal Problems*. Malibu, CA: Daniel Spencer, 1982.

Thayer, R. H. Tutorial: *Software Engineering Project Management*. Los Alamitos, CA: IEEE Computer Society Press, 1990.

Tompkins, B. G. *Project Cost Control for Managers, An Inside Track to Eliminating Cost Overruns*. Houston: Gulf Publishing, 1985.

Wideman, R. M. (ed.). *Project Management Body of Knowledge (PMBOK) Glossary of Terms*. Drexel Hill, PA: Project Management Institute, 1987.

10
Marketing and Quality

GEORGE F. KLEITZ
George Kleitz and Associates, Inc.
Wheaton, Illinois

I. INTRODUCTION

Once upon a time, the owner of an advertising agency was making a sales presentation to the president of a medium-sized company. The agency owner's goal was to secure the company's advertising and sales promotion account.

The president of the company was agreeable enough, and he smiled, nodding his approval as the agency owner went on giving the presentation, stressing the creative ideas that could help sell more business. She was right "on track"; or so she thought.

When she turned to the subject of marketing, however, she saw her prospective client's attitude change. Instead of a smile, his well-tanned face became grim and that grimness intensified as he leaned forward. Almost completely out of his chair he thundered "Marketing. . . . ? That may be great for General Motors or General Foods, but we don't have time for it. We have to concentrate on selling." The ad agency owner was stunned. The concept of marketing was the foundation that would make her selling ideas come to life.

Just at that point, the president's secretary popped into the office: "Mike (the company's star salesman) is on the phone . . . you'd better talk to him. Now. It sounds serious."

As the president listened to Mike, his grim expression deepened, then turned to a look of angry puzzlement. "How could they (the company's largest customer) cancel that order? How could they give that order to . . . (the company's biggest competitor). It's just not possible." But the customer did just that . . . a story that is repeated all too often.

II. THE CAUSE

What happened? The company lost the customer because the quality hadn't been there. Even though the customer had brought the problems to the company's attention any number of times, no one had been listening. Although there was enough blame to spread around, it was the lack of belief in marketing that was really at fault.

III. REAL MARKETING

Originally, the definition of "to market" was simply to bring one's product to the marketplace; a form of selling. In the late 20th century, however, "marketing" has come to mean much more than mere selling.

Marketing does imply successfully selling a product or service through personal effort, advertising, and promotion, but first it means that the seller has found out a great deal of information about the market. This information includes determining what customers want—or, more likely, demand—in the product or service. As such, marketing is closely tied to quality. In fact, quality and marketing should work in tandem, the joint goal being that of securing customer acceptance of the company's products and services, at a profitable price, again and again, without reservation.

Perhaps the strongest example of the difference between "just selling" and the marketing/quality team is the plight of the U.S. automakers—the "Big 3"—who, in the early 1980s, simply ignored the U.S. customer. It was not until later in that decade, after drastic market share losses, that the Big 3 discovered that marketing involves the customer—and that you must listen, and act to satisfy that customer.

IV. THE MARKETING FUNCTION

Probably more than any other activity area in a firm, marketing has the closest, most frequent contacts with customers. These contacts may be through sales, sales promotion, advertising, merchandising, or customer service.

Thus the marketing department, or the marketing function, if the department is referred to by a different name, should be the eyes and ears of the company in terms of finding out what the customer's requirements are.

Of course, the aim of the marketing function, in the long run, is to convince the customer that the company's products are just what the customer is looking for, but that won't happen unless the company first finds out what the customer wants.

A modern marketing function is integrated, all along the way, with the other functions of the company. Along with quality assurance, the marketing function—and the difference between marketing and selling—must be recognized by top management as being vitally important. Otherwise, as was presented at the beginning of this chapter, the company cannot fulfill its capabilities.

To be successful with marketing, the company must seek out and employ skilled practitioners who have a track record of accomplishment in the various phases of marketing. Too often, critical marketing jobs are relegated to the new, untried employee, or, as seems to be increasing, to those with a speaking acquaintanceship with desktop publishing.

There is also an "underside" to marketing that does not fit the popular perception of the well-dressed person working with glamorous media people and directing the expenditure of millions of dollars on television advertising. Such executives are a small portion of the cadre, the "tip of the iceberg." Behind them is a group of talented, hardworking "unsung" people who determine who the customer is; what it is that the customer wants; and what it will take to keep them sold. This is what successful marketing is built on.

V. THE MARKETING PLAN

Where do successful marketers start? With a real Marketing Plan. Surprisingly, many small and mid-size companies do not have a marketing plan, believing—as did the company president mentioned at the beginning of this chapter—that marketing plans are nonsense. The reasoning in these companies is that selling, not marketing, is what is really required.

As this chapter was being written, a typical example was occurring. A company in the automotive aftermarket, one that had grown by its reliance on personal selling, decided to go into an allied market niche. The company had been in this market previously, but didn't do well. The current idea was that some "new" advertising would fix the previous problem. It didn't. The new ad campaign didn't work, because when the officers of the company were asked who their customers and prospects would be, they hadn't any idea. Neither did they know who the competition was nor what the competition was doing!

A good marketing operation is very much like the intelligence operations that were so important during World War II. Everything was organized; everyone had an assignment and concentrated on doing it well. After all, lives were at stake! Modern professional football teams also follow a format except that it is called a "game plan."

Essentially, the marketing plan is very much like the military intelligence operation. Consisting of seven sections, or chapters, it contains information on the market and deductions as to courses of action. It is written, monitored, and continuously updated so as to be of constantly high value to the company.

VI. THE PLAN OUTLINE

A general outline for a marketing plan would contain the following sections:

A. The Company—History and Background

How does the company see itself?
What is its reason for being in business?
What specific markets is the company targeting?
What products and services does the company emphasize?
From which market segments will the company derive the most profit?
How does the company handle production; sales; research?
What is the company's definition of quality?
What are the company's competitive advantages and disadvantages?
What are the company's sales goals? Now? Three years from now?
What strategies does the company pursue to achieve its ends?
What is the relationship between pricing and performance?
What is the relationship between sales and performance?

What new techniques, products, services, technologies, are projected? In the near term? In the long run?

B. The Industry and the Marketplace

People often assume that they really know the specifics of the industry and the marketplace they are in. Most often, they don't, simply because they are too busy with day-to-day matters to stay well informed. Is the industry mature but stable? Growing? Contracting? The specifics will greatly affect the marketing operations plan.

The marketing plan takes an in-depth, dispassionate look at the industry, its history, and its make-up in terms of important trends, competition, and functions. The plan also seeks to establish past and current market share breakout.

C. Customers and Prospects

It is important to know exactly who your customers and prospects are and what quality levels each is looking for. What are they saying and doing? What is their impression of your company?

Since, typically, some 20% of the companies account for 80% of the market opportunities in a given industry, how do any of the smaller players go after a bigger bite?

Who are more likely, given the company's character and all of the aspects of the market, to be the company's best prospects for new business?

D. The Competition

Who are the company's competitors, ranked by size and threat to market share?

How does the company rank in comparison to the competition?

Have any benchmarking studies been done? Planned?

What are the characteristics of each competitor in terms of product quality, customers held, services and products offered, and technological capabilities?

What is the competition's apparent sales strategy?

On what does the competition put its marketing emphasis?

How does the competition operate in the field?

How does the competition communicate with customers and prospects?

How does this company compare with our competitors in all respects?

E. Evaluation

After recording and distilling all of the information noted in sections A through D above, what logical conclusions can be reached, generally and specifically, as to our overall marketing presence?

What pitfalls exist; i.e., lack of quality, low technology level, poor communication with the customer, and the like?

What opportunities exist?
How can we effectively mount an attack that will successfully reach our long- and short-term objectives?

F. Strategies and Tactics

Define the position that the company should take with regard to the market both in general and specifically.
Define the methods the company chooses to use to overcome pitfalls and maximize opportunities.
Define the tools and policies that the company will use both internally and externally.
Define the company's marketing themes and concepts.

G. Plans of Implementation

A plan of implementation is most important since even the best made, best laid plans are worthless unless they are implemented.

Implementation steps:

What, when, and how will we proceed both internally and externally.
How will we budget-in those tools and logistics that we will need to meet whatever requirements we find are necessary to achieve our goals?
Lay out the actual plan, based on a real calendar with exact timing, cost, and review periods.[1]

VII. THE PARTICIPANTS

The Marketing Plan requires the cooperation of every department head in the company, otherwise it is soon of little worth. The required participation is not always easy, because the plan requires facing up to reality. The plan is only as good as it is accurate.

What is being done more and more often to overcome the problem of eventual disinterest and/or intracompany barriers is to select a qualified outside consultant. The outside person or group can generally get the work done not only faster, but more accurately and more completely as well.

The marketing plan result is typically most effective when it is requested by the highest level of management but derived under the control of the marketing department.

Many knowledgeable companies today recognize the need for one group to take the major responsibility for finding out what the customer wants. That group, by whatever title and of whatever size, is the Marketing department. These companies also recognize that one group has the major responsibility for translating the

[1] Refer to Chapter 9, "Project Management." The techniques of project management are a powerful way of implementing a marketing plan.

customer's needs to materials and machines or means of providing the service. That group, by whatever title and of whatever size, is the Engineering department. Such companies also recognize that one group has the major responsibility for using the materials and machines to produce effectively those things or services that the customer wants. That group, of course, is Production or Operations. Finally, these same companies recognize that one group has the major responsibility to measure, test, evaluate and report. That group is Quality Assurance or Quality Control.

VIII. CONCLUSION

Marketing is an ongoing process. It doesn't stop once the plan of action is put into effect. Listening to customers—listening seriously—is crucial. Certainly one function of the sales force is to report accurately what is happening—what customers really feel—and what they are doing. And it is important to get this information to top management, production, research, and quality assurance quickly and accurately.

This approach is more than academic. A consistent, high-quality product made to fit the customer's needs and priced competitively is the key to ever greater sales. And when the entire company is engaged in putting its best foot forward every day, its marketing efforts can be considered a success.

FURTHER READING

Crosby, P. B. *Quality Without Tears*, New York: McGraw-Hill, 1984.
Grenier, R. W. *Customer Satisfaction Through Total Quality Assurance*. Wheaton, IL: Hitchcock, 1988.
Levin, T. *Marketing Myopia*.

11

The Malcolm Baldrige National Quality Award

CURT W. REIMANN and HARRY S. HERTZ
National Institute of Standards and Technology
Gaithersburg, Maryland

I. INTRODUCTION

The Malcolm Baldrige National Quality Award (MBNQA) is an award for quality management and achievement presented annually by the President of the United States. Created by Public Law 100-107 in 1987, the award is intended to serve as a focal point that U.S. companies can use for performance improvement.

The Secretary of Commerce and the National Institute of Standards and Technology (NIST), an agency of the Department of Commerce, were given responsibilities to develop and manage the Award with cooperation and financial support from the private sector. Although currently open only to for-profit companies, the award program, including award criteria, has been designed for the involvement of all U.S. organizations. (Pilot programs were conducted in education and health care in 1995.)

Early experiences in the award program show that award winners and contenders have demonstrated improvement in quality of products and services, a reduction in operating costs, and numerous other operating improvements. Organizations throughout the United States are moving to learn from the successful strategies made known through the award program. In addition, the award criteria have evolved to capture the lessons learned, and they represent a body of knowledge available to improve performance practices in companies throughout the United States.

Malcolm Baldrige

Malcolm Baldrige was Secretary of Commerce from 1981 until his death in a rodeo accident in July 1987. Baldrige was a proponent of quality management as a key to the U.S. prosperity and long-term strength. He took a personal interest in the quality improvement act that was eventually named after him and helped draft one of the early versions. In recognition of his contributions, Congress named the award in his honor.

II. AWARD MANAGEMENT

A. The Foundation

The Foundation for the Malcolm Baldrige National Quality Award was created to foster the success of the Program. The Foundation's main objective is to raise funds permanently to endow the Award Program.

Prominent leaders from U.S. companies serve as Foundation Trustees to ensure that the Foundation's objectives are accomplished. Donor organizations vary in size and type, and they are representative of many kinds of businesses and business groups.

B. Department of Commerce

Responsibility for the Award is assigned to the Department of Commerce. The National Institute of Standards and Technology (NIST), an agency of the Department's Technology Administration, manages the Award Program.

NIST's goals are to aid U.S. industry through research and services; to contribute to public health, safety, and the environment; and to support the U.S. scientific and engineering research communities. NIST conducts basic and applied research in the physical sciences and engineering and develops measurement techniques, test methods, and standards. Much of NIST's work relates directly to technology development and technology utilization.

C. Board of Overseers

The Board of Overseers is the advisory organization to the Department of Commerce. The Board is appointed by the Secretary of Commerce and consists of distinguished leaders from all sectors of the U.S. economy.

The Board evaluates all aspects of the Award Program, including the adequacy of the criteria and processes for making awards. An important part of the responsibility is to assess how well the Award is serving the national interest. Accordingly, the Board makes recommendations to the Secretary of Commerce and to the Director of NIST regarding changes and improvements in the Award Program.

D. Board of Examiners

The Board of Examiners evaluates award applications; prepares feedback reports; and makes award recommendations to the Director of NIST. The Board consists of business and quality experts primarily from the private sector; its members are selected by NIST through a competitive application process.

For 1995, the Board consists of about 270 members. Of these, 9 serve as Judges, and approximately 50 serve as Senior Examiners. The remainder serve as Examiners. All members of the Board take part in an Examiner preparation course.

In addition to their application review responsibilities, Board members contribute significantly to information transfer activities. Many of these activities involve the hundreds of professional, trade, community, and state organizations to which Board members belong.

III. AWARD PURPOSES

The Malcolm Baldrige National Quality Award was established for three basic purposes:

1. To promote awareness of the importance of quality and performance improvement to U.S. competitiveness
2. To recognize successful U.S. companies for their performance management and strategies and performance improvement results
3. To promote sharing of successful quality strategies among U.S. organizations

IV. ELIGIBILITY

Public Law 100-107 establishes the three eligibility categories of the Award: manufacturing, service, and small business. Any for-profit business located in the United States or its territories may apply for the Award. Eligibility for the Award is intended to be as open as possible.

Minor eligibility restrictions and conditions ensure fairness and consistency in definition. For example, publicly or privately owned, domestic or foreign-owned, joint ventures, incorporated firms, sole proprietorships, partnerships, and holding companies may apply. Not eligible are local, state, and national government agencies; not-for-profit organizations; trade associations; and professional societies.

Award Eligibility Categories

Manufacturing

Companies or subsidiaries[1] that produce and sell manufactured products or manufacturing processes and those companies that produce agricultural, mining, or construction products.

Service

Companies or subsidiaries that sell services.

Small Business

A small business is considered to be a complete business with not more than 500 full-time employees. Business activities may include manufacturing and/or service.

---

[1] For purposes of applying for the Award, a subsidiary is taken to mean an actual subsidiary, business unit, division, or like organization. In the manufacturing and service categories, subsidiary units of a company may be eligible for the Award. Small businesses must apply as a whole; subsidiary units of small businesses are not eligible.

A small business must be able to document that it functions independently of any other businesses which are equity owners.

For example, a small business owned by a holding company would be eligible if it can document its independent operation and that other units of the holding company are in different businesses.

If there are equity owners with some management control, at least 50% of the small business's customer base (dollar volume for products and services) must be from other than the equity owners or other businesses owned by the equity owners.

The proper classification of companies that perform both manufacturing and service is determined by the larger percentage of sales.

Up to two awards may be given each year in each of the three eligibility categories. The removal of the limit of two awards per category has been proposed by the Award Program's Board of Overseers and forwarded to Congress.

V. APPLYING FOR THE AWARD

In order to participate in the award evaluation process, companies are required to submit application packages. An application package consists of basic information about the company; approved eligibility forms; and responses to the 24 examination items that make up the award criteria. There is a 70-page limit for these responses.

The award criteria are shown in Figure 1. Fees are charged to cover some of the costs of review. All applications are confidential and applicants are not expected to provide proprietary information about products or processes.

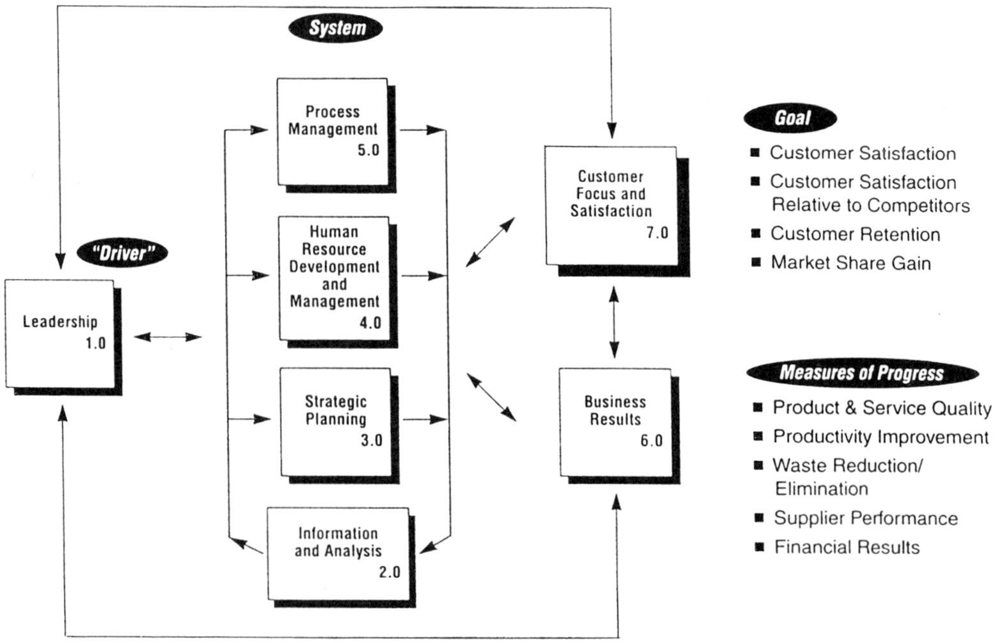

Figure 1 Baldrige Award Criteria framework: dynamic relationships.

A. Consultants

The Award places primary emphasis on facts and data. Although a company is free to hire consultants to advise on total quality assurance and the Baldrige Award itself, there are no "secret" answers or even right or wrong answers to the Baldrige Award application. The Award cannot be won by hiring someone to "fill in the blanks."

Instead, the company must show through facts and data that it has a world-class performance management system in place and that it is continually looking for ways to improve.

As a final check before recommending winners, members of the Board of Examiners visit the more outstanding candidates for the Award. During these site visits, examiners interview employees and review pertinent records and data. The objective is to verify the information provided in the application and to answer questions raised during the Board's review. A company that hired someone to fill out its application would never make it through this rigorous review without the internal processes and results to warrant receiving the Award.

B. Application Review

Applications are reviewed and evaluated by members of the Board of Examiners in a four-stage process:

Stage 1: Independent review and evaluation by at least five members of the Board

Stage 2: Consensus review and evaluation for applications that score well in Stage 1

Stage 3: Site visits to applicants that score well in Stage 2

Stage 4: Judges' review and recommendations

Board members are assigned to applications taking into account the nature of the applicants' businesses and the expertise of the Examiners. Assignments are made in accord with strict rules regarding conflict of interest.

Applications are reviewed without funding from the U.S. government. Review expenses are paid primarily through application fees; partial support for the reviews is provided by the Foundation for the Malcolm Baldrige National Quality Award.

C. Feedback

All applicants for the Award receive a written feedback report that is based upon a rigorous evaluation of the applicant's responses to the requirements of the Award criteria. Considered to be the most important element of the Baldrige Award system, an applicant company's feedback reports typically contain over 150 areas of strength and areas that need improvement. A tool for continuous improvement, feedback reports are often used by companies for strategic planning with a focus on improvement of processes for delivering customer value and for improving productivity and asset utilization.

D. Award Recipient's Responsibilities

Award recipients are required to share information on their successful performance and quality strategies with other U.S. organizations. However, recipients are not required to share proprietary information, even if such information was part of their Award application. The principal mechanism for sharing information is the annual Quest for Excellence Conference.

Award recipients in the first seven years of the Award have been very generous in their commitment to improving U.S. competitiveness, and manufacturing and service quality. They have shared information with hundreds of thousands of companies, educational institutions, government agencies, health care organizations, and others. This sharing far exceeds expectations and Program requirements. Award winners' efforts have encouraged many other organizations in all sectors of the U.S. economy to undertake their own performance improvement efforts.

VI. AWARD CRITERIA

The Malcolm Baldrige National Quality Award Criteria are the basis for both making Awards and providing feedback to applicants. In addition, the criteria can be used to help evaluate quality standards and expectations; to facilitate communication and sharing between and within organizations of all types based upon common understanding of key quality requirements; and to serve as a working tool for planning, training, assessment, to list only a few of the many uses.

The Award Criteria are simultaneously directed toward two results-oriented goals: that of providing key requirements for delivering ever-improving value to customers; and that of maximizing the overall effectiveness and productivity of the delivering organization. To achieve these results-oriented goals, the Criteria are built upon a set of values that together address and integrate the overall customer and company performance requirements.

A. Core Values and Concepts

The Award Criteria are built upon 11 core values and concepts:

1. Customer-driven quality
2. Leadership
3. Continuous improvement and learning
4. Employee participation and development
5. Fast response
6. Design quality and prevention
7. Long-range view of the future
8. Management by fact
9. Partnership development
10. Corporate responsibility and citizenship
11. Results orientation

These core values and concepts are described in depth in the Malcolm Baldrige National Quality Award Criteria booklet.

B. Criteria Framework

The core values and concepts are embodied in seven categories:

- 1.0. Leadership
- 2.0. Information and analysis

3.0. Strategic planning
4.0. Human resource development and management
5.0. Process management
6.0. Business results
7.0. Customer focus and satisfaction

The framework connecting and integrating the categories is shown in Figure 1. The framework has four basic elements: driver, system, measures of progress, and goal.

Driver

Senior executive leadership acts as the driver by setting directions; creating values, goals, and systems; and guiding the pursuit of customer value and company performance improvement.

System

The system comprises the set of well-defined and well-designed processes for meeting the company's customer and performance requirements.

Measures of Progress

Measures of Progress provide a results-oriented basis for channeling actions to delivering ever-improving customer value and company performance.

Goal

The goal—i.e., the basic aims—of the system are the delivery of ever-improving value to customers and success in the marketplace. The seven Criteria Categories shown in Figure 1 are subdivided into examination items and areas to address.

C. Examination Items

Under the Criteria framework are 24 examination items, each focusing on a major requirement. The Items consist of sets of areas to address (Areas). Information is submitted by applicants in response to specific requirements of these Areas. Item titles and point values are shown in Table 1.

Key Characteristics of the Award Criteria

The Criteria Are Directed Toward Business Results

The Criteria focus principally on seven key areas of business performance. Business performance is a composite of:

1. Customer satisfaction and retention
2. Market share and new market development
3. Product and service quality
4. Financial indicators, productivity, operational effectiveness, and responsiveness
5. Human resource performance and development

Table 1 1995 Award Examination Criteria—Item Listing

1995 Examination Categories/Items		Point Value
1.0	Leadership	90
	1.1 Senior Executive Leadership	
	1.2 Leadership System and Organization	
	1.3 Public Responsibility and Corporate Citizenship	
2.0	Information and Analysis	75
	2.1 Management of Information and Data	
	2.2 Competitive Comparisons and Benchmarking	
	2.3 Analysis and Use of Company-Level Data	
3.0	Strategic Planning	55
	3.1 Strategy Development	
	3.2 Strategy Deployment	
4.0	Human Resource Development and Management	140
	4.1 Human Resource Planning and Evaluation	
	4.2 High Performance Work Systems	
	4.3 Employee Education, Training, and Development	
	4.4 Employee Well-Being and Satisfaction	
5.0	Process Management	140
	5.1 Design and Introduction of Products and Services	
	5.2 Process Management Product and Service Production and Delivery	
	5.3 Process Management: Support Services	
	5.4 Management of Supplier Performance	
6.0	Business Results	250
	6.1 Produce and Service Quality Results	
	6.2 Company Operational and Financial Results	
	6.3 Supplier Performance Results	
7.0	Customer Focus and Satisfaction	250
	7.1 Customer and Market Knowledge	
	7.2 Customer Relationship Management	
	7.3 Customer Satisfaction Determination	
	7.4 Customer Satisfaction Results	
	7.5 Customer Satisfaction Comparison	
Total Points		1000

6. Supplier performance development
7. Public responsibility corporate citizenship

Improvements in these seven results areas contribute to overall company performance, including financial performance. The results areas also recognize the importance of suppliers and community and national well-being.

The use of a composite of indicators helps to ensure that strategies are balanced—that they do not trade-off among important stakeholders or objectives. The composite of indicators also helps to ensure that company strategies bridge short-term and long-term goals.

The Criteria Are Nonprescriptive

The Criteria provide wide latitude in how requirements are met. Accordingly, the Criteria do not prescribe:

- Specific tools, techniques, technologies, systems, or starting points
- That there should or should not be within a company a quality or a planning department
- How the company itself should be organized

The Criteria do emphasize that these and other factors be regularly evaluated as part of the company's improvement processes. The factors listed are important and are very likely to change as needs and strategies evolve.

The Criteria are nonprescriptive because:

1. The focus is on results, not on procedures, tools, or organizations. Companies are encouraged to develop and *demonstrate* creative, adaptive, and flexible approaches to meeting basic requirements. Nonprescriptive requirements are intended to foster both incremental and major ("breakthrough") improvement.
2. Selection of tools, techniques, systems, and organizations usually depends upon many factors such as business size, business type, the company stage of development, and employee capabilities.
3. Focusing on common requirements within a company, rather than on specific procedures, fosters better understanding, communication, and sharing while supporting creativity in approaches.

The Criteria Are Comprehensive

The Criteria address all internal and external requirements of the company, including those related to fulfilling its public responsibilities. Accordingly, all processes of all company work units are tied to these requirements. New or changing strategies may be readily adapted within the same set of Criteria requirements.

The Criteria Include Interrelated (Process Results) Learning Cycles

There is dynamic linkage among the Criteria requirements. Learning—and action based upon that learning—takes place via feedback among the process and results elements.

The results have four clearly defined stages:

1. Planning, including design of processes, selection of indicators, and deployment of requirements
2. Execution of plans
3. Assessment of progress, taking into account internal and external (results) indicators
4. Revision of plans based upon assessment findings

The Criteria Emphasize Alignment

The Criteria call for improvement (learning) cycles in all parts of the company. To ensure that these cycles—carried out in different parts of the company—support one another, overall aims need to be consistent or aligned. Alignment in the Criteria is achieved via connecting and reinforcing measures derived from overall company requirements.

These measures tie directly to customer value and to operational performance. The use of measures thus channels different activities in agreed-upon directions. The use of measures often avoids the need for detailed procedures or centralization of decision making or process management. Measures thus provide a communications tool and a basis for deploying consistent customer and operational performance requirements. Such alignment ensures consistency of purpose while at the same time supporting speed, innovation, and decentralized decision making.

The Criteria Are Part of a Diagnostic System

The Criteria and the Scoring Guidelines make up a two-part diagnostic (assessment) system. The Criteria are a set of 24 basic, results-oriented requirements. The scoring guidelines spell out the assessment dimensions—approach, deployment, and results—and the key factors used in assessment relative to each dimension. An assessment thus provides a profile of strengths and areas for improvement relative to the 24 requirements. In this way, the assessment directs attention to actions that contribute to the results composite described above.

VII. ORGANIZATIONAL IMPROVEMENT

Nonprescriptive, results-oriented Criteria and key measures and indicators focus on *what* needs to be improved. This approach helps to ensure that improvements throughout the organization contribute to the organization's overall objectives. In addition to fostering creativity in approach and organization, results-oriented Criteria and key measures and indicators encourage "breakthrough thinking"—that is, openness to the possibility for major improvements as well as to incremental ones. However, if key measures and indicators are tied too directly to existing work organizations and processes, breakthrough changes may be discouraged. For this reason, analysis of processes and progress should focus on the selection of and the value of the measures and indicators themselves. This will help to ensure that measure and indicator selection does not stifle creativity that may lead to entirely new approaches.

Benchmarking may also serve a useful purpose in stimulating breakthrough thinking. Benchmarking may lead to significant improvements based on adoption or adaptation of current best practice.

In addition, benchmarks help encourage creativity through exposure to alternative approaches and represent a clear challenge to "beat the best," thus stimulating the search for major improvements rather than only incremental refinements of existing approaches. As with key measures and indicators, benchmark selection is critical, and benchmarks should be reviewed periodically for appropriateness.

A. Business Strategy and Decisions

The focus on superior offerings and lower costs of operation means that the Criteria's principal route to improved financial performance is through requirements that seek to channel company activities toward producing superior overall value. Delivering superior value—an important part of business strategy—also supports other business strategies such as pricing. For example, superior value offers the possibility of price premiums or competing via lower prices. Pricing decisions may enhance market share and asset utilization, and thus may also contribute to improved financial performance.

Business strategy usually addresses factors in addition to quality and value. For example, strategy may address market niche, alliances, facilities location, diversification, acquisition, export development, research, technology leadership, and rapid product turnover. The Criteria support the development, deployment, and evaluation of business decisions and strategies, even though these involve many factors other than product and service quality.

Examples of applications of the Criteria to business decisions and strategies include:

- Management of the information used in business decisions and strategy—scope, validity, and analysis
- Requirements of niches, new businesses, and export target markets
- Use of benchmarking information in decisions relating to outsourcing, alliances, and acquisitions
- Analysis of factors—societal, regulatory, economic, competitive, and risk—that may bear upon the success or failure of strategy
- Development of scenarios built around possible outcomes of strategy or decisions, including risks and consequences of failures
- Lessons learned from previous strategy developments—within the company or available through research.

B. Financial Performance

The Criteria address financial performance via three major avenues: (1) emphasis on requirements that lead to superior offerings and thus to better market performance, market share gain, and customer retention; (2) emphasis on improved productivity, asset utilization, and lower overall operating costs; and (3) support for business strategy development, business decisions, and innovation.

The Criteria and evaluation system take into account market share, customer retention, customer satisfaction, productivity, asset utilization, and numerous other factors that contribute to financial performance. The Criteria *do encourage* the use of financial information, including profit trends, in analyses and reporting of results derived from performance improvement. However, companies are encouraged to demonstrate the connection between operational performance improvement and financial performance.

C. Innovation and Creativity

Innovation and creativity are important aspects of delivering ever-improving value to customers and of maximizing productivity.

Examples of mechanisms used in the Criteria to encourage innovation and creativity include:

- Nonprescriptive criteria, supported by benchmarks and indicators, encourage creativity and breakthrough thinking as they channel activities toward purpose, not toward following procedures.

- Customer-driven quality places major emphasis on the "positive side of quality," which stresses enhancement, new services, and customer relationship management. Success with the positive side of quality depends heavily on creativity—usually more so than steps to reduce errors and defects which tend to rely more on well-defined techniques.
- Continuous improvement and cycles of learning are stressed as integral parts of the activities of all work units. This encourages analysis and problem solving everywhere within the company.
- Strong emphasis on cycle time reduction in all company operations encourages companies to analyze work paths, work organizations, and the value-added contributions of all process steps. This fosters change, innovation, and creative thinking in how work is organized and conducted.
- A focus on future requirements of customers, customer segments, and customers of competitors encourages companies to seek innovative and creative ways to serve needs.

Examples of specific process management mechanisms to improve new product and process innovation include:

- A strong emphasis on cycle time in the design phase to encourage rapid introduction of new products and services derived from company research. Success requires stage-to-stage coordination of functions and activities ranging from basic research to commercialization.
- Requirements for research and development units that address:
    1. A climate for innovation, including research opportunities and career advancement
    2. Unit awareness of fundamental knowledge that bears upon success
    3. Unit awareness of national and world leadership centers in universities, government laboratories, and other companies
    4. Shortening the patenting cycle; effectiveness of services to research and development by other units including procurement, facilities management, and technical support
    5. Key determinants in project success and project cancellation
    6. Company communication links, including internal technology transfer
    7. Key technical and reporting requirements and communications
    8. Key measures of success—such as problem-solving effectiveness, responsiveness, and value creation—for research and development units

VIII. SCORING SYSTEM

The system for scoring applicant responses to examination items and for developing feedback is based upon three evaluation dimensions: (1) approach, (2) deployment, and (3) results.

A. Approach

Approach refers to how the applicant addresses the item requirements—i.e., the *method*(s) used. The factors used to evaluate approaches include the following:

- The appropriateness of the use of methods to meet the requirements
- The effectiveness of the methods used
- The degree to which the approach is systematic, integrated, and consistently applied
- The degree to which the approach embodies evaluation and improvement cycles and is based upon data and information that are objective and reliable evidence of innovation (this includes significant and effective adaptations of approaches used in other applications or types of businesses).

B. Deployment

Deployment refers to the *extent* to which the applicant's approach is applied to all item requirements. The factors used to evaluate deployment include the following:

- Use of the approach in addressing business and item requirements
- Use of the approach by all appropriate work units

C. Results

Results refers to *outcomes* in achieving the purposes given in the item. The factors used to evaluate results include the following:

- Current performance levels
- Performance levels relative to appropriate comparisons and/or benchmarks
- Rate, breadth, and importance of performance improvements
- Demonstration of sustained improvement and/or sustained high-level performance

D. Relevance and Importance as Scoring Factors

The three evaluation dimensions described above are all critical to the assessment and feedback processes. However, evaluations and feedback must also consider the relevance and importance to the applicant's businesses of *improvements* in these dimensions. The areas of greatest relevance and importance are addressed in the business overview. Of particular importance are the key customer requirements and key business drivers.

IX. NATIONAL AND INTERNATIONAL QUALITY THRUSTS

The globalization of markets; increasing quality requirements; tough competition and supplier pressures have led to two parallel and visible "quality" thrusts, nationally and internationally. These thrusts are reflected in the Malcolm Baldrige National Quality Award in the United States and the ISO 9000 series of international standards.

- The focus of the Baldrige Award is that of enhancing competitiveness. Its central purpose is educational—to encourage sharing of competitiveness learning and to "drive" this learning, creating an nationally evolving body of knowledge. The content of the Baldrige Award criteria reflects two parallel key competitiveness thrusts: (1) delivery of ever-improving value to customers, and (2) systematic improvement of company operational performance.

- The focus of ISO 9000 registration is that of ensuring conformance quality. Its central purpose is to provide a common basis for an independent and transportable supplier qualification system. It is directed toward reducing audit costs and helping to assure buyers that specified practices are being followed; both of these are important to enhance and facilitate trade. The content of the ISO 9000 requirement standards reflects a key conformity thrust; consistent practice in specified operations, including proper documentation of such practice.

Overall, the Baldrige Award Criteria provide an integrated, results-orientated framework for designing, implementing, and assessing a process for managing all operations. The ISO 9000 series are standards used in implementing a compliance system and assessing conformity in company-selected operations.

Despite their numerous, major differences, there is now much confusion regarding the Baldrige Award and ISO 9000.[2] Commentary, surveys, and advertising are often inaccurate or misleading. Two common misconceptions stand out: (1) the Baldrige Award and ISO 9000 registration are similar—they cover similar requirements in the same way; and (2) the Baldrige Award and ISO 9000 registration both address improvement, both rely on high-quality results, and hence both are forms of recognition. Based upon these widely held beliefs, some conclude that the Baldrige Award and ISO 9000 are equivalent rather than distinctly different instruments that can reinforce one another when properly used.

The misconceptions are cause for concern not merely because they reflect confusion between award and registration requirements. More importantly, the confusion relates to basic business management issues: quality system definition, operational excellence requirements, and how quality is managed. For example, whereas competitiveness issues tend to marshal executive involvement, compliance and conformity matters more often are delegated, much as inspection and quality control were in the past. If all "quality" is delegated and focuses primarily on conformity and documentation, there is likely to be a "disconnect" between quality and closely related competitiveness requirements, reversing a trend toward their integration. Within companies, this may result in fragmentation of effort, slow

[2] Although the standards are referred to as ISO 9000 in this article and in the general literature, ISO Standards 9001, 9002, and 9003 are the requirements documents. Conformity to ISO 9000 standards refers to the use of one of these three standards in third-party registration or other assessment schemes. The ISO 9000 standards also include guidance documents, 9000 and 9004 (with subparts). ISO 9004 provides guidance for developing internal quality systems and is not used in third-party registration. Some confusion relating to the Baldrige Award/ISO 9000 comparison derives from failure to distinguish between ISO 9004 and ISO 9000 registration requirements in ISO 9001, 9002, and 9003. Specific comparisons are with ISO 9001 and 9002 and third-party registration to these requirements standards.

response, and weak productivity growth. Nationally, the focus on improving competitiveness could be diminished.

Strengthening U.S. competitiveness depends upon organizations in all sectors pursuing overall operational excellence. Success depends on effective integration of all requirements. Conformance quality is only one of *many* such requirements. The existence of two visible, partially overlapping but very different thrusts creates new impediments and costs, frequently confusing efforts to strengthen national competitiveness. Shown in Table 2 are the most important Baldrige Award criteria requirements and how they relate to ISO 9000 registration requirements. This summary should assist those who have launched ISO 9000 efforts to integrate these efforts into the Baldrige Award competitiveness improvement initiative.

Key Differences Between the Baldrige Award and ISO 9000 Registration

The Baldrige Award and ISO 9000 standards differ significantly in focus, purpose, and content as shown in Table 2. Despite their major differences, they are often confused and depicted as equivalent. This occurs because of (1) a misconception that they cover the same requirements; and (2) a perception that they both recognize high quality. The comparison assists companies which have launched ISO 9000 efforts to integrate these efforts with competitiveness improvement. In highlighting the differences between the Baldrige Award and ISO 9000 registration, however, the intent is not to portray these instruments as mutually exclusive: To the contrary, the authors wish to encourage U.S. companies which need, or elect, to pursue ISO 9000 registration also to take part in the national effort to promote improved competitiveness. There are many routes to participation in the Baldrige process or in related state and local award processes (Bemowski, 1993). An increasing number of excellent companies are using the Baldrige Award and ISO 9000 compatibility, sometimes simultaneously and sometimes sequentially. Either approach requires understanding of their important differences.

Despite the importance of Award and registration efforts, most of the millions of "business units" in companies, health care organizations, schools, government agencies, and nonprofit organizations in the United States are unlikely to seek either ISO 9000 registration or the Baldrige Award. Nevertheless, these units face the same basic requirements as those who do and may benefit equally from available knowledge and information sharing. Major purposes of the Baldrige Award Criteria are to promote education, self-assessment, and sharing.

X. EDUCATION

Although the Baldrige Award Criteria are the basis for Awards, the Criteria booklet is designed primarily as an educational instrument. Its principal use is as a "daily working tool." The Criteria define a basic integrated framework spelling out and relating key business requirements. The creation of an integrated framework for all requirements represents an important stage in the evolution of quality—from being largely separated from other operational requirements to being fully inte-

Table 2 Comparison of Criteria for the Baldrige Award and ISO 9000 Registration

	Baldrige Award Program	ISO 9000 Registration
Focus	Competitiveness—criteria reflect two key competitiveness thrusts: (1) delivery of ever-improving value to customers, and (2) improvement of overall company operational performance.	Conformity to practices specified in the registrant's own quality system.
Purpose	Educational—to encourage sharing of competitiveness learning and to "drive" this learning nationally. It fulfills this purpose by (1) promoting awareness of quality as an important element in competitiveness, (2) recognizing companies for successful quality strategies, and (3) fostering information sharing of lessons learned.	To provide a common basis for assuring buyers that specific practices, including documentation, are in conformance with the providers' stated quality systems. Some organizations use the ISO 9000 standards to bring basic process discipline to their operations.
Meaning of "Quality"	Customer-driven quality—addressing total purchase, ownership, relationship experience—concerned with all factors that matter to customers. Conformity issues are included in criteria, under Process Management, which addresses other key operational requirements.	Conformity of specified operations to documented requirements.
Improvement Results	Depends heavily on results—"Results" are a composite of competitiveness factors: customer-related, employee-related, product and service quality, and overall productivity. "Management by fact," tied to results, is a core value. Trends (improvement) and levels (comparisons to competitors and best performance) are taken into account. Results play dual role: (1) representing business improvement indicators needed to demonstrate a successful quality strategy, and (2) representing indicators that drive improvement.	Does not assess outcome-oriented results or improvement trends. Does not require demonstration of high quality, improving quality, efficient operations, or similar levels of quality among registered companies.
Role in the Marketplace	A form of recognition. Despite its heavy reliance on results, it is not intended to be a product endorsement, registration, or certification. Award winners may publicize and advertise their recognition and must share quality strategies with other U.S. organizations. The winners' role is public education and inducement for others to improve. Award winners adhere to a voluntary advertising guideline that prohibits attributing their Awards to their products.	Provides customers with assurances that a registered supplier has a documented quality system and follows it. In some cases, registration will reduce the number of independent audits otherwise conducted by customers. Although some registrars encourage advertising registration as a market advantage, some also prohibit advertising that registration signifies a product evaluation or high quality. ISO

Table 2 Continued

	Baldrige Award Program	ISO 9000 Registration
		9000 registration does not translate meaningfully into a Baldrige Award assessment score.
Nature of the Assessment	Involves a four-stage review conducted by volunteer private-sector Board of Examiners. Applications reviewed by 5–15 board members, depending on application progress. Final contenders receive site visits (2–5 days) by a team of 6–8 examiners. Focus is the customer and the marketplace. Evidence of pervasive improvement, backed by results, must be in place. Improvement includes customer-related and operations-related factors. It may include relevant financial indicators provided they are tied to the other indicators. Conformity and documentation are addressed as part of process management. Assessment is not an audit or conformity assessment.	Evaluates organization's quality manual and working documents, a site audit to ensure conformance to stated practices, and periodic reaudits after registration. Focus is on documentation of a quality system and on conformity to that documentation.
Feedback	Applicants receive feedback covering 24 items in seven categories. Feedback is diagnostic, highlighting strengths and areas for improvement in overall competitiveness management system. Three scoring dimensions: approach, deployment, and results.	Audit feedback covers discrepancies and findings related to practices and documentation. Feedback takes the form of major and minor nonconformities. Assesses organization's documented quality system requirements and deployment of these requirements.
Criteria Improvement	Criteria booklet is revised annually to capture lessons learned from each cycle. Since 1988, the document has undergone seven cycles of improvement, becoming more focused on business management. Major changes include a greater results orientation; more emphasis on speed, competitiveness, productivity; improved integration of quality and other business management requirements; greater emphasis on human resource development; and better accommodation to service organizations' requirements.	Revisions of ISO 9001, 9002, and 9003 were issued in 1994, with a focus on clarification of the 1987 documents, themselves based on the first commercial quality system standard, BS 5750, developed in 1979 (Sawin and Hutchens, 1991). The roots of BS 5750 trace to MIL-Q-9858A, established by the U.S. Department of Defense in 1959. Since 1987, additional guidance documents have been and are being developed.
Responsibility for Information Sharing	Award winners are required to share nonproprietary information on their successful quality strategies with other U.S. organizations.	Registrants have no obligation to share information with others.

Table 2 Continued

	Baldrige Award Program	ISO 9000 Registration
Service Quality	A principal concern in guiding criteria evolution has been compatibility with service excellence. Criteria and supporting information are evaluated to improve compatibility with requirements for service excellence. Criteria are relevant to service organizations. The most important "process" item (Customer Relationship Management) is a principal concern of service organizations, but also a major concern for manufacturers which seek competitive advantage via service.	ISO 9000 standards are directed toward the demonstration of a supplier's capability to control the processes that determine the acceptability of product or service, including design processes in ISO 9001. ISO 9000 standards are more oriented toward repetitive processes, without an equivalent focus on critical service quality issues such as relationship management and human resource development.
Scope of Coverage	Criteria address all operations and processes of all work units to improve overall company productivity, responsiveness, effectiveness, and quality. Approach offers wide latitude in developing customer-focused cost-reduction strategies, such as reengineering of business processes. Owing to broader nature of the Baldrige Criteria and assessment, a rigorous audit of a printed quality manual and compliance with its procedures do not occur during an assessment.	ISO 9001 registration covers only design/development, production, installation, and servicing. Registration covers parts of several items in the Baldrige Award Criteria (primarily parts of Process Management). ISO 9001 requirements address less than 10% of the scope of the Baldrige criteria, and do not fully address any of the 24 criteria items. All ISO 9001 requirements are within the scope of the Baldrige Award.
Documentation Requirement	Criteria do not spell out ongoing documentation requirement. Criteria imply that documentation should be tailored to fit requirements and circumstances, including internal, contractual, or regulatory requirements. Some analysts confuse application report with an ongoing documentation requirement. Assessment relies on evidence and data, but this does not define or prescribe a documentation system.	Documentation is a central audit requirement. Documentation requirements are ongoing, meaning that documents are a permanent part of the quality system needed to maintain registration.
Self-Assessment	Principal use of criteria is in self-assessment of improvement practices. Inclusion of a scoring system and evaluation factors allows companies to chart their own progress. Some companies correlate their progress in Baldrige Criteria self-assessment with changes in financial indicators.	ISO 9000 standards are used primarily in "contractual situations" or other external audits. Additional registrar-developed audit checklists define actual criteria/requirements for registration. Registration by external assessor is needed to fulfill most contractual

Table 2 Continued

Baldrige Award Program	ISO 9000 Registration
	requirements (i.e., self-assessment is generally not accepted). Aside from benefits of self-assessment while pursuing registration, it is not clear that ISO 9000 self-assessment after registration leads to operational improvement because standards do not address continuous improvement or competitiveness factors.

grated with them. The process-results linkages, coupled with a scoring system, create the basis for a results-based assessment.

There are three important points to note regarding the Baldrige Award Criteria as an educational tool:

1. The Criteria framework integrates two major thrusts: delivery of ever-improving value to customers, and systematic improvement of company operational and business performance.
2. The criteria framework addresses generic requirements that are equally applicable to organizations of all types—manufacturing and service businesses and not-for-profit organizations. This is intended to promote and facilitate interest, communication, and sharing among all organizations. Nationally, sharing and communications are built around harmonization of requirements.
3. The framework subordinates tools, techniques, and procedures to key customer and performance requirements.

   To support the larger educational purposes of the Baldrige Award Program, the Award Criteria booklet includes five interrelated and reinforcing elements:

   a. An introduction outlining the core values, concepts, framework, and characteristics of the award criteria—a key characteristic of the Criteria is that they are nonprescriptive. The Criteria do not prescribe tools, techniques, or organizations, thus fostering innovation. The introduction also addresses four important topics:
      i. Incremental and breakthrough improvement
      ii. Business strategy and decisions
      iii. Financial performance
      iv. Innovation and creativity
   b. Criteria that focus on 24 key requirements (examination items) among the seven categories—these items are detailed, giving the users many ideas on improving practices.
   c. Item notes—some quite detailed—that illustrate, clarify, and educate regarding the item requirements.

d. The scoring system commentary which gives users a perspective on the factors that are considered in evaluating progress and achievement. The scoring system uses three evaluation dimensions—approach, deployment, and results—which aid in developing diagnostic feedback.
e. Instructions to applicants that help all users of the criteria to understand how to organize information to facilitate internal and external assessment.

Continuing Education—The Evolution of the Baldrige Award Criteria

To fulfill its educational role, the Award program must itself be a rapid-cycle-time, learning organization. The primary vehicle to capture and disseminate learning is the annual Award Criteria booklet. Since 1988, the booklet has undergone seven cycles of revision. With the exception of the seven-category framework, all other educational materials are either entirely new or incorporate major learning.

Lessons learned are derived from many sources—from the strengths and weaknesses in U.S. organizations, for example—and from well-publicized deficiencies in TQM (total quality management) or other management thrusts. In aggregate, the changes from 1988 to 1995 are numerous and significant. They reflect better integration of all business management and competitiveness factors and stronger results orientation. Many other important changes respond to the requirements of service organizations.

XI. SELF-ASSESSMENT

Few organizations seek either awards or registration, but most wish to have non-costly means to learn how well they are doing and how they could get better. The most cost-effective means to achieve these purposes is through self-assessment. A realistic and thorough self-assessment has three components: (1) understanding of all requirements, what influences them, and how they are changing; (2) how requirements are deployed throughout the organization; and (3) how well requirements are being met.

An effective self-assessment instrument should have four characteristics: (1) educational value, (2) completeness—addressing all requirements and how they are deployed, (3) an integrated way to collect information so that it may be meaningfully evaluated, and (4) results indicators that address how well requirements are met. Answering "how well?" types of questions depends upon trends (Are we making progress?) and levels (How do our results compare with others?). Together, trend and level information provide a good basis for improvement actions.

The Baldrige Award booklet contains the key elements needed for self-assessment. Case studies prepared for examiner training, and subsequently available to all organizations, provide information to support the assessment learning process.

XII. FURTHER INFORMATION

For applications or information, write or call: Malcolm Baldrige National Quality Award, National Institute of Standards and Technology, Administration Building,

Room A537, Gaithersburg, Maryland 20899 (tel.: 301-975-2036; fax: 301-948-3716; e-mail oqp@nist.gov).

FURTHER READING

Bemowski, K., *The state of the States,* Quality Progress, 26(5), 1993, 27–36.

ISO 9000-1987, *Quality Management and Quality Assurance Standards—Guidelines for Selection and Use.*

ISO 9001-1987, *Quality Systems—Model for Quality Assurance in Design/Development, Production, Installation and Servicing.*

ISO 9002-1987, *Quality Systems—Model for Quality in Final Inspection and Test.*

ISO 9003-1987, *Quality Systems—Model for Quality Assurance in Final Inspection and Test.*

ISO 9004-1987, *Quality Management and Quality System Elements—Guidelines.*

(These standards are available in the United States as ANSI/ASQC Q90-1987 through Q94-1987, from the American National Standards Institute, New York, N.Y, and the American Society for Quality Control, Milwaukee, Wis.)

1995 Award Criteria, Malcolm Baldrige National Quality Award, U.S. Department of Commerce, Technology Administration, National Institute of Standards and Technology, Gaithersburg, MD 20899. The criteria booklet is available free of charge. Telephone: 301-975-2036.

Sawin, S.D., and Hutchens Jr., S., *ISO-9000 in Operation,* 1991 ASQC Quality Congress Transactions, Milwaukee, Wis., 915–916.

12

Electronic Product Design

RICHARD Y. MOSS
Hewlett-Packard Company
Palo Alto, California

I. INTRODUCTION

During the design of a new electronic product, obvious features such as price, package, and performance often get the most attention. But it is during this same period that the quality and reliability of the product are determined, and although they may not have the same visibility, it is those characteristics which can, in the end, "make" or "break" the product.

Analysis of the successes and failures of a large number of electronic products has led to two important observations:

1. About three-fourths all warranty repairs to the electronic hardware studied involved the *replacement of electrical parts*. Finding the causes and cures of component failures is, therefore, the key to achieving major improvements in reliability.
2. High failure rates in electronic components are most often caused by a combination of inadequate design margins, excessive stress on components and materials, unnecessarily high part counts, and poorly controlled or stressful manufacturing processes.

The above observations lead to the conclusion that high quality and reliability must be *designed into* a product from the start. Given that conclusion, it is never too early in the design cycle to begin reliability improvement!

II. INVESTIGATION PHASE

The design of an electronic product is an evolutionary process, which can be divided into two phases (Fig. 1). The first phase is called the Investigation phase, during

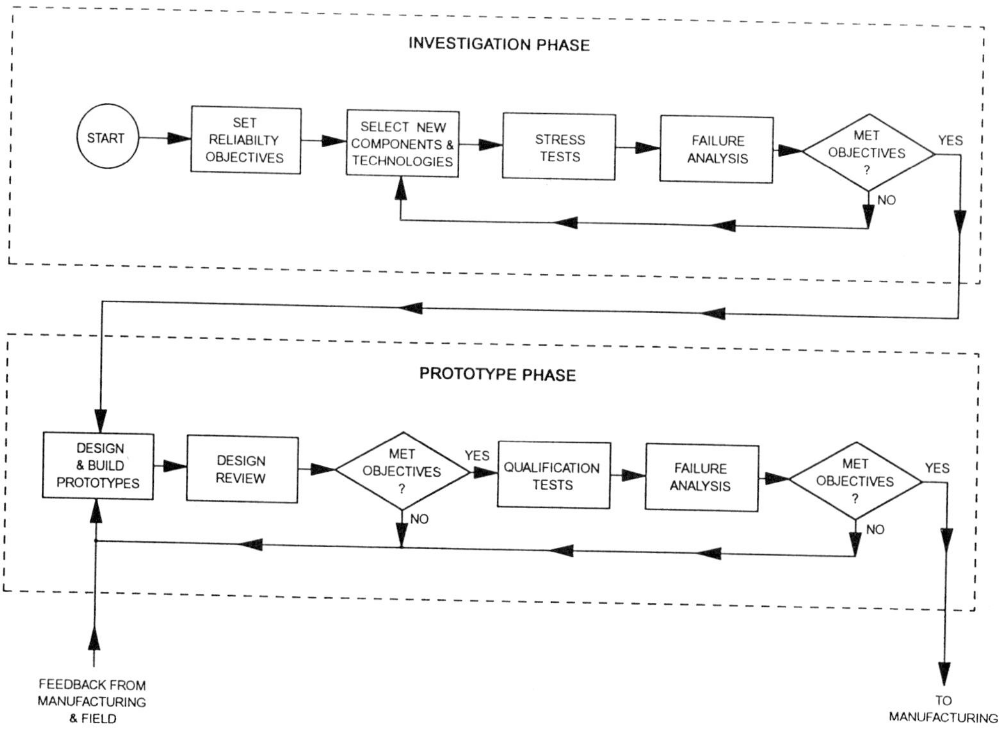

Figure 1 Product reliability evolution in design.

which a new product idea is studied for technical and market feasibility. Product specifications, development cost estimates, and schedule proposals are generated in the second phase, and this almost always entails some development and testing of critical new circuits or a new fabrication process. Two activities take place which have a significant bearing on the product's eventual reliability:

1. Setting of quality and reliability objectives
2. Selection and testing of new technologies and processes

A. Setting Objectives

The process of setting performance objectives for a new product exemplifies the essence of the design engineering process: trying to find the optimum trade-off between performance, cost, and schedule in a rapidly changing technological environment. Objectives need to be set in several areas: electrical and mechanical specifications, manufacturing cost, quality, and reliability. To be effective, these objectives must be challenging but believable, and they must be objectively measurable. The kinds of quality and reliability objectives that have been successfully used in electronic design projects fall into five general classifications: safety and regulatory requirements, reliability, service life, state of the art, and quality costs. Of course, not all of these objectives are necessarily applicable to any one project.

Safety and Regulatory Requirements

The design is obviously affected by safety (e.g., electrical shock hazard) and electromagnetic compatibility (EMC) considerations, which are discussed further in the chapter on packaging (Chapter 13). In addition, circuits which operate at high voltages can generate ozone or x-rays which can be harmful; high current or high-power circuits can become hot enough to start a fire under some fault conditions; and an electronic product which fails at a critical moment (particularly in medical electronics or process control) can result in a product liability lawsuit which will be expensive and damaging to defend, even if the defense is successful.

Management's responsibility here is to state and enforce a clear policy that no product may be released to manufacturing or delivered to a user which is known to contain a preventable safety hazard, or which has not been thoroughly tested to verify that it is free of discoverable hazards. One possible method of discovering such problems is to do a potential problem analysis (PPA) of the foreseeable consequences of a failure or misapplication, and then attempt to design the product to be safe even under those conditions (1). Techniques such as FMEA (failure modes and effects analysis) can be very helpful in discovering serious failure modes before the fact. Experience has shown that reducing the failure rate of a product tends to reduce the total cost of liability by about the same ratio, so that the return on the extra engineering investment can be quite large.

In recent years, the product regulations climate has become much more pervasive. In addition to regulating electromagnetic emissions and susceptibility, government and private regulators are specifying acoustic noise, chemicals contained in or used in the manufacture of the product, reusability or recyclability of the shipping containers, and disposal of the product itself at end of life. Thus the design of a product is no longer dictated solely by cost and utility, but now includes environmental and political issues as well.

Reliability

High reliability, like any other performance feature, is more likely to result when a challenging objective is set and met than it is by "happy accident." One measure that is commonly used to specify reliability is MTBF (mean time between failures) or its reciprocal, average failure rate [λ (t)]. What is puzzling is that although many managers are familiar with the reliability bathtub curve (Fig. 2), which clearly shows that failure rate is a variable with time, they fail to recognize that MTBF also varies with time (the age of the product). Thus, if an attempt is made to measure the MTBF of a new design using a short duration test, a pessimistic value corresponding to the early failure region of the bathtub curve will be obtained. A better practice is to specify *reliability*, which is the probability of survival without failure for a specified period of time and under specified conditions. This makes it impossible to "play games" with the results by making the test longer or shorter to influence the outcome. A still better practice is to keep an accurate record of the actual times to failure as testing progresses, and then fit these to a statistical life distribution model (2). By plotting a graph of the failure rate versus time of your product, you can expose early failure or wear-out behavior in the product and make it easier to find the causes of failure, as well as more accurately predict the expected operating costs and support needed.

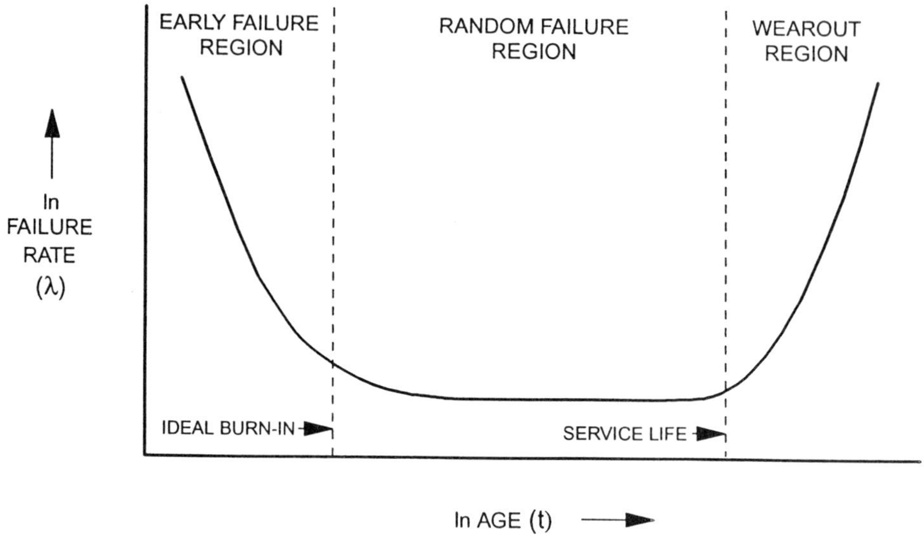

Figure 2 Failure rate bathtub curve.

Service Life

A second quantity which is often overlooked in specifying the reliability of a product is the service life, which is the useful life expectancy of the product before wear-out makes it uneconomical to repair and maintain. Although the market life of a particular design may be only a year because of rapid technological change, it is not unusual for a customer to expect a service life of 10 years from the product. Designers of solid-state electronic products are often unaware that the components they use have wear-out failure mechanisms, and hence finite life expectancies. These wear-out mechanisms must be eliminated or at least minimized wherever encountered, or the components made easily replaceable where there is no alternative. (Incandescent lamps and batteries are examples of components with limited life expectancies, and which are made easily replaceable as compensation.)

State of the Art

A popular rule of thumb in electronics is that it takes 5 years to develop a complex fabrication process, such as those used in integrated circuit manufacturing, to the point that it is well understood and is producing its highest quality and reliability. Designers are not able to wait that long before adopting new technologies and, in fact, if new technologies weren't widely used before that, the pressure to improve would be less and the mature state might take even longer to achieve. At the other extreme, using obsolete processes or technology can also increase costs, as well as produce poor quality and reliability, so sticking to old processes is not the solution either. The optimum strategy seems to be to take a calculated risk, with as much data as possible obtained by accelerated stress testing and failure analysis, the catastrophic problems eliminated, and a plan for long-term improvement relentlessly pursued. It is obviously the job of management to make sure this happens.

Quality Costs

Reduction of quality costs is among the most effective kind of goals, because these costs are readily measurable and are related to other management objectives, such as expense control and profits. Ideally, an accounting system should be set up to collect all the quality costs—those related to appraisal and prevention, as well as failure and corrective action costs. If total quality costs cannot be determined, warranty cost as a percentage of sales is a useful measure, so long as it is realized that this one item is only about 20% of the total quality cost. Warranty cost has the advantage (or disadvantage) that it is a function of both failure rate and repair cost per unit, and some managers prefer to track each factor separately. In any case, it is reasonable to set a goal of 2:1 or 3:1 improvement on a new product compared to the previous model. (This corresponds to 25–35% per year improvement for a typical product life cycle of 3–5 years).

The first time that total quality costs are computed they usually shock management, since it is not unusual to discover that they total 20–25% of the sales dollar. This shock is especially painful in a company that thinks that its quality is pretty good, because the obvious quality costs are warranty and scrap, which total less than 5% of sales (3). When you realize that 25–50% of your payroll, facilities, and inventory are also contributing to the cost of quality because they are employed in redesign, retest, troubleshooting, touchup and rework, rescheduling, servicing, and generally producing and repairing defects, it is easy to realize what huge gains in productivity and profit can be made by improving the quality and reliability of the goods and services provided. It is the responsibility of management to determine and control the *total* cost of quality, not just the "tip of the iceberg" items like warranty and customer returns.

B. Technology Selection

Technical performance and cost are normally the primary criteria for selecting a new technology or process. There are usually several alternatives, each with a different combination of performance, cost, and reliability attributes. A good strategy is to *first eliminate the unreliable choices* by accelerated stress testing and failure analysis, and then decide between the others—if more than one survives—on the basis of cost and performance. Usually a new technology or process is not totally new but has evolved from existing ones. Something is known about what weaknesses and failure mechanisms to look for, and there will be a previous level of performance for comparison. An obvious but useful bit of advice is to select the component or technology which is least susceptible to the stresses that are expected in the product's defined environment. For example, it would not be smart to select a temperature-sensitive part to be used in a product aimed at wide temperature range applications. Similarly, delicate parts cannot be specified where there is a stressful manufacturing process.

It is also worthwhile to remember that if the product is going to be manufactured for 3–5 years and have a service life of 10 years, then the components used in it will have to be available for the next 15 years! That means that the newest technologies—which haven't yet been proven successful or promising enough to attract competing suppliers (and may never succeed)—and old technologies that are no longer competitive represent the biggest risks in terms of future availability.

In the case of electronic components for general-purpose use, there are usually many alternatives from which every designer will make different choices. Standardization is the key to combating this, and one successful way to implement it is to set up panels or "sounding boards" consisting of designers, materials engineers, and reliability engineers to select and designate *preferred* parts for new designs. An additional benefit of standardization will be savings by reducing both manufacturing and service inventories; lower costs because the preferred parts can be purchased in higher volumes; and a greatly enhanced ability to monitor the quality history of each part and each supplier, because there are fewer to track.

C. Stress Testing

Qualification testing of *all* new component and technology types is vital to disaster prevention. The test philosophy at the Investigation phase should be to *test to failure*—that is, to design the tests so that the stress level or duration is increased in predetermined steps until *multiple* failures occur and a *pattern* of failures is established. In the world of the reliability engineer, failures are information, and "testing for success" is a waste of time, because no failures yield little or no information. Although much can be learned from even a single failure, even more can be gained from the discovery of a *pattern* of related failures. Even when failures occur randomly in time, virtually all turn out to have nonrandom causes, which can be discovered and eliminated.

Nearly all failure mechanisms are accelerated (i.e., the failure rate is increased) by elevated stresses, which can be grouped into four classifications: electrical, thermal, chemical, and mechanical. Table 1 lists some examples of these four types,

Table 1 Failure Mechanisms Accelerated by Elevated Stress

Stress type	Stress	Failure mechanism	Component type
Electrical	Voltage	Dielectric breakdown	Capacitors, high voltage insulation, MOS devices
		Avalanche breakdown	Semiconductor junctions
	Current	Electromigration, fusing	Thin films, integrated circuits
		Current hogging	Switching semiconductors
Thermal	Heat	Chemical reaction	Batteries, electrolytic capacitors
		Ionic contamination	Semiconductor devices
		Intermetallic growth	Interconnections
Chemical	Moisture	Corrosion, dendrites	Thin films, plating
		Ionic contamination	Nonhermetic semiconductors
		Leakage resistance	High impedance and high voltage circuits
Mechanical	Temperature cycles	Differential expansion	Seals, power circuits, polymer encapsulation
		Condensation	Cavity packages
	Shock and vibration	Fatigue	Interconnections
		Loosening, movement	Mountings, adjustments
		Conducting particles	Cavity packages

the most common failure mechanisms they activate, and some component types frequently affected.

The basic electrical stresses are voltage and current. The product of the two is, of course, power which results in thermal stress. But even by themselves each stress can do quite a bit of damage. The most common voltage-induced failure mode is breakdown, either exceeding the dielectric strength of an insulator or avalanching a reverse-biased semiconductor junction. High temperature tends to increase the susceptibility of most devices to voltage breakdown, and unless current is limited, damage or degradation is virtually certain. The source of the damaging voltage may be designed-in, such as a circuit operating too close to its specified breakdown, or may be externally generated by electrostatic discharge (ESD), power surges, or operator error. Thin dielectrics, such as the silicon dioxide layers on integrated circuits, can be fatally damaged in less than a microsecond, so voltage-limiting circuits must be capable of very fast responses. There are several current-activated failure mechanisms, and all of them are strongly accelerated by temperature. Electromigration is a particularly dangerous mechanism, since it is like the slow erosion of a riverbank by the rapid flow of water, and the longer the erosion continues the weaker it becomes, so the failure rate increases with age. Thin aluminum films, used to make the interconnections in large-scale integrated circuits, are particularly prone to electromigration failure, and it is becoming one of the most serious limitations to very large-scale integrated circuit (VLSIC) technology. (As circuit feature sizes shrink, current density increases.) Junction "current hogging" is another mechanism that has been known for a long time, but has become a more serious concern recently as switching power supplies have come into vogue. In this failure mode, the entire junction area does not switch on (or off) simultaneously, so the current briefly concentrates in a small fraction of the junction, leading to overheating of that small area, further concentration of the current in the hot spot, and so on until failure results.

Heat is the arch enemy of reliability. Chemists have long known that (chemical) reaction rates increase exponentially with temperature, so electrical components that depend upon chemical systems, such as batteries or electrolytic capacitors, are greatly affected by temperature. Electronic design engineers are not always so aware, however, that elevated temperatures greatly increase failures such as high leakage currents caused by increased mobility of ionic contaminants, and high temperatures also can promote the growth of intermetallic compounds such as "purple plague," a brittle gold-aluminum compound which can lead to wire bond failures inside a transistor or integrated circuit package. Finally, heat tends to accelerate most of the other failure mechanisms listed in Table 1. Consequently, elevated temperature is nearly always a component of an accelerated stress test plan, and is a major contributor to field failures of electronics.

Chemical stress, particularly when the chemicals are dissolved (ionized) and transported by water, is a serious threat to reliability, because the failure mechanisms are mostly of the irreversible, wear-out variety. Corrosion and growth of metal "whiskers" (dendrites) are major reliability problems in severely contaminated (and humid) environments, such as petroleum refineries, paper and wood pulp processing, marine climates, or large metropolitan areas with their higher concentration of exhaust fumes and smoke. Copper, silver, and aluminum all corrode rapidly when exposed to more than a few parts per billion of chlorine or sulfur, especially if moisture and voltage are also present. Gold plating of connections and contacts has been one traditional answer, but gold is expensive and not effective in really severe environments—corrosion can occur through a tiny pore or scratch or around the edges of the gold plating. Currently, tin-plated interconnection systems are employed in the most severe environments, with the requirement that the contact pressure is high enough to create a gas-tight joint at the point of contact. Finally, water containing any dissolved impurities is an electrical conductor, so leakage paths between high-impedance or high-voltage circuits must be controlled by encapsulation; lengthened leakage paths (like the ridges on a power line insulator); use of insulator materials such as Teflon, which tend to inhibit the information of moisture films; or by reducing humidity using environmental control. Since there is no such thing as a truly hermetic plastic, polymer encapsulation is not a very good solution where humid environments must be tolerated. Plastic has the advantage that it doesn't corrode, but the disadvantage that water will diffuse into it over a few days or weeks, carrying contaminants in solution to attack the circuits inside.

Mechanical failures occur in electronic parts, even so-called solid-state ones with no moving parts. The biggest effect of temperature variation on many products is that it causes expansion and contraction of the various metals and plastics from which the components are constructed. The mismatch of their coefficients of thermal expansion (CTE) leads to tension, compression, and shear stresses on the interfaces between different materials, and this results in cracked seals, broken wire bonds, and even cracked semiconductor chips. Components encapsulated in polymers, particularly high-power devices which heat and cool repeatedly, tend to fail the most often owing to this phenomenon. Another effect of temperature cycles is to cause humidity cycles, particularly as temperature drops below the dew point and moisture condenses on or inside a component. Finally, there are the traditional failure mechanisms induced by shock and vibration: open circuits due to wear or cracking of repeatedly flexed interconnections, loosening and changes of settings of adjustable or critically aligned parts, and momentary short circuits caused by loose conducting particles such as tiny solder balls or wire scraps. Because of their transient nature, these mechanisms can be very difficult to find in failure analysis. Calculating the acceleration factor (the ratio of failure rate at elevated stress to that at normal stress) of accelerated stress tests seems to be an obsession with many designers and reliability engineers, particularly when confronted with the necessity of doing stress tests for the first time. In many cases, calculating this acceleration factor is difficult or impossible, which leads to wildly speculative assumptions. If a calibrated acceleration factor is needed, the established method is to replicate the tests at three or more different stress levels, determine a parameter of the failure distribution such as median life at each stress, and then try to model mathematically the behavior of this parameter versus stress. An example of this is shown in Figure 3, which shows that the acceleration factor of temperature stress on this integrated circuit fits a mathematical model called the Arrhenius relationship, after the Swedish chemist of the same name. In this model, median life is an exponential function of absolute temperature (°kelvin). Although it is useful to know that testing at 175°C *decreases* the median life by a factor of 26 compared to 125°C, the really important result of these tests was to discover that mobile charge migration was the root cause of failure, and then take steps to eliminate it!

There are two major pitfalls of elevated stress testing which must be avoided. The first is the possibility of raising the stress so high that a new failure mechanism,

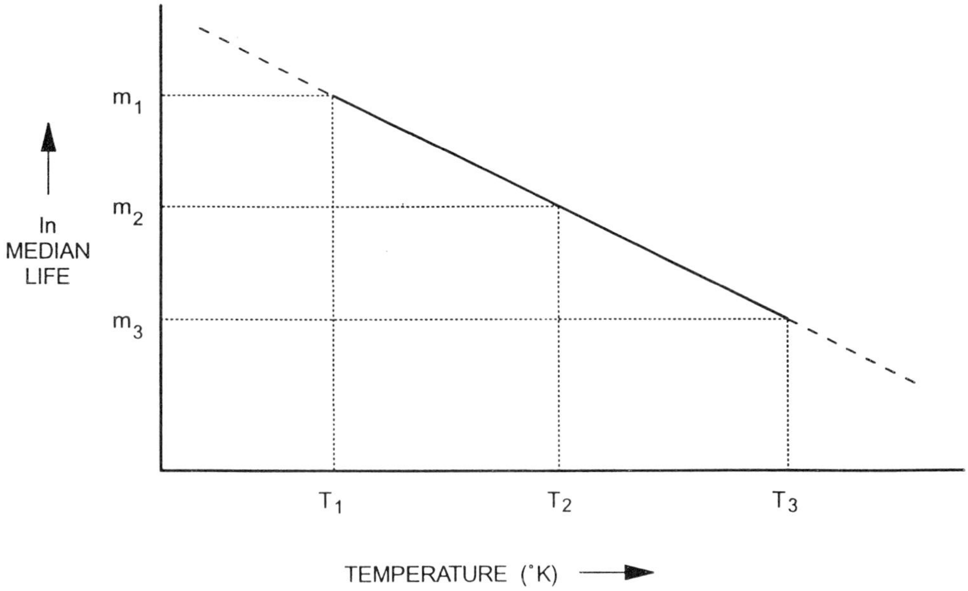

Figure 3 Arrhenius relationship.

one not found in real life, occurs because of a change of physical or mechanical state in the material. For example, it is not a valid accelerated test to raise the temperature of a component to the point where part of it, such as a solder connection, melts. The occurrence of this would usually be signaled in the mathematical model by a sudden change of slope parameter. The second pitfall is more subtle, and could occur in the case where a part under test has two important failure mechanisms, one of which is accelerated by the test stress and the other not. As a result of the stress test, and especially the calculations of the acceleration factor, one is tempted to divide the high failure rate at elevated stress by a large acceleration factor and conclude that the reliability at normal stress will be very good, whereas in reality, the undiscovered failure mechanism will dominate the statistics at normal stress and will produce unacceptably poor reliability in the field. An example of this is shown in Figure 4. There are two ways to avoid this pitfall:

1. By testing with a variety of stresses, particularly using several stresses simultaneously, so that you are less likely to overlook a failure mechanism
2. By avoiding extrapolation in cases where the acceleration factor is large (>20) and a small error in estimating its slope will produce a big error in estimating the reliability

To do conservative testing requires larger sample sizes and longer tests in order to get multiple failures and accurately establish an acceleration factor and dominant mechanism.

D. Failure Analysis

Failure analysis is a *vital* part of the design process, since it provides clues as to the root cause of failure, so that the failure mechanism can be minimized or elimi-

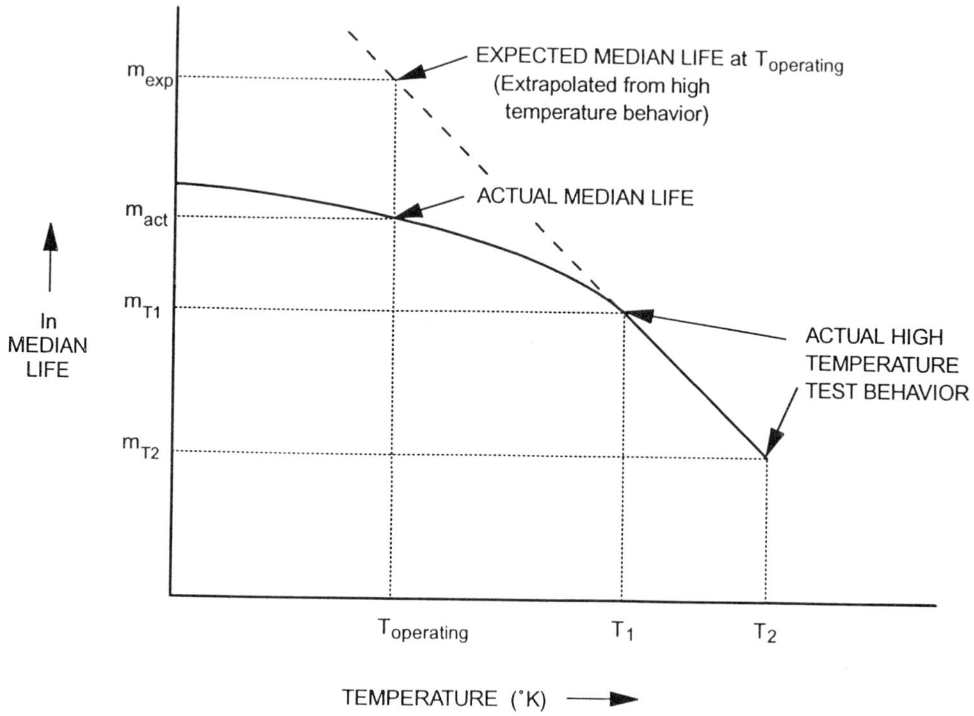

Figure 4 Pitfalls to accelerated testing.

nated. There are two distinct parts to failure analysis—statistical analysis and physical examination. Statistical analysis can provide important clues to the nature of the failure mechanism, as well as helping to determine the relative seriousness of the problem and the most economical cure. For example, a failure mechanism which exhibits a *decreasing* failure rate with time might be most economically eliminated by subjecting all the parts to an aging process such as burn-in, whereas a failure mechanism which exhibits an *increasing* failure rate behavior, such as is caused by corrosion, metal migration, or mechanical fatigue, must be designed out, since aging will only make it worse. The statistical analysis should also give a measure of the magnitude and cost of the problem, so that the most serious problems are attacked first. Simple statistical techniques such as graphs, histograms, and process control charts can be taught to workers at nearly all levels of skill and education, and with the assistance of computer programs even more complex procedures can be widely used. Teaching which ones to use, writing the programs or procedures, and teaching their use is best accomplished by a statistician. This fact has been demonstrated by the success of the American statisticians Dr. Joseph R. Juran and Dr. W. Edwards Deming in Japan (4).

The physical examination part of failure analysis provides objective evidence with which to identify the mechanism causing the failures, and hence allows insight into ways to eliminate it, as well as better tests to detect that particular type of failure. Most importantly, it allows the determination of who is responsible for fixing the problem, which is not a trivial concern where large, decentralized corporations are both the supplier and the user, and each is pointing a finger at the other. Physical failure analysis cannot always find the exact cause of every failure, but the "batting average" is high enough that problems can nearly always be solved. The failures caused by construction or processing defects seem to be the ones which are most amenable to analysis in the laboratory, whereas failures caused by momentary overstress or changes of an unspecified parameter are the hardest to resolve. In such cases, the failure analyst must also be a very good applications engineer or else work closely with one. It is also a very good idea in such cases to "prove" the analyst's hypothesis by intentionally recreating identical failures. A small but surprisingly effective failure analysis laboratory can be created with a capital investment of about $50,000 for equipment, and it can be kept very busy by a company or site electronic manufacturing at a level of less than $20 million per year. As volume and experience grow, the facility can be expanded to match the needs, since failure analysis is of immense value to production as well as product development. The analysis of parts from qualification testing, incoming inspection, production failures, and field returns is possible, and statistical analysis will help determine the priorities (5).

III. PROTOTYPE PHASE

The second phase in a new product's evolution is the prototype phase. Often there are two or more iterations of this phase, during which the complete product is designed and packaged, the performance characterized, and the production documentation such as schematics, material lists, and test procedures produced. Events which affect the quality and reliability of the product happen all through the cycle, but for discussion purposes, they can be grouped under three headings: design rules, design review, and qualification testing.

A. Design Rules

For some reason, design rules often exist for such activities as integrated circuit or printed circuit layout but seldom for electrical circuit design. The growth rate in many electronics companies is such that 50% of the designers have less than 5 years' experience, and they are currently working on their first design project. Typically, they learned the theory but not the practice of design in school, and it is a foolish management which allows each of them to learn the latter by trial and error, when the errors are so costly. Design rules should particularly address those areas where formal engineering education stops short, such as:

1. *Thermal management*: Just because the vendor's data sheet says that the absolute maximum junction temperature of an integrated circuit is 150°C (302°F) doesn't mean that it will operate reliably for 10 years at that temperature. The failure mechanisms which plague semiconductor devices are accelerated by a factor of 500 at 150°C compared to 85°C (185°F), so a device with a median life of 1000 years at 85°C will last only 2 years at 150°C. Teaching thermal management to the designers may well be the single most effective improvement measure that management can employ (6).

2. *Electrical derating and margin*: If the manufacturer of an electronic component is practicing statistical quality control (SQC), the data sheet specification limits of the components will probably be set at the 3 σ (sigma) points of the population. That means that only about 0.27% of the parts are likely to be outside the data sheet limits initially. However, the realities of life when employing commercial-grade components are that you are lucky if the data sheet limits reflect the 2 σ (sigma) points of the population. That means that about 5% of the parts can be expected to be outside the data sheet limits initially, and probably more after the effects of age and stress are considered. A consistent program of derating and extra margin in interpreting data sheet parameters of components will have a large payoff in product quality and reliability, second only to good thermal management.
3. *Regulatory compliance—safety and EMC*: As was mentioned previously, many countries have made compliance with their regulations a requirement for selling products there, and the penalties include both fines and imprisonment. The frequency and cost of civil liability are also increasing, so it is essential that management develops and enforces a set of design rules which will assure safe and legal products.
4. *Process compatibility*: One of the quickest routes to disaster is to release a design which cannot be manufactured by the processes available. This is the area that is usually addressed by design rules for integrated circuit masks, printed circuit hoards, and mechanical parts. A more subtle type of compatibility problem is the one where an electrical component is chosen which cannot withstand the stresses of automated handling, soldering, washing, and test. Qualification testing of new components must include exposure to these process stresses, since a part damaged by the manufacturing process can have a 10-fold higher failure rate in the field. This is particularly true of surface mount technology (SMT), because the stresses during reflow soldering are potentially damaging.
5. *Standardization*: Although management is sometimes afraid that imposing restrictions on component choices may stifle creativity or limit performance, just the reverse is true. Most design engineers find that complete freedom really is the worst tyranny, since it takes much longer to evaluate all the choices and increases the risk of a bad choice. A well-managed standardization program can be a boon to engineering creativity and productivity rather than a hindrance. Letting the design engineers participate in the standardization process is a good way to minimize complaints and mistakes.

B. Design Review

It is shocking to see how seldom new product designs are subjected to a serious electrical design review. Perhaps managers are afraid they will bruise the egos of their engineers. More likely, they (the managers) are afraid to expose their own ignorance. Whatever the excuse, it is not sufficient to justify the result. A design review, properly executed, is a valuable tool for assuring the quality and reliability of a new product. There should be a minimum of one design review during each phase of product development, and the following subjects should be examined:

1. *Design rule compliance*: Make sure there are no misapplications, overstress, incompatible or nonstandard components, nor violations of policy regarding safety and regulatory compliance.
2. *Compatibility*: Is the design upward, downward, and laterally compatible with other products being developed or currently marketed? In this age of the computer, hardware and software incompatibilities are quickly evident to the customer, and they are both expensive and embarrassing to the manufacturer.
3. *Performance*: Are the latest reliability, electrical performance, and cost estimates converging on the product's final objectives? It is unwise but common human nature to assume that the problems in the prototype version of the product will somehow magically disappear when a cleaned-up version is built. They will not, and you will probably gain some new problems as well.

An approach to design review, particularly of computer software, which has been gaining favor in some circles is called peer review. Here engineers of approximately equal organizational rank review each other's designs in an atmosphere presumably less threatening than a management review. There are undoubted benefits of such a system, but it should be a supplement rather than a substitute for a thorough review by a smart manager who knows how to ask the right questions, even if he or she doesn't know all the answers. And, as we shall see shortly, the best questions are often asked simply by testing the design.

C. Qualification Testing

Included under this heading are the physical tests made on the prototype hardware, and generally repeated on first production units. Three types of tests are performed: margin tests, life tests, and field trials. As with previously mentioned, design margin plays major role in determining the frequency of failure due to parameter shifts in components, and the margin tests are intended to be a check of that. A design standard should be established which sets the minimum ranges of environmental stress over which every product must operate satisfactorily, and still more severe ranges over which they must survive operation or storage without damage. These ranges should be wider than the real world in order to compensate for that fact that the tests are performed on a very small sample, perhaps only one or two prototypes, and passing such a small number doesn't give much confidence about the whole population. For example, available data show that even in the most severe tropical climates, the temperature rarely exceeds 35°C (95°F) at 95% relative humidity (RH), but to give reasonable confidence, the design margin should be tested by stressing prototypes at 40°C (104°F), 95% RH.

A typical way of documenting a design standard is to organize it into several environmental classes, such as outdoor portable environment, industrial/commercial environment, and home environment. Military specification MIL-T-28800 is an example of a document which divides the electronic test equipment environments into seven classes with gradually decreasing severity (7). Conditions for operating and nonoperating temperature, humidity, vibration, mechanical shock, altitude, and power fluctuations are specified for each class. While conducting these types of tests, it is a good idea to vary electrical parameters such as supply voltages, output loading, input signal levels, and timing to try to discover previously unsuspected interactions or marginal conditions. This same design standard can also set test conditions and performance requirements in the areas of electromagnetic compatibility (EMC), transient and electrostatic discharge (ESD) susceptibility, and electrical leakage currents (safety). These latter requirements will need to be reviewed and updated frequently, as a result of the growing body of international laws governing safety and EMC.

The complete series of tests typically requires several weeks to perform even if enough prototypes are available so that some tests can be run in parallel. Specialized facilities are necessary to run the more technically difficult tests such as EMC, but many of the tests which are most valuable from the point of view of impact on quality and reliability can be run using inexpensive sources of heat, cold, moisture, and vibration. A test using an oven, a deep freeze, a humidifier, and a homemade "shaker" can uncover lots of problems to be designed out, and private test labs can be employed to do EMC or safety certification tests where acquiring in-house facilities would be too expensive to justify.

Reliability verification tests (RVT) differ from margin tests in several important ways. First, a much larger sample size and much longer duration are required in order to be able to assign reasonable confidence bounds to the results. Sample sizes of 10–100 units are typical, as are durations of from 1000 to 4000 hours. Test conditions are tailored to the particular type of product being qualified, and generally involve one or two accelerated stresses. A typical RVT will be run at elevated temperature and with maximum duty factor to achieve a high acceleration factor, and with sufficient units and hours to demonstrate the reliability goal with 80 or 90% confidence. There are many statistically designed test plans available to help management select the particular combination of units, duration, and confidence desired. A popular example is the sequential test plan found in MIL-STD-781, where the duration of the test is lengthened or shortened automatically as a function of how much better or worse the product is than the goal (8). The unwary should be warned, however, that there are two serious limitations to this and other similar plans:

1. An exponential life distribution is assumed; that is, it is assumed that the failure rate is constant, and that there are no early failures or wear-out. Experience tends to show that these are dangerous and unwarranted assumptions, especially in a new design that has not yet been "shaken down."
2. Constant-stress tests, with their beguilingly well-documented acceleration factors, can lead to a fool's paradise in which a serious failure mechanism is completely overlooked because it wasn't accelerated by the chosen constant stress.

A variation on these constant stress life tests that is currently being adopted by military and commercial manufacturers alike involves the use of multiple, cyclical stresses. For example, the ambient temperature would be cycled between a low temperature (−40°C; −40°F) and high one (60°C; 140°F) at 1- to 4-hour intervals, with temperature ramp rates of up to 10–20°C per minute during the transitions. AC power may be simultaneously cycled on and off and short intervals of vibration introduced (9). This type of testing is known as STRIFE (for *str*ess l*ife*) or HALT (highly accelerated stress testing). New products which have been subjected to 10,000 or more unit-hours (1000–4000 hours per unit, 3–10 units) of such testing and the resulting pattern failures designed out, have enjoyed warranty failure rates dramatically less than those of similar products which had not been so tested, even when the comparison products were "mature" designs with several years of manufacturing history. Two key points have emerged from testing to date:

1. The value of the testing is lost unless the pattern failures are eliminated, so the earlier in the design cycle it is begun, the better
2. Low temperatures (below room temperature) are as important as high temperatures

There are several possible reasons for item 2 above, but the increase of humidity as the test chamber air is cooled is certainly a major factor—in fact, if the dew point is reached and moisture allowed to condense on the product, problems will almost certainly be found in a new design.

Field trials are the third way in which a new product can be evaluated. In a large corporation, it may be possible to conduct the field trial internally, using the "next bench" or next department as the test site. This is most successful when the new product is in final form, so that it is safe to use and the user need not be more sophisticated or better trained than the intended customer. Problems with the human interface, documentation, and abuse resistance of the product are often uncovered in this kind of trial.

The design rules, design reviews, and qualification testing which are employed during the prototype phase of the design cycle are all vital to achieving a reliable product, but they are rarely enough. There are just too many degrees of freedom in a complex electronic product to be sure of getting it exactly right on the first try. The design nearly always continues to evolve after it is "released" to manufacturing, and management needs to assure that the evolution is toward a more reliable product, not just one that is cheaper to manufacture.

IV. MANUFACTURING PHASE

A. Transfer of Knowledge

The first problem to be solved is the transfer of knowledge from development engineering to production engineering. Good documentation is vital, but not all the important knowledge in the designers' heads is going to get documented, much less be absorbed by production. One solution is to transfer one or more design engineers to production along with the new design, as a sort of human repository of undocumented knowledge. An obvious problem with this approach is that while the designer is knowledgeable about the design, he or she may not be similarly expert in the manufacturing process, and may even feel that this assignment is a dangerous detour in his or her career path. Another approach has been to assign a production engineer to the product at the start of the prototype phase to absorb all he or she can about the design, to see that it is thoroughly documented, and to try to influence the design so that "manufacturability" and compatibility with current manufacturing processes is assured. With today's fast-changing technology, the problem is that manufacturing processes are also evolving, and the designers find themselves trying to hit a moving target.

B. Process Design

A new buzzword in manufacturing management jargon is "concurrent design," in which a new product design is supposed to evolve in parallel with a new process design. It seems to me that this approach will be most successful where computer-aided design (CAD) is employed extensively, because every time the process is changed, the design rules must also change, and hence a complete design review and some changes may be needed. When it comes to producing a high-quality, high-reliability product, there is no substitute for a mature process that is under control and which has well-tested and documented design rules (10). This is also the kind of process that can be automated successfully; automating a "flaky" design or process will only produce "flaky" products faster.

C. Stress Testing

Just as margin testing, stress testing, and failure analysis were vital to the design phases, the same kinds of testing on production units can lead to further improvements. Quality can be improved by testing of components, subassemblies, and finished product, and reliability can be improved by incorporating stress into production tests to weed out weak but functioning parts and discover patterns of failure which can be designed out. Multiple, cyclical stresses have been found to be more effective than single or constant stresses, so production tests with temperature, power, and sometimes vibration cycles have evolved (11). This type of testing is known as ESS (environmental stress screening) or HASS (highly accelerated stress screening). In an experiment to demonstrate the greater effectiveness of cyclical stresses, 1300 production units of a tabletop digital computer were randomly divided into two groups as they were produced and their serial numbers recorded. One group (A) of 650 computers was subjected to 48 hours of simultaneous temperature and power cycling, with temperature alternating between 0°C (32°F) and 40°C (104°F) once every hour, and with the 120 volt AC power turned off for 30 seconds once each hour. The units ran a digital self-test continuously, and results were monitored by a computer system. The second group (B) of 650 computers was allowed to remain at room temperature for 48 hours, but was subjected to the same power on/off cycling and computer-monitored self-test as group A. The results were dramatically different for the two groups! Of those in group B (which ran at room temperature), 24% recorded some kind of failure during the 48-hour "heat run," whereas 55% of those in group A (which were temperature cycled) had a failure. All 1300 units were then thoroughly retested, repaired, or readjusted as needed, and shipped to users without any attempt to segregate group A or B units. Warranty failures or complaints from users were recorded as received through normal field service channels for the first 90 days of use. Here group A had a much lower failure rate, about 12%, compared to group B's, 19%. Since the cost of a field service call is usually more than 20 times that of a manufacturing line repair on this type of product, the 7% improvement in field failures repaid the 31% increase in factory repairs several times over, even including the cost and energy consumption of the temperature cycling facility. Moreover, the increased level of factory repairs resulted in a study of the causes, and as a pattern of failure emerged from the data, steps were taken to eliminate the most frequent modes. Because design and production engineers could get trustworthy data and examine the failed units themselves, they were able to take action much more quickly than when the failure reports dribbled in from scattered and widely differing field sites. As these patterns of failure were removed from the product and process, both the factory and field failure rates improved significantly (>3 : 1)!

D. Design Control

It is important to assure that design control be maintained in production, and that the data from production and the user be fed back to the designers. One reason is to try to avoid creating new problems with changes, but a more important one is to help the designers improve their design rules and practices. A closed-loop system can produce a much more accurate result than an open-loop one, whether it is an electronic or human system.

One approach to this has been called design defect tracking (DDT). A simple tracking report, such as a spreadsheet, is used to record every failure, suspected bug, or other unexplained perturbation as a new product is tested. An "owner" is assigned to every entry, and the entries are classified into five states (0–4):

State 0. Cause Unknown

A defect or unexplained phenomenon has been observed and an engineer assigned responsibility to investigate it.

State 1. Cause Understood

The root cause has been found.

State 2. Solution Designed

A solution has been proposed and peer group reviewed.

State 3. Solution Implemented

The proposed solution has been temporarily implemented for testing.

State 4. Solution Verified

The proposed solution has been tested, cures the problem, and causes no new problems.

Using a system like DDT benefits everybody; problems don't get "forgotten" in the rush to market; management can check status at any time; no one is confused about who "owns" a problem or when it was first discovered; and engineers can point to the list of unresolved problems to fend off undue pressure to "ship it."

V. CONCLUSIONS

Reliability is not something that is added by inspection or a burn-in step in manufacturing or by requiring it in a specification. It is designed in, a little at a time, as each part or process is chosen and incorporated into the design, and problems are discovered and solved. Therefore, every person who makes a design decision has a hand in determining that design's reliability. Training and feedback are the tools to achieve improved reliability—training appropriate to each level in the organization, so that each person knows how to contribute, and feedback so that they can see the effects of their actions. Testing, begun early in development and continued through manufacturing, with cyclical stresses that condense years into weeks or even days, provides the signals for the feedback loops. A failure should never be ignored or written off as random; it may occur at a random time or place, but it rarely has a random cause. Find the cause, and you will probably be able to eliminate or minimize it.

The way to make sure that a reliable product evolves is for management to get involved in setting reliability objectives, measuring the progress toward them, and making sure that everyone understands that they are the highest priority. The more actively management is involved and the more deeply it is committed, the better the results will be! In 1982, a Japanese-American joint venture company won the Deming prize, Japan's highest honor in the field of statistical quality control. Five years earlier, this same joint venture's quality and reliability were average by U.S. standards, and probably worse than average by Japanese standards. Every employee worked hard to make this improvement happen, but the unanimous opinion was that this happened primarily because the company's president became the personal leader and champion of the effort.

REFERENCES

1. C. H. Kepner, B. B. Tregoe, *The Rational Manager*. McGraw-Hill, New York, 1955.
2. R. Y. Moss, Modeling variable hazard rate life data. *Proceedings 28th Electronic Components Conference*, April 24–26, 1978, IEEE 78CH1349-0 CHMT, pp. 16–22.
3. P. B. Crosby, *Quality is Free*. New York: McGraw-Hill, 1979.
4. W. E. Deming, *On The Management Of Statistical Techniques For Quality & Productivity*, seminar handout, 1981.
5. *Microcircuit Failure Analysis Techniques Procedural Guide*, Reliability Analysis Center RADC/RBRAC, 1981.
6. G. N. Morrison, et al., *RADC Thermal Guide For Reliability Engineers*, RADC-TR-82-172, Hughes Aircraft Company, Rome, NY: Rome Air Development Center, June 1982.
7. MIL-T-28800, *General Specification For Test Equipment for Use With Electrical & Electronic Equipment*, Naval Electronic Systems Command.
8. MIL-STD-781, *Reliability Tests: Exponential Distribution*, Naval Air Systems Command.
9. R. A. Bailey, R. A. Gilbert, Strife testing. *Quality*, November: pp. 53–55, 1982.
10. E. R. Ott, *Process Quality Control*, New York: McGraw-Hill, 1975.
11. A. E. Saari, R. E. Schafer, and S. J. VanDenBerg, *Stress Screening Of Electronic Hardware*, RADC-TR-82-87, Hughes Aircraft Company, Rome, NY: Rome Air Development Center, May 1982.

FURTHER READING

Amerasekera, E. A., and D. S. Campbell, *Failure Mechanisms in Semiconductor Devices*, Wiley, New York, 1987.

Condra, L. W., *Reliability Improvement With Design of Experiments*, Marcel Dekker, New York, 1993.

Fuqua, N. B., *Reliability Engineering for Electronic Design*, Marcel Dekker, New York, 1987.

Ireson, W. G., and C. F. Coombs Jr., *Handbook of Reliability Engineering and Management*, 2nd ed., McGraw-Hill, New York, 1996.

Kececioglu, D., *Reliability Engineering Handbook*, Vols. 1 & 2, Prentice-Hall, Englewood Cliffs, NJ, 1991.

Klinger, D. J., Y. Nakada, and M. A. Menendez, *AT&T Reliability Manual*, Van Nostrand Reinhold, New York, 1991.

Nelson, W., *Applied Life Data Analysis*, Wiley, New York, 1982.

Nelson, W., *Accelerated Testing*, Wiley, New York, 1990.

O'Connor, P. D. T., *Practical Reliability Engineering*, 3rd ed., Wiley, New York, 1991.

13
Product Package Design

RICHARD Y. MOSS
Hewlett-Packard Company
Palo Alto, California

I. INTRODUCTION

In this chapter, the term *package* means the cabinet, enclosure, case, or body of a system, product, or component, which is present while the product is being used—NOT the carton or packing material used only to transport it to the customer. That a good package is an essential ingredient of a reliable product is scarcely controversial, yet arriving at that goal is a process which is often accompanied by conflict. Several possible reasons for this come to mind:

1. Package design is much more difficult than is generally recognized, since it involves poorly understood areas of design such as high frequency and magnetic shielding, heat transfer, mechanical engineering for strength and weight, and human factors.
2. The package design is often almost completely original rather than being simply an original arrangement of standard parts, as much of the electrical design is.
3. The objectives and acceptance criteria for the package design are usually not clearly defined and recorded, and involve compromises between function and style or engineering and art. Worse yet, it is often not clearly established who will have final approval authority over the package design, whereas there is seldom any argument over authority for approving the electrical design. The result is that upper management can get embroiled in deciding what color to paint the package while a crucial electrical design decision made at a much lower level goes unchallenged.
4. Whether or not a particular design is even feasible may well be a function of the availability of expensive tooling or a specific manufacturing process. Thus it is necessary to include the tooling and fabrication experts in the

package design reviews. The management challenge is much greater when a whole committee must be pleased!

II. LEVELS OF PACKAGING

A. Component

Just as there are different levels of product complexity, ranging from components to systems, so there are different considerations at each level of packaging. At the component level, packaging tends to emphasize the protection of fragile or sensitive internal parts, and may even have to provide a special "environment," such as excluding light or maintaining a vacuum. Often, component packages must conform to industry standards, so the only freedom allowed the designer is in the choice of materials and manufacturing techniques. The contribution of component packaging to reliability is crucial. Package-induced component failure modes such as overheating, electrical leakage due to contamination, corrosion, and mechanical fatigue due to coefficient of thermal expansion (CTE) mismatch dominate the failure statistics of electrical components. The rapid increase in the use of epoxy and silicone encapsulants, and reduced usage of corrosion-resistant but expensive precious metal platings, have both contributed to this increase. Today's component packaging requires materials and process control expertise, and is definitely no place for amateurs!

B. Subassembly

The next level of packaging is the subassembly. The emphasis at this level is on the mounting and interconnection of components which are functionally grouped for convenience of testing and field maintenance, and to allow optional features to be added or removed easily. In many electronic products, the subassembly level usually employs printed circuit technology to accomplish both the mounting and interconnection functions. Reliability of the interconnections is of prime importance at this level of packaging, although the contribution to cooling and vibration resistance is more important than is usually recognized. The majority of small electrical components accomplish more than 75% of their cooling by conducting the internally generated heat through the metal electrical leads to the printed circuit board, from which it is dissipated into the environment by air convection.

Physical flexing, especially at resonant frequencies, of the printed circuit substrate (usually fiberglass-reinforced epoxy resin) is the major source of vibration damage to small electrical components. This usually takes the form of plastic or glass component package cracking, and is of major importance in product reliability.

C. Product

In many cases, the final product is the highest level of packaging, and often has the greatest variety of demands placed on it. The product package provides an interface between the operator and the equipment, and is supposed to protect each from the other. This means that there are legal considerations as well as practical;

for example, compliance with regulations dealing with safety and electromagnetic compatibility (EMC).

Another problem is that the product package is often the most original, least standardized level of packaging, and its appearance sets the "style" of the company's products. This tends to generate controversy, since decisions of "artistic taste" tend to get made (and changed) by upper management.

The interface functions of the package deserve additional discussion. The product package generally supports the control and display functions of the product, with digital and graphical displays dominating. Most of the labeling and instructions which are required as a part of the product are mounted on, or are part of, the product package. Connectors and power cables generally are part of the package, which typically provides cable routing and strain relief. Finally, aids to transportation and mounting, such as handles, wheels, and mounting brackets, are usually considered part of the package design.

D. System

The highest level of assembly which will be discussed in this chapter is the system, which is often a collection of products interconnected so as to optimize the performance of a specific application. The function of the system-level packaging is to "integrate" and interconnect products which may or may not be particularly compatible in size or appearance, and to provide system convenience features, such as a desk or console, aimed at a particular application. Improved efficiency, the reduction of operator errors, and future expandability are benefits of a well-designed system package. Unfortunately, there are few roads, much less maps, in this part of the "desert," so that the system design ends up being the user's idea in one case, a particular designer's in another, and a compromise designed by a committee in a third.

Endless variety is possible, and production quantities of large systems are often so low that there is little incentive to employ extensive tooling or design standardization. Vertically integrated corporations can be the exception to this rule, if they have the foresight and discipline to define a system packaging plan at the start, and then make the product packaging a modular subset of that. Obviously, management attention is necessary to make that happen!

III. FUNCTIONS OF PACKAGING

The primary function of a package is that of a protective container; protection against the physical forces of gravity and rapid acceleration, and a container to keep out (or in) heat, light, acoustic noise, foreign materials such as solids, liquids, and gases, and to act as a barrier (insulator and shield) to electrical signals. Beginning with the physical forces, each of these functions will be discussed in turn.

A. Withstanding Physical Forces

Physical forces usually result from the interaction of acceleration and mass, according to the simple equation $F = m \cdot a$. (Force equals mass times acceleration.) The acceleration may be the result of gravity, rotation, or complex motion such as occurs in vibration or physical shock. The application of enough force to a component, subassembly, or product will result in deformation of some part; if this deformation is small, there may not be any malfunction; if it is large, failure will result. Many people are under the illusion that if the elastic limit of the material is not exceeded, no damage will result, but this is not always true. Buckling is an example of a common failure mode which can be initiated while stresses are below the elastic limit; for instance, when a long, thin member with insufficient stiffness is compression loaded. High cycle fatigue also occurs within the elastic range after long periods of cyclical loading (1).

Plastic deformation and creep are examples of failure modes occurring when the elastic limit is exceeded. The difference is that plastic deformation either ceases after a short time or, if the stress is cyclical, ends in low-cycle fatigue. This is what happens when you break a piece of wire, such as a metal coat hanger or paper clip, by bending it back and forth past its elastic limit. Creep tends to continue as long as the stress is present, until the material fractures, either in sheer or cleavage.

Another whole family of mechanical failure modes are loosely classed as wear, either adhesive or abrasive. Adhesive wear is probably the most common, and occurs when microwelds form and then shear, at the tiny points of contact that are the result of microscopic roughness of the load-bearing surfaces. Galling, scuffing, scoring, and seizing are all examples of adhesive wear. Abrasive wear is a self-descriptive term, and occurs when hard particles are present between two surfaces that are rubbing together, so that the surfaces are gouged or chipped away rapidly. If the particles come from the rubbing surfaces themselves, the term *fretting* may be applied. Other similar failure modes are pitting, where high-contact stresses result in cracking away of pieces of a surface, and cavitation, caused by pressure pulses in a fluid in contact with the surface. [This is the basis of ultrasonic cleaning (2,3)].

As was previously mentioned, the forces to activate all these failure modes generally come from the acceleration of gravity or motion. Vibratory motion is particularly destructive when it contains frequencies which excite resonance in a structure. (That's what happened to the Tacoma Narrows bridge!) In resonance, the forces are increased as a function of the "Q," or lack of damping, of the structure, so that the chances of exceeding the elastic limit are greatly increased. Examples of this type of failure are seen in electronic products where heavy components are mounted and supported by their electrical leads, and fail by cracking and breakage of those leads when subjected to vibration. This can be avoided by securing more massive components by their cases, and by supporting the printed circuit (PC) assemblies and making them more rigid so that damaging resonances are minimized. Thus vibration testing, including searches for resonances, is very important to assure the reliability of a new package. It is also obvious that the units tested must be exactly like the final design if the results are to be valid.

B. Heat Transfer

The next most important function of the package is the control of heat. Heat is transferred from a warm object to a cooler environment by a combination of three means: conduction, radiation, and convection. Evaporative cooling is an associated phenomenon.) Conductive cooling occurs when there is direct contact between a source of heat and a cooler mass, or "sink." This is the primary means by which most components transfer heat to the printed circuit board, and also contributes to the further transfer of heat from inside a product or system cabinet to outside, as you will learn the hard way if you change from a good thermal conductor, like metal, to a poor one, like plastic. The conductive heat flow is a function of the temperature difference, cross-sectional area, path length, and coefficient of thermal conductivity of the material; most metals and some ceramics are very good conductors of heat, whereas most plastics are poor conductors of heat.

Radiation is the means by which heat escapes from a warm object to a cooler surrounding space without any contact. Radiation cooling can happen in a vacuum, and is a function of the emissivity of the hot surface (nonglossy black paint has high emissivity, whereas a mirror has low emissivity) and the temperature difference between the object and its surroundings. In most electronic products, the temperature difference is only a few degrees, so radiation makes almost no contribution to their cooling.

Convection, the third means of cooling, requires the presence of a gaseous or liquid medium in which to set up convection currents. Convection is the main means by which the heat from electronic products is dissipated into the ambient air; unfortunately, the rate of cooling is a function of the orientation (vertical vs horizontal) and dimensions of the hot surface, and of properties such as the density and velocity of the convective medium. This means that the temperature of a product which is cooled by air convection is affected by the altitude and humidity, the rate of airflow, and which side is up. Forced air convection is far more effective at cooling than natural convection, providing that an air blockage (dirty filter) or fan failure does not occur. It is a good idea to provide some kind of automatic thermal cutout or warning indicator in products where failure or blockage of the fan could result in damaging temperatures (4,5).

C. Light

The function of the package as a container of light is most obvious in products with intense light sources, such as projectors or photocopiers; perhaps less obvious is the need to shield certain types of components from ambient light. Not only are traditional photographic materials light sensitive, but so are many devices with semiconductor junctions. Low-leakage diodes and transistors must nearly always be protected from light to prevent photocurrents that can cause a malfunction.

D. Acoustics

Vibrations which are coupled to the surrounding air result in acoustical noise, which is an annoyance in quiet environments and dangerous if it exceeds specified levels. It is becoming increasingly common for vibration-damping and noise-absorbent materials to be designed into new products and systems, and more and more companies are building their own sound level measurement facilities. One approach that has been employed successfully is to combine the functions of an electromagnetically shielded "screen room" with those of an acoustically quiet (anechoic) chamber, thus saving capital funds and floor space by combining two functions in one space. To do this, it is generally only necessary to make the dimensions, particularly the ceiling height, slightly larger than would be required for the screen room alone. This is simple and inexpensive to accomplish if planned in advance (6,7).

E. Contaminant Barrier

An obvious function of the package is to keep contaminants out or, occasionally, in. At the component level, nearly all technologies must be protected from substances which cause leakage or corrosion, particularly those which easily dissolve in water. Water, in liquid or vapor form, and soluble compounds of chlorine or phosphorus are particularly troublesome around semiconductors and reactive metals such as aluminum, copper, or silver. The contamination problem has become more acute as component packaging has become more reliant on plastic encapsulation or polymer sealing rather than the more expensive hermetic seals using glass, ceramic, and metals with matched coefficients of thermal expansion. There are no plastics which are capable of making a perfectly moisture-impervious seal, so component technologies which are packaged in plastic must be limited to ones that are insensitive to water and its attendant impurities. A promising approach to making plastic-encapsulated semiconductors reliable is the use of a moisture-impermeable passivation, such as silicon nitride, on the semiconductor surface itself. This protects all but the bonding pads from corrosion or contamination.

At the subassembly, product, or system level, it is less common to have a sealed package, so airborne contaminants usually circulate freely around and through the product. Splash-proof products are fairly common, but immersion-proof ones are not, particularly as they get larger. In addition to the difficulty of achieving a reliable seal on a large product, there is the problem of building the package strong enough to withstand the enormous pressure differences that occur at even moderate depths. The pressure at 10-feet depth in water is 624 lb/ft^2, and at 50 feet the pressure is over 3000 lb/ft^2! Most of us are not in the business of building submarines, so submersible products larger than a wristwatch or camera are unusual. Airborne dust and corrosives are a common problem in all sizes of package, particularly sulfur compounds. Corrosion failure of thin metal films, such as those used in electrical connectors, is a major cause of field failures. Gold plating has been the traditional solution to this problem, but the rising price of gold coupled with the impossibility of covering all exposed metals with sufficient thickness to resist the ever-increasing concentrations of corrosives in the environment have prompted an intense search for substitute metals. No easy or inexpensive solutions have been found to date, so it is wise to continue to use quality connectors where they are required, and to minimize the number of connectors in new designs.

F. Electrical Insulation and Shielding

Another "protective container" function of the package is that of an electrical insulator and shield. Not only are the electrical signals inside the product a potential safety hazard, but with the trend toward increased density of electronics in our homes and everyday lives—television and video games, microwave ovens, digital calculators and home computers, and cellular telephones, to name only a few—the management of the electromagnetic "environment" has become an urgent problem.

The regulatory agencies of most industrialized nations are leaping into the breach, proposing sweeping regulations with a sometimes tenuous basis in theory and a large bureaucracy to enforce them. Even if the technical considerations are separated from the economic and political, the trend is inexorably toward more regulation. Management's best strategy is to establish a product regulations organization within the company whose mission is to become expert in these regulated areas, to formulate company strategy for compliance and keep everyone informed of current and proposed rules and to try to influence the regulatory agencies to act reasonably. A list of important regulatory agencies is shown in Table 1.

Taking safety considerations first, there are six categories of hazard to be prevented by packaging:

1. *Electrical shock*: Preventive measures include double insulation, shielding and grounding, interlocks which cut power when the package is opened, fuses, circuit breakers and ground current interrupters, current limiters, cover panels requiring special tools for removal, proper size and location of enclosure openings, automatic discharge of energy-storage devices at power down, strain relief and elimination of sharp radii in wiring, and warning labels.
2. *Mechanical hazards*: Preventive measures include elimination of sharp edges, protective grilles over rotating parts such as fans (adequate to keep out hair as well as fingers), designs with wide bases and low centers of gravity to prevent tipover, and shields against explosion or implosion of breakable components such as cathode ray tubes.
3. *Fire and burns*: Preventive measures include the use of fire-retardant materials, designs which limit the maximum temperatures of exposed surfaces, and packages which will contain burning or molten material generated by an internal failure rather than allowing it to run out an opening.
4. *Radiation*: Intense visible, ultraviolet, or laser light, or objectionable sound must be contained or eliminated, and hazards such as x-ray or microwave radiation must be prevented. Cathode ray tube (CRT) displays, such as computer displays and television sets, are currently the subject of controversy and conflicting studies concerning the possible health effects of radiation they generate.
5. *Hazardous chemicals and materials*: The best bet is not to use them at all; where they are an intrinsic part of the component, such as acid electrolyte in a battery, they should be packaged so as to be leakproof, and labeled to warn against opening the package.
6. *Labeling*: The risk is that the warning or instructional labels will be missing, illegible, or not readily visible, and thus fail to perform their intended function. Labels must last as long as the product, which is many years in most cases, and must be kept simple and few.

The second area of regulatory agency interest is electromagnetic compatibility (EMC). The regulators are often interested only in the interference which the product generates, but your customer is also interested in the *susceptibility* of your product to electromagnetic interference (EMI), since he does not want the product to malfunction when subjected to EMI such as lightning or a radio transmitter.

Table 1 World Product Standards and Regulatory Organizations

Abbreviation	Country	Agency name
AFNOR	France	French National Standards Organization
ANSI	USA	American National Standards Institute
BRH	USA	Bureau of Radiological Health
BSI	UK	British Standards Institute
CEBEC	Belgium	Belgian National Safety Agency
CEN	Europe	European Committee for Standardization
CENELEC	Europe	European Committee for Electrotechnical Standardization
CISPR	International	International Special Committee on Radio Interference
CPSC	USA	Consumer Product Safety Commission
CSA	Canada	Canadian Standards Association
DEMKO	Denmark	Danish Electrical Material Control Organization
DIN	Germany	Deutches Institut für Normung e.V. (German National Standard Institute)
EI	Finland	Finnish Electrical Inspectorate
EIA	USA	Electronic Industries Association
ETSI	Europe	European Telecommunications Standards Institution
EZU	Czech Republic	Czech Republic Test House
FCC	USA	Federal Communications Commission
FDA	USA	Food & Drug Administration
FIMCO	Finland	Finnish Electrical Approval Agency
IEC	International	International Electrotechnical Commission
IEEE	USA	Institute of Electrical & Electronic Engineers
IMQ	Italy	Instituto Italiano del Marchio de Qualita
ISO	International	International Organization for Standardization
JEIDA	Japan	Japan Electronic Industry Development Association
KEMA	Netherlands	Keuring Van Elektrotechnisch Materialen
MITI	Japan	Ministry of International Trade & Industry
NEMKO	Norway	Norges Elektriske Materiell Kontroll
NFPA	USA	National Fire Protection Association
NOM	Mexico	Norma Obligatoria Mexicana
OSHA	USA	Occupational Safety and Health Administration
OVE	Austria	Osterreichischer Verband fur Elektrotechnik
SAA	Australia	Standards Association of Australia
SEC	Australia	State Electricity Commission of Victoria
SEMKO	Sweden	Svenska Elektriska Materiel Kontrollanstalten
SEV	Switzerland	Schweizer Electrotechnischer Vereinigung
TUV	Germany	Technischer Uberwachungs-Verein (Testing Org.)
UL	USA	Underwriter's Laboratories Inc.
UTE	France	Union Technique de L'Electricite
VCC	Japan	Voluntary Control Council for Interference
VDE	Germany	Verband Deutscher Elektrotechniker

The measures necessary to control interference generation and susceptibility are mostly the same: shielding by enclosure in high-conductivity metals, avoidance of slot radiators by using conductive gaskets at joints and round holes with diameter much less than the shortest wavelength to be attenuated, shielded cables and connectors, and electrical filter circuits on connectors to unshielded wiring such as the AC power mains (8). In the case of low-frequency magnetic field containment, the shielding metal must have a high magnetic permeability, such as certain steels or special alloys, and wiring must either be coaxial or twisted pair, so that external magnetic fields are either minimized or canceled.

Electrostatic discharge (ESD) is a common electrical phenomenon that is not yet widely regulated, but is none the less serious. History is full of examples of catastrophes caused by a static spark in the presence of a flammable or explosive material such as textile lint, gunpowder, rocket fuel, petroleum fumes, anesthetics, or hydrogen gas, to name only a few. More recently, it has been realized that ESD is responsible for a significant fraction, probably 10–30%, of all the electronic component failures in production and field use. Several degrees of failure can be described: In the most severe cases, the energy of a static discharge can cause immediate, catastrophic failure of a semiconductor device or other sensitive component. A second, more subtle kind of failure results when a static discharge <1 kV, too low in voltage to generate a visible spark or to be felt by a human, damages or degrades a part so that its probability of failing later is greatly increased. This is probably the most common situation, where the initial damage occurs during the manufacture of components and subassemblies, but it is the product or system which subsequently fails. The third kind of failure is a "soft" failure, or error, which most often plagues digital computer systems. Here it is usually the EMI generated by the static discharge which changes one or more bits in a digital signal, resulting in data errors or even system crashes.

The prevention of ESD problems is simple in principle, and affects the package in several ways:

1. Products and systems must be designed to withstand ESD at the worst-case levels encountered in the field. This generally means that the outer package must be a "Faraday cage" (a complete, conducting box), and all wiring which enters or leaves the enclosure must incorporate protective circuitry.
2. All ESD-sensitive components and assemblies must be stored and transported in static-protective containers. The best protective container is, again, a Faraday cage. Conductive (carbon-loaded) plastic, metalized plastic, and metal foil are examples of such containers which are readily available (19).
3. Precautions against ESD damage must be observed where sensitive items are handled—in manufacturing and also service and support. This means maintaining static-free work stations, and grounding all possible static charge generators, including humans.

G. Maintenance Accessibility

The last, but not least, function of the package is the role that it plays in maintenance. Probably every one of us has had a personal experience with a product that was unserviceable because of poor access or lack of adjustability. This problem most often occurs where the subassembly is tested and calibrated separately at the factory, but must be serviced while installed in the product or system. Not only must there be easy access to test points and adjustments, but also subassemblies or components with the highest probability of maintenance should be the easiest to access or remove. A simple example is the air filter in front of a cooling fan; the product can fail due to overheating if the filter is not easy to see and clean. In planning the partitioning of components into subassemblies, items requiring frequent replacement should be mounted so that they can be individually replaced; if they are mounted on a larger assembly which must also be replaced, the only result is to raise the average repair cost.

IV. THE DESIGN PROCESS

A. Objectives

Setting objectives is just as important in package design as it is in electrical design. The objectives should be written down and approved at the appropriate management levels, not just informally discussed or understood. There are five major factors to be considered in setting objectives for the package design.

Legal Requirements

Creating a safe product, which complies with the regulatory agency requirements in the countries and market sectors chosen is certainly top priority. Although it is a legitimate management decision to exclude a certain country or market segment, and hence to avoid the need to comply with that country's regulations, it is often difficult to implement that decision. Products have a way of being resold to a third party, who may not observe the restrictions. Or the products might be transferred from the country of purchase to one where the product is not legal. It is far safer to meet or exceed all the worldwide regulations that apply to that type of product, so long as they are not mutually contradictory. This may cost extra in design or even in manufacture, but it may well be cheaper in the long run because of reduced liability.

Customer's Expectations

The customer has a right to expect a product which is of high quality, reliability, and serviceability. In the package design, that translates into a package that meets all its specifications, is long lived, and permits easy access for maintenance or upgrades. Quantifying these objectives is difficult: Quality might be specified as the percentage with no defects, reliability would be measured by the survival under extreme conditions or the survival percentage in warranty, and serviceability could be measured in terms of the average time to perform certain benchmark repairs.

Environment

The objective is to meet the stated quality and reliability goals over a wide range of environmental conditions, including temperature, humidity, altitude, vibration, and contaminants. Not only must the package itself survive, but also its supporting functions must be performed successfully; adequate cooling for the electronics must be provided, even at maximum altitude and temperature, for example. A new kind of environmental issue has recently surfaced: the disposal of the product at end of life (when it is no longer useful) without impacting the environment. This means "design for disassembly," and use of materials that are safely recyclable.

State of the Art

As is the case with electrical design, we often wish to gain performance or reduce cost with state of the art technology, but it is also usually true that the newest technique is not the most reliable until after several years of evolution. Jumping into a new packaging technology too soon can cost an unaffordable price in redesign, retooling, and warranty replacements, but so can the use of an obsolete approach. A successful strategy from a quality viewpoint is to be neither the first nor the last to adopt the newest technology, but be the best! Setting tough but realistic objectives for design, tooling and fabrication costs, schedules, and yields is a good way to start.

Quality Costs

The bottom line is, after all, profit or loss. Scrap, rework, inspection, and replacements all reduce profits, so setting objectives for reducing these is appropriate. Over the long haul, warranty costs will tend to be the most painful, because they represent a dissatisfied customer who has an increased probability of taking his or her business elsewhere next time. Studies at two large corporations have shown that the result is a multiplication of the effect of warranty costs on profit; each 1% of the sales dollar spent on warranty is associated with a 5–6% reduction in pretax profits.

B. Project Management

Since most package designs are accomplished concurrently with other design activities, such as electrical or software design, the management challenge is to promote communication between the separate parts of the project, and to keep them all on schedule and aimed at the same target. The bigger the project, the more meetings there must be (10); in this situation, it is particularly important that there be planning of the purpose of each meeting, and that each participant comes fully prepared. As the project moves toward completion, there will be more conflicting details to be resolved, and larger expenditures to be authorized. The project manager should be sure to review all the appropriate data in the categories of product design, stress analysis, stress testing, regulatory conformance, manufacturability, and marketability.

Product Design

In package design, more can be accomplished using models and mockups than is the case in electrical design. One reason is that three-dimensional objects are difficult to visualize corrrectly from two-dimensional drawings. Three-dimensional mockups are a much safer means of reviewing reaction to subjective factors such as styling and color; to assess human factors such as convenience and placement of elements; and to study accessibility for maintenance, subassembly mounting, and interconnection problems—and even some aspects of cooling such as provision for airflow. Computer-aided design (CAD) and computer-aided artwork (CAA) are valuable in this area, as well as allowing the use of design rule and tolerance-checking programs to catch errors early; thus greatly reducing the number of iterations before an error-free design is achieved.

Stress Analysis

The effects of the "four horsemen" of stress—electrical, mechanical, thermal, and chemical—must be tested and reviewed repeatedly during the evolution of the package design (see Chapter 12). Especially important is the margin between maximum stresses to be encountered and the minimum strength of the design. Usually the designer knows, or thinks he or she knows, this margin; it is important to document this information, and then to verify it by testing. This raises the delicate issue of the designer whose ego would be bruised by the implication that you don't believe or trust him or her, and hence must verify his or her design's quality by testing. If you are the type who would travel on an airplane that had never before been flown, then go ahead and release a product which has not been tested. Otherwise, plan on stress tests for each stage in the evolution.

Stress Testing

There are two distinctly different types of stress testing: step-stress testing to determine the design margin, and stress-accelerated life tests to verify reliability. Thermal mapping, environmental testing, and abuse testing are all ways of verifying that the design has sufficient margin, and confirms the designer's expectations. In step stress testing, it is important to continue to increase the stress until a consistent pattern of failures emerges, since this tells us much more about the design than a test with no failures—the position of the threshold of failure is a sensitive barometer of change. A design where the failure threshold is 50% above the ratings inspires much more confidence than one where the operating margin is only 5%, or one where the margin has deteriorated since the previous test! Stress-accelerated life tests generally subject the design to a fixed stress, or combination of stresses, and it is the duration of the test that is increased until a consistent pattern of failures emerges. (The stresses may be applied cyclically, but the limits and rates are fixed.) Early in the design, the intent is to find the dominant failure mechanisms and attempt to design them out; at the end, the intent of the test may well be to establish that a specified minimum reliability has been achieved.

Regulatory Conformance

Since creating a safe and legally conforming product is the top priority, it follows that review of the results of tests is an important project management function. If achieving reliability without testing is difficult to imagine, then convincing a regulatory agency (or, in the worst case, a jury) that you have a safe and conforming product is impossible. It is also nearly impossible to achieve such a design as an afterthought or in a hurry. Product safety and electromagnetic compatibility must be part of the design from the start, and they must be verified by testing.

Manufacturability

There are many activities related to the eventual manufacture of the new design which must be planned and initiated long in advance of the start of production, and many of them depend upon the design team for their initiation, if not completion. Tooling must be designed and built, and items fabricated with that tooling tested as outlined above. Documentation (such as drawings and material lists) must be completed and reviewed, and long-lead materials ordered, received, and incoming acceptance tests performed. (It is especially important to perform incoming tests on new materials or on those from new vendors.) In addition to tooling, new or reassigned capital equipment needed for fabricating, finishing, or testing the new design must be budgeted, purchased, and installed. With sophisticated computer-controlled or numerically controlled machine tools often needed in the fabrication of package designs, it is important that accuracy and calibration be checked when the equipment is first installed and regularly thereafter. Last, but certainly not least, it is generally necessary to provide training for the manufacturing personnel in the peculiar or critical aspects of assembly and test procedures. Since these concerns are generally known only to the designers in the beginning, it is the responsibility of management to make sure that they get communicated and documented.

Marketability

You may well ask how marketing affects the quality and reliability of a package design. The answer is: in several ways. If the product is announced, shown to customers, or introduced at an industry show before it has completed its design and testing, the result can be a temptation to skip some tests or ship product before the tests are completed and all the major problems solved. This certainly affects the quality of the product. Also, if operation and service manuals are not finished, or if sales and service personnel are not trained and inventories of spares and special test equipment available, the customer's expectations will not be satisfied, and a fiasco can result. The excitement generated by receiving a new product soon fades, but the bad taste of poor performance lives on and on!

V. CONCLUSIONS

Successful package design requires strong management skills as well as originality and competence in a broad range of engineering skills.

Today's complex products are usually designed by a team rather than a single individual, so a high degree of cooperation and coordination is required. The traditional pitfalls of subjective judgment, controversy, and the resulting redesign can be avoided by careful planning and defining goals so that objective decisions are made at each project checkpoint. The levels of management approval that should be obtained are roughly parallel to the hierarchy of levels of packaging; that is, the system and product levels require the highest levels of management approval and components the lowest (unless components are the final product, of course). When it comes to quality and reliability, however, none of the levels of packaging can be considered less important than the others. A weakness at any level can result in a system or product failure; the chain is truly only as strong as its weakest link!

REFERENCES

1. P. H. Wirshing, J. E. Kempert. A fresh look at fatigue. *Machine Design*, May 20:120–123, 1976.
2. C. Lipson. Basic course in failure analysis—lesson 1; how parts can fail. *Machine Design*, October 16:146–149, 1969.
3. A. D. S. Carter. *Mechanical Reliability*, 2nd ed. New York: Wiley, 1986.
4. A. W. Scott. *Cooling of Electronic Equipment*. New York: Wiley, 1974.
5. D. S. Steinberg. *Cooling Techniques for Electronic Equipment*, 2nd ed. New York: Wiley, 1991.
6. L. L. Beranek. *Noise and Vibration Control*. New York: McGraw-Hill, 1971.
7. A. P. G. Peterson, Gross Jr. E. E. *Handbook of Noise Measurement*, 9th ed., Concord, MA: General Radio Company, 1980.
8. H. W. Ott. *Noise Reduction Techniques in Electronic Systems*, 2nd ed., New York: Wiley, 1988.
9. *Electrostatic Discharge Training Manual*, NAVSEA SE 003-AA-TRN-010, Naval Sea Systems Command, September, 1980.
10. F. P. Brooks, Jr. *The Mythical Man-Month*. Redding, MA: Addison-Wesley, 1975.

FURTHER READING

General Packaging

Dally, James W., *Packaging of Electronic systems, A Mechanical Engineering Approach*, New York: McGraw-Hill, 1990.

Manzione, L. T., *Plastic Packaging of Microelectronic Devices*, New York: Van Nostrand Reinhold, 1990.

Morris, J. E., *Electronics Packaging Forum*, Vols. 1 & 2, New York: Van Nostrand Reinhold, 1990.

Seraphim, D. P., R. Lasky, and Che-Yu Li, *Principles of Electronic Packaging*, New York: McGraw-Hill, 1989.

Sloan, J. L., *Design and Packaging of Electronic Equipment*, New York: Van Nostrand Reinhold, 1985.

EMC

Morrison, R., *Grounding and Shielding Techniques in Instrumentation*, 3rd ed., New York: Wiley, 1986.

Paul, C. R., *Introduction to EMC*, New York: Wiley, 1992.

Williams, T., *EMC For Product Designers*, Oxford, England: B-H Newnes, 1992.

ESD

Bhar, T. N., and E. J. McMahon, *ESD Control*, Carmel, IN: Hayden, 1983.

Boxleitner, W., *ESD and Electronic Equipment*, New York: IEEE Press, 1988.

Dangelmayer, G. T., *ESD Program Management*, New York: Van Nostrand Reinhold, 1990.

Kolyer, J. M., and D. E. Watson, *ESD From A to Z*, New York: Van Nostrand Reinhold, 1990.
McAteer, O. J., *ESD Control*, New York: McGraw-Hill, 1989.

Standards and Regulations

Cargill, C. F., *Information Technology Standardization—Theory, Process, and Organizations*, Maynard, MA: Digital Press, 1989.
Kreuger, H., *Product Safety in Information Technology Equipment*, Germany: VDE-Verlag, 1992.
Spring, M. B., "Improving the Standardizing Process," Department of Information Science, Pittsburgh, PA: University of Pittsburgh, May 31, 1994.
Safety and EMC Newsletter, Surrey, United Kingdom: ERA Technology.

Thermal and Vibration Analysis

Lau, J. H., *Thermal Stress and Strain in Microelectronics Packaging*, New York: McGraw-Hill, 1989.
Steinberg, D. S., *Vibration Analysis for Electronic Equipment*, 2nd ed., New York: Wiley, 1988.

14
Pharmaceutical Product Quality Management

JAMES RICHARD SANTOS
Fidus Medical Technology Corporation
Fremont, California

I. INTRODUCTION

In 1962, the drug thalidomide caused severe birth defects in infants whose mothers had ingested it during pregnancy in the United States and abroad. In response, President John F. Kennedy recalled Congress to pass the *Kefauver-Harris* Amendments to the Food, Drug and Cosmetics (FD&C) Act. These amendments resulted in the Good Manufacturing Practices regulations, and, further, required drug developers to prove product efficacy.

Before the 1962 Kefauver-Harris Amendments, there was no legal obligation for pharmaceutical manufacturers to show the effectiveness of any drug. From then on, the effectiveness of proposed new drugs had to be demonstrated before the U.S. Food and Drug Administration (FDA) would grant approval.

In 1972, the FDA reviewed drugs that had been on the market from 1938 to 1962 to evaluate their effectiveness. The FDA found that 512 of the many drugs brought to the market during that time were "over-the-counter" (OTC)—that is, nonprescription drugs. Of these OTC drugs, 300 lacked substantial evidence of effectiveness of one or more of their ingredients.

Many prescription drugs were found to be in the same problem state. It had become clear that the FDA would have to correct the situation. Because of the FDA's limited resources, not all drugs could be evaluated immediately. Thus prescription drug review was given precedence over the OTC drugs.

II. PRESCRIPTION DRUGS

Part of the requirements for prescription drugs is based upon the Drug Efficacy Study Implementation (DESI) (1), which was initiated by the FDA in 1968. The purpose of the study was to assure that every prescription drug offered to the American public is both safe *and* effective. The study reviewed all prescription drugs on the American market from the 1938 FD&C Act to 1962 (Kefauver-Harris Amendments).

A. DESI Results

The results of the DESI study were as follows:

1. Hundreds of ineffective drugs were removed from the market.
2. Thousands of drugs were relabeled for more accurate therapeutic quality information.
3. Physicians have more reliable and objective label information to base their prescriptions on.
4. New criteria for combination drug marketing were developed. Some combination drugs proved to be more of a hazard than a help to consumers.

Because of a defect of the 1906 Food and Drug Law, products covered by the law had to cause injury before the FDA could seek their removal from the market. The 1938 FD&C Act rectified the situation by requiring that manufacturers *prove* the safety of their food and drug products *before* they could be marketed. New drugs had to be submitted to, and approved by, the FDA before they could reach the market.

The 1962 Kefauver-Harris Amendments required prescription drug manufacturers to prove not only safety but effectiveness. Thalidomide, although effective for its intended purpose as a sedative and hypnotic caused unacceptable side effects in fetuses which forced tighter controls.

B. Research Assistance

In 1963, the FDA asked the National Academy of Sciences (NAS) and its subsidiary the National Research Council (NRC) to help in the initial part of the drug efficacy study review. NAS provided many top scientific experts with proven abilities. Thirty panels—each consisting of six experts in specific fields such as antibiotics or heart drugs—were formed.

There were, at the time of the study, some 3000 drugs on the market. In addition, however, some 16,000 claims had been made for the beneficial results of using these drugs. Beyond that, for every 1938–1962 approved prescription drug, five others, known as "me-too" versions, were sold. These were chemically identical copies of the previously approved prescription drugs.

Each manufacturer had to provide evidence of their drugs' effectiveness. Each claim or purpose was reviewed.

NAS/NRC submitted their reports to the FDA in late 1967 through mid 1969. In January of 1968, the FDA formed a task force to put the NAS/NRC reports into effect. The Drug Efficiency Study Initiative was underway.

DESI evaluated data submitted by manufacturers trying to support the claims made for their prescription drugs as well as material contained in NDA files and the medical literature. DESI also coordinated physicians, pharmacologists, and chemists to study the NAS/NRC findings and recommendations and other evidence.

C. Study Guidelines

The following effectiveness guidelines were used:

Effective: The evidence was adequate to justify the claims made for the drug; that is, the drug was effective in doing what was claimed for it.

Effective with reservations: The claims made for the drug might be accurate, but some change had to be made, such as in labeling or ingredients.

Probably effective: More evidence was needed to support the claim(s) made for the drug, but the likelihood was that the evidence could be gathered.

Possibly effective: The drug needed further study. However, the chances that evidence could be gathered to prove the drug's effectiveness were only moderate.

Lacking substantial evidence of effectiveness: There was no scientific evidence to support the claim(s) made for the drug.

Ineffective as a fixed combination: Although each of several ingredients in a combination might be effective, combining them did not make the drug anymore useful than if one or more of the ingredients were used separately.

Manufacturers whose drugs fell into one of the following categories were required to take corrective action:

Lacking in substantial evidence of effectiveness: The manufacturer had to withdraw the product from the market unless the manufacturer could supply positive data within 60 days.

Possibly effective: Six months' grace was allowed for providing positive supporting data. Extensions could be granted if bona fide studies were underway.

Probably effective: One year was given to provide supporting data. Reasonable extensions could be granted.

D. An Overview of Human Testing of Drugs

An *investigational new drug* (IND) (2) application may be submitted for one or more phases of an investigation. 21 CFR (Code of Federal Regulations), Part 312, Subpart B describes in detail the requirements for the IND. The clinical investigation of a previously untested drug generally consists of three phases conducted in sequence. They may, however, overlap. The three phases of an investigation are as follows.

Phase I

Phase I includes the initial introduction of an investigational new drug into humans. Phase I studies are usually closely monitored, and they may be conducted using patients or volunteer subjects. These studies are designed to find the metabolism and pharmacological actions of the drug in humans. Side effects associated with increasing doses are studied. If possible, early evidence on effectiveness is sought.

During Phase I, sufficient information about the drug's pharmacokinetics (how drug reacts within the body) and pharmacological (the effects of the drug on the targeted problem) effects should be obtained to permit the design of well-controlled, scientifically valid Phase II studies. The total number of subjects and patients included in Phase I studies varies with the drug. It is usually between 20 and 80.

Phase I studies also include studies of drug metabolism, structure-activity relationships (i.e., the way in which chemically active sites on a drug-molecule structure relate to the mating structure of a targeted disease molecule), and the mechanism of action in humans. Studies also are included which investigational new drugs are used as research tools to explore biological phenomena or disease processes.

Phase II

Phase II includes controlled clinical studies. These are conducted to evaluate the effectiveness of the drug for a particular indication or indications in patients with the disease or condition under study. Controlled clinical studies are also done to learn the common short-term side effects and risks associated with the drug.

Phase II studies are well controlled and closely monitored. They are conducted on in a relatively small number of patients; usually no more than several hundred.

Phase III

Phase III studies are expanded controlled and uncontrolled trials. They are done after obtaining preliminary evidence suggesting effectiveness of the drug. The trials gather additional information, including the effectiveness and safety information needed to evaluate the total benefit-risk relationship of the drug. This information is also used to provide an adequate basis for physician labeling.

Phase III studies usually include from several hundred to several thousand subjects. The drug in Phase III must be the same as the drug proposed to be sold to the market (3) in:

1. Dosage form
2. Potency
3. Ingredients
4. Bioavailability—that is, the amount and rate at which the active ingredient(s) are absorbed into the human body

Phase I, II, and III Protocols

A protocol is required for each planned study. Generally, Phase I studies are less detailed and more flexible than Phase II or III study protocols. Phase I protocols should primarily provide an outline of the investigation. The outline is:

1. An estimate of the number of patients to be involved
2. A description of safety exclusions (i.e., those aspects of a patient's health that would cause them to be excluded from a study)
3. A description of the dosing plan, including duration, dose, or method to be used in determining dose

The outline should specify in detail only those elements of the study that are critical to safety. An example of this would be monitoring of vital signs and blood chemistries. Modifications of the experimental design of Phase I studies that do not affect critical safety assessments are required to be reported to the FDA only in the annual report.

Detailed protocols describing all aspects of the study must be submitted for Phases II and III. Such protocols should be designed to allow some deviation from the study design, if necessary, as the investigation progresses. Alternatives or contingencies should be built into the protocols at the beginning, if the sponsor anticipates potential deviations. For example, a protocol for a short-term study might include a plan for an early crossover of nonresponders (persons taking part in a pharmaceutical study who show no response to the drug being tested) to an alternative therapy.

E. IND Rewrite

In 1989, under CFR Part 312, *Investigational New Drug Application* (Revised), the FDA revised its regulations governing the submission and review of IND applications (4). The new regulations were called the IND Rewrite. They were changed to ensure the FDA's ability to monitor carefully the safety of patients who take part in clinical investigations. The regulations were also designed to help the development of promising new drug therapies. The improvements in the regulations were intended to help sponsors of clinical investigations prepare and submit high-quality IND applications and to let the FDA review the applications more efficiently and with the least delay.

Two changes (5) were made to the Code of Federal Regulations (CFR). The first concerned 21 CFR 312.34 (*Treatment use of an investigational new drug*). The other involved 21CFR, 312.7(d) (*Sale of an investigational new drug*). The focus of the changes by the FDA was the reproposing of procedures to the pharmaceutical industry to make investigational new drugs available to desperately ill patients before general marketing began (6).

These procedures serve to help the availability of promising new drugs to patients as early as possible. They apply to patients with immediate life-threatening or other serious diseases for which no satisfactory alternate therapies exist.

For example, the procedures would apply to advanced cases of acquired immune deficiency syndrome (AIDS); to certain uncontrollable cardiac arrhythmias; and for other serious diseases such as Alzheimer's disease and multiple sclerosis.

The FDA also reproposed conditions to the pharmaceutical industry under which drug manufacturers may sell investigational new drug products. With the revolution in biotechnology, it was important for the FDA to recognize the need to provide sufficient incentives for the rapid development of drugs. This is more so when no satisfactory alternate therapy is available.

F. Institutional Review Boards for Clinical Investigations

The FDA amended the revised regulations of July 27, 1981, for informed consent and those applicable to institutional review boards (IRBs), effective August 19, 1991. This was done to conform both regulations to the Federal Policy for the Protection of Human Research Subjects. Existing FDA regulations then governing the protection of human subjects shared a common core with the federal policy and carried out the fundamental principles of that policy (7).

G. Drug Test Safety Requirements

The FDA (8) is charged by statute with ensuring the protection of the rights, safety, and welfare of human subjects who participate in clinical investigations. This applies to those investigations subject to sections 505(i), 507(d), or 520(g) of the FD&C Act. Included also were clinical investigations that support applications for research or marketing permits required for the FDA-regulated products. This involved drugs and biological products for human use.

In the *Federal Register* (FR) of January 27, 1981, the FDA adopted regulations governing two areas. One was informed consent of human subjects (21 CFR Part 50, 46 FR 8942). The other consisted of the regulations establishing standards for the composition, operation, and responsibilities of IRBs that review clinical investigations involving human subjects (21 CFR Part 56, 46 FR 8958). Simultaneously, the Department of Health and Human Services (HHS) adopted regulations protecting human research subjects (45 CFR Part 46, 46 FR 8366). The FDA and HHS regulations share a common framework.

In December 1981, the President's *Commission for the Study of Ethical Problems in Medicine and Behavioral Research* issued its first biennial report (9). The commission recommended that all federal departments and agencies adopt the HHS regulations (45 CFR Part 46).

In May 1982, the President's science advisor, the Office of Science and Technology Policy (OSTP), appointed an ad hoc Committee for the Protection of Human Research Subjects, which served under the Federal Coordinating Council for Science, Engineering, and Technology (FCCSET). This was done to respond to the recommendations of the commission. The committee developed responses to the commission in consultation with OSTP and the Office of Management and Budget (OMB).

The committee agreed that uniformity of federal regulations on human subject protection was desirable for two reasons: (1) to eliminate unnecessary regulations and (2) to promote increased understanding by institutions that conduct federally supported or regulated research.

The committee developed a model policy that OSTP, however, later revised. OSTP, with the concurrence of affected federal departments and agencies such as the FDA and the Centers for Disease Control, published the revised policy as a proposal in the *Federal Register* of June 3, 1988 (51 FR 20204). More than 200 comments were submitted by the pharmaceutical development and manufacturing community in response to the proposal. The final ruling was published in the *Federal Register* of June 3, 1988 (56 FR 28025).

The FDA concurred in the final rule. In the *Federal Register* of November 10, 1988 (53 FR 45678), the agency proposed to amend its regulations in 21 CFR, parts 50 and 56, so that they would conform to federal policy stated in the final ruling, with two qualifications. The first qualification dealt with the unique requirements of the FD&C Act. The second qualification was the fact that the FDA is a regulatory agency that rarely supports or conducts research under its regulations.

H. Substantive Changes

Substantive changes to the 21 CFR Part 56, became effective on August 19, 1991 (10). The FDA's IRB regulations were revised to conform to the language of the

Common Rule. For the most part, the changes to the FDA regulations are not substantive. The substantive changes are presented in Table 1.

III. OVER-THE-COUNTER DRUGS

In 1972, the Food and Drug Administration established the *Over-the-Counter (OTC) Review.* As a result, over-the-counter drugs (10) then, now, and in the future must meet standards of safety, effectiveness, and adequate labeling.

At the time, two choices existed for the OTC review:

1. Review each product individually (about 300,000 products)—this would take decades to complete.
2. Do rule-making by active ingredient—about 1000 active ingredients existed within the therapeutic classes.

Table 1 Substantive Changes—*Federal Register* of June 3, 1988

Subsection	Change description
56.104(d)	Taste and food quality evaluations and consumer acceptance studies are exempt from IRB requirements. That is true only if any additive is at or below the level found safe or approved by the FDA, Environmental Protection Administration, or the Department of Agriculture.
56.107(a)	***Consideration*** is given to the selection of ***gender*** to membership of an IRB.
56.107(a)	***Consideration*** is given for the inclusion of person(s) to membership of an IRB to represent ***vulnerable*** populations. VULNERABLE POPULATIONS: • Fetuses. • Neonates. • Children. • Elderly persons. • Emergency room patients. • Women of childbearing age. • Pregnant women. • Mentally hindered persons. • Physically disabled persons. • Economically disadvantaged persons. • Educationally disadvantaged persons. • Persons with AIDS. • Persons with inoperable cancer. • Prisoners.
56.107(c)	The IRB must include at least one member whose primary concerns are in scientific areas.
56.108(b) (2)	An IRB must follow *written* procedures. These are for prompt reporting to the FDA and appropriate institutional officials. The reports involve serious or continuing noncompliance with these regulations or the requirements or determinations of the IRB.

The second option was chosen as the most effective and fairest approach of the two possible methods.

The first phase of the OTC review used advisory panels to provide a broad perspective. Panel members consisted of academic, medical profession, pharmacy profession, and biological and physical scientists concerned with each category of drug. One expert nominated by a consumer organization was also on each panel. Nonvoting members included representatives of industry and consumer groups. The professionals supplied specialized knowledge on particular issues to augment the FDA's limited staff. Consumer and industry members gave practical information about their side of the story.

The OTC review panels classified OTC drug ingredients into three categories:

Category I: Ingredients are generally recognized as safe (GRAS) and effective and not misbranded.
Category II: Ingredients are NOT GRAS/effective or *are* misbranded.
Category III: Available data are insufficient to allow final safety/effectiveness classification at the time of the panel's review.

The final report of each review panel contained a detailed explanation of reasons for the panel's recommendation. These recommendations included specification of the time during which products with ingredients or claims in Category III could continue to be marketed pending completion of further testing. Most reports contained detailed guidelines for such testing.

There were 27 categories of ingredients for which 17 panels of nongovernmental experts were appointed to help the FDA review the scientific literature and clinical studies. Some panels looked at more than one category. Results from the first 271 active ingredients reviewed showed the following:

50% were Category I, Safe and Effective
20% were Category II, NOT Safe or Effective
30% required additional studies

Category II classed ingredients problems were mostly due to:

1. Apparent lack of effectiveness of a single ingredient
2. Apparent lack of effectiveness of specific combinations of ingredients
3. Insupportable labeling claims

In cases where a panel found significant safety problems with certain ingredients, the FDA took effective action to remove the products from the market. Some ingredients included the following: hexachlorophene, tribromsalam, and zirconium in aerosols and antiperspirants.

IV. GOOD MANUFACTURING PRACTICE

Good manufacturing practice (GMP) or, more recently, *current* good manufacturing practice (cGMP), can be compared to a less-structured ISO 9000 system. GMP is a quality-based documentation system incorporating plain good business sense. Pharmaceutical companies must know: what is being done; what has been done; are the manufacturing processes used to make the products effective and efficient?

The pharmaceutical industry is, by its nature, heavily regulated. In the United States, the Food and Drug Administration is the primary regulator. Adherence to pharmaceutical GMP can be said to include the following elements.

A. Clean It

Facilities must be kept clean and free of contamination and controls must be in place to ensure that different products do not cross contaminate one another.

B. Validate It

Process validation means that when a process is set up, a key part of the setup is to prove by testing that products made by the process (and any intermediate processes) actually meet specifications and perform consistently (i.e., the processes are capable and "in control").

C. Verify It

Inspection, the old standby technique of quality control, comes into play here. It is important that the ingredients, the process, and the final product are tested on an ongoing basis to verify compliance to quality, purity, and sterility requirements.

D. Document It

Each of the following should be documented:

- The master production batch record
- The individual production batch record
- The bulk production batch record
- The product part numbers
- The batch lot numbers

Each batch must, at a minimum, be recorded properly, filed appropriately, and archived safely according to documentation requirements.

Traceability is the key to this world of documentation. For example, a GMP product record must include information such as production batch records, quality control records, sterilization charts, move tickets, and all other critical batch-associated documentation.

Distribution records must contain, at a minimum, enough data to access all customers should a recall become a necessity. The record system must also ensure that all recalled product is returned to the manufacturing facility.

Written and approved procedures must ensure that proper corrective action is taken on every complaint filed, including a proper assessment of the accuracy of the complaint. Records must contain complete information, including actions taken and notifications made. Also, records must be filed using a system that will ensure the safety and retrievability of the records.

E. Let the Inspector In

State or federal inspections can be performed with or without the manufacturer's consent. Should any manufacturer choose not to submit to an inspection, a court-issued warrant could be used to gain access to the manufacturer's facility.

F. Recall It

Should something really go wrong, a voluntary recall is better for your company image than an FDA-enforced recall. Generally, a voluntary recall can be done at any time for any reason by any manufacturer or distributor.

REFERENCES

1. "Desi Who?", from *FDA CONSUMER,* DHEW Pub. Number FDA 73-3013, Food and Drug Administration, Rockville, MD, October, 1972.
2. 21 CFR Part 312 (revised as of April 1, 1989). Subpart B—Investigational New Drug Application, Subsections 312.21, 312.23.
3. Author's research notes for FDA Speaker's Bureau talks while a chemist for the San Francisco District Laboratory of the U.S. Food and Drug Administration (1975–1980).
4. *Federal Register*, Vol. 52, No. 53, Thursday, March 19, 1987, Rules and Regulations (p. 8798).
5. 21 CFR Part 312 (revised as of April 1, 1989).
6. *Federal Register*, Vol. 52, No. 53, Thursday, March 19, 1987, Proposed Rules (p. 8850).
7. *Federal Register*, Vol. 58, No. 117, Tuesday, June 18, 1991, Rules and Regulations (p. 28025).
8. *Federal Register*, Vol. 58, No. 117, Tuesday, June 18, 1991, Rules and Regulations (p. 28025).
9. *First Biennial Report on the Adequacy and Uniformity of Federal Rules and Policies, and their Implementation, for the Protection of Human Subjects in Biomedical and Behavioral Research, Protecting Human Subjects.*
10. Substantive Changes. Orientation Meeting Information Packet for New Employees and Compliance Personnel (Section Informed Consent 50 & IRB 56). Palo Alto, CA: Syntex Corp. Bioresearch Compliance (obtained December 1, 1992).
11. From a speech entitled "Food and Drug Administration's OTC Review" by Cynthia Leggett, U.S. FDA, Atlanta District Office, Atlanta, December 5, 1977.

FURTHER READING

DeSain, C., *Documentation Basics that Support Good Manufacturing Practices*, Aster, Eugene, OR, 1993.

Willig, S. H., and Stoker, J. R., *Good Manufacturing Practices for Pharmaceuticals: A Plan for Total Quality Control*, 3rd ed., Marcel Dekker, New York, 1996.

Whitmore, E., *Product Development Planning for Health Care Products Regulated by the FDA.* ASCQ Quality Press, Milwaukee, WI, 1997.

15
Healthcare Quality Management

JAMES RICHARD SANTOS
Fidus Medical Technology Corporation
Fremont, California

I. INTRODUCTION

For the healthcare industry, continuous quality improvement, or CQI, is becoming the quality method of choice, As the name implies, CQI is an ongoing effort aimed at what has been often termed "never-ending improvement." The process involves major cultural, and possibly organizational, changes. The extent of these changes depends on the level at which the organization is starting. Continuous quality improvement affects every process and every person in the organization; it is a radical change.

II. HEALTHCARE QUALITY REQUIREMENTS

Healthcare organizations have successfully approached CQI in many different ways (1). There is no set pattern for the final structure for any organization launching into a sustained effort of quality improvement. Four key elements: customer focus, continuous improvement, total organizational involvement, and the appropriate organizational setting, however, always exist in every CQI success story.

A. Customer Focus

The needs, desires, and requirements of the customer must be understood and acted upon if CQI is to succeed. This customer focus is the driving force for all CQI elements, and it must extend to both internal and external customers.

Internal customers are those within the organization. These include physicians, nursing and technical staff, employees, administrators, and others. External customers are those outside the organization. Such customers include patients, government, insurance companies, and the surrounding community. Patients, of course, are the most important customers. All parts of a successful organization must know how they affect their customers. Also, motivation of the members of the organization must occur to improve the service or product to meet, and exceed, their customer's needs.

B. Continuous Improvement

Continuous improvement has the premise that good enough is not. Anything can be better, whether it is a procedure, product, or service. All the "good enoughs" will be obsolete before too long. An organization must follow two avenues in its quest for continuous improvement. The organization must improve on the existing good—that is, it must be innovative and embrace new methodologies. On the other hand, it must also seek out and destroy the causes of waste, rework, and inefficiency.

C. Total Organizational Involvement

Without top management's involvement in the CQI process, quality improvement results are small. Significant amounts of management's time must be put into the process improvement program. Only then is larger success possible. Maximum results only occur if everyone in the organization is involved and empowered to make improvements. Unfortunately, many organizations go forward with only a small bit of their total capability. The involvement of the total organization is the real energy source behind any quality improvement process.

D. Organizational Setting

The organizational setting is the link that holds the quality improvement effort together. This includes:

Top management's involvement and leadership that actively model the values and ideas of continuous quality improvement

Long-term goals and short-term objectives to bring CQI into normal work functions

Systematic planning to change goals and objectives into linked efforts that cross functional boundaries

Setting up methods that turn plans into effective actions

Information systems that allow teams to monitor, control, and upgrade processes

Measurement, reporting, and actions that keep up the focus on critical areas

Communication to connect the goals and efforts of individuals and teams

Recognition, rewards, and celebration to encourage and reward progress along the path of continuous quality improvement

III. ACCREDITATION

Hospitals and other healthcare facilities must meet certain criteria for method of operation, construction, and safety. One such criterion is accreditation. The Joint

Commission on Accreditation of Healthcare Organizations provides the certification and quality improvement requirements for hospitals and other healthcare facilities.[1]

Originally founded in 1951 as the Joint Commission on Accreditation of Hospitals, the organization changed its name in 1987 to the Joint Commission on Accreditation of Healthcare Organizations, which is the current name.

The charter of the Joint Commission (2) is to improve the quality of healthcare given to the public. The Joint Commission develops standards of quality together with health professionals and others. It stimulates healthcare organizations to meet or exceed the standards in two ways; it offers accreditation and it provides instruction in quality improvement concepts. The Joint Commission evaluates and accredits more than 5400 hospitals and 3600 other healthcare organizations and programs.

A. Categories of Accreditation

Four categories of accreditation exist (3):

Accreditation with commendation Organizations receiving aggregate scores of 90 or above on the accreditation grid and meeting other criteria are accredited with commendation.

Accreditation Organizations that are in substantial compliance with the standards of the Joint Commission are accredited.

Conditional accreditation Organizations that are not in substantial compliance with the Joint Commission standards but have filed an acceptable corrective action plan within a specified period are conditionally accredited. Such organizations must undergo a follow-up survey within six months. Then, the organization must show enough improvement to merit an accreditation award.

Denied accreditation Organizations that fail to meet the standards of the Joint Commission are denied accreditation and considered not accredited.

B. Organizations

Accreditation is available for the following healthcare organizations:

- General and psychiatric hospitals
- Home care organizations; Substance abuse treatment programs
- Community mental health centers
- Programs for the mentally retarded and developmentally disabled
- Adult and adolescent psychiatric programs
- Outpatient surgery centers; Urgent care clinics
- Group practices
- Community health centers
- Managed care organizations
- Nursing homes and other long-term care facilities

---

[1] Joint Commission on Accreditation of Healthcare Organizations, 1 Renaissance Boulevard, Oakbrook Terrace, Illinois 60181 (tel.: 630-916-5600).

Healthcare organizations seek Joint Commission accreditation because accreditation:

- Stimulates improvement in patient care
- Strengthens community confidence
- Supports staff education
- Enhances medical staff recruitment
- May substitute for federal certification
- Fulfills licensure requirements in many states
- Speeds up third-party payments

C. Commission Board

A 24-member Board of Commissioners that includes public members governs the Joint Commission. Members of the Board represent:

The American College of Physicians
American Hospital Association
American College of Surgeons
American Medical Association
American Dental Association
The American public

D. Commission Surveys

The Joint Commission employs over 500 people to do its surveys (4). These people include physicians, nurses, healthcare administrators, medical technologists, and psychologists. Also among these 500 people are respiratory therapists, pharmacists, durable medical equipment providers, and social workers. Each surveyor is a veteran healthcare professional with excellent credentials and substantial experience. This experience must be in hospitals, mental health services, clinics, nursing homes, home care agencies, or laboratories.

Surveyors are either employees or contractors. They conduct on-site surveys across the entire country. As a traveling consultant, each surveyor must do a thorough evaluation of standards compliance. Also, each surveyor must help the organization under survey in both achieving standards compliance and improving performance.

E. Quality Healthcare Resources

Quality Healthcare Resources Inc. (QHR) is a nonprofit subsidiary of the Joint Commission that provides on-site technical help to healthcare organizations (5). This aid focuses on monitoring, evaluating, and improving the quality of patient care. On-site technical help focuses on practical ways to assess and improve quality. The techniques are tailored to the specific organization with the goal of developing totally manageable systems.

QHR provides services in all types of healthcare organizations. These services address medical staff, clinical support services, plant technology and safety management, and department-specific consultation needs.

The technical assistance provided is fully separate from the accreditation activities of the Joint Commission. No healthcare organization that requests services will derive any special advantage nor receive any special treatment in the accreditation process because it has retained QHR.[2]

F. The Joint Commission and QCI

One major change that impressed the Joint Commission and the healthcare industry during the late 1980s was introduction of continuous quality improvement ideas (6). The Joint Commission introduced quality improvement into its internal operations in 1987. Since 1992, a Quality Improvement Steering Committee oversees the activities of many Quality Action Teams. These teams have addressed a wide range of issues, including survey report turnaround times and the carrying out of their *Agenda for Change.* Quality improvement has become an essential part in the standards development process, and it is expected to lead to standards that promote continuous quality improvement within healthcare organizations.

G. Joint Commission Training Program

The Joint Commission conducts about 300 educational programs (7), including national, international, and satellite video conferences. The programs are for accredited organizations and for those interested in the accreditation process. Programs are often cosponsored by state, national, and international organizations. The programs provide important communication forums for the healthcare community. They also serve to further the interests of all participants in promoting quality patient care.

A variety of publications supplement the educational programs. Included are basic texts such as standards manuals and scoring guidelines, topical materials, and a series of periodicals. *Joint Commission Perspectives* is the official newsletter of the Joint Commission.

IV. FEDERAL AND STATE REQUIREMENTS

Federal certification of hospitals is the responsibility of the Healthcare and Financial Administration (HFA), headquartered in Baltimore, Maryland. The HFA is on the same level of government as the U.S. Public Health Service. The main document used in the federal certification[3] of hospitals is the *Health Insurance Manual* Number 7, which is also known as the State Operations Manual. The manual is available from the U.S. Government Printing Office.

---

[2] Quality Healthcare Resources, Inc. Suite 1100, 1 Renaissance Boulevard, Oakbrook Terrace, Illinois 60181 (tel.: 630-916-4333).

[3] The federal regulation covering this certification is 42 CFR Part 488. Refer to the *Federal Register* for this regulation for more information. Statutory authority comes from the Social Security Act, Section 1864 (Certification) and Section 1865 (Use of Accreditation for Public Certification Purposes).

For further information, contact the following federal office: Chief of Acute Care Services Branch, Office of Standards and Certification, Health Standards and Quality Bureau, Healthcare and Financial Administration, Baltimore, Maryland 21207 (tel.: 410-786-3000).

Individual states may have their own licensing and certifying agencies. For example, California has the State Department of Health Services Licensing and Certification Division. That division reviews the care and services provided by the state's health facilities. It covers hospital and skilled nursing facility licensing and certification. Barclay's California Code of Regulations (22 CCR inclusive) is the specific regulation for California healthcare facilities.[4]

V. SAFETY CODES AND OTHER REQUIREMENTS

Key regulations and requirements[5] apply to several areas in hospitals and other healthcare facilities. These include flammable gas, liquids, electrical safety, and other specialized area criteria.

A. Uniform Codes

The Uniform Fire Code (UFC) was developed by the International Conference of Building Officials[6] and the Western Fire Chief's Association.[7] As they developed the UFC, the two organizations routinely compared their work with the provisions of the Uniform Building Code (UBC) to prevent conflicts between the two standards.

The Uniform Fire Code and the Uniform Building Code are only two codes of a total package. Along with them, three other codes, the Uniform Mechanical Code (UMC), Uniform Plumbing Code (UPC), and the Uniform Electrical Code (UEC), exist. Thus, cities, counties, and states have a complete group of model codes. Each code is compatible with the others.

When a governing body adopts the UFC for its use, it adopts by reference all listed references in UFC Appendix VI-A. When the UFC is silent on a subject area, the governing body may adopt a nationally recognized standard of good practice. National Fire Protection Association (NPFA), Underwriters Laboratories (UL), Factory Mutual (FM), American National Standards Institute (ANSI), and the U.S. government list standards that are examples of recognized standards of good practice.

---

[4] A good place to look for both federal and state healthcare regulations is a local law library. Universities that have a law school are also good places to check.

[5] Interested parties should contact their city clerk to check on the codes and other standards.

[6] International Conference of Building Officials (ICBO), 5360 South Workman Mill Road, Whittier, California 90601.

[7] Western Fire Chief's Association, Palm Brook Corporate Center, 3602 Inland Empire Boulevard, Suite B-205, Ontario, California 91764.

An example of a state fire code is the *California Fire Code*. Listed under California Code of Regulations (CCR) Title 24, Part 9, the code adopted the 1991 edition of the Uniform Fire Code by reference and, in addition, added necessary state amendments.

B. Uniform Fire Code; Part VII Special Subjects

Article 74—Compressed Gases

Article 74 of the UFC covers the storage, use, and handling of oxygen and other compressed gases and gas systems (8). The following items should be noted:

1. Related provisions appear in UFC Articles 49, 75, and 80.
2. An exception is liquified petroleum (LP) gas, which is addressed in UFC Article 82.
3. Provisions for oxygen also apply to nitrous oxide.
4. The permit to store, use, or handle compressed gases is described in UFC Section 4.108.

Division II of Article 74 includes medical applications. It covers flammable anesthetic and nonflammable medical gases at hospitals and similar facilities. Fixed installations of nonflammable medical gases intended for sedation in which the patient is not rendered unconscious are also included. These installations include, but are not limited to, analgesia systems for dentistry, podiatry, veterinary, and similar uses.

Article 75—Cryogenic Fluids

Article 75 presents general information on cryogenic fluids and defines requirements for their storage, handling, and transportation (9). Included are the following:

1. Design, construction, and testing of containers and other vessels
2. Pressure-relief devices
3. Pressure-relief vent piping
4. Requirements for handling, storage, and unloading areas

Article 79—Flammable and Combustible Liquids

Article 79 describes requirements for the use, dispensing, mixing, and handling of flammable and combustible liquids (10).

Article 80—Hazardous Materials

Article 80 describes requirements related to the prevention, control, and mitigation of dangerous conditions that may be involved with hazardous materials (11). Included are discussions of criteria that relate to storage, dispensing, use, and handling of hazardous materials and information needed by emergency response personnel. Also given is classification by hazard.

Article 80 divides hazardous materials into categories, including materials regulated under article 80 or elsewhere in the UFC. The materials listed are classified as *physical or health hazards*. A material with a primary classification as a physical hazard can also present a health hazard. Physical hazards include the following:

1. Explosives and blasting agents, regulated elsewhere in the UFC
2. Compressed gases, regulated in this article and elsewhere in the UFC
3. Flammable and combustible liquids, regulated elsewhere in the UFC
4. Flammable solids
5. Organic peroxides
6. Oxidizers
7. Pyrophoric materials
8. Unstable (reactive) materials
9. Water-reactive solids and liquids
10. Cryogenic fluids, regulated under this article and elsewhere in the UFC

The materials listed in the article also include those classified as *health hazards.* A material with a primary classification as a health hazard can also present a physical hazard. Health hazards include the following:

1. Toxic materials, including toxic compressed gases
2. Radioactive materials
3. Corrosives
4. Carcinogens, irritants, sensitizers, and other health hazards

Uniform Building Code

The Uniform Building Code (12) (UBC) is dedicated to the development of better building construction and greater safety to the public. Founded on broad-based performance principles, the Code provides uniformity in building laws. Enacted by the International Conference of Building Officials in 1927, the UBC is revised every seven years.

Chapters 10, 10A, 10B, and 10C of the Uniform Building Code, apply to healthcare facilities. Chapters 10A, 10B, and 10C are also California state-specific requirements and were drafted to give the reader an understanding of state-imposed requirements. (One should check with the state in which their healthcare organization is conducting operations.)

Chapter 10 provides *Requirements for Group I Occupancies.* These requirements apply to those found in Table 1.

Chapter 33, *Section 3320(a)—Exiting from Group I Occupancies.* Stafford (9) emphasized that this section is "over and above" those general exiting requirements for every building.

Table 1 Requirements for Group I Occupancies

Hospitals, sanitariums, nursing homes with nonambulatory patients, and similar buildings.
Healthcare centers for ambulatory patients receiving outpatient medical care that may render the patient incapable of unassisted self-preservation.
Mental hospitals, and mental sanitariums.
Other occupancy situations are included, but not presented here.

National Fire Codes (13)

The *National Fire Codes* are a compilation of National Fire Prevention Association recommended practices and guides. The complete set contains the codes, standards, recommended practices, and guides that have been developed by the technical committees of the Association according to the NFPA Regulations Governing Committee Projects. The NFPA may be contacted at the following address: National Fire Protection Association, 1 Batterymarch Park, Quincy, Massachusetts 02269-9101.

The NFPA was organized in 1896. Its purpose is to promote the science and improve the methods of fire protection. Also included are fire prevention, electrical safety, and other related safety goals. These improvements and science promotion are done to obtain and circulate information on these subjects. They are also done to secure the cooperation of its members and the public in establishing proper safeguards against loss of life and property. The Association is an international, charitable, technical, and educational organization. Its membership includes over 100 national and regional societies and associations, and over 55,000 individuals, corporations, and organizations.

The *National Fire Codes* are annual compilations of the Codes, Standards, and Recommended Practices and Guides prepared by Technical Committees organized under NFPA sponsorship. Only those documents adopted by the NFPA are included in the National Fire Codes. Users of the *National Fire Codes* should also consult applicable federal, state, and local laws and regulations.

The codes are living documents. Revisions to them continue on a three-year cycle. Approximately one third of them are revised each year.

Master Index to the National Fire Codes

Published by the NFPA, the Master Index to the National Fire Codes (14) contains examples specific to healthcare, hospitals, and medical applications as shown in Table 2.

NFPA 70—The National Electrical Code

NFPA has acted as sponsor of the *National Electrical Code* (15) since 1911. The original code document was developed in 1897 as a result of the unified efforts of

Table 2 Master Index to National Fire Codes (Examples)

Ambulatory healthcare centers	**101**:12-6, **101**:13-6, **99**:C-13, and **99**: Chap. 13
Anesthetizing locations	**99**:12-4.1, **99**:13-4.1, **99**:A-12-4.1.1.4 thru A-8-12-4.1.5, and **99**:C-12
Apparatus, gas anesthesia	**99**:8-5.1.2.1
Gas systems, healthcare facilities	**99**:4-4.11, **99**:4-4.2.9, **99**:4.5.1.4, **99**:4-5.2.4, **99**:A-4-4.1.2(d) thru A-4-4.1.1.3(d)
Health care facilities	**99**
Healthcare occupancies	**101**:12-3.4, **101**:13-3.4
Hospitals	**99**:A-12, **99**:C-13, and **99**:Chap. 12 (see also Health care facilities; Occupancies; Health care; and Anesthetizing locations.)
Vacuum systems, medical-surgical	**99**:4-8.1.1.5, **99**:4-8.1.1.7, **99**:4-8.1.1.8, and **99**:4-9.1.4

Note that the chapters are in boldface type.

various insurance, electrical, architectural, and allied interests. NFPA publishes the current *National Electrical Code Handbook.*

Specific chapter names are found in Table 3. Article 517, Health Care Facilities is of special interest for healthcare organizations.

NFPA 99—The Standard for Health Care Facilities

The origins of the *Standard for Health Care Facilities* (16) standard began with a series of explosions in operating rooms during surgery in the 1930s. The problem was traced to static discharges of sufficient energy to ignite the flammable anesthetic agents being used. A series of recommendations followed to moderate the hazard. In addition, a Committee on Hospitals was created with responsibility for periodically reviewing and revising the recommendations.

Over the years, the concerns of the Committee have come to encompass other fire and fire-related hazards related to the delivery of healthcare services. As a result, the Committee has enlarged its scope to include all healthcare facilities.

The major concerns of the Committee are those hazards linked with operating a healthcare facility, treating patients, and operating laboratories. Areas covered include electrical systems, gas systems, medical equipment, and environmental conditions peculiar to healthcare facilities operation. Also covered is the management of a facility in case disasters disrupt normal patient care.

The scope of NFPA 99 is to establish criteria to reduce the hazards of fire, explosion, and electricity in healthcare facilities. The criteria include performance, maintenance, testing, and safe practices for facilities, material, equipment, and appliances, in addition to hazards associated with the primary hazards.

NFPA 99 applies to all healthcare facilities. Construction and equipment requirements apply only to new construction and new equipment, except as modified in individual chapters. Chapters 12 through 18 specify the conditions under which the requirements of Chapters 3 through 11 apply in Chapters 12 through 18.

NFPA 99 serves those persons involved in design, construction, inspection, and operation of healthcare facilities. It is also useful in the design, manufacture, and testing of appliances and equipment used in patient care areas of healthcare facilities.

Table 3 NFPA 70—National Electrical Code

1. General
2. Wiring and Protection
3. Wiring Methods and Materials
4. Equipment for General Use
5. Special Occupancies
    - 517 Health Care Facilities
        - Part A. General
        - Part B. Wiring and Protection
        - Part C. Essential Electrical Systems
        - Part D. Inhalation Anesthetizing Locations
        - Part E. X-Ray Installations
        - Part F. Communications, Signaling Systems, Data Systems, Fire Protective Signaling Systems, and Systems Less than 120 Volts, Nominal
        - Part G. Isolated Power Systems

Table 4 NFPA 99—Standard for Healthcare Facilities

1. Introduction
2. Definitions
3. Electrical Systems
4. Gas and Vacuum Systems
5. Environmental Systems
6. Materials
7. Electrical Equipment, Healthcare Facilities
8. Gas Equipment (Positive and Negative Pressure), Healthcare Facilities
9. Manufacturer Requirements
10. Laboratories
11. (Reserved)
    Chapters 3 through 11 contain requirements, but do not state where they are applicable.
12. Hospital Requirements
13. Ambulatory Healthcare Center Requirements
14. Clinic Requirements
15. Medical and Dental Office Requirements
16. Nursing Home Requirements
17. Limited Care Facilities
18. (Reserved)
    Chapters 12 through 18 are "facility" chapters. They list the requirements from Chapters 3 through 11 applicable to specific facilities. These chapters also include additional requirements specific to that particular facility.
19. Hyperbaric Facilities
    Chapter 19 contains safety requirements for hyperbaric facilities, whether freestanding or part of a larger facility.
20. Referenced Publications
    Chapter 20 lists required references made in Chapters 1 through 19.

Appendix A. Explanatory Notes to Chapters 1-19
  This appendix contains nonmandatory information keyed to the text of Chapters 1 through 19. For example, A-5-4.2 is explanatory information on 5-4.2.

Appendix B. Referenced Publications (these are informatory references)

Appendix C. Additional Explanatory Notes to Chapters 1-19
  More explanatory information on chapters 1 through 19, though it is not keyed to specific text.

Annex 1. Health Care Emergency Preparedness
  Disaster planning information.

Annex 2. The Safe Use of High-Frequency Electricity in Healthcare Facilities
  Self-explanatory.

Cross-Reference to Previous Individual Documents

Index

This publication is set up for teaching the reader. For example, it can tell what the electrical system requirements are for a hospital:

- First, turn to Chapter 12 (Hospital Requirements).
- Next, turn to Subsection 12-3.3 (Electrical System Requirements).

The general chapter format of the code is given in Table 4. Subsequent revisions may add to or delete some sections. It is set up to help a reader interested in learning.

NFPA 801—Recommended Fire Protection Practice for Facilities Handling Radioactive Materials (17)

The Committee on Atomic Energy (CAE) was organized in 1953. The CAE was formed for the purpose of giving the fire-protection specialist certain basic information on radioactive materials and handling. It also served to give designers and operators of such laboratories some guidance on practices necessary for fire safety. The first edition of NFPA 801 was originally limited to laboratories handling radioactive materials. The 1955 meeting of the National Fire Protection Association adopted it.

In 1970, NFPA 801 was revised to reflect current thinking and practices. It then applied to all locations, except nuclear reactors, where radioactive materials are stored, handled, or used.

The 1980 edition clarified the presence of and levels of radiation. It added cautionary statements about assumption of risks by the fire officer. It spoke to the importance of training in the handling of radioactive materials by fire department personnel. This edition also clarified variations of the intensity of a radiation field.

A total revision came with the 1991 edition, which reorganized all the chapters in the publication. Significantly, the American National Standards Institute approved the 1991 edition.

This recommended practice covers fire-protection practices intended to reduce the risks of fire and explosion at facilities handling radioactive materials. The recommendations are applicable to all locations where radioactive materials may be stored, handled, or used. This includes hospitals, laboratories, and industrial properties.

The object of NFPA 801 is to reduce personal hazards, property damage, and process interruption resulting from fire and explosion. Radioactive contamination may or may not be a factor in these risks.

The general layout of NFPA 801 is as follows:

Chapter 1. Introduction
Chapter 2. Administrative Controls
Chapter 3. General Facility Design
Chapter 4. General Fire Protection Systems and Equipment
Chapter 5. Special Radiation Facilities and Equipment
Chapter 6. Referenced Publications
Appendix A
Appendix B. Sources of Radiation—The Nature of the Fire Problem
Appendix C. Referenced Publications
Appendix D. Additional Publications
Index

REFERENCES

1. Brown L. Continuous Quality Improvement in Healthcare Roundtable. Conference Proceedings, QC'92 and ASQC Western Regional Conference. Santa Clara, CA: Santa Clara Valley Section, ASQC, 1992, pp. 1-67 to 1-68.

2. Facts About the Joint Commission. *Blue Book A Customer Service Directory 1992–1993.* Oakbrook Terrace, IL: Joint Commission of Accreditation of Healthcare Organizations.
3. Accreditation Categories. *Joint Commission Corporate Report 1992.* Oakbrook Terrace, IL: Joint Commission of Accreditation of Healthcare Organizations, p. 11.
4. Surveyors—Teachers and Evaluators in the Health Care Setting. *Joint Commission Corporate Report 1992.* Oakbrook Terrace, IL: Joint Commission of Accreditation on Healthcare Organizations, p. 20.
5. Gallagher K. Quality Healthcare Resources, Inc. *Blue Book A Customer Service Directory 1992–1993.* Oakbrook Terrace, IL: Joint Commission of Accreditation of Healthcare Organizations, 1992, p. 15.
6. A New Philosophy: Continuous Quality Improvement. *Joint Commission Corporate Report 1992.* Oakbrook Terrace, IL: Joint Commission of Accreditation of Healthcare Organizations, p. 7.
7. Educating Through Programs and Publications. *Joint Commission Corporate Report 1992.* Oakbrook Terrace, IL: Joint Commission of Accreditation of Healthcare Organizations, p. 18.
8. Uniform Building Code (1991 Edition). International Conference of Building Officials, Whittier, CA, Article 75, Cryogenic Fluids, pp. 213, 216, 217.
9. Uniform Building Code (1991 Edition). International Conference of Building Officials, Whittier, CA, Article 79, Flammable and Cryogenic Fluids, pp. 255, 263, 285, 307.
10. Uniform Building Code (1991 Edition). International Conference of Building Officials, Whittier, CA, Article 80, Hazardous Materials, pp. 361, 362, 366, 367.
11. Uniform Building Code (1991 Edition). International Conference of Building Officials, Whittier, CA, Article 74, Compressed Gases, pp. 203–204.
12. Uniform Building Code and California Building Code (1991 Edition). International Conference of Building Officials, Whittier, CA, pp. iii, 95, 100.21, 100.45, 100.49.
13. National Fire Prevention Association; Quincy, MA, Vol. 5, pp. ii–iv.
14. Master Index to National Fire Codes (No Date). National Fire Prevention Association; Quincy, MA, pp. 5, 6, 66, 69.
15. NFPA 70 National Electrical Code. National Fire Prevention Association; Quincy, MA, 1993, pp. 70-i, 70-v to 70-vi.
16. NFPA 99. Standard for Health Care Facilities, Vol. 5, 1990, pp. 99-1, 99-11 to 99-13, 99-15 to 99-16, 99-101 to 99-102. Quincy, MA, National Fire Codes, National Fire Prevention Association.
17. NFPA 801. Recommended Fire Protection Practice for Facilities Handling Radioactive Materials, pp. 801-1, 801-3 to 801-5. Quincy, MA, National Fire Prevention Association.

FURTHER READING

All of the following are available from the Joint Commission of Accreditation of Healthcare Organizations, Oak Brook, IL:

Striving Toward Improvement.

Defining Nursing Care in Quality Assurance in a Hospital Setting.

Transition from Quality Assurance to Quality Improvement Performance in Mental Health Organizations.

16

Total Quality Assurance

ROBERT W. GRENIER
Consultant
Springfield, Illinois

I. INTRODUCTION

Today, there is a myriad of quality systems, each of which is considered by someone to be the ultimate quality system. Some of these systems are The Malcolm Baldrige Award, ISO 9000 (series), MIL-Q-9858, Quality Function Deployment, and Taguchi techniques. There are many others. If any or all of these systems are, in fact, the ultimate system, the word has not yet gotten around.

One reason for the lack of real quality accomplishment of so many U.S. companies is because their quality programs and efforts are directed predominantly, or completely, to the manufacturing area. Of the four typical "functions"*—that is, manufacturing, marketing, management, and design engineering—manufacturing often has the *least* impact on the way the company is perceived, overall, by customers. Every other major area (i.e., function) of a company affects the company in the eyes of customers at least as much as does manufacturing. As examples:

- If management fails to implement a quality policy because of lack of commitment, the resulting quality problems will be felt by the customer.
- The majority of product field problems are due to problems in design engineering.
- A marketing decision that results in inadequate field access to spare parts and thus results in delayed repairs can result in significant customer dissatisfaction.

---

* In this chapter, the word *function* will be used to refer to a major segment of a company. The "typical" functions discussed in this chapter are manufacturing, marketing, design engineering, and management. Functions in other companies might be accounting, shipping, molding, finance, etc.

- Customer service or other sales personnel who are curt or rude can result in lost business as surely as the shipment of defective products.
- Order entry errors that result in the receipt of the wrong parts are as big a problem to the customer as defective parts.

II. THE TQA SYSTEM

There is, however, a quality system that can be used to evaluate not only manufacturing but each of the other functions that are significant contributors to the company's image and performance. Known as Total Quality Assurance, or TQA, the system is used to measure quantitatively the performance of each function and the company overall at the start of the program. Then, measurements of performance are periodically made over time to assure that gains are actually occurring during the quality improvement cycle.

A significant difference between TQA and, for example, ISO 9000 (series), is that TQA provides measurements of customer satisfaction. Thus it provides a measure of external performance, as viewed by customers, as well as a measure of internal compliance to procedures. The TQA system provides internally generated, numerical, quantitative performance measurements that do not require third-party auditors or certifiers.

A TQA system is structured by rating the performance of the selected functions and the degree of customer satisfaction with the company's products. The system must be driven by the desire for customer satisfaction.

- Each function's performance measurements are weighted in order of impact on customer satisfaction.
- The system must be results oriented rather than systems or procedures driven.

III. TQA RESPONSIBILITIES

A. Management Responsibilities

As regards quality, top management's number one priority must be commitment. How many times have we read that a company is "quality oriented," but when it comes down to action to back up the words, it is so often found to be just lip service. How many times does top management really get involved to make certain that the quality systems are integrated; that every function is committed to quality; that everyone knows what their basic responsibilities are in the quality loop?

How many times at management meetings is the comment made that "we have a quality problem," when, in fact, the translation is "we have a design problem," or "we have a marketing problem." The words *quality problem* imply that there is a problem in the quality department, and this typically is not a reflection of the real world. The quality department does not design, make, or sell the product. If there are design, manufacturing, or marketing problems, then that is what they should be called.

Another example of top management's responsibilities and how they can impact customer acceptance is when a company establishes a deadline for the introduction of a new product and will not allow the date to slip even when there are substantial problems. If design engineering is not satisfied with the progress being made during prototype testing, then management must grant additional time until the design is ready for production.

What, then, should the quality function do? Initially, the main charge of the quality function can be that of assisting management in establishing the total quality assurance system. This assistance can come in the form of personnel training and counsel for each of the other functions. Once the quality system is in place, then continuous auditing of the system becomes one of quality's major responsibilities. As a result of the audits, recommendations in the form of an annual quality plan are made to correct present problems and to make continuous improvements in the overall system.

B. Marketing Responsibilities

The one function that constantly demands quality, especially when there is a buyer's market, is marketing. Yet, how often does the marketing department acknowledge that they have quality responsibilities? Too often, marketing considers itself divorced from quality responsibilities.

What are some of marketing's responsibilities? A major marketing responsibility would be to assure that when the company introduces a new product into the market, spare parts and optional equipment are available at service centers. All printed materials such as parts books, service manuals, and the like must be completed, proofread, printed, and distributed prior to the launch. Reliability and life cycle cost targets have been established and verified. Training of company, dealer, and customer service personnel must be completed.

Do marketing personnel communicate clearly and objectively to design engineering, quality, and manufacturing when they report customer or user problems? Marketing must be the driving force in the timely correction of field problems. Ask yourself, do your company's marketing personnel ever admit that they are part of the quality team? Because of marketing's direct contact with customers and users, they can exert a powerful impact on product acceptance.

C. Design Engineering Responsibilities

Many, if not most, product problems in the field are due to design deficiencies. Design problems can be measured in terms of reliability or performance goals not being met. Often, catastrophic failures are seen in early product life that are, in fact, design failures and not manufacturing deficiencies.

What are some of the reasons for these design-related problems?

- Perhaps the major reason is inadequate prototype testing. This occurs when the length and total number of device-test hours are dictated by a market introduction schedule date rather than the necessities of the product.
- It is not unusual for a catastrophic failure to occur toward the end of prototype testing and result in a design change. It is also not unusual for the change not to be adequately tested, and, as a result, cause long-term problems. The customer is the one who will be affected.

- Some companies do not hold design reviews. Some companies hold reviews that are, in fact, engineering department PR presentations rather than design reviews.
- Are defect classifications—that is, critical, major, minor, and other—identified on blueprints and drawings? Defect classifications are necessary to communicate to manufacturing and quality assurance what design criteria must be controlled more closely than other characteristics.

D. Manufacturing Responsibilities

Manufacturing's main responsibility is conformance to requirements. Unfortunately, manufacturing quality (i.e., conformance) equates to overall quality assurance in the minds of too many people. While manufacturing conformance is mandatory, it is only one part of an overall quality program.

IV. THE RATING SYSTEM

The TQA system discussed in this chapter can be applied regardless of the size of the company, its geographical location, or the existence of "overseas" divisions. Each functional area in each divisional location is rated independently. In addition, the customer satisfaction rating is determined independently. If a product is manufactured at two different sites, each is rated separately, so that the results become the catalyst for annual quality plans specific to each location.

A certified total quality assurance system should be the company's objective. The rating system is used to gage if the division or company can be considered to have achieved a sufficiently good quality system or if improvements must be developed. The rating system must clearly define the strengths and weaknesses of the operation and point out in what specific areas additional work is required to maintain growth and achieve and maintain certification.

The various work elements and features that make up each of the functions are used to evaluate the operation. The overall system would be expected to score at least a total of 850 points out of a possible 1000 points (i.e., 85%) for initial certification.

A. Classifications

Two major rating system classifications are discussed in this chapter: work required and acceptable. The work required classification indicates that the level of achievement of an individual work element, or subsystem, is less than what is considered appropriate.

The work required classification can be further divided into four categories: poor, weak, minimal, and fair. The evaluation process is sometimes simplified by using only three categories: poor, weak, and fair under the work required classification.

A classification of acceptable indicates that the elements or subsystems meet or exceed the minimum level deemed appropriate for certification. The acceptable classification is divided into two categories: qualified and outstanding.

The sum of all of the scores determines the rating of the overall system. Thus, certain elements may be rated at less than would be required by the overall system, but other elements may be rated higher than the minimum required.

B. Subsystem Rating

In evaluating an operation's overall quality system each of the functional areas (management, marketing, design engineering, and manufacturing in this chapter) are assessed individually. Each function is initially assigned work elements, and "weights" are assigned to each of the elements in accordance with how important the element is to the overall function. The weighted elements are thus, ultimately, a reflection of customer satisfaction (Fig. 1).

During the initial evaluation, each work element is rated against predefined criteria and is assigned a rating of 0 (complete lack of activity) to 10 (essentially perfect performance). This assigned numerical rating is then entered into the appropriate column on the evaluation work sheet, as shown in Figure 1. The product of the numerical rating and the weight is the score for each work element.

After all of the work elements have been scored, the weighted scores are added to derive the total score for the function. This score is then divided by 1000 to derive a percentage score. Total scores shown as percentages are easier to discuss and work with than are total weighted numerical scores.

In assessing and rating each subsystem, certain predefined questions must be answered as each functional area is being evaluated. The answers to these questions are used to assist the auditor(s) in arriving at the rating.

C. Product Line Evaluation

Customer satisfaction must drive the overall quality system. The measurement of customer satisfaction has to be through an evaluation of the company's products rated by those people who use the products. To determine the effectiveness of the overall system, a significant and representative segment of the market must be surveyed. From the response to the survey, the company should be able to evaluate the degree of product acceptance (i.e., satisfaction) and determine any possible areas of needed improvements.

Quantitative measurement is a better trend indicator than is a qualitative rating. From the response to the satisfaction survey, the company should be able to determine any areas where improvements are needed.

User response to product acceptance surveys can be achieved in several ways. These include:

- Direct mail of a survey form to the user
- Direct personal contact between the user and company employee survey personnel via, e.g., visits, phone calls
- Studies conducted by internal company marketing personnel
- Hiring an independent consulting group

In addition, it is usually desirable to have company marketing personnel or other employees rate the product lines as well.

Total Quality Assurance Evaluation Worksheet

Operation: **Product Line Evaluation** **Work Elements:**	**RATING (R)** Work Required (0 ◄) Poor	Weak	Minimal	Fair	Acceptable (► 10) Qualified	Outstanding	**WEIGHT (W)**	**SCORING (R × W)** Score
■ QUALITY								
1) Perception of Quality								
• By dealer from factory								
• By user from dealer					8.5		14	119
• By user from factory								
■ PRODUCT LINE ACCEPTANCE								
2) Versatility				8.0			2	16
3) Reliability				7.5			10	75
4) Durability				7.9			12	95
5) Serviceability					8.5		15	128
6) Parts Availability				7.6			10	76
■ MANUFACTURER'S RESPONSIVENESS								
7) Time Taken to Correct, and Adequacy of Correction of, Component Failures			6.0				11	66
8) Time Taken to Correct, and Adequacy of Correction of, Performance Problems			5.0				11	55
■ ECONOMICS								
9) Owning and Operating Costs				8.1			5	41
10) Productivity				7.1			5	36
11) Warranty Terms					8.5		5	43
Rating				X			100	750
					Score (%)			75

Figure 1 Evaluation worksheet.

A representative Product Evaluation Work Sheet is shown in Figure 1. This worksheet has been divided into four categories and 11 work elements. Although the *Parts Availability* work element was rated as "outstanding," two elements, *Time Taken to Correct . . . Problems, and Productivity* were rated as "minimal" because of the company's slowness in responding to customer problems, and because the equipment being evaluated was inefficient by current standards. The overall rating of customer satisfaction with the company's products was 76%, or a rating of "fair."

While this worksheet is based on an evaluation of an actual company, the categories and work elements listed are not intended to apply to every company in every market. If the elements listed are not appropriate for your company, tailor a worksheet to your requirements. In any case, the bottom line is: Use feedback from the people who use your products to create a product line acceptance rating score.

D. Management Evaluation

A representative Management evaluation worksheet is shown in Figure 2. This worksheet has been subdivided into three subcategories—*System Prerequisites, Quality Organization, and Operations*—and a total of 10 work elements. *Quality Planning* has been given the most weight (15 points), because it has been determined that quality planning will have a substantial impact on the company in the next several years. Because of the company's "personality," it has been determined that the specific reporting level of the quality organization is relatively uncritical; thus this work element has been given a weight of 5 points. The remainder of the work elements have been weighted equally at 10 points.

The rating (0–10) shown for each element is the result of an audit/evaluation of the entire management function. Note that *Quality Systems Audits* and *Employee Involvement Plan* have been rated as "weak," indicating that the person (or team) performing the evaluation has determined that management has put very little effort into the accomplishment of these elements. On the other hand, the company's Quality Cost System has been rated as "outstanding," indicating that quality costs are effectively analyzed, and the results of the analysis are used by management to guide the company.

The overall rating of the management function is 68%, which is considered to be only a "fair" rating. Contributing to the low evaluation is a rating of 4.0 for *Quality Systems Audits* and 3.0 for *Employee Involvement Programs(s)*. While there had been some initial discussion, management essentially did not believe in formally auditing the various company operations or in using production time to involve employees in programs such as problem solving teams or quality circles.

E. Marketing Evaluation

A representative marketing evaluation worksheet is shown in Figure 3. This worksheet was divided into three subdivisions and 12 work elements. Because of the type of products and markets that this company was involved in, *Service Training Programs and Effectiveness* was considered to be the most important work element and thus was weighted at 15 points. *Design Review Participation*, both work elements under *Competitive Product Analysis*, and both work elements dealing with spare

Total Quality Assurance Evaluation Worksheet

Operation:

Function: Management

Work Elements	RATING (R)						WEIGHT (W)	SCORING (R × W)
	Work Required				Acceptable			
	0 ⟵					⟶ 10		
	Poor	Weak	Minimal	Fair	Qualified	Outstanding		Score
■ SYSTEM PREREQUISITES								
1) Attitude toward and Support of the Quality System				8.0			10	80
2) Integration				8.0			10	80
■ QUALITY ORGANIZATION								
3) Reporting Level of the Quality Operation			5.0				5	25
4) Structure of the Quality Operation					8.8		10	88
■ OPERATIONS								
5) Quality Planning				7.0			15	105
6) Quality Operating Manual				7.8			10	78
7) TQA Training			6.0				10	60
8) Quality Systems Audits		4.0					10	40
9) Quality Cost System						9.5	10	95
10) Employee Involvement Program		3.0					10	30
Rating				X			100	681
					Score (%)			68

Figure 2 Management function evaluation worksheet.

Total Quality Assurance Evaluation Worksheet

Operation:

Function: Marketing

Work Elements:	RATING (R)						WEIGHT (W)	SCORING (R × W)
	Work Required				Acceptable			
	0 ⟶ 10							
	Poor	Weak	Minimal	Fair	Qualified	Outstanding		Score
■ PLANNING SYSTEMS								
1) Product Plan Document				8.0			10	80
2) Problem Reporting System				7.0			10	70
3) Design Review Participation			6.0				5	30
4) Launch Procedure					9.0		10	90
■ COMPETITIVE PRODUCT ANALYSIS								
5) Laboratory Evaluation				8.0			5	40
6) Marketplace Evaluation				7.0			5	35
■ OPERATIONS								
7) Quality of Publications				8.0			10	80
8) Service Training Programs and Effectiveness				7.0			15	105
9) Service Quality Guide			5.0				10	50
10) Parts Availability at the Manufacturer				8.0			5	40
11) Parts Availability at the Dealers				8.0			5	40
12) Attitude toward Quality		3.0					10	30
Rating				X			100	690
					Score (%)			69

Figure 3 Marketing function evaluation worksheet.

and replacement parts under the category of *Operations* were considered to be less critical and were weighted at 5 points.

The overall rating for the Marketing function was 69%, a rating that was considered to be only "fair." Low rating scores were given to *Design Review Participation*, the *Service Quality Guide* (an important document for service and repair persons), and the function's attitude toward the company's quality program.

F. Design Engineering Evaluation

A representative design engineering evaluation worksheet is shown in Figure 4. This functional evaluation was divided into four subsystems or categories and 11 work elements.

Two work elements *Design Standards* and *Approval of Add-on Products* were weighted at 5 points. All the rest of the work elements were weighted at 10 points. Except for the *Product Safety Program*, which was rated at 8.3, or "fair," all of the work elements in this function were rated as "qualified" or "outstanding." The overall rating for the Design Engineering function was 89%, which is classified as "qualified."

G. Manufacturing Evaluation

A representative manufacturing evaluation worksheet is shown in Figure 5. This function worksheet is divided into three categories and 26 work elements. Three work elements—*Operator-Control Quality Plan*, *Measurement Equipment Calibration and Control*, and *Management Feedback Information*—were assigned a weight of 6. Note that the sum of all the weights assigned to a function's work elements must total 100.

The total score for the manufacturing function was 758, or 76%, when rounded off; a classification of "fair." Ten of the work elements, however, measured 85% or more, indicating that a quality plan would find a good foundation.

Operator-Control Quality Plan

It has become more and more accepted that it is the individual operator around whom the quality plan for conformance quality is designed. The operator-control quality plan must be supplemented by a climate that motivates the individual operator to use the system to help produce a quality product. This motivation is largely supplied by the actions and deeds of the immediate supervision and upper management. Unless upper management shows continued interest in product quality by deed as well as by word, not much will happen in any area.

Before the operator-control concept can be completely implemented, several things must be present. These include adequate, calibrated measuring tools and a stable, capable process. The operator must also have proper instruction, adequate time in which to do the work, and, of course, the proper skills. In other words, the operator requires a reasonably good oppportunity to do a good job on his or her own.

Cooperation by manufacturing supervision is essential in the early stages of the program's development and implementation, because it is in this area that the preliminary studies are often made and the first results achieved. A product-by-

Total Quality Assurance Evaluation Worksheet

	RATING (R)							SCORING (R × W)
	Work Required				Acceptable			
	0 ⟵					⟶ 10		
Operation:								
Function: Design Engineering								
Work Elements:	Poor	Weak	Minimal	Fair	Qualified	Outstanding	WEIGHT (W)	Score
■ PLANNING PROCESSES FOR NEW DESIGNS								
1) Quality in Design					9.0		10	90
2) Design Standards					9.0		5	45
3) Reliability Application					8.5		10	85
4) Design Review						9.5	10	95
5) Product Safety Program				8.3			10	83
6) Prototype Testing					9.2		10	92
■ QUALITY OF CURRENT PRODUCT DESIGN								
7) Classification of Characteristics					8.8		10	88
8) Engineering Change Control					9.0		10	90
■ DESIGN EFFECTIVENESS REVIEW								
9) Review of Warranty Data						9.5	10	95
10) Review of Life Cycle Data					8.5		10	85
■ OPERATIONS								
11) Approval of Add-on Products					8.7		5	44
Rating					X		100	892
					Score (%)			89

Figure 4 Design engineering function evaluation worksheet.

Total Quality Assurance Evaluation Worksheet

Operation:

Function: Manufacturing

Work Elements:	RATING (R): Work Required (0 →)				RATING (R): Acceptable (→ 10)		WEIGHT (W)	SCORING (R × W)
	Poor	Weak	Minimal	Fair	Qualified	Outstanding		Score
■ SUPPLIER QUALITY ASSURANCE								
1) Evaluating the Quality of Purchased Materials				7.0			4	28
2) Supplier Quality Levels			6.0				4	24
3) Incoming Inspection Plan				8.0			5	40
4) Supplier Surveys					9.0		5	45
5) First Article Complete Inspection					8.5		5	43
6) Nonconforming Material Disposition and Routing				7.5			4	30
7) Supplier Rating			6.5				5	33
8) Supplier Surveillance Program				7.0			4	28
9) Communication with Suppliers				7.0			4	28
■ MANUFACTURING QUALITY ASSURANCE								
10) Measuring Performance				8.0			4	32
11) Process Flow Planning					8.5		3	26
12) Manufacturing Quality Level				8.0			2	16
13) Process Capability					9.0		3	27
14) Operator Control Quality Plan			5.0				6	30
15) Statistical Process Control					9.0		4	36
16) First Article Complete Inspection					8.5		5	43
17) Nonconforming Material Control					8.5		2	17
18) Measurement Equipment Calibration and Control				7.0			6	42
19) Special Manufacturing Evaluation Tests				7.0			3	21
20) Packaging and Shipping			6.0				2	12
21) Product Evaluation Test		4.0					2	8
22) Storage of Material					9.2		2	18
23) New Capital Equipment Qualification/Existing Jig and Fixture Control					8.5		3	26
24) Operator Rating Program					9.0		3	27
■ QUALITY INFORMATION FEEDBACK								
25) Management Feedback Information				8.0			6	48
26) Product Quality Promotion Information				7.5			4	30
Rating				X			100	758
							Score (%)	76

Figure 5 Manufacturing function evaluation worksheet.

product or area-by-area approach toward implementation should be used rather than trying to install the procedure in the whole plant at one time.

Operator and supervisor training is an important task that must be done early on. It does no good to develop effective quality plans if they are not followed or used. It is the responsibility of manufacturing management to assure that the plans are understood and followed.

V. CERTIFICATION

The key to success in the implementation of a total quality assurance system is in the certification process.

The first step toward certification is an evaluation of the operation's current quality system in terms of the functional areas with their appropriate work elements. After reviewing its needs as determined from the initial evaluation, each operation must develop a *program plan* that, when implemented, will result in the operation being "certified."

VI. THE TQA PROGRAM PLAN

The program plan contains a brief description of each element that must be improved or added to the quality system and an identification of the functions responsible for making the improvements.

As implementation progresses, a quality manual should be developed and grow in parallel with the system improvements. Since most elements of a TQA system can be described by procedures, the procedures in the quality manual—and their effectiveness—demonstrate to a great extent the health of the quality system.

An important part of the program plan is the schedule of implementation. In developing the plan, each element is evaluated for priority and the length of time required to:

- Develop the steps that are to be taken and the way that they will be accomplished
- Write and coordinate the required procedures
- Accomplish sufficient implementation to assure the permanence of the new procedures

The completion date of the last elements to be implemented determines the length of time needed to develop the total quality assurance system. The length of the plan schedule will depend on the initial state of the operation's quality program, management's attitude, how much needs to be done, and the dedication with which employees approach the tasks. In general, 2–3 years is considered average.

It should be recognized that for any program to be successful, it must be fully coordinated and integrated by each function and by the highest level of management.

After the program plan is properly introduced, the implementation process is monitored through periodic progress evaluations. Quality should review each area and provide a revised rating for the operation on a periodic basis. The time between evaluations should be a maximum of 1 year.

VII. ACHIEVING TQA CERTIFICATION

There are two main requirements that an operation must satisfy to achieve certification of its quality system. These are:

- A high degree of quality system implementation; 85% is considered to be the minimum composite average with no function below 60%.
- Evidence that the effect of the quality system appears in the product; i.e., a customer satisfaction (acceptance) rating of 85% or higher.

When the plan is completed and progress evaluations indicate successful implementation, quality assurance should conduct an in-depth evaluation of the system. Emphasis should be placed on the adequacy of the various elements of the system as represented by quality manual procedures. The following questions must be answered affirmatively:

- Are the procedures routinely followed in the conduct of the business?
- Are the procedures effective and "living" procedures?
- Are the operation's products well received by customers?

The degree of system integration that has been achieved should also be carefully examined. System integration is determined through discussion with the function managers and by evidence of their department's participation and action.

Further indications of a well-integrated system are the degree to which operator responsibility for quality has been accepted and how this concept is reflected in the control methods that are employed. Also to be sought is evidence that engineering has identified and communicated design quality information. How well the quality system reacts to this communication, follows through with audits, and provides evidence of control including product liability protection, is also important.

Another important indicator required for certification is the operation's quality cost performance. Where quality cost history has been poor, there must be adequate demonstration of significant improvement. In operations where quality cost performance has been relatively good, there must be evidence of a continued satisfactory level.

In addition to the requirements noted above, before certification can occur there must be evidence that the effects of the quality system can be seen in the product. Unless this happens, the effort to implement a new quality system has been wasted.

The products must have no significant after-sales quality problems and there must be assurance that the product currently being manufactured does not have latent manufacturing or reliability deficiencies.

When an operation has complied with all of the requirements, quality assurance should present a recommendation for certification to top management. A recommendation form such as that shown in Figure 6 is useful in recommending certification. When certification is approved, a formal certificate signed by the chief executive officer is presented to the operation. This certificate becomes tangible evidence of the implementation of the TQA system.

TQA CERTIFICATION STATUS WORKSHEET

Initial Certification ☐ Recertification ☐

Company, Division or Operation ______________ Date ______

		Yes	No
1. Quality System Rating.	☐		
Is there a function rating below 60 percent?		☐	☐
2. Customer Acceptance Rating	☐		
Is there a product line rating below 50 percent?		☐	☐
3. Average of 1 and 2		☐	
4. Quality Cost Performance	☐		
Total Composite Score		☐	

CERTIFICATION STATUS

RECOMMENDATION		DISPOSITION	
Certify System	☐	Certification Approved	☐
Deny Certification	☐	Certification Denied	☐
Recertify System	☐	Recertification Approved	☐
Deny Recertification	☐	Recertification Denied	☐

Final Status Undetermined (Held Open) ☐ Executive Quality Review Board

Signed ______ Date: ______ Signed ______ Date: ______

Quality Assurance Chairman, Executive Quality Review Board

Figure 6 Certification worksheet.

Certification Validation/Withdrawal

After initial certification, the operation should be subjected to annual recertification using the same general procedures used for the initial certification. Recertification evaluation serves three purposes:

- It discovers quality system weaknesses or slippage
- It measures system growth and effectiveness
- It measures the growth of customer satisfaction in the company's products

The certification procedure should include a procedure for certification withdrawal if the operation slips to an unacceptable degree. Indications of such slippage would be:

- A significant increase in quality costs
- Significant problems in new designs
- A decreasing level of customer satisfaction with the company's products

FURTHER READING

Grenier, R. W., *Customer Satisfaction through Total Quality Assurance*, Hitchcock, Wheaton, IL, 1988.

17
Error Cause Removal

DAVID C. CROSBY
The Crosby Company
Glen Ellyn, Illinois

I. INTRODUCTION

Every product defect or improper performance of a service that is attributable to a person can be traced to one or more of three causes: attitude, ability, and environment. Problems in the work environment are often referred to as "bugs in the workplace."

Attitude affects quality through the thought habits of the people who manage the company as well as those who produce the work. Ability—or lack of ability—of course, speaks for itself. It is this third cause, bugs in the workplace, that we will deal with here. Bugs are such things as hard to reach controls, poor lighting, foggy instruction, buggy software, parts that don't fit together properly: the physical problems people must deal with to produce a product or service.

This chapter will discuss an error cause removal (ECR) system known as the Bug Program.[1] There are many ways to conduct an ECR program. The Bug Program is one that works. The concept is simply to ask employees what prevents them from doing their job right every time they do it. Listen to them, then get rid of that problem.

---

[1] The title "Bug Program" is a trademark of The Crosby Co., Glen Ellyn, IL, for a specific error cause removal program.

II. THE ROOTS OF THE BUG

A formalized concept of error cause removal was developed in the early 1960s as a part of the Zero Defects[2] (ZD) movement. Believed to be the brainchild of General Electric in Boston, error cause removal was created to help employees identify problems in their work area. ECR had its share of successes and failures. There are a couple of problems when using the ECR technique as a continuing program. First, supervision usually does not want it known that they have problems in their area. Second, it is murder to keep employees interested day after day, year after year. The "what's buggin' you" idea takes care of both problems. It takes the seriousness out of things: The bug is the problem, not the supervisor.

Since the first-line supervisor is the key person in the ECR program, much of the program material must be designed with this person in mind. Always remember the first law of quality: *People Perform to the Standards Set (or Accepted) by Their Leader.*

The first-line supervisor is the leader, so if he or she thinks that an error cause removal program such as the Bug Program is a good idea, the program is a good idea. Otherwise, forget it. The idea that you must get over to the first-line supervisor is that bugs are bugs, and we've all got 'em, so let's get rid of them. Simple.

The second problem, that of keeping everyone informed, motivated, amused, and amazed, is solved because the Bug Program only runs 30 working days—about six weeks. You don't need to sustain it day after day after day after day. The first two weeks are spent bringing all levels of management up to speed on the program and training supervisors in handling bugs. Four weeks are spent on an all-out drive to, "get the bugs out of your job." There is a little teaser campaign two weeks before you start the four-week drive. *But* at the end of the six weeks, you can pack all the promotional stuff up and toss it out. The program is a snap. It's easy, it's fun, it works.

If you follow the methods and techniques presented here, you will have a successful error cause removal campaign. You will discover that the people in your organization have about 2.5 bugs each. Of course, some people will not report a single bug, and some people will make you a little crazy. If you do everything just as it is described, the employees in your organization will respond, and you will put many bugs to rest. Read this guideline. Develop a plan and follow it. You will be successful; you will gain the attention of your leaders.

---

[2] Created by Philip B. Crosby, Zero Defects is a performance standard under which defects are not permitted. The rule is: Meet the quality standard every time the job is done. Starting around 1961, Zero Defects was turned into a quality improvement program. It was backed by the U.S. Defense Department and used by all military services and most defense contractors. For the most part, the ZD movement was misunderstood. Since a lot of promotional material was used to promote the program, it was taken by many (including the American Society for Quality Control) to be motivational. Thousands of ZD programs were started and soon failed because of neglect and insufficient planning and commitment. The U.S. Army ended their Zero Defects program around 1970 when it took the reporting requirement out of the Army Regulation.

The program outlined here is complete and should work for almost any organization. But feel free to make changes to better fit your situation. If you are in the food or hotel business, you might not want to use the word "Bug." So come up with something new. Try BIPs, for built in problems. "Get the BIPs out of your job!" It works. Just stick with the rules and the basic concept.

III. THE BUG PROGRAM

A. What Is It?

Quality needs to be improved. Quality will always need to be improved. The Bug Program is a powerful tool for management and supervision to help each employee improve quality by getting the bugs out of their job. The concept of the Bug Program is very simple. Each employee is asked by his or her leader, "Tell me what it is that prevents you from doing your job right every time, and I'll get rid of it." It doesn't get more complicated than that. The material and ideas presented here help you execute a well-planned, personalized advertizing program that is designed to capture the attention of your employees and move them to action.

B. How Does It Work?

The Bug Program that you will create will make the employees of your organization think about their jobs in a new light. They are asked, "What is it that prevents you from doing your job right every time?" And then they are told, "Tell me and I'll fix it." They tell you, you fix it. Simple.

C. The Mechanics

Bugs, or problems, or ECRs, are written out on an Extermination Request. The completed form is deposited in a box by the employee. You check and empty the box(s) every day, and route and control each bug until it is gone! Exterminated! Put to rest!

D. Bugs and Suggestions

Are bugs suggestions? One major difference between the Bug Program and the typical suggestion program is that in the Bug Program the employee need only identify a bug, not offer a solution to a problem. It's up to supervision and management to find the solution to the problem (or kill the bug). That's why management gets paid all that money!

You may find that by submitting a bug the employee has, in fact, made a suggestion. You will do both the employee and your organization a favor if you review each bug for possible suggestion material. Remember, if a problem is stated clearly enough, the solution is obvious. Another thing that makes the bug program different from a suggestion program is that after thirty days of operation, you toss out all the promotional material and get on with your life. You do not need to spend your time trying to keep interested all year long. That job is almost impossible. When the Bug Program ends, when everyone has had their chance to get rid of a

Figure 1 Example of promotional artwork.

bug or two, it's time to add up the score, let everyone know what a great job they did, and pass out a little recognition for a job well done! You can do it again next year—"The Bug Is Back!!!!"

E. What You Will Do

The material in this chapter is written for the person who will plan the program, create or purchase camera-ready artwork for publicity, and, in general, coordinate the whole thing. If that sounds like you, then here's what you will be doing:

You should sit down and read through this whole chapter. It's important that you have a general understanding of what you will be doing so that you can plan every detail. Then:

- You will make up the overall implementation plan; the master plan.
- You will decide how to obtain artwork,[3] and which additional devices and events you will use to conduct the program. An example of promotional artwork is shown in Figure 1.
- You will determine exactly how much material is needed, arrange for printing, and perhaps order supplies from specialty advertising suppliers.

---

[3] You will need artwork to communicate with your people. The artwork shown in this chapter is protected by U.S. copyright laws. Do not adopt the bug character as a mascot for your little league team. The bug character itself, and the artwork, are the property of David Crosby. A complete set of artwork can be purchased from The Crosby Co., P.O. Box 2433, Glen Ellyn, Illinois 60138.

- You will arrange to store your advertising material away from prying eyes, and distribute the material when it is time to use it.
- You will make a formal presentation to the big boss that describes the program, estimates its cost, and predicts the results.
- You will make a formal presentation (perhaps several) to all levels of management to tell them about the program, how it works, and how they will be involved.
- You will present the same information to all levels of supervision; plus, you will train them in how to handle the program kickoff, and how to handle questions and the bugs themselves.
- You will collect bugs from the box(es), log them, send them to the proper supervision, and track them to conclusion.
- You will contact supervisors who are delinquent in action on an Extermination Request, and offer help if it is needed.
- You will arrange for recognition of employees who identify bugs and of employees who killed a bug.
- You will collect performance data about the Bug Program and prepare a report that can be presented to the big boss.
- You will probably help the big boss prepare information that can be presented to all employees about the success of the Bug Program.
- You will collect and destroy all the promotional material used in the Bug Program when the program has been completed.
- You will be successful.
- You will be tired.

F. Using These Instructions

Up until the day the Bug Program is kicked off, it must be kept a secret from your employees. Can you keep a secret? You must now become security minded. You must take security precautions to keep the program material from getting into the wrong hands. No, not some distant enemy or even your No. 1 competitor. They can buy this book too! What we are talking about is all those wonderful employees out there. You must keep it a secret from them for now. For example, if you must work through a purchasing agent when dealing with printers, and ordering supplies, be sure to explain that the material is secret. Think about how material (printing, publicity, devices, etc.) will be received. You don't want a receiving inspector to yell out, "Hey, look at this great poster." Will word leak out? You will also need a safe place to store supplies. Find out who has keys to the storage area. As they say in the spy business, "Let's get paranoid!"

You can't do this job alone. You will need the help and understanding of all department heads, all levels of supervision, and eventually all employees. If you do not have a Quality Improvement Team (QIT) or committee, you should organize one. The QIT may not do much actual work for you. If you do an expert job of training them, informing them, and listening to them, at least they won't fight you, and they will know what they are expected to do. You should work with your committee to open up communications lines, but in the back of your mind you should keep the old saying, "Search your parks; you will find no statues dedicated to a committee."

IV. WHY PEOPLE MAKE MISTAKES

Mistakes don't just happen, they are caused. It's true! All mistakes are caused by one or more of three things: attitude, ability, or environment.

A. Attitude

An attitude is a way of thinking—a habit. The way someone thinks about the importance of their work, their company, or the product pretty well determines how well their work will be done. Many people seem to have a dual standard about work being done right. They have one standard when it comes to work done for them . . . their paycheck, medical treatment, the car they buy, and so on. Everything must be perfect. But when it comes to mistakes in their own work, the story changes. A few errors in their work is "only human." The Bug Program will help improve attitude in several ways. First, it communicates to every employee that the company is interested in quality. It gives employees a chance to get involved, and it also takes away some of the excuses for not doing work right.

B. Ability

It is necessary to have the physical skills to perform work right. Someone could have a terrific attitude about the importance of the work but not have the skills (or the capacity to develop the skills) to get the work done right. Witness the Olympic Games. How we all would like to be able to perform like those wonderful athletes. Most of us do not have, have never had, and never will have that kind of ability.

C. Bugs on the Job

The lights, the noise, the tools, workbenches, desks, typewriters, instructions, specifications, software, cranky computers, etc. Here is where most of the bugs reside. While it is very possible to improve quality by improving attitude or by training to improve ability, it is not very likely that an employee will turn in a bug to improve their own attitude or ability. They will, however, turn in bugs about their work environment, and that includes training. Almost everyone is willing to point out a problem in their own work environment. Those problems are bugs.

V. GETTING RID OF MISTAKES

The best way to get rid of a problem is to remove its cause. This is especially true when the cause is the tooling or the process—something that actually prevents the employee from doing the job right each time. What is needed is a system that lets people tell their leader what is wrong. The Bug Program does just that. If you could eat dinner with your employees, you would find out about many bugs in their work. That's one of the things most of us talk about at dinner—"They want me to get the work out, but the machine breaks down every hour. . . . " Through meetings, advertising, and action, the program helps employees identify problems. Problems, not solutions. A key difference between the Bug Program and a suggestion program is that the Bug Program does not require a solution to the problem from the employee.

It's management's job to come up with a solution. "What are the bugs in your job?" "What prevents you from doing your job right every time?" Experience shows that you will receive about two or three bugs from each employee. At least 85% of the bugs can be corrected by the first-line supervisor. These are problems dealing with light, instructions, material handling, etc. While these bugs might seem minor to some old pros like you, understand that a very small bug can become a huge irritation. They must be taken care of fast.

You'll probably get at least one bug that says, "The @#$%&*@#$ Bug Program bugs me!!" If you can find out who that person is, go talk with him or her. Somebody out there is angry, and needs someone to talk with. Maybe they will turn in a real bug and feel better if they get help. You will get some bugs that must be handled by another department. Sometimes work sent to an employee from another department is incomplete or wrong. And there is always the second shift. Who knows, maybe that old excuse will go away. If instructions, methods, or specifications must be changed, another department could be involved. For this reason, you must log in all bugs and keep a follow-up file so that you know where they are and when they are due back to you.

About 2 or 3% of the bugs will work their way up to the big boss. Most of these problems will require the spending of a lot of money. Some could affect the business plan. They could point out major problems and require a lot of work.

VI. THE STEPS TO SUCCESS

The Bug Program can be successfully run for 10 employees or 10,000. There are only a few steps to a successful program, and the whole thing takes only thirty working days, or about six weeks. Of course, you will have to work a little longer than six weeks. You must plan the implementation of the program, arrange for promotional materials to be produced, conduct briefing/training sessions for managers and supervisors, hold hands and help some of those managers and supervisors get through the six weeks, and then wrap everything up when its over. Oh yes, you must make up a report for your leader, and your leader should report back to everyone else. Your six weeks will be more like 14 or 16 weeks if you count the time required for the printer to print and the last bug to clear the system. But still, it's a part-time job, and you'll find it to be one of the most rewarding jobs you have ever performed. It's also a learning experience.

Step-By-Step Instructions

Discussed below are the step-by-step instructions to plan and implement your own Bug Program. It's assumed that you are the one who will coordinate the whole program. It's also assumed that you are running the Program in addition to your normal job. Don't worry. Almost all the detail work has been done.

Step 1. Learning

Step 1 is for you to finish reading this chapter. Maybe read it twice. Make certain that you understand just what you will be doing. Obtain or create the artwork that you need. Now is the time to decide if there is anything that needs to be changed. Spend a lot of time with your flow chart.

Step 2. Planning

Now that you have read everything, and have become familiar with all of the material, it's time to make a plan and a schedule. You will need to determine how much promotional material to produce, and how it will be used. You will have to discuss lead-time with the printer, and think about a secure storage spot for all of your supplies. Make up a master planning sheet, a media planning sheet, and a flow chart showing how a bug is processed.

Let's gather some important data first. Your media planning sheet is used to record materials used to promote the program, plan locations, and determine quantities and costs. It will also serve as a kind of check sheet so that you don't miss anything. Use the master planning sheet to plan and schedule events, and to keep track of progress.

The following are the things you will be planning.

Printing

If you are in a very small company, you can use a copy machine to produce your advertising material. But with all the copy shops around today, it's a simple task to have your material printed. Many big organizations have their own print shop; however, remember the security aspect. Go forth and talk with your printer. If you have the opportunity, go talk with several printers. Show him or her the artwork, discuss paper stock, colors, folding, cutting, packaging, and delivery. Of course you will want to discuss price. If you talk with several printers, make certain that you show each one exactly the same work plan, and ask the same questions. Once the printer gives you a firm delivery date, you should add about ten days to that date for insurance.

Everything must be printed before you kick off the program. It is the only way to be certain that posters, table tents, etc., will be ready when you need them. An example of a table tent is shown in Figure 2.

It would be very bad for the health of your job to get into the third week of the program and find out that your printer's shops burned down, or that she or he won the lottery and moved to Hawaii. Since you are going to have everything printed in advance, that means you will need a dry, cool storage area. This storage area must also be secure, and it should have shelves so that you can organize your material for easy distribution.

Specialty Advertising Items

You may want to include advertising material that your printer doesn't handle, such as pins, T-shirts, balloons, badges, pens, or plaques. Some companies have their own in-house catalog of goodies to be used as "give-aways," incentives, etc.

Figure 2 Table tents are placed on tables in the lobby, cafeteria, and elsewhere.

Just because you have not seen a catalog does not mean that one doesn't exist. Check with the sales or marketing manager. Ask the big boss. Usually, you can "buy" things and the money never shows up as a direct cost. If your organization does not have its own catalog, you can purchase a wide variety of specialty items. Look in the yellow pages under Advertising Specialties. These folks seem to be pretty good on delivery, but don't take a chance. Add another ten days for insurance, anyway. Remember to talk with the purchasing and receiving people about keeping things a secret.

Events

Make a list of all the events you want in your program. These are things like management orientation, supervisor orientation, kickoff, and the release of each advertising item. Look at the example, and select the items you want to include in your program. Later, there is a list of recommended projects and events and a list of fifty nifty ways to plug the bug.

Dates and Schedules

Use your planning sheet to make up an actual schedule of events. First, select the kickoff date. Get out a good business calendar that shows holidays and any special days important to your company. It's not wise to kickoff a major event while all the employees are off celebrating Bastille Day (July 14th) or something. Here's a short list of special days that you may want to take note of: New Year's Day, Martin Luther King's Birthday, Washington's Birthday, President's Day, Lincoln's Birthday, Good Friday, May Day, Passover, Memorial Day, Independence Day, Labor Day, Chinese New Year, Yom Kippur, Columbus Day, Election Day, Armed Forces Day, Thanksgiving Day, Christmas, Rosh Hashanah, Hanukkah. If you're in the United Kingdom, Japan, Australia, or Germany, it's okay to kickoff on the 4th of July, but you better make up your own important date list.

Kickoff Date

To find the best kickoff date, first select a general time period that looks good. If you live in Florida, you might kickoff with a picnic in January. If you live in Chicago, that's not such a great idea. The one thing that you'll find to be true is that there is no perfect time to run the program. In the summer, everyone is on vacation. In the fall, everyone is busy with new models and start-up operations. In the spring, nobody wants to do anything. Maybe there is a special day that means something to your organization. The founder's birthday? The anniversary of a new building? An important date in your town? Your birthday? (It will be our little secret.) A Monday is the best day of the week to kick off. Using the master planning sheet and your best intuition, pick the date. Get the boss to approve it. Then, count back ten boxes from the right-hand edge of the sheet and write "Kickoff" on that line. Place a large X in the space. Each block is one week. Put the date of the kickoff on the date line on top. You can now enter the dates for each week. Use the date of each Friday, because that is the best day to change table tents and posters. You'll do that every Friday evening while everyone else is out having fun. That way, everyone will be surprised and delighted when they come to work on Monday. By the way, the reason you used ten weeks as a benchmark is because it takes six weeks to run the program and about four weeks for the wrap-up, reports, etc. Also remember that if a bug is turned in on the last day of the program, it could take three weeks or so to clear the system. All of your effort now is to reach that kickoff day with everything in place, including your nervous system, and your job. You can now count back from the kickoff and determine when you must place the order for printing; when to make a presentation to top management, to middle management (if you have any left), and to supervision. Just count backward from the kickoff date. Set aside your master plan for now, and let's figure out how much material and money you will need. This next exercise will also let you put little X's on your master plan. Make a few copies of the blank Media Planning Sheet. You'll probably use up a few just trying to figure out what you will need. Just keep in mind that all of this planning will guarantee a successful program and convince your boss that you're just the bright light your company needs. If you're the boss, you already know that.

What you are going to do next is figure out exactly where you are going to place your posters, table tents, etc., and how many copies of each you will need to carry out your plan. They say that Dwight D. Eisenhower was promoted over all the other generals because he planned outstanding war games. This exercise should give you the total number of units required, and thus you should be able to make up your printing order. You're doing great! It's time to clean up your paper work, make a few overhead slides, and educate management and supervision. Isn't it funny how we "brief" managers and "train" supervisors?

Step 3. Management Orientation

Step 3 is almost as important as the planning effort you have put into the program. All of your planning work must be complete and approved before you take this step. You must make sure that each and every manager in the organization understands why the program is being conducted, how the program will be conducted, their role in the scheme of things, and how they will benefit. Try to keep in mind that this is not your program—this is the boss's program. You are the coordinator. If you are the boss, you still need to conduct the orientation, but you have a slight advantage. So, the first thing you do, even before you present the program to all managers, is present the program to the boss. It's recommended that you make this presentation as formal as possible. Complete with slides and handouts. This will give you a chance to see how your presentation to the other managers will play, and of course the boss can inject his or her wisdom into things. What do you present? You present the concept, and you present your plan.

First, discuss why people make mistakes, then present the concept of error cause removal: We ask all employees to identify the things in their job that prevent them from doing the job right each time (the bugs in their job), and we promise to get rid of those bugs.

Next, introduce the concept of using the bug: Using the bug relieves the supervisor of guilt, gives us a good theme to publicize the program, and gets us off the hook in thirty days. Be sure to have a lot of slides, but it is a good idea not to hand out material. At this point in time, the program is a deep, dark secret. In fact, you must tell these managers that it is a secret, and warn that they must play along to prevent security leaks. If you hand out material, you risk the secret leaking out. Papers have a way of being left around. By the way, did you leave your masters in the copy machine?

In addition to a discussion of why people make mistakes and an overview of error cause removal, your presentation should cover at least the following:

- An overview of the whole program
- Your plan
- The materials to be used
- Management's role
- Questions

The time it takes to get this orientation job done depends on the size of your organization. The key is to make certain that each manager understands the program, knows what is expected of him or her, and knows that the material is secret. Every manager must be involved.

Step 4. Supervisory Orientation

Supervision will prove to be the determining factor of the success of the program. If the direct supervisors—those first-line leaders of most of the employees—are sold on the program, it can't fail. Since you are going to conduct a formal presentation for all supervisors, you will be a success. Why? In the first place, ECR and this program are good solid ideas. Second, you have created a professional plan to implement and conduct the program. And finally, you have taken the time to make certain each and every supervisor understands the program and how it works. Has anyone ever called them in for a formal presentation before? Don't count layoffs. Tell the big boss that he or she must be there to say a few words to the supervisors and to answer questions. The statement, "Yes, we will make every effort to get rid of every bug," means more when it comes from the boss! Use the same presentation you used for the managers, but cover the flow chart in depth. During the presentation, ask a couple of the supervisors to act out the role of talking to their employees. Maybe all of them should try this. It would be great experience for them. Keep it light, have a good time. Remember that the fear of speaking in public is number one on the phobia list. Some of these folks have never talked to all of their people at the same time as a group. Supervisors will want to know how to handle questions from the employees who report to them. For example, what if it is going to take six months to correct the problem? Tell the employee who identified the bug exactly why it is going to take six months. Convince the supervisors that management wants to get rid of these bugs, and that they will really take action. About the only time you can get trapped in an awkward situation is if someone stumbles onto secret information. Let's say, for example, that a product line is about to be changed, or dropped, or added, and it is important that this information does not leak out just now. Sooner or later, you must reveal the truth, but you may have to delay the answer until it is safe.

Step 5. Artwork

It's time to make up your artwork for the printer. Before you start on your artwork, you should talk with the printer again. Take your plan with you. The idea is to find out how the printer wants the artwork laid out. A printer with a large press may be able to print all your table tents on one sheet of card stock and cut them up. The same thing goes for posters and other artwork. Ask the printer about folding your table tents. You may just want them scored so that you can fold them yourself. You should know how the printer wants this noted on the artwork. One technique used to indicate locations and instructions on artwork is the light blue pencil. Check this out with your printer too. Make notes! About all you need to do is paste up the finished art for the printer. If you want the name of your organization on the material, you can paste up a letterhead or use press-type. If you have never done this kind of work, don't worry, make a few copies of the artwork and practice on that. There are a few rules you must observe. First, keep your artwork clean. What you see is what you get. If you're not careful, you could end up with the best-known fingerprints in the organization! Paste up your finished, camera-ready artwork on light illustration board, and protect it with a cover of clean paper.

Paste Pots and Scissors

To make your own custom artwork, you can paste-up bits and pieces to make the camera-ready art for the printer. The term "paste-up" could be misleading. You will probably use rubber cement, not paste. On the other hand, paste will work too. Rubber cement can get messy. The idea is to paste things to the board without getting finger prints and excess cement hanging around the edge of the work. The printer's camera sees excess rubber cement as just so much artwork.

Press-Type

In case you have not heard of press-type, it is a sheet of ready-made letters mounted on a sort of waxed paper. You can transfer these letters to your artwork by placing the letter over the position where you want it, and gently rubbing the back of the paper with a pencil or a special burnishing tool. The letter is transferred to your artwork! Press-type is called different things in different areas of the world, so you may have to resort to a little sign language to convince the clerk that such a thing exists. Visit your local stationery store or art supply shop. Ask for press-type, run-on letters (rub off?), or transfer letters. They should have a large selection of types and sizes.

Hand-Printed Letters

Forget it! If you are an expert in calligraphy, save it for certificates. It's great, but not on posters, pamphlets, and the like. In case you don't want to do all the paste-up, rub-on stuff, you might ask the people in the art supply store to recommend a good starving artist.

Dates on Artwork

It's a good idea to place a very small date or date code on the bottom right-hand side of every piece of artwork. It's the date the artwork is to be displayed. This may not be necessary if your organization is not very large, and only you will be distributing publicity material. However, if you are the coordinator in a large organization, you may have to depend on others. These others can cause you great grief if they put out the wrong poster or table tent. It would be a good idea to provide these helpers with a printed sheet of dates and description with a space for check mark. Also, don't give your helpers material in advance. If you do, it's guaranteed that a poster or some other gadget will be distributed early because someone thought it was cute. Stay in control. Work your plan.

Step 6. Publicity

The very first thing to appear is the teaser material. Two posters and two table tents; one each week before kickoff. Of course, only you and your insiders know what these mean. Everyone will have to stonewall the questions. Did you cover that in their orientation? The key to a successful program now is to follow your master plan. If you made a mistake or two in your plan, no one will know but you. Anyway it's too late to change the plan now.

Promotional Material and Its Use

The following is a list of all the suggested promotional material and a short statement of how and/or when it is to be used. You are in charge. You can delete any material that doesn't fit in with your operation and/or add items that will help you communicate the bug idea to all the people. For example, you may want to include pens, T-Shirts, or plaques. You may want to give each employee a new Thunderbird. That's up to you.

Posters

The posters should be 11 × 17 inches. They are suitable for use in both office and shop areas. Table tents carry the same message as the poster, and they should be used at the same time

Buttons

Buttons can be ordered from a specialty advertising house. Buttons would carry similar messages to those of the posters and table tents.

"What's Buggin' You?" buttons could be worn by supervisors during kickoff.

Collection Boxes

Collection boxes for collecting extermination forms. Use artwork to paste on the front of cardboard or wooden box.

Fifty Nifty Ideas

Here are fifty idea joggers that might give you an idea or two on how to plug the bug. Read over the list and your mind will race with ideas. T-shirts—"NO BUGS"; coin or token pens & pencils; frisbees; Name The Bug contest; pay envelope stuffers; color the bug contest; lady bug radios; bumper stickers; countdown calendar; stickers on telephone; a help line; score board; sticker for hard hats; shipping label; door mat; bug hot line; message on clock; pocket knives; fax cover sheet; bug ear rings; radio spots; key chain; wooden nickels; bug napkins; bug picnic; raffle a "bug" (VW); get bug on TV show; pocket guards; bug web page; coffee mugs; business cards; invent a mixed drink; bug getter club; float in parade; drink cups; display of bugs found; rent billboard; parking space; bug getter day off; bug getter flag; bug toys; matchbooks; baseball caps; telephone answering machine; shoulder patches; "bug mobile"; bug costume.

Rules for Using Advertising Material

1. Place all advertising material out per your plan. Always on time, always in the proper location.
2. Remove and discard outdated material. Never, never, never leave a poster out to get dirty. Your employees will want to keep some of the material. Let them, but request that they take it home for now. The danger is that they use our artwork for their bowling league, or to make a political statement. Their politics and bowling problems are their business, so they can get their own artwork. Talk with them; they will understand.
3. If you see publicity material that is damaged, but still within its display period, remove it. Replace it if you have extra copies. Didn't the printer overrun the job?
4. When the program is over, collect all the promo stuff, everything, and get it in the trash. This includes your supplies, left over forms, pamphlets, everything. It's a great idea to make up a scrap-book to help you plan next year's activity, and to let everyone know what a wonderful job you have done.
5. If you are a division of a corporation, you can be assured that the corporate seagulls (they fly in, make lots of noise, eat your food, make a mess, and fly away) will come to visit when they find out what a great job you have done. They will know, because the balance sheet will show the improvement, and because the big boss will leak out the word. Show them everything you did, use the slides, your scrap book, the works.

Step 7. Handling Bugs

Make up a flow chart of how bugs will be handled. The first rule in handling bugs is promptness. Promptness is next to godliness. Promptness in getting them into and out of the system. Promptness says you care. Most supervisors have enough paperwork to keep them busy. Most supervisors do not like paperwork and do not want, or need, any more. For this reason, use a box to collect your bugs. You can purchase cardboard boxes of the size you want and paste on the artwork. These boxes should be placed in areas that are convenient to the employees. You might want to number the boxes and keep a tally of the number of bugs from each box. These boxes should be checked at least once a day. From there on out it is a matter of following the flow chart. If there is a cardinal rule in the Bug Program, it's "don't let a bug get lost." Follow your flow chart. Assign a number to each bug you receive. Log it in. It's a good idea to purchase an automatic numbering stamp to assign the numbers. Enter each bug in a 15-day tickler file. Follow-up if it doesn't show up on time. The employees must trust this system. Lose a few, and word of mouth advertising will do you in. Use a follow-up log. If a bug must be handled by another department, be sure that information is entered into your control log, and that a new tickler date is assigned. Also make sure the number is cleared from the first supervisor's record.

Step 8. Wrap-Up and Report

It will take you an extra two or three weeks to clean everything up and close the books on the program. Actually, it could take just a little longer as the last few bugs get killed off. But for the most part, you're off the hook. On the last Friday of the program, after everyone has gone for the day, you should make the rounds and gather up all advertising material and throw it out. Make certain that you get it all. The program is over, the bug is dead. Is that a tear in the corner of your eye? Now it is time for you to make a report to your leader. What do you report? Report how many bugs were received by department, group, or section. Report how many extermination requests resulted in a dead bug and any known or estimated savings. Report the total cost of running the program. And give some case histories of unusual or interesting bugs. Two or three pages should do. Just keep in mind that the big boss has a bigger boss, who of course has a still bigger boss. Each boss needs material to impress the next bigger boss.

Step 9. Reports and Recognition

It is important that everything about the program be closed out in a formal manner. You made a big deal in kicking the thing off, you should make a big deal in ending it. Several things should happen. First, you must report the results to the big boss. These data should be broken down in a way that makes sense to your organization. The boss should gather all of his or her underlings and let them know what happened and what a wonderful job they did. This is a great time and place for the big boss (or even the big, big boss) to present an award. Second, all the supervisors who gathered their people to introduce the program and hand out the "what's buggin' you" pamphlets should once again gather their people to let them know how they did. Each supervisor should report how many bugs were gathered, how this group did in relation to the rest of the groups, and which individuals did an outstanding job. Are awards in order? It would be nice to recognize a supervisor in front of his or her people. Third, while those employees who turned in a bug did receive a badge, those whose bugs were killed should receive something else. A pin, a plaque, a certificate, a T-shirt, a pat on the back. It would be okay for the big boss to attend these little meetings between the supervisor and their people and pass out a few thanks—maybe even the awards.

Step 10. Take a Rest

Well, it's over. You did a great job, and you deserve a rest. Make sure you give everyone who helped you a pat on the back—they deserve it.

18

Reliability, Accessibility, and Maintainability

JOHN JOURDAN HELDT
Free Lance Reliability Service
San Jose, California

I. INTRODUCTION

The curve shown in Figure 1 is representative of the operating performance of devices such as transistors and bearings—and human beings. The curve is usually called the *bathtub curve* because of its distinct shape.

The left-hand—or early life—portion of the curve is usually referred to as the early failure or infant mortality phase. In this phase, the failure rate of devices can be very high. In the center portion of the curve, called *useful life*, the failure rate for a typical device is both low and very constant; usually over a long period of time. The portion of the curve on the right-hand side is called the *wear-out* phase.

A. Early Failure

In products, early life failures, or infant mortalities, are most often due to workmanship errors, inadequate design, or marginal components. Typical examples of workmanship errors are cold solder joints, bent or misaligned pins of components, or diodes and capacitors installed with wrong polarities.

Failures due to overvoltage or overheating are usually design failures, simply caused by the designer's failure to anticipate the effects of applied voltages or other stresses.

Marginal components are usually due to manufacturers who have not followed their own procedures for fabrication of parts or the inadvertent use of wrong material in the assembly phase.

Many schemes have not proposed to force early wear-out failure of weak parts before the product gets into the customer's hands. Most of these plans include aging each of the individual components or subassemblies. The aging is usually expedited by operation at higher than normal temperatures or through the applica-

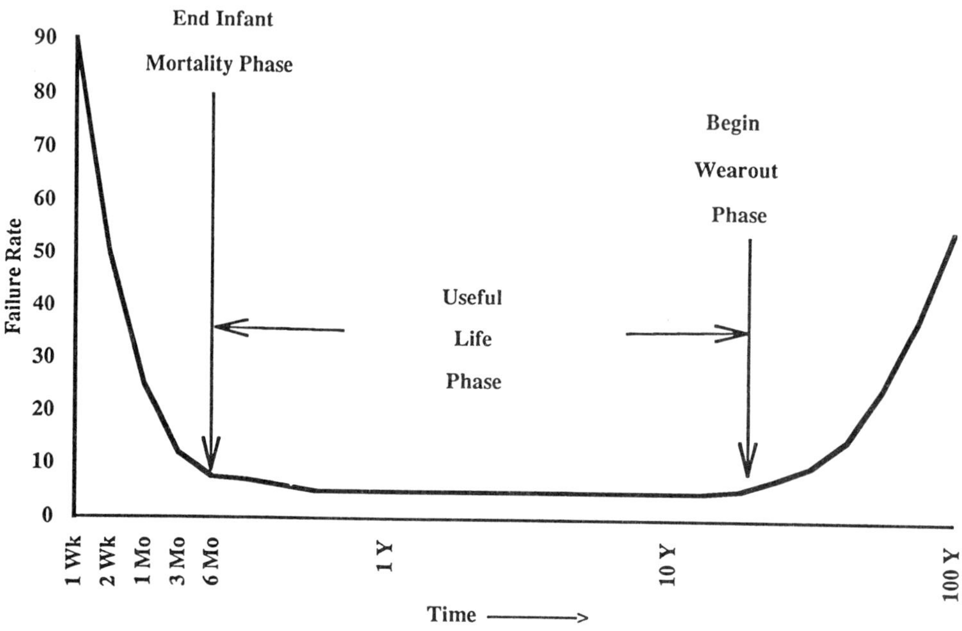

Figure 1 Bathtub curve.

tion of higher voltage or current levels than normal. Sometimes the part is vibrated while the thermal and electrical stresses are being applied. Most plans are aimed to make a week of stress operation look like 6 months of real time to the parts or assemblies. However, it is a good idea to chart the effect of accelerated aging, since almost all of the early failures are removed in the first few hours.

B. Useful Life

The basic formula for time-related reliability is shown in the following equation[1]. This formula is valid only when the failure rate is relatively constant; (i.e., the middle part of the bath tub curve). This formula is generally used for electronic and electromechanical parts.

---

[1] These formulas are the basis for "exponential" or "time-related" reliability. The importance of these relationships will become evident when we study MTBF determination in the sections that follow. $R = e^{-\lambda t}$ is the same formula as the Weibull probability function, when:

$$\alpha \text{ (scale parameter)} = 1$$
$$\beta \text{ (shape parameter)} = 1$$
$$\lambda \text{ (location parameter)} = 0$$

$R = e^{-\lambda t}$ is also the same formula as the Poisson law of small numbers, when

$$r \text{ (number of occurrences)} = 0$$

$R(t) = e^{-\lambda t} = e^{-t/MTBF}$

Where:

R(t) = probability that the item will operate without failure for a given period of time, "t" (usually expressed in hours), under stated operating conditions.

e = 2.7182, the base of the natural logarithms.

λ = Greek letter lambda, the symbol for the item failure rate. The failure rate is usually expressed in failures per million hours (or failures per million cycles), or, recently, as *FITS*[2]. Since the failure rate is level on the useful life part of the curve, it is assumed to be constant for this formula.

MTBF = reciprocal of λ (the failure rate). That is, MTBF = 1/λ.

The reciprocal of the failure rate is called the mean time between failure (MTBF). Sometimes this is called the MTTF, which stands for mean time *to* failure. These are the same value. The use of MTBF implies that the item can be repaired and put back into operation. MTTF is used where the failed part or assembly cannot be repaired and must be replaced. MTTF can be used to describe the life of a space probe outside the solar system, where a failed unit could not be repaired. Therefore, MTBF = 1/λ and λ = 1/MTBF.

C. Wear Out

The basic formula for time-related reliability is valid only throughout the useful life period. Almost all well-designed products are sold with a warranty period and in actual practice outlast their warranties by at least twice. Degraded product, such as appliances that have been exposed to humid temperatures or excessive voltage, may experience early wear out. In most cases, this kind of wear out can be avoided by standard operating procedures and routine maintenance.

II. RELIABILITY ASSESSMENTS

Time-related reliability assessments are easy, since the reliability of each component and subassembly is exponential in nature. This means that we find the failure rate for each component in MIL-HDBK-217[3], and then add the failure rates of all of the parts in the subassembly to find the subassembly's failure rate. Likewise, we

---

[2] More robust productions methods have brought failure rates for common parts such as resistors and capacitors to very low figures such as 0.0012 failures per million hours (i.e., 0.0012×10^{-6} failures). This has brought about a new designation called a FIT, which is used mostly in Europe and by Bell Systems. FIT is the acronym for 'failures in time' and means that the multiplier is 10^{-9}. In other words, 0.0012×10^{-6} is the same as 1.2 FITs.

[3] See the current amendment of MIL-HDBK-217—*Military Handbook, Reliability Prediction Of Electronic Equipment*. Bell Systems Information Publication IP 10475, *AT&T Reliability Manual*, is another failure rate source.

can add the failure rates of all of the subassemblies to find the failure rate for the entire assembly.

A. Exponential Math (Formulas)

Figure 2 demonstrates multiplication by adding exponents. Each of the blocks in the *single-thread reliability* model represents a unit that would cause the assembly to fail should that block fail. This means that the overall reliability for the assembly is the multiplier of all four of the block reliabilities, and reliability of the assembly is lower than the reliability of any block.

$$R_{overall} = R_1 \times R_2 \times R_3 \times R_4$$

Substituting the exponential value.

$$R_{overall} = e^{-(\lambda 1)t} \times e^{-(\lambda 2)t} \times e^{-(\lambda 3)t} \times e^{-(\lambda 4)t}$$

When the failure rates are grouped together

$$R_{overall} = e^{-(\lambda 1+\lambda 2+\lambda 3+\lambda 4)t}$$

The summed failure rate formula follows,

$$R_{overall} = e^{-(\Sigma\lambda)t}$$

B. Exponential Math (Using Real Numbers)

Figure 3 demonstrates multiplication by adding exponents, with real numbers in place:

$$R_{overall} = 99.9\% \times 99.6\% \times 99.8\% \times 99.7\%$$

Substituting the exponential value,

$$R_{overall} = e^{-0.001} \times e^{-0.004} \times e^{-0.002} \times e^{-0.003}$$

When the failure rates are grouped together

$$R_{overall} = e^{-(0.001+0.004+0.002+0.003)}$$

$$R_{overall} = e^{-0.010} = 99.005\%$$

C. An Exponential Math Example

When the exponential value ($-\lambda t$) is very small, the reliability value is very close to "$1 - \lambda t$". For example, when

$$-\lambda t = 0.01, \quad e^{-\lambda t} \approx 1 - 0.01 \approx 99.0\%$$

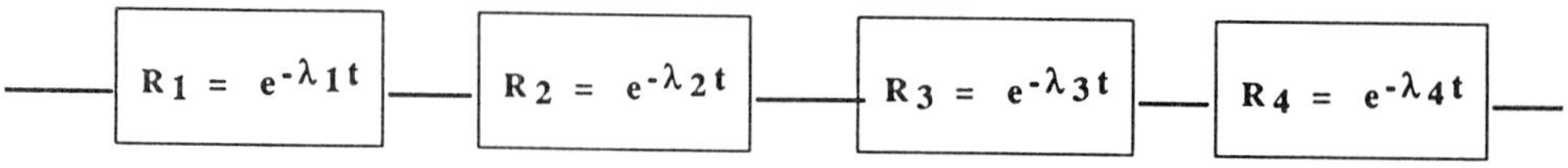

Figure 2 Single-thread reliability model: general values.

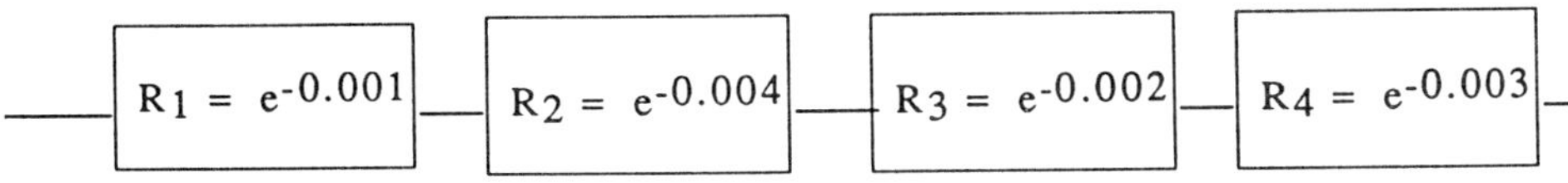

Figure 3 Single-thread reliability model: specific values.

As "$-\lambda t$" becomes larger, the approximation of the reliability figure loses more and more accuracy, but it is still close enough for many practical purposes.

D. Reliability Allocation Example

A conceptual design review is scheduled for a new and larger disk drive for Leptron's* current computer. Leptron's reliability engineer has been asked to establish the reliability goal for each of the disk drive's subassemblies that will result in 97% overall reliability for the new disk drive.

The major elements of the disc drive are spindle, signal processing card, read-write system, and chassis and wiring.

$$R_{overall} = 1 - Q_{overall} = 1 - 0.003 = 0.997$$

In words, this states that the allocated reliability is equal to one minus the unreliability. Q is the symbol usually used for unreliability. Also, Q closely approximates the exponent (λt)

Q allocation		(Q)	Allocated reliability (1 − Q)		
Q_{spin}	=	0.001	R_{spin}	=	0.999
Q_{spc}	=	0.0005	R_{spc}	=	0.9995
Q_{rs}	=	0.001	R_{rs}	=	0.999
Q_{mf}	=	0.0005	R_{mf}	=	0.9995
$Q_{overall}$	=	0.003 (Qs added)	$R_{overall}$	=	0.9970032 (Rs multiplied)

Allocating the unreliability goal of the assembly into reasonable unreliabilities for each of the subassemblies is a simple operation. Any allocation, simple or complex, can be made using this method. Guidelines follow:

1. Determine those components which could cause failure of the entire assembly. Show these components in the single-thread reliability model.
2. Find the unreliability (Q) of the reliability goal by subtracting the reliability goal from 1. Apportion this unreliability among the subassemblies. The total of the subassembly unreliabilities must equal the unreliability of the assembly. Some judgment may be needed here, since you may need

* A fictitious company.

to allocate more unreliability to the complex subassemblies and less for the simple components.

3. Multiply the resulting reliability allocations to verify the accuracy of your apportionment.

E. Failure Rate Allocation Example

The actual failure rate allocation for each subassembly can be determined from each element's allocated reliability by using the exponential formula:

$$R = e^{-\lambda t}$$

The allocated reliability for spindle is 99.9%, thus

$$e^{-\lambda t} = 0.999$$

From this we find $\lambda t \approx 0.001$ (i.e., $\lambda t \approx 1 - R$).

Divide 0.001 (λt) by 173.3 hours (the number of hours in a single shift operation for one month) to find:

$$\lambda(\text{spindle}) = \frac{\lambda t}{t} \approx \frac{0.001}{173.3}$$

$$\approx 0.00000564 \text{ failures per hour } (5.64 \times 10^{-6})$$

Note: $-\lambda t$ is the Napierian logarithm of the reliability. However, λt is approximately equal to 1 minus the reliability. This approximation approaches the calculated value as the reliability becomes closer to unity (100%).

Once the reliability is apportioned and the failure rate allowed for each subassembly has been determined, it still remains for the designer to propose components for the unit which have failure rates that will add up to the failure rate allocated to each subassembly. In other words, the sum of the failure rates of all of the components that go into the subassembly must not be more than the allocated failure rate. Some of the factors that result in parts with lower failure rates are listed here:

1. Parts that are more robust have lower failure rates. Select parts that are designed to be more rugged or parts which have been screened to demonstrate freedom from workmanship errors.
2. Use parts that have at least 50% higher temperature or power ratings than required. One rule of thumb is that a part operating at half power will last at least four times longer than a part operating at full power. (This means that 50% derating reduces the failure rate by four.)
3. Design redundancies into the unit. An appropriately added component, a subassembly or even a standby assembly, are often-used techniques that can result in increased reliability.

F. Selected Exponential Values

Exponential values can be found using most hand-held calculators.[4] For the calculator that has an e^x function, one enters the exponential value $-\lambda t$ and presses the e^x key to find the corresponding reliability. To find the $-\lambda t$ corresponding to any

---

[4] Using spreadsheet software, personal computers can also be used to calculate $e^{-\lambda t}$ reliabilities using the formula "exp(−x)." Extensive $e^{-\lambda t}$ tables can be developed with this software.

—x	Exp x
-0.001	0.999000
-0.002	0.998002
-0.003	0.997004
-0.004	0.996008
-0.005	0.995012
-0.006	0.994018
-0.007	0.993024
-0.008	0.992032
-0.009	0.991040
-0.010	0.990050

—x	Exp x
-0.010	0.990050
-0.020	0.980199
-0.030	0.970446
-0.040	0.960789
-0.050	0.951229
-0.060	0.941765
-0.070	0.932394
-0.080	0.923116
-0.090	0.913931
-0.100	0.904837

Figure 4 Negative exponential values.

reliability, one enters the reliability value and presses the inverse key then the e^x key. The lnx key may also be used. Enter $-\lambda t$ and press the inverse key then the lnx key to find the corresponding reliability. To go from reliability to $-\lambda t$, enter the reliability value and press the lnx key.

Caution: Whenever λt is entered, the negative sign (−) must be entered by pressing the *change sign* (±) before proceeding. Refer to Figure 4 if a scientific calculator isn't available. Figure 4 may also be used to verify the correct use of your calculator.

III. LIFE TEST FOR RELIABILITY

A. Estimation of the MTBF

Much time is used predicting reliability of the product and a great deal of creativity goes into improving its reliability, but in the end, it all boils down to answering the question: How well will our product perform in the job for which it was intended?

One way to answer this question is to determine the mean time between failures (MTBF).

Life testing is one way to find the MTBF. Life test theory presumes that all of the parts or assemblies are within the useful life portion of the bath tub curve. This means that 10,000 unit test hours (UTH) can be accumulated by putting 10,000 units under test for 1 hour as well as testing one unit for 10,000 hours.[5] In either case, the 10,000 test hours can be divided by the number of failures for a rough estimate of the MTBF.

When the expected value (np′) of the Poisson distribution associated with the number of failures is available, the MTBF is found as shown[6]:

---

[5] There are other ways to accumulate 10,000 unit test hours (UTH) such as: operate 100 parts for 10 hours at an elevated temperature, which accelerates the aging by 10 (i.e., $10 \times 10 \times 100 = 10{,}000$ equivalent hours).

[6] The formula, MTBF = Unit Test Hours (UTH)/Expected Value (np′) is valid, because the expected value can be rewritten as "n = UTH" and "p′ = λ." Thus, the formula becomes

$$\text{MTBF} = \frac{\text{UTH (actual)}}{\text{Unit Test Hours}_{(\text{Poisson})} \times \lambda_{(\text{Poisson})}} = \frac{1}{\lambda}$$

$$\text{MTBF} = \frac{\text{Unit Test Hours (UTH)}}{\text{Expected Value } (np')}$$

Use of the Poisson distribution has a side benefit, since the MTBF can be calculated at different confidence levels. Calculating the MTBF at the 10% confidence level and again at the 90% confidence level shows the 80% envelope for the actual MTBF. In other words, there is an 80% probability that the actual MTBF of the unit will fall within this envelope. The 60% MTBF envelope can be calculated using the MTBFs at 20 and 80% confidence levels[7].

B. Examples of Estimation of the MTBF

The Poisson table shown in Figure 5 is tabulated[8] to make it easy to use for life test calculations. The numbers along the left side of the chart correspond to the number of failures in the life test. Usually the number of failures is identified as a lower case c. In this table, the column is identified as "Failures" as well as "c." The numbers in the body of the table are "Expected Values" (np′). n is the number of samples tested and p′ (pronounced p-prime) is the actual failure rate. In this table, the columns are headed by the confidence level (CL)[9] associated with the np′ values listed in the body of the table.

For life test purposes, the total test hours[10] (TTH) is divided by np′, which yields the MTBF.

1. Example 1: Estimate the MTBF at a 60% confidence level, based on the following scenario: 24 units were under test for a total of 13,000 hours. One unit failed at 12,100 hours and was not replaced. (i.e., TTH = 23 × 13,000 = 299,000 plus 12,100 = 311,100 unit test hours).

- When the life test is stopped[11] on a specified number of hours, as in this scenario, the line number used is the actual failure number (i.e., "c = 1" line).
- Go down the CL = 60% column to find the "expected value" (in this case, 2.022 on the "c = 1 line" is outlined in Fig. 5).

---

[7] Because of the shape of the Poisson distribution, $\text{MTBF}_{(\text{CL} = 50\%)}$ comes about at about 37% of the time line. For this reason, many people use $\text{MTBF}_{(\text{CL} = 60\%)}$, since this is closer to 50% on the time line.

[8] Special tabulations like this one are easily done with the use of your personal computer.

[9] CL is the complement of P_a (i.e., CL = 1 − P_a). Thus all of the values shown across the top of our table are actually the P_a value subtracted from 1.

[10] The total test hour designation is used to include unit test hours (i.e., hours accrued when many parts are under test at the same time) and equivalent test hours (i.e., hours amassed by time acceleration or aging).

[11] When life test is ended on a specific number of failures, the failure number used to find np′ is the actual number of failures at test end minus one. When a test ends on a specific number of hours, as this one, the actual failure number is used.

Failures (c)	Confidence Levels (CL) 95%	90%	80%	70%	60%	50%	40%	30%	20%	10%	5%
0	2.996	2.303	1.609	1.204	0.916	0.693	0.511	0.357	0.223	0.105	0.051
1	4.744	3.890	2.994	2.439	2.022	1.678	1.376	1.097	0.824	0.532	0.355
2	6.296	5.322	4.279	3.616	3.105	2.674	2.285	1.914	1.535	1.102	0.818
3	7.754	6.681	5.515	4.762	4.175	3.672	3.211	2.764	2.297	1.745	1.366
4	9.154	7.994	6.721	5.890	5.237	4.671	4.148	3.634	3.090	2.433	1.970
5	10.513	9.275	7.906	7.006	6.292	5.670	5.091	4.517	3.904	3.152	2.613
6	11.842	10.532	9.075	8.111	7.343	6.670	6.039	5.411	4.734	3.895	3.285
7	13.148	11.771	10.233	9.209	8.390	7.669	6.991	6.312	5.576	4.656	3.981
8	14.435	12.995	11.380	10.301	9.434	8.669	7.947	7.220	6.428	5.432	4.695
9	15.705	14.206	12.519	11.387	10.476	9.669	8.904	8.133	7.289	6.221	5.425
10	16.962	15.407	13.651	12.470	11.515	10.669	9.864	9.050	8.157	7.021	6.169
11	18.208	16.598	14.777	13.548	12.553	11.668	10.826	9.972	9.031	7.829	6.924
12	19.443	17.782	15.897	14.623	13.589	12.668	11.790	10.896	9.910	8.646	7.690
13	20.669	18.958	17.013	15.695	14.624	13.668	12.755	11.824	10.794	9.470	8.464
14	21.886	20.128	18.125	16.765	15.658	14.668	13.721	12.754	11.682	10.300	9.246
15	23.097	21.292	19.233	17.832	16.690	15.668	14.688	13.686	12.574	11.135	10.036
16	24.301	22.452	20.338	18.898	17.722	16.668	15.657	14.621	13.469	11.976	10.832
17	25.499	23.606	21.439	19.961	18.752	17.668	16.626	15.558	14.367	12.822	11.634
18	26.692	24.756	22.538	21.023	19.782	18.668	17.596	16.496	15.269	13.671	12.442
19	27.879	25.903	23.634	22.082	20.811	19.668	18.567	17.436	16.172	14.525	13.255
20	29.062	27.045	24.728	23.141	21.839	20.668	19.539	18.377	17.079	15.383	14.072
21	30.240	28.184	25.819	24.198	22.867	21.668	20.511	19.320	17.987	16.244	14.894
22	31.415	29.320	26.909	25.254	23.894	22.668	21.484	20.265	18.898	17.108	15.719
23	32.585	30.453	27.996	26.308	24.920	23.668	22.458	21.210	19.810	17.975	16.549
24	33.752	31.584	29.082	27.361	25.946	24.667	23.432	22.157	20.725	18.844	17.382
25	34.916	32.711	30.166	28.414	26.971	25.667	24.407	23.104	21.641	19.717	18.219

Figure 5 Poisson table of expected values and confidence levels.

- Divide the life test hours (TTH) by the expected value (np′) to find the MTBF, as shown below:

$$\text{MTBF}_{(\text{CL}\,=\,60\%)} = \frac{311100}{2.022} = 153{,}857.567 \text{ hours}$$ [12]

2. Example 2: About 15 years ago, the Megawatt Bulb Company* installed 20 of their new fluorescent lamps in an auxiliary warehouse. Someone thought it would be a good idea to get some life test data on these new bulbs, so an elapsed time meter was installed on the lighting circuit. This year, the safety committee noted that the warehouse had at least half the lights out. The electrician who was sent to replace the bulbs found that the elapsed time meter showed that the lights had been lit for 128,631 hours, and that 15 bulbs were no longer functional.

Can you estimate the demonstrated MTBF and confidence levels for this model bulb?

Note: Use 128,631 hours each for the five functional bulbs. Use half the elapsed time (i.e., 128631/2 = 64,315 hours) for each of the 15 failed units.

[12] In actual practice, this figure would be rounded up to MTBF = 154,000 hours.
* A fictitious company.

- The total test hours (TTH) is estimated as follows:

  5 working bulbs: 5 × 128,631 = 643,155 hours;
  15 failed units: 15 × 64,315 = 964,725
  TTH = 1,607,880 hours

- Go down the "CL = 10%," "CL = 90," and the "CL = 50" columns to the "c = 15 line" to find the expected values:

  $np'_{(CL = 10\%)} = 11.135$
  $np'_{(CL = 90\%)} = 21.292$
  $np'_{(CL = 50\%)} = 15.668$

  Note: These values are identified on Figure 5.
- Peform all three divisions:

$$MTBF_{(at\ 10\%\ CL)} = \frac{1607880}{11.135} = 144{,}400 \text{ hours}$$

$$MTBF_{(at\ 90\%\ CL)} = \frac{1607880}{21.292} = 75{,}520 \text{ hours}$$

$$MTBF_{(at\ 50\%\ CL)} = \frac{1607880}{15.668} = 102{,}622 \text{ hours}$$

- Conclusions: 102,622 hrs is the fluorescent bulb's MTBF; 80% of the MTBF's for the entire universe of this model fluorescent bulb will fall between 75,520 and 144,400 hours.

C. Poisson Table and the Life Test

The Poisson table can also be used to design an MTBF verification[13] test. One of the by-products of the MTBF verification test is that units that have poor construction, poor workmanship, or inadequate materials also result in a lower MTBF. This means the verification test is also a guard against early failure symptoms. Intelligent interpretation of the MTBF test will also give signals that the time used for burn-in of units might be lowered.

Steps for development of the life test:

1. Estimate the desired MTBF for the product and choose the number of failures to be used for acceptance. The number of failures to use is arbitrary; but the number of the unit test hours required for the test becomes larger as the acceptance number rises; and the test become less discriminatory as the number of unit test hours goes lower.
2. Use Figure 5 to find the expected value (np′) at the intersection of the number of failures line and the 20% confidence level column. CL = 20% corresponds with P_a (probability of acceptance or, as it is sometimes called, probability of success) = 80%. CL + P_a = 1. P_a = 80% is also an arbitrary

[13] The MTBF verification test is often referred to as a life test.

choice: P_a = 80% was chosen because it means that there is an 80% probability of acceptance when the actual MTBF is equal to the calculated MTBF.

3. The number of unit test hours is found by multiplying your desired MTBF by the expected value that you found in step two.
4. Go across the "c line" in Figure 5, to each of the other confidence level values. Divide the total unit test hours by each of the corresponding nps. This gives an MTBF for each of the P_as as shown in Figure 6a. Plot the operating characteristic (OC) curve for the test as P_a versus MTBF. In other words, each CL needs to be converted into P_a, using the "$P_a = 1 - CL$" relationship.
5. Plot each of the MTBFs against its corresponding P_a. The resultant graph is a life test OC curve like the curve shown in Figure 6b.
6. The OC curve can be used to estimate the probability of acceptance, according the life test, for any MTBF shown at the bottom of the graph.

D. Life Test Example

The concept of this example is that it is desired to design a life test and use the results to accept or reject the statement: "The units under test have a mean time between failure (MTBF) of not less than ____ hours."

Assume that the following information is given:

Desired MTBF = 6000 hours
Desired test confidence = 80%

np'(c=5)	23,400/np'	Pa = 1 -CL
2.613	8955	95%
3.152	7424	90%
3.904	5994	80%
4.517	5180	70%
5.091	4596	60%
5.670	4127	50%
6.292	3719	40%
7.006	3340	30%
7.906	2960	20%
9.275	2523	10%
10.513	2226	5%

(a)

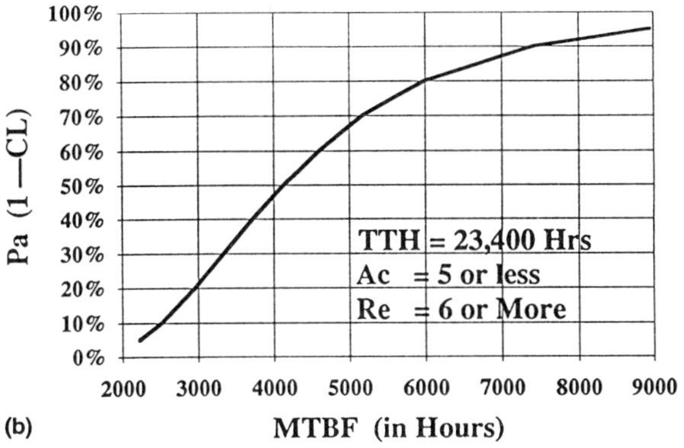

(b)

Figure 6 (a) *np'* values; (b) OC curve for life test.

Choose a number of unit test failures that will be permitted during the test duration. Test failures beyond this number will result in a conclusion that the units on test do not possess the desired MTBF. While there is a certain arbitraryness about choosing the "accept" number, in general, a larger "accept" number results in longer test time but improves the test result accurancy. For this example, assume the decision is made to "accept" on five and "reject" on six unit failures.

Using Figure 5, go across the C = 5 row to the 20% column. Note that the probability of acceptance is equal to {1 − the confidence level}, so that {1 − 20%} equals an 80% probability of acceptance.

Note the value at the intersection of C = 5 and 20%, which is 3.904. The total number of unit test hours, then, in this example, is equal to the MTBF goal (6000) multiplied by 3.904, or 23,424 unit hours. Since we chose an "accept" on five failures, we have to test at least five units for 4685 hours (in which case, all five units would have to be operating at the end of the test) or any mix up to 23,424 units for 1 hour (in which case, 23,419 units would have to be operating at the end of the hour). If all five units were not operating at the end of the 4,685 hour time or only 23,418 units were operating at the end of the hour, we would have to conclude that the unit MTBF was less than 6000 hours. Eighty percent test probability of acceptance means that five or fewer failures will occur during 23,424 unit test hours (as in this example) at least four out of five times when the units do, in fact, have an MTBF of 6000 hours. As in any statistical estimate (i.e., when there is less than perfect information), there is a possibility of reaching an incorrect conclusion.

Referring to Figure 5, it can be seen that if the MTBF is 3000 hours rather than 6000 hours, there will be only a 20% chance that there will be five or fewer failures during this life test.

Sometimes it is desirable to devise a test that will result in rejecting (or failing) the statement "The units under test have a mean time between failure of ____ hours," rather than to pass it 80% of the time. The following examples refer to life tests with probabilities of acceptance (P_a) of = 20% for MTBFs equal to 3000 hours. Remember to use the CL = 80% column for P_as = 20%.

Accept number	np′ × MTBF	Total unit test hours
1	2.994 × 3000	8982
5	7.906 × 3000	23,718
10	13.651 × 3000	40,953

Generic life test operating characteristic curves have been developed by the U.S. Department of Defense. These are shown in MIL-STD-781—*Reliability Design Qualification and Production Acceptance Tests: Exponential Distribution*. These curves have two sets of MTBF values (designated as θ_0 and θ_1 for the "X" axis: when the top line (labeled θ_0) is used, each θ_0 is multiplied by the desired MTBF. The life test defined by the θ_0 OC curve is essentially the same as the OC curve for the test developed in the Poisson illustration above (i.e., there is an 80% probability for acceptance when the true MTBF is equal to the desired MTBF). When the θ_1 "MTBF" axis values are used, the product has an 80% probability of failure when the true MTBF equals half of the desired MTBF. (The life test and OC curve developments can also be done using a chi square table.)

IV. MECHANICAL DEVICES FAILURE PREDICTION

A. The Normal Distribution

When the strength of a material is normally distributed about an average and stress on that material is also normally distributed, failures will occur if the actual stress exceeds the material strength. The probability of overlap can be predicted when the averages and standard deviations are known. The mechanical part will have optimum reliability when the area of the overlap is at 2σ or more.

As seen in Figure 7, the probabilities of the strengths of material and the material stresses are shown as areas under normal distribution curves. The probability of failure is designated by the area where these distributions overlap. Abnormally low strength units will fail mechanically when subjected to an abnormally high stress. The reliability of the material is designated as the probability that the strength of units will not fall within the overlap region at the same time the unit is stressed into the same overlap region. Specifically, it is the probability of the strength being greater than the applied stress based on the two distributions.

B. Normal Distribution Example

In Figure 7, the probability of failure is estimated for the event where the two distributions intersect. The 2 sigma (σ) point overlap means that the probability of failure is equal to the area under each tail multiplied by the other. The area below

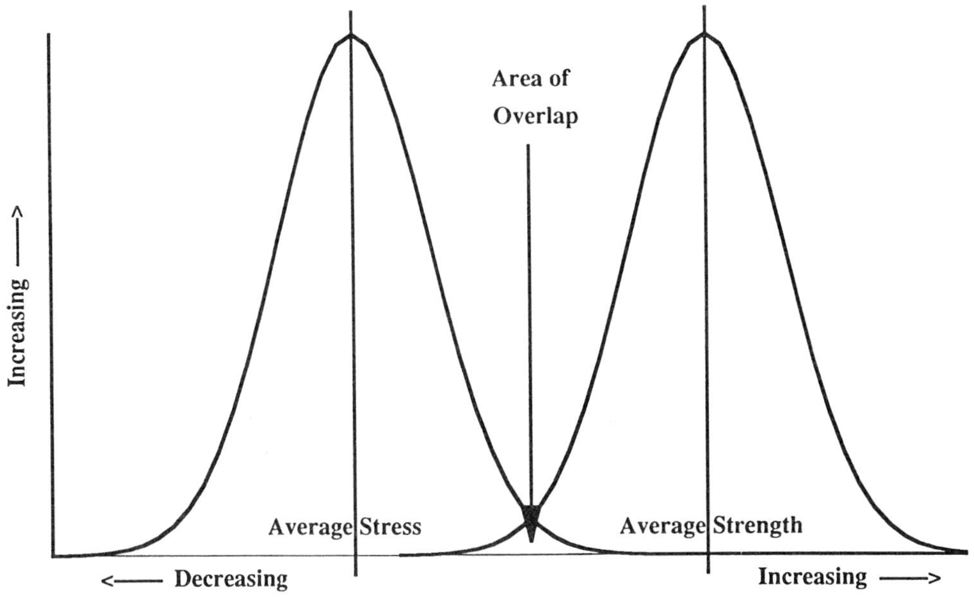

Figure 7 Stress and strength reliability.

the -2σ point on the strength curve indicates that 2.28% of the material will fall below this strength. The $+2\sigma$ point on the stress curve indicates that 2.28% of the time, the material is stressed enough to fall within the lowest strength of the material.

Failure can occur when both happen at the same time. This probability of failure is:

$0.0228 \times 0.0228 = 0.00052$ (i.e., 0.052%)

Reliability for this case = 100% − 0.052% = 99.948%

R = 99.948% means that there are 52 chances of failure out of 100,000 strength/stress occurrences.

C. Logarithmic Ruled Paper

Figure 8 shows how to estimate reliability based on production parts that have been tested to destruction. The scenario for the reliability estimate follows:

Ten sample radar antennas have been weathered in a 95% salt spray atmosphere for 1 year while being stressed at 110% load conditions. Test results to date are:

Six antennas, still operating
First failure, 6200 hours
Second failure, 6700 hours
Third failure, 7800 hours
Fourth failure, 8000 hours

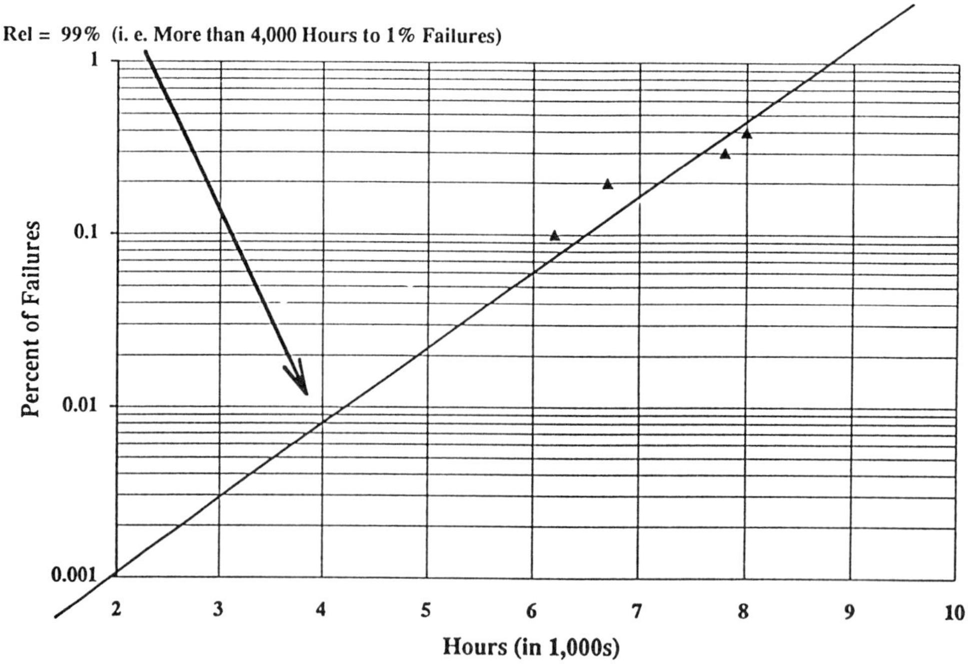

Figure 8 Graphical reliability.

The reliability goal is R = 99% for 6 months at sea. Test results of R = 99% for more than 4000 hours of operation will indicate a successful demonstration for the antenna design. The steps are outlined here:

1. Failures are plotted as logarithmic reliability versus linear time.
2. A best-fit line is drawn to fit these data.
3. The best-fit line is extrapolated to find where it crosses the 1% line (i.e., Where r = 99% is indicated).

D. Weibull Ruled Paper

Graphs using the Weibull probability distribution instead of the logarithmic distribution are often used (Fig. 9). Failure patterns of mechanical assemblies often follow different distributions. Since the Weibull function varies greatly depending on the numerical values of the shape parameters, most mechanical failure rate shape can be graphed on one or the other of the Weibull ruled papers. When Weibull ruled paper is used, the steps are the same as used for the logarithmic example:

- Plot Weibull reliabilities versus failure times.
- Draw the best-fit line.
- Extrapolate the best-fit line to find the indicated reliability.

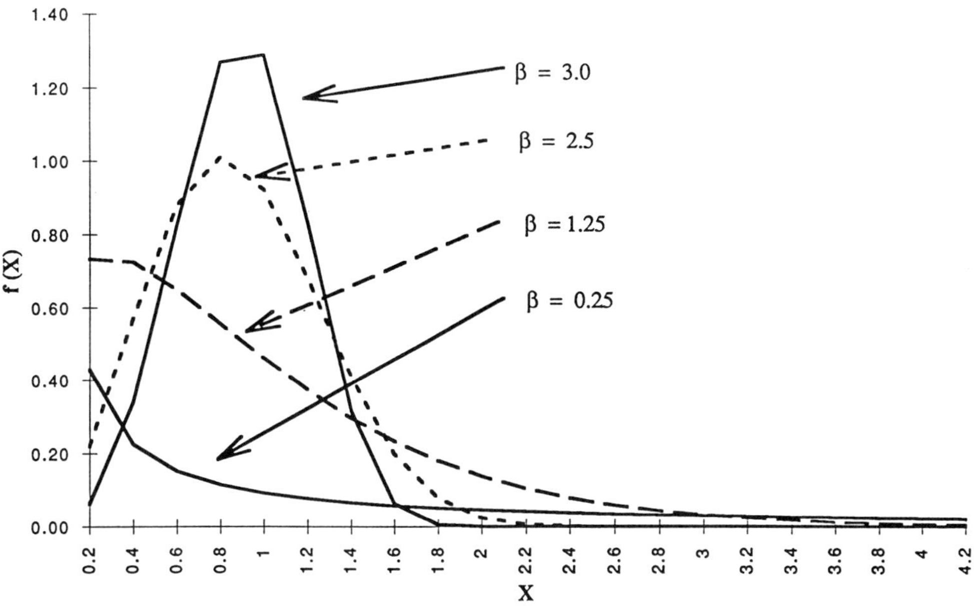

Figure 9 Selected Weibull distributions.

It isn't necessary to do the Weibull arithmetic, since predictions are made directly from Weibull probability paper.[14] The solution is to plot the data on the paper. When the points fall on an approximately straight line, the best-fit straight line takes in most of the points. This straight line can be interpolated or extrapolated for probability predictions.

V. IMPROVING DESIGN RELIABILITY

A. Derating

About 100 years ago, a Swedish chemist named Arrhenius developed an empirical model that described the effect of temperature on the rate of inversion of sucrose. This model coined the term *experimental activation energy* (E_a). E_a is the factor which determines the slope which describes the accelerating effect that temperature has on the rate of reaction.

This Arrhenius reaction rate model is generally accepted as the basis for time-temperature failure rate effects[15] often used in life test data. This means that stress due to higher temperatures is universally accepted to cause parts to fail sooner. The Arrhenius reaction rate model is broadly extended to include "greater loading causes higher failure rates." Theoretical development of the Arrhenius equation is as yet empirical; however, detailed failure rate data recorded on stress-temperature matrices are available in MIL-HDBK-217. Use of these data show the high level of verification for all Arrhenius-based assumptions.

Based on the Arrhenius principles, qualification testing, either of parts or units, usually includes operation at higher temperatures and operation under high- and low-voltage application. Testing of components in this manner will expose anomalies within the manufacturer's processing and often indicates the superiority of one manufacturer's product over another's. Testing of the assembly at temperature extremes will reveal any unforeseen operating or performance problems; accelerate any drift or damage due to aging; and serve to verify the designer's derating expertise.

B. Improvement Revue

Reliability can be improved by careful design and by continuing measures of derating and prevention.

1. *Quality failure*: quality failures are not related to stress, nor are they time related. Generally, they are related to poor workmanship. In normal

---

[14] Weibull probability paper is available from Team Graph Papers, Box 25A, Tamworth, NH 03886 (tel.: 603-323-8843).

[15] In general this interpretation of Arrhenius' reaction rate model is the basis for almost all of the time-acceleration efforts used in the *aging* process used for the accelerated elimination of infant mortality in electronic, electromechanical, and mechanical parts alike.

circumstances, we will be able to find this type of failure by inspection and remove it from the system. Accelerated aging (burn-in) can be used to reveal potential quality failures that have been missed.

2. *Reliability failure*: reliability failures are stress dependent, such as parts that are substandard due to material deficiencies or parts that have been degraded by poor processing. Usually, this kind of failure is removed by screening.
3. *Wear-out failure*: wear-out failures are time dependent and consist of parts that have been degraded by time or over stressing. Wear-out failures can be eliminated by predicting the average length of time before failure and replacing them when they reach the failure average minus 3 σ.
4. *Design failure*: design failures may be stress related, time dependent, or both. This type of failure can be circumvented by proper derating of components and correct application of parts.

VI. RELIABILITY PREDICTIONS

A. MIL-HDBK-217

Over the years, the Department of Defense (DoD) has gathered and standardized the latest information pertaining to reliability prediction procedures. The DoD has always made this information available to government suppliers and commercial enterprises alike by publishing this information in "MIL-HDBK-217.[16] This handbook details two methods of reliability assessments:

Stress analysis: each component of the assembly is evaluated to determine the effects of the operating conditions; mainly temperature and environment. The generic failure rate of each component is multiplied by factors based on these operating conditions to yield the best estimate for the parts failure rate in the specified environment. The total of all of the individual components' failure rates is used as the failure rate for the entire assembly.

Parts count analysis: in this case, like components are grouped together: all like resistors are grouped; all like capacitors are grouped; all like semiconductors are grouped, and so forth. The number of components in each group is multiplied by that component's generic[17] failure rate and

---

[16] MIL-HDBK-217 is available from COMMANDER, Rome Laboratory, AFSC, ATTN: ERSS, Griffiss Air Force Base, NY 13441-5700. An ancillary reliability source is RAC Reliability Analysis Center, P.O. Box 4700, Rome, NY 13442-4700. The Reliability Analysis Center is a DoD Information Analysis Center sponsored by the Defense Technical Information Center. In addition to electronic reliability information, RAC dispenses design, mechanical, and system software reliability data.

[17] Generic failure rates are furnished for each environmental condition. These rates are based on the component's operation at the expected environmental temperature and 50% of normal load. Based on the author's past experience, the total difference between the assembly's stress analysis failure rate and the failure rate calculated as parts count is always less than 1.5%.

that component's quality factor. The total of all of the component groups' failure rates are added together to be used as the failure rate for the assembly.

B. TR-NWT-000332

More recently, Bell Communications Research, Inc. (Bellcore) has evolved a technical reference, TR-NWT-000332—*Reliability Prediction Procedure for Electronic Equipment.*[18] This document was originally patterned after MIL-HDBK-217B, empirically augmented with data based on the telephone company's long history of reliable operations.

VII. MAINTAINABILITY

A natural outgrowth of the reliability prediction is the maintainability prediction. Once the assessment is made concerning how long a circuit or system will operate (on the average) before it needs to be repaired, we can predict how big the task will be to put it back in operation. From this, we can determine how many and what kinds of spare parts we need and where to keep them (logistics).

If we are smart, we will make it easy to replace the components or subassemblies that are expected to fail most often. Or we might redesign the system to make it more robust.[19] In either event, the maintainability prediction will have been the mechanism that pointed out what was required to make our maintenance task easier and our reliability better.

A. Maintainability Analysis

From Table 1 and the text that follows, the estimated downtime per million hours of the radio frequency (rf) amplifer is 390.9 hours. We can use the basic formula to calculate the operational availability (A_0) for the assembly:

$$A_0 = \frac{\text{Total Hours} - \text{Downtime}}{\text{Total Hours}}$$

$$A_0 = \frac{1{,}000{,}000 - 390.9}{1{,}000{,}000} = 99.96\%$$

The MTBF is 1,000,000/311.94 = 3206 hours.

Also, from Table 1 and the calculated MTBF, it can be estimated that, on the average, once every 3206 hours, the amplifier will be down for 1 hour and 15[20] minutes while it is being repaired.

---

[18] The public contact for this (and other) documents is Bellcore, Customer Services, 60 New England Avenue, DSC 1B-252, Piscataway, NJ 08854-4196 (tel.: 1-800-521-CORE; for foreign calls, use 908-699-5800).

[19] *Robust* is the term used to describe product that is more reliable; i.e., a higher MTBF.

[20] This time to repair or time to restore (TTR) includes the diagnosis time to find the subassembly or component that failed, the time to replace the unit, the time to check out the assembly, and the start-up time needed put the assembly back into operation.

An alternate method of calculating A_0 is:

A_0 = MTBF/(MTBF + MTTR)
A_0 = 3206/(3206 + 1.25)
A_0 = 99.967

B. Availability Guideline

A vital rule of thumb based on the principle of operational availability is: When a subassembly can be replaced within an hour with a spare unit following a failure signal, the failure rate assessed for the subassembly can be divided by two (i.e., the MTBF is considered to be doubled).

One reason we can expect our phone calls to go through the first time we dial is that the phone companies take advantage of this reliability enhancement by providing a failure-detection circuit that illuminates a failure light on the front of any failed unit. A maintenance person quickly replaced the failed unit, which immediately puts the circuit back into operation. This is especially effective when there are banks of similar units. The failed unit can then be repaired by the maintenance person, thus making it available if another system failure occurs.

C. Maintainability Calculations

A hypothetical rf amplifier contains six circuits. The failure rates and average (mean) time to repair the circuit in the event of a failure have been calculated from field experience and maintenance department input.

Table 1 Hypothetical rf Amplifier

Circuit	Failure rate (λ) (1)	Mean time to repair (MTTR) (2)	(λ) × MTTR (3)
1 Chassis	136.08	1.3	176.9
2 Filter/driver	46.05	0.7	32.2
3 Power amplifier	84.01	1.5	126.0
4 Input/output circuit	10.71	0.8	8.6
5 Controller	4.72	1.0	4.7
6 Power supply	30.37	1.4	42.5
Average (mean) value	311.94	—	390.9

The *average* value for column 1 is average number of amplifier failures per million hours of operation. The average value for column 3 is the average number of hours of amplifier downtime per million hours of operation.

The overall mean time to repair for the amplifier cannot be determined by averaging the MTTR column (column 2), because that does not take into account the different number of failure occurrences for each circuit. Instead, the mean time to repair is determined by dividing column 3 by column 1:

$$\frac{390.9}{311.94} = 1.25 \text{ hours}$$

That is, whenever something in the amplier fails, on the average, it will take 1 hour and 15 minutes to repair it and put the amplifier back in operation.

VIII. ANALYSIS OF FAULTS

A. Types of Faults

In general, there are four types of failures caused by faults:

1. *Quality failure*: a quality fault is not usually related to time or stress. In most cases, a potential quality failure is most often removed from assembly by inspection.
2. *Reliability failure*: a reliability fault is stress dependent. It is usually due to early failure and can usually be removed by screening.
3. *Wear-out failure*: a wear-out fault is time dependent. A wear-out fault can be eliminated by replacement of the component before wear out is reached.
4. *Design failure*: a design fault may be time or stress related. This kind of fault can only be avoided by using components that are more robust or by using a fail-safe design.

B. Failure Avoidance

The first three types of failures can usually be avoided. Quality failures have been brought into the parts per million range through intelligent use of statistical methods. Likewise, reliability failures are almost wholly removed when power aging[21] methods are used. Wear-out failures are best eliminated by removing the part before it has a chance to wear out. Probably, automobile tires are the best example of this principle. Tires have a *wear tread* molded in between the *bearing treads*. When the 80% of the bearing tread's thickness has worn off, the wear treads are the same height as the bearing treads. This indicates that it is time to replace that tire, since further wear could reduce the cross section of the tire's body to the point where catastrophic failure (like a blowout) might occur.

C. Bottom Up Fault-Reduction Tools

Two major fault-reduction tools are effective in reducing design failures to their optimum levels. Both are simple "bottom up" failure analyses that start by making an analysis of the lowest level components to investigate how failures of basic components affect the next level of assembly. The failure path is followed up through the assembly until the failure effect of each the lowest level parts on the function of the top assembly is known. Specific bottom up analyses are known as failure modes effect and analysis (FME&A) and failure modes effect and criticality analyses (FME&CA).

---

[21] Power aging to get rid of the effects of infant mortality is sometimes called burn-in, shake and bake, etc.

The FME&A notes the effect of the upward effect of each of the failure mode of each component's failure. In addition to the effect on the top assembly, the probability of the event is also assessed.

The FME&CA operates in much the same way as the FME&A except that all degradation is evaluated according to its criticality. This answers the question as to how serious is the danger to the top assembly when the individual part fails.

Note: Both methods review the components used in the assembly and assess their fitness for use. Risks exposed by this kind of assessment are usually removed or reduced by a change in the design of the assembly such as replacing the part with a more robust component or by adding redundant parts.[22] In other words, both methods assess the probability level of the failure occurrence and look for methods of reducing or eliminating the failure cause entirely.

2. The second is a top down system that investigates a top-level failure by looking at the immediate cause of the system degradation and then following down through the lower tier causes down until it is no longer profitable to continue the investigation.

D. Top Down Fault Reduction

The fault tree analysis is known as a *top down analysis*.[23] This method is essentially a vertical flow chart with special symbols.[24] It uses the same kind of reasoning analysis as the FME&A and FME&CA except that it does not require as much detail, since it addresses a top-level failure and investigates it only until the failure cause is defined and failure reduction methods can be defined and implemented.

---

[22] Adding a redundant part usually indicates that both parts have to fail before the function of the assembly is impaired.

[23] *Top down* means that the top assembly is first analyzed to determine which (and how much) subassemblies contribute to potential failures. Then each subassembly is analyzed to determine which of the subassembly's component are likely contributors to those failures. This method allows the investigator to stop digging when further work shows indicates little chance of finding areas for design improvement.

[24] Fault tree symbols are defined in Figure 2-12 of *Reliability Design Handbook*, Catalog No. RDH 376, IIT Research Institute, 10 W. 35th Street, Chicago, IL 60616.

19
Supplier Surveillance

RAY A. KLOTZ
World Class Quality Consulting Company
Escondido, California

I. INTRODUCTION

Business today requires that a company's quality organization operate almost as though it were a profit and loss function. This has come full circle from the days when quality groups recommended and implemented process improvements to capture cost-reduction methods.

Today, the quality function is required to lead the company into the realm of quality that is expected to fit all sizes of business and directly affect the bottom line. More and more companies, from the garage shop[1] to the multinational combine, have now embraced the ISO 9000 (series) of quality systems. Government agencies as well as professional societies—namely, the American Society for Quality Control[2]—(ASQC) have been busily preparing everyone for the international[3] approach.

---

[1] The term *garage shop* is the vernacular for a shoestring business which often begins or is run from a spare room or a garage. The term has come to mean any new or low-volume business operating more or less out of the founder's hip pocket. (For example, the Apple Computer Co. started as a garage shop.)

[2] The American Society for Quality Control has been instrumental in preparing the American equivalents of the ISO 9000 series. Specifically, these equivalents are: ANSI/ASQC 090; ANSI/ASQC 091; ANSI/ASQC 092; ANSI/ASQC 093; and ANSI/ASQC 094.

[3] Australia, Austria, Belgium, Canada, Denmark, Finland, France, Germany, India, Ireland, The Netherlands, Norway, South Africa, Spain, Switzerland, United Kingdom, Yugoslavia, and the European Community among others are included in those who have prepared or are preparing ISO 9000 equivalent documents.

So where does supplier surveillance come into play? It is pretty well known that auditing, or surveillance, is being urged as an important requirement in the not so gradual transfer from MIL-Q-9858 (and MIL-I-45208) to the ISO 9000 series requirements. The watchword is *prevention*. This movement is being pushed within the defense Contract Management Administration (DCMAO), the U.S. Department of Defense (DoD), and other services to step up to the ISO requirements and away from MIL-Q-9858.

II. COMPARING STANDARDS

Comparisons of MIL-Q-9858 and ISO 9001 are available from many sources and can be used to enhance your understanding of the changes in attitude that will probably be required. In general, the ISO documents include everything that is covered in MIL-Q and MIL-STD-45662 and more. The Department of Defense (DoD) seems to be using an interim strategy of implementing total quality management (TQM) on their way to the goal of a universal quality system.

It is important to understand that the relationship of the contractor and supplier must be strong—almost like a marriage. Like a marriage, there must be complete honesty and it must contain a give and take attitude. Essentially, this is the gray area of resolution that separates true ISO 9000 operation from past MIL-Q-9858 practices.

As a good review of upcoming DoD quality management, we look to the contract management accorded to aerospace and defense suppliers. Thus, we review the two government quality specifications—MIL-Q-9858 and MIL-I-45208—whose precepts have been in force for as long as we can remember. The bulk of this chapter will examine supplier surveillance from the point of view of one of the toughest customers as far as quality is concerned—the U.S. government.

MIL-Q-9858, or MIL-Q, as it is often called, is the military specification that outlines the fundamentals for a quality assurance system. MIL-I-45208, or MIL-I, as it is often called, is the standard for the fundamentals of an inspection system.

Most large suppliers prepare their own quality system standards that often include substantial parts of MIL-I or MIL-Q. These "specs" are usually defined as *meet or exceed* documents.[4]

III. SUPPLIER QUALITY MANUALS

Many suppliers—large and small—supplying both to the government and to general industry—publish their own quality manual.[5] The manual is usually intended to

---

[4] A "meet or exceed" quality document includes a statement such as one of the following: "This document meets or exceeds the requirements of MIL-Q-9858" or "This document meets or exceeds the requirements of MIL-I-45208."

[5] In the past, the Small Business Administration has distributed a generic quality manual, Technical Aid No. 91: *A Tested System for Achieving Quality Control*, to help small suppliers. Software quality manuals which can be customized are also readily available.

outline the company's projected level of quality and tells how the company intends to achieve this level. Under most supplier evaluation programs, the quality manual is one of the major areas to be evaluated to see whether or not the supplier's stated quality expectations meet those of the customer.

Knowledgeable suppliers have often constructed their quality manuals to meet the requirements of MIL-I or MIL-Q.[6] Suppliers to the Department of Defense qualify to MIL-Q—mainly because it is required. But more and more suppliers of off-the-shelf (i.e., commercial) products create quality manuals not only to show prospective customers that they can expect quality merchandise, but also to demonstrate to their own suppliers the level of quality expected of them.

IV. GOVERNMENT QUALITY GUIDELINES

In the late 1980s, the U.S. government began to establish quality guidelines aimed at simplifying the interpretation of government quality specifications. The overall aim was to improve contractor quality performance. The result is that DoD is still evolving from an "inspection" mode, where the integrity of the product required blanket proof via intensive inspection, to a quality *management* mode. Toward this end, the Department of Defense issued a memo entitled *DoD Posture on Quality*[7] (Fig. 1).

The basic premise of the DoD philosophy is simply stated: Every member of every supplier's organization must be trained and oriented to find out what the customer—the Department of Defense, in this case—needs and to participate in such a way that the customer will receive the items and services it's paying for.

V. TOTAL QUALITY MANAGEMENT

The Department of Defense apparently intends to spread the total quality management (TQM) strategy throughout the defense industry. This is evidenced by its support of DoD 5000.51 *Total Quality Management; A Guide for Implementation.* "A Guide" appears to be the key words in the title, because it is DoD's aim to install this change of thinking and as improvements to the (now previous) quality assurance and inspection systems.

Perhaps the Foreword to 5000.51 (Fig. 2) gives the best synopsis of the government's posture. In general, this document seems to be based on the automobile industry's successful supplier quality programs. These include the Ford Motor Company's *Total Quality Excellence*; the General Motor's *Targets for Excellence*; and the Chrysler Corporation's *Quality Excellence.*

---

[6] Most suppliers to the government aim at the MIL-Q-9858 level. A company which has a MIL-Q quality system in place is deemed to be qualified as a MIL-I-45208 contractor.

[7] The DoD Posture on Quality memo was issued March 30, 1988, by Frank Carlucci, Secretary of Defense. This is the first government document to sanction the DoD version of total quality management.

DoD POSTURE ON QUALITY

- Quality is absolutely vital to our defense, and requires a commitment to continuous Improvement by all DoD personnel.
- A quality and productivity oriented Defense Industry with its underlying industrial base is the key to our ability to maintain a superior level of readiness.
- Sustained DoD wide emphasis and concern with respect to high quality and productivity must be an integral part of our daily activities.
- Quality Improvement is a key to productivity Improvement and must be pursued with the necessary resources to produce tangible benefits.
- Technology, being one of our greatest assets, must be widely used to Improve continuously the quality of Defense systems, equipment and services.
- Emphasis must change from relying on Inspection, to designing and building quality into the process and product.
- Quality must be a key element of competition.
- Acquisition strategies must include requirements for continuous improvement of quality and reduced ownership costs.
- Managers and personnel at all levels must take responsibility for the quality of their efforts.
- Competent, dedicated employees make the greatest contributions to quality and productivity. They must be recognized and rewarded accordingly.
- Quality concepts must be ingrained throughout every organization with the proper training at each level, starting with top management.
- Principles of quality improvement must involve all personnel and products, including the generation of products in paper and data form.

Figure 1 Secretary of Defense Frank Carlucci's memo.

Specifically, TQM principles encompass:

- *Commitment*: Company management is to provide constant support of the TQM principle through example and guidance.
- *Forethought*: TQM strategy is one of continuous process improvement[8] involving everyone in the organization—managers and workers alike.
- *Posture*: The TQM management model for continuous process improvement focuses on providing leadership, training, and motivation to bring about that improvement—internally and with suppliers.

---

[8] Continuous process improvement is not restricted to production work. Management administration and other ancillary functions are also expected to continue to do better.

FOREWORD

Government and Industry have come to understand that previously acceptable norms of goods and services are no longer acceptable.

Customer satisfaction, reliability, productivity, costs, and for Industry, market share, profitability, and even survival are directly affected by the quality of an organization's products and performance.

Therefore it becomes essential to develop attitudes and systems—at all levels of the organization—that promote and implement continuous improvement of procedures, processes, products, and services. Those attitudes and systems are the focus of Total Quality Management (TQM).

This guide supports the implementation of DoD Directive 5000.51 on Total Quality Management and is designed to provide a basic understanding of TQM. Executives and managers may find the guide to be particularly useful in this regard. Also its use is encouraged to support training.

The implementation of TQM in any organization must take into consideration such factors as the organization's unique product or service, culture, customers, level of knowledge, and experience.

This guide, therefore, must be tailored to its specific application. The guide provides one approach, but others are possible. Innovative approaches are encouraged.

This guide is not to be contractually mandated (reference: DoD Directive 5000.43, Acquisition Streamlining).

If you have comments to contribute to the continuous improvement of this document, please forward them to:

Office of the Deputy Assistant Secretary of Defense for Total Quality Management
OASD(P&L) TQM
Pentagon, Washington, DC 20301

Figure 2 Department of Defense acquisition guidelines.

- *Essentials*: TQM includes additional guidance on the use of tools and techniques required to bring about continuous process improvement.

TQM and DoD acquisition systems are not wholly compatible. Thus, it is imperative that TQM concepts, practices, tools, and techniques be rapidly integrated.

VI. DoD QUALITY LEVELS

Too often in the past, it has been easier for the Department of Defense to ask a prospective contractor to perform to a higher level quality system than needed; at least in part because the higher level allowed for more involvement by the Defense Contract Management Administration Office—even when it wasn't needed. Needless to say, this involvement was an added aggravation—and cost—to the contractor; and in the long run, to the government.

The following is a guide for determining the general quality level required for various products.

Level 1: The highest level of DoD quality. In everyday language this could be called a planned and systematic pattern of actions necessary to provide adequate confidence that the item or product conforms to established technical requirements. To prepare for this level of quality, the supplier must have a quality plan (product assurance manual) and is usually audited to make sure that the company is following the plan. In government talk, this level meets MIL-Q-9858A Quality Program Requirements.

Level 2: Level 2 also requires a written plan. It is almost as stringent as Level 1,[9] except that its stated aim is to see that the product conforms to the drawings and all of the provisions of the contract are met in general. A supplier can accept a Level 2 contract without question when the supplier already meets the requirements of Level 1. This leads many suppliers to aim their quality systems at Level 1, since it is little more bother than Level 2 and they feel that it makes them more versatile. This level of quality corresponds with MIL-I-45208A inspection system requirements.

Level 3: For Level 3 quality, the supplier is required to "maintain an inspection system acceptable to the customer." Customers reserve the right to witness any inspection or test if they want, or they can perform their own inspection or test. This last reservation is usually exercised only when customers have doubts about their suppliers or when they want correlation between their own test equipment and the suppliers' test gear.

Level 4: Quality Level 4 simply makes the supplier responsible for the inspection and test of products before offering them to the customer. Most of the time, evidence of this is the inspector's acceptance stamp on the product or the accompanying paperwork.

Level 5: At Level 5, there is no specific quality requirement in the contract. Instead, the customer relies on the supplier's internal control of quality. Product made at Level 5 is usually designated at "off-the-shelf."

VII. PRODUCT CATEGORIES

The following are generally accepted product category definitions:

Commercial. Products made using commercial drawings and/or standards. Generally listed in commercial catalogs.

Common. Products that have multiple applications—e.g., screws.

Complex. Products that have characteristics that can not all be measured in the finished item.

---

[9] The major difference between Level 1 and Level 2 is that Level 1 requires a quality cost system and Level 2 does not.

Critical. Products whose failure could injure people or jeopardize entire tasks.

Customer PQA (Customer Procurement Quality Assurance). The customer determines that the quality and quality requirements are met.

Military-Federal. Products made using federal or military drawings and/or specifications.

Noncomplex. Products that can be completely tested by relatively simple end item testing.

Off-the-Shelf. Products that are produced and placed in stock before sale of the item.

Peculiar. Products that have only one application.

Table 1 expands on the discussion of quality levels.

VIII. SUPPLIER EVALUATION CHECKLISTS

Much effort has gone into the development of quality checklists. Government contractors and suppliers pore over DoD requirements to visualize all of the features that need to be present in an acceptable quality system.

Table 1 Simplified DoD Matrix: DoD Quality Level Required

Technical description	Kind of item	Application	Quality level
Commercial	Noncomplex	Noncritical Common	5
Commercial	Noncomplex	Noncritical Peculiar	5
Commercial	Noncomplex	Critical	3
Commercial	Complex	Noncritical Common	5
Commercial	Complex	Noncritical Peculiar	3
Commercial	Complex	Critical	2
Military-federal	Noncomplex	Noncritical Common	3
Military-federal	Noncomplex	Noncritical Peculiar	3
Military-federal	Noncomplex	Critical	2
Military-federal	Complex	Noncritical Common	3
Military-federal	Complex	Noncritical Peculiar	2
Military-federal	Complex	Critical	1
Off-the-shelf	All	Noncritical	5
Off-the-shelf	All	Critical	3

This table is a guide for selecting the DoD quality level. Special circumstances may warrant quality levels greater or lesser than indicated.

The contractors usually put this in the form of an audit checklist so they can make sure that their suppliers can furnish quality material to them and maintain competitiveness.

Lower tier suppliers develop checklists not only for use in the audit of their suppliers but also to determine all of the questions that might be asked of them by their customer so they will know how to answer them and how to use the data and information gathered.

DoD has long known that such efforts go a long way toward letting the suppliers know what is expected of them.

The following self-audit checklist was originally prepared by a major aerospace company for distribution to all of its suppliers. It might appear that *self*-audit would be doomed to self-destruct, because the element of surprise is lost. The intention, however, is not to "catch" suppliers but to make them better suppliers. Enlightened customers have long pushed for enlightened suppliers. Indeed, Secretary of Defense Frank Carlucci's TQM implementation is aimed solely at speeding up the improvement process and making that speed up an ongoing process.[10]

Following the self-audit checklist is a listing of common deficiencies that are regularly found during audits.

IX. SELF-AUDIT CHECKLIST

Instructions: Enter *1*, *2*, *3*, or *N/A* in the space provided at the left of each of the elements below according to the following schedule:

- *1* means a written policy is in place and the element is under complete control in your facility.
- *2* means that the element is being adequately controlled but documentation or implementation may be somewhat lacking.
- *3* means the element is out of control.
- *N/A* indicates that this element is not pertinent in your case.

____ Segregation and control of incoming materials
____ Indication of inspection status on all materials
____ Inspection status entries on process traveler documents
____ Adherence to documented inspection instructions and test procedures
____ Compatibility of shop drawings and specifications with shop orders
____ Unauthorized red-lining of acceptance criteria
____ Adequacy of work order documentation of material in-process
____ Recording of inspection and test results on proper forms (i.e., not scratch paper)
____ Accuracy of entries on inspection and test tags and forms
____ Availability of acceptance standards and procedures in working areas
____ Validity of written procedural coverage of routine activities
____ Currency of parts lists and technical documents in use

[10] We can add the Baldrige Award Principles, ISO 9000 precepts, and Benchmarking, among other things, to this list.

____ Validity of special process *personnel* certifications
____ Validity of special *process* certifications
____ Availability of calibration traceability to bureau of standards
____ Validity of calibration decals
____ Adequacy of protection of inspection and test equipment
____ Maintenance of procedure and standards manuals
____ Availability of contractually imposed quality program documentation
____ Currency of engineering and production change orders
____ Enforcement of material preservation & handling standards
____ Segregation and identification of non conforming materials
____ Conduct of training programs
____ Imposition of quality requirements on procurement orders
____ Timeliness of remedial action to prevent recurrence of deficiencies
____ Adherence to statistical sampling plans
____ Timeliness and accuracy of control charts
____ Availability of quality rejection records
____ Identity and storage of limited life items
____ Adequacy of final inspection and shipping inspection

X. COMMON DEFICIENCIES FOUND IN THE EVALUATION OF QUALITY PROGRAMS

Note: This is an empirical government checklist originally formulated by the Defense Contract Department. It is an excellent checklist for the auditor and the *auditee* alike.

A. Management Organization

The scope of the responsibility and authority of the quality organization cannot be determined.

Quality decisions are being compromised by the production or engineering departments.

Quality personnel and operations do not agree with the documented organization chart.

Operating departments do not adhere to procedures and override quality decisions.

B. Receiving

Inspection and tests are being performed without written instructions.

Inspection and test instructions with inadequate information to perform the operation, or personnel don't follow established instructions.

Unauthorized, marked up, illegible, or defaced prints or documents in the operating areas.

Inspection and test equipment overdue for calibration.

Inspection and test equipment is improperly tagged as to use.

Inspection and test equipment and personally owned tools are not in the calibration system.

Sampling inspection plans are not adequate.
Sampling sizes taken vary from what table indicates.
Inspection and test set-ups do not agree with procedures.
Obsolete drawings in the inspection folders, files and in working area.

C. Stores

No inspection status identification on stored items.
Procedure for controlling shelf-life items expiration not followed for limited-life items.
Overage materials in stores.
Material not physically protected from damage or corrosion.
Raw material not fully identified.
Rejected, obsolete, or nonconforming materials in storeroom area.
Chemical and physical test certifications not available or incomplete.
Materials awaiting test and rejected material not separated from accepted material.
Material in stores does not meet inspection criteria.
The first in, first out (FIFO) principle of stores handling is not adhered to.

D. Manufacturing In-Process Controls

Work instructions are not available.
Work instructions are inadequate to perform the operation.
Inspection instructions are not available.
Inspection instructions are inadequate or incomplete to perform the inspection.
Tools and equipment in use are not authorized or are misused.
Inspection stages are inadequate, skipped, or omitted.
Test equipment is not included in the calibration system.
Inspection and test equipment overdue for calibration.
Operators are not certified.
Obsolete drawings or documents are in use in the manufacturing operation.
Plating solutions do not meet specifications.
Unauthorized changes appear on drawings.
Unauthorized changes are found in inspection instructions.
Unauthorized changes are found in manufacturing work instructions.
Motors and gauges that control processes are not calibrated.

E. Final Assembly and Test

Assembly and inspection instructions are not provided.
Assembly and inspection instructions are inadequate to perform the task.
Test equipment in use does not comply with that specified by the test procedure.
Test procedures are not adequate to test the product.
Inspectors do not have sufficient criteria for inspection judgments.
Unauthorized changes have been made to cost procedures.

Assembly operators and test personnel do not follow approved instructions.
Test equipment is overdue for calibration.
Test equipment is not serviced within the calibration due dates.

F. Nonconforming Material

Material review area is not adequately segregated from the acceptable material.
Operating procedures are not available.
There is no authorized personnel list.
Recurrent deficiencies are not adequately recorded.
Standard rework and repair procedures and instructions are not available.
Standard rework and repair procedures and instructions are not adequate.
Rework and repair inspection instructions are not available.
Rework and repair inspection instructions are not adequate.
Decisions to use discrepant material are made by the supplier beyond his or her delegated authority.

G. Preservation, Packaging, Packing, Storage, and Shipping

Packaging or handling procedures or instructions are nonexistent.
Work instructions are incomplete.
Work instructions do not comply with contractual requirements.
There are no inspection instructions.
Inspection instructions are incomplete.
Preservation methods are inadequate.
Rust or corrosion can be detected on stored items.
Wrong packaging material is used.
Containers do not specify special environment.
Provisions for protecting the item during storage are inadequate.

H. Procurement and Supplier Control

Quality assurance does not review the purchase request or orders.
Quality assurance review of purchase request or orders has been bypassed.
Purchases are made from suppliers not on the approved supplier list without authorization by the quality control manager.
An approved supplier list is not generated.
Purchasing does not have a current approved supplier list.
No evidence of corrective action is shown in the file on poor-quality suppliers.
Quality requirements are not included on some purchase orders.
Purchase order does not include a revision or indicate the correct print revision.
Inadequate quality requirements are placed in purchase orders.

I. Change Control

The drawing change control system does not provide for the orderly disposition and removal of obsolete drawings.

The document change control system does not provide for the disposition and removal of obsolete documents.
Obsolete drawings are found in the operating area.
Obsolete documents are found in the operating area.
The system does not provide for the quality function to review necessary change requests before the fact.
The supplier makes changes which have not been authorized by the purchasing activity.
The supplier does not maintain a record of print recipients.
The supplier does not maintain a record of document recipients.

J. The Corrective Action Program

There is no established program corrective action system.
There is no established product corrective action system.
The program corrective action system is not fully documented.
The product corrective action system is not fully documented.
Corrective action due dates are not required.
Corrective action due dates are often delinquent.
The action taken does not always correct the problem, it merely corrects the specific instance.
The supplier does not have an adequate follow-up system for delinquent or unacceptable replies to corrective action requests.
Program corrective action does not involve the entire organization.

K. Quality Assurance

Quality planning is not adequate.
There is no adequate program for defects analysis.
Defect status reports are not made available to operating personnel.
Quality cost data are not collected.
Quality cost data that are collected are not complete.
Quality does not review all change orders.
Quality's review of change is made after the fact.
The audit function is not performed.
Audits are performed only on quality functions.
Replies to quality audit reports do not go to management higher than the quality director.

L. Measurement and Test Equipment Standards

All test equipment is not in the calibration system.
Expired calibration dates are found on test equipment in use.
Calibration instructions are not available.
Calibration instructions are inadequate to perform the required task.
Complete calibration data are not maintained.
Calibration intervals are revised upward without justification.
Instruments are in use which have no calibration indications.

Instruments are in use for which the calibration laboratory has no record.
Recorded laboratory temperature and humility limits do not meet those specified.
The test equipment does not adequately measure the product's characteristics.
Similar items of test equipment do not indicate the same result.

M. Personnel Training and Certification

Classroom training courses are not provided.
Personnel are performing special operations beyond their certification date or capability.
Personnel are performing special operations without adequate certification.
Training programs do not provide a quantitative means of determining the proficiency of the trainee.
Training courses are not documented as to exact content.

Note: When this Common Deficiencies List is used to prepare a checklist for any particular vendor or for self-audit, those items which are not applicable can be so marked or simply dropped from the list. Another source for checklist material is Handbook 50, *Evaluation of a Contractor's Quality Program* or Handbook 51, *Evaluation of a Contractor's Inspection System* (these handbooks extend MIL-Q-9858A and MIL-I-45208A, respectively).

FURTHER READING

Deming, W. E., *Out of the Crisis*, Massachusetts Institute of Technology, Center for Advanced Engineering Study, Cambridge, MA, 1986.
Feigenbaum, A. V., *Total Quality Control*, McGraw-Hill, New York, 1983.
Gavin, D. A., *Managing Quality*, Free Press, New York, 1988.
Harrington, H. J., *The Improvement Process*, McGraw-Hill, New York, 1987.
Imai, M., *Kaizen*, Random House, New York, 1986.
Ishikawa, K., *What Is Total Quality Control*? Prentice-Hall, Englewood Cliffs, NJ, 1985.
Juran, J. M., *Managerial Breakthrough*, McGraw-Hill, New York, 1964.
Managing Quality and Productivity in Aerospace and Defense, Defense Systems Management College, Fort Belvoir, VA, 1988.
Scherkenbach, W., *The Deming Route to Quality and Productivity*, Cee Press, Washington, DC, 1986.
Schonberger, R. J., *Japanese Manufacturing Techniques: Nine Hidden Lessons in Simplicity*, Free Press, New York, 1982.
Townsend, P. L., *Commit to Quality*, Wiley, New York, 1986.

20
Managing Color

C. S. McCAMY
Consultant in Color Science
Wappinger Falls, New York

I. INTRODUCTION

It is very likely that human color vision evolved to a high degree because it was useful in the struggle for survival. Today, we use color not only to judge the quality of foods and other natural products but also to satisfy our innate love of color for its own sake. Thus color is a valuable physical property often having more commercial importance than linear dimensions, area, volume, or mass. Good managers use good design, process control, and quality assurance as management tools to control physical variables as part of the effort to maximize their return on investment. Color is managed in much the same way as other physical properties. The management of color demands an understanding of the nature of color, the way it is measured, and the methods of maximizing productivity and minimizing cost.

II. THE LANGUAGE OF COLOR

The normal human being can discriminate over 10 million different colors. Specifying colors in such a vast assortment is made possible by the fact that we perceive or can generate orderly relationships among colors.

The most generally recognized color-order system is that defined by A. H. Munsell in 1905. Munsell was a commercial artist who had suffered from the lack of a precise way of describing color. He devised a method and published it in his small booklet *A Color Notation* (1). He observed that the colors of surfaces could be described by three attributes. The first of these he called *hue*. Hue designates the attribute of color we characterize by such names as red, yellow, green, blue, or purple. Munsell chose the hue names just given as the five principal hues. He then interpolated five hues between these. He designated them yellow-red, green-yellow, blue-green, purple-blue, and red-purple. We may blend paints or we may merely imagine colors starting with red and gradually becoming yellower until the color is yellow, then blending to green, blue, purple, and finally back to red. It is in the nature of human color vision that this hue series returns to its starting point; that is, we may construct a hue circle. Munsell proposed to divide that circle into 100 equal parts. Thus he assigned 10 such parts to each of the 10 hues he had named.

The second attribute of color is *value*. The value of a color merely describes how light or dark it is. Given two surfaces, an observer can decide whether they are of the same lightness or whether one is lighter or darker than the other. Munsell proposed that we imagine a scale of lightness ranging from the blackest possible black to the lightest possible white and that we divide that scale into 10 parts that appeared to be equal. This established his value scale.

The third attribute of color was called *chroma*. Chroma is the difference from gray. We may imagine a very pure red color having the same value as a gray. We may then imagine the addition of very small amounts of the red to the gray, causing a gradual departure from the gray. The more red that would be added, the greater would be the departure. In this way, we can establish a chroma dimension. The chroma increases toward the pure chromatic color; red in the example given.

These three attributes of color may be regarded as the three dimensions of a color space, as illustrated in Figure 1. The space is represented by a cylindrical coordinate system. Value is shown along the vertical axis; hue is represented by the angular displacement around the cylinder, in much the same way that the equator of the earth surrounds the earth; and chroma is measured outward from the value axis. Black is at the bottom of this space and white is at the top. All neutral colors (i.e., the various grays) are distributed between black and white along the axis. The hues are arranged radially about the axis and chroma increases laterally.

Munsell developed a notation for color. He used the notation H for hue, V for value, and C for chroma. He indicated the value, to the necessary number of decimal places, on the scale of 10 mentioned earlier. The chroma scale was visually divided so that vermilion (a bright red pigment consisting of mercuric sulfide) had a chroma of 10. The hue was designated either by the scale of 100 for the whole

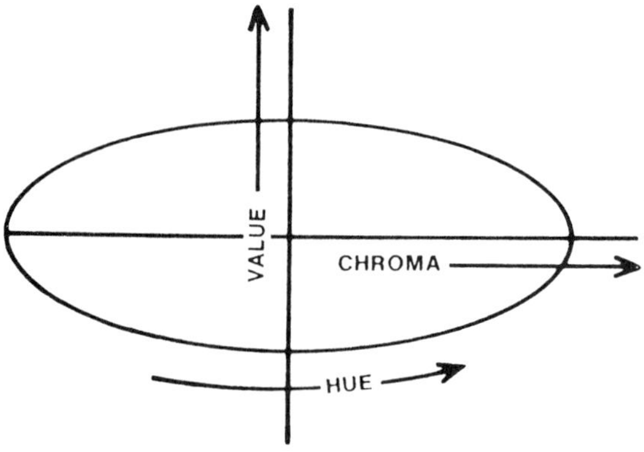

Figure 1 Diagram of the three-dimensional color space defined by Munsell as hue, value, and chroma.

circle or, more often, by the hue and the step on a 10-step scale within the hue. The hues were symbolized by the initial letters of the names of the hues: R, YR, Y, GY, G, BG, B, PB, P, and RP. His notation for the red at the center of the 10 steps assigned to red was "5R." He annotated the hue, value, and chroma in the form H V/C. The notation "7.5R 4/8" meant that the hue was seven and a half units toward yellow in the hue region assigned to red, the value was four tenths of the way from black to white, and the chroma was eight steps away from gray.

In 1918, Munsell founded the Munsell Color Company to provide physical standards in the form of painted-paper "chips." This operation, now a part of the Macbeth Division of Kollmorgen Corporation, is to this day the principal supplier of such standards to colorists in business, science, and industry. Each standard is identified by a Munsell notation.

There have been a number of dictionaries of color that listed the names of colors and associated them with printed color patches. The colors were defined by exhibiting a sample. Very often, the names were taken from natural objects such as "rose," "peach," and "lemon yellow." Sometimes there was a reference to the pigment or dye used. Very many color names are of unknown origin and many have little apparent connection with the color.

The Inter-Society Color Council and the National Bureau of Standards (NBS; now National Institute of Standards and Technology, NIST) developed a simplified language of color based on the Munsell system. The Munsell space was divided into regions bounded by specified values of hue, value, and chroma. Each of these was given a name. The names were the Munsell hue names, or such designators as pink or brown, as appropriate. A standard set of adjectives was adopted. These included pale, dark, light, moderate, brilliant, strong, deep, and vivid. The National Bureau of Standards published *Color, Universal Language and Dictionary of Names*, NBS Special Publication 440 (2). With this reference book, a standard name can be chosen for any Munsell notation. Conversely, the Munsell notation can be determined if the general color name is known. In addition, all of the major previous color dictionaries were consulted and the standard names were given for all the names previously listed. (The list is worth perusal for amusement's sake alone. We find "dream fluff," "happy day," "French nude," "Paris mud," "heart's desire," "pearly gates," "wafted feather," and "intimate mood," together with the more prosaic "battleship gray.")

To permit visualization of the 267 different colors for which standard names were chosen, the National Bureau of Standards issued the ISCC-NBS Centroid Color Charts. These are color charts showing painted paper samples of the color at the centroid of each of the regions of Munsell space that has been given a standard color name. This publication is a supplement to NBS Special Publication 440, referred to as Standard Reference Sample 2107 (3).

The Munsell system provides the best-known basis for visual color-tolerance standards. The ideal color is specified by the Munsell notation and the acceptable tolerances are then given in terms of deviations in hue, value, and chroma.

In some cases, a colored product may vary in hue only, in value only, or in chroma only. Such a variation is one dimensional. In the case of a product varying in value only, we need only one standard representing the ideal color, one for the lighter limit, and one the darker limit. We may encounter products that vary in hue, value, and chroma but find that these variations are interrelated in such a way that the variation is essentially along a single line, possibly a curved line, in Munsell space. In such cases, it is still possible to use only three standards; one for the ideal, one for the variation in one direction, and one for the variation in the other direction. This is regarded as a one-dimensional color-tolerance set even though the colors representing the tolerances differ from the ideal in all three of the Munsell dimensions.

The colors of some products vary independently in two dimensions, such as hue and value. In such cases, the color tolerance set must include a desired color, two standards for the variation in one dimension, and two standards for the variation in the second dimension. When there is a two-dimensional variation, it is generally advantageous to set the tolerances on the basis of the Munsell dimensions of hue, value, or chroma. These dimensions are visual in nature and are more readily communicated, visualized, and understood than dimensions made up of combinations of the Munsell dimensions.

When the color may vary in any way about the desired color, it is necessary to have a three-dimensional color-tolerance set. A three-dimensional color-tolerance set has the desired color, two tolerance standards for the variation in hue, two for value, and two for chroma. These colored standards are affixed to a card having openings so that any of the standards can be compared directly to the underlying surface.

Even when colors are measured and tolerances are specified numerically, it is useful to have a color-tolerance set made. Then the color tolerances can be shown to designers, formulators, managers, customers, and others involved in the business. If, on seeing the tolerances, someone thinks they are not right, they can be redefined. Numerical tolerances are not so directly comprehended.

III. LIGHT AND COLOR

When we compare colors, we must control the illumination if we are to make reliable and consistent judgments. Suppose that we compare two paints, A and B, in yellow light. We might find that they match and their color is kelly green. We might compare two other paints, C and D, and find that they match and they are peach colored. Viewing the same four paints in white light, we may find that A is blue, B is green, C is pink, and D is yellow. Although most people have a vague notion that colored things look different under different illuminants, it is not generally known that two colors can match under one illuminant and not match under another. When two colors match under one illuminant but fail to match under another, it is because they reflect light of various wavelengths differently. We say that they are a metameric match under the first illuminant, or we refer to them as metamers. The phenomenon of matching under one illuminant but not another is referred to as metamerism. If color judgments are to be consistent from time to time and place to place, they must be made under standard illumination.

In 1666, Sir Isaac Newton did a famous experiment that revealed the relationship between light and color. A beam of white light, in his case sunlight, was passed through a triangular prism. The beam was deviated to a different direction and at the same time dispersed into a spectrum. He observed that the spectrum had colors, which he called violet, indigo, blue, green, yellow, orange, and red, in that order.

He observed that all these hues were components of white light. He then passed the colored spectrum light through a prism that deviated the light back in the original direction. When he did so, the components of the spectrum recombined to make white light. Thus he showed that white light could be separated into component colors and that the component colors could be recombined to make white light.

Suppose that we let the spectrum light fall on a white surface and we place colored glass in the beam. When red glass is placed in the beam, it transmits the red light but absorbs the blue, green, and much of the yellow. When blue glass is placed in the beam, it absorbs the yellow light but transmits the blue light. When green glass is placed in the beam, it absorbs the blue, yellow, and red light but transmits the green light. When yellow glass is placed in the beam, it transmits the yellow and red but absorbs the blue. On the basis of such observations, Newton concluded that the colors of objects are caused by the fact that the objects reflect or transmit more or less light of different parts of the spectrum.

Many years later, it was found that light is an electromagnetic wave phenomenon. A light beam is a series of waves of electric and magnetic fields rippling through space. The distance from the top of one wave to the top of the next is called the wavelength. In the light of the visible spectrum, the wavelength increases from the shortest wavelength for violet light to the longest wavelength for red light. It is this difference in wavelength that causes the light of different colors to leave the prism at different angles. These wavelengths are extremely short distances, so they are measured in nanometers. (About 25 million nanometers equals 1 in.) The visible spectrum has wavelengths ranging from 380 nm at the violet end to 780 nm at the red end.

The International Commission on Illumination[1] has standardized illuminants for color science. CIE Illuminant A is an ordinary 100-watt tungsten lamp. There are many CIE fluorescent illuminants, designated F1, F2, etc., and several kinds of daylight. Daylight has various phases, from the reddish light of sunrise to the white light of noon, and back to the reddish light of sunset. The various phases of daylight nearly match the colors of the inside of a glowing furnace at various temperatures. Thus, the phases of daylight can be specified very simply by specifying a temperature on the Kelvin scale, known as the "correlated color temperature." The CIE selected the temperatures 5500 K, 6500 K, and 7500 K and uses the symbols D_{55}, D_{65}, and D_{75} for these phases of daylight. D_{55} is direct sunlight, D_{65} is average daylight, and D_{75} is light from a lightly overcast north sky (in the Northern Hemisphere). The CIE recommends that D_{65} be used unless there is a reason to do otherwise. The CIE specified these illuminants by giving the relative amount of light in each small wavelength interval across the spectrum based on many measurements of natural daylight.

Natural daylight varies with cloud conditions, altitude, and other atmospheric conditions from day to day and even from minute to minute. Natural daylight is not available at night nor in interior rooms of buildings. For these reasons, in 1915,

---

[1] The International Commission on Illumination is generally referred to as the CIE, from the initials of its name in French: Commission Internationale de l'Eclairage.

an illuminating engineer, Norman Macbeth, founded a company to make equipment to provide artificial daylight for judging color. It is an ongoing business to this day. A lighting booth has various kinds of standard lighting in it and switches to select the desired kind. Most booths today have artificial daylight, incandescent lamplight, and cool white fluorescent light. All are important in business, science, industry, and art. It is desirable to have colors match under all three kinds of light.

Color judgments must be made by qualified observers. About one man in 20 has defective color vision and would be considered color-blind. Only about half that proportion of women have this defect. An ophthalmologist can determine whether or not a person has normal color vision. A more complete test is the Farnsworth-Munsell 100-Hue Test, available from the Munsell Department of Macbeth Division, Kollmorgen Corp., Newburgh, NY. Even among so-called normal observers, there is as much variation in perceived color as an individual might find when changing illumination from incandescent lamplight to daylight.

There are many physical and psychological factors involved in making judgments of small color differences. The conditions of observation have been standardized by the American Society for Testing and Materials (ASTM) (4).

Sometimes we see objects so brightly colored that they appear to glow. They appear to be reflecting more light than the amount of light shining on them. This appearance is usually caused by fluorescence. Fluorescent materials absorb energy in the ultraviolet or short-wavelength visible region and then reemit it in the longer wavelengths in the visible spectrum. Viewing booths have ultraviolet lamps and a switch so that the ultraviolet can be turned on or off. In this way, one can readily see whether or not a surface is noticeably fluorescent.

IV. THE EYE

A light source may be characterized by the amount of light emitted in each part of the spectrum. When the light is transmitted through a material or reflected from a surface, the selective nature of the transmission or reflection determines the kind of light going to the eye. The color perceived depends, of course, on the nature of the human visual system.

The human eye is insensitive to radiant energy in the ultraviolet region of the spectrum at wavelengths shorter than about 350 nm. At these short wavelengths, the sensitivity of the eye begins to increase with increasing wavelength. It reaches a maximum at 555 nm, and then it falls off until we reach the red end of the visible spectrum. Beyond that is the infrared region, to which our eyes are not sensitive. This relationship of sensitivity to wavelength is called the human luminous efficiency function.

If we have three beams of light colored blue, green, and red, and we can mix these three kinds of light in any desired proportion, we can obtain light of any hue. The blue and green lights combine to give all the hues from blue through blue-green to green; green and red light combine to give the hues from green to green-yellow to yellow to yellow-red or orange and finally, to red. Red and blue combine to give all the various purples. Mixing the third color with a mixture of the other two reduces the chroma. The three colored lights may be mixed in the right proportions to make white light. This process is called additive mixture, because we start with darkness and add light until eventually, when all three are mixed in ample proportions, we have white light. The red, green, and blue lights are called the "additive primaries." Incidentally, the face of a color TV tube has numerous small areas of phosphors, the phosphors being of three kinds, emitting blue, green, and red light. Electron guns excite these dots in various proportions to produce all the desired colors. In color photography and color printing, dyes and inks are used to absorb light that would otherwise be reflected from the white paper. A dye that absorbs red light is bluish-green, a color called "cyan." One that absorbs green is a purplish color called "magenta," and one that absorbs blue is yellow. Because these dyes absorb rather than add light, yellow, magenta, and cyan are called "subtractive primaries."

In 1801, Thomas Young observed that if we can get all hues by mixing the three primary colors of light, the human eye must contain three kinds of receptors. Researchers have now been able to identify three spectral sensitivity bands for the eye. Modern research has shown that there is very much more to it than that, but the basic idea is right.

The facts of color mixture and this insight into the nature of color vision suggest a method of measuring color. Suppose that we have a white screen divided by a black partition, so that we may illuminate the two halves of the screen with different light sources. Let one half be illuminated by blue, green, and red lights. Let us have some provision for adjusting the amounts of these three independently and some means of knowing how much of each is used. Let the other half of the screen be illuminated by light of some unknown color. We have only to adjust the three colored lights until we obtain a visual match. Then we record the amounts of the three lights required and we have characterized the unknown color by three numbers. When this is done, we find that there are some colors that cannot be matched with the given three colored lights. However, in all cases, a match can be obtained if one of the colored lights can be added to the other side of the screen. When we do that, we regard that amount of light as a negative quantity. If this sort of experiment is set up in various places with different red, green, and blue sources, we must expect to get different sets of three numbers for the same unknown light source. If enough unknowns are measured on two instruments, a mathematical relationship can be found for predicting the values that will be obtained on one instrument once they are known for the other. This is known as a transformation.

To provide a basis for obtaining the same numbers in different laboratories, the CIE standardized three primaries by stating the amounts of each required to match the light of the spectrum at each wavelength. The red primary was called $\bar{x}$ the green $\bar{y}$ and the blue $\bar{z}$. The $\bar{x}$ function has two lobes. These three functions are called *CIE spectral tristimulus values* or *CIE color-matching functions.*

V. COLOR MEASUREMENT

Suppose that we have a standard light source, and the light is reflected from a surface to be measured. Let that light pass through one of four different filters to a photosensitive sensor. Let each filter and sensor combination be designed so that its spectral sensitivity is equivalent to one of the tristimulus value curves standardized by the CIE. (We need four instead of three filters, because the $\bar{x}$ function has two lobes. The values measured with these two filters must be combined.) The three values resulting from measurements through these filters characterize the color of the reflecting surface. Such a device is called a filter colorimeter. This illustrates the fundamental principle of all color measurement.

This system is very simple, but it is not without its problems. Such devices have been built, but it is very difficult to design a light source and appropriate filters to produce a standard illuminant. Similarly, it is difficult to design filter and sensor combinations that closely approximate the desired sensitivity functions. Lamps, filters, and photosensors have notorious variation from one lot to the next, so even if an instrument can be designed and a production run is satisfactory, variation in components may make the next run unsatisfactory. The components may change with time.

We get around these problems by setting up a system to measure the amount of light reflected in each of a number of narrow wavelength intervals. A light source is used to illuminate the specimen to be measured. The light reflected (or transmitted) from the specimen is directed to a prism or grating that disperses the light into a spectrum. The amount of light reflected in each narrow wavelength interval is directly compared to the amount of light reflected from a standard white surface. In this way, the spectral emittance of the standard lamp does not have to match the CIE curve exactly. There are no filters in the receiver, and the variation in spectral sensitivity of the photosensor is canceled out when we make such ratio measurements. A device that does this is called a spectrophotometer. The spectrophotometer provides a curve or table of values showing the percentage of light reflected by the specimen in each wavelength interval in the spectrum. This percentage is called the spectral reflectance factor.

Once the spectral reflectance factor of the specimen is known, it is possible to compute the three values that would have been obtained with a perfect filter colorimeter. These three numbers are called tristimulus values and are given the symbols X, Y, and Z. This computation takes into account the nature of the illuminant in which we are interested, the spectral reflectance (or transmittance) factor of the specimen which was measured with a spectrophotometer, and the spectral sensitivity of the standard observer, as represented by CIE spectral tristimulus values $\bar{x}$, $\bar{y}$, and $\bar{z}$. These three tristimulus values X, Y, and Z represent the color of the specimen under the desired illuminant. Since $\bar{y}$ was transformed to match the sensitivity of the eye to light of various wavelengths, the tristimulus value Y is a direct measure of the lightness of the color.

For many purposes, we are interested in the chromatic nature of a color without regard for its lightness. We compute two numbers: $x = X/X + Y + Z$ and $y = Y/X + Y + Z$, called *chromaticities.* If we plot y vs. x, we call them *chromaticity coordinates.* Such a plot is shown in Figure 2.

A plot of y with respect to x is called a CIE chromaticity diagram. On this diagram, we may plot the colors of the light of each of the wavelengths in the spectrum. We find that the points plot along a horseshoe-shaped curved line. This is known as the spectrum locus. The chromaticity diagram has a very interesting property: If we have lights represented by two points on this diagram and we mix these two lights in all proportions, we find that the chromaticities obtained lie along the straight line joining the two points. Therefore, the chromaticity diagram is a light mixture diagram.

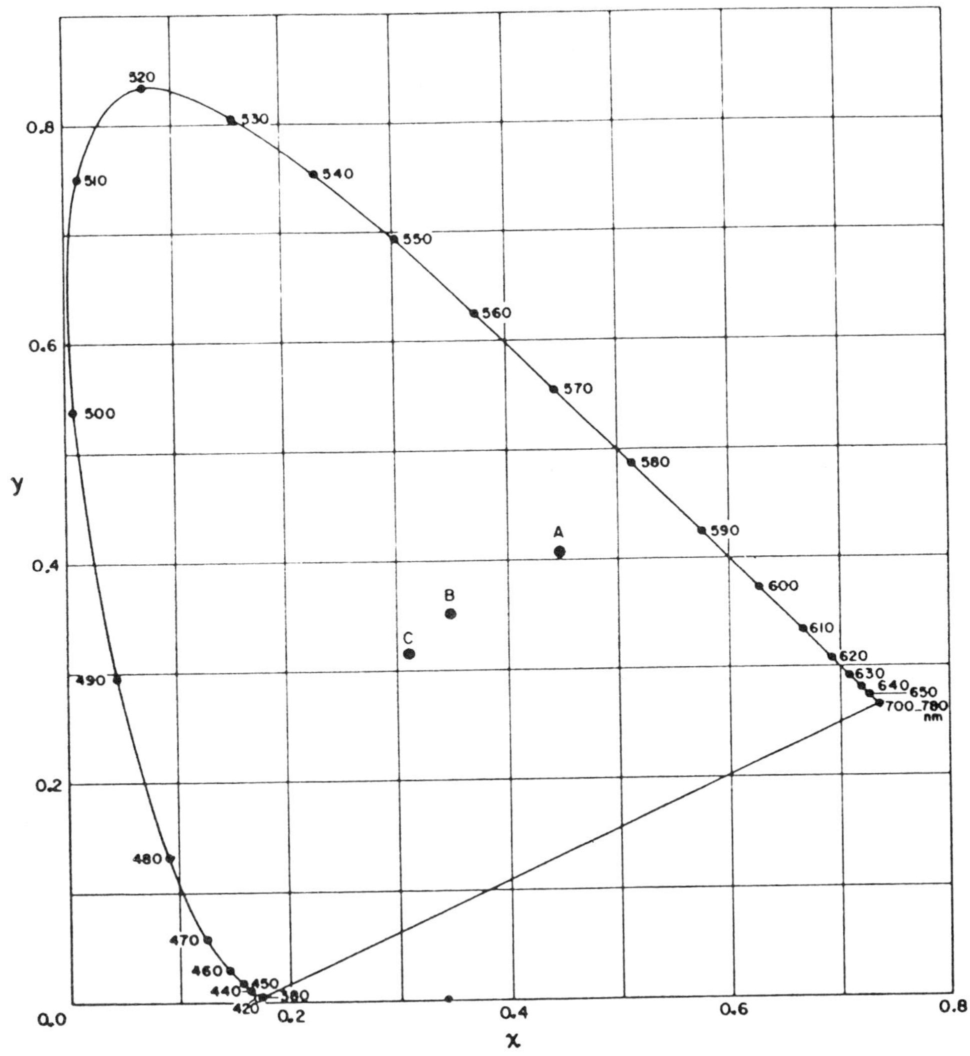

Figure 2 CIE chromaticity diagram. The chromaticities of all colors are in the area bounded above by the spectrum locus and below by the straight purple limit. The chromaticities of CIE illuminants A, B, and C are shown.

It follows from this that if we join the two ends of the spectrum locus with a straight line, it will represent the colors obtained when we mix the light from the two ends of the visible spectrum. Light of any color must be made of some mixture of light of various wavelengths in the spectrum, and therefore the chromaticities of all colors are inside that closed region.

We may plot the chromaticity of the illuminant on this diagram. The illuminant might be, for example, illuminant A. That point also represents the chromaticity of a white surface reflecting light exactly like that coming from the illuminant. Therefore, that point is called the illuminant point or the white point.

There is a triangular region at the bottom of the diagram defined by the illuminant point, the violet end of the spectrum locus, and the red end of the spectrum locus. The chromaticities in that triangle correspond to purple colors. It is important to note that purple does not appear in the spectrum. Purple is mostly a mixture of light from the two ends of the spectrum.

There have been studies of the exact relationship between Munsell hue, value, and chroma and the CIE chromaticities. Diagrams and tables have been published showing this relationship, so if we know the CIE chromaticity coordinates and Y, we can determine the Munsell hue, value, and chroma (5).

In the early 1940s, David MacAdam performed a series of experiments on the repeatability of color matching. He used a visual colorimeter, presented a color, and asked observers to adjust the amounts of red, green, and blue light to achieve a match. In some parts of the chromaticity diagram, the data clustered very tightly. This meant that people could detect very small differences in chromaticity. In other parts of the chromaticity diagram, the data spread quite widely, meaning that rather large chromaticity differences were inperceptible to those who were trying to make the match. The data systematically spread out more in one direction than in the perpendicular direction.

This proved that, for small color differences, the chromaticity diagram is perceptually nonuniform. We would like a space in which the distance between two chromaticities is proportional to the perceived color difference. Although it is known that no mathematical transformation of the space can be uniform, some approximate transformations have been adopted by the CIE (6). The most widely used is the transformation from tristimulus values X, Y, and Z to rectangular coordinates L^*, a^*, and b^*. L^* represents lightness; it is on a vertical scale with black at the bottom and white at the top. The positive a axis is in the red direction; the negative a axis is in the green direction. The positive b axis is in the yellow direction; the negative b axis is in the blue direction. This color space is called CIELAB (pronounced "sea lab"). It is illustrated in Figure 3.

Given these rectangular coordinates, it is a simple matter to transform to cylindrical coordinates L^*, C^*, and h, where L^* is the lightness, corresponding to Munsell value; C^* is chroma, corresponding to Munsell chroma; and h is the hue angle, corresponding to Munsell hue. (These correspondences are good, but do not imply equality.) The significance of these numbers is readily comprehended, because they correspond to the psychological attributes of color named by Munsell. People are more critical of variations in hue than variations in lightness or chroma. If tolerances are given in terms of L^*, C^*, and h, the tolerance in h can be made tighter and the others looser. This ability to relax two of three tolerances can have a big impact on the cost of producing a color.

It seems natural that the color of an object would depend on the spectral quality of the light in which it is seen, but most people do not realize that the color may depend on the direction of the illumination and the angle from which the object is observed. We speak of these conditions as geometric conditions. Satin, velvet, and crushed velore look the way they do because their appearance varies with the angles of illumination and angle of view. Polished wood can display a very considerable change in appearance as these angles are changed.

For these reasons, it is necessary to standardize the geometric conditions of measurement. There are two general types of geometry, one being highly directional,

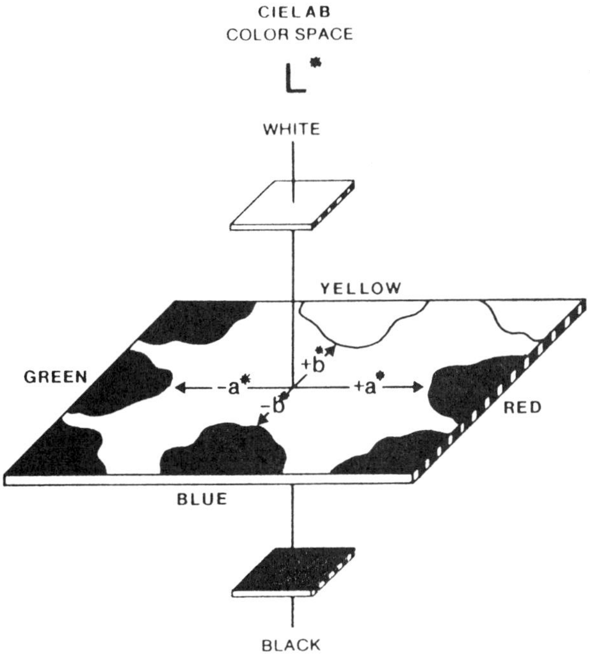

Figure 3 Diagram of an approximately uniform color space recommended by the International Commission on Illumination. The three dimensions are L*, a*, and b*. The abbreviated name of the space is CIELAB.

the other being nondirectional. The highly directional mode can be of two types: Either we illuminate the sample at 45° to the perpendicular and view along the perpendicular, or we reverse that, illuminating along the perpendicular and viewing at 45° to the perpendicular. If we interchange the illuminating beam and viewing beam, ray for ray, we measure the same value of the reflectance factor, so these two methods of measurement are interchangeable.

To measure reflectance factors with a nondirectional system, we use an integrating sphere, which is a hollow sphere painted flat white inside. Again, we may proceed in either of two ways, which are interchangeable. The first is to have the light enter the interior of the sphere and strike it on one side. The light striking the side of the sphere is reflected in all directions many times, so the interior of the sphere is uniformly bright. If a specimen is placed at an opening in the sphere, it is illuminated uniformly at all angles. The light reflected from the specimen is measured through a port on or near the perpendicular to the specimen. The alternative method is to illuminate the specimen along the perpendicular to the sample, let the sample reflect light to the sphere, let the sphere integrate the light reflected in all directions, and then measure the light reflected from the wall of the sphere.

Fluorescent specimens must be measured by illuminating them with light closely matching a standard illuminant, including the right amount of radiant power in the ultraviolet part of the spectrum. The ultraviolet excites the material to glow in the visible part of the spectrum, so it can greatly affect the observed and measured color.

VI. APPLIED COLORIMETRY

Now for the payoff. How is all this used to make a profit?

First, color science makes it possible to specify colors precisely in terms that are understood internationally. A designer may toss a painted swatch on the table and say, "That's the color." The swatch may be measured or compared to the Munsell book of color. The Munsell or CIE notation can be used as a basis for communicating the color specification from place to place and recording the specification for future use. This ability to communicate clearly and precisely is essential to efficient buying and selling in any business involving color.

The Munsell chips can be used to explore the range of variation that the designer considers tolerable. In setting tolerances, the designer must bear in mind that tight tolerances are expensive, because greater care must be exercised in producing the color and colors out of tolerance represent a loss. Single part items that need not match each other or anything else exactly can be assigned larger tolerances than items made of parts that must match or items that must match each other or some other thing. Munsell papers can be used to make color tolerance sets for use by inspectors. Careful management of the selection of tolerances optimizes return on investment by making the product no more precisely than is necessary to satisfy the manufacturing and marketing requirements.

Small businesses, those employing only simple operations with a few colors, and those for which color is not a critical factor may find a visual system of color specification and control entirely satisfactory. However, wherever color is a major concern and the business is of any size, it is likely that color measurement will more than pay for itself. Colorimetry is objective, repeatable, and much faster than visual interpolation. The measured values can be directly communicated from the spectrophotometer or colorimeter to computers, process controls, inventory control systems, and so on, to facilitate highly automated operations.

In "color businesses" such as paints, plastics, textiles, cosmetics, and packaging, it is necessary to manage the design, production, and marketing of many colors that are often changed. The methods used may be illustrated by following a color through the whole cycle in a textile firm.

A designer selects a color. A swatch of the color can be measured on a spectrophotometer. The spectral reflectance factor at each wavelength is transmitted directly to a computer. The computer computes a dye recipe that will produce that color. In fact, the computer may be designed to come as close as possible to matching the spectral curve at each wavelength. If that is done, the resulting product will match the designer's sample under any illumination. If that is not needed, other design considerations, such as fastness and cost, may prevail. Obviously, the computer must be programmed to do all these things, and it must be supplied with the data resulting from the measurement of many dyed samples, fastness data, cost data, and so on. This procedure is known as colorant formulation. It is a key step in the successful management of color. It tells you how to make a desired color.

Once a formula has been computed, a batch of the colorant is mixed and a trial specimen is made. The trial is usually close but not quite on color. Colorants may have aged since the calibration data were taken, new shipments of colorants may be slightly different, the process of application of color may be a little different, or the substrate may not take color as before. The trial specimen is measured and the computer decides what would be the best way to correct the batch to make the color right. This procedure is known as batch correction. Many an amateur interior decorator has finally corrected the color of the paint but wound up with three times as much paint as was needed. This and other pitfalls must be avoided by well-designed batch correction systems. The entire profit may depend on wise batch correction or the decision to set a bad batch aside for use in some other way.

Once a good color is in production, the next concern is to see that production is uniform. This is called color process control. The color of the product is measured. If the process is continuous, as it is in the textile and paper industries, measurement must be made on-line; that is, equipment is positioned in place or scanned from place to place to measure the color of the product as it moves past, often at high speed. Piece goods are measured off-line. In either case, measured colors and measured color differences may be shown on video displays so that machine operators can control the process and similar displays may be made available in remote locations, such as the quality assurance department. Data may be displayed in the form of control charts, which are plots of color variation from the ideal, as functions of time. Usually, three plots are used to display the three dimensions of color. Tolerance limits are displayed as lines above and below the ideal line, so it is immediately obvious if tolerances are exceeded or there is a trend toward going out of tolerance. Alarms may be signaled if these conditions exist. In a few cases, the measured and computed data are automatically transmitted to mechanisms controlling conditions affecting the process to control the process automatically. This procedure is called closed-loop operation. As of this writing, most processes are monitored but not controlled automatically. The profitability of modern process manufacturing industries depends on good process control and carefully chosen tolerances.

Incidentally, one should not overlook the fact that spectrophotometry and colorimetry are processes subject to the very kind of process control described above. Periodic measurements of stable colored standards are displayed on a control chart to reveal the condition of the measurement process in historic perspective. You cannot keep a production process in control with measuring instruments out of control.

By the time we have reached this stage, it would appear that little remains to be done in the way of color management. Not so; one of the most profitable techniques of color management comes in here. After all the good color design, colorant formulation, batch correction, and process control, suppose that there remains so much variation in the dyed fabric that your customers, who are apparel manufacturers, complain that the coat does not match the pants. You have done your best, but the variation of natural materials prevents you from achieving the uniformity demanded. Suppose, for simplicity, that a blue tends to range from greenish to purplish. The range is unacceptable. Maybe even half the range is unacceptable, but suppose that a third of the range would be all right. Measure the product and sort it into three lots that we might call "green," "blue," and "purple." Then organize order entry and shipping schedules so that a given customer and his or her subcontractors always receive material from the same lot. This technique is called shade sorting. The three-dimensional color space surrounding the centroid color is divided into shades, as many as necessary to achieve satisfactory uniformity within one shade designation. The customer gets matching goods and the manufacturer gets top price for all the product. That is color management.

REFERENCES

1. A. H. Munsell, *A Color Notation*, Munsell Department, Macbeth Division, Kollmorgen Corp., Newburgh, NY.
2. K. A. Kelly, and D. B. Judd, *Color: Universal Language and Dictionary of Names*, NBS Special Publication 440, U.S. Government Printing Office, Washington, DC, 1976.
3. *ISCC-NBS Centroid Color Charts*, Standard Reference Material 2107, National Institute for Standards and Technology, Gaithersburg, MD.
4. ASTM Standard D 1729-89, *Standard Practice for Visual Evaluation of Color Differences of Opaque Materials.* American Society for Testing and Materials, Philadelphia, 1989.
5. ASTM Standard D 1535-89, *Standard Test Method for Specifying Color by the Munsell System.* American Society for Testing and Materials, Philadelphia, 1989.
6. *Colorimetry*, 2nd ed., CIE Publication No. 15.2 (1986). Central Bureau of the CIE, Vienna, Austria, 1986.

FURTHER READING

Billmeyer, F. W., Jr., and M. Saltzman, *Principles of Color Technology*, 2nd ed., Wiley-Interscience, New York, 1981.

Committee on Colorimetry, Optical Society of America, *The Science of Color*, Optical Society of America, Washington, DC, 1963.

Hunter, R. S., and R. W. Harold, *The Measurement of Appearance*, 2nd ed., Wiley, New York, 1987.

Judd, D. B., and G. W. Wyszecki, *Color in Business, Science, and Industry*, 3rd ed., Wiley, New York, 1975.

Wyszecki, G. W., and W. S. Stiles, *Color Science*, 2nd ed., Wiley, New York, 1982.

21

Electronic Test Tactics and Techniques

JON TURINO
Integrated Measurement Systems
Beaverton, Oregon

I. INTRODUCTION

Testing is the key element of most quality control programs. Testing is the method used to verify product conformance to specifications and the workmanship of the manufacturing processes up to the moment when the test units are gathered for testing. A major fringe benefit of testing is the isolation and detection of circuit faults; a process that substantially helps to eliminate recurring failures at their source.

Mechanical, visual, and similar evaluations of products—usually called inspections—are of course, very important. Typically, these functions are passive in the sense that the equipment is not on or running when the measurements are made. The word *testing* is generally considered to be a dynamic term referring to mechanical or electrical tests made on operating equipment. In this chapter, only electrical testing will be presented.[1]

Subassemblies, such as printed circuit board assemblies (PCBAs), hybrid microcircuits, cabinet wiring, and wiring or harness assemblies are normally tested after assembly. When these subassemblies are meshed into a final assembly, each unit is again tested as a part of the overall system.

II. FAULT DETECTION

One of the major goals of any testing tactic is to remove faults as early in the building process as possible. In part, this is because rework and repair are much

[1] In industry, testing almost always relates to electrical analysis of some type. Physical measurements are indeed an important part of testing, but as stated, they are essentially measurements, which are covered in another section of this handbook.

Table 1 Typical Performance Characteristics of Continuity Testers

- High-speed testing of point-to-point wiring in cables, harnesses, wire-wrap panels, and multilayer PCBs
- High-pot testing for isulation resistance verification to 1.5 KV
- Resistance measurements from 0.01 ohm to 1000 megohms
  - Applied voltages from 5 to 1500 VDC
  - Accuracy at ±5%
- DC voltage measurements
  - 0.1 millivolts to 1500 VDC
  - Accuracy at ±1%
- Incremental delay and dwell times for high-stability measurements
- Programmable parameters for
  - Go/No Go testing of unit under test (UUT)
  - Measurement of embedded resistors

less expensive when there are fewer installed parts to work around. Finding a faulty part after it has been installed in the subassembly costs about 10 times as much as it does to find the bad unit at the component test level.

That same bad part can cost 1000 times more to find and replace once the assembly is in the hands of the customer. As an example: It costs about 15¢ to find a bad transistor at component test, but it costs about $15 to find and replace that same part at PCBA test![2]

Manufacturers need to consider all available testing options. Which options a company should choose depends on the volume of products to be tested and the mix of components, printed circuit boards, and systems.

III. CONTINUITY TESTING

Several test equipment manufacturers provide testers that are used primarily for cable backplane and bareboard continuity testing. Most of these are high-speed testers. Table 1 (above) lists some of the characteristics of continuity testers.

IV. LOADED BOARD OPENS AND SHORTS TESTING

Opens and shorts testing for loaded boards is based on the principle that most PCB defects are due to problems related to opens and short circuits. Most opens and shorts are due to workmanship problems in the assembly and solder operations.

[2] Electronic components like integrated circuits, transistors, capacitors, and resistors are usually sampled when received. Rejected lots are usually screened to remove all defective units. Enlightened managers apply the "100 times rule": i.e., order "100% screening" only when the best estimate of the lot, based on the sample, indicates that the lot is more than 1% defective. It is cost effective to *not* remove the rejects from the subassemblies whenever the lot is less than 1% bad.

Table 2 Benefits and Limitations of Opens and Shorts Testing

Benefits
- Process-induced shorts and opens are isolated at an early stage
- A small amount of programming effort is required
- Less skill and training is required for operator know-how

Limitations
- Little or no parametric capability is provided
- No functional test capability is used

Testing for opens and shorts as a first step increases productivity and lowers cost. Removing any faults before further work is done reduces testing time required later by the more complex automatic test equipment (ATE). Opens and shorts testing can also reduce or eliminate the situation of "smoking" a complex and expensive printed circuit board.

Fixturing for opens and shorts is accomplished using a device called a "bed-of-nails." A bed-of-nails is a specialized test fixture that holds spring-loaded probes—called nails—that make electrical contact with specific portions—usually called "nodes"—of the printed circuit board under test. The probes connect the nodes to the test equipment through cabling. Bed-of-nails test fixtures are generally used only to detect short circuits in the assembly. This testing is extremely cost effective, because 90% of the faults found in plated through hole printed circuit boards are typically short circuits.

To fixture a bed-of-nails to test boards for both opens and shorts requires a test nail at both ends of every trace. This makes testing for opens almost twice as costly as simple shorts testing, since twice as many nails are used in the fixturing.

For surface mount technology (SMT)* boards, opens are the more common failure mechanism. However, only one nail per node is required for testing open circuits on these boards as long as all nodes are available on the bottom side of the board (i.e., all nodes can be reached by the probes in the bed of nails).

Open circuits will be detected along with any component defects when tested. Depending on the refinement of the tester, it may even be able to detect opens related to digital ICs. It may still be necessary to use in-circuit functional testing to detect open circuits associated with digital ICs.

The objective of loaded board opens and shorts testing is to isolate simple faults (e.g., those points that are connected and should not be; and those points that should be connected together and aren't). This testing technique differs from bare board testing in that fixturing is typically connected to one side of the board only, and this technique picks up workmanship shorts and opens that were not present when the bare board was tested.

Table 2 (above) lists some of the benefits and limitations of opens and shorts testing.

---

* See glossary.

V. MANUFACTURING DEFECTS TESTING

There is a class of testers called *manufacturing defects testers* (MDTs) or *manufacturing defects analyzers* (MDAs). These testers are either enhanced opens and shorts testers or stripped-down in-circuit testers. MDTs have emerged because they are less expensive than full-capability in-circuit testers. On occasion, an MDT has evolved into an in-circuit tester with functional capability. MDTs typically perform opens and shorts testing while testing for missing or wrong components. They also measure resistance, capacitance, and inductance while looking for component insertion faults. Typically, they do not check digital device truth tables. This job still needs to be done by an in-circuit tester.

An MDT usually replaces an opens and shorts tester and an in-circuit tester. The MDT finds almost all of the manufacturing-type defects, but it does not exercise* the functions of the digital or analog components on the board under test. This means that a few functional faults will need to be removed at a later, or higher level, test stage.

VI. IN-CIRCUIT BOARD TESTING

When each component on a printed wiring board is tested on an individual basis, without regard to its intended function in the circuit, the technique of testing is called in-circuit testing. In-circuit testing philosophy is based on the following logic:

1. When all components are correctly installed and are the correct value, the printed circuit board assembly will function correctly.
2. The majority of defects on printed circuit boards are due to poor workmanship or mishandling.

In-circuit testers are excellent for detecting and isolating manufacturing-induced faults on a PCB. Digital ICs are tested for their basic truth tables at the same time resistors, capacitors, and inductors are measured.

When possible, electrical guarding is put into use when discrete components are tested on an assembled but unpowered PCB. The principle is to guard against any effects induced by the other components. It is often very difficult to get a good measurement of the device under test due to "sneak" current paths and components in parallel.

While there is little need for making contact at each end of every trace—except on critical bus lines—a nail from the bed-of-nails fixture should contact each node. Typical input and output data for in-circuit testers is shown in Table 3. The benefits and limitations of in-circuit testing are shown in Table 4.

VII. DIGITAL FUNCTIONAL TESTING

There are several types of digital functional test protocols* that are widely used today. The reasoning, methods, technology, rewards, and flaws of each test type are outlined here.

---

* See glossary.

Table 3 Typical Input and Output Data for In-Circuit Test Equipment

- Programmable short detection from 5 to 20 ohms
- Resistance measurements from 1 ohm to 9.99 megohms
  Accuracy ±1% for resistors from 1 to 10 ohms
  Accuracy ±0.5% for resistors greater than 10 ohms
- Capacitance measurements from 10 to 999 pF:
  Accuracy ±1% for capacitors from 10 to 1000 pF
  Accuracy ±2% for capacitors greater than 1000 pF
- Inductance measurement range of 1 mH to 10 H:
  Accuracy ±5% for inductors less than 10 mH.
  Accuracy ±3% for Inductors from 10 mH to 10 H.
- Constant current supply: 10 nA to 99.9 mA, with ±5% accuracy
- Constant voltage supply: 0.1 to 99.9V, with ±0.5% accuracy
- DC voltage measurement range: 0.1 to 99.9 V with ±0.5% accuracy
- Capable of in-circuit differential voltage measurements in a powered circuit without altering circuit performance (e.g., the ability to measure DC voltages in the presence of an AC or DC common mode voltage level)
  Measurement range: 0.1 to 15 VDC with a common mode voltage of $\pm 15V_{p\text{-}p}$ maximum
  Accuracy +0.5%

A. Static Functional Testing

Static functional testing involves testing a printed circuit board assembly as an entity.* Static tests involves sequencing the test vectors* slower than the normal operating speed of the unit under test (UUT). This technique is based on the concept that all components on the board are correct and correctly installed and that there are no process-induced problems such as short circuits.

Table 4 Benefits and Limitations of In-Circuit Testing

Benefits
- One-pass diagnostics
- Easy to program
- Easy board handling
- Low operator skill requirements

Limitations
- Some components are untestable (in circuit)
- Tolerance of tester versus tolerance of UUT
- Component interaction cannot be checked
- Limited functional test capability

* See glossary.

Table 5 Static Functional Testing

Benefits
• Verifies the functional integrity of the assembly
• Usually provides higher next assembly yields than previous methods
Limitations
• Does not test for timing-related faults
• Has a slow test execution with long test patterns
• Finds only one fault per pass

The unit under test is usually connected to the automatic test equipment (ATE) by its edge connector. Some static functional testers use an ancillary bed-of-nails fixture to provide increased access to the circuitry under test.

Input stimulus vectors are designed to verify suspected faulty circuit paths when the unit is exercised. This protocol will demonstrate that the board will perform properly when installed in the next assembly. When a fault has been verified, the test equipment will let the operator know which node to probe so the fault can be located. Some automatic test equipment systems have a fault dictionary which explains which node to investigate and what outputs to look for. In this mode, the automatic test equipment provides the operator with a fault reference number to be used with the fault description dictionary. Usually, a separate dictionary is put together for each board type.

The speed of test execution is limited by the digital I/O (input/output) rate of the equipment for a Go/No Go analysis. For diagnostics, the limiting factor becomes the fault isolation time and the number of manual probing operations required. For analog testing, the ATE is limited by the instrument conversion rate on the general-purpose instrumentation bus (GPIB) or the voltage measurement indicator (VXI) for the instrumentation bus. All tests are done at discrete stimulus/response setups, and this technique requires only medium skill level for operation. The benefits and limitations of static functional testing are shown in Table 5 (above).

B. Dynamic Functional Testing

Dynamic functional testing involves stepping through* the test vectors at the operating speed* (or higher) of the unit under test. Dynamic testing is most often used for the types of board which contain read-only memory (ROM), random access memory (RAM), microprocessor, and other ancillary large-scale and very large-scale (LSI and VLSI, respectively) chips.

At the board test level, speeds in excess of 20 MHz* are usually referred to as dynamic, based on ATE manufacturer's definitions. Real dynamic testing usually means tests are performed at actual (i.e., normal) operating speeds of the unit under test.

* See glossary.

Table 6 Dynamic Functional Testing

Benefits	
	• Verifies functional and speed-related integrity of assembly
	• Executes quickly
Limitations	
	• High cost of equipment
	• Finds only one fault per pass

The gap between these definitions is one of practice and terminology. Some devices cease to operate at speeds below about 800 kHz,* and dynamic testing is mandated. In addition, higher test rates reduce output test times. *Soft errors*, such as pattern sensitivity, are more likely to be detected by using dynamic testing whether or not it is ATE dynamic or real dynamic.

The internal design of the dynamic functional tester is similar to that of a static functional tester except that memory behind each pin is used to store patterns for a rapid broadside* of stimulus and response vectors.

Signature analysis is sometimes used to reduce large amounts of unit under test data to a single four- to eight-digit number and thus compress the large amount of response data which must be analyzed by the ATE.

Bed-of-nails fixtures are not normally used, because they induce electrical noise and high capacitance loading. New fixtures to work around these problems are, however, now being introduced. These new testers provide high-speed functional and in-circuit capabilities using the same test gear.

High-speed functional ATE has gained popularity in order to bypass problems in finding more complex types of soft errors. At the same time, the in-circuit capabilities approach true dynamic testing conditions. But ATE has some limitations. Good design practice means using mature (i.e., slow) components in the design of an test system. Meantime designers are bringing out new products which need to be tested using very high-speed state of the art technology. Typical benefit and limitations of modern dynamic function testers are shown in Table 6.

C. In-Circuit Emulation

In-circuit emulation is particularly useful in development applications. This method is often used for production and field service testing. It requires good software design and, usually, a high level of skill for diagnostics.

This technique mimics the end-item operation via functional stimulus programs which are input to the board from the microprocessor or ROM socket. Loopback board design can be employed to provide a good test of peripheral and random logic. The technique also uses signature analysis to gather and analyze test data where the loopback technique cannot be used. The emulator provides a captive microprocessor of the same family as that used on the printed circuit board assembly,

* See glossary.

Table 7 In-Circuit Evaluation

Benefits
• Provides for full-speed operation
• Verifies functional and speed integrity
• Can use existing self-test programs
• Can be used in field service test
Limitations
• Minimal diagnostic capabilities
• Microprocessor must be removed during test
• Requires high skill level in production test

running at full speed, in what is called a *personality module.* The printed circuit board assembly under test can run either its own programs or special diagnostic exercises stored in read-only memory. Test programs can then be structured to display repeatability of signatures. In-circuit emulation has the advantage of allowing the transportation of engineering programs to the testing functions provided proper care has been taken. The capital cost for an in circuit emulator relatively low, but usage requires careful attention to design.

Stimulus speeds of more than 40 MHz are possible. Connection of the tester to the printed circuit board assembly under test requires a socket for the microprocessor. An RS-232C serial input/output port can interface with host computers. Large development systems are usually available with this type of tester for added flexibility in program generation and editing (Table 7, above).

D. Dynamic Reference Testing

Dynamic reference testing is a technique whereby input stimuli are provided to the printed circuit board assembly under test and to a known good board simultaneously. Outputs from the two boards are monitored* for differences. This type of ATE usually has generators to provide the patterns for difference monitoring as well as the ability to use stored outputs from units known to be good. Unlimited response vector lengths are theoretically possible, because the reference board acts like an infinite-length read-only memory for storage of diagnostic nodal data. Pattern speeds in excess of 50 MHz are attainable with this type of ATE, and signature analysis techniques are used to verify that the known good board has not degraded (Table 8).

E. Signature Analysis

Signature analysis is not a stand-alone test strategy, unless the input stimulus vectors are provided from a source external to the board being tested. Signature analysis is often used as a partner to several other test methods, including in-circuit and functional testing. The basic value of signature analysis is as a data compression technique.

Data compression is possible because the signature analyzer has a built-in 16-bit feedback shift register. When data are entered, this shift register can encode

* See glossary.

Table 8 Dynamic Reference Testing

Benefits
- Large numbers of test vectors can be applied at high speeds
- No LSI modeling required
- Verifies functional integrity

Limitations
- Requires manual test program generation
- Requires reference board for diagnostics

and display 65,536 (i.e., 16^4) distinct, separate states on four indicators. Any change in the node under test will produce a different signature, which is the indication of a possible circuit malfunction. Measurement intervals exceeding 16^4 clock cycles will still produce repeatable signatures because of the compression algorithms chosen.

As long as enough patterns have been circulated through the shift register, the signature display uniquely defines the output from the node. Input stimulus vectors can be provided by on-board software or from an external source (Table 9).

F. Dedicated Testers

A dedicated tester is commonly an in-house built tester designed expressly for testing one type of circuit board. This kind of equipment can be designed to test either digital or analog boards. Usually, this kind of test gear is designed by an in-house test engineering department. Typically, the diagnostic capabilities are not as good as those developed by the ATE manufacturers, who have all invested millions of dollars in research and development.

Dedicated testers are best used when product volume is great and high throughput is needed. Alternatively, dedicated testers are often used when production volumes are too low to justify commercial, general-purpose automatic test equipment (Table 10).

G. Hot Mockup Testers

Hot mockup testers are a form of a dedicated tester in that they are usually an in-house design and is dedicated to testing one type of circuit board. A hot mockup tester is a complete final system with the exception of the unit to be tested. All of

Table 9 Signature Analysis

Benefits
- Many thousands of tests can be applied at high speeds
- Fast program generation in many cases
- Large amounts of response data can be compressed
- Can be used for field service

Limitations
- Requires careful consideration in the design for testability
- Diagnostic resolution is poor in feedback loops and bus structured boards

Table 10 Dedicated Testers

Benefits	
	• Tests can be tailored to the specific task
	• Can be inexpensive if designed properly
Limitations	
	• Limited diagnostic capabilities
	• High skill levels are typically required
	• Improper design can be very expensive

the components of the assembly are known to be good except for the unit under test. The test system exercises the unit under test and monitors responses.

Hot mockups can be used for quality assurance purposes and for test program improvement providing that good records are kept. Most of the capital equipment cost is buried, because inventory surplus can be used to build the tester. In most cases, this technique is highly labor intensive owing to limited diagnostics, but it does provide the ultimate in functional testing (Table 11).

VIII. ANALOG PRINTED CIRCUIT BOARD TEST EQUIPMENT

Analog printed circuit board test equipment ideally is structured for use within the company. An example of such equipment might be an automatic analog and dynamic logic test station. This system can be used to test PCBs and assemblies which are either completely analog or some combination of analog and digital. It is usually put together using some variety of commercial test equipment, a UUT interface panel, and a computer with disk memory.

In operation, the stimulus and measurements are controlled by the computer. The test procedure should be entirely automatic; thus the PCB under test should not require extensive adjustments during the test. The UUT interface panel, typically containing circuitry such as loads, special stimulus circuits, and interconnections, must be designed specifically for the individual PCB under test. Elements that might be included in the stimulus and measurement module include:

- DC voltage sources
- Signal sources
- Digital stimulus

Table 11 Hot Mockup Testers

Benefits	
	• Requires little design effort
	• Inexpensive in terms of capital costs
Limitations	
	• Needs little or no diagnostic capability
	• Requires high skill level operators
	• Usually has long test times

- Switching matrix
- Digital measurement
- Bus programmable DMM measurements
- Bus programmable counter measurements
- Bus programmable wave form analyzer

It is possible to design manually operated test stations for things like radio frequency/intermediate frequency (RF/IF) circuit testing. Such a system can be used to test, align, troubleshoot, and tweak RF and/or IF assemblies to the component level. This tester can operate at frequencies as high as 1000 MHz. It can be put together using separate pieces of commercial test equipment and an interface panel.

Typically, the interface panel will some mix of switchable attenuators, RF amplifiers, RF power dividers, mixers, and the like. A well-designed interface can be adapted to all UUTs. A station like this is capable of making measurements that include:

- Gain
- Bandwidth
- Phase
- Standing wave ratio
- Insertion loss
- Swept frequency response
- Pulse characteristics
- Noise figure

Specially designed automatic test equipment may be installed and used to test, align, troubleshoot, and tweak active and passive microwave assemblies, as well as components and networks. It is composed of various modules of commercial test equipment and a computer section capable of controlling all phases of the testing. This control extends to the stimulus and measurement test equipment as well. The types of measurements this kind of equipment can make include:

- Voltage standing wave ratio (VSWR)
- Reflection (real)
- Insertion loss (dB)
- Reflection (imaginary)
- Phase (transmission)
- Z magnitude (mhos), angle (degrees)
- Z angle (degrees)
- Phase deviation (degrees)
- Y magnitude (mhos)
- Loss deviation (dB)
- R (ohms)
- Flatness deviation from
- X (ohms); mean (dB)
- Gain
- Group delay (nsec)
- R/Z_o
- Isolation (dB)
- X/Z_o

- Reflection magnitude
- G (mhos)
- Return angle (degrees)
- B (mhos)
- Return loss (dB)
- G/Y_o
- Transmission (real)
- B/Y_o
- Transmission (imaginary)

IX. COMBINATIONAL TESTERS

Combinational testers are high-speed, high-performance test systems that include both in-circuit and functional (sometimes called performance) test capability. Combinational testers use a bed-of-nails fixture to overcome an inherent lack of testability features in the UUT and to drive test patterns through individual ICs (or clusters of components that cannot be isolated for individual testing) on the board.

The advantage of a combinational tester is that it can perform both in-circuit testing (i.e., test for manufacturing defects) and functional testing of the board under test in one operation. Since the bed-of-nails tester accesses almost all of the nodes, testability flaws go away.

Major limitations of combinational testers are their high initial cost, the expensive test programming, and the lengthy test times.

X. CHOOSING A TEST STRATEGY

The selection of the best test strategy for a particular situation should not be a guessing game or a matter of opinion. There are rigorous systems for matching the correct strategy to the situation. This section outlines the most direct method for matching a test strategy to the job.

Information is needed in order to develop a test strategy that reinforces the overall strategy of the business. Thus the test developer needs information on the following criteria in order to develop the best strategy:

- *Design criteria*: What technology will be used? The design may limit the number of choices or available testing options.
- *Marketing criteria*: How many items will be sold? This has crucial impact on money available for automated versus manual testing.
- *Quality criteria*: What level of quality shall the product have? How good does the testing have to be?
- *Support criteria*: What product service plan will be used? How will field returns be handled? What is the plan for swapping good units for bad?
- *Application criteria*: What is the product ambient environment? That is, what are our environmental stress screening options?
- *Financial criteria*: How much money is available to finance the test strategy?

Data can be gathered from past experience, from consultants, from magazines, at trade shows, and even from competitors or noncompeting manufacturers of similar type and complexity products. Early estimates should be logged for reference in case revisions are needed.

A. Incoming Inspection

The strategy at this point is relatively straightforward. Simply calculate how much it costs to find a defective component at board level. Then, determine how much it costs to test an individual component at incoming inspection. The cost ratio, (Cost at Board Test)/(Individual Test Cost), sets the incoming inspection criteria. When the cost ratio is 100, it shows that the break even point for that component is 1%. In other words, it costs as much to test 100 parts at the component level as it costs to remove one bad unit at board test. At this point, it may be good inspection strategy to sample all incoming lots using an acceptance quality level (AQL) of 0.001% (10 defects per million).

However, it may be cost effective to inspect to an AQL = 0.1%,[14] since the cost of sampling at the higher AQL may well outweigh the cost of finding a bad component on one board out of a thousand.

B. At Board Level

Here diagnostics and rework enter the picture and things are a little more complicated. First, ascertain or estimate the board level fault spectrum that is anticipated. Figure 1 shows the Pareto analysis of a typical spectrum of failure modes for printed circuit boards (with plated through holes) manufactured in the United States.

The objective of the test strategy is to detect defects as early in the process as possible where they are typically easier to detect and cheaper to correct. But *all* faults are hard to find at *any* level. We would like to think that all bare board faults will be found at incoming inspection, but a few bare board faults will usually survive to be found in later test. Some are even found at the final system level.

Functional interactive faults, on the other hand, can only be found at board or system level.

---

[14] When the test cost ratio is 100, a good strategy with regard to sampling of electronic parts is as follows. Sample incoming lots at an AQL = 0.25%. (This means that on the average accepted lots would have less than 1 bad part out of 400.) For failed lots, use the percentage of bad parts in the sample as the best estimate of the percentage of defects in the entire lot: More than 1%, Return the lot to vendor. Less than 1%, Use as is. Exactly 1%, This is the break-even point. Do what is most convenient! When your disposition is use 0.25% as is, it is good incoming inspection policy to take your supplier to task for not delivering the quality that you have paid for!

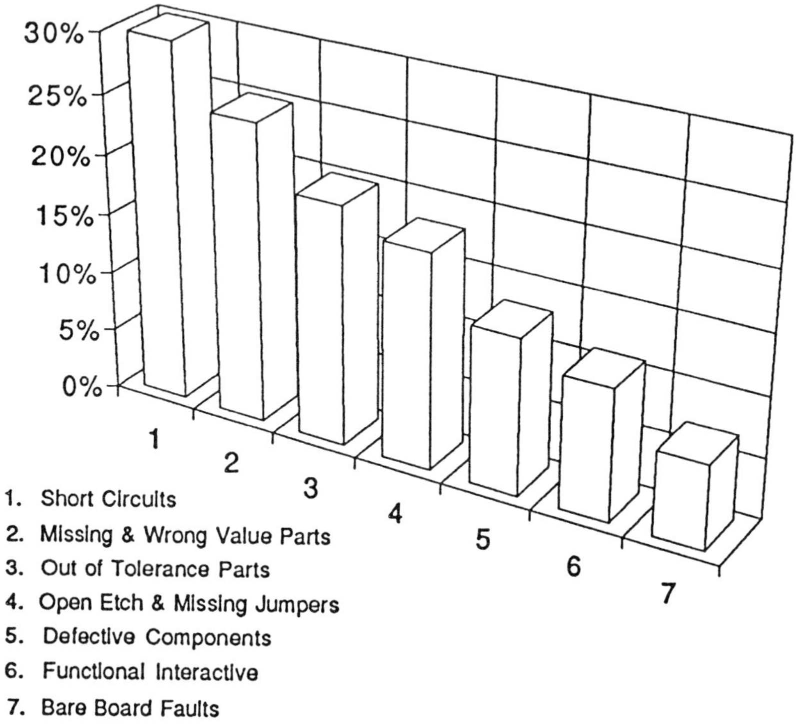

Figure 1 Through-hole printed circuit board failures distribution. (From Turino, 1990.)

The major elements to be considered, at each level of testing, in the test strategy calculations include:

- Capital equipment cost
- Test programming cost
- Test fixturing cost
- Test operation cost
- Diagnostic and rework cost
- Fault coverage impact on next level of test

At board level *diagnostic and rework cost* and *fault coverage impact on next level of test* are the most important on an ongoing basis. An in-circuit tester can diagnose multiple faults in one pass. A functional tester typically finds only one fault at a time, so, depending on the number of faults estimated per board, diagnostic time will have a large impact on the testing strategy. Figure 2 illustrates the effect of multiple faults on costs using both strategies. One important thing to note is the cost of diagnosing faults. Using either method, the testing cost at *zero defects* is zero dollars.

Is it possible to calculate the cost of each possible test strategy? The answer is "yes," and the method is easy. Make a flowchart showing each step in the plan, create a formula for that flowchart, then string the formulas for the various steps together to get the formula for the total scheme.

As an example, we examine a functional board test. To calculate the operating

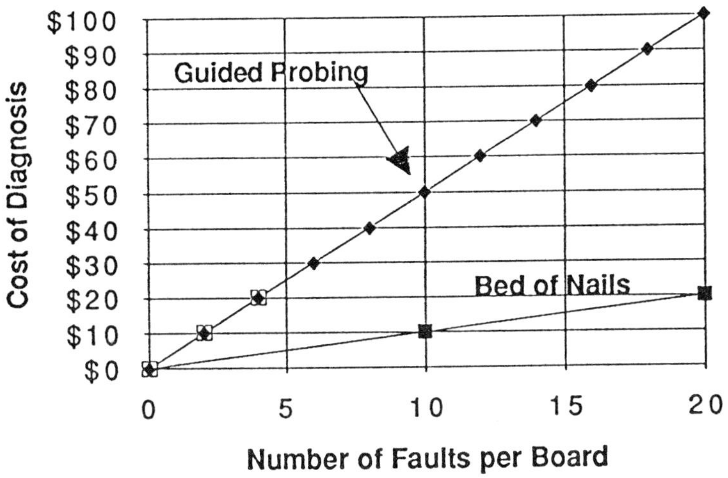

Figure 2 Comparison of diagnostic costs. Bed-of-nails fixtures reduce diagnostic time, compared to guided probing techniques, even when the functional testers are employed. (From Turino, 1990.)

cost of testing, we construct the flowchart shown in Figure 3. From the flowchart, we create the formula that correctly calculates the cost of the operation:

$$TC = M\{(HT + TT) + (DT + RT + HT + TT)(F1) + (DT + RT + HT + TT)(F2) + \ldots\}$$

Where:

TC = Total cost
M = $/hour
HT = Handling time

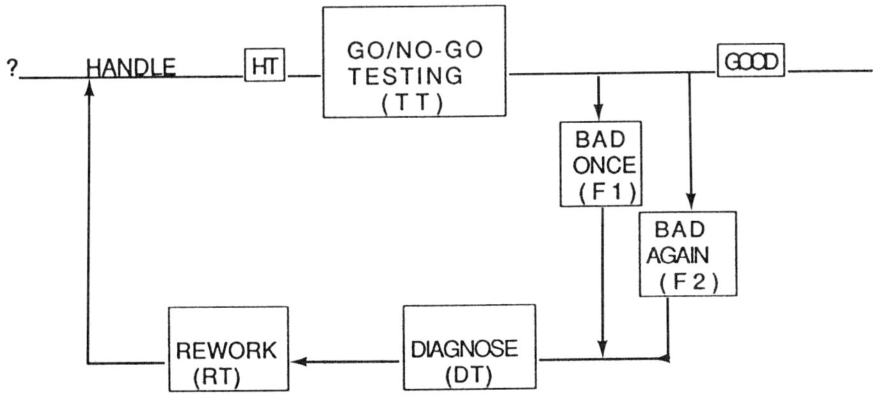

Figure 3 Flowchart of test operation. Each operation can be flowcharted and its costs determined or estimated. Large factors, such as F1 and DT, should be looked at first before looking at such things as HT.

TT = Test time (Go/No-Go)
DT = Diagnostic time (per fault)
RT = Rework time
F1 = Portion of defective boards first time through
F2 = Portion of defective boards second time through

Anything done to reduce the F1 value will have a major impact on overall operating costs. In other words, defect prevention in the manufacturing process is like money in the bank.

The next big factor is diagnostic time (DT). Design for testability and anything else that reduces DT will lower costs dramatically. Handling time and test time are usually the least fruitful elements for improvement efforts.

A similar flowchart and formula can be constructed and developed for system-level testing. This should include engineering evaluation, as required, of boards that pass board test but fail at system test.

The F1 and F2 rates at system test are governed by the fault coverage (test effectiveness) at the board test level. F1 will be lower at system level following *high-speed functional board test* than it will be when only *manufacturing defects testing* has been performed on boards prior to their assembly into the system.

Test coverage may thus be simplified to the ABC level:

Test Coverage = (A)(B)(C)

Where

A = Fraction of defects detectable in a design
B = Fraction of defects detectable by test equipment
C = Fraction of defects detected by test software

Consider the following illustrations.

Example One

1. We use a design with 100% detectable faults.
2. Our manufacturing defects tester can detect a maximum of 60% of the faults that could occur.
3. The tester is programmed such that 90% of the faults that it can detect actually get detected.

Since the fault coverage is 100 × 90 × 60 = 54%, 46% of the boards going into the system will have faults on them after test.

Example Two

1. We use a design with 100% detectable faults.
2. Our manufacturing defects tester can detect a maximum of 95% of the faults that could occur.
3. The tester is programmed such that 90% of the faults that it can detect actually get detected.

In this example, the fault coverage is (100 × 95 × 90) = 85.5%. Thus only 14.5% of the boards going into the system may still have faults on them after test.

From these examples, we can see that the board test coverage has a big impact on the system level F1 figure. Considering that diagnosis is at least 10 times more expensive at system level than at board level, the better job we do at board level, the lower our system testing costs. That is why all of the formulas for the various stages of a strategy must be added together, and the overall cost of the system must be examined as an entity.

Thus we construct a formula for the complete strategy as follows:

Total Cost = (incoming inspection cost) + (board test cost)
+ (unit test cost) + (system test cost)

Cost, as used in this formula, includes:

- Capital equipment cost
- Test programming costs
- Test fixturing costs
- Testing and troubleshooting costs on an ongoing basis

The next two examples will illustrate the use of this approach and point out the importance of looking at the whole process.

Example Three

1. The database for this example includes a system composed of 10 printed circuit boards.
2. Each of the 10 printed circuit boards is made up of 100 components.
3. A total of 1000 systems per year will be produced.

   The manufacturing process yields 2.5 faults per board (one component fault and 1.5 workmanship defects).

   The system yield is 85% after boards are tested on an in-circuit tester (with 95% system yield after functional board test).

The major cost elements are added and compared in Table 12. Clearly, the right strategy in this case is in-circuit board test. As shown in Table 12, in-circuit testing cost is roughly half the price of the functional test strategy.

Example Four

1. What happens if we change the scenario?

   Raise the number of systems to be built each year from 1000 to 5000.

Table 12 Test Strategy Trade-off

Example 3: cost	In-circuit ($)	Functional ($)
Capital equipment	250,000	500,000
Test fixturing	35,000	5,000
Test programming	30,000	80,000
Board-level test	25,000	125,000
System-level test	37,500	12,500
Total	377,500	722,500

Table 13 Test Strategy Trade-off

Example 3: cost	In-circuit ($)	Functional ($)
Capital equipment	250,000	500,000
Test fixturing	175,000	25,000
Test programming	150,000	400,000
Board-level test	125,000	500,000
System-level test	1,875,000	625,000
Total	2,575,000	2,050,000

Increase the complexity of the systems from 10 boards each to 50 boards each (this raises the cost of system level testing and diagnostics).

2. Hold all other data the same.

The results of this scenario are shown in Table 13.

Functional testing is now clearly the strategy of choice. When one looks only at the cost of the board test operation, ignoring the impact of board test fault coverage on system-level testing and diagnostic costs, one could come to an incorrect choice. When system-level costs are ignored, it looks like the selection of the in-circuit test strategy is the course to take. We can see that this would be a half million dollar mistake, because system test costs are very real.

A difference of only 10% fault coverage at board level, in this instance, makes a three to one difference in system level testing costs. Example four illustrates how critical is the need to consider the entire manufacturing process when selecting test strategies rather than looking at each step in isolation.

The same formula and flowchart approach can be used to play *what if*: What if we could raise our board level yield from 85% (the average in the United States) to 98% (the average in Japan)?

This may not be possible, but we can calculate the financial impact. Just lower the F1 number from 0.15 to 0.02 and recalculate the test cost.

The flowchart and formula approach is incredibly powerful. You can calculate the cost of testing with and without testability features. The difference in test costs is the amount of money available to you for added parts costs to implement the testability.

You can "what if": Yields, fault distributions, fault coverage figures, and production quantities to see whether changes in conditions will have a major impact on your test strategy recommendations and decisions. You can find out what it is costing you to test, and then you can benefit when you dig the gold out of those costs.

XI. GLOSSARY

ATE Automatic test equipment.

Bare board The printed circuit board is stuffed and soldered to become a printed circuit board assembly (see also PCB).

Bed-of-nails A specialized fixture used to detect shorts in an assembly.

Broadside Term for the application of a great number of stimuli or vectors simultaneously or in a very short time period.

Chip Integrated circuit, transistor, capacitor, resistor, and almost any other electronic component prior to encapsulation. Example: Hybrid circuits use surface mount technology (SMT) chips in their circuitry.

Clock rate Another term for the unit's operating speed.

DUT Device under test.

Error, hard In the retrieval of data, or information, a hard error occurs when data *cannot be retrieved* after a specified number of retrieval tries.

Error, soft In the retrieval of data, or information, a soft error occurs when data *can be retrieved* within a specified number of retrieval tries.

Exclusive or test When the output from a unit under test is compared to the output from a unit that is known to be good, the comparison of the two outputs is known as an exclusive or test. In other words, the two outputs must be the same for the UUT to pass the exclusive or test.

Exercise A unit under test (UUT) is said to be "exercised" when repetitive "stimulis" are applied to the UUT and the UUT output is monitored for sequential acceptable outputs. Also referred to as "exercising the unit under test."

GPIB General purpose instrumentation bus: used as the automatic test equipment interface for analog testing.

Gold standard In the exclusive or test, the "known to be good" unit is referred to as the gold standard.

Hot mockup tester A dedicated tester that is usually designed in-house and dedicated to testing one type of circuit board.

Hybrid circuit Hybrid circuits use surface mount technology (SMT) chips (usually mounted on ceramic substrates with plated on gold circuitry). One or more of these ceramic substrates are connected together before the hybrid circuit is hermetically sealed.

IC Integrated circuit or microcircuit.

IF Intermediate frequency.

In-circuit board testing A test of each component on a printed wiring board is tested on an individual basis without regard to its intended circuit function.

In-circuit testing, benefits One-pass diagnostics; easy to program; easy board handling; low operator skill requirements.

In-circuit testing, limitations Some components are untestable, in circuit; tolerance of tester versus tolerance of UUT; component interaction cannot be checked; limited functional test capability.

I/O, In-Out Usually refers to the digital in out rate of the equipment for a Go/No Go analysis.

KHz Abbreviation or symbol for kilohertz. For example, 800 KHz is the same as 800 thousand cycles per second, or 800 thousand operations per second.

LSI Large-scale integrated circuit. For example, microprocessor and other large-scale chips (see also VLSI).

MCM Multichip modules. The MCM technique is used in SMT applications and hybrid circuits.
MDA Manufacturing defects analyer.
MDT Manufacturing defects tester. An MDT usually replaces an opens and shorts tester and an in-circuit tester.
MHz Abbreviation or symbol for megahertz. For example, 20 MHz is the same as 20 million cycles per second, or 20 million operations for second.
1 percent rule It costs about 15¢ to find a bad transistor at component test, but it costs about $15 to find (and replace) that same part at PCBA test.
Opens and shorts testing, benefits Process-induced shorts and opens are isolated at an early stage; a small amount of programming effort is required; less skill and training is required for operator know-how.
Opens and shorts testing, limitations Little or no parametric capability is provided; no functional test capability is used.
PCB The printed circuit board before it becomes a PCBA (same as PWB).
PCBA Printed circuit board assembly (same as PWBA).
PTH Plated through hole.
PWB Printed wiring board (same as PCB).
PWBA Printed wiring board assembly (same as PCBA).
RAM Random access memory.
RF Radio frequency.
ROM Read-only memory.
Signature analysis A method used to code large amounts of UUT data to a single byte to compress large amounts of ATE response data. Within the signature analyzer is a 16-bit feedback shift register into which the data are entered. In all, there are 65,536 possible states to which the register can be set during a measurement window.
SMT Surface mount technology. Examples are multichip modules (MCM) and hybrid circuits.
Sneak paths Usually due to parallel components in the circuit. An example of a configuration that cannot be effectively guarded from sneak paths during circuit testing is a parallel bypass capacitor array on a power bus.
Stepping through Common term for following through a unit under test's test sequence.
Stimulus Term used to indicate the signal applied to the unit under test (UUT) so the output can be monitored to determine that the UUT is functioning properly.
Testing The method used to verify conformance to specifications and the good workmanship of the assembly processes up to the point of test.
Testing an entity Term used for testing the assembly as a functional unit, as opposed to making measurements of each of the entity's components.
Test protocol Term used for the program or the sequencing of the test stimuli provided to the unit under test for testing or exercising.
UUT Unit under test.

Vector Term used when referring to test patterns or stimulus patterns applied to the unit under test. Static test protocol is the sequencing of the various Vectors.

VSWR Voltage standing wave ratio.

VLSI Very large-scale integrated circuit. For example, microprocessor and other very large-scale chips (see also LSI).

FURTHER READING

Business Process Re-engineering with CE/IPD, 5-Part Videotaped Course, Logical Solutions Technology Inc. (LSTI), Campbell, CA, 1994.

Concurrent Engineering Management Session, Logical Solutions Technology Inc. (LSTI), Campbell, CA, 1990–.

Concurrent Engineering Seminar, Logical Solutions Technology Inc. (LSTI), Campbell, CA, 1990–.

Turino, J., *Design to Test*, 2nd Ed., Van Nostrand Reinhold, New York, 1990.

Turino, J., *Managing Concurrent Engineering*: *Buying Time To Market*, VNR, 1992.

22
Heat Treating

DAN D. ASHCRAFT
Oklahoma State University
Stillwater, Oklahoma

I. INTRODUCTION

According to the American Society for Metals (ASM), heat treatment is defined as "heating and cooling a solid metal or alloy in such a way as to obtain desired conditions or properties." Such a broad, generic definition requires further explanation to provide meaning and be technologically useful. Heat treatment can be applied to a number of commercial metals with a variety of results. Rather than delving into the science of metals, the focus of this chapter will be on the practical side of this area of physical metallurgy. Furthermore, discussion will be limited to changes in mechanical properties (e.g., strength, hardness, ductility, toughness) brought about by the effects of heat treatment.

Most applications of engineering metals, as dictated by the design requirements, are based on the economics of selection and processing techniques available. Seldom does a single metal or alloy have all the attributes needed to satisfy the engineering specifications even though it may be possible to enhance a desirable mechanical property by heat treating. Sometimes there are competing requirements. The necessity of changing mechanical properties to fit the performance criteria often results in a compromise of properties. For instance, strength and hardness are usually gained at the expense of ductility. Also, there are those situations requiring the opposite effect in a material; specifically, the reduction of certain mechanical properties. This occurs frequently in the processing of metals, where it becomes necessary intentionally to decrease strength and increase ductility to accommodate further processing.

Classically, there are three methods by which metallic properties can be altered. The first technique is to control the chemical composition of the metal by adding and deleting certain chemical elements. The second technique is to change the shape of the metal mechanically by plastically deforming the metal; that is, to

Table 1 Common Mechanical Properties

Property	Definition	Measurement units[a]	Alterable by heat treating
Strength (ultimate tensile)	Ratio of maximum load to original cross-sectional area	psi MPa	Yes
Hardness	Resistance of a metal to plastic deformation, usually by indentation	HB HRC	Yes
Ductility	Ability of material to deform plastically without fracturing	% e % RA	Yes
Toughness	Ability of a metal to absorb energy and deform plastically before fracturing	lbf-ft N-m	Yes
Stiffness	Ability of a metal or shape to resist elastic deflection (modulus of elasticity)	psi GPa	No

[a] psi, pounds per square inch (tensile test); MPa, megapascal (tensile test); HB, hardness Brinell (Brinell hardness test); HRC, hardness Rockwell C scale (Rockwell hardness test); % e, percent elongation (tensile test); % RA, percent reduction of area (tensile test); lbf-ft, pounds force-foot (impact test); N-m, newton-meter (impact test); GPa, gigapascal (tensile test).
Source: Ref. 1.

apply sufficient pressure under controlled conditions to cause a permanent change in shape—a technique typically referred to as "cold working." Finally, the thermal technique of heat treatment can also have a pronounced affect on mechanical properties of metals. Not only does the temperature itself have an affect, but of equal importance is the control of the thermal cycle. It is the combination of the rate of heating to a certain temperature, the holding time at temperature, and the cooling rate that will determine the resulting properties. Table 1 indicates which mechanical properties can be affected by heat treatment.

To better understand the process of heat treating of metals and their alloys, a brief introduction will describe metals and the principles behind the internal phenomena that bring about the changes in mechanical properties. Following this will be a presentation of common heat treatment processes used for both ferrous and nonferrous metals. Although it is far from the intent of this chapter to be an exhaustive treatise, it will provide the necessary terminology and basic understanding of heat treatment. The concepts addressed here, as well as those beyond this chapter, are covered in many excellent references on heat treating; some of which are listed at the end of this chapter.

II. STRUCTURE OF METALS

Most commercial engineering metals are not single-element, pure metals, but are commonly alloys, containing two or more elements. In some alloys, the presence of an element or elements is intentional. In others, they are by-products of the ore-extracting and ore-processing operations. Although the cost for complete removal is prohibitive, by-products are kept to a minimum to reduce deleterious effects.

In the transition of a metal from the molten liquid to the solid state, clusters of metal atoms arrange themselves into a regular and repeated geometric pattern in a crystalline structure called a "lattice." As solidification continues, more atoms are added and this arrangement grows three dimensionally until it abuts an adjacent lattice or exhausts the surrounding liquid. This structure is commonly referred to as a "crystal." Material properties are largely dependent on the arrangement of this lattice structure. Commercially used metals (Fig. 1) are aggregates of small, irregularly shaped, randomly oriented crystals called *grains.* The grain size depends on the number of original nucleation sites present during solidification and can subsequently be altered by plastic deformation and heat treatment.

Grains grow at such a rapid pace during solidification that perfect alignment of all atoms in the lattice structure is prevented. This less than perfect structure occurs within individual grains as well as at the boundary between adjacent grains. The resulting crystal imperfections are important in two areas. First, in the deformation of metals, these imperfections can be moved when sufficient stress is applied,

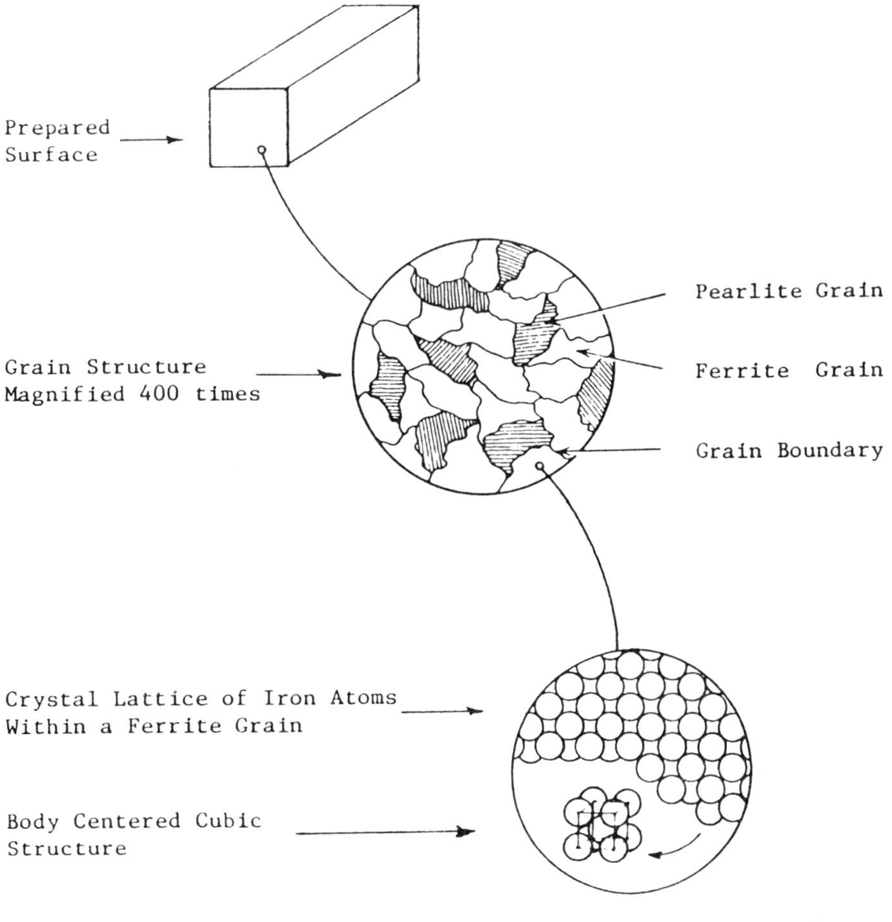

Figure 1 Schematic representation of an annealed 1040 steel at the microscopic and atomic levels.

allowing metals to deform plastically. Plastic deformation capitalizes on the ability of metals to shift their atomic bonds without rupture or cleavage. Planes of high atomic density, or packing, slide past one another in a compressed ripple effect. Commonly referred to as "slip", this phenomenon makes possible many manufacturing operations, such as forging, extrusion, drawing, forming, and cutting. To improve the mechanical properties of strength and hardness, it is only necessary to impede the movement of dislocations or other crystal imperfections. Locking these imperfections in place can be accomplished by alloying with certain elements. This alloying restricts dislocation movement or distorts the lattice structure. Higher stress levels are then required for plastic deformation, thus improving strength and hardness. Cold working or heat treating can also distort the atomic structure to such a degree as to alter mechanical properties.

The second area where imperfections are helpful is in solid-state diffusion or migration of atoms. At elevated temperatures, for example during heat treating, atoms are more dynamic and vibrate so actively that they can move from one lattice position to another. This is all the more facilitated by "vacancies" or other imperfections and is responsible for transformations that occur within metallic structures.

Commercial uses of alloyed metals are so numerous and varied that it is necessary to use general terms to describe their metallic structures in order to discuss their heat treatment and the resultant transformations. The basic structure—that which is physically distinct and mechanically separable—is called a *phase.* The first of two possibilities are the homogeneous phases, consisting basically of *compounds* and *solid solutions.* Chemical compounds are unique substances, the characteristics of which can be used to enhance mechanical properties. Solid solutions consist of alloying atoms randomly substituting for atoms in the lattice structure (substitutional solid solutions) or fitting in the spaces between the solvent atoms (interstitial solid solutions). Both solid solutions and uniformly distributed compounds cause distortion of the lattice structure in the area surrounding the solute atoms, with a resulting change in properties.

The structure known as a *mixture* constitutes the second phase possibility. Mixtures consist of any combination of two or more homogeneous phases. The dominant and most continuous phase determines a material's properties. Heat treatment makes it possible to alter these phases.

III. HEAT TREATMENT PRINCIPLES AND STAGES

Changing the mechanical properties of a metal by altering its metallurgical structure is the main function of heat treatment. Since this involves controlled heating and cooling of the metal, it is important to understand the solid-state transformations or changes that occur during this cycle. The key to successful heat treatment is knowing that phases are present for a given material at a certain temperature and, after a prescribed cooling cycle, being able to predict properties based on resulting structures.

Some metals, notably iron and its alloys, can change lattice structures readily. This polymorphic (many shapes) phenomenon has a pronounced effect on the heat treatability of these materials. A metal that changes from one lattice structure to another is said to undergo a phase transformation. Some metals, such as aluminum, do not change lattice types.

Three stages of heat effect can be considered to occur during a heating cycle. A single-phase solid solution will be used to illustrate these stages. The first is a low-temperature process called recovery. Assuming that the metal was previously cold worked, the lattice would be distorted and grains elongated. As the metal heats up, thermal energy begins to relax the strained lattice. The bonds between metal atoms, which were stretched and held into that position by internal stress, are now relieved and elastic recovery has been established. During this springback of the elastically displaced atoms, there is only a slight change in strength and hardness. If no previous distortion of lattice exists while heating through this region, no alteration in microstructure takes place.

The next stage, recrystallization, occurs as heating continues beyond the recovery range. At the grain boundaries, a strain-free lattice begins to form from the plastically distorted lattice. The resulting small grains are typical of those formed upon solidification and have the same composition and lattice as the original structure. More minute grains form within the boundaries of each original grain until all the distorted lattice has recrystallized. The temperature at which this begins is dependent on the amount of prior cold working; the more severe the deformation, the more residual energy that is present, and a lower temperature will initiate the process.

Further heating, or holding for long periods after recrystallization, results in grain growth. Some of the new grains just formed will begin to enlarge in size by absorbing other less inclined grains. Since mechanical properties are closely linked with grain size, it is important to control the grain growth. As a rule, the smaller grains provide increased strength, hardness, and toughness. When compared to a cold-formed metal, a recrystallized metal is soft, has the lowest strength, and is the most ductile—prerequisites for further plastic deformation. In most cases, after recrystallization grain size will be refined, but the constituents of the grains can vary dramatically with cooling rate.

The final stage of any heat treatment involves the cooling of the metal. The rate of cooling will determine the grain structure and, subsequently, the mechanical properties. Some metals respond by an increase in mechanical properties when cooled rapidly, whereas others may show a decrease. Very slow cooling produces yet another effect. Rates of cooling between these can result in a variety of structures. Different metals typically do not exhibit the same response to a specific heat treatment; the resulting change in properties being dependent on the metal treated. Discussing the heat treatment of specific classes of metals will clarify this concept.

IV. HEAT TREATMENT OF FERROUS METALS

Ferrous, or iron-based, metals constitute the largest tonnage of metals heat treated. Included in this category are plain carbon steels, alloy steels, cast irons, tool steels, stainless steels, and heat-resistant alloys. To simplify this topic, only plain carbon steels are discussed here in any detail. The term *plain carbon* is a qualifier for steels used to designate those that have only carbon as a major alloying ingredient. The term *alloy steel* generally refers to an alloy of iron and an element such as nickel or chromium. Carbon may also be included in this category. The other categories are discussed in a number of references listed at the end of the chapter.

It is important to note that the fundamental heat treatment principles apply regardless of which type of ferrous metal is treated. The main emphasis is the control of heat treatment parameters and the special precautions necessary to produce a satisfactory outcome. Various steels respond differently to heat treatment cycles; the major cause being their distinctive chemical compositions. Knowing at what temeperatures certain phase(s) exist and when phase changes can occur, as a result of a particular cooling rate, is indispensable to heat treating. Two types of transformation diagrams have been developed which provide this information graphically: the equilibrium diagram and the isothermal transformation diagram.

A. Iron Carbon Equilibrium Diagram

The classic iron carbon equilibrium diagram, a portion of which is shown in Figure 2, is used to illustrate the phase(s) in a plain carbon steel. The phase(s) present, at a given temperature, is dependent on the carbon content of the steel when the metal is held at that temperature long enough for equilibrium. Equilibrium is a solid-state condition of stability or balance among atoms so that there is no driving force for change. Detailed discussions of these concepts are contained in the references. However, to apply the basic heat treatment concepts, it will be necessary to have a brief exposure to the transformations that take place within this diagram.

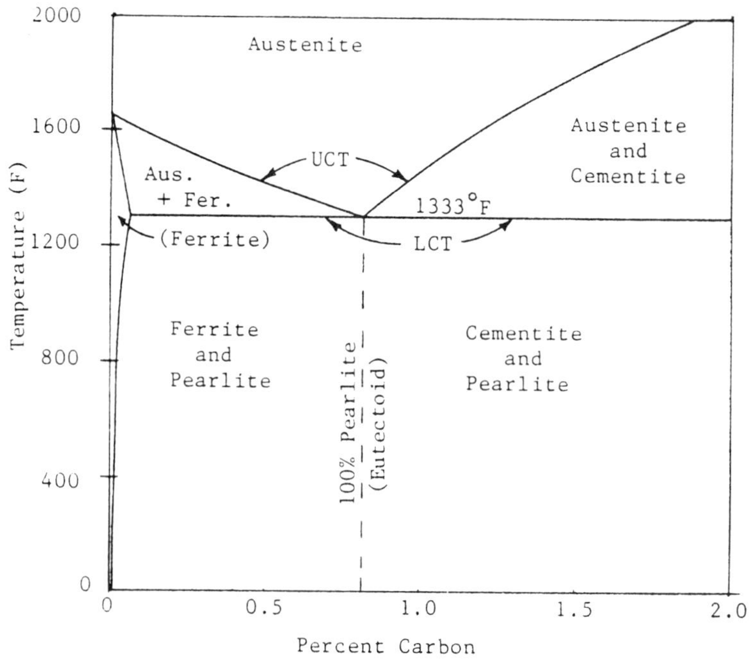

Figure 2 The portion of the iron-carbon equilibrium diagram applicable to heat treatment of plain carbon steels. UCT and LCT indicate upper and lower critical temperatures.

The fact that steel can exist in many phases is among its most important characteristics. For the purpose of discussing heat treating of steels, compositions of less than 2% carbon at temperatures below 2000°F will encompass the most frequently encountered phases (see Fig. 2). In this area, there are regions separated by critical temperature lines which indicate that a change of phase will occur when a line is passed or crossed.

AISI 1040 (American Iron and Steel Institute designation for plain steel with 0.4% carbon) will serve as an example. Since the grain structure of this metal determines many of its mechanical properties, it will be a major focus of this discussion. The heating cycle will be used to illustrate the constituents present and the phase changes that occur at different temperatures. A typical heat treatment cycle begins with the steel at room temperature. A carbon steel, with less than 0.8% C, when slowly cooled to room temperature after solidification, exists as a two-phase structure of ferrite and pearlite (see Fig. 1). The relative amount of each depends on the carbon content: The further the percentage is below 0.8%, the more ferrite; the closer the percentage is to 0.8%, the more pearlite. Ferrite is a solid solution whose body-centered-cubic (BCC) structure has virtually no solubility of carbon. As a result, it is the softest, the lowest strength, and the most ductile phase of steel. The pearlite grain, on the other hand, is a mixture composed of ferrite and cementite that forms during the eutectoid reaction at 1333°F. This transformation generates a laminated grain consisting of alternating layers, or platelets, of ferrite and cementite. Cementite, an iron carbide compound, is very hard and wear resistant, constituting the hardest phase in the iron carbon equilibrium diagram. Because of the laminar cementite, the pearlite grain structure is much stronger, less ductile, and harder than ferrite.

Heat treating the 1040 steel, which can now be considered a polycrystalline aggregate of ferrite and pearlite grains, begins by heating to the lower critical temperature (see Fig. 2). At this point, the pearlite grains begin to transform. Cementite becomes unstable at about 1350°F and starts to dissociate. Simultaneously, the laminar ferrite allotropically transforms by shifting its BCC structure into a new face-centered-cubic (FCC) structure called austenite. Austenite has the ability to absorb interstitially more carbon than ferrite; as much as 2%. Acting somewhat like a sponge, austenite absorbs all the carbon atoms dissolving from cementite until the supply of carbide is exhausted. The microstructure is now the original ferrite grains and the newly formed austenite grains. Interestingly, the reaction is a form of recrystallization and the size of the old pearlite grain has been reduced. Each old pearlite grain is composed of several small austenite grains.

As the temperature rises further, the remaining ferrite grains also begin to transform to austenite. Again, the transformation takes place primarily at the grain boundary, and the ferrite grain is refined into smaller austenite grains. When the upper critical temperature is reached, the resultant 100% austenitic grain structure undergoes an important reaction. The previously pearlitic austenite is high in carbon, 0.8%, and the austenite from the old ferrite is very low. If the temperature is raised 100°F above upper critical and held there, the carbon atoms will diffuse or migrate from areas of high concentration to areas of low concentration. The resulting microstructure will be more homogeneous with respect to carbon distribution and more likely to have uniform properties, particularly when cooled quickly.

Austenite is a key structure in steels. Not only is it desirable for hot forging owing to the ease with which FCC plastically deforms, but austenite signals completion of the first and second stages of heat treatment: those of solution treating and homogenization. When a steel has been completely austenitized, it is ready for the final stage of heat treatment—cooling. A metal incompletely austenized will not reach full potential when cooled, particularly when a single-phase structure is desired as a final structure. Thus metals are held at austenizing temperatures for an hour per inch of the largest cross section to ensure complete transformation.

B. Isothermal Transformation Diagram

The rate at which austenite is cooled determines the final structure of the metal and thus its resulting properties. The diagram most useful in determining the correct rate of cooling is called an isothermal transformation diagram, or more commonly a time-temperature transformation (TTT) diagram. Figure 3 illustrates schematically the transformation products as a function of the cooling rate of austenite for a specific steel. Although developed by studying isothermal cooling effects, the diagrams can be superimposed with continuous cooling curves, resulting in a conser-

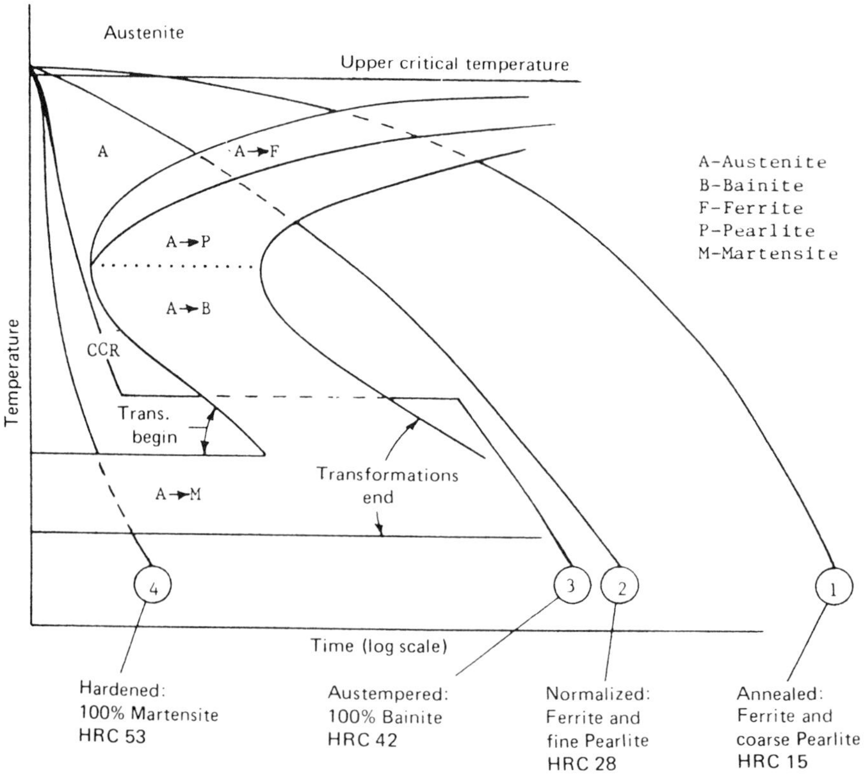

Figure 3 Hypothetical isothermal transformation diagram for AISI 1040 steel superimposed with cooling curves, 1 through 4, illustrating typical heat treatments. Dashed lines indicate transformations.

vative estimate of transformation products. This is normally sufficient for most engineering applications. When more stringent requirements prevail, the continuous cooling diagrams are used.

C. Common Heat-Treatment Processes

The majority of heat-treatment processes require the same first and second stages, differing only in temperature and soaking time, depending on alloy composition. These processes include full annealing, normalizing, and hardening. The exceptions are those restricted to heating temperatures below critical transformation ranges, such as in stress relieving and tempering.

Annealing

The term *annealing* generally refers to the heat-treatment cycle that will impart softness and ductility to a metal in any other previous condition (i.e., cold worked or hardened). When the term is used without qualification, it is considered to mean full annealing. The process involves fully austenitizing the metal and cooling slowly at 50–100°F per hour (see curve 1, Fig. 3), usually controlled by cooling in the heat-treating oven or furnace.

It is important when austenitizing that metals be heated slowly and uniformly. With rapid heating, a temperature gradient can exist; the outside reaches the critical temperature while the inside is still below it. This condition can promote stress at the surface that could cause warpage. Another deleterious outcome of large thermal gradients is that the higher temperature outside will be austenitic for a longer period. There is a possibility that grain growth will occur in this region.

After annealing, the microstructure, representing steel in its lowest strength condition, consists of recrystallized ferrite and pearlite grains. Typical uses for annealing are for improved machinability, for further processing requiring plastic deformation, or for completely removing any internal stresses present from operations, such as casting, welding, forging, and so on.

Stress Relieving

Generally considered as a special category of annealing, stress relieving is a heat treatment for reducing the internal stress. After various manufacturing operations requiring plastic deformation, residual stresses are produced. Even castings or weldments can have induced internal stresses due to the nonuniform cooling normally involved. If these locked-in stresses remain in a material, they can cause premature failure, cracking, or excessive distortion during processing or while in service, rendering the part unfit for use.

The process of stress relief is one of subcritical heat treating; that is, the workpiece is heated to a temperature below the lower transformation or recrystallization temperature. The exact temperature depends on the amount of prior cold working and chemical content. For plain carbon steel, the temperature range of not greater than 1200°F is generally satisfactory. However, should this temperature exceed the recovery range, the metal will begin to recrystallize. Although the stress will certainly be relieved, other desirable mechanical properties may be reduced.

The idea is to reduce only the internal stress without sacrificing strength, hardness, or other properties intentionally induced.

The actual process of stress relieving consists first of heating to the required temperature. The workpiece is held at temperature until uniformly heated; it is then removed from the furnace and allowed to cool in still air or simply furnace cooled. As a rule, parts are placed in a warm furnace and heated to temperature rather than having cold parts thrust into a hot furnace. The nonuniform heating can cause distortion, and the higher the heating temperature, the more important it is to control the heating rate.

Normalizing

Similar to annealing, parts to be normalized are fully austenitized and held at temperature until homogeneous. But instead of the slow furnace cool of annealing, the part is removed from the furnace and cooled by ambient air (normal cooling) at a slightly quicker rate, 200°F per hour (see curve 2, Fig. 3). The faster cooling generates smaller grained ferrite and more abundant pearlite upon transformation from austenite. Additionally, the cementite and ferrite platelets are closer together, resulting in a finer pearlite that that of the coarse-grained annealed pearlite. This contributes greatly to the increased strength and hardness of normalized steels compared to those annealed. It is important to point out that some high alloy/medium carbon steels may respond to air cooling by substantially hardening. Normalizing, with its subsequently refined grain size and corresponding increases in mechanical properties, is often a final heat treatment. Normalizing can also be used to improve machinability characteristics, particularly in low-carbon steels, which when annealed are often too soft and "gummy."

Hardening

One of the most interesting and useful characteristics of carbon steel is its ability to attain extreme hardness and high strength under controlled conditions. Three conditions are essential for successful hardening. The first condition is that sufficient carbon, the hardening element, be present. Steels with carbon contents of 0.3% or above show appreciable response to hardening heat treatment, reaching a high hardness of Rockwell C65 at about 0.8% carbon without significant improvement in hardness as content increases further.

The second condition, as in all heat treatments, is to heat to the correct temperature range and hold to form a homogeneous fully austenitic steel. The final condition is that the steel be cooled at the required rate to prevent the formation of other softer transformation products (i.e., ferrite and pearlite). If an austenitized steel is cooled very rapidly by being plunged into a cold quenching medium, such as water, the normal equilibrium transformation reactions do not have enough time to take place. The carbon normally expelled when slowly cooled does not precipitate out of solution with the iron and form carbides. Instead, the FCC lattice of austenite shifts so abruptly that the carbon is trapped. The FCC iron tries to transform to a BCC structure, but because of the excess carbon atoms present interstitially, a highly strained body-centered tetragonal (BCT) system is formed. This so distorts the atomic lattice structure that very high hardness ensues. The hardness is a function of how severely distorted or strained the lattice becomes as a result of the amount of carbon present in the BCT. This supersaturated solid solution of carbon in BCT iron is called martensite. Martensite does not all form at the same temperature. Transformation begins at a temperature dependent on the chemical composition of the steel and ends several hundred degrees lower. The main objective of quench hardening is to form a fully martensitic structure. The minimum cooling rate that will accomplish this is called the critical cooling rate (CCR). The controlling factor in determining the CCR is again the chemical composition of the steel. The I-T diagrams shown in Figures 4 and 5 illustrate this effect. Note that the nose, or extreme left-hand portion, of the curve is shifted farther to the right in the alloy (4340) than in the plain carbon (1050).

Hardenability is defined as the ability of a steel to be hardened uniformly from the surface of the part to the center. For a given cooling rate, the part size can influence the transformation products. Since the mass effect of large sections causes them to take longer to cool than do thin sections, the result may be softer structures at the center. Alloy ingredients added to steels enhance response to heat treatment and tend to make the material easier to heat treat, particularly when thicker cross-sectioned parts require through hardening. By essentially moving the nose of the TTT curve to the right, the cooling rate at the center of the part may be sufficient for full hardness. Another aspect of this concept is that a slower cooling rate can produce the same result as a faster rate in plain carbon steel. Therefore, slower quenching mediums, such as oil, may be used. If sufficient alloy content is present, the CCR may slow to a point requiring only air cooling while still producing a very hard structure of martensite uniformly throughout its thickness. Metals with this characteristic are said to have high or good hardenability.

Hardenability is commonly measured by a technique called the end quench test. A comparison of the end quench (E-Q) hardenability curves of Figures 4 and 5 shows that both steels have high hardness, HRC 60–55. This is due to the fact that the hardness of a critically quenched, fully martensitic steel is primarily a function of carbon content. But the hardness drops abruptly in the 1050, indicating low hardenability, whereas the 4340 maintains this hardness and thus has high hardenability, the depth to which the hardness exists being largely a function of alloy content.

Tempering

With its high hardness, martensite is correspondingly brittle; usually not a desirable mechanical property. It is therefore common practice to remove or reduce the brittleness by process of tempering. By reheating a quenched, hardened steel to a temperature some point below the lower transformation, additional adjustments in mechanical properties are possible. As a martensitic steel is heated, the lattice structure—which is very distorted as it tightly grips the carbon—begins to release some carbon from solid solution and relaxes. This decomposition of martensite is a function of temperature and holding time. Basically, the higher the temperature and the longer the time, the more pronounced the effect.

When, depending on material design requirements, it may be desirable to maintain high hardness and wear resistance while reducing brittleness, a low tempering temperature, 200–400°F, would be chosen. At the other extreme, if improvement in toughness were required, a higher tempering temperature, 1200°F, would sacrifice

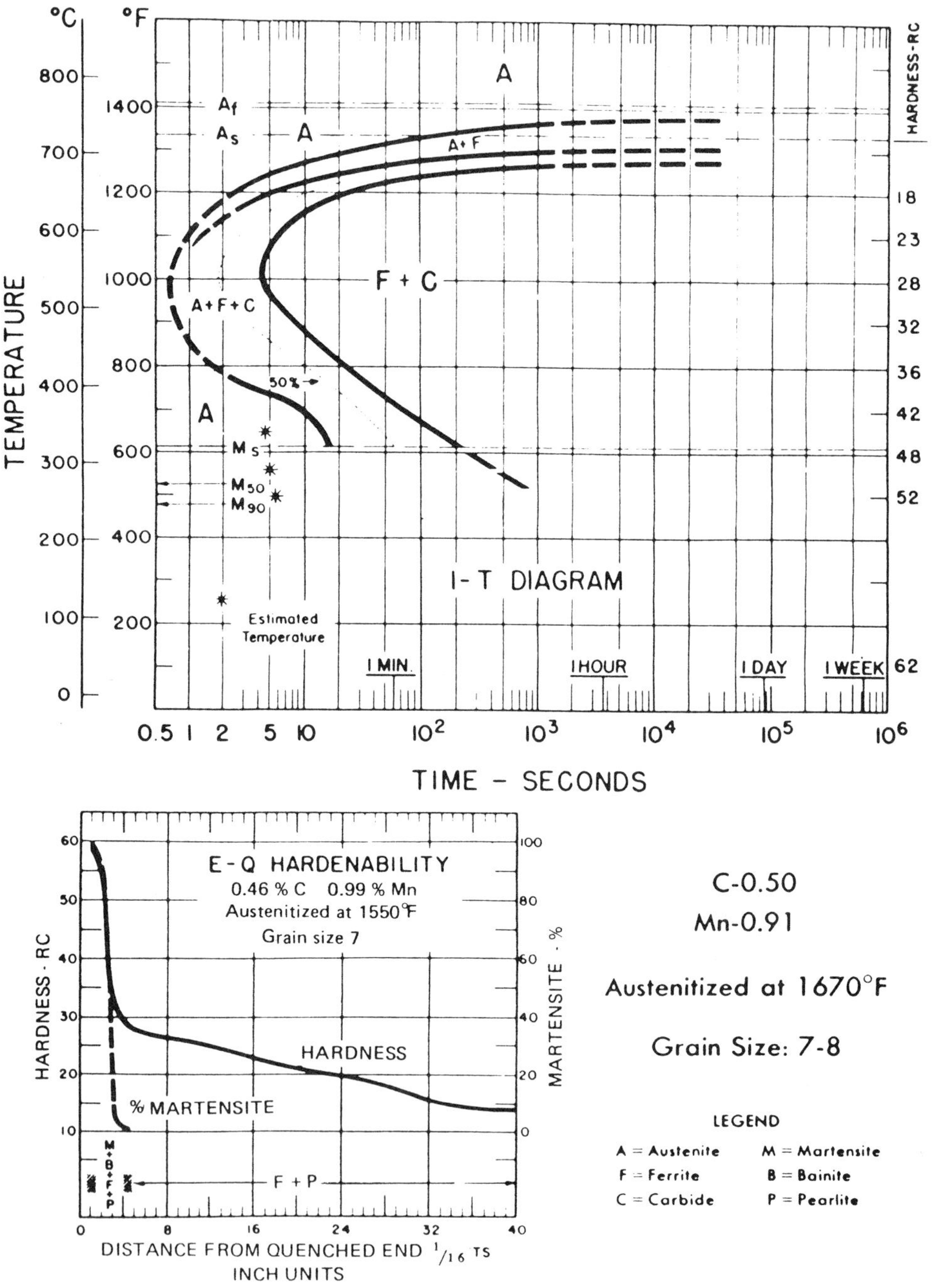

Figure 4 Isothermal transformation (I-T) diagram and end quench (E-Q) hardenability chart for AISI 1050 plain carbon steel. (By permission from Ref. 2. © American Society for Metals.)

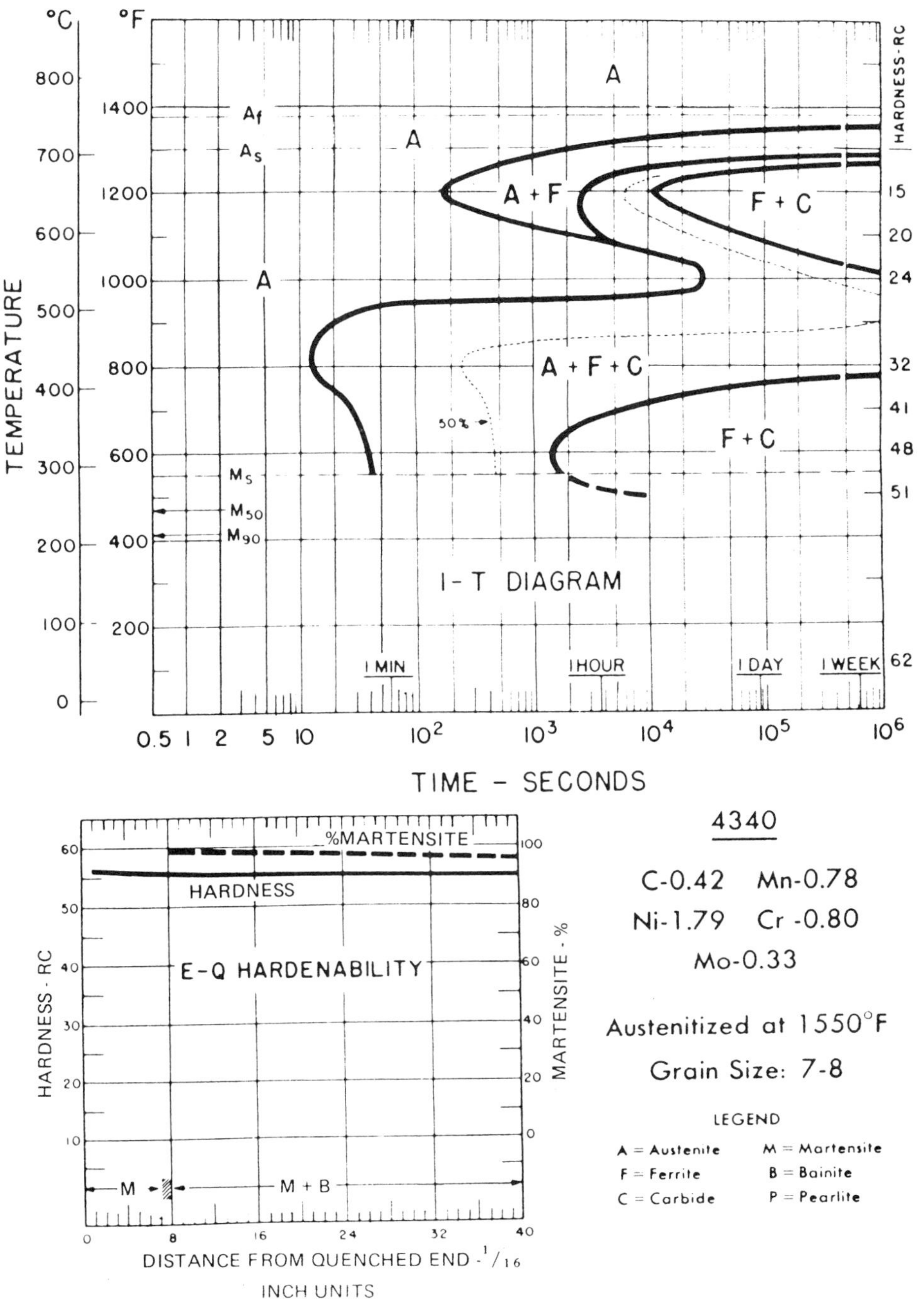

Figure 5 Isothermal transformation (I-T) diagram and end quench (E-Q) hardenability chart for AISI 4340 alloy steel. (By permission from Ref. 2. © American Society for Metals.)

the high hardness for better impact properties. Tempering between these two extremes represents a compromise of hardness and strength for ductility and toughness, as illustrated in Figure 6.

Austempering

An intermediate transformation product known as bainite is the last member of the "ite" family. Full bainitic structure can be formed only by an interrupted quenching process. Austenitized steel is cooled critically, generally into a liquid salt bath, to a temperature above the martensite start temperature. It is then held there isothermally until the complete austenite to bainite transformation occurs and then cooled normally to room temperature (see curve 3, Fig. 3). Metals cooled in this fashion must be of small cross section to be cooled critically. When such a transformation results in 100% bainite, the heat treatment process is called austempering. While having similar strength and hardness of equivalent steels that were quench hardened, an austempered steel will exhibit increased ductility and a significant increase in toughness; very good for lawn mower blades, shovels, and so on. Another advantage of austempering is that it is a complete heat treatment, whereas quench hardening is normally followed by secondary tempering treatment.

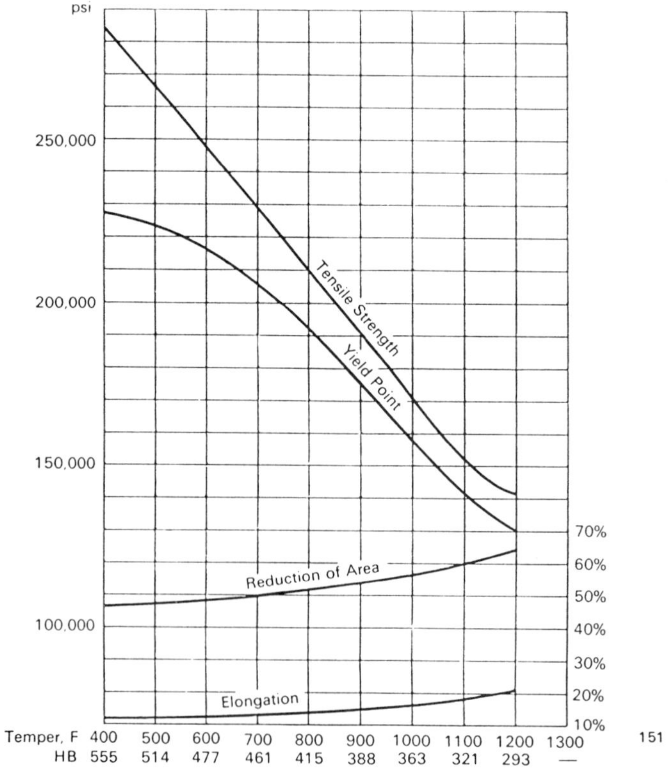

Figure 6 Tempering temperature effect on mechanical properties of an oil quenched 4340. (From Ref. 3.)

Surface Heat Treatment

In some cases, it is desirable to increase the mechanical properties at the surface of a metal only. Two techniques make it possible to treat the surface of steels. First, when the carbon content in a steel is below that necessary for substantial hardening, it can be increased by one of several carburizing processes. Parts to be carburized are austenitized in the presence of a carbonaceous medium, causing carbon diffusion into the surface. With sufficient carbon at the surface, quenching will harden the case while the core, with insufficient carbon, remains softer. It should be pointed out that carburizing is a reversible reaction. Therefore, in heat treatment processes at elevated temperatures, the furnace atmosphere must be controlled to prevent surface decarburizing and scaling. Other surface alloying processes are used to diffuse a variety of elements, often nitrogen, into the surface of steels.

In the second technique, called selective hardening, a hardenable steel (steel with enough carbon to quench harden) is heated on the surface only and rapidly cooled, forming a hard martensite case. The two common processes are known as flame and induction hardening, which refer to the method used to heat the surface. The main advantage of selective hardening is that it is a much faster process and more easily automated than surface alloying. Currently, laser and electron-beam technologies are being used to provide finer control and are particularly well suited to selectively harden parts having complex geometry.

V. HEAT TREATMENT OF NONFERROUS METALS

Whereas ferrous alloys are used for more engineering applications than are nonferrous metals, nonferrous metals have a wider variety of useful characteristics and properties. Most widely used of the nonferrous group for structural applications are the aluminum alloys. Among their important properties is their light weight, coupled with the ability to improve their mechanical properties. These attributes often make aluminum alloys useful as a substitute for the lower strength steels.

Mechanical properties of aluminum alloys can be enhanced by some of the same techniques that are used for steel. For most commercial uses, the pure form of a metal lacks the required mechanical properties. Three mechanisms exist for altering properties of aluminum: alloying, strain hardening, and heat treatment. The first of these creates the most common form of aluminum. Alloyed aluminum can be further enhanced, with respect to mechanical properties, by the other two mechanisms, which categorize aluminum alloys as either non–heat-treatable or heat-treatable alloys.

A. Non–Heat-Treatable Aluminum Alloys

Beyond the initial alloy strengthening, the non–heat-treatable aluminum can be strain hardened. In fact, cold working is the only method possible for increasing the strength and hardness of these alloys, as they do not respond to strengthening heat treatments. Table 2 shows the increases in mechanical properties of typical aluminum alloys that have been work hardened. Notice the temper heading. This temper designation, one of four such developed by the Aluminum Association, does not have the same connotation as tempering heat treatment used for steels.

With respect to aluminum alloys, it refers to additional operations performed on the alloy beyond the initial alloying process. In this case, the series suffix "H" indicates degrees of strain hardening (e.g., 3003 H14).

B. Heat-Treatable Aluminum Alloys

The heat-treatable aluminum alloys can be subjected to thermal processing, which will enhance mechanical properties. This is possible because the elements used in the alloys of this series exhibit increased ability to absorb more alloying elements at high temperatures and a decrease in this solubility at lower temperatures—prime prerequisites for the process known as precipitation hardening. The process involves heating to elevated temperature until alloy ingredients dissolve into solution and become homogeneously distributed among the aluminum atoms. After solution heat treatment, the alloy is rapidly cooled, generally by quenching in water. This traps the elements. The metal is soft and ductile and easily workable immediately after cooling. However, the unstable supersaturated structure soon begins to precipitate naturally. The alloying elements come out of solution in the form of fine compounds which are uniformly distributed. The subsequent distortion of the lattice structure improves strength and hardness of these alloys. This natural aging begins to strengthen rapidly within the first hour and reaches a maximum in 4–5 days, depending on the alloy. The aging process can be accelerated by heating above room temperature. Referred to as artificial aging, this technique is quicker but does not attain as high a strength. Heat-treatable alloys can also be cold worked if done immediately after solution treating, and appreciable increases in strength and hardness result when naturally or artificially aged. Typical mechanical properties are shown in Table 2.

Two additional heat treatments, annealing and stress relief, can be employed regardless of whether the aluminum alloy is categorized as heat treatable or not.

Table 2 Representative Mechanical Properties of Typical Wrought Aluminum Alloys

	Non–heat-treatable						Heat-treatable					
Alloy	1100		3003		5005		2024		6061		7075	
Temper[a]	0	H18	0	H14	0	H18	0	T4	0	T4	0	T6
Tensile strength (ksi)	13	24	16	22	18	29	27	69	18	35	33	83
Yield strength (ksi)	5	22	6	14	6	28	11	57	8	21	15	73
Elongation (% in 2 in)	35	9	30	8	30	6	20	10	25	22	17	11
Hardness (BHN)	23	44	28	40	30	51	47	125	30	65	60	150

[a] O, annealed; H14, strain hardened, half hard; H18, strain hardened, full hard; T4, solution treated, naturally aged; T6, solution treated, artificially aged.
Source: Ref. 4.

Annealing of aluminum or its alloys requires heating to the recrystallization range, which will produce a smaller strain-free grain structure. This softening is particularly important when cold forming, where the reduction is so severe that the metal becomes hard and brittle. The cooling rate after recrystallization is generally slow. The compounds that precipitate during this cooling generally coalesce into large groups and do not impart substantial strengthening to the alloy that is commonly produced by finely dispersed precipitate.

When a manufacturing process induces deleterious stresses on an aluminum alloy it is often recommended that stress relief treatment be employed. The alloy is heated to below its recrystallization temperature, allowed to soak, and then cooled at any rate that will not reintroduce stress. No appreciable loss in strength will occur and harmful stresses will be reduced.

C. Heat Treatment of Copper Alloys

Another, but no less important, category of nonferrous metals is that of the copper alloys. Copper has as its most useful properties high electrical and thermal conductivity. Coupled with high ductility for formability and corrosion resistance for longer environmental life, copper is suitable for a number of commercial applications. When strength and hardness of unalloyed copper are less than required, they can be increased only by cold working, as unalloyed copper does not respond to strengthening heat treatments.

Copper, however, is commonly alloyed with a variety of different elements to meet necessary design requirements. Classified commercially as brasses and bronzes, these copper alloys can be further modified, with respect to mechanical properties, by either plastic deformation (cold working) or heat treatment or both. Each copper alloy has a response to heat treatment unique to its chemical composition.

Brasses

Brass is a copper alloy with zinc as the major alloying element. Brasses can vary in zinc content up to 45% and are commonly called red or yellow, depending on their color. Strength and hardness of brasses increase as a result of solid solution strengthening up to about 40% zinc. Beyond this amount, strength and ductility deteriorate owing to the presence of a brittle second-phase structure which has the possibility for heat treatment, although it is rarely performed. The majority of brasses are of the single-phase type and therefore are not strengthenable by heat treatment. The common strengthening procedure is cold working. To soften a work-hardened brass, an annealing heat treatment is required. This is particularily useful when drastic reductions in cross section, requiring repeated deformations, and typically referred to as severe cold working are needed.

As described previously, distortion of grain structure from cold working sets up a severely strained atomic structure. These residual stresses are relieved in the first stage of annealing. Brasses, especially yellow, are susceptible to the phenomenon of stress-corrosion cracking. This occurs where a stressed metal exhibits an accelerated corrosion at a local surface stress concentration. The localized corrosion causes the formation of a crack which is further propagated by the existing stress, often ending in a premature failure.

By heating to stress-relieving temperatures, but below the recrystallization point, the material will retain cold-worked mechanical properties without stress-corrosion cracking. If reduction in strength and hardness is desired, the recrystallization stage of annealing must be reached. The cooling rate from these temperatures has no effect on single-phase brasses.

Bronzes

Bronze is a generic name applied to most other alloys of copper, with the exception being the copper-zinc brasses. The name implies a better class of alloy than brass with respect to strength and corrosion resistance. Among the numerous bronzes, the most common commercially used are those where copper is alloyed with tin, aluminum, silicon, or beryllium.

Tin bronze has higher strength and hardness and wear resistance due to the presence of a relatively brittle copper-tin compound which forms upon solidification. Stress relieving and annealing are the most common heat treatments for bronze.

Aluminum bronze, unlike tin bronze, can be quench hardened for superior strength, hardness, and wear resistance. The resulting microstructures are similar to that found in steel. When tempered, aluminum bronze will be additionally strengthened owing to a small amount of precipitation hardening.

Silicon bronze, similar to tin bronze, can only be annealed or stress relieved, as it does not respond to quench hardening. However, silicon bronze has the distinction of being the strongest of the work-hardenable copper alloys, with comparable mechanical properties to mild steel and corrosion resistance of copper.

Beryllium bronze (also called beryllium copper) enjoys commercial importance due in part to acquired strengths as high as those of alloy steel. This most dramatic improvement in mechanical properties, coupled with good conductivity and corrosion resistance, make beryllium bronze widely used in electric contacts, switches, and so on, where high hardness, fatigue strength, and wear resistance are needed. These enhanced properties are obtained by precipitation hardening similar to that used in aluminum alloys (see Section V.B).

All copper alloys can be obtained from the mill or supplier in the desired form and condition (e.g., strip, bar; solution annealed, hardened). Of particular importance is specifying the condition particularly when additional fabrication is necessary, such as the punch press operations of piercing, blanking, and forming. The ease with which these operations are performed is related to their preforming properties and should be matched accordingly.

The precipitation hardening of beryllium bronze is a two-stage thermal treatment. The solution heat treatment can and should be supplied by the mill. The end user then need only perform the hardening heat treatment after forming to desired configuration. The age hardening is accomplished simply by heating to the recommended low temperature, causing a fine dispersion of precipitated beryllium-copper compounds. When severely deforming beryllium bronze for special operations, such as deep drawing, it may be necessary to soften and restore preforming properties. This would normally be accomplished by annealing; however, if beryllium bronze is fully annealed, it will not respond to precipation hardening. Therefore, the work should be solution annealed by heating to above the recommended critical temperatures for a sufficient time to allow complete decomposition of precipitate followed by critical quenching. This will produce a workable product by preventing the precipitate from forming. The ease of prehardened fabrication and advantage of age hardening to high strengths make beryllium bronze one of the more commercially important copper alloys.

VI. QUALITY IN HEAT TREATMENT

Heat-treatment parameters must be considered and properly orchestrated to produce the desired effect of the heat-treatment cycle. Obviously, the process must be carefully controlled using the proper equipment, correct and accurate temperatures, suitable quenching media, and prescribed procedures for consistently satisfactory performance. In assuring the quality of heat-treated parts, process control requires being able to determine when parts do not meet engineering specifications. A number of destructive and nondestructive tests are available to ascertain the change in mechanical properties due to heat treatment. Among the most important of these tools are hardness testing and metallographic analysis (see Chapter 37).

Hardness testing after heat treatment can indicate if a part is within hardness specifications. Hardness of a metal is generally a good indication of the microstructure present. The Rockwell and the Brinell hardness tests can be performed with relative ease and speed; therefore lending themselves well to process control.

If parts are not within tolerance, the cause should be investigated. Hardness testing, eddy current testing, and other such testing techniques will indicate only when an out-of-tolerance condition exists, not why it exists. Clues to the cause(s) are often revealed in the microstructure and can be examined metallographically. This involved process prepares the surface of a sample part for microscopical viewing and subsequent interpretation. Grain structures are clearly visible, making it possible to determine what phases are present. For example, incorrect phases in a quench-hardened steel (e.g., ferrite and pearlite together with the desired martensite) may reveal improper heat treatment and indicate a possible cause to investigate.

VII. CONCLUSION

The focus of this chapter has been to introduce the concept of heat treatment as a common manufacturing process which can drastically modify metallic properties and thus extend the usefulness of engineering metals. For the process to be effective, it must be correlated with other manufacturing processes so that economical use of metals can be optimized.

Heat treatment was defined as controlled heating and cooling of metals purposefully to obtain desirable mechanical properties. Several manufacturing operations involve incidental heating and cooling cycles in their processing (i.e., casting, hot forging, and welding) but are not considered to be heat treating. Only when thermal treatment is for the express purpose of predictably altering mechanical properties is it considered heat treating. However, this is not to say that the heat effect and rate of cooling will not alter mechanical properties with these other operations.

In review, there are several keys or factors to successful heat treatment of metals. First, the content of the alloy to be treated must be known as well as its prior processing history. Cold working or hot working can affect the metal's response to a particular heat treatment. Also, the quality of the chemical content must be controlled as well as the cleanliness, freedom from segregation, and grain size. Next, the required heating temperature and rate of heating needs to be determined. The proper elevated temperature is crucial in modifying mechanical properties. After reaching the desired temperature, a soaking period is necessary so that the heat affects the microstructure equally. The metal is held at temperature until uniformly homogeneous or until the desired condition is obtained. It is important that furnace atmospheres be controlled so that deterioration of the part surface is reduced. The final stage, and perhaps the most crucial, is the cooling rate. When metals are heated to above their critical temperatures and cooled at rates varying from drastic water quenching to milder furnace cooling, a variety of mechanical properties are possible.

It is recommended that those interested in heat treatment consult the list of references at the end of this chapter. The one single source for comprehensive and in-depth information regarding heat treatment and process control is the American Society for Metals. Many other useful references are also listed, ranging from theoretical to the practical. Major manufacturers and suppliers of commercial metals can also be excellent sources of information.

REFERENCES

1. American Society for Metals, *Metals Handbook,* Vol. 1, Properties and Selection, ASM, Metals Park, OH.
2. American Society for Metals, *Atlas of Isothermal Transformation and Cooling Transformation Diagrams,* ASM, Metals Park, OH, 1977.
3. Bethlehem Steel Corp., *Modern Steels and Their Properties*, Handbook 3310, Bethlehem, PA.
4. Aluminum Association, Inc. *Aluminum Standards,* Washington, DC, 1979.

FURTHER READING

Aluminum Association, Inc., *Publications Guide*, AAI, Washington, DC.

American Society for Metals, *Metals Handbook*, 9th ed. Vol. 4, Heat Treating, ASM, Metals Park, OH, 1982.

American Society for Metals, *Heat Treater's Guide*, ASM, Metals Park, OH, 1982.

American Society for Metals, *Principles of Heat Treatment of Steel*, ASM, Metals Park, OH, 1980.

Avner, S. H., *Introduction to Physical Metallurgy*, McGraw-Hill, New York, 1974.

Climax Molybdenum Co., *Literature on Molybdenum Bearing Steels*, Greenwich, CT.

Flinn, R. A., and P. K., Trojan, *Engineering Materials and Their Applications*, Houghton Mifflin, Boston, 1981.

Machine Design, Materials Reference Issue, Penton/IPC, Inc., Cleveland, OH, 1983.

MIL-H-6875F, *Process for Heat Treatment of Steel*, and MIL-H-6088F, *Heat Treatment of Aluminum Alloys*, Naval Publications and Forms Center, Philadelphia, PA.

Republic Steel Corp., *Alloying Elements and Their Effects*, Cleveland, OH, 1976.

United States Steel, *The Making, Shaping and Treating Steel*, 8th ed. Pittsburgh, PA, 1964.

23
Grinding Technology

K. SUBRAMANIAN and SRIHARI NANDYAL
Norton Company
Worcester, Massachusetts

I. INTRODUCTION

Grinding as a surface generation process may be defined as a material removal process to achieve surfaces of desired geometry, tolerances, and surface characteristics. These surface generation methods are used usually to achieve certain functional characteristics of the surface. The processes addressed in this chapter are occasionally used to achieve a better appearance or improve the cosmetic features of surfaces, even though these could occur as secondary benefits.

Most grinding methods are carried out as part of a series of industrial processes used to produce or manufacture parts or components. As an example, a cast product may need snagging or cutting off of gates and risers using grinding wheels before it is shipped for use or sent to the next component fabrication department. A forged shaft may be cut or finished by grinding wheels before it becomes an industrial component for use in an assembly process. Bearing or gear components after heat treatment may be ground to desired tolerances and surface quality before they are assembled into finished bearings or transmissions, respectively. Computer parts such as microchips or magnetic heads may be ground, lapped, or polished before they are sent to assembly operations. Jet engine blades may have their surfaces ground to geometry and tolerances prior to their use in the assembly process to manufacture jet engines. Grinding technology plays a key role in a wide range of industries.

In this chapter, an overview of the various grinding processes is presented followed by a detailed description of the constituents of grinding wheels and their effect on grinding performance. A "systems approach" to apply our knowledge of the grinding systems effectively is also addressed in detail.

II. OVERVIEW OF GRINDING PROCESSES

Grinding is a precision material removal process which utilizes hard abrasive particles as the cutting medium. Material is removed from the workpiece by the mechanical action of abrasive grains. These abrasives may be used in loose form or in a rigid or flexible backing depending on the precision and quality requirements of the manufacturing operation. In general, grinding processes can be classified into rough, precision, and ultraprecision operations. Figure 1 details the various abrasive processes.

A. Rough Grinding

Rough grinding, which involves high stock removal, is usually carried out with coarse abrasives held on a rigid (bonded) or semirigid (coated) matrix. In these operations, the surface quality of the workpiece, although important, is not the main concern. Cutoff, snagging, and steel conditioning are typical rough grinding operations. *Cutoff* operations involve cutting off of blanks from a rod and utilize rubber or resinoid bonded wheels. These wheels are usually reinforced in order to withstand the high pressures involved in the operation. *Snagging* operations are performed using a portable grinding machine and are used for removing risers, gates, and flashings on semifinished castings. Figure 2 shows a typical snagging operation. *Steel conditioning* operations involve very high stock removal rates and utilize high-speed, high-horsepower grinders to remove defects, such as, for example,

ROUGH GRINDING

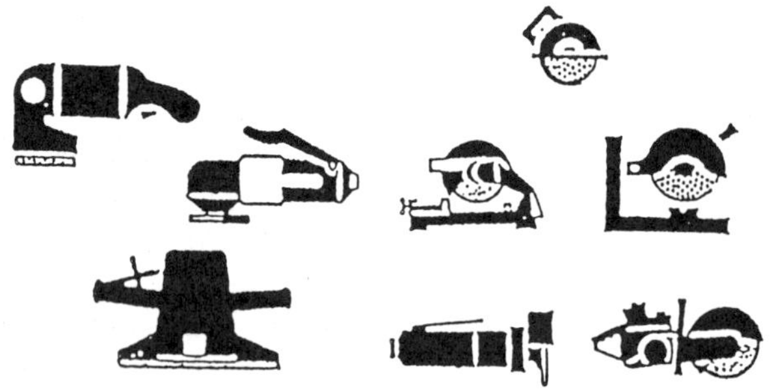

END USE/APPLICATIONS

- Foundry Snagging
- Billet Conditioning
- Cut off bars, rods, etc.
- Stone sawing
- Concrete Cutting

Figure 1 Overview of abrasive machining processes.

PRECISION GRINDING

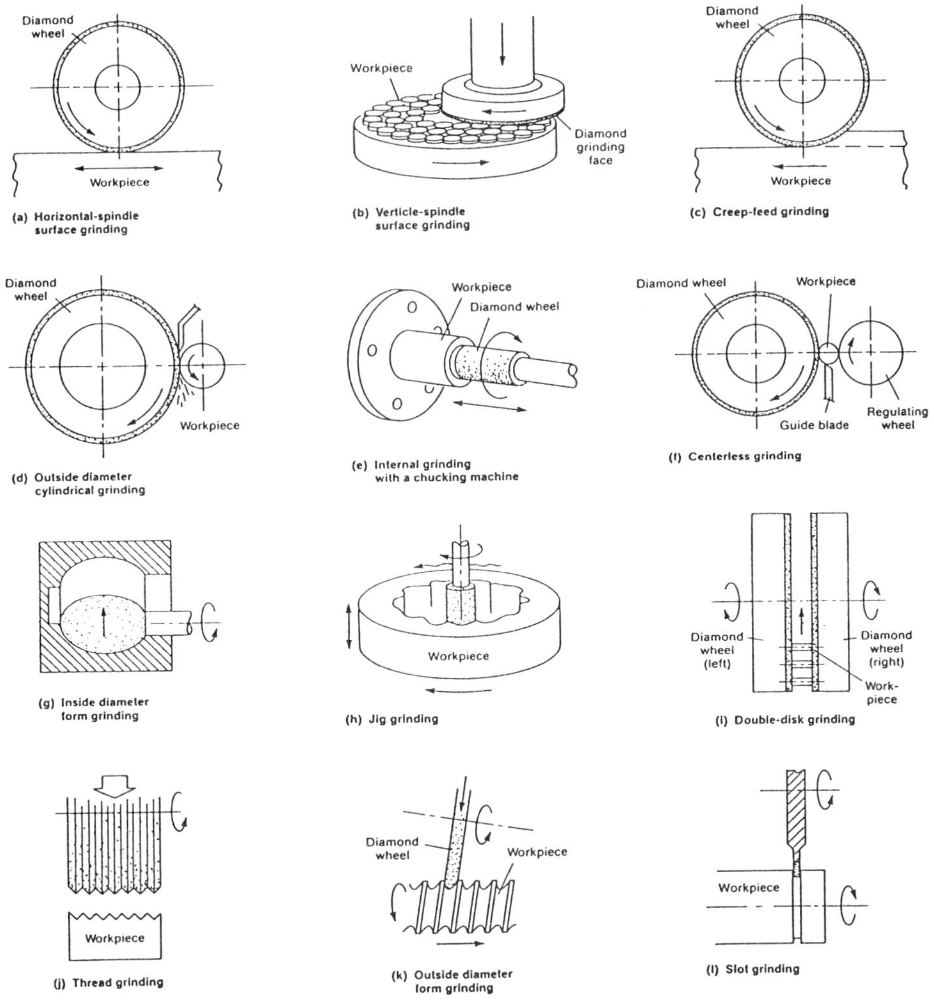

END USE/APPLICATIONS

- Precision Components Finishing
- Automotive
- Aerospace
- Tool Production
- Wear Parts
- Bearings
- Electronics
- Optics

Figure 1 Continued

ULTRA-PRECISION PROCESSES

Lapping

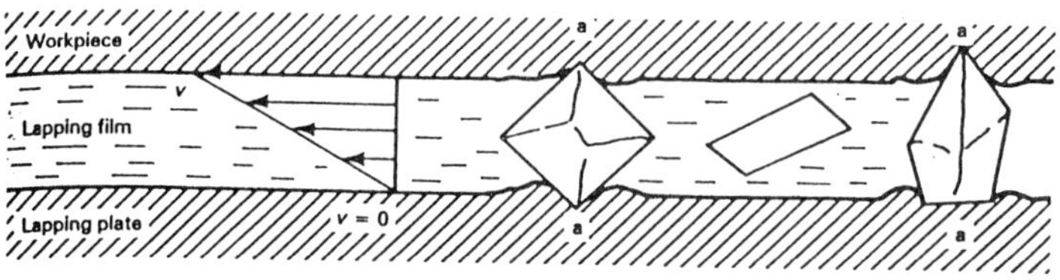

Honing

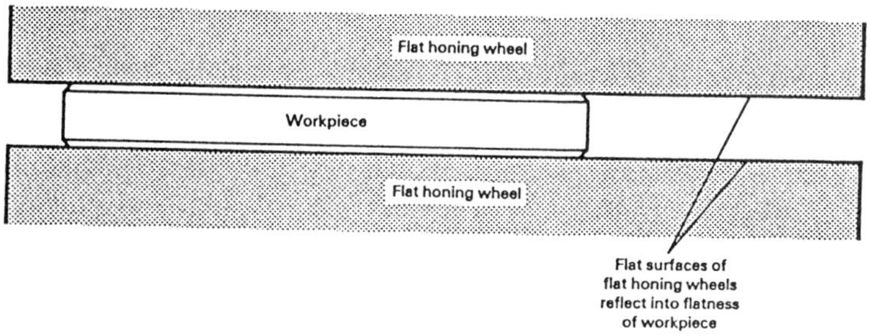

Polishing

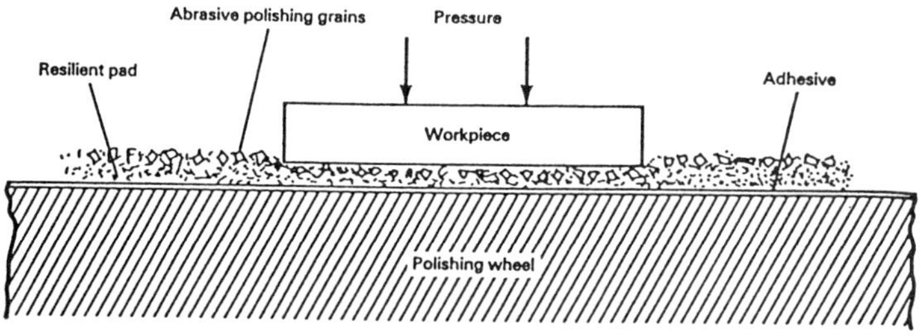

END USE/APPLICATIONS

- Precision Mating Surfaces used in a variety of tribological applications (e.g.)
    - Hydraulic Cylinders
    - Piston/Cylinder
    - Bearings
    - Magnetic Heads
- Precision surfaces used for optics, such as lenses

Figure 1 Continued

Figure 2 Snagging operation using an offhand grinding machine.

scabs, seams, or cinder patches, from semifinished steel that have been hot rolled from steel ingots and allowed to cool.

B. Precision Grinding

Precision grinding with bonded or coated abrasive tools involves low to high material removal rates using fine to coarse abrasive grains. The operations involving bonded grinding wheels utilize stiff, precision grinding machines and are the most commonly used finishing operations in the industry today. Horizontal spindle surface grinding, internal grinding, cylindrical grinding, creepfeed grinding, centerless grinding, and form grinding are typical precision grinding operations. Figures 3a and b are examples of external cylindrical grinding and dry tool and cutter grinding operations, respectively. Typical coated abrasive operations include coil grinding, abrasive machining, conveyor grinding, and sheet and plate dimensioning.

C. Ultraprecision Operations

Ultraprecision operations such as lapping, honing, and polishing mainly use micron-sized abrasive particles in bonded and loose forms. These processes involve very

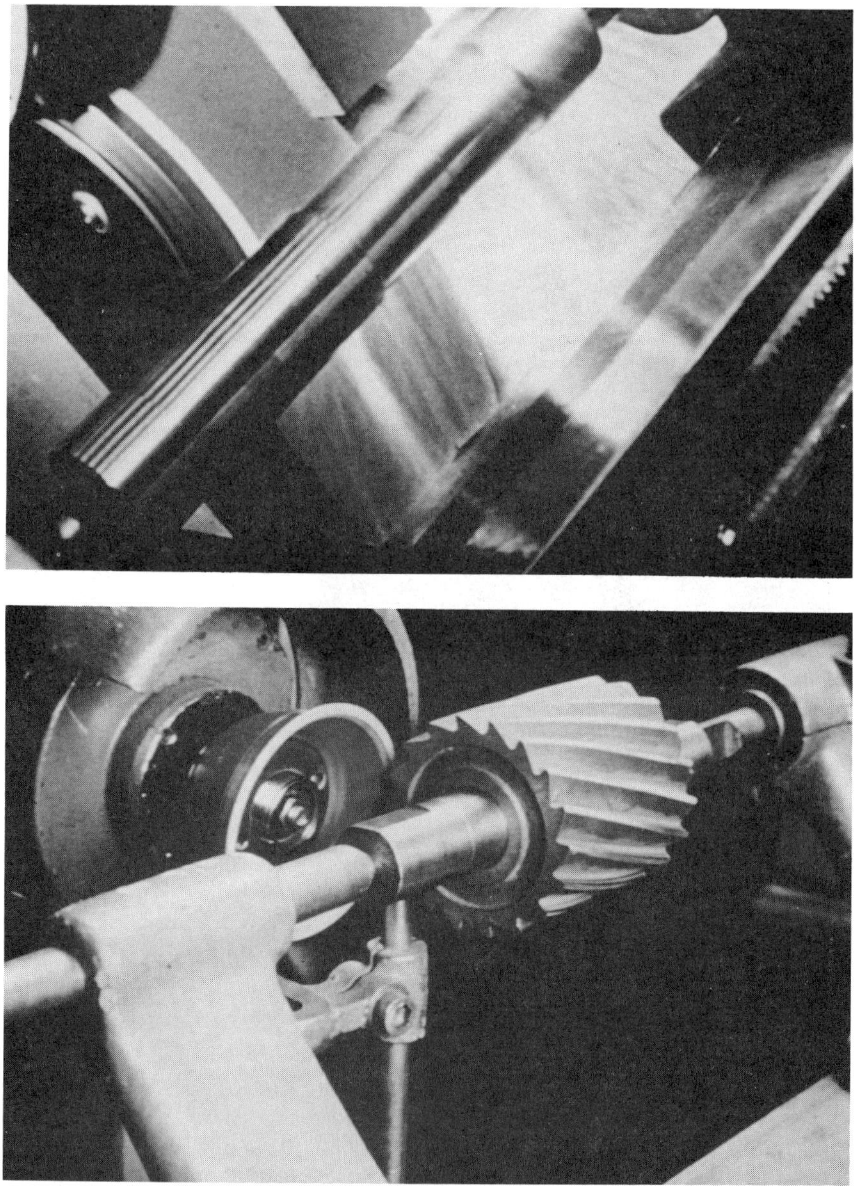

Figure 3 Examples of precision grinding operations: (top) external cylindrical grinding; (bottom) dry tool and cutter grinding.

low material removal rates and can generate parts with extremely fine surface finishes. Honing, which utilizes fine grit bonded abrasive sticks, is a controlled, low-speed, surface finishing process. It can produce surface finishes in the range of 32–8 μin (0.8–0.2 micron) and is used to finish gear teeth, valve components, cylinder bores, and races for ball and roller bearings. Figure 4 shows some of the

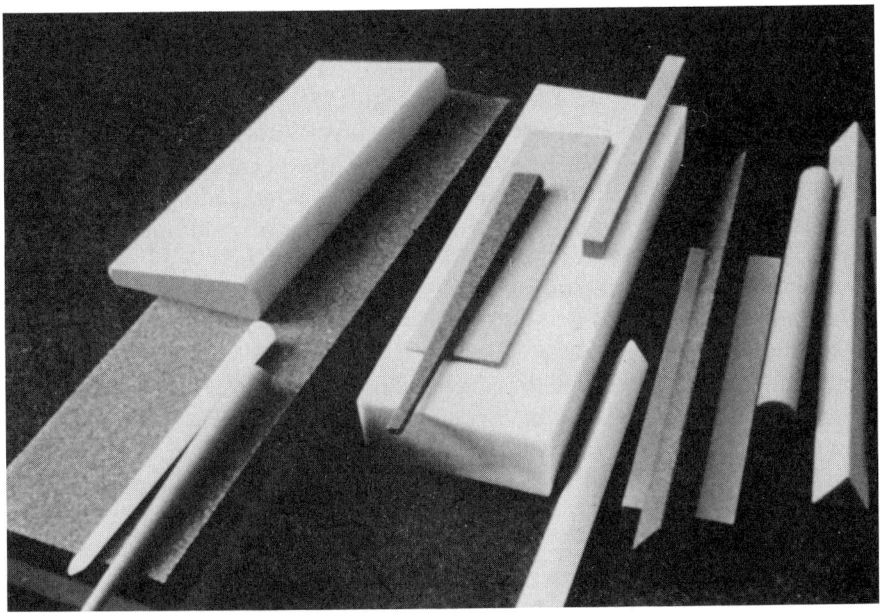

Figure 4 Fine grit honing sticks.

commonly used bonded honing sticks. Lapping and polishing are low-speed, low-pressure abrading operations which involve material removal with the help of loose abrasives carried in a liquid medium. They are commonly used for refinement of surface finish, to generate extremely close fit between mating surfaces, and to remove damaged surface and subsurface layers generated by previous machining operations.

In this chapter, emphasis will be given to precision grinding operations as they are the most widely used finishing operations used in the industry today.

III. CONSTITUENTS OF A GRINDING WHEEL

The most important tool used in a grinding operation is the grinding wheel. A grinding wheel consists of hard abrasive particles held together in a rigid matrix. The matrix, or bond, which may or may not be porous, is made of materials such as glass, resin, or metal depending on the type of abrasive being used. In this section, the composition of grinding wheels will be considered in order to understand how grinding wheels are marked and selected.

A. Wheel Marking System

In order to identify the composition of a grinding wheel, simple marking systems have been developed by wheel manufacturers. These marking systems indicate the following:

1. Abrasive type and size in the wheel
2. Hardness of the wheel as represented by the grade

3. Structure or the amount of abrasive content
4. Bond type
5. Bond modification which varies for different manufacturers

The standard marking system used for conventional abrasive wheels is presented in Figure 5 (1). The letters A or C indicate the abrasive type. A is for aluminum oxide and C is for silicon carbide. The number preceding these letters refers to a modification of the abrasive grain, as several types of aluminum oxide and silicon carbide are available with different chemical compositions and properties. Recently, a new aluminum oxide abrasive called "seeded-gel" has been introduced in the market and is designated as SG. Other "sol-gel" abrasives are also being introduced by abrasive manufacturers.

The number to the right of the abrasive type indicates the size of the abrasive grain in terms of the grit number, which correlates to the number of linear holes per inch of a sieve through which the abrasive grains can pass. The greater the grit number, the finer the abrasive particles in the wheel. Table 1 lists the typical grains sizes for a given mesh or grit size (1).

The grade, or hardness, refers to the strength of the grinding wheel. It is a measure of how strongly the abrasive grains are held together in the grinding wheel and indicates the extent of the bond as compared to the abrasives. Thus a harder grade wheel will have more bond than a softer grade wheel for the same abrasive content. Wheel grades are generally designated by a letter from A to Z; A being the softest and Z being hardest.

The structure number is an indication of the volumetric concentration of the abrasive grains in the wheel and is also a measure of the distance between the abrasive particles. The structure number usually varies from 0 to 25; a higher number indicating fewer abrasive grains and a more "open" wheel.

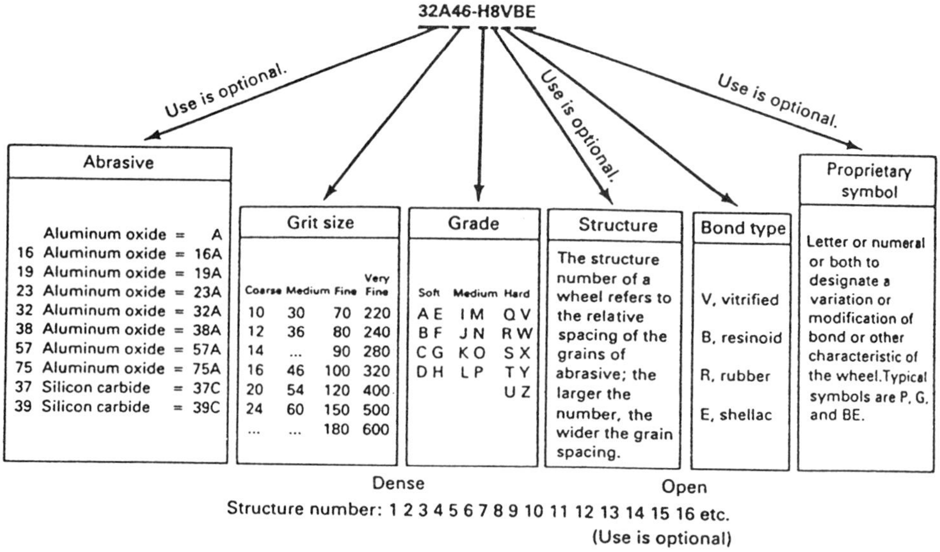

Figure 5 Standard marking system for conventional abrasive grinding wheels.

Table 1 Mean Particle Sizes for Grits Used in Conventional Abrasive Grinding Wheels

	Particle size (mean)	
Grit size	μm	in
4	6848	0.2577
6	5630	0.2117
8	4620	0.1817
10	3460	0.1366
12	2550	0.1003
14	2100	0.0830
16	1660	0.0655
20	1340	0.0528
24	1035	0.0408
30	930	0.0365
36	710	0.0280
46	508	0.0200
54	430	0.0170
60	406	0.0160
70	328	0.0131
80	266	0.0105
90	216	0.0085
100	173	0.0068
120	142	0.0056
150	122	0.0048
180	86	0.0034
220	66	0.0026
240	63	0.0024
280	44	0.0017
320	32	0.0012
400	23	0.0009
500	16	0.0006
600	8	0.0003
900	6	0.0002
Levigated alumina	3	0.0001

Note: Grit size varies indirectly with particle size.
Source: Ref. 1.

The bond type used is represented by a letter: V for vitrified bonds, E for shellac bonds, and R for resinoid bonds. The symbols after the bond type are usually the manufacturer's designation, which helps to identify the constituents and nature of the bond.

Figure 6 shows the wheel-marking system for superabrasive wheels: The system is similar to the conventional system, with the difference being the manner in which the abrasive content is designated. The concentration number refers to the volumetric concentration of the superabrasive grits in the wheel. Thus a wheel with a 100 concentration would contain ≈25 volume percent of abrasive grains. Diamonds are designated by the letter D, whereas cubic boron nitride (CBN) is designated by the letter B. The

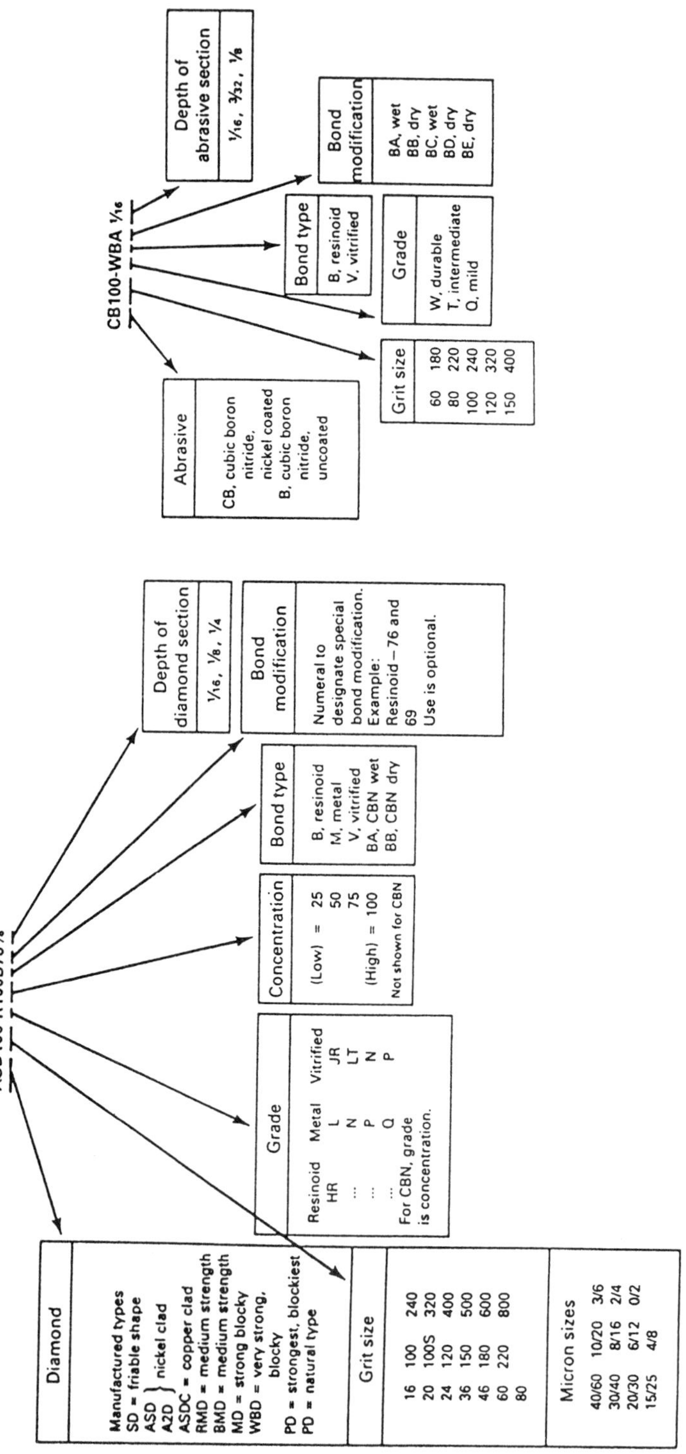

Figure 6 Standard marking system for superabrasive grinding wheels.

fraction number at the end of the markings refers to the thickness of the diamond or CBN section. Superabrasive wheels, which are usually costly, consist of a core material to which the diamond or CBN rim is cemented. Superabrasive wheels are also available in metal bonds and such bonds are represented by the letter M. There are also superabrasive wheels made of electroplated or brazed layers of abrasives on a steel preform. These are designated as EP or MSL, respectively.

B. Types of Abrasive Grains

Grinding is a material removal operation involving chip formation and abrasion and in order to remove material the abrasives used should be at least twice as hard as the material being ground. For this purpose, extremely hard abrasives are utilized in a grinding wheel. Abrasives are usually classified as *conventional* abrasives and *super*abrasives. Aluminum oxide, silicon carbide, and zirconia-alumina are conventional abrasives, whereas cubic boron nitride (CBN) and diamond are classified as superabrasives. The essential difference between the two abrasive classifications is evident from Figure 7, which compares the Knoop hardnesses of the various abrasives against some common work materials. The conventional abrasives, SiC and aluminum oxide, have much lower hardnesses than diamond and CBN superabrasives. Figure 8 is a chart of the relative usage of the various abrasives. Aluminum oxide and CBN are used mostly to grind ferrous metals, whereas SiC and diamond are used for grinding nonferrous metals, ceramics, and carbides. Diamond, in spite

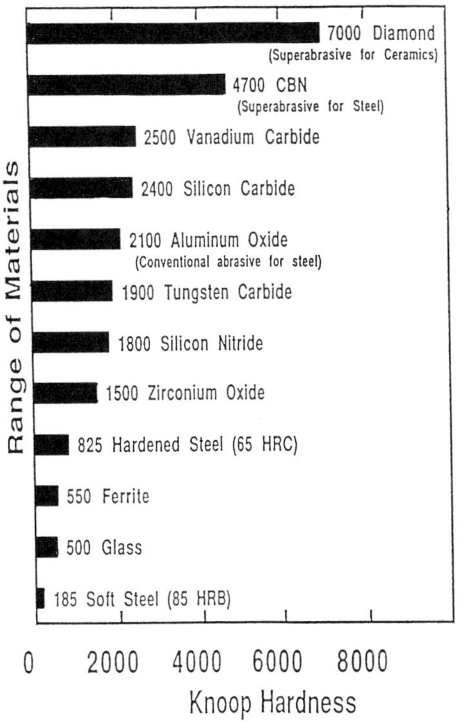

Figure 7 Comparison of hardness of various abrasives and some common work materials.

WORK MATERIAL	ABRASIVE PRODUCTS			
	CONVENTIONAL ABRASIVES		SUPER ABRASIVES	POWDERS, SLURRIES, COMPOUNDS (CONVEN./SUPER ABRASIVES)
	BONDED	COATED		
PLASTICS	LOW	LOW	LOW	NONE
COMPOSITES	LOW	HIGH	HIGH	LOW
METALS	HIGH	HIGH	HIGH	HIGH
STEEL	HIGH	HIGH	HIGH	HIGH
GLASS	LOW	HIGH	HIGH	HIGH
CARBIDES	HIGH	LOW	HIGH	HIGH
CERAMICS	LOW	LOW	HIGH	HIGH
WOOD	LOW	HIGH	HIGH	NONE
STONE	NONE	LOW	HIGH	LOW
MINERALS	NONE	NONE	HIGH	NONE

RELATIVE USAGE: HIGH (filled box), LOW (shaded box), NONE (empty box)

Figure 8 Relative usage of various abrasive products.

of its extreme hardness, has been found to be uneconomical for the grinding of ferrous materials owning to graphitization and carbon diffusion into the iron causing excessive diamond wear.

Conventional Abrasives

Fused aluminum oxides are manufactured with bauxite as the main raw material by three different methods (2). Bauxite may be fused directly with coke and iron in an electric furnace; it may be processed to form *Bayer* process alumina, which is then fused; or it may be sintered after pressing.

Regular or brown aluminum oxide produced by fusing calcined bauxite with a small amount of coke and iron has lower hardness and higher toughness than the purer white and monocrystalline varieties. This tough abrasive is used in a wide variety of operations from heavy-duty to rough and semifinish grinding.

Monocrystalline aluminum oxide is manufactured by the fusion of a mixture of bauxite, coke, iron sulfide, iron filings, and alkaline metal oxides. This material is purer than brown aluminum oxide and contains small amounts of metallic oxides as alloying elements. This type of aluminum oxide is used extensively in precision grinding operations.

White aluminum oxide is manufactured by the fusion of Bayer process alumina. This material is very pure and contains small amounts of residual sodium oxide which volatilize during melting giving rise to a porous structure. Although these abrasives have low impact strength, they find wide applications in precision grinding.

Sintered alumina is manufactured by sintering at 1500°C granules obtained by crushing a pressed or extruded form of finely milled (1–5 micron) calcined bauxite. This abrasive has a very fine crystal size and high strength and is suitable for conditioning and other heavy duty grinding operations.

Zirconia-alumina is manufactured by the fusion of a mixture of calcined bauxite, zircon sand, petroleum coke, and iron filings in an electric arc furnace. Rapid solidification of the molten crude on water cooled steel plates results in a eutectic of alumina containing 40% zirconium. Such a material has high impact strength and toughness, which are the desirable characteristics for an abrasive to be useful in heavy grinding conditions.

The most recent development in aluminum oxide synthesis is an unfused variety called seeded-gel (or sol-gel) alumina. This is a very pure form of the aluminum oxide and is made by a ceramic process in which submicron particles are sintered to form microcrystalline abrasive grits. These abrasives are strong and provide dramatic performance improvements compared to fused aluminum oxides. Other forms of sol-gel alumina abrasives are also commercially available with superior performance over monocrystalline alumina abrasives.

Silicon carbide is manufactured by the reduction of sand with excess coke in an electric arc furnace. It is available in two varieties, green and black, with the green silicon carbide being purer and slightly harder than the black variety. Silicon carbides are used in the precision grinding of nonferrous metals, ceramics, and cast irons. They are not suitable for grinding of most ferrous metals because of their reactivity with iron in the steels (3).

Superabrasives

Diamond and cubic boron nitride (CBN) are the two superabrasive materials used in grinding applications. Diamond, available in both natural and synthetic forms, is the hardest known material. Cubic boron nitride, which is the second hardest, is a manmade material and does not exist naturally. Both the superabrasive materials are synthesized at high pressures and temperatures in the presence of molten catalyst solvents.

Synthetic diamonds are available in a wide range of shapes and crystal structures. Weak, friable, polycrystalline diamonds are used for grinding tough and brittle materials like ceramics and carbides, whereas stronger, well-formed crystals find applications in the grinding of concrete, stone, and glass.

The most popular form of CBN, used for grinding of most steels and nonferrous high-strength alloys, is a monocrystalline variety. A tougher, microcrystalline form of CBN is also available.

Diamond and CBN are also available with a coating of nickel metal. These metal-coated abrasives are predominantly used in resin bonded wheels and increase the grinding wheel life two to three times that of wheels with uncoated diamonds. There are several advantages in using thick metal coatings on the superabrasives in resin bond systems (4):

- High strength and large surface area reinforces retention of abrasive.
- Relatively low thermal conductivity metal buffers degradable resin from grinding heat.
- Higher surface area and superior wettability increase bonding strength over uncoated abrasive.

Diamond abrasives are also available with copper coatings and are predominantly used in dry grinding applications.

C. Types of Bond

In order to grind effectively the large range of materials required by industry, a variety of bonding systems is available to hold the abrasives in the wheel. The most common bonding materials are resin (phenolic), ceramic (glass), or metal (bronze). Other bond types such as shellac, oxychloride, rubber, and silicate are also available but are not commonly used. Figure 9 is a schematic of the various abrasive products and the bonding medium used to hold the abrasives together. The superabrasive wheels are also available in a metal-single-layer (MSL)–type specification which consists of a single layer of the abrasives held together chemically with the help of a metallic braze. Other superabrasive products with single layer of abrasives electroplated to the preform are called

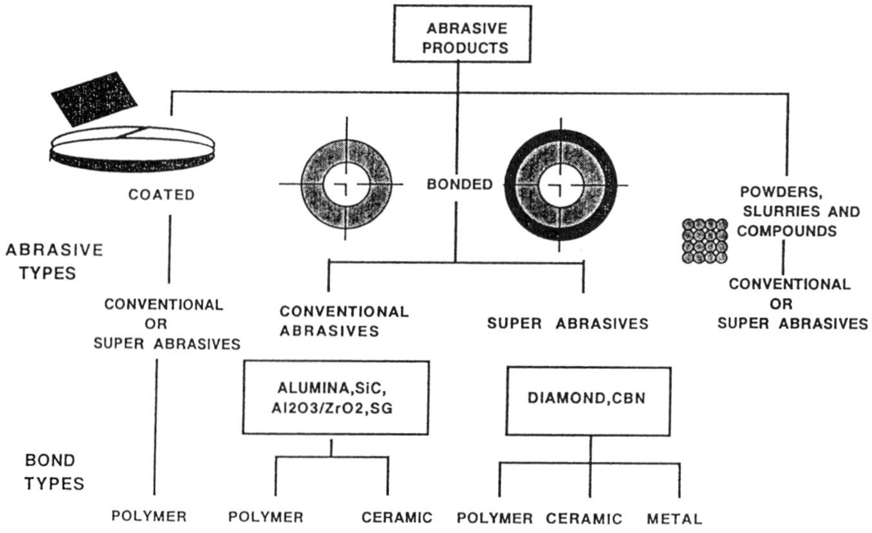

Figure 9 Abrasive products and the various bonding materials used.

plated or "E-process" wheels. Table 2 lists the characteristics of the four bond types available for grinding wheels.

Vitreous bonds made of clay, feldspar, and a glass frit are the most commonly used bonds for conventional abrasives and are becoming popular with superabrasives. They are rigid, free cutting, and have very good form retention. In addition, the porosity in these bonds can be controlled for more chip clearance and better coolant application at the grinding zone. These bonds, however, do not have high impact resistance and are not used in heavy-pressure operations such as foundry snagging or steel conditioning.

Resinoid bonds are made of thermosetting polymers; usually phenol formaldehyde or epoxy resins. These bonds are resilient, have good impact resistance, and are very free cutting. Their rigidity can be varied by adding fillers such as glass fibers. Resin bonds are used in most rough grinding operations such as snagging,

Table 2 Advantages of Abrasive Bond Types

Resin bond
Readily available
Easy to true and dress
Moderate freeness of cut
Applicable for a range of operations
First selection for learning the use of diamond wheels
Vitrified bond
Free cutting
Easy to true
Does not need dressing (if selected and trued properly)
Controlled porosity to enable coolant flow to the grinding zone and chip removal
Intricate forms can be crush formed on the wheels
Suitable for creepfeed or deep grinding, inside diameter grinding, or high-conformity grinding
Potential for longer wheel life than resin bond
Excellent under oil as coolant
Metal bond
Very durable
Excellent for thin slot, groove, cutoff, simple form, or slot grinding
High stiffness
Good form holding
Good thermal conductivity
Potential for high-speed operation
Generally requires high grinding forces and power
Difficult to true and dress
Layered products
Single abrasive layer plated on a premachined steel preform
Extremely free cutting
High unit-width metal removal rates
Form wheels, easily produced
Form accuracy dependent on preform and plating accuracy
High abrasive density
Generally not truable
Generally poorer surface finish than bonded abrasive wheels

weld grinding, and cutoff. With superabrasives, resin bonds are extremely popular and find extensive usage in tool and cutter grinding, grinding of ceramics, carbide drill fluting, and glass beveling applications. The resilience of these bonds results in reduced chippage of brittle workpiece materials.

Metal bonds are commonly used with superabrasives and are made from sintered bronze produced by powder metallurgy methods. Of late, cast iron and aluminum bonds are also becoming popular. These bonds are very durable, have excellent form-holding characteristics, and have high stiffness. They require rigid machines, however, in order to withstand the high forces generated. Metal bonds are very popular in geological drilling, in asphalt and concrete cutting, and for cutoff wheels in precision electronic applications. Metal bonds generally use strong, blocky diamonds as abrasives.

D. Wheel Shapes

Grinding wheels are used in a variety of operations such as tool and cutter grinding, form grinding, and ball grinding. All these different geometric configurations entail that suitable grinding wheel shapes are also available in order to perform these operations successfully. Hence, grinding wheels are made in several different shapes. Figure 10 shows the standard wheel configurations for conventional abrasive wheels. Superabrasives wheels are available in wider varieties and shapes as shown in Figure

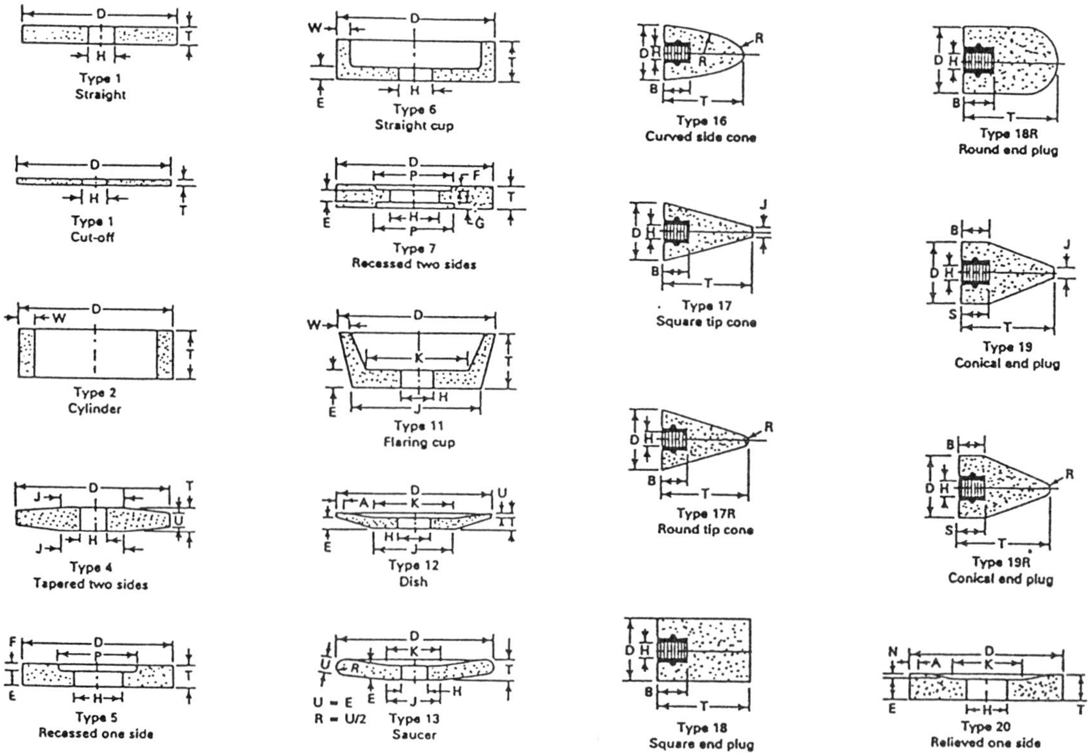

Figure 10 Standard wheel configurations for conventional abrasive grinding wheels.

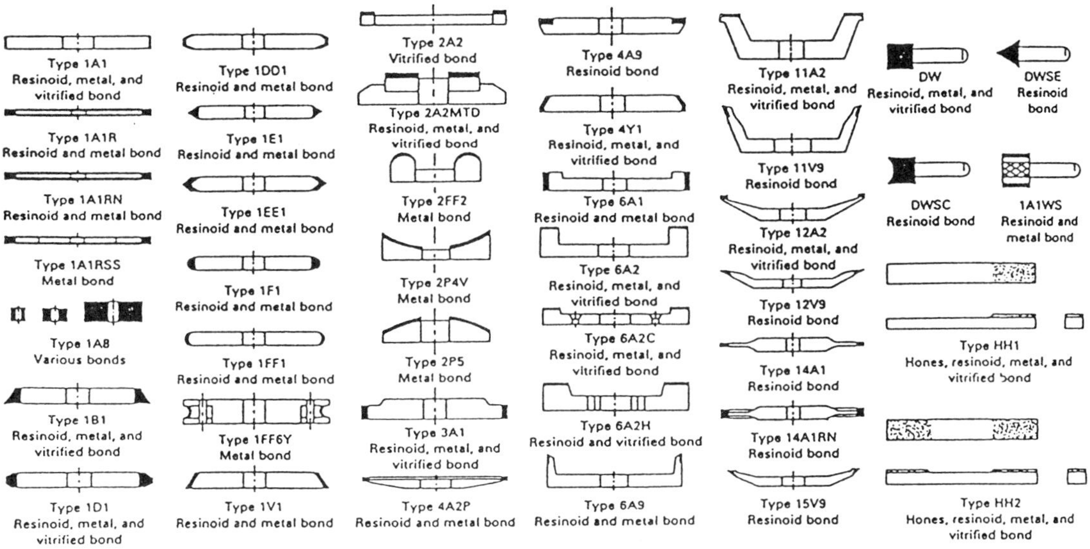

Figure 11 Standard wheel configurations for superabrasive grinding wheels.

11. In general, conventional abrasive wheels offer some flexibility in shape, as they can be readily "machined" to other desired shapes. This is generally not true with superabrasive wheels, and hence they are almost always made to near net shape, as required prior to their use. In Figure 12a and b, some commonly used conventional and superabrasives wheels are shown.

IV. TRUING AND DRESSING OF GRINDING WHEELS

The tolerances and surface finishes produced on workpiece surfaces and the forces developed during grinding depends to a great extent on the manner in which the grinding wheel was prepared for operation. Preparation of the grinding wheel generally involves two operations, *truing* and *dressing*, which are required to maintain the uniformity of the grinding wheel surface and also to keep the wheel open and free cutting.

Truing refers to the process of generating a geometrically correct wheel surface in order to grind with minimum or no chatter. The successful use of grinding wheels requires that the wheel be concentric and free of lobes (Fig. 13) (5). Conventional abrasives are most commonly trued by feeding a single point or multipoint diamond dressing tool across the rotating wheel surface. Superabrasive wheels are trued with a vitrified silicon carbide truing wheel mounted on a brake-controlled truing device. Diamond truing rolls driven by hydraulic motors and diamond crush rollers are some of the other truing devices used today.

Dressing is the process of opening up the wheel surface after a truing operation or after grinding in order to remove grinding "swarf." After truing, the wheel surface is generally very smooth without much exposure of the abrasive grits. The bond adjacent to the grits will have to be eroded away in order to expose the grits for efficient grinding. Occasionally, during grinding the chips tend to fill up the

(a)

(b)

Figure 12 (a) Conventional abrasive and (b) superabrasive grinding wheels.

pores and clog the wheel resulting in an inefficient cutting action. The wheels then have to be dressed in order to remove the chips from the wheel surface and expose the grits. Dressing is usually accomplished by rubbing an abrasive stick against the wheel surface. Recently, in-process dressing of metal-bond diamond wheels using electrodischarge machining (EDM) has been introduced. Electrolytic inprocess dressing (ELID) is another emerging technology for dressing fine abrasive metal bond diamond wheels.

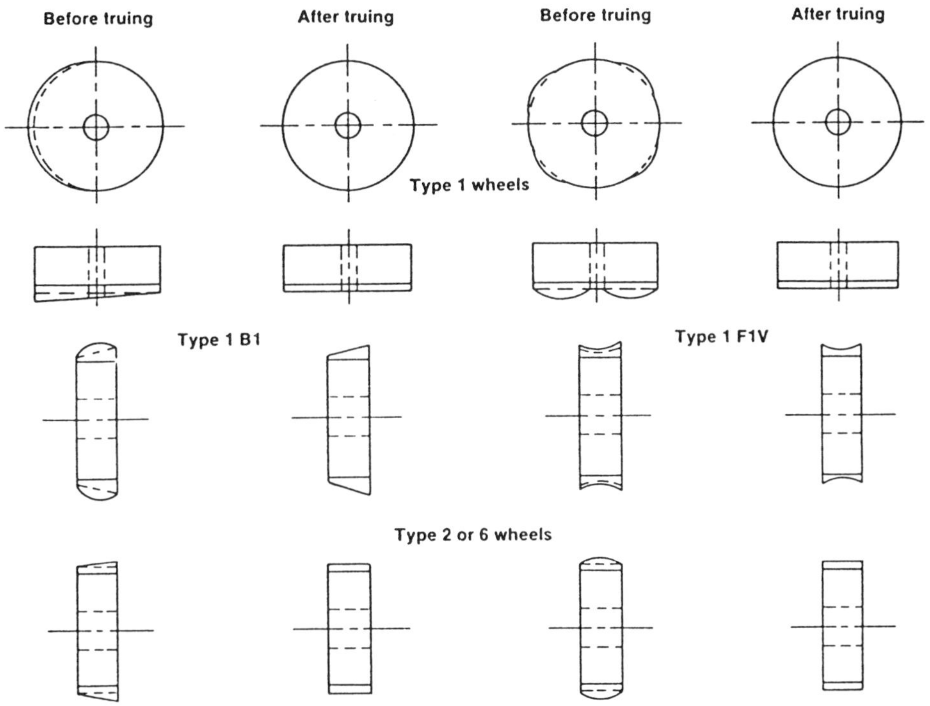

Figure 13 Typical examples of conditions that require truing.

V. GRINDING FLUIDS

In grinding, the conditions are very severe and high pressures, temperatures, and friction prevail at the grinding zone. Temperatures of the order of 1000°C or higher have been reported. These high temperatures can cause large-scale residual stresses, burn, and thermal damage to the workpiece and also reduce the life of the grinding wheel. Hence, in order to minimize these deleterious effects, proper cooling and lubrication should be provided at the grinding zone.

Grinding fluids or coolants are the most effective way of providing lubricity and cooling action at the grinding zone. Several types of coolants are available and are classified into four main categories:

- Straight oils
- Water-soluble oils
- Synthetic fluids
- Plain water with additives

Straight oils when combined with extreme pressure additives, such as chlorine, sulfur, and phosphorus, provide the best lubricating action for very severe grinding applications such as form grinding, gear grinding, and flute grinding. However, they have low flash points, which makes them easily combustible and they are considered as health hazards.

Water-soluble oils are the most popular coolants used in the industry today. They provide a good combination of cooling and lubricating properties and are

Table 3 Relative Rating of the Four Types of Grinding Fluids on the Basis of Their Properties

	Grinding fluids				
Fluid property	Petroleum-base and mineral-base cutting oils	Water-soluble oils	Synthetic fluids	Semisynthetic fluids	Water plus additives
Cooling	D	B-C	A	B	A
Lubricity	A	B-C	B-C	B-C	D
Rust protection	A	B-C	B-C	B-C	A
Cleanliness	D	C-D	A	B	A
Stability	A	C-D	A	B-C	C
Tolerance to contamination	A-B	C-D	A-B	B-C	C
System life	A	B-C	A	B-C	C
Health and safety	C-D	C	A-B	B	B
Disposal	B-C	C	A-C	B-C	C
Fire hazard	D	A-B	A	A	A

A, excellent; B, very good; C, good; D, poor.
Source: Van Straaten Corporation.

accepted by most operators because of their improved cleanliness and reduced fire and health hazards. Water-soluble oils are used in most light-, moderate-, and heavy-duty grinding operations.

Synthetic fluids are non–petroleum-based coolants, and they are gaining in popularity. They are miscible with water and usually contain lubricating and rust-preventing components and are much safer alternatives to oil containing products.

Table 3 is a comparison of the characteristics of the various grinding fluids (1).

VI. SYSTEMS APPROACH FOR GRINDING

Since grinding is often selected as a final finishing operation because of its ability to satisfy stringent requirements of surface roughness and tolerances, it is imperative that involved personnel understand the effects of grinding process variables on surface quality.

The precision grinding process is an interaction between a number of variables which can be grouped into four major categories—machine tools, wheel selection, work material, and operational factors—as shown in Figure 14. A clear and systematic understanding of each of these variables is critical to successful grinding results. The influence of these input variables, through process variables such as grinding forces and energy, result in the output of the grinding process (i.e., part geometry, tolerances, retained strength, surface quality). This input/output model and its use

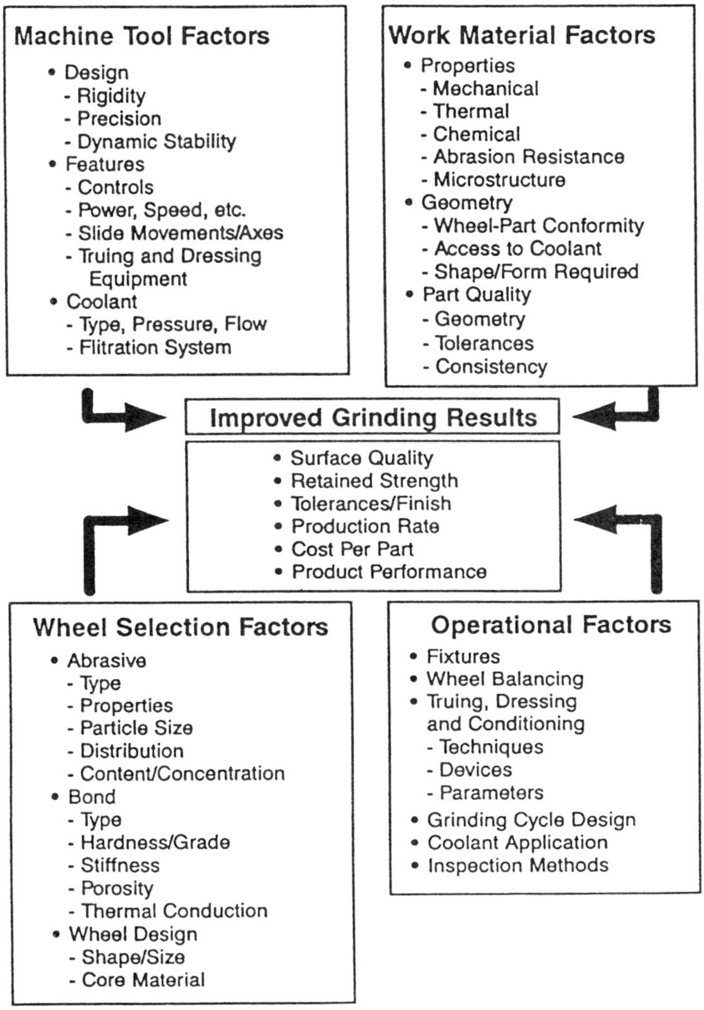

Figure 14 Selected variables influencing the abrasive grinding system.

to optimize the grinding process is called the systems approach (6). Figure 15 is a schematic representation of a typical grinding system.

Figure 16 shows an input/output model to be considered when pursuing the systems approach (7). This readily shows that the precision ground product is the output of a large number of variables which can be grouped in four categories. Each of these four primary input categories consists of a number of specific factors, as shown in Figure 14. However, more often than not, most of the input variables contribute to very few, well-defined microscopic process interactions. In the case of grinding processes, these may be grouped into cutting, plowing, and sliding. These can be measured or monitored using several well-defined parameters such as power, forces, energy, and temperature. The resultant output can be described in technical terms such as wheel wear, surface quality, and material removal rate, or in economic terms (system output) such as production rate, cost per part, and

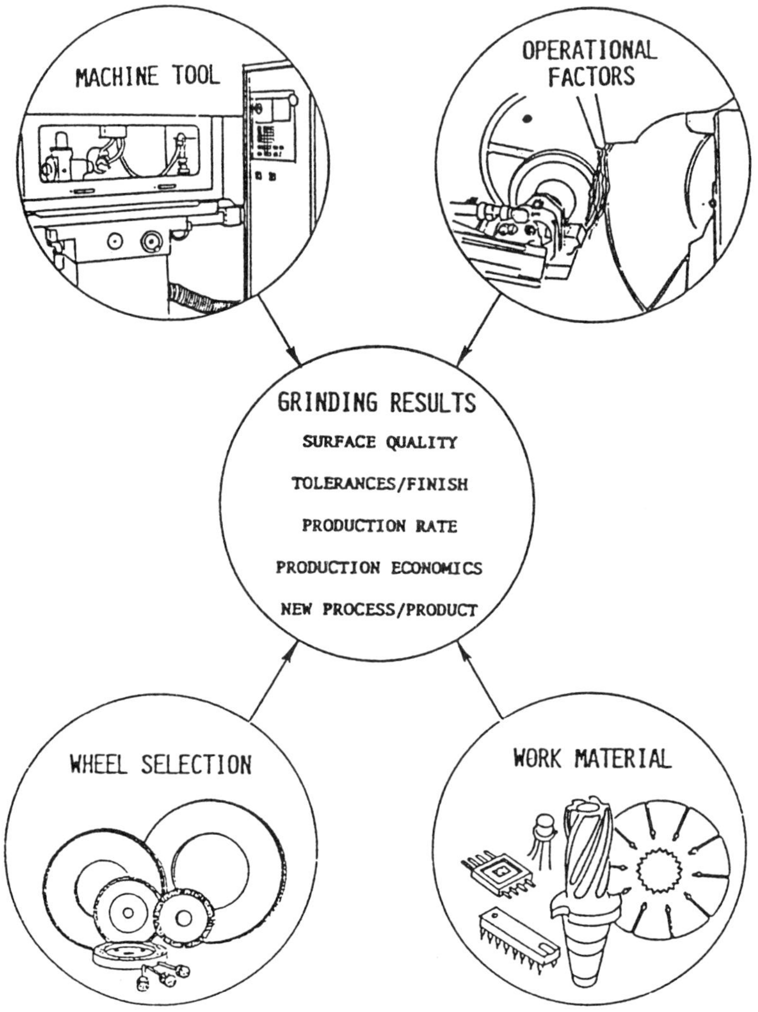

Figure 15 Schematic representation of the production grinding system.

yield. Thus, grinding as a manufacturing process is an input/output process with well-defined causal relationships with intrinsic or microscopic interactions. It is perceived as an empirical or stochastic process when we fail to study or understand the microscopic interactions or their causal relationships. Hence the science of grinding may be defined as:

> Modulate selected few input parameters among four input groups—that is, machine tool, wheel factors, work material, and operational factors—to maximize the cutting or chip formation process while minimizing the frictional or tribological components caused by plowing and rubbing interactions at the grinding zone. This optimization is achieved at the minimum expenditure of forces and energy while meeting the output requirements.

When this systems approach is pursued, we can vary one input variable at a time. This usually results in small or marginal improvements in quality or output

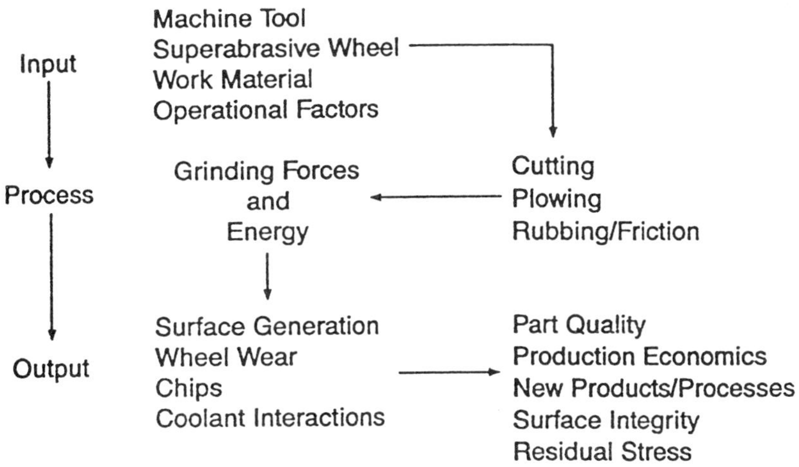

Figure 16 An input/output model of the precision grinding process.

of the systems. Sometimes this may be described as continuous improvement. However, when all the four input groupings are simultaneously varied, while achieving the constraints described above, the results are usually quantum improvements in productivity, quality, or total cost.

A. Machine Tool

The following machine tool parameters and operational factors predominantly determine the product quality of precision ground parts (8):

- Rigidity/stiffness
- Vibration level
- Coolant systems
- Precision movements and slides
- On-machine dynamic balancing
- Truing and dressing systems
- Multiaxis CNC capability
- Materials handling system

Both static and dynamic stiffness in the spindle, wheel, work fixture, and table assembly must be carefully considered in order for the grinding wheel to produce the required straightness, flatness, or similar surface requirements. Stiff machines are specially required while grinding ceramics and when CBN is used as an abrasive. Low levels of vibrations are crucial to prevent chatter marks on the finished surface of the part and while grinding brittle materials in order to avoid chippage. The direction, pressure, and flow of coolant applied are very critical in order to avoid grinding burn and thermal damage to the ground parts. Water-soluble oils are excellent cooling agents, whereas straight oils provide effective lubrication.

Tolerances and finishes are becoming tighter. With the advent of new materials, such as ceramics, that have superior thermal stability and hardness, obtaining precision tolerances and finishes requires machine tools capable of precision movements, with a high degree of repeatability, stability, and positioning accuracy.

Imbalance in the grinding wheel will result in vibrations and parts with improper finish. It is essential that the grinding wheel be dynamically balanced during operation to minimize vibration. This on-machine balancing operation should be applied as part of the truing process prior to the use of the grinding wheel in the grinding operation.

One of the key requirements of production grinding economics is the decrease in the total cost of grinding. Setup time, machine to machine movement time, as well as in-process inventory costs contribute heavily to the total cost of fabrication. These costs can be significantly decreased if the part can be fabricated on one machine using a single setup. This approach is being used successfully via the application of multiaxis CNC grinding systems specifically developed for the use with CBN wheels to grind steel parts.

B. Wheel Selection Criteria

The grinding wheel is perhaps the most important tool used in the grinding operation and the manner in which it is selected can influence the finish and tolerances produced on the workpiece. Figure 17 shows the schematic representation of a grinding process. It is imperative to maximize the abrasive/work interactions leading to chip generation and grinding efficiency and to minimize the rubbing interaction at the bond/work, chip/bond, or chip/work interfaces.

The selection of proper wheel specification is a key input to the grinding system and should not be approached in isolation. Wheel selection should be considered in conjunction with the other three inputs to the grinding system:

- Grit size
- Grain content
- Wheel grade.

Effect of Grit Size

The size of abrasive used in a grinding wheel can have a pronounced effect on the G-ratios (work removed per unit of wheel volume consumed), material removal rate (MRR), and the surface finish. Figure 18 shows the general effect of grit size on these output variables. Although the wheel life is reduced with the use of finer grit sizes, the surface finish is improved considerably.

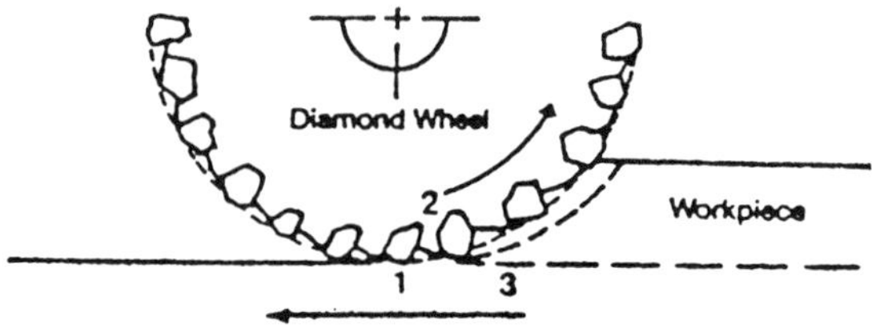

Figure 17 Schematic representation of the grinding process: (1) abrasive/work interaction; (2) chip/bond interaction; (3) chip/work interaction.

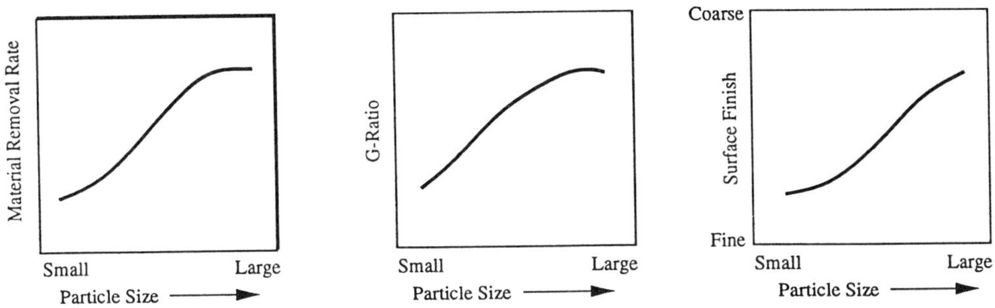

Figure 18 Effect of grit size on material removal rate, G-ratios, and surface finish.

Effect of Grain Content or Concentration

The amount of abrasive in a wheel relates to the number of active cutting points. The higher the abrasive content, the more will be the active cutting points and hence smaller the chip thicknesses. This will result in smoother surface finishes. An increase in abrasive content also means a less porous wheel and consequently higher forces. Larger abrasive content also implies longer lasting wheels or higher G-ratios. Figure 19 shows the general trends of the effect of abrasive concentration on some output variables.

Effect of Grade or Hardness of the Wheel

The grade of a wheel indicates how strongly the abrasive grains are held in a grinding wheel. Harder grade wheels do not release abrasive grains readily and therefore draw higher power and forces as compared to softer grade wheels. The surface finish produced by the harder grades is generally better than softer grades. Figure 20 shows the variation of MRR, G-ratios, and finish with the wheel grade.

C. Work Material Considerations

Figure 8 demonstrates some of the common work materials and the abrasives used to grind them. This is a very simplified version, as each grouping of work material represents a range. The thermal, chemical, and mechanical properties of these work

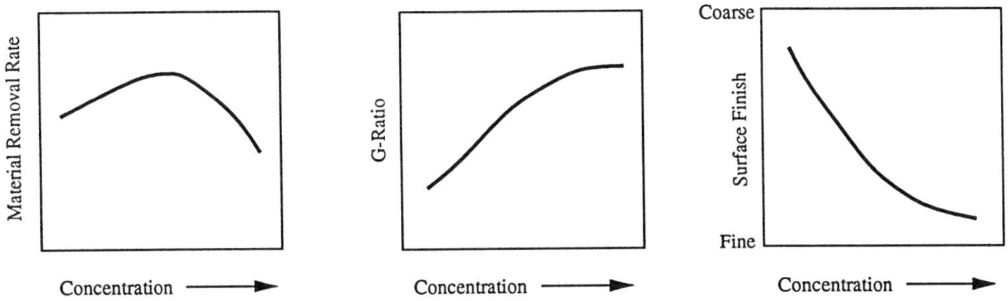

Figure 19 Effect of abrasive concentration on material removal rate, G-ratios, and surface finish.

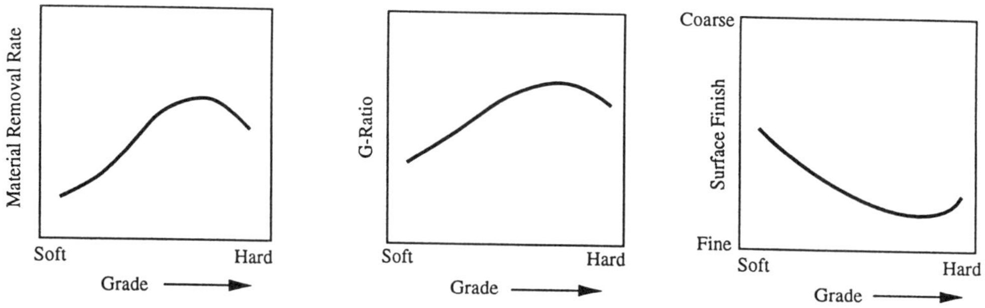

Figure 20 Effect of wheel grade on material removal rate, G-ratios, and surface finish.

materials largely influence the grinding process results. As a general rule, lower hardness and more ductile materials are easier to grind, but if there is insufficient chip clearance, they tend to "load" the grinding wheel leading to excessive chip/bond and chip/work interactions. This is often minimized by using large abrasive grit sizes, bond types which prevent adhesion of chips, and also copious coolant flow that washes away the swarf material.

Materials of higher hardness are difficult to grind, but if they are homogeneous and without hard secondary phase materials such as carbides (e.g., 52100 steel), then grinding difficulty is much less than that for tool steels containing secondary carbide particles. Materials of poor thermal conductivity and high temperature strength such as stainless steel and aerospace alloys are very difficult to grind. Such grinding situations are improved using oil coolant, which improves lubricity thus minimizing frictional interactions in the grinding zone.

The grinding of ceramic materials is often described as a process of crushing or pulverizing. This is often not the case. Recent research has shown that ceramics can be successfully ground under plastic deformation conditions resulting in extremely complex shapes and close toleranced parts (9).

Plastics and composites which produce long stringy chips are often ground using abrasive products with grains of large exposure which allow for adequate chip clearance.

D. Operational Factors

When the work material is subjected to a grinding process, the interactions can be represented by the following parameters:

- *Material removal rate* (MRR) is the volume of material removed in a unit time.

  MRR = Work Speed (v_w) × Depth of Cut (d) × Width of cut (b_w)

- When grinding takes place uniformly along the entire width of the wheel, the MRR can be normalized using unit width MRR or the specific material removal rate (MRR′):

  MRR′ = MRR/Width of Cut (in^3/min, in)

- Grinding processes of various configurations can be normalized for ease of comparison using equivalent diameter (D_e), where D_w is the work diameter and D_s is the wheel diameter. Figure 21 is a schematic representation of the concept of equivalent diameter (10).

OD grinding	De = Dw × Ds/Dw + Ds
ID grinding	De = Dw × Ds/Ds − Ds
Surface grinding	De = Ds

- The size of the chip produced by the abrasive grain during a grinding process can be estimated as

$$h = \left(\frac{v_w}{v_s}\right)^{1/2} \times \left(\frac{d}{D_e}\right)^{1/4} \times \left(\frac{1}{kC}\right)^{1/2}$$

  where k = 1 to 20 and C = number of grains/square inch of grinding wheel surface. C is larger for fine abrasive grains and small for large abrasive grains.
- *Grinding force* is the force exerted between the grinding wheel and the work material. These forces can be normal to the work surface (F_n), tangential to the wheel (F_t) and occasionally in a transverse direction (F_z).
- *Specific energy* is the ratio of grinding power to the material removal rate. It is a measure of the energy input per unit volume of material removed. As a general rule, specific energy decreases as the chip thickness is increased.
- *G-ratio* is the ratio of the volume of work removed per unit volume of abrasive product consumed. It is a measure of the life or durability of abrasive product for a given application. In many applications, the wheel may be trued or dressed frequently or sometimes continuously. In those situations, the G-ratio may need to reflect the abrasive consumed during these nongrinding operations.
- One measure of the ease or difficulty in grinding is termed as "*grindability*" or *grinding system performance index* (GSPI). This is defined as:

$$\text{GSPI} = \frac{\text{G-Ratio}}{\text{Specific Energy}}$$

  Ranking of GSPI for a variety of work materials in precision external cylindrical grinding is shown in Table 4.

The grinding of performance is measured in terms of the above parameters and can be represented or analyzed as follows. Figure 22a and b show the variation of normal force and power with material removal rate in a typical grinding operation. In general, the forces and power increase with increase in material removal rate and a minimum force is required to initiate cutting. This minimum value of force or power is termed the threshold force or power. Thus,

$$F_{total} = F_{cutting} + F_{threshold}$$
$$F_{Total}/MRR = F_{cutting}/MRR + F_{threshold}/MRR$$
$$= 1/WRP + F_{threshold}/MRR$$

Figure 22c is a plot relating the material removal rate to the normal force. Such data can be easily obtained on a grinding machine using force measuring equipment and can be effectively used to categorize the "sharpness" or cutting

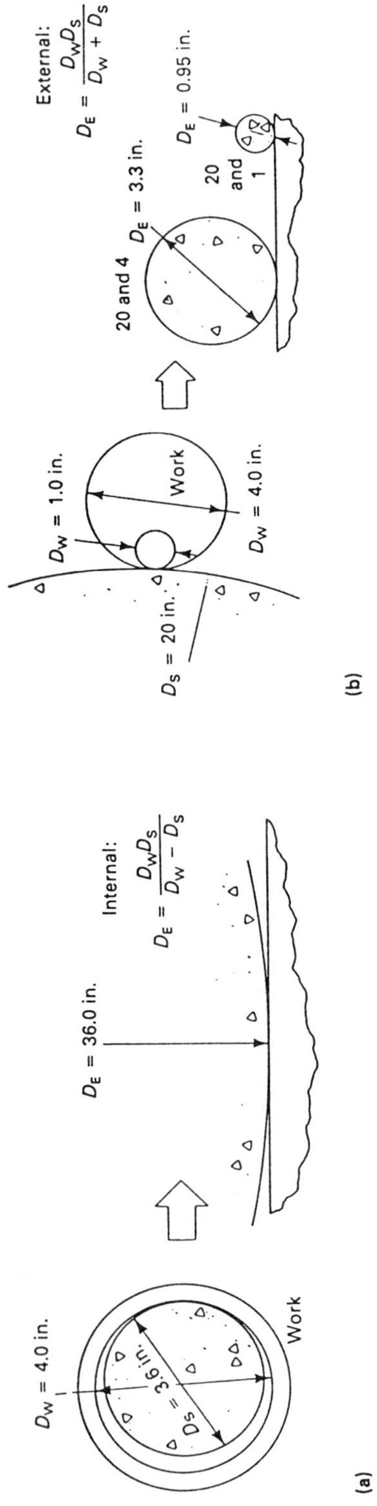

Figure 21 Relating (a) internal and (b) external cylindrical grinding to surface grinding using equivalent diameter. (From Ref. 10.)

Table 4 Grindability of Work Materials (Wet External Grinding)

Work material	Relative grindability
52100 bearing steel	72
4340 structural steel	75
1020 low carbon steel	44
M7, D2 tool steel	3
440SS stainless steel	1.2
304SS stainless steel	1.1
INCO718 aerospace alloy	1

efficiency of grinding wheels. The slope of the MRR versus F_n line is called the work removal parameter (WRP). A steep slope indicates a sharp wheel and low forces while a shallow slope means a dull wheel and high forces.

Similar analysis with power equations leads to:

$$P_{total} = P_{cutting} + P_{threshold}$$

or

$$P_{total}/MRR = P_{cutting}/MRR + P_{threshold}/MRR$$

that is

$$\text{Specific Energy} = \text{Specific Power} + P_{threshold}/MRR$$

Specific power is the slope of the power versus MRR curve and is another important parameter used to analyze grinding results. It can be considered to represent the cutting component of power. The threshold power may be approximated as the power component required to overcome friction and material deformation effects. Thus if these effects are absent:

$$\text{Specific Energy} = \text{Specific Power}$$

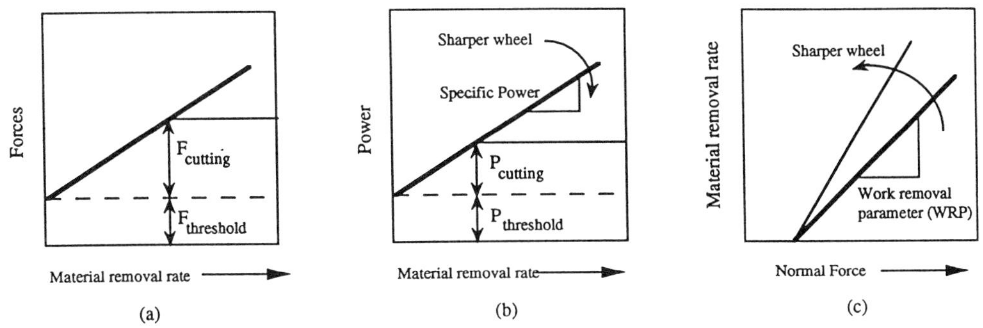

Figure 22 Effect of material removal rate on (a) forces and (b) power; (c) work removal parameter.

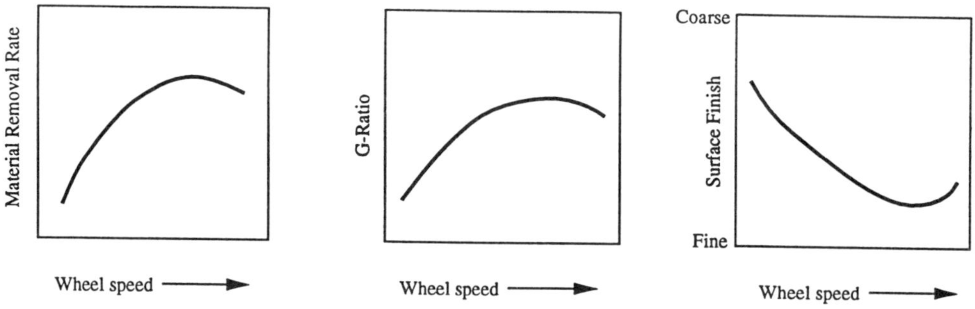

Figure 23 Effect of wheel speed on material removal rate, G-ratios, and surface finish.

Effect of Wheel Speed

Figure 23 shows the effect of wheel speed on the MRR, surface finish, and G-ratios. Increasing the wheel speed increases the MRR and G-ratios and improves the surface finish. The power drawn, however, increases with the wheel speed.

Effect of Coolant Types

Coolants can play a significant role in grinding. Table 5 compares the forces, G-ratios, and finishes obtained while grinding with water-soluble oil and straight oil. The beneficial effects of using straight oil is clearly evident. Grinding with straight oils as coolant will reduce forces, improve G-ratios, and improve the surface finishes, particularly on difficult to grind materials such as tool steels.

Output of the Grinding System

Grinding technologists should strive to maximize the output of the grinding system, to achieve parts with close tolerances and consistency keeping in mind the importance of reducing the total cost per part. Proper understanding of the effect of input variables and the application of systems approach should help in achieving these goals effectively. In the following, the significance of the various outputs of the grinding system are described:

Tolerances

Tolerances in generic terms imply permitted deviation from the nominal value. Whatever the "tolerance" referred to, it is clear that one of the driving forces for

Table 5 Comparison of Coolant Types

	Water-soluble oil	Straight oil
Forces	High	Low
G-ratios	Low	High
Surface finish	Good	Very good

grinding methods is to achieve improved or closer tolerances. As an example, finished parts of closer tolerances in a hydraulic power steering pump help to achieve higher pump efficiency, which results in improved power steering performance thus enhancing the quality of an automobile. Closer tolerances in gears enables them to roll against other gears more readily while reducing sliding friction. This results in, for example, better gear efficiency, increased torque-carrying capability, and lower noise level.

Consistency

For a given tolerance of a finished part, it is critical that we achieve them consistently over a batch of components. Consistency is established through the confidence limits arrived at through Statistical Process Control. Although tolerances may be viewed as a "set point" or location in geometric space, consistency defines our ability to stay close to the set point. Grinding methods constantly strive to improve the consistency of their output, i.e., generate surfaces of required geometry and surface characteristics.

It is crucial for the reader to recognize the role of consistency in the output of the grinding methods. Imagine an improvement in the form accuracy or tolerances of a gear surface achieved through a new or improved grinding method. This could result in significant improvement in the performance of the transmission in which the gear is used. This new process will become successful only when two conditions are met; that is, the new process delivers the improved tolerances consistently and all other components in the transmission are also manufactured consistently. If either of the consistency requirements are not met, the new transmission does not perform to the higher performance levels and the new finishing method developed may not see the light of the day. Conversely, constant improvement in consistency is a key technology driver that closely follows improvement in tolerances of the finished components.

Surface Quality

Grinding results in generation of surfaces. Through this process, the surface layers of the work material is altered. This alteration may result in changes in surface characteristics such as:

- Surface finish
- Fatigue strength
- Residual stress
- Retained strength

All finishing methods constantly strive to improve upon these and other characteristics. It is suffice here to say that surface quality and its improvement is a key technology driver for grinding methods.

Grinding of New Materials

The advent of new materials such as silicon aluminum alloys, fiber reinforced composites, metal matrix composites, ceramics, ceramic coatings, particle boards, and metallic glass, constantly pose challenges for both innovation and improvements in grinding methods. Such new materials are constantly introduced to achieve lower weight, higher strength/weight ratio, high temperature resistance, operation under severe environments of corrosion, erosion, fatigue, for example, or to achieve other unique performance improvements. These in turn influence the choice and improvements necessary in finishing methods.

Productivity

Although we have thus far stated technology drivers that relate to technical aspects of grinding methods, it is obvious that such results are to be achieved at lower cost, anywhere in the world. This requirement simply translates into improved productivity. All finishing methods seek to achieve higher productivity through either incremental or small improvements in productivity or achieve quantum improvements through new processes. Productivity in grinding processes may be measured in terms of such factors as total cost/part, cycle time improvements, or improved yield.

VII. SUMMARY

In this chapter, the various aspects of grinding technology has been described. The wheel-marking systems for both conventional and superabrasives grinding wheels has been explained in detail in order to identify and apply the constituents of a grinding wheel easily. The effect of several variables such as abrasive size, concentration, wheel grade, wheel speed, and coolant type on the output of the grinding process has also been highlighted.

The importance of a "systems approach" that attempts to consider all aspects of the grinding process to achieve optimum results has also been addressed. This approach integrates the input variables of machine tool, wheel, work material, and operational parameters, while minimizing the process variables of grinding forces and energy. This results in the desired output variables: part geometry, quality, low cost/part, and process economics.

REFERENCES

1. Ault, W. N., "Grinding Equipment and Processes," *ASM Metals Handbook*, Vol. 16, 9th ed., 1989.
2. Coes, L., *Abrasives.* Springer-Verlag, New York, 1971.
3. Komanduri, R., and Shaw, M. C., "Attritious Wear of Silicon Carbide," ASME J. of Engrg. for Industry, Vol. 98, pp. 1125, 1976.
4. Krar, S. F., and Ratterman, E., *Superabrasives—Grinding and Machining with CBN and Diamond.* Glencoe/McGraw-Hill, 1990.
5. Subramanian, K., "Make the Best Use of CBN Wheels by Proper Truing, Dressing, and Conditioning," *Proceedings of Superabrasives '85*, Chicago, April, 1995.
6. Subramanian, K., "Superabrasives for Precision Production Grinding—A Case for Interdisciplinary Effort," *Proceedings of the Symposium on Interdisciplinary Issues in Materials and Manufacturing,* Vol. 2, pp. 665–676, 1987.
7. Subramanian, K., Redington, P. D., and Ramanath, S., "A Systems Approach for Grinding of Ceramics," *Proceedings of the International Conference on Machining of Advanced Materials*, NIST, Gaithersburg, MD, 1993.

8. Subramanian, K., and Ramanath, S., "Machine Tool Developments Required for Precision Production Grinding of Ceramics," *Proceedings of Third Biennial International Manufacturing Technology Research Forum*, Tokyo, 1989.
9. Subramanian, K., and Ramanath, S., "Principles of Abrasive Machining Processes," Sect. 5, Vol. 4, Ceramics and Glasses, *ASM Handbook*, 1991.
10. Lindsay, R. P., "Chapter 2: Principles of Grinding," *Handbook of Modern Grinding Technology*, 30.

24
Foundry Technology

PAUL J. MIKELONIS
Consultant
New Berlin, Wisconsin

I. INTRODUCTION

Metal casting is the most direct method of producing a desired shape and is accomplished by pouring a molten metal into a form that retains the metal as it solidifies. The form containing the shape or casting configuration is usually produced from sand that has been bonded with clay, water, and other additives to give it strength for handling and also for holding the molten metal. Forms (molds) can also be produced from other materials, such as graphite, rubber, plaster, metal, or refractory mixtures of alumina, silicates, gypsum, and so on. Sands can be bonded with materials such as sodium silicate, furfurals, phenolics, cement, thermosetting resins, or oils (linseed, cottonseed, or fish oils are used). Since the major tonnage of castings is produced from clay-bonded sand molds and ferrous metals (which include gray cast iron, steel, ductile cast iron, and malleable cast iron), the practices in this chapter will relate to these materials and processes. A great many of the practices are adaptable to all metal-casting operations.

The metal-casting industry is one of the largest industries in the United States in dollar value added by manufacturing. Metal casting or founding dates back to 3100 BC, as verified by cast bronze sculpture and statuary found by archeological teams in Asia and Africa. The Chinese were casting iron in a variety of shapes several hundred years before the birth of Christ. The area that is now called Tanzania, in eastern Africa, shows evidence of irons and even a form of steel being melted and cast somewhere around AD 800. Between AD 1000 and 1500, metal casting evolved from an expression of art form to the casting of military hardware, print type, and even some engineering shapes.

With over 20 million tons of ferrous and nonferrous castings produced in a year, every other manufacturing industry is touched in some manner by metal castings as a component of the product they produce, the product itself, or in the equipment or machines used in production.

The intricacies and multitudinous variables that enter into a metal-casting operation can be appreciated by looking at a generalized flowchart of a foundry operation in Figure 1. The variety of materials and processes that enter into the foundry work require a great deal of process control to produce a cast product that will fulfill the casting users' requirements and specifications.

The assurance of quality through control of processes and quality control procedures in metal casting is a growing science. In the early days of metal castings, when primarily art castings were being produced, a single artisan performed all the work. Artisans would create a pattern, produce the mold, melt the metal, pour the molten metal into the mold, and then clean the casting to a level of quality that would meet their level of perfection. Since the means of inspection were only sensory appraisal techniques, the approach to quality was relatively simple.

A great number of foundries were in operation in the United States by 1900. In 1897, E. H. Putnam, publisher of the magazine *The Trades Man*, in commenting on the quality of casting production stated: "Without abating quantity, quality must be worked up to the highest degree and when this is reached, the hard work is just begun, for any abatement of vigilance will be followed by a corresponding decline in quality." Since liability suits on product failure were evident as early as 1875, the evaluation of casting material quality was adopted and standardized at the start of the century. Material specifications for pig iron and other foundry materials appeared in 1907. By 1915, inspection procedures for assessing casting quality were advocated together with defect reporting and analysis by forward-looking foundrymen, who contended that properly organized inspection departments were necessary for the success of the metal-casting industry.

In 1942, an industry paper by H. H. Fairfield entitled "Statistical Methods as an Aid to Foundry Operations" was a first study and report on quality control techniques in the foundry. The American Foundryman's Society published its handbook *Statistical Quality Control for Foundries* in 1952. The use of control methods, including procedures to know the costs of quality, are more prevalent as castings customers become more sophisticated in their purchases and as castings producers realize the need to generate integrity in their castings.

As foundries progress from simply assessing defects to plotting inspection results in some form, various forms of analysis are used to define sources of defects and levels of quality. As a first approach to decision making, because it is the most economical one, previous experience is explored to determine a course of corrective action. In this sense, quality assurance is attained through the art of experience and judgment. When this action is further based on statistical evaluation of current as well as previous quality levels, the quality assurance discipline becomes more successful in a remedial way to sell the part.

II. DESIGNING A QUALITY ORGANIZATION

A. The Quality Assurance Manual

A formal quality assurance program cannot be adequately installed or disciplined, once in operation, without documenting what and how it is all about. The balance of this section contains much information that should be part of the foundry quality

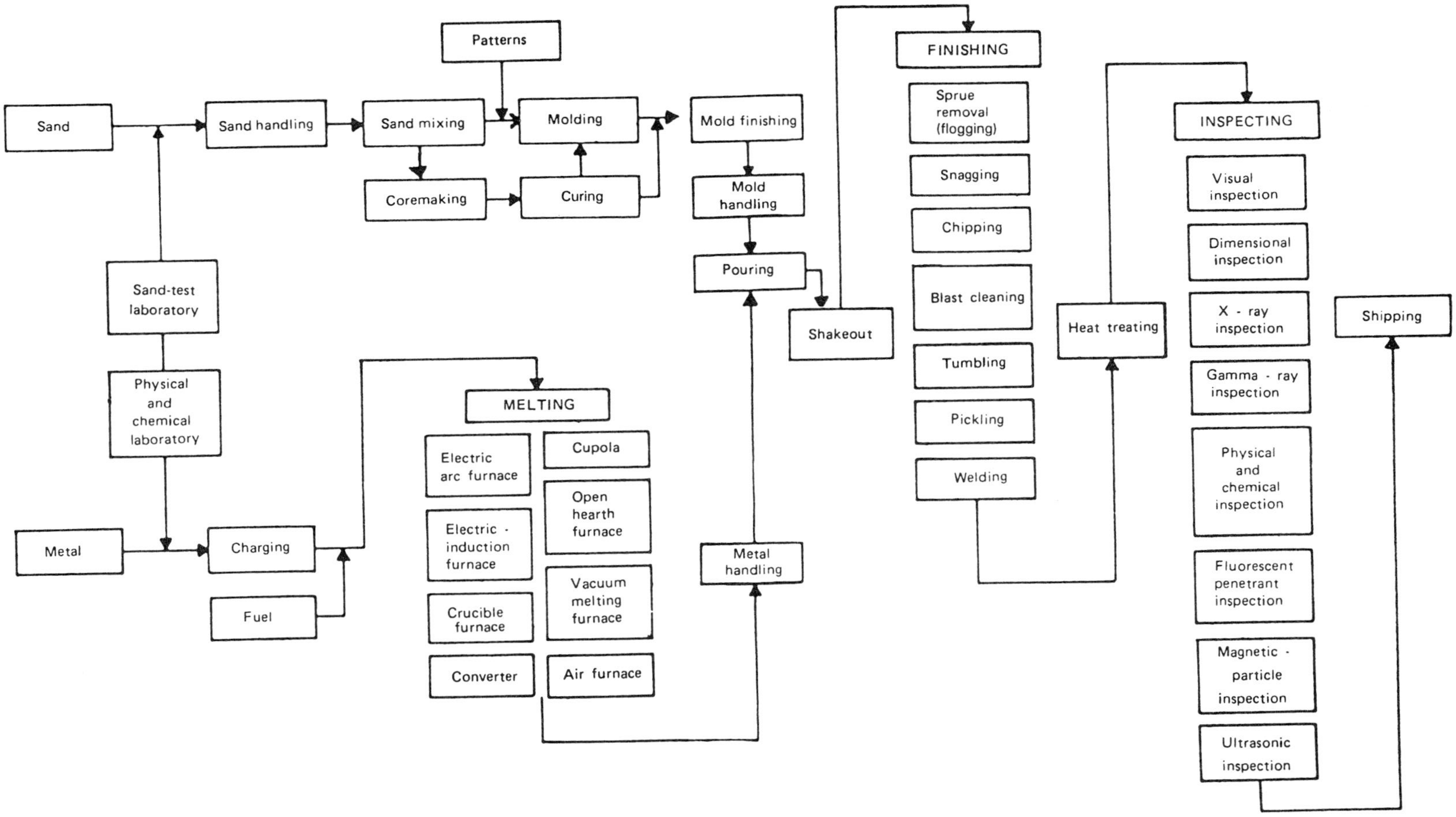

Figure 1 Flowchart of a typical foundry operation.

manual. Staffing, responsibilities, process procedures, job instructions, process auditing, materials control, and inspection procedures should be covered in writing.

B. Staffing

The staffing of a foundry for a quality assurance program will be a function of the size of the physical plant, the size and number of castings produced, and the complexity or criticality of these castings. Since a comparison of the percentage of employees involved in the quality program to total employees in the plant is not necessarily a good guide when the foregoing variables are considered, it is best to use dollars expended in the quality program as a percentage of sales dollars for a guide. This type of association permits better control of personnel and facilities with a direct comparison of the value of the work performed as reflected by the plant financial statement. A good guide for this is the manual *Quality Costs—What and How* published by the American Society for Quality Control. See the Bibliography at the end of this chapter.

A review of singular and multiplant staffing will give the reader an opportunity to appraise his or her facilities for personnel needs. One must think in terms of the prevention and appraisal aspects of process and quality control to best estimate staffing requirements.

A 100-melt-ton-per-day operation producing 50 tons of good, relatively intricate cored castings in gray iron would need the following personnel based on a one-shift operation.

For the prevention aspects of the program:

Quality assurance manager	1
Process engineer	1
Process control observer	1

For the appraisal aspects of the program:

Chief inspector	1
Inspection supervisor	1
Line inspector	3
Audit inspector	1
Layout inspector	1
Laboratory technician	1

Some personnel would be required who would spend only part of their time in each of the foregoing functions; these would be:

Technical director	1
Plant metallurgist	1
Quality assurance supervisor	1
Process control supervisor	1
Laboratory supervisor	1
Clerk and secretary	1

In this example, we have 11 full-time employees plus 6 part-time employees who will be spending 10–30% of their time on quality assurance. A semiautomatic, intricately cored casting operation would then be utilizing the equivalent of 14 people, or 4.7%, of a possible 300 total employment.

In the case of multiplant operations, plant staffing would remain as above and a corporate staff would consist of:

Director of quality assurance	1
Staff quality assurance engineer	1
Staff quality assurance technician	1

These personnel would aid and audit the plant programs as well as provide direction to conform to the corporate philosophy on quality.

An organization chart for personnel at the plant level is shown in Figure 2. At the corporate level in the multiplant operation, the director of quality assurance would report directly to the president and the staff personnel would report to the director.

Responsibilities of the full-time plant quality assurance personnel are defined as follows:

1. *Quality Assurance Manager*
   a. Establishes and monitors process control procedures
   b. Initiates scrap reduction projects
   c. Maintains quality records and customer specifications
   d. Maintains plant quality assurance manual
   e. Establishes and maintains incoming materials control
   f. Maintains instrument surveillance for accuracy and calibration
   g. Schedules and monitors internal quality audits

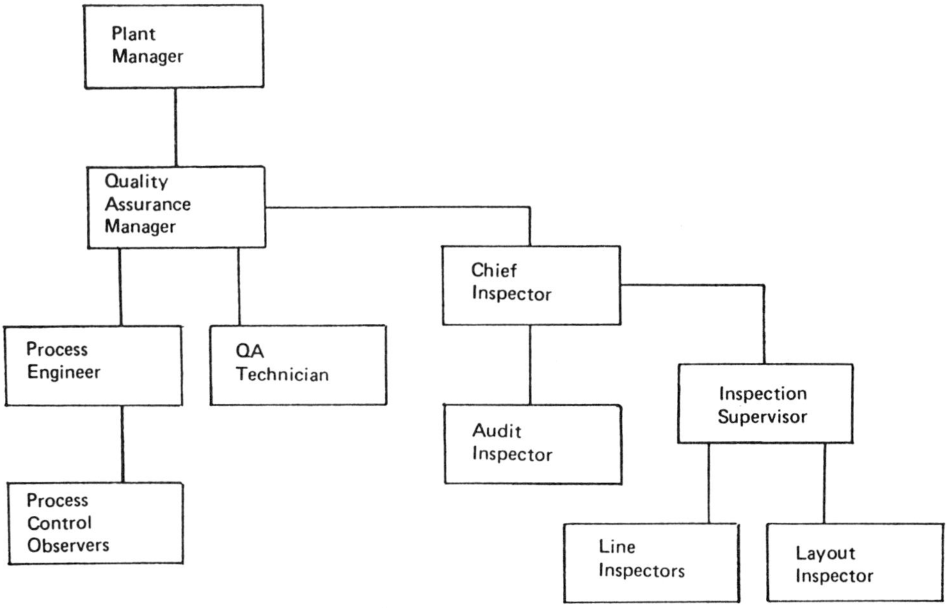

Figure 2 Plant-level quality organizational chart.

2. *Process Engineer*
   a. Aids line supervisors in writing process control procedures
   b. Supervises process control observer
   c. Defines process procedures in conjunction with line supervisors
   d. Documents process procedures
3. *Process Control Observer*
   a. Observes and audits compliance of process procedures to documented procedures
   b. Aids line foremen in detecting process deviations
   c. Aids in associating casting defects with process deviations
4. *Chief Inspector*
   a. Supervises all sensory inspection procedures
   b. Supervises all nondestructive evaluations of castings
   c. Aids in defining and documenting defects for correction
   d. Acts as foundry-customer liason on casting problems
5. *Audit Inspectors*
   a. Audit casting quality beyond line inspection

III. PURCHASED MATERIAL CONTROL

All materials used in the foundry should be purchased to materials specifications mutually agreed on by the materials supplier and the foundry. Materials received should be accepted on the supplier's inspection and test results or, preferably, the foundry's inspection and/or test results.

Base materials such as sand, metal, and alloys should be easily controlled, as standard specifications and testing have been available and used for many years. Other materials such as sand binders and additives have been caught up in a rapidly changing product technology and close cooperation between the producers and users is vital in designing acceptance tests and specifications.

A. Material Specifications

Specifications for foundry materials have been developed by a number of organizations involved in standards writing. Materials that are common to both ferrous and nonferrous casting operations are listed in Table 1 with the availability of specifications. Typical material specifications are shown in Figures 3–5.

B. Purchasing to Suppliers' Test Results

The reasons for purchasing to suppliers' test results include situations where costly and sophisticated testing is required; where a supplier has established itself as qualified and reliable; or where a foundry is too small to have all the required test equipment and personnel to operate the equipment. The test results should be sent prior to or with the material to define acceptance or rejection of the material prior to use in the foundry. One person in the foundry should be responsible for receiving and taking action on the material test results as well as maintaining a file of the test records for a period of a year after the material is used. A time-versus-data plot

Table 1 Material Specification Sources

Material	Specification source[a]
Sand	AFS, supplier
Sand additive carbonaceous	Supplier
Sand additive clay	SFSA, supplier
Sand binders	Supplier
Sand castings	Supplier
Resin-coated sand	AFS, supplier
Chaplets	Supplier
Pig iron	ASTM, supplier
Scrap charge metals	ISRI, supplier
Primary melting alloys	AA, supplier
Secondary melting alloys	AA, supplier
Ladle alloys	ASTM, supplier
Ladle fluxes	Supplier
Melting fluxes	Supplier
Abrasive shot	SAE, SFSA, supplier
Abrasive grinding wheels	Supplier

[a] AFS, American Foundrymen's Society; SFSA, Steel Founders Society of America; ASTM, American Society for Testing and Materials; ISRI, Institute of Scrap Recycling Industries; AA, Aluminum Association; SAE, Society of Automotive Engineers.

for purchased material is useful to maintain a record of the supplier's performance on material properties.

C. Receiving Inspection and Testing of Materials

Where the size of the plant will justify it, a receiving inspector, reporting to the quality assurance manager, can be charged with inspecting, sampling, and having material tests performed. These control procedures can be performed with guidance from the department general supervisor whose materials are involved. The sampling and tests to be used should be established in conjunction with the supplier. If sampling and testing differ from those of suppliers, the results can be questionable. Test procedures are sometimes provided by the national standards writing bodies noted previously; if this is the case, both supplier and foundry should use the standard tests.

D. Auditing Supplier's Facilities

The need to visit new and existing suppliers to view their quality programs is an important part of a total quality effort. An audit sheet can be used to assure compliance of all important controls. Noting items such as incoming materials control, record keeping, instrument inspection and calibration, identification and routing of defective material, documentation of test results, and the use of a formal quality program will aid in establishing supplier integrity. If the suppliers have current ISO or QS9000 certification, this may be sufficient evidence of a qualified quality program to accept them without further audits.

DATE: NO. M2

CHANGE NO. 3 PAGES 1 of 1

MATERIAL: Purchased steel scrap
Cut structural and plate scrap, 2 feet and under

DESCRIPTION: This is clean steel, such as plate structural shapes, shearings, punchings, and crop ends.

USE: Used as a low carbon, low silicon, free from alloys charge material for steel, gray iron, and ductile iron.

REQUIRED SPECIFICATIONS:

1. CHEMISTRY -

Carbon	0.10—0.40%
Silicon	0.10—0.30%
Manganese	0.30—0.90%
Phosphorus	0.04% maximum
Sulfur	0.04% maximum

Free of all other alloys, except in residual percentages.

2. SIZE: Dimensions shall be a maximum of 24" long × 12" wide, and not less than 1/4" thick, with a minimum weight of one pound. Not more than 5%, by weight, of material less than one pound unit weight, will be acceptable in any one shipment.

3. GENERAL: All material shall be free of all attachments, dirt, excessive rust or corrosion. There shall be no nonferrous materials or foreign material of any kind.

SAMPLING & ACCEPTANCE: Incoming material shall be visually inspected by authorized personnel from the melting department before unloading commences who will acknowledge acceptance of the material by signing the weigh ticket.

Where a material does not meet the required specifications, terms for acceptance or rejection will be negotiated between foundry plant personnel and the seller.

Chemical analysis may be performed on any suspect material, using ASTM standard analytical methods.

Figure 3 A typical purchase specification for scrap metal.

DATE:	NO. S1
CHANGE NO. 2	PAGES 1 of 1
MATERIAL: Lake sand	

DESCRIPTION: A subangular sand from dunes and bottom deposits in the Great Lakes area. This material may be relatively free of AFS clay, but may have up to 8.0% non-silica materials. Screen distribution is usually on three screens, peaking on the 70 screen. AFS grain fineness can be 45 to 70.

USE: This sand is used primarily in the manufacture of cores in the oil, furan, or resin coated mixes.

REQUIRED SPECIFICATIONS:

1. SCREEN ANALYSIS

Screen Number	Minimum %	Maximum %
30	0.0	1.0
40	4.0	7.0
50	25.0	35.0
70	45.0	55.0
100	15.0	20.0
140	0.0	2.0
GFN	48	52

2. PHYSICAL AND CHEMICAL PROPERTIES:
   a. Acid Demand Value = 35 maximum
   b. AFS clay = 1.0% maximum
   c. Moisture as rec'd. = 0.5% maximum
   d. Sand temp. as rec'd. = 100°F maximum

SAMPLING & ACCEPTANCE: Incoming sand shall be tested at a frequency designated by the purchasing plant. Where vendor reliability has been established, acceptance shall be based on the supplier's laboratory report.

All tests will conform to the latest standard testing procedures as outlined in the AFS Sand Handbook.

REJECTION: All material failing to meet specifications will be subject to rejection.

Figure 4 A typical purchase specification for core sand.

DATE:	NO. C51
CHANGE NO. 6	PAGES 1 of 1
MATERIAL: Core oil	

DESCRIPTION: A material formulated from drying oils, resins, and solvents in proportions at the discretion of the supplier to their chemical properties ranges. The composition is not a part of the specification.

USE: Used as a binder in the production of core sands.

REQUIRED SPECIFICATIONS:

1. The oil will be suitable for producing oil sand core mixes used in cores at curing temperatures of 400°F to 460°F for a time of 3.0 to 6.0 hours, dependent on core configuration.
2. The oil will be homogeneous and free from all contaminants including water, dirt, and sediment. It shall not separate during storage and must be compatible with cereal binders, iron oxide and other core sand additives.
3. TECHNICAL PROPERTIES
   a. Non-volatiles (calculated) = 80—85%
   b. Specific gravity @ 60°F = 0.945—0.947
   c. Viscosity @ 100°F = 270 ssu
   d. Baked tensile strength (400°F baking temperature) of standard core sand mix.

Baking Time	Tensile in PSI
30 min.	105
45 min.	310
60 min.	300
90 min.	275
120 min.	255
180 min.	255

   Test procedures will be those found in the latest edition of the AFS Foundry Sand Handbook.
4. The oil must fulfill all foundry production requirements as well as meet technical properties.

SAMPLING & ACCEPTANCE: Acceptance shall be based on suppliers' test results where suppliers' reliability has been established. If the material does not meet the properties and requirements specifications, it shall be subject to rejection as agreed on by plant personnel and the foundry Purchasing Department.

Figure 5 A typical purchase specification for core binder oil.

IV. PROCESS CONTROL

Foundry process control is an effort expended to prevent the production and use of nonconforming materials, cores, core equipment, molds, mold equipment, metal, and so on, to aid in producing a reliable and usable casting. Process control in the metal-casting industry, through inspection and auditing in the various processing areas, is a more economically sound means of producing good-quality castings than any attempts to inspect quality into the final cast product. Both process audits (Fig. 6) and job instruction sheets (Fig. 7) can be instrumental in defining the working of a process control system.

A. Control of Core-Making Processes

All core-sand mixes will have process write-ups giving the material quantities, sequence of material additions, and mixing instructions regarding times and other special instructions. Required properties and frequency of testing of the core sand should also be stated. Instructions for disposal of sand or cores not meeting minimum required properties should also be given in the process write-ups.

Core coatings will have a process write-up stating mixing instructions, required test properties, and frequency of testing. Posting of test results at the coating workstations permits easy auditing.

Equipment for producing cores, after being accepted initially as accurate and usable, should be inspected prior to and after production runs. In the case of long-running production work, the inspection of equipment prior to each production shift will reduce the chance of poor-quality cores being produced.

The core maker, core inspector, and process control observer should be part of a system to inspect all cores for correct surface contour, soundness, proper curing,

Process Check	Location	Time	Comments/Action
Sand Preparation			
1. Sand mix posted			
2. Additives added in proper sequence			
3. Muller mixing at proper cycle times			
4. Sand tests taken from muller and stations as required			
5. All sand tests are documented			
6. Sand test equipment cleaned and calibrated regularly			

Process audit for sand preparation, core sand or mold sand.

Figure 6 Process audit.

Customer ______________________ Part No. ____________

Original Issue Date ______________

Material Specification ______________

Mechanical Properties ______________ Hardness Range ________

Chemical Requirements ______________

Molding

Flask ______________

Chaplets: Cope: Size ____ No. ____ Drag: Size ____ No. ____

Ram-up Core ____ Chills ____ Nail Up Core ____ Vents ____

Exothermic Riser Sleeve: Size ____ No. ____

Other Requirements ______________________

Core

Cores/Casting ________ Type: Shell ____ Oil ____ Insocure ____

Production By: Bench ____ Small Blower ____ Shell ____ U180 ____

Hot Box ____ Sand Composition ______________

Blow Pressure ____ Invest Time ____ Cure Time ____ Box Temp ____ °F

Venting ______________ Core Weight ______________

Pour and Shift

Metal Temperature: Max. ________ Min. ________ Pour Time ________

Mold Weights: No. ________ Size ________ Shakeout Time ____

Riser Hot Top ______________ Other Requirements ________

Cleaning and Heat Treating

Grinding ________ Gauging ________ Straightening ________

Heat Treat Cycle ______________________

Pieces/Tray ______________

Inspection

% Hardness Tested ______________ Test Location ______________

Ultrasonic ______________ Eddy Current ______________

Figure 7 Job instruction sheet.

and dimensional accuracy. Intricate cores may need fixtures for gaging. Using assembly fixtures for multicomponent core assemblies will assure dimensional accuracy.

B. Control of Molding Processes

Molding-sand mixes are documented just as the core-sand mixes in the preceding section. Since most molding and mixing operations perform a number of sand properties tests, plotting of test results at that station can be an aid to tracking results and anticipating necessary changes as trends develop.

Good, well-maintained pattern equipment is an essential starting point for producing good molds. Inspection of patterns prior to startup of production for proper contour, cuts or gouges, properly anchored gates and risers, and so on as listed in Figure 8 can be used to check molding equipment by process control observers, molding supervisors, or the molders.

Molds containing cores or complicated core assemblies should have core-setting fixtures to help establish good dimensional control and also avoid some core-related casting defects.

Closing and handling of molds on storage lines is also important. Abuse of the molds can result in casting surface defects.

C. Control of Melting, Pouring, and Shakeout Processes

Quality cast metal starts with control and inspection of charge materials used to produce the metal. Once the correct material is available and used, the accurate weighing and placing of the charge materials in the melt unit must be followed. Most metals do not offer a great latitude in the chemistry of their composition, so charge materials should be purchased to a specified chemical analysis, which may be certified by the supplier or accepted on the basis of user test results.

Good melting control is dependent on measuring and controlling metallurgical qualities, metal temperature, metal chemistry, and even gas content of the metal as the melting process progresses. Many forms of process control equipment and testing—such as molten metal pyrometers, spectrometers, thermal analysis equipment, chill tests, and microstructure examination—are means of assessing metal quality. The use of these means to get data for making melting adjustments, where necessary to meet required specifications, adds to the metal quality.

Batch melting, compared to continuous melting methods, affords a better opportunity to make adjustments prior to pouring. However, the judicious and programmed use of metal testing in continuous melting can be useful in reducing property ranges and improving quality.

Many metals require ladle additives such as alloys, inoculants, degassers, or slag coagulants. Too often, the mechanics of adding these additives is ignored. Each material has an optimum addition rate and time as well as point of entry into the molten metal. These procedures should be documented and followed to assure the proper response of the metal to the additive.

Once the metal is ready for pouring, the handling and postmelting treatment of the metal require disciplined control. Metal temperature control is important to both ferrous and nonferrous metals. Many casting defects, ranging from gas porosity

Molding	Performed	NA
1. Check pattern equipment for condition		
2. Check mold procedure		
3. Check that peen boards and core setting gauges are used		
4. Check scrap trend against molding (ramoff, drops, etc.)		
5. Proper cope off procedure (one of 1st five molds)		
6. Check core fit and defective cores		
7. Correct number, size and quality of cores, chills and chaplets		
8. Correct diameter and condition of down sprue		
9. Flask is opened and closed properly		
10. Tight flasks are clamped properly		
11. Proper length of closing pins		
12. Proper mold vents and size		
13. Proper blow out procedure		
14. Check that mold hardness is taken regularly		
15. Check for defective or worn equipment		
16. Check for loose or worn pins and bushings		
17. Proper push off procedure		
18. Correct sand or facing used		
19. Correct number of flask bars and gaggers		
20. Check for cracked molds		
21. Proper codes and test bars		
22. Check molds for cleanliness		
23. Inspect interior of mold		

Figure 8 Mold process inspection list.

to surface inclusions, can be related directly to metal temperature control in handling and pouring the molten metal. Maintaining a documented system of job cards that identify the proper metal temperature pouring range as well as optimum pouring times for the given mold becomes the basis for a process control audit on pouring practices.

Mold shakeout control after pouring off the molds is another process area requiring documentation of job requirements as to time from pour to shakeout and possibly special handling instructions. Because all metals are section sensitive to some degree, metallurgical control that can affect mechanical properties is also a function of the optimum time that castings remain in the mold prior to shakeout. Observance of this discipline is another step in meeting the required customer specification.

D. Control of Cleaning and Heat-Treating Processes

The cleaning department has the prime function of preparing the raw castings to a condition and configuration suitable for the customer's processing or use. Cleaning operations define the physical appearance of the casting, and the user's incoming inspection makes its initial judgment of casting acceptance based on its sensory qualities. The control of abrasive shot blasting, grinding, and chipping affects the surface finish, cosmetic appearance, and dimensions of the castings. The use of a job process card to guide the work performed on the casting and serve as an audit inspection guide is worthwhile documentation. A photograph of the casting referencing critical cleaning areas can be attached to the process card for further direction.

Heat treatment is required for many ferrous as well as nonferrous metals. Correct heat treating requires adherence to proper time-at-temperature cycles. Heat treatment results can be affected by the amount and geometry of loading the castings into the heat treatment furnace. Improper loading can result in missing the required metallurgical or mechanical properties, warpage or distortion of the castings, and wasting of heat energy. The job process card showing loading characteristics and optimum load weight for a given furnace will serve to identify proper control of heat treatment. Temperature and time at temperature should also be part of the job process card information. Castings that have been heat treated should be qualified by some means, such as hardness inspection or microstructure evaluation (see also Chapter 37) on a sampling basis. Heat treatment lots are identified and associated with the sample inspection and documented accordingly.

V. INSPECTION

Inspection is the act of evaluating some characteristic of the casting as compared to a standard to determine if the part conforms to a specification. The level of inspection required will be derived from historical data on the same or similar castings, performance of the part regarding machining by the customer, service life, and the producing foundry's process capabilities. Specifications may not cover all the characteristics pertinent to the casting being inspected, and evaluation requires that a preliminary judgment be made if the casting is fit for its intended use. Obviously, the customer makes the final judgment. The use of an actual casting for a Go/No Go standard can be a supplementary aid to an inspection job process card. The use of this actual standard reduces questionable judgment calls that may be made on marginal quality castings and helps to preserve the marketability of the casting.

A. Sensory Inspection Characteristics

An inspection of sensory characteristics, which includes all surface-related items such as discontinuities, inclusions, and not-to-contour surface, can be performed at many processing points in the cleaning room. Work areas such as shakeout, gate and riser removal, abrasive cleaning, grinding, and chipping can be informal inspection stations where nonconforming castings can be removed from the main flow of work. In all cases, though, the inspection department still makes the final decision on casting integrity prior to shipment.

B. Internal Sensory Inspection Characteristics

Castings such as cylinder heads, compressor housings, valve bodies, and so on that have complex internal configurations may require optical aids to assess internal surface quality. Dental mirrors, fiberoptics, and special viewing lights allow for a more critical appraisal of surface conditions.

C. Dimensional Inspection Characteristics

Inspection for conformance to dimensional specifications can be accomplished by use of a *checking fixture* that accepts the contour of the casting or picks up certain gage points. When high production is obtained from a given pattern, a complete dimensional layout should be performed at a frequency to assure that customer-required dimensions are maintained to the drawing tolerance. A number of items will determine the frequency of layout inspection required: material used for the pattern equipment, molding process, type of molding, and number of molds produced.

D. Gaging Inspection

Assurance that castings will be compatible with machining fixtures and with other components in assembly can best be determined by using *gaging fixtures*. Gages may range from simple contour forms such as are used to define the opening in a manifold to more sophisticated forms as used for complicated castings such as heads, blocks, or housings. The latter type of fixture may not only define contours but also assess machining fixture points to the extent that they are machined to a given dimension.

E. Hardness Inspection

The mechanical hardness test is probably the most extensively used means of evaluating casting quality, since hardness relates to mechanical properties as well as the microstructure.

Hardness testing measures a metal's resistance to permanent deformation by a certain sized or shaped indenter under a given load. The most common of these tests are the Brinell and the Rockwell test. Other tests for hardness include the Vickers, Knoop, and the scleroscope (see also Chapter 41).

The Brinell hardness test is performed by applying a constant load, usually 3000 kg, on a hardened steel ball, called the indenter, that is 10 mm in diameter. ASTM Standard E10 describes the complete testing procedure, as well as the calibrating and standardizing methods for Brinell test equipment. Loads of 500 or 1500 kg can be used for softer materials or thin sections, where cracking may occur.

If the hardness is expected to be over 400 HB (indentation diameter of 2.90 mm or less), a tungsten carbide ball should be used rather than a hardened steel ball.

The quality and reliability of Brinell testing can be adversely affected by both the operator and the machine. Operator-induced problems that can contribute to inaccurate results are:

1. Ground surface on test piece is too rough.
2. Surface of test piece is on an angle.
3. Hardness indentation is too close to the edge of the casting.
4. Test piece temperature is much higher than room temperature [should be less than 50° F (differential)].
5. Measuring scope tipped when reading indentation diameter.

An example of the influence of these five factors on accuracy is shown in Table 2, where sections of a standard steel bar were tested by five inspectors under different conditions.

Machine-induced conditions can also result in erroneous readings. Machine variations include:

1. Indenter ball flattened.
2. Machine does not hold pressure.
3. Machine pressure readings are in error.

Hardness test measurements should be made on a number of castings rather than on individual castings when differences occur. It is found that the frequency distribution and average of hardness readings will be quite comparable despite differences on individual readings.

Another widely used quality test procedure is the Rockwell hardness test. This test develops a number based on an indentor's penetration depth under a relatively light load beyond which it has been driven by a heavy load. The light load of 10 kg is applied followed by the major load, which is applied and removed with a direct reading being obtained on the indicator dial of the Rockwell test

Table 2 Hardness Measurement Variations Due to Differences in Test Accuracy

Inspector	#1	#2	#3	#4	#5
Indentation diameter (mm)					
4.70					xxxxx
4.65		x			
4.60	xx	xx	xx	xxxxx	xxxxx
4.55	xx	xx	xxxxxx		
4.50	xxxxxxxxxxxx	xxxxxx	xxxxx	xxxxxxxxxx	xxxxxxxxx
4.45	xxx	xxxx	xxxx	x	
4.40	xxxxxx	xxxx	xxxxx	xxxxxxxx	xxxx
4.35	x	xxx	xxx		
4.30	x	xxxx	x	xxxx	xxxx
$\overline{X} =$	4.485	4.448	4.468	4.459	4.511

machine. A number of different scales are used, the B and C scales being the most common for metals. The B scale uses a 1/16-in-diameter hardened steel ball (penetrator) and a major load of 100 kg. The C scale uses a 120° diamond cone penetrator requiring a 150-kg major load.

The same precautions used to assure reliable readings in the Brinell hardness test are also common to the Rockwell hardness test. The quality of the surface finish, angled or nonparallel surfaces, rounded surfaces, and surface scale are critical items to be controlled for accurate test results.

F. Sampling Plans

Acceptance inspection by the foundry in assessing product quality or by the user in incoming inspection can be advantageously accomplished by use of sampling plans. Assuming that inspectors follow a prescribed plan for sampling and the inspection is performed accurately to conformance requirements, some economy is gained. From time to time it is useful to perform destructive testing on the samples. Good sampling is dependent on randomness and avoidance of any bias. Any bias such as previewing the lot and selecting castings as to good and bad quality levels, avoiding inconvenient pieces or always sampling from the same location in a container will void a good sampling plan. Good supervision and auditing are necessary to avoid these problems by assuring that the agreed-on sampling plan is being followed.

Many casting producers and consumers use sampling plans without realizing that there is a necessary discipline in executing the plan and that there are risks, such as rejecting usable lots and accepting poor-quality lots. It is recommended that people responsible for supervising inspection sampling utilize standard sampling plans such as MIL-STD-105E or MIL-STD-414. These two standards define sampling for inspection by attribute (casting is good or bad) and by variable (a specific quality characteristic is measured).

It is not too unusual for foundries to use an acceptable quality level (AQL) of 1.5%. This is the maximum percentage defective that can be considered satisfactory as a process average for acceptance of casting lots for a defect or a group of defects. This tells the supplier that the plan will regularly accept the lots provided that the process average level (of defectives) shown is less than 1.5 per 100. Thus the AQL relates to an average for a number of lots rather than a single lot regarding the expectations of nonconformance.

A 100% inspection plan is used for critical castings, large castings, or small lots. In the case of small castings, consideration should be given to determine if the service application justifies a 100% inspection where large lots are involved.

With any type of sampling plans where a large number of characteristics are to be inspected on each casting, human failing is reduced if one person checks one characteristic per casting, rather than checking a number of characteristics on each casting before moving to the next.

VI. TESTING

The determination of a casting's quality and reliability is often best performed by testing. Testing entails means other than the immediate sensory determination of a casting's fitness for use. Mechanical and electronic aids have been technologically developed. These testing and evaluation tools include x-ray and gamma-ray radiography, ultrasonic and sonic testing, eddy current testing, magnetic particle testing, and liquid penetrant testing together with older, established procedures such as hardness and leak testing. All these test methods can be used to determine one or more of a casting's properties such as soundness, surface or subsurface condition, mechanical properties, or a metallurgical property, which in turn reflects the quality and reliability of the casting being tested.

Advances have been made in other forms of testing and the near future will see expanded use in casting applications. Methods such as acoustic and optical holography, acoustic emission, infrared testing, and sophisticated forms of radiography will be available for qualifying casting integrity. An aid to selecting test methods is shown in Table 3.

A. Internal Soundness

The internal soundness of castings can be determined to any degree of accuracy by only two methods: radiography or ultrasonics. Radiography is a means of providing a two-dimensional picture of the intensity distribution of radiation that has passed from a source through the casting onto a film. The solid material attenuates the intensity of the radiation, and internal discontinuities or changes in material section size will attenuate the radiation at varying intensity to produce a shadow image on the film. X-rays have been used successfully on castings since the Coolidge x-ray tube was developed in 1912. Present-day equipment usually consists of a high-voltage power supply or transformer and the x-ray tube and x-ray type photographic film specifically selected for the application.

Table 3 Nondestructive Test Comparison

Methods	Flaws detected	Advantages	Limitations
Eddy-current	Cracks, seams, and microstructure	Moderate cost, portable, readily automated	Conductive metals only, shallow penetration, geometry sensitive, reference standards of help
Magnetic particle	Cracks, seams, voids, inclusions, porosity	Simple, inexpensive, senses shallow surface flaws	Ferromagnetic materials only, operator dependent, can give irrelavent indications
Liquid penetrant	Cracks, laps, seams, porosity	Inexpensive, portable, easy to use	Operator dependent, irrelevant indications can occur, flaw must be open to accessible surface
Ultrasonic	Cracks, voids, porosity, laps	Excellent penetration, good sensitivity, requires access to one side only	Requires mechanical coupling to surface, manual method is slow, operator dependent
X-ray radiography	Porosity, voids, inclusions, cracks	Detects internal flaws, portable, gives permanent record, can be used on all metals	Costly compared to other methods, possible health hazard, may be insensitive to some laminar flaws

Gamma rays (for radiography) are a form of electromagnetic radiation just like x-rays, but differ in wavelength. Gamma rays are emitted from radioactive atomic nuclei. The most common gamma-ray sources are cobalt-60 and iridium-192, although other sources, such as thalium-170, cesium-137, and radium, are available. Gamma-ray testing equipment includes the radioisotope encapsulated in a lead-shielded storage safe. A cable or pneumatic drive handling system is used to move the radioisotope from its storage safe to the radiographic position.

Radiography can be used to detect a variety of discontinuities and flawed conditions in castings that may not be visible externally. Such conditions as cold shuts, cracks, gas porosity, shrinkage, misruns, unfused chaplets, core shifts, nonmetallic inclusions, and segregation can be detected and a film record obtained. To assess correctly the film record for casting condition, standard reference radiographs are used.

Radiographic inspection has a number of advantages and limitations compared to other testing methods. The advantages are that

1. It provides a record of the inspection.
2. It is nondestructive in nature.
3. It reveals the internal nature of the casting to determine its serviceability.
4. It can be relatively inexpensive for small castings where the internal quality and soundness is vital.

The limitations of radiography are as follows:

1. It can be relatively costly for large, intricately shaped castings requiring a large number of film exposures.
2. Laminar-type discontinuities may not be discovered.
3. Equipment, facilities, and operation costs may be high if few radiographs are to be taken.

The use of ultrasonics to detect internal discontinuities is common for castings. Sound waves traveling through a casting will be reflected at any interface, such as a flaw. These reflected waves are analyzed for the distance they have traversed to isolate and determine the extent of an internal defect. Gray cast irons, however, may be difficult to interpret because of the free graphite in the metal which scatters the ultrasonic energy, thus diffusing the sound beam.

The advantages of ultrasonic testing for flaw detection are:

1. Economical for single shapes for production lots.
2. Castings can be tested from one side.
3. Has good sensitivity in detecting small flaws.
4. Has good comparative accuracy in defining the size and depth of flaws.
5. Allows for automated procedures in a Go/No Go basis.

Limitations of ultrasonic testing are:

1. An experienced operator is needed to perform the inspection competently.
2. This is a small-area-coverage technique, and large castings require a great deal of time.
3. Requires good coupling of transducer and casting, which is sometimes difficult.
4. For metals such as cast iron, expert interpretation of readouts is necessary.

B. Surface Discontinuities

Flaws that are at the surface or slightly subsurface may or may not be visible. To detect this type of discontinuity, several techniques are available. These include magnetic particle for ferromagnetic materials, liquid penetrant, or eddy current testing.

Magnetic particle inspection is based on inducing a magnetic field or flux into the ferromagnetic material. The flux will be interrupted by any break such as a crack in the casting. The magnetism at the surface is increased by the field bypassing the defect. On applying a fine magnetic powder, dry or suspended in a liquid, the powder will build up around the discontinuity and define it. When the defect is at right angles to the magnetic field, good definition is obtained. Magnetic fields can be induced by using permanent magnets, electric currents, or electromagnetic yokes, coils, or prods. This method can be used for large or small castings, although it may be a slow procedure for small castings.

Liquid penetrant inspection is used to define discontinuities of a porous nature that are internal but open up to the surface of the casting. Color contrasting and fluorescent dyes are used in the liquid penetrant to enhance the visibility of the penetrant drawn through the flaw by capillary action, onto a developer that is placed on the surface of the casting.

Eddy current techniques can be used to detect surface or slightly subsurface defects such as cold shuts, cracks, inclusions, and porous areas. By using a source of magnetic field which induces eddy currents, a sensor picks up changes in the magnetic field caused by the eddy currents. These changes in the magnetic field are then read on a meter where the reading is proportional to the magnetic field change, or on a display where the readings are proportional to the phase magnitude or modulation of the magnetic field. This method of detecting flaws is not as reliable as the ultrasonic procedure. For surface cracks or seams, however, the eddy current procedure can be used at a relatively fast testing rate.

C. Mechanical Properties

The most common mechanical-properties tests are those for determining strength and hardness. The use of separately cast test bars, or "coupons," that represent the metal in the casting has been the standard specification procedure for obtaining data on mechanical properties. In the last few years, as liability problems have increased, there has been a growing trend to section tensile specimens from a casting at some critical section to determine actual properties in the casting. This trend will continue and standard specifications will probably be written accordingly.

Strength can be determined through measurement of other properties, such as hardness, acoustical resonant frequency, ultrasonic energy velocity, and electrical or magnetic properties. These various properties are influenced by the metallurgical makeup of the metal, which in turn can be associated with the strength of the metal.

Hardness has an established relationship with the metal's tensile strength and is referenced for some metals in a number of ASTM and SAE material specifications.

The acoustical property *resonant frequency* can be used to determine certain mechanical properties. By creating mechanical energy that causes vibrations in a test piece and measuring the amplitude of these mechanical vibrations, it is found that the amplitude is at a maximum level of the resonant frequency of the cast iron under test. It was found that an empirical relationship exists between the resonant frequency and the tensile strength in cast irons.

High-frequency energy, where sound waves have a frequency greater than 20,000 hz, has been useful in defining graphite form in cast irons, which in turn relates to tensile strength. An example of the relationship of velocity to tensile strength of cast irons is shown in Figure 9.

Eddy current testing of ferromagnetic materials using induced electric currents can also be used to determine the mechanical properties of metal castings. Test equipment usually consists of an amplifier, cathode ray display, and two sets of coils, one for encircling a standard piece and a second to encircle the unknown piece. A correlation exists between the magnetic properties of the metal and the metallic matrix, which in turn relates to hardness and tensile strength.

D. Metallurgical Properties

Many factors such as the chemical makeup of the metal, postmelt treatment of the molten metal, cooling rate of the casting as related to design and shakeout time,

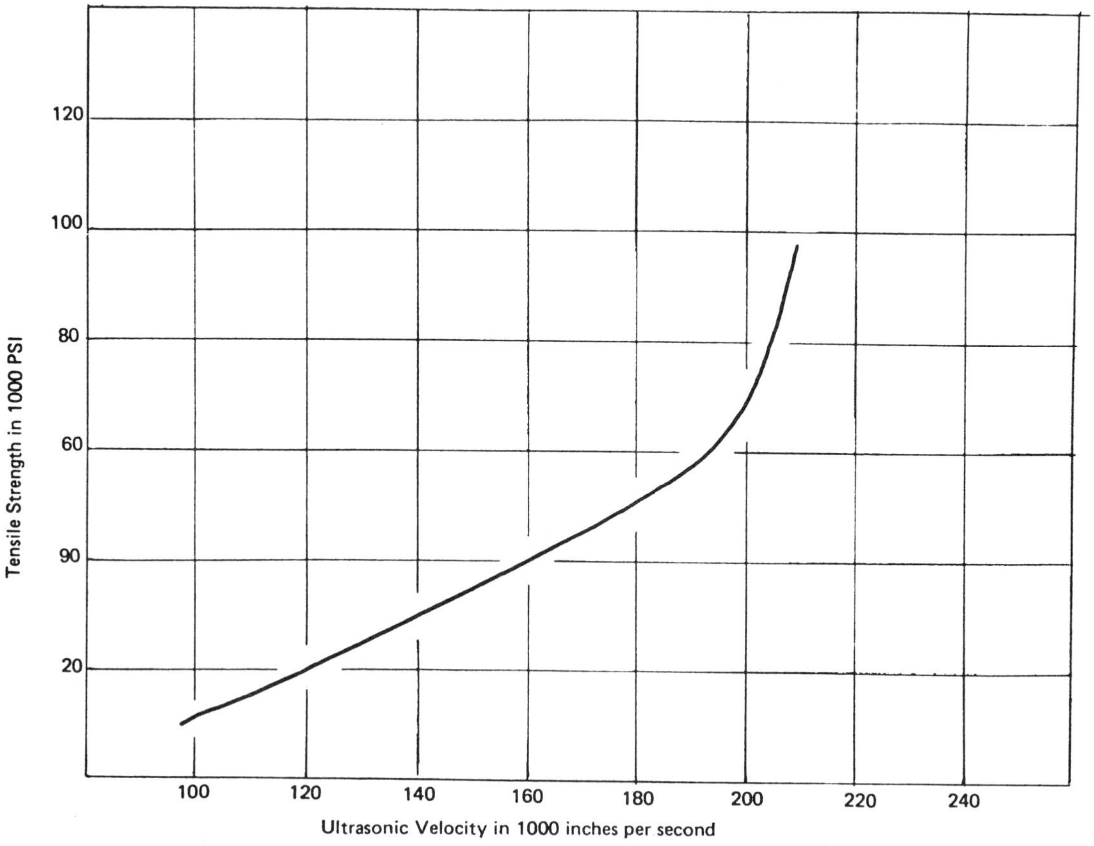

Figure 9 Tensile strength versus sound velocity in cast iron.

and heat treatment affect metallurgical quality. In most metals, the metallurgical quality is predictable, depending on the known control of the foregoing factors. However, with metals such as gray cast iron and its section sensitivity, ductile and compacted graphite cast iron and its control through ladle or mold treatment, assurance of quality can be gained only through microstructural tests. These tests are performed on separate cast specimens, lugs attached to the casting, or sections of the casting itself. Steel microstructural control is quite predictable, depending on its chemistry and heat treatment control.

VII. QUALITY AUDIT SYSTEMS

Quality audit systems are designed to supplement normal process control procedures and to act as an early warning system to assure production of quality castings. These systems can also act as an aid to supervisors in maintaining a high level of quality for the work performed in their specific process area. Materials, processes, and product are covered in the quality audit with a great deal of emphasis placed on "appraisal" and "prevention" to assure material and process conformance and thus to minimize product nonconformance.

The manning of a quality audit system may be handled in several ways. In a single plant operation, the process control observer would be responsible for the audits of materials and processes, whereas the audit inspector would be responsible for the casting (product) quality audit. In a very small plant, the quality assurance manager would be responsible for all audits. In multiplant operations, the director of quality assurance and his staff would have the responsibility for performing quality audits. It is not unusual in multiplant operations to have plant audit teams exchange audits with other plants. Frequency of the audits, which can cover the entire plant or only certain processing areas, should be determined by the quality of performance.

A. Materials

Material audits cover materials produced in metal-casting operations and also purchased materials. Audit points should include:

1. Do purchase orders adequately define material requirements including standard specifications?
2. Are purchased materials tested in-house at a frequency that is properly defined?
3. Are materials purchased on suppliers' tests properly accepted or rejected on test results? Are tests on file for review?
4. Are materials properly segregated and identified? Are they used on a first-in first-out basis? Are they stored per the manufacturer's specified environmental conditions?

B. Processes

A good process control system requires that job process cards be maintained. This documentation becomes the basis for a quality audit of processes. Audits of the various process areas should be cognizant of the following items:

1. Core-making
   a. Are core boxes and associated equipment maintained properly?
   b. Is core sand mixed and prepared to documented recipes?
   c. Are cores made to proper contours and free of defects?
   d. Are stored cores inspected and properly identified as fit to use?
2. Molding
   a. Are patterns, flasks, and associated equipment accurate and properly maintained?
   b. Is molding sand prepared to documented recipes and delivered to the molding floor to stated specifications?
   c. Is molding performed to stated standards?
   d. Is the closing and handling of molds performed with care?
3. Melting, Metal Handling, Mold Shakeout
   a. Is the melt charge makeup properly weighed and delivered to the melt unit?
   b. Is the melt unit operated properly regarding energy input, time of melting, fluxing, and tapping from the melt unit?
   c. Is the molten metal qualified for chemical and metallurgical properties before being dispatched to pouring stations?
   d. Does the metal get the proper ladle treatment?
   e. Is the molten metal of the proper temperature for pouring, and is it poured at the required rate?
   f. Are molds shaken out at the required time and the castings properly handled to avoid damage?
4. Cleaning, Heat Treatment, Inspection
   a. Are the castings cleaned to the specified contour?
   b. When required, are castings heat treated to the specified time and temperature cycle and so identified?
   c. Are castings appraised for sensory characteristics throughout the cleaning process?
   d. Are castings inspected to customer requirements as identified on job inspection process cards?
   e. Are critical inspection tests such as leak and soundness documented and castings marked as inspected?

Product quality audits of the castings can be performed in the shipping area from shipments ready for the customer. Sample lots should be selected with randomness and audit inspected for all customer-specified characteristics such as hardness, dimensions, soundness, or microstructure. Frequency of inspection audits should be determined by the foundry's past performance on the given job, as judged by tabulation of customer complaints and returned nonconforming castings.

VIII. STATISTICAL APPLICATIONS FOR METAL-CASTING QUALITY PROGRAMS

There are a great number of statistical techniques available as aids in controlling metal-casting processes, assessing material or process capability, tabulating defects and process problems, solving defects where multivariable effects are present, and data gathering and appraisal. With the use of low-cost computer time sharing, personal computers, and programmable calculators much of the work can be performed quickly. However, a great deal of basic and even advanced statistical work can be performed by hand calculations or with the help of a hand-held calculator.

Effective quality control starts with knowing what you are doing in all your processes. This requires the collecting of accurate process data in statistical fashion to properly appraise the process capabilities. Typically, any bias in accumulating data can be detrimental to data evaluation. Randomization of data collection is necessary to avoid conclusions that are slanted and possibly erroneous.

A. Control Charts

Control charts are produced by plotting a series of observations which form a pattern. In a foundry, the observations are measurements taken of conditions within the process being studied. By applying statistical tests, control limits for the process can be established. Any fluctuation of the observations beyond the control limits indicate that the pattern is unnatural and the process is out of control. Combining past job experience and knowledge with the information on control charts can provide confidence as well as answers to many production problems.

Two types of control charts are used in foundry operations. One is the average and range or X-bar and R chart and the other is the p chart or proportion defective chart.

The average and range chart is a very sensitive control chart for tracking performance and identifying causes for unnatural fluctuations. The use of this chart is extensive regarding sand and metal properties variations. A typical application of this form of plotting is shown in Figure 10. This chart is used as an aid in controlling the final sulfur content of steel through monitoring the sulfur content of incoming steel scrap. The sulfur analysis is taken on each of five loads of scrap steel received daily and the daily average, $\overline{X}$, is plotted together with the range, R, of the daily sulfur analyses. Near the end, a definite upward trend is noted. Five consecutive readings above or below the $\overline{X}$ or average line indicates a trend. The range, R, shows the daily range to be low, and therefore the scrap steel shipments as a group are indicating similar, but higher, sulfur contents. Action should be taken to maintain control. Getting another form of steel scrap or going to another supplier might be in order.

The p, or proportion defective, chart represents the proportion of defective pieces as a percentage of the total. Thus where we check for only bad and good pieces, we are performing an attributes measurement. The p charts are frequently used in foundry applications to assess performance of various production items. An example of a p chart is shown in Figure 11. In this case, a pipe core for an aluminum manifold was being checked for cracks prior to leaving the core room for the molding area. The investigation of the results over the upper control limits indicated that a substitute machine operator had not been instructed on procedures for removing the core from the core box and storing the cores properly in storage racks, resulting in an unusually high number of defective pieces.

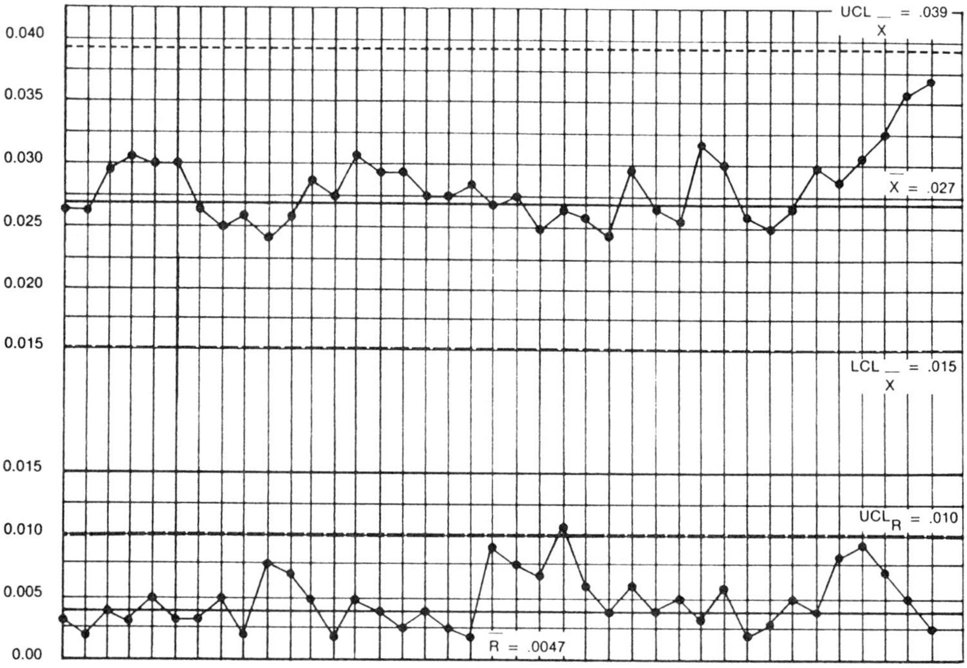

Figure 10 Typical X-bar and R chart.

Data that have been developed for the control charts can also be presented in other forms. The frequency distribution plot and the normal probability plot can be developed from the base data. Both of these plotting techniques can be useful tools to determine job or process capabilities.

The grouped frequency distribution (graphically displayed as a bar chart) represents a set of observations showing the frequency of occurrence of the values of the variable in ordered classes. A good example of this formal plotting is shown, using Brinell hardness readings, in Figure 12. The plotted data show the Brinell reading hardness spread from 163 to 241 with the mode (the highest point) of the distribution at 217. There is a skewing of the readings to the left, on the lower hardness end, although the overall distribution appears relatively normal. This frequency distribution not only gives us information about the population of castings being inspected but can be used in other ways. This plot should be compared with plots for the same castings produced at different times to determine if changes have occurred in the hardness distribution. If there are changes, it may be necessary to check the processes that affect hardness to see if they are in control. Comparing the plot with plots for other Brinell test operators on similar lots of castings can also indicate if operators are showing any bias or if the test procedure is properly followed.

Another technique utilizing data as compiled for frequency distribution studies or control chart plotting is called a normal probability plot. This type of plot can be useful in process capability studies. The following study demonstrates the use of a frequency distribution and a normal probability plot to evaluate dimensional control and possible action to improve this control in a

Plant ABC Co.

Department No. core	Part Name pipe core	Part No. 32684 B
Machine No. B4	Oper. No. & Description core making - blower machine	
Pcs. per Hr. 100		
Subgroup Size		
Sample Size 20		

NO.	Number Defective	Fraction Defective	Reasons for Reject: CRACK					Remarks
1	2	0.10	2					
2	0	0	0					
3	1	0.05	1					
4	0	0	0					
5	1	0.05	1					
6	1	0.05	1					
7	3	0.15	3					
8	0	0	0					
9	0	0	0					
10	1	0.05	1					
11	1	0.05	1					
12	2	0.10	2					
13	0	0	0					
14	0	0	0					
15	0	0	0					
16	1	0.05	1					
17	0	0.0	0					
18	2	0.10	2					
19	1	0.05	1					
20	0	0	0					
21	4	0.20	4					
22	3	0.15	3					
23	5	0.25	5					
24	6	0.30	6					
25	1	0.05	1					

Figure 11 p data sheet and p chart.

large (700-lb) casting. The study was undertaken to determine what the variation in wall thickness would be in the cope and drag (top and bottom walls as cast) of the casting. Since a very large body core formed the inside cavity of the casting, a number of variables are under consideration: dimensional accuracy of the core, tightness of core prints, possible core movement in the mold, mold

Plant ABC co

Department No.	Part Name pipe core		Part No. 32684B
Machine No. B4	Oper. No. & Description core making – blower machine		
Pcs. per Hr. 100	Subgroup Size	Sample Size 20	

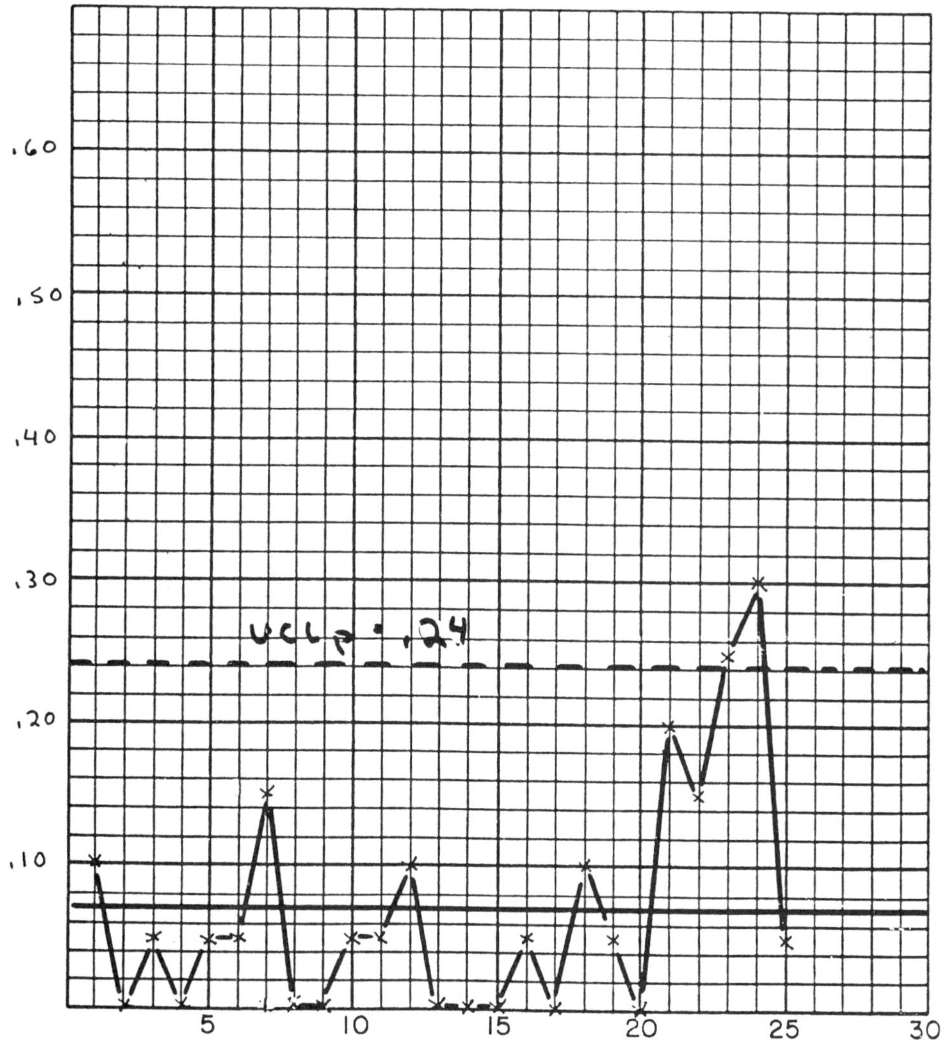

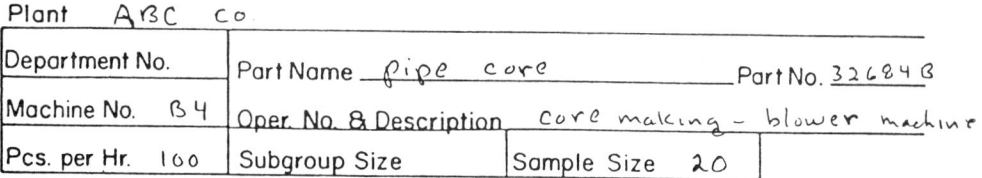

Figure 11 Continued

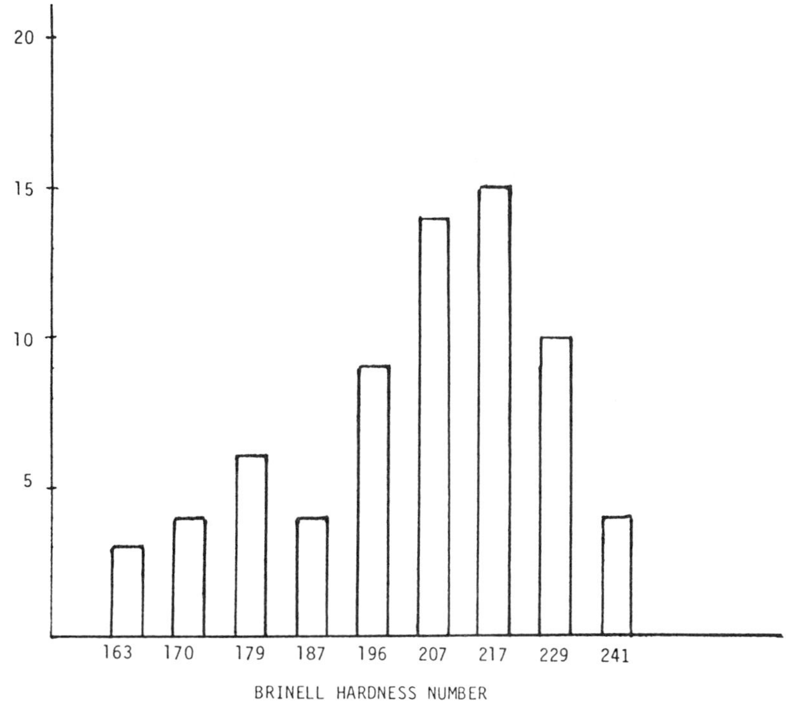

Figure 12 Frequency distribution of casting hardness.

wall movement, and possibly others. The frequency plot in Figure 13 shows that wall thickness varied from 0.48 to 0.75 in in the cope, whereas the variation was 0.56 to 0.72 in in the drag. This told us that the large core tended to be pushed upward (thinner walls on the cope side) as a result of molten metal pressure as the casting was being poured. By tightening up the core prints and placing reinforcing bars in the cope flask to minimize core and mold wall movement, wall thickness became more equal and the variation reduced.

The data taken from the cope and drag wall measurements were used to develop a probability plot. Table 4 shows the readings given in 0.02-in increments.

The midpoint percentage is plotted against the wall thickness on normal probability paper as shown in Figure 14. The best-fit line is placed through the plotted points. The process average is at the intersection of the 50% line and is shown to be 0.625. The 3σ upper limit is at the intersection of the rightmost vertical line and the best-fit line and is found to be 0.81. The 3σ low limit is at the intersection of the leftmost vertical line and the best-fit line and reads 0.44. The process capability—that is, the "spread" of the process—is the upper 3σ limit minus the lower 3σ limit, or 0.37. An estimate of the standard deviation is the process capability divided by 6 or 0.062. Other information can be determined from the plot, such as prediction of percentages above and below a given value, comparison with specification values, and obviously a determination of process or machine capability under the conditions studied.

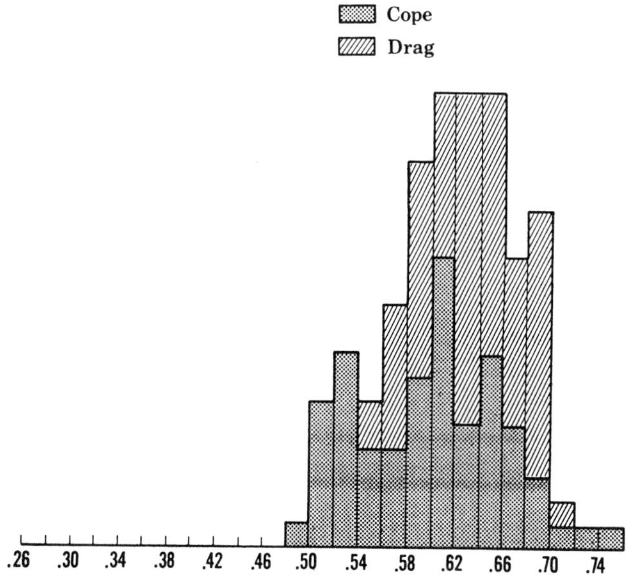

Figure 13 Frequency distribution of wall thickness.

The techniques mentioned in this section on control charts can be used in many areas of foundry processing, utilizing the data that have been gathered through routine testing and checking of sands, cores, molds, metal, and castings. Proper and thorough analysis of the plots, whether control charts, frequency distribution studies, or probability plots, will provide statistically sound information for decision making. A number of references listed at the end of this chapter should be studied to enhance one's knowledge in these areas.

Table 4 Casting Wall Thickness Measurements

Dimension	No. of readings	Cumulative readings	Cumulative (%)	Midpoint (%)
0.49–0.50	1	1	0.8	0.4
0.51–0.52	6	7	5.2	3.0
0.53–0.54	8	15	11.2	8.2
0.55–0.56	6	21	15.7	13.3
0.57–0.58	10	31	23.1	19.4
0.59–0.60	16	47	35.1	29.1
0.61–0.62	19	66	49.3	42.2
0.63–0.64	19	85	63.4	56.3
0.65–0.66	19	104	77.6	70.5
0.67–0.68	12	116	86.6	82.1
0.69–0.70	14	130	97.0	91.8
0.71–0.72	2	132	98.5	97.7
0.73–0.74	1	133	99.3	98.9
0.75–0.76	1	134	100.0	99.6

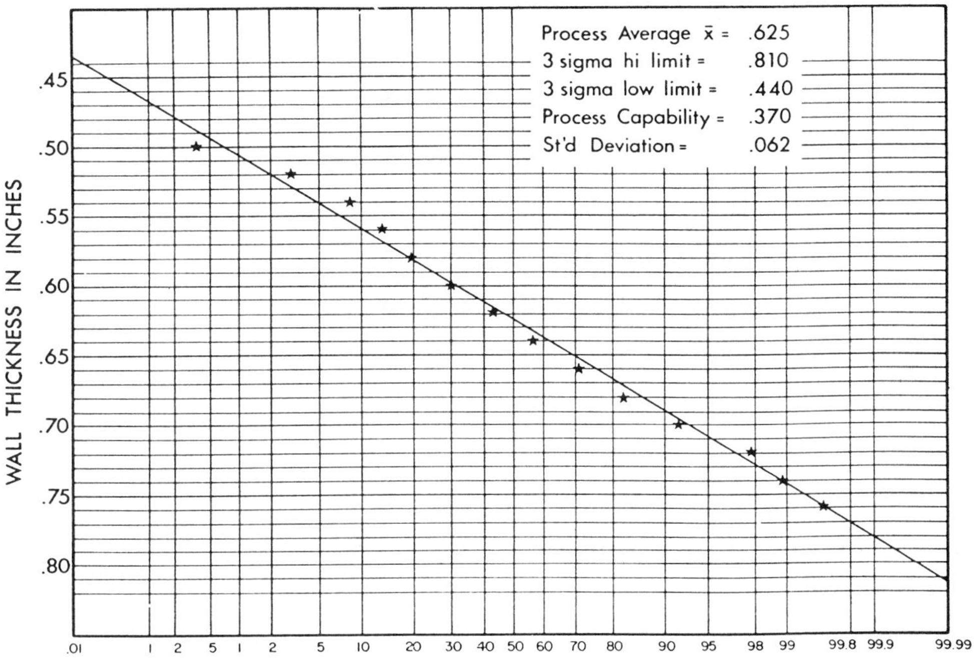

Figure 14 Process capability study.

B. Defect Studies

Statistical techniques for defect studies are used extensively. Among those used are the ones covered in the section on control charts. The control charts of metal and sand properties data can be associated with defects having a direct relation to those properties. Using attribute control charts by job, by defect, or by defect for the individual job is useful in recognizing the occurrence of trends. Distribution studies of defect types in Pareto chart form indicate where the effort to reduce defects should be stressed. It usually shows that four or five defects are the major contributors to the overall defect total and by concentrating the defect reduction effort on these few will net the greatest economic gain. Probability plotting will determine the expected results for material or machine performance, and this can relate to defect occurrence if the process related to material or machine is "out of control."

Whenever possible, casting defects should not be rated on a Go/No Go basis. Defects occur at some level of severity, and to judge properly the effects of the variables in the casting process, defects should be rated to a given severity level or number. For example, by rating from 0 or perfect to 5 or exceedingly bad, judgment on the effect of a process change can be evaluated and statistical procedures more easily applied. When working with small castings, visual standards can be selected to allow sensory judgment of the defect. By selecting the standards to represent quality between the six standard ratings of 0, 1, 2, 3, 4, and 5, judging of test castings can be performed by comparing with the standards as better or worse than the standard and a whole number assigned as to defect severity.

Defect records should be informative enough to provide clues to the source of the problem. If the casting has more than one type of defect, record all visible defects, since one defect may be associated with another and point out some process deficiency. On a single-type defect study, the location of the defect should be recorded. A heavy concentration of the same defect in a given area of the casting can lead to the cause of the problem.

Once the cause is found, it should be tested to see if it is controllable and actually is the root cause for the defect. If you cannot create the defect by varying the cause, you do not have the root cause, and further study is required. A number of causes and effects, like a series of chain links, must be studied until the controllable cause is established. Most defects are a result of a multivariance effect resulting in the need for the study above (see also Chapter 25).

It is also true that a great deal of defect solving can be accomplished by using past experiences in similar situations as a guide for corrective action. Certainly, this is the most economical and the most expedient approach to defect solving. When this fails, the use of statistical methods should be advocated. Most foundries generate and collect a substantial amount of data in their processes. This data can be used in statistical studies and the results used to solve problems.

C. Problem Solving

Many years ago, Frederick Taylor advocated a scientific approach to problem solving and decision making. At the time Taylor professed his philosophy, quality control techniques were virtually unknown and tools such as calculators and computers were nonexistent. However, even with all the progress made in statistical aids and equipment, a review of Taylor's precepts is in order.

1. Properly identify the problem. Probably the most difficult step and as some unknown author once stated, "If you can't define the problem in one sentence, you don't know the problem."
2. Gather all the needed facts and evaluate them.
3. Determine the various solutions and recommend the most likely solution.
4. Once the solution has been proven, act on it.

The first three steps can be performed by using various statistical and analytical techniques, some of which we will demonstrate. These methods will speed up the decision-making process and verify any conclusions reached.

Many problems in metal casting relate to a decision in a material or process change, the relationship of process variables to the quality of the product, or perhaps deciding which of two materials or methods will produce the best product after limited testing is performed. Several test procedures can be used to establish the better of two items. One of these procedures is the *t test*, a statistical test to determine if a significant difference exists between the averages of two methods. The data obtained from the two methods are considered as coming from a normal population and the variances of the two methods are considered equal. An example of the t test is shown in Table 5. Values are shown in Table 6. This is a foundry experiment where the effects of two gating systems on surface inclusions were studied. The castings from the two methods were rated as to defect severity in the manner described earlier in this chapter.

Table 5 Results of the Defect Severity Using Gating Systems A and B[a]

Gating system A			Gating system B		
2	1	0	2	2	1
1	2	2	3	3	3
3	2	2	3	3	2
2	3	3	2	5	1
4	1	1	4	2	3

[a] 0, perfect; 1, very slight; 2, slight; 3, poor; 4, very bad; 5, horrible.

From Table 6, we substitute the known quantities in the equation

$$\sigma^2 = \frac{\Sigma(X_A^2) - [(\Sigma X_A)^2/n_1] + \Sigma(X_B^2) - [(\Sigma X_B)^2/n_2]}{n_1 + n_2 - 2}$$

$$= \frac{71 - [(29)^2/15] + 117 - [(39)^2/15]}{(15 + 15 - 2)}$$

$$= 1.09 \text{ or } \sigma = 1.04$$

Testing for t gives us

$$t = \frac{\overline{X}_A - \overline{X}_B}{\sigma} \frac{n_A \times n_B}{n_A + n_B}$$

$$= \frac{1.93 - 2.60}{1.04} \frac{15 \times 15}{15 + 15} = 1.76 \text{ with 28 degrees of freedom}$$

Table 6 Squaring the Defect Severity Ratings of Table 5 (X_A^2 and X_B^2)

Gating system A	X_A^2	Gating system B	X_B^2
2	4	2	4
1	1	3	9
3	9	3	9
2	4	2	4
4	16	4	16
1	1	2	4
2	4	3	9
2	4	3	9
3	9	5	25
1	1	2	4
0	0	1	1
2	4	3	9
2	4	2	4
3	9	1	1
1	1	3	9
$\Sigma X_A = 29$	$\Sigma X_A^2 = 71$	$\Sigma X_B = 39$	$\Sigma X_B^2 = 117$
$\overline{X}_A = \frac{29}{15} = 1.93$		$\overline{X}_B = \frac{39}{15} = 2.60$	

Checking Table 7, we see that t at 1.76 is not significant, and no real difference exists between the results of the two gating systems.

Another procedure that can be used to provide a measure of significance between sets of results from two methods is called the endpoint count. The simplest form of this significance testing would require (from the previous example) six castings produced with gating A and six castings produced with gating B. Again, we are to determine the effects of the gating system on the surface inclusion defect. The 12 molds are produced, 6 of each gating system, and then poured in a randomized manner to avoid any bias regarding the other processing variables. After the castings are cleaned up, they are ranked with regard to the inclusion defect from the worst to the best casting and the results set up as in Table 8.

The castings above the upper line (before a change of rank as related to gating system occurs) and below the lower line constitute an end count of 5. The balance of the castings overlap between the two gating systems in regard to defect severity.

Table 7 Factors for t Tests

	t_1 (%)		
Degrees of freedom	5	1	0.1
1	12.71	63.66	636.62
2	4.30	9.93	31.60
3	3.18	5.84	12.94
4	2.78	4.60	8.61
5	2.57	4.03	6.86
6	2.45	3.71	5.96
7	2.37	3.50	5.41
8	2.31	3.36	5.04
9	2.26	3.25	4.78
10	2.23	3.17	4.59
11	2.20	3.11	4.44
12	2.18	3.06	4.32
13	2.16	3.01	4.22
14	2.15	2.98	4.14
15	2.13	2.95	4.07
16	2.12	2.92	4.02
17	2.11	2.90	3.97
18	2.10	2.88	3.92
19	2.09	2.86	3.88
20	2.09	2.85	3.85
21	2.08	2.83	3.82
22	2.07	2.82	3.79
23	2.07	2.81	3.77
24	2.06	2.80	3.75
25	2.06	2.79	3.73
27	2.05	2.77	3.69
28	2.05	2.76	3.67
29	2.04	2.76	3.66
30	2.04	2.75	3.65

Table 8 Example of an Endpoint Test

	Rank	Gating system A	Gating system B		End count
Worst	12	x			
	11	x		=	4
	10	x			
	9	x			
	8		x		
	7		x		
	6		x		
	5		x		
	4	x			
	3		x		
	2	x			
Best	1		x	=	1
					5

The end count determines the significance of the difference between the two methods:

End count less than 7; there is no difference
End count 7–9 would indicate a significant difference
End count 10 or more indicates a proven difference

Since the end count in this exercise was 5, it is concluded that neither gating system, A or B, has a significant effect on the inclusion defect.

Other tests of significance are available. For example, the F or variance ratio test can be used where three or more groups of data are compared. The sophistication of statistical methods for appraising control of processes, finding causes for lack of control, defect solving, and so on, goes well beyond the scope of the information presented in this section of the chapter. It is not advisable to attempt using all the available methods until the basic ones are mastered well. Certainly, if one starts a statistical program with control charts, they should be set up with good data, plotted accurately and analyzed carefully. Just this one technique can be tremendously valuable in attaining process control that will be economically advantageous. After all the possibilities are exhausted in control chart applications, time can be spent in learning and using other statistical aids. A number of reference books are presented below to aid in designing your program.

IX. CERTIFIED FOUNDRY QUALITY

In recent years, an increasing number of manufacturing firms, including foundries, have revised their quality systems to comply with ISO 9000 (series) standards. Originally drafted by the International Organization for Standardization (see Chapters 7 and 8), the standards were revised by Ford, General Motors, Chrysler, and the Automotive Industry Action Group (AIAG) under the QS9000 Quality System Requirements now largely used by the automakers and their suppliers. Among other changes, QS9000 adds requirements for production part approval, continuous improvement, and manufacturing capability to the ISO 9000 standards.

BIBLIOGRAPHY

American Foundrymen's Society, *Statistical Methods as an Aid to Foundry Operations*, Transactions of the American Foundrymen's Society, Des Plaines, IL, 1942.

American Foundrymen's Society, *Statistical Quality Control for Foundries*, AFS, Des Plaines, IL, 1952.

American Foundrymen's Society, *AFS Metalcasters Reference and Guide*, AFS, Des Plaines, IL, 1972.

American Society for Quality Control, *Quality Costs—What and How*, 2nd ed., ASQC, Milwaukee, WI, 1971.

American Society for Quality Control, *Guide for Reducing Quality Costs*, ASQC, Milwaukee, WI, 1977.

American Society for Testing Materials, *Manual on Presentation of Data and Control Chart Analysis*, STP 15D, ASTM, Philadelphia, 1976.

Boyde, D. W. Sr., Milwaukee, WI, personal communication.

Department of Defense, MIL-STD-105D and MIL-STD-414, U.S. Government Printing Office, Washington, DC, 1963 and 1957.

Freund, J., and I. Miller, *Probability and Statistics for Engineers*, 2nd ed., Prentice-Hall, Englewood Cliffs, NJ, 1977.

Grant, E. L., and R. Leavenworth, *Statistical Quality Control*, 4th ed., McGraw-Hill, New York, 1972.

National Aeronautics and Space Administration, *Nondestructive Testing*, NASA SP-5113, U.S. Government Printing Office, Washington, DC, 1973.

Rowe, C. A., Milwaukee, WI, personal communication.

Western Electric Company, *Statistical Quality Control Handbook*, 2nd ed., WEC, New York, 1958.

FURTHER READING

American Foundrymen's Society, *Iron Castings Handbook*, 3rd ed., AFS, Des Plaines, IL, 1981.

American Foundrymen's Society, *Metalcasters Reference and Guide*, AFS, Des Plaines, IL.

American Foundrymen's Society, *Metal Casting and Molding Processes*, AFS, Des Plaines, IL.

American Foundrymen's Society, *Steel Castings Handbook*, AFS, Des Plaines, IL.

Sanders and Gould, *History Cast in Metal*, American Foundrymen's Society, Des Plaines, IL.

Taylor, Flemings, and Wulff, *Foundry Engineering*, 2nd ed., American Foundrymen's Society, Des Plaines, IL, 1959.

25
Isolating the Key Variables

ROBERT W. TRAVER
Traver Associates
Averill Park, New York

I. INTRODUCTION

In recent years, there has been a great deal of emphasis on getting the hourly workers involved in solving manufacturing problems. Statistical process control (SPC) has been emphasized in many companies and training has become a key element. A current emphasis is on ISO 9000 as a foundation for a good quality system. Notwithstanding all of these current initiatives, there are many instances where reducing variability is the crucial goal: the customer's major desire.

II. COMPARING APPROACHES

Often, the traditional engineering approach solves the problem of excessive product variation or at least helps to reduce variation. Yet, almost every company has at one time or another faced a problem of excessive variability which has resisted all attempts at isolation and reduction.

In such a situation of apparently unsolvable problems, management often resorts to:

Accepting reduced quality
Purchasing more expensive manufacturing equipment
Selecting a different method of manufacturing
Eliminating the product from the product line
Designing around the problem
Subcontracting the job

Any of these options may be costly and usually are chosen because the real cause of the product problem—that is, the key process variable (or variables)—has not been identified.

This traditional engineering approach to problem solving is based on someone guessing at the answer or at least guessing the right variables based on past experience, knowledge, or hunches and is characterized by the following four steps.

1. Observe
2. Think
3. Try
4. Explain

Using these steps an investigator will first *observe* the problem. Then, he or she will *think* of what process variable(s) might be causing the problem. For example, the cause might seem to be that the pressure is too low, the temperature is too high, the time is too short, or the raw material is marginal. The investigator will then *try* changing one of these variables at a time and then *explain* the reason for success or lack of success.

The traditional engineering approach is an efficient and effective method for solving many types of problems. There is nothing more effective than the person who knows the answer. This method is so effective that people use the method (keep guessing at variables) even when the method fails to work. The approach may fail because there are four major difficulties with this traditional approach:

- The root cause of the problem must be on the list of suspect variables
- The practical impossibility of holding all other variables constant
- Interactions cannot be found
- Testing one factor at a time takes too long

As a result of these difficulties with the traditional approach, something different may be required. It is in these instances where *key variable isolation* is most appreciated. Key variable isolation has been described as engineering judgment, manufacturing know-how, common sense, and some statistical methods all woven together. It is not just one of the above but all four aspects interwoven.

The structure for using key variable isolation is outlined in the nine-step problem-solving process (this technique was developed by the author during on-the-shop-floor problem analyses) below.

1. Provide focus—*determine importance of problem*
    1. Quality costs
    2. Pareto analysis
    3. p charts
    4. Check sheets
    5. Histograms
    6. Corrective action reports
    7. Customer surveys
    8. Customer complaints
    9. Warranty costs
2. Get close to the problem—*see what is really going on*
3. Quantify the output—*develop a variable scale (if none exists)*
    1. Measurement capability
    2. 0–5 Scale
    3. Jury panels

4. Run multi-vari studies—*let the process do the talking*
   1. Multi-vari studies
      a. Visual analysis
      b. Patterns of variation
      c. Nonrandom patterns and differences
   2. Measles charts
5. Design experiment(s)—*test the variables for causes and effects*
   1. Full factorials
   2. Variables search
   3. Component search
   4. Parallel path experimentation
   5. Random and multiple balance
   6. Fractional factorials
6. Turn the problem on and off—*prove you have the answer*
7. Optimize—*run tests with proven variables*
   1. Designed experiments—multilevel
   2. EVOP
   3. REVOP
8. Install process controls on key variable(s)—*use most cost-effective tool*
   1. Precontrol
   2. Median charts
   3. X bar ($\overline{X}$) and R charts
   4. p Charts
9. Measure before and after results—*quantify continuous improvement*
   1. Dollars
   2. Quality costs
   3. Scrap costs
   4. Rework costs
   5. Customer complaints
   6. Warranty costs

In contrast to the traditional approach, the key variable isolation strategy begins with a stratified sampling of the products being produced by the process. This step in the process is called a multi-vari study, because it looks at multiple *types* of variation. The results are then plotted in different ways, so that graphical analyses can be made. By reviewing the data in a multi-vari format, you use the product to determine the type of variation. *Listen to the process; let the process do the talking.* Types of variation may be within-piece, piece-to-piece, time-to-time, side-to-side, end-to-end, assembly-to-assembly, day-to-day, shift-to-shift, crew-to-crew, and so on. Figure 1 shows the three most general types of variation. These plots can suggest or rule out possible sources of variation; the step-by-step identification of types of variation is used progressively to reduce the field of variables.

This strategy may seem elementary but is in fact powerful. Many problems go unsolved for long periods of time, because in the "think" step of the traditional approach, an incorrect assumption is made and the investigation goes off in the wrong direction. By answering the questions of within-piece, piece-to-piece, or time-to-time variation, we rule out about two thirds of all possible variables that can be important. Further investigations are thereafter confined to variables that fit the clues.

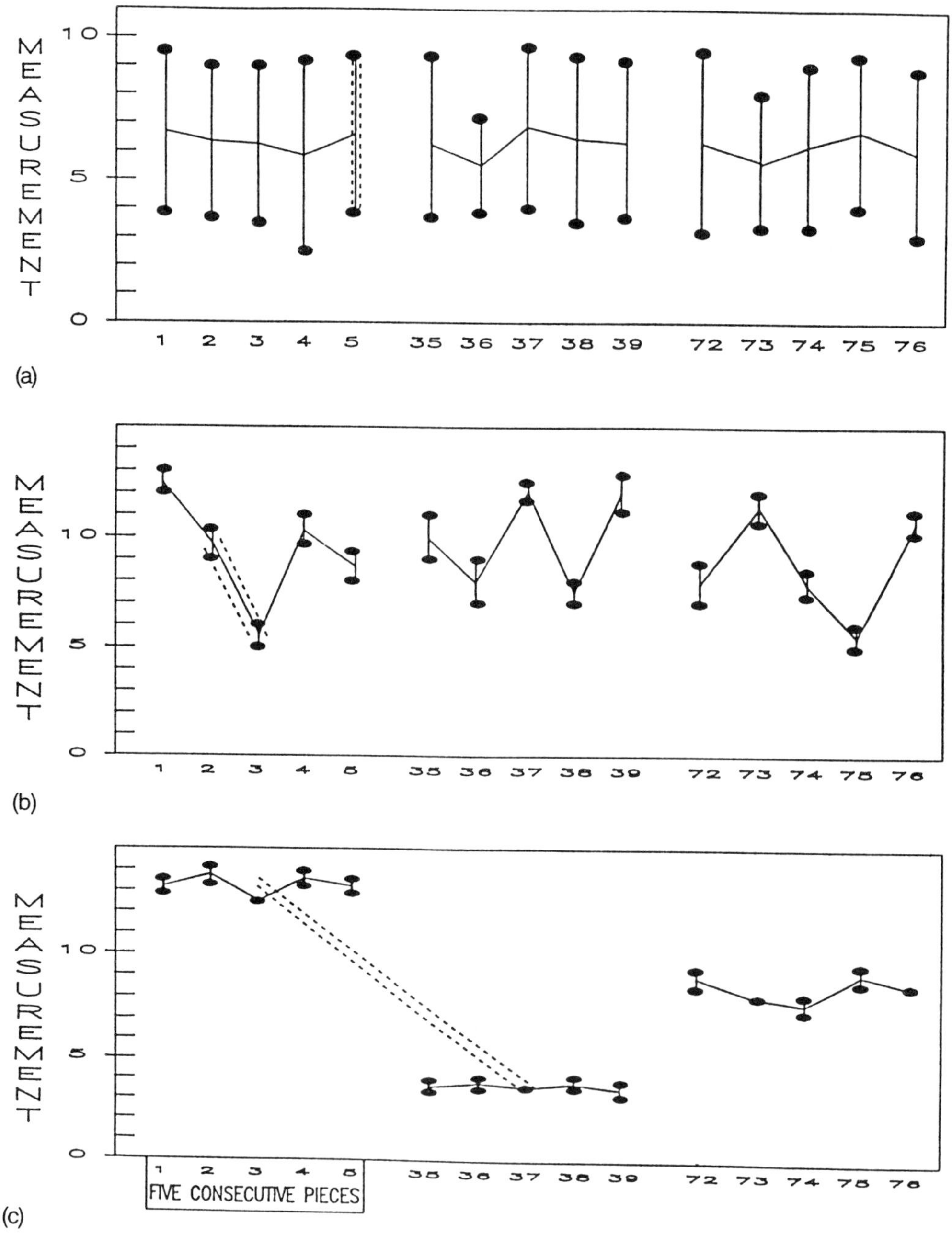

Figure 1 Multi-vari charts. The three main types of variation: (a) within-piece variation; (b) piece-to-piece variation; (c) time-to-time variation.

The investigator using key variable isolation has the advantage of no preconceived notions about the source of the problem and therefore no biases.

There are also significant differences between key variable isolation and the classic statistical approach to problems of product variability. For example, in the classic treatment, the investigator must be content with his or her colleagues' ability to *think* of the potential variables and then try to determine whether these variables are indeed part of the problem, doing so by the use of standard statistical techniques. As might be expected, many designed experiments fail because the right variables are not included in the experiment. In contrast, the key variable isolation investigator is able to design a program that leads him to the key types of variations from the start. Subsequently, the field of potential variables is narrowed to the extent that, in most manufacturing situations, the key variables become apparent.

The key variable isolation strategy often incorporates design of experiments into the study. The difference is how the variables of the designed experiment are selected. In the classic approach, the variables are often selected by brainstorming (guessing), whereas in key variable isolation studies, the results of the multi-vari study lead the investigator toward variables that fit the patterns of variation and exclude variables that do not fit the clues.

However, it must be remembered that regardless of how the variables are selected, all properly conducted designed experiments contain certain key elements. For practical analysis, they should be *balanced.* To be valid, they must be *randomized.* To give a good estimate of repeatability, there should be more than one *replication* (tested more than once at each set of conditions). The use of attribute data (pass or fail) is inadequate and is a major reason for many problems going unsolved. Instead, an agreed-upon *variable measurement* is necessary. Some variables already have scales such as ohms, inches, or pounds, but on other occasions, a variable scale must be devised. The importance of these variable scales in key variable isolation studies cannot be overemphasized. An example of a variable scale developed for an appearance characteristic is:

Numerical Rating Scale for Defect Score
0 = No defect
1 = Barely noticeable defect
2 = Small defect
3 = Large defect
4 = Gross defect
5 = As bad as it ever gets

Using the multi-vari technique to identify the process variables that are most likely to be causing product variation problems may be compared to playing the game of 20 questions. This game starts with one person thinking of some physical thing. Another person then has 20 questions to determine what that thing is. The first question is whether the item is animal, vegetable, or mineral. If the answer is animal, two thirds of the "universe" has been ruled out. The questioner then asks a further narrowing question such as, does it have four legs? If the answer is yes, humans, birds, fish, and insects are ruled out, and the search is confined to a four-legged animal. The next question would not be whether it is a mouse, tiger, or elephant but rather another narrowing question such as, is it a Western Hemisphere animal? If it is a Western Hemisphere animal, the search is further restricted to a four-legged animal found in the Western Hemisphere. The game continues with questions asked that further reduce the field of possible answers.

Both 20 questions and key variable isolation using the multi-vari technique work in this narrowing fashion. This approach contrasts sharply with the method of continuing to guess randomly at the answer. The continued guessing method might be compared to learning in 20 questions that the item had been narrowed down to a Western Hemisphere four-legged animal and then guessing artichoke. The answer, however, cannot be artichoke, because it does not fit the clues. Using the multi-vari technique, the investigator can set aside the investigation of those process variables that are not likely to be causing problems and concentrate on those that are most likely to be key variables.

III. MULTI-VARI COMPARED TO DESIGNED EXPERIMENTS

It is important to distinguish between the multi-vari technique and designed experiments. In a multi-vari study, the manufacturing process is not interrupted. A stratified sample is selected from an ongoing process to represent what happens during regular production without disturbing that process; that is, products manufactured during a multi-vari study are typical products.

In a designed experiment, an investigator creates a special experiment in which process input variables are intentionally adjusted to determine which variables may be causing the problem. A designed experiment invariably interrupts production. Multi-vari studies often solve the problem or may only narrow down the number of process variables suspected of causing product problems. The investigator could then design a specific type of experiment to test those remaining variables and determine which of them is the key variable.

To see how this problem-solving process might apply to a manufacturing problem consider the following example. An iron foundry was experiencing a serious visual defect on a casting for its key customer. The customer was threatening to change suppliers if the foundry did not reduce by a considerable amount the frequency and severity of this visual defect.

There was a great deal of pressure put on the technical people by plant management to improve the quality. With the spotlight on the problem, many tests were run to find the cause or causes of the defect. Almost every day some tests were run. Although the people involved were good foundry personnel, they had not been well trained in running statistical experiments to solve problems.

As a result of brainstorming to find the causes they experimented by running one day using Arizona sand and the next day using New Mexico sand and comparing the percentage of scrap. Or, on two days they would run with different percentages of aluminum, copper, or other additives and compare the results of these two days.

The studies were run varying one thing at a time while trying to hold everything else constant. However, in a foundry or almost any other real manufacturing process, things vary whether or not you want them to do so.

There were so many experiments being run, many of which had a negative impact on the results, that production was suffering even more.

There were many theories abounding, and it seemed that the popularity of any given theory waxed and waned depending on who suggested the theory, how emphatic he or she was of the potential validity, or how high the promoter was on the organizational chart. Also, there were those who claimed that the problem really was worse in the castings from the two outside cavities of the four cavities in the mold; that is, castings numbered 1 and 4 were thought to be bigger problems than castings numbered 2 and 3. With all the pressure, emotion, and anxiety, no one had even taken the time to count the percentage of defects by cavity to confirm or refute this theory.

Management doubted that anyone outside of the foundry could possibly help, but they were desperate. As a last resort, management agreed to bring in an industrial investigator who specialized in solving problems of excess product variation.

The investigator convened a cross-functional problem-solving team composed of a metallurgist, a foundry engineer, a quality control specialist, and the production supervisor. The team showed the investigator the process and then they collected castings with varying degrees of the visual defect and developed the 0–5 numerical rating scale shown in Section II. When analyzing results using a numerical rating scale, the score is determined by totaling the defects. That is, a casting with a 3, two 2s, and seven 1s would have a score of 14.

With a method of quantifying the degree of defectiveness on a variable scale now available, the investigator described to the team how to run a multi-vari study. The team considered variation within-piece by exact location including side-to-bottom of the casting, cavity-to-cavity within the mold (four cavities), mold-to-mold for five consecutive molds, the beginning and end of a pouring ladle of molten metal, three times within a day, and day-to-day for two days.

Then they drew a picture of the four individual castings within a mold and separated the side and bottom results. The picture of each casting was shown looking down into the product at the bottom with the two sides "opened outward" so the defects could be drawn on the inside surface of the sides of the casting. Also, five consecutive molds were shown to compare results from mold-to-mold. This picture of the product would be used to plot the location and severity of the visual defects on the casting using a red marker. This type of plot is called a *measles chart*, because the picture with the red spots used to denote the location and severity of defects might look like it had measles (see Fig. 2).

The first measles chart plotted was for the results of the four castings in the mold with the defects for the side and bottom separated and five consecutive molds from the beginning of the pouring ladle. The same type of data sheet was used for recording the score from the end of the pouring ladle, and two more sheets each were used for the second and third similar sets of samples later in the day. Another six sheets were used for the identical study on the second day. (The defects plotted on Figure 2 are the composite of all twelve sheets.) Some people prefer to call this pictorial representation a concentration diagram instead of a measles chart.

With these twelve individual sheets, defect scores can be totaled by day, showing day-to-day variation. The sheets can be totaled to show results from the beginning to the end of the pouring ladle and the time-to-time variation within the day. Also, totals are completed to show variation from mold-to-mold for the five consecutive molds. Totals over the twelve sheets are done as shown in Figure 2 for the four castings, so there are totals for casting number 1, 2, 3, and 4. Totals are then done by the side of all castings versus the bottom of all castings. By making

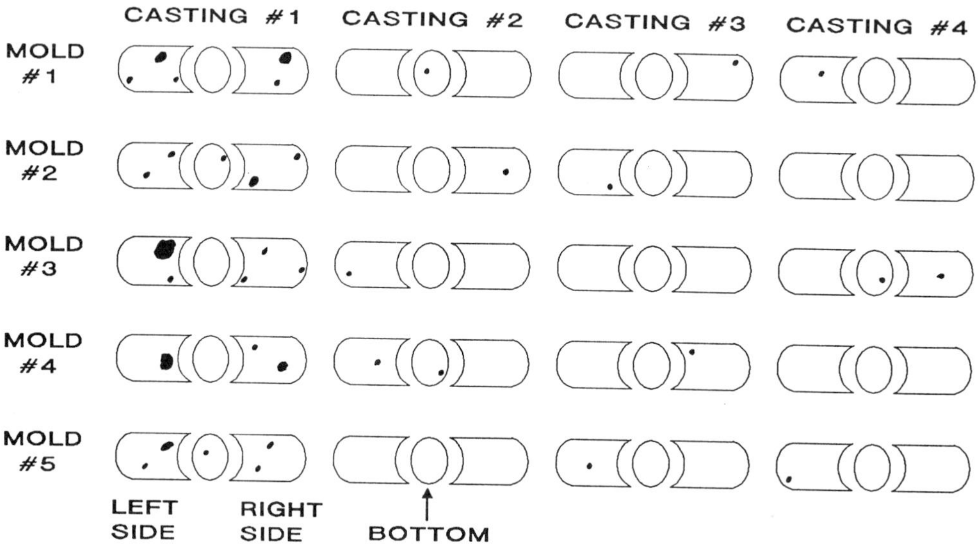

Figure 2 Measles chart of visual defect study results.

transparent copies of each of these twelve measles charts, a quick visual analysis can be made of the location of defects within the cavity, including side-to-bottom, cavity-to-cavity, mold-to-mold, beginning to end of ladle, and so on. The sheets themselves can be compared, but transparencies make it a little easier to see the "measles."

When the data were analyzed in this fashion, the results showed that there was a very large difference in the defect score from the beginning to the end of pouring ladles (Fig. 3a). This pattern was true both days and there was a negligible difference from day-to-day, thereby ruling out variables that vary from day-to-day (Fig. 3b). There was little variation over the three times within the day ruling out those variables that vary during the day. Mold-to-mold variation of the five consecutively poured molds was also negligible.

Before the study, it was suggested that castings from the outside two cavities (1 and 4) were worse than castings from the middle two cavities (2 and 3). The data showed that castings from the left (outside), cavity #1, was by far the worst but not the right (outside) casting, cavity #4 (Fig. 4a). This difference in casting cavity results was most pronounced at the end of the pouring ladle (Fig. 4b).

When the defect score in the side of all castings was compared to the defect score in the bottom of all castings, the side was much worse than the bottom (Fig. 5a). However, by looking at Fig. 5b, it is obvious that this result was caused mostly by the high defect score on the *side* of *casting #1*. There was not a significant increase in defects on casting #1 on the *bottom.* Further analysis showed this higher defect score on the *side* of casting #1 held true for castings poured from both the beginning and end of the ladle (Fig. 6). The real problem was concentrated on the *side* of *casting #1* and was worse at *the end of the ladle.*

Armed with these pictorial representations, the team concentrated its effort to find what caused the molten metal from the end of the pouring ladle to give a

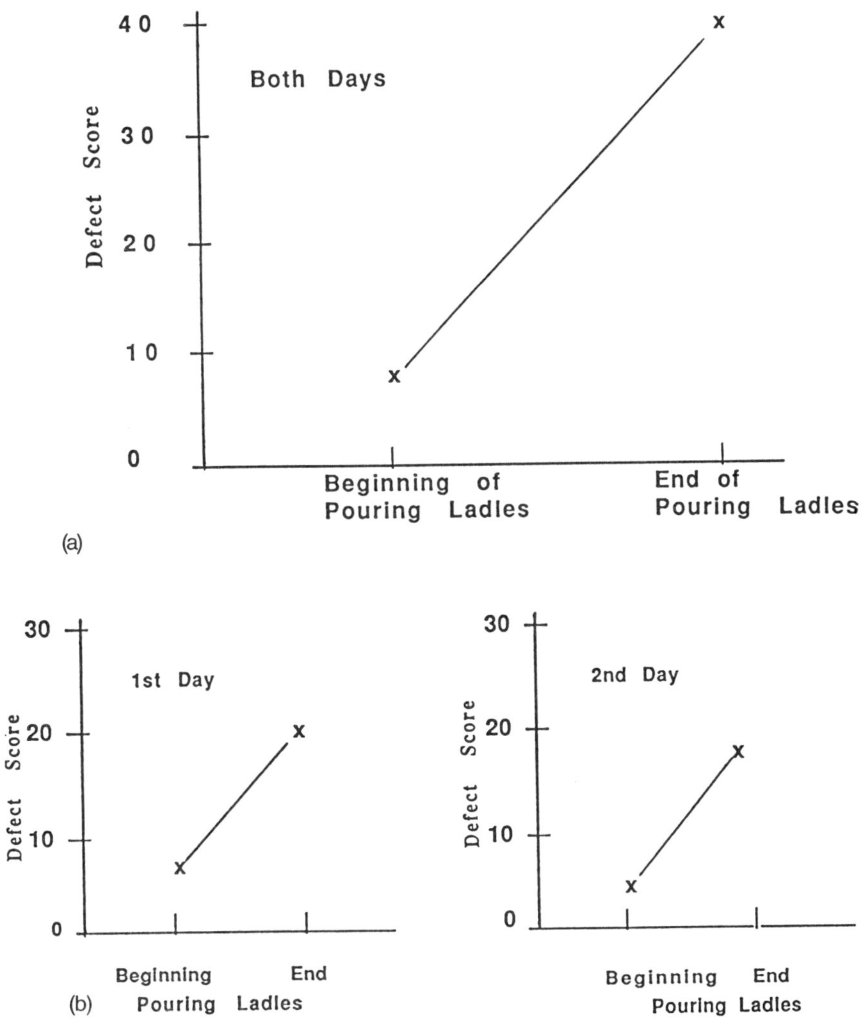

Figure 3 Visual defect study results: within pouring ladle.

higher defect score than the molten metal from the beginning of the pouring ladle. Now that they had the spotlight on the right thing, they were able to pinpoint the cause of the difference. They instituted a different pouring procedure.

When four supposedly identical castings in a mold provide very different results, it cannot be the sand or chemical properties of the metal or other frequently guessed variables. All casting cavities have the same sand around them and the same metal in them. The team looked at what might be different about the side of cavity #1 and the pattern[1] used to make this cavity. Once they concentrated their

---

[1] A pattern is the model of the product that is built, often from wood, to form the cavity in the mold where the molten metal is poured to create the product.

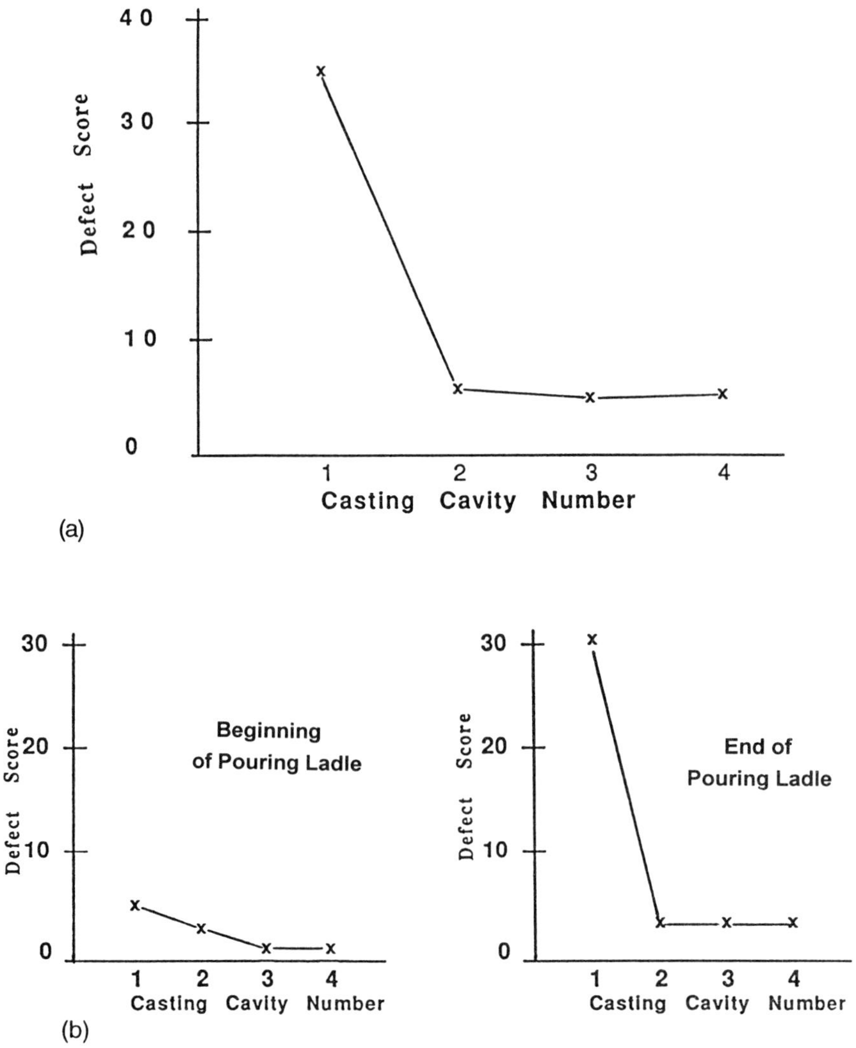

Figure 4 Visual defect study results: casting cavity-to-casting cavity.

attention, they found some subtle differences in the pattern in the gate[2] area that allowed a slightly different flow characteristic for the molten metal for casting #1 that affected the side of the casting but not the bottom. They corrected the pattern.

There was a tremendous reduction in defects as a result of the two corrective actions:

1. Procedural changes in the pouring operation to reduce the problem caused by products poured from the end of the pouring ladle
2. The changes in the gate area of the pattern for casting #1

[2] The gate is the area through which the molten metal flows into the casting cavity.

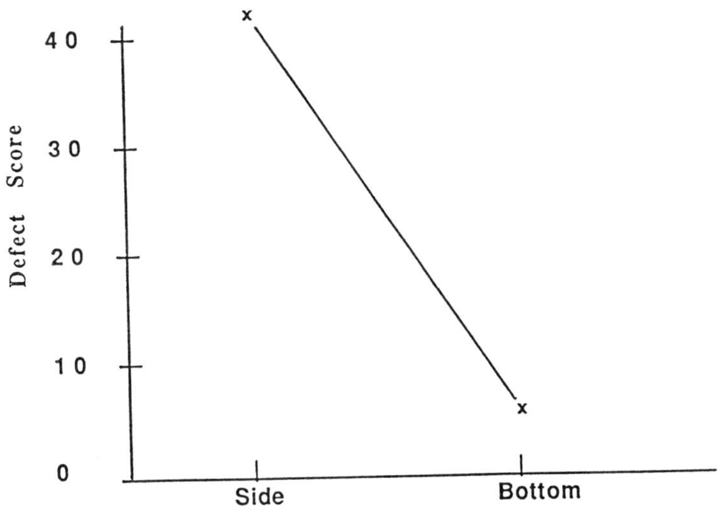

(a)

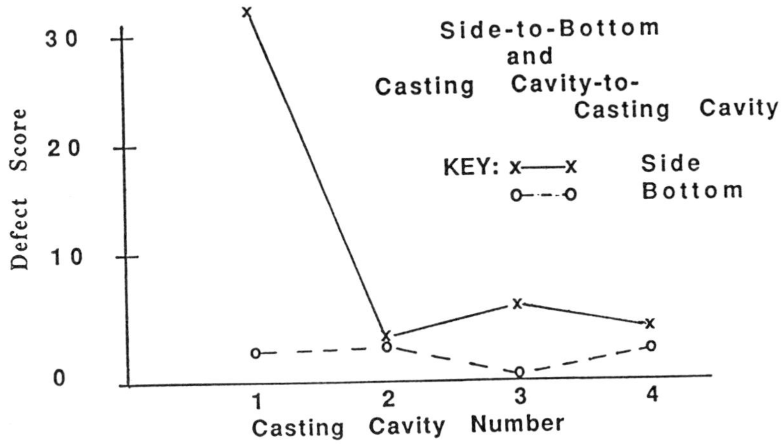

(b)

Figure 5 Visual defect study results: side-to-bottom.

The steps taken to solve this stubborn problem are compared to the nine-step problem-solving process.

1. Provide focus—*determine importance of problem*
   The top item on the Pareto chart of scrap was a casting which had high losses for a visual defect. This key customer threatened to change suppliers.
2. Get close to the problem—*see what is really going on*
   The cross-functional problem-solving team that was organized to solve this problem showed the investigator the operation.
3. Quantify the output—*develop a variable scale (if none exists)*
   The team developed a 0–5 scale for the degree of the defect by selecting a sample of defective castings that covered the range of defectiveness.

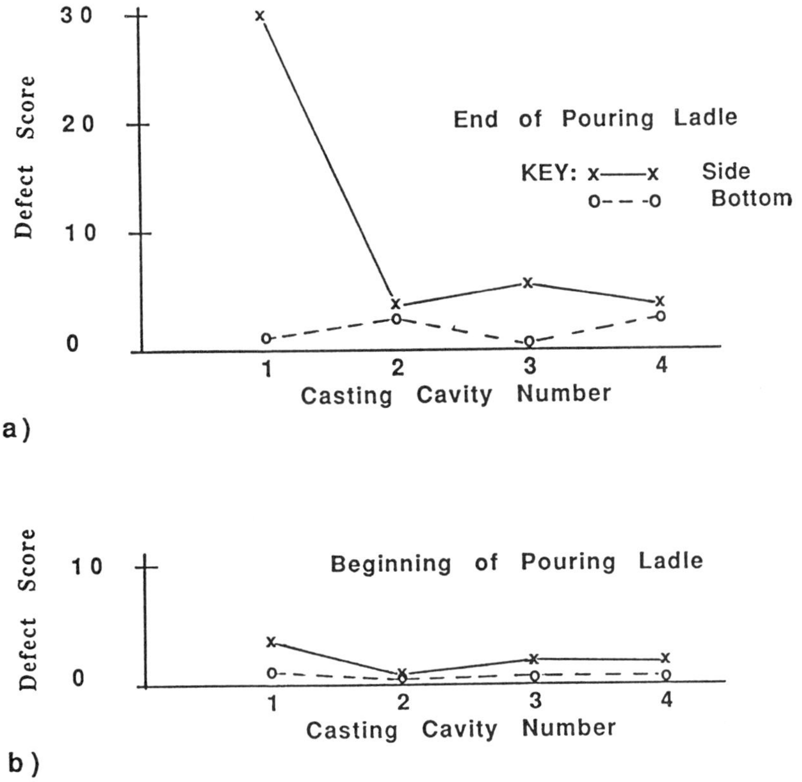

Figure 6 Visual defect study results: casting cavity-to-casting cavity and side-to-bottom, beginning to end of pouring ladle.

4. Run a multi-vari study—*let the process do the talking*
   A multi-vari study ruled out many of the suspected variables and showed which patterns of variation stood out.
5. Design experiment(s)—*test the variables for causes and effects*
   A designed experiment was not needed, because the multi-vari study solved the problem.
6. Turn the problem on and off—*prove you have the answer*
   By changing back to the original pouring procedure and using an uncorrected pattern, the problem could be turned on and off.
7. Optimize—*run tests with proven variables*
   Optimizing was not necessary because the key variables were isolated and eliminated.
8. Install process controls on key variable(s)—*use most cost-effective tool*
   The patterns were corrected so the only process control was to implement the new pouring procedure.
9. Measure before and after results—*quantify continuous improvement*
   Before the study, this casting was the top item on the scrap list. After implementing the corrective actions indicated by the results of the study,

the casting was near the bottom on the scrap list and the customer was quite satisfied. The key customer was saved.

After a heretofore unsolvable problem gets solved, there are often those who feel it was really rather simple. This is usually the way it is with detective work; after the mystery is solved, the solution seems obvious. However, during the investigation, the causes(s) of the problem are elusive mysteries. Although the key variation isolation concept seems easy to apply, most people cannot simply study it and go out and apply it successfully. Instead, it is best learned in an internship mode, in which an investigator works with an experienced practitioner who provides coaching and guidance in the nine-step problem-solving process.

Too often, problems of high losses persist for a long time. The technical people involved are usually quite reluctant to request outside help. They think the request would reflect badly on them. Instead, failing to call for help puts the company in jeopardy. It is up to management to determine when the company's best interests are served by requesting outside help.

Managers need not become proficient in the use of the nine-step problem-solving process, but they certainly should know when their technical people are spinning their wheels and would benefit by using this process. Over the last thirty years, the author has worked with numerous companies to isolate the key variables, reduce variability, and reduce losses in manufactured products. Using this problem-solving process, product lines have been saved, plants have been saved, expensive equipment purchases or overhauls have been avoided, and even companies have been rescued.

It is important to point out that experienced practitioners do not work by themselves. They are most effective when they receive good support from technical people who are familiar with their own products and processes. It is a team effort. The plant personnel get the credit for the results.

FURTHER READING

Traver, R. W., *Manfacturing Solutions for Consistent Quality and Reliability*, Amacom, Averill Park, NY, 1995.

26

Geometric Dimensioning and Tolerancing

LOWELL W. FOSTER
Lowell W. Foster Associates, Inc.
Minneapolis, Minnesota

I. INTRODUCTION

Geometric dimensioning and tolerancing, or geometrics, is a method of establishing controls for the geometry of a piece of hardware. It can be defined as a means of dimensioning and tolerancing a drawing with respect to the actual function or relationship of part features which can be most economically produced. The language of geometrics conveys design requirements more clearly on the drawing. It provides a uniform communication to the manufacturing operations, and this melds all phases of the technical processes required to produce an electromechanical product:

- It saves money by providing maximum producibility of parts through maximum production tolerances. It provides "bonus" or extra tolerances in many cases.
- It ensures that design requirements are clearly stated and thus carried out with less misunderstanding.
- It adapts to, and assists, computerization techniques in design and manufacture.
- It assures interchangeability of mating parts at assembly.
- It provides maximum tolerances to production.
- It provides a uniform and universal drawing language.
- Older methods of defining products no longer suffice on much of today's sophisticated design.
- It is rapidly becoming the "spoken word" in engineering communication around the world.

Geometrics is an effective management tool. It facilitates communication between design, manufacturing, and all technical personnel involved. It further encourages valuable manufacturing and inspection techniques and utilizes good standardization practices. It enhances the disciplined structure of engineering management by maximizing employee output in the technical areas and broadening its capabilities.

Geometrics should be used in the following situations:

- When part features are critical to function or interchangeability
- When datum references are necessary to ensure consistency between design, manufacture, and inspection operations
- When computerization techniques in design and manufacture are desirable
- When functional gaging techniques are desirable
- When standard interpretation is not already implied

Geometrics does not necessarily replace conventional or coordinate dimensioning and tolerancing. One method supplements the other, and should be used in combination for best advantage for the situation.

Geometrics can be successfully introduced in many ways: gradually, to replace confusing or archaic notes or through a more expansive program to introduce the system as a new tool on a new project. Whichever method is chosen, do not expect broad success without a commitment to some form of education on the subject. This education can vary from introduction to shop personnel via shop meetings, pocket references, wall charts, bulletin boards, short classroom sessions, and so on, to in-depth seminars or courses for technical and professional personnel.

Finally, geometrics, or geometric dimensioning and tolerancing, in its most modern form, is based on national and international standards. The American National Standards Institute, the top authority in the United States for voluntary consensus standards, provides ANSI/ASME[1] Y14.5M-1994, Dimensioning and Tolerancing. The International Standards Organization (ISO) provides ISO 1101, Tolerances of Form and Position and numerous other standards on the subject (ISO 5458, 5459, 2692, 3040, 1660, 406, 129, 10578, 10579, 8015, etc.).

II. GEOMETRICS SYMBOLS

Geometrics symbols (Fig. 1) can be used in place of notes on a drawing. Although the symbolic language must be learned, it has numerous advantages:

- Symbols have a uniform meaning.
- Symbols can be placed on the drawing more directly where the relationship applies.
- Symbols can be drawn quickly and neatly with templates or conventional drawing tools.
- Symbols adapt to computer-aided design or drafting.
- Symbols are an international language.

Figure 2 compares a drawing using symbols versus notes. Figure 3 illustrates some additional symbols. The following terms are also used:

---

[1] This standard was developed under the standards writing body, The American Society of Mechanical Engineers (ASME), and thus carries that identity.

⏥	Flatness
—	Straightness
○	Roundness (Circularity)
⌭	Cylindricity
⊥	Perpendicularity (Squareness)
∠	Angularity
//	Parallelism
⌓	Profile Of A Surface
⌒	Profile Of A Line
↗	Circular Runout
⌰	Total Runout
⌖	Position
◎	Concentricity
⌯	Symmetry

Figure 1 Geometric characteristics and symbols.

Resultant condition: collective effect of all tolerance variations on a feature in reverse from the virtual condition.

Actual size: general term for the size of a produced feature; includes the actual mating size and actual local size.

Actual mating envelope (size): the dimensional value of the counterpart of the smallest size which can be circumscribed about an external feature or inscribed within an internal feature.

Actual local size: the dimensional value of any individual distance at any cross-section feature.

Regardless of feature size: the term used to indicate that a geometric tolerance or datum reference applies at any increment of size.

Note: This chapter uses the U.S. customary or inch system as a basis. However, it must be clearly understood that the metric system base (the millimeter) can be used to no prejudice of the system; metric values of proper nomenclature are simply inserted in place of the inch value. With the gradual, yet steady, trend toward greater metric use, this option must be recognized.

III. FUNDAMENTAL PRINCIPLES

A. Maximum Material Condition Principle

Maximum material condition (MMC) provides the designer or drafter with a mechanism to calculate tolerances realistically and also provides advantages to production and inspection. By definition, maximum material condition is that condition of the part feature wherein it contains the maximum amount of material permitted by the stated size tolerance on the drawing (e.g., minimum hole size and maximum shaft size). Figure 4 illustrates the maximum material condition as it applies to representative mating parts.

The MMC principle is applicable only when:

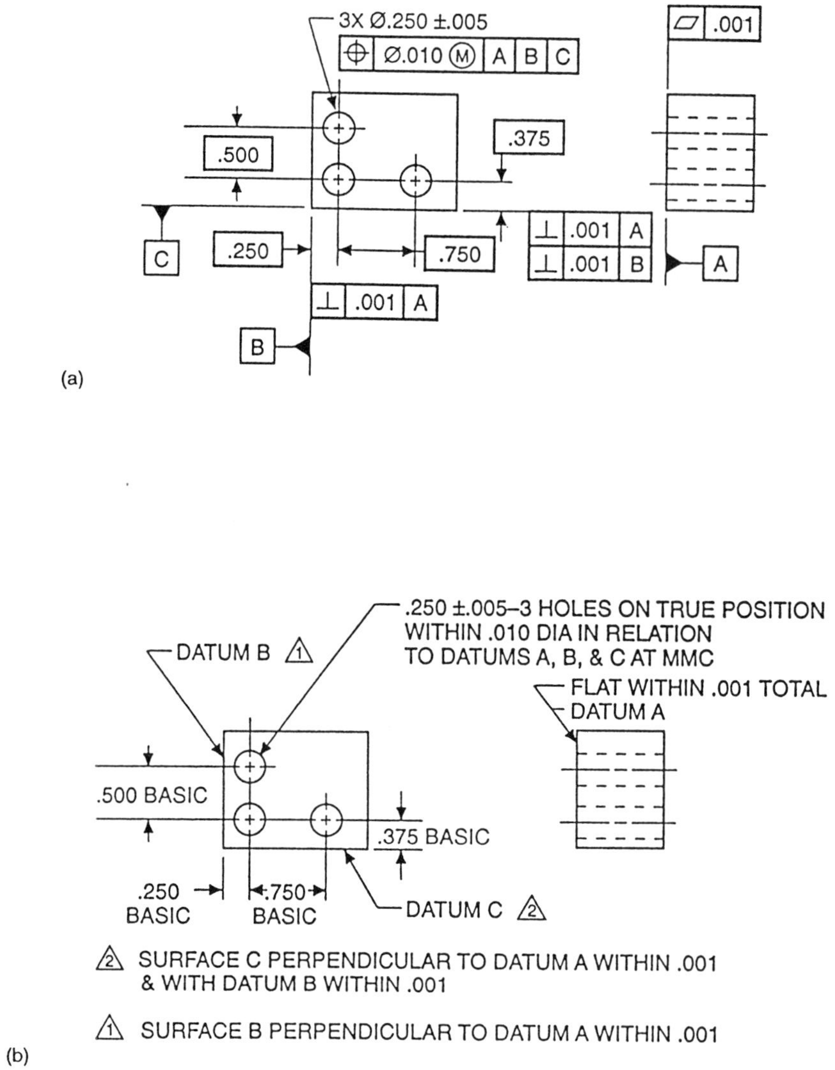

Figure 2 Using symbols and notes.

1. The feature controlled has some geometric relationship with another feature (e.g., a hole and an edge or surface, two or more holes in a pattern, a shaft and its axis.)
2. The feature, or features, is a "size" feature (e.g., a hole, pin, slot) which has a centerline, axis, or center plane.

Note that the MMC principle is applicable in two orders of magnitude; that is, it has two separate and distinct meanings:

- One meaning is to indicate the "worst condition" (or other appropriate descriptors) that the part mating features will present to one another at assembly.

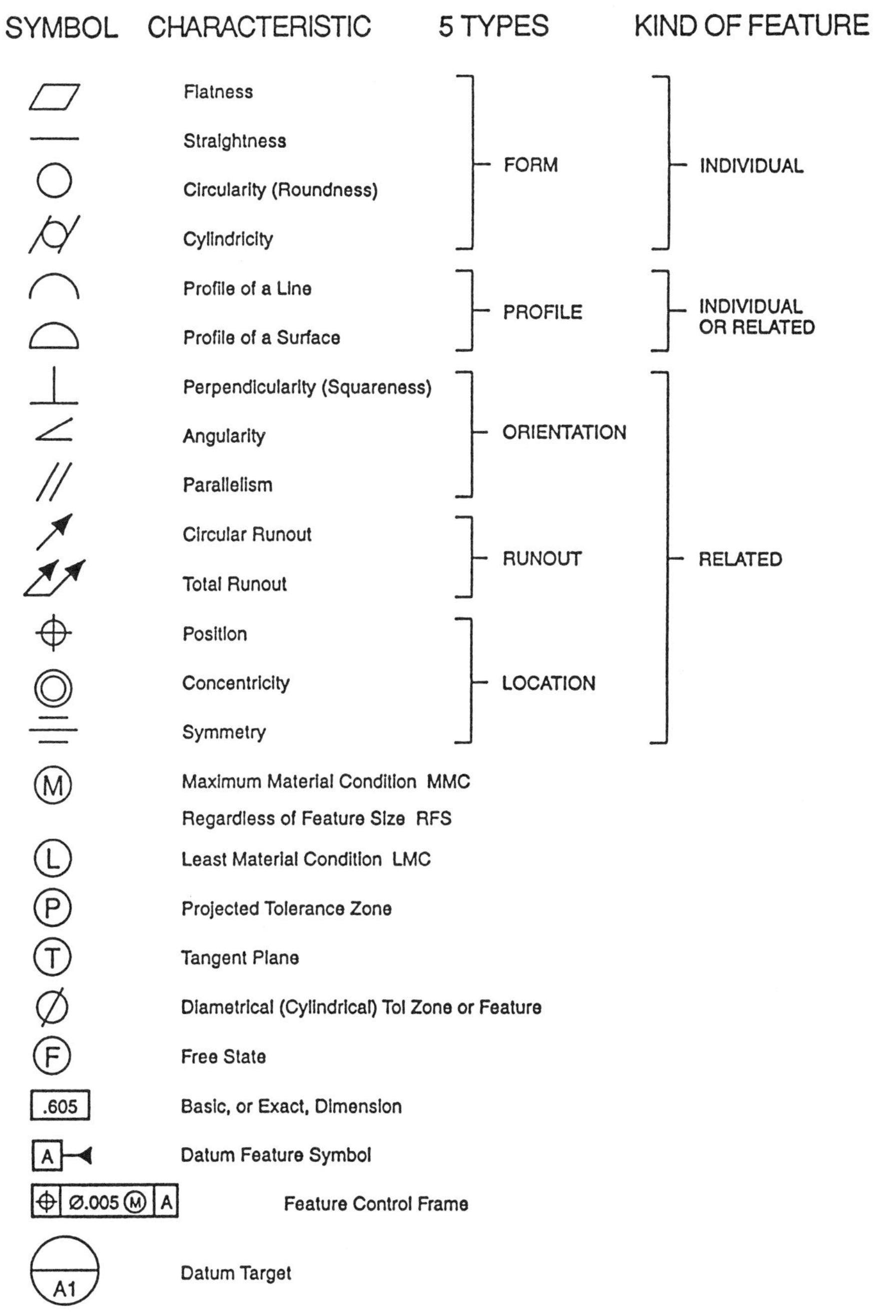

Figure 3 Other symbols.

- The second and most important meaning is the principle or concept invoked by the symbol "circle M" (Ⓜ). This captures the subtlety of the effect of size deviation upon location. That is, as the part pin, for example, is produced at some actual mating size smaller than its MMC, that feature acquires an

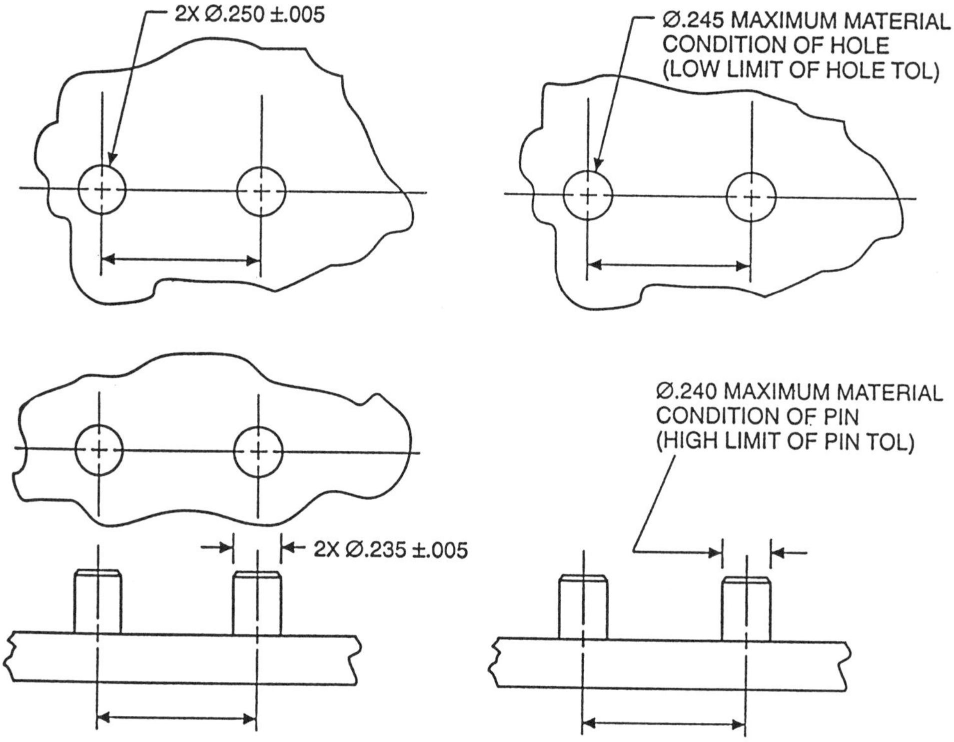

Figure 4 Maximum material conditions principle.

added form or positional tolerance in the amount of that departure from MMC. The part function is assured and more production tolerance is available, and if desired, functional gages can be utilized.

B. Regardless of Feature Size

Another important principle is known as *regardless of feature size* (RFS). This simply means that the effect of the size of the features, such as those shown in Figure 4 (if, of course, RFS is invoked on the drawing and not MMC) will not have an effect on the form or positional tolerance. The stated form or positional tolerance is only to that amount (maximum) regardless of the size to which the features are produced. Regardless of feature size means that the tolerance of form or position must be met regardless of where the feature lies within its size tolerance.

As discussed later (i.e., according to Section IV.B [rule 2]), RFS is automatically invoked on all geometric tolerances (where applied to a size feature relationship) according to rule 2.

As may be deduced, the RFS principle is usually applied to more precision–oriented parts; that is, where the effect of size departure from the MMC of the controlled feature (or features) cannot be permitted to increase the form or location tolerance.

IV. GENERAL RULES OF GEOMETRICS[2]

Also essential to a good understanding of geometrics are the general rules of application. In addition to being very fundamental to the system, the general rules provide "handles" for the user. The general rules are self-explanatory, as shown in Figures 5–9 and described in the text.

A. Rule 1

Rule 1 deals with the relationship of size to form controls (see Fig. 5). Unless otherwise specified, the limits of size of a feature prescribe the extent within which variations of geometric form, as well as size, are allowed. This control applies solely to individual features of size.

Individual Feature of Size

The limits of size of an individual feature prescribe the extent to which variations in its geometric for as well as size are allowed, when only a tolerance of size is specified.

Variations of Size

At any cross section, the actual local size of an individual feature shall be within the specified tolerance of size.

Variations of Form (Envelope Principle)

The form of an individual feature is controlled by its limits of size to the extent prescribed in item 1 in the list below.

1. The surface or surfaces of a feature shall not extend beyond the envelope of perfect form MMC. This boundary is the true geometric form represented in Figure 5. No variation in form is permitted if the feature is produced at its MMC limit of size.
2. Where the actual local size of a feature has departed from the MMC toward the LMC, a variation in form is allowed equal to the amount of such departure.
3. There is no requirement for a boundary of perfect form at LMC. Thus a feature produced at its LMC limit of size is permitted to vary from true form to the maximum variation allowed by the boundary of perfect form at MMC.

Interrelated Features

The form control provision of Rule 1, applies only to individual features and not to the interrelationship of features. Where a boundary of perfect form at MMC

---

[2] The symbols for maximum material condition Ⓜ and least material condition Ⓛ are also known as "modifiers"; that is, they are sometimes used to modify or change the implications of the following rules.

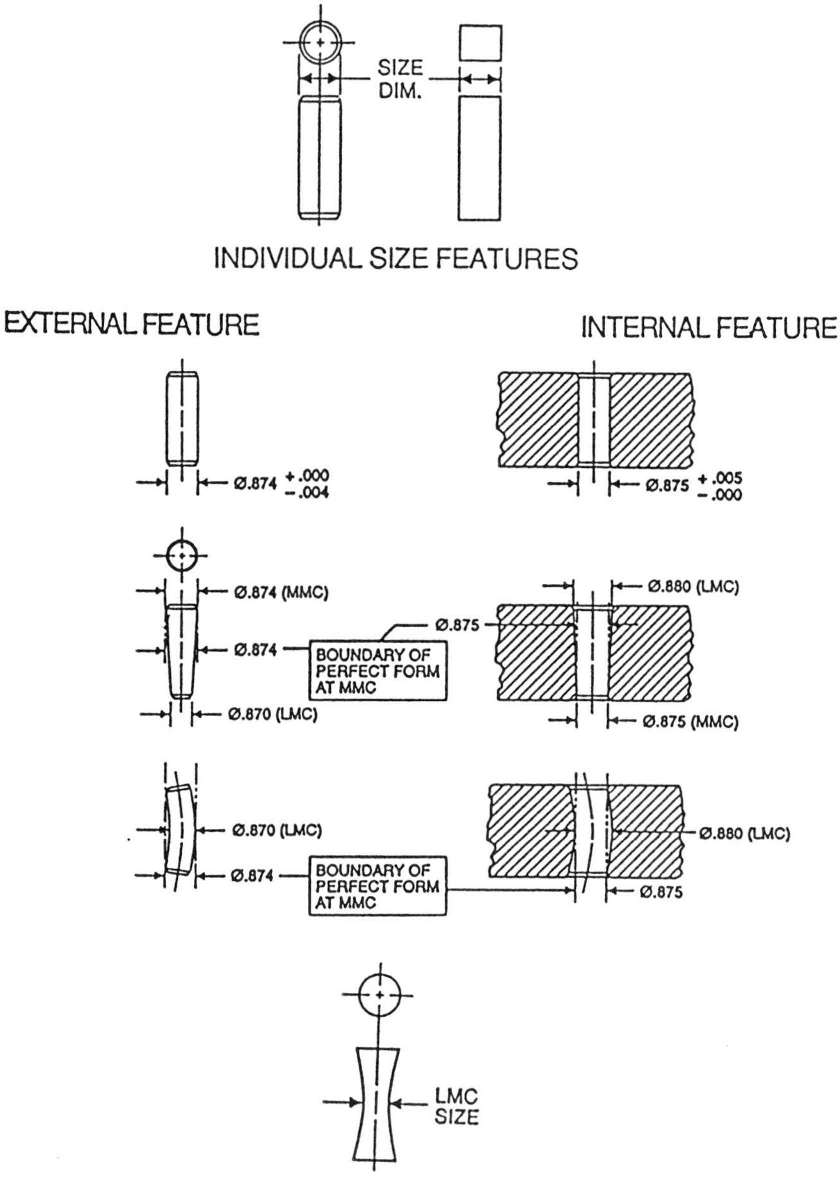

Figure 5 Rule 1—limits of size rule.

control of interrelated features is necessary, one of the following methods should be used to the extent dictated by the design requirements:

1. Specify a zero form orientation or positional tolerance at MMC for the features and datums (if applicable). (See Fig. 6)
2. Indicate this control for the features involved by a note such as "perfect form at MMC required for interrelated features."
3. Relate the dimensions to a datum reference frame or by a local or general note indicating datum precedence.

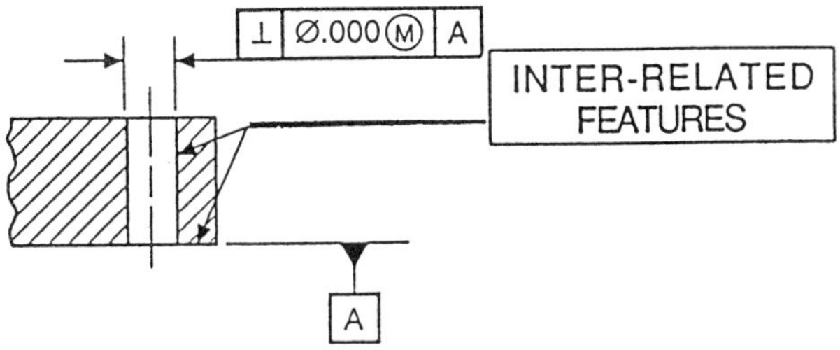

Figure 6 Interrelationship of features.

Perfect Form at MMC Not Required

Where it is desired to permit a specified tolerance of form to exceed the boundary of perfect form at MMC, this may be done by adding to the drawing the suitable form tolerance and a note specifically exempting the pertinent size dimensions from the perfect form rule 1 requirement. A suitable note might be "PERFECT FORM AT MMC NOT REQUIRED."

B. Rule 2

Rule 2 deals with the symbology required to state MMC, RFS, or LMC conditions on applicable geometric tolerancing (Fig. 7).

RULE 2 - **MATERIAL CONDITION RULE**

For *all* applicable geometric tolerances, RFS applies with respect to the individual tolerance, datum reference, or both, where no modifying symbol is specified. MMC, Ⓜ, or LMC, Ⓛ, must be specified on the drawing where it is required. Such as:

(MAY BE SPECIFIED ON SIZE FEATURES AS APPROPRIATE)

NOTE The below characteristics and controls are *always* applicable at RFS and due to the nature of the control cannot be applied at MMC or LMC.

RULE 2A - **ALTERNATE PRACTICE RULE**

For a tolerance of position RFS may be specified on the drawing with respect to the individual tolerance, datum reference, or both, as applicable. (Past Practices) (Not compatible with ISO Practices.)

Figure 7 Rules 2 and 2A.

Each tolerance of orientation or position and datum reference specified for a screw thread applies to the axis of the thread derived from the pitch cylinder. Where an exception to this practice is necessary, the specific feature of the screw thread (such as MAJOR ∅ or MINOR ∅) shall be stated beneath the feature control frame or beneath the datum feature symbol, as applicable.

Each tolerance of orientation or location and datum reference specified for gears, splines, etc., must designate the specific feature of the gear, spline, etc., to which it applies (such as PITCH ∅, PD, MAJOR ∅, or MINOR ∅). This information is stated beneath the feature control frame or beneath the datum feature symbol.

Figure 8 Pitch diameter rule.

C. Rule 2A

Rule 2A deals with alternate use of RFS symbol Ⓢ (Fig. 7).

D. Pitch Diameter Rule

The pitch diameter rule deals with the symbology required to clarify geometric application to screw threads (Fig. 8).

E. Datum Virtual Condition Rule

The datum virtual condition rule deals with the inference of virtual condition as it results from the interrelationship of size, form, orientation, and location tolerances. This rule will not be clearly understood until good knowledge of the fundamentals of geometrics is acquired. Although referenced in a feature control frame at MMC or LMC, a datum feature of size controlled by a separate tolerance of form, orientation, or position applies at its virtual condition. Where it is not intended for the virtual condition to apply, a zero tolerance at MMC should be specified for the appropriate data features.

- Size + form, orientation, or position tolerance = shaft MMC virtual condition

A virtual condition exists for a datum feature of size where its axis or center plane is controlled by a geometric tolerance. In such cases, the datum feature applies as its virtual condition even though it is referenced in a feature control frame at MMC or LMC.

Figure 9 Datum/virtual control rule.

- Size − form, orientation, or position tolerance = hole MMC virtual condition
- Size − form, orientation, or position tolerance = shaft LMC virtual condition
- Size + form, orientation, or position tolerance = hole LMC virtual condition

(See Figures 9 and 10).

Datums

The use of datums plays an important role in the application of geometrics. Datums provide a mechanism to capture design requirements (i.e., how the part mounts or the relationship of features) and ensure uniformity in fixturing, manufacturing, and inspecting the part.

Datum points, lines, and planes are assumed to be exact for purposes of computation or reference and from which the location of related features of a part may be established. Datums are established by, or are relative to, actual part features or surfaces. Datum surfaces and datum features are actual part surfaces or features used to establish datum and which include all the surface or feature irregularities and inaccuracies. The datum feature symbol is shown in Fig. 11.

Establishing Datums from Datum Surfaces and Datum Features

As seen in the foregoing definitions of datums, datum surfaces, and datum features, the datum as an exact entity is established by contact of the actual part feature with a simulated datum feature as represented by a surface plate, gage pin, precision

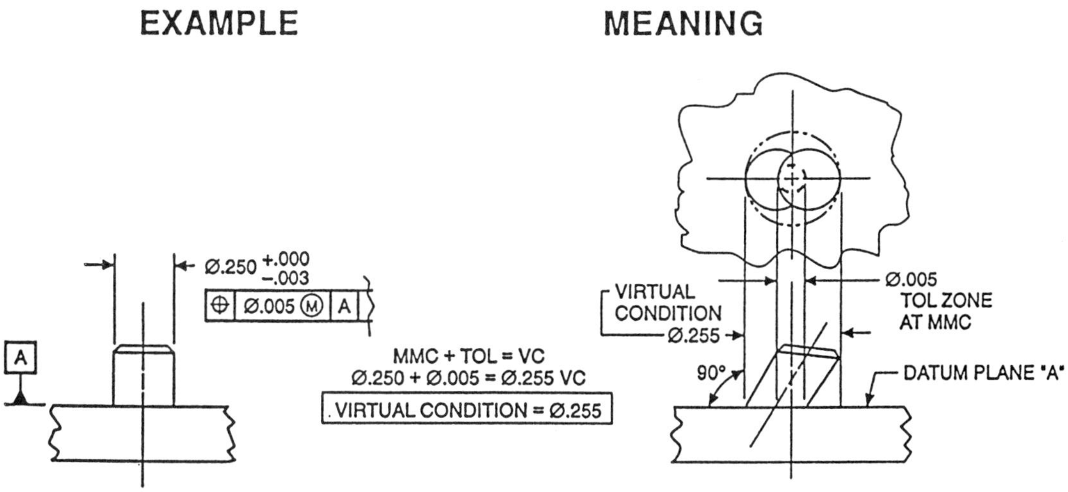

Figure 10 MMC Virtual Condition. In the case of an internal feature such as a hole, the virtual condition is determined by: MMC − TOL + VC; e.g., hole MMC minus the stated orientation or position tolerance equals the hole virtual condition. The virtual condition of a hole is always a "constant value" and can be referred to as the "inner boundary (locus)" in the worst-case analysis calculations.

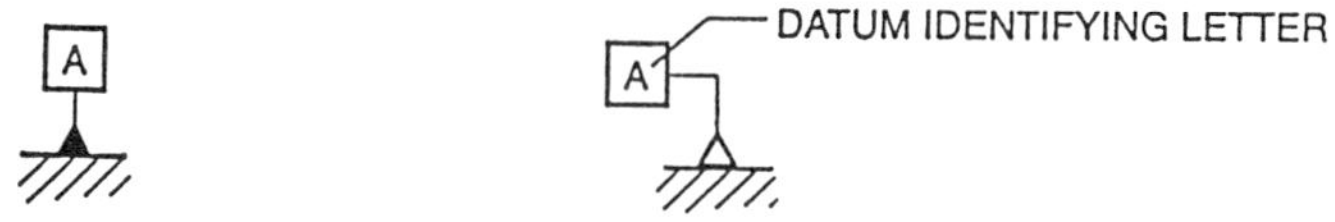

The datum feature symbol consists of a capitol letter enclosed in a square frame, a leader line extending from the frame to the concerned feature and terminating with a triangle.

Figure 11 Datum feature symbol. To identify a feature as a datum, this symbol is used. (The datum feature triangle may be filled or open. The leader may be appropriately directed to a feature.) Each datum requiring identification is assigned a different reference letter. Do not use letters I, O, or Q. If the single letter alphabet is exhausted, double letters may be used, i.e., AA, AB, etc. Where datum feature symbol is repeated to identify the same feature in other locations of a drawing, it need not be identified as a reference.

collet, and so on. The datums are assumed to exist in the datum simulators; that is, the associated processing equipment which is of high quality and thus simulates datum planes, axes, and so on, for purposes of verification. See Figures 12 and 13 for examples of establishing datums from datum features and datum surfaces.

Datum Reference Plane (The Three-Plane Concept)

The three mutually perpendicular planes are fundamental to everyday life: to mathematics, geometry, machine movement, and engineering design. Without the three planes, engineering would be very hard pressed to define its product and inspection to measure it. The three planes have always existed; we have simply used them by intent in geometrics to identify and establish feature relationships. The three planes are, by geometric definition, mutually perpendicular. Figure 14 illustrates the establishment of the three datum planes and how the system invokes datum priority while reading the feature control symbol letter symbols left to right: that is, primary, secondary, and tertiary datums.

Feature Control Frame

Figure 15 illustrates the make up of the feature control frame and the sequence in which the elements of the system are stated.

V. APPLICATION OF TOLERANCES OF FORM, ORIENTATION, PROFILE, RUNOUT, AND LOCATION

The following series of illustrations and related text presents typical (and hypothetical) examples of the application of each control. The text explanation is minimized in favor of the pictorial illustrations. As described in the introductory paragraphs, a brief yet reasonably complete study of the major aspects of the subject is presented here. Should questions arise, or there is need for in-depth information, refer to one of the text references listed at the end of the section or the standard authority, ANSI/ASME Y14.5 (latest issue).

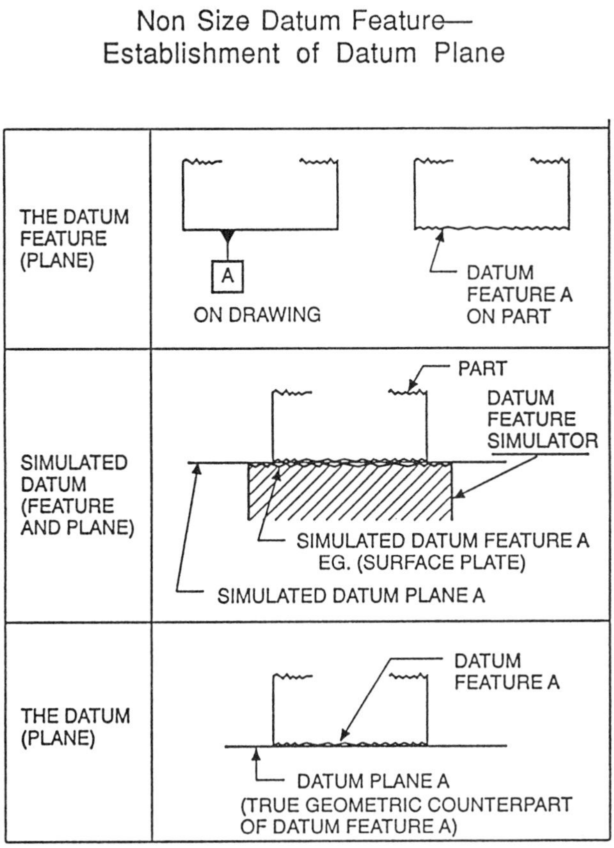

Figure 12 Establishing datum planes from datum surfaces. The simulated datum feature is the surface of manufacturing or verification equipment. The simulated datum plane is derived from the datum feature simulator. Note: Simulated datum features are used as the practical embodiment of the datums during manufacture and inspection.

A. Application to Individual or Related Features

It is helpful in learning, or applying, geometrics to identify more clearly certain geometric characteristics for certain uses. That is, the more elementary controls (i.e., flatness, straightness, circularity, and cylindricity) are applied only to "individual" features, whereas perpendicularity, for example, is applied in a "related" feature requirement. Figures following will explain the reasoning of each. An individual feature can be clearly identified as one that does not relate to a datum reference (i.e., there is no relationship other than to itself), although, of course, an individual feature can be a datum if desired. Related features always have datum relationships; it is their nature. Figure 16 will assist in identifying the characteristic types or categories as they are applicable to individual or related feature requirements.

B. Tolerances of Form, Orientation, Profile, and Runout

Tolerances of form, orientation, profile, and runout state the permissible variation of actual surfaces, axes, or centerplanes of features from the ideal as implied

Size Datum Feature

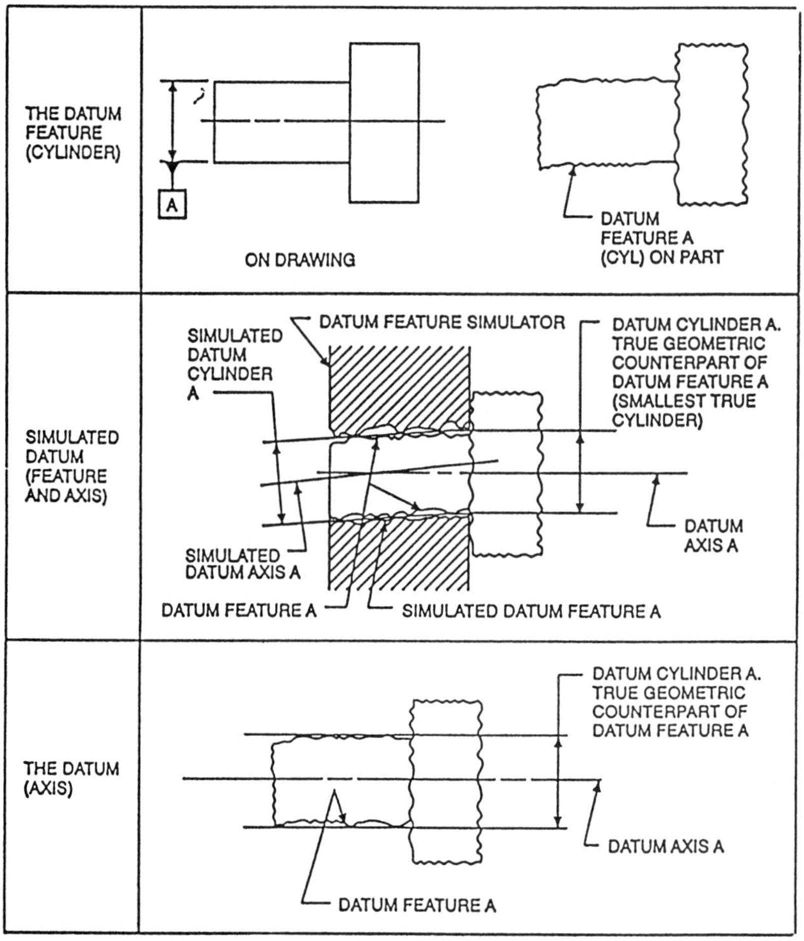

Figure 13 Establishing datum cylinders from datum features. The simulated datum feature is the surface of manufacturing or verification equipment. The simulated datum axis is derived from the datum feature simulator. Note: Simulated datum features are used as the practical embodiment of the datums during manufacturing and inspection.

by the drawing. Form tolerances, refer to flatness, straightness, circularity, and cylindricity; orientation tolerances, refer to parallelism, perpendicularity, and angularity; profile of a surface, profile of a line, and circular and total runout tolerance are unique variations of form and orientation tolerance and are considered as separate types of characteristics.

These tolerances should be specified for all features critical to the design requirements:

1. Where workshop practices cannot be relied on to provide the required accuracy

ESTABLISHING DATUM PLANES FROM DATUM SURFACES/FEATURES— THREE PLANE CONCEPT— DATUM REFERENCE FRAME

EXAMPLE

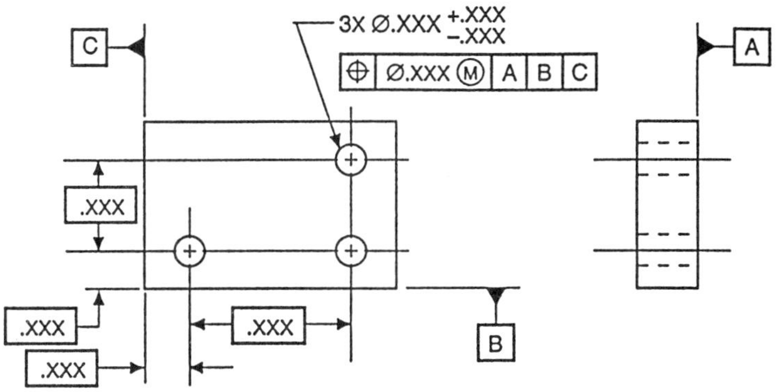

ESTABLISHING THE DATUM PLANES

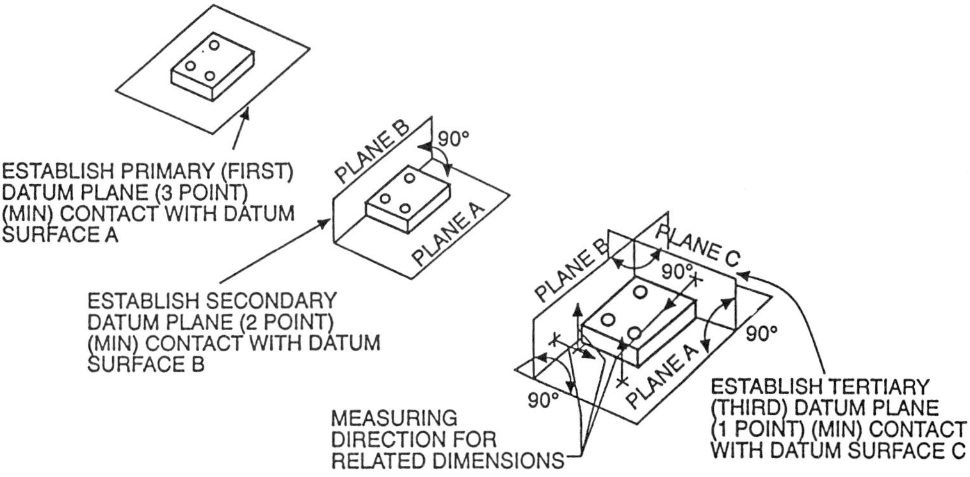

Figure 14 Three-plane concept.

2. Where documents establishing suitable standards of workmanship cannot be prescribed
3. Where tolerances of size and location do not provide the necessary control

The following series of form, orientation, profile, and runout tolerance examples are presented for purposes of explanation of the basic principles. Symbolic notation of these tolerances is, of course, recommended and emphasized throughout this section.

EXAMPLE

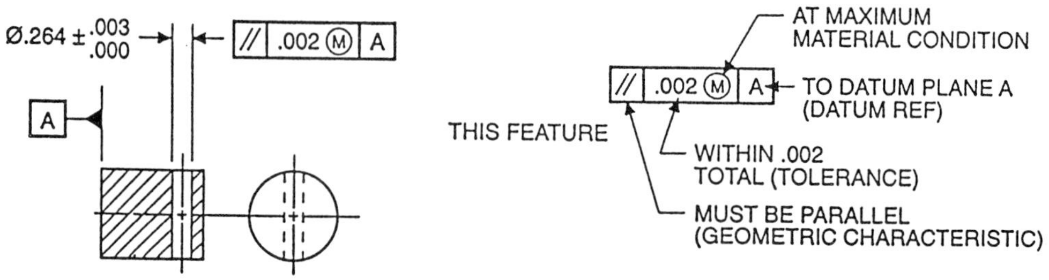

Figure 15 The feature control frame consists of a box containing the geometric characteristic symbol, datum references, tolerance, and the material condition symbol (e.g., for MMC) if applicable. This example shows this feature control frame as used on a part drawing.

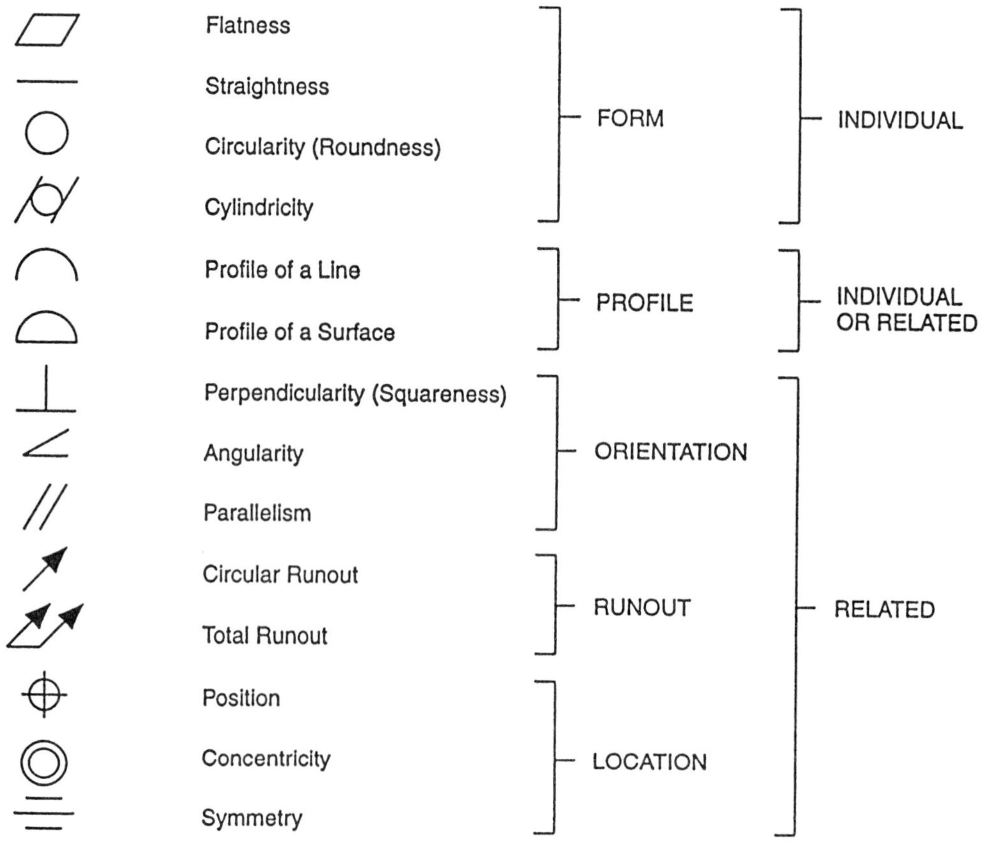

Figure 16 Characteristics related to individual or related feature requirements.

Tolerances of Form, Individual Features

Tolerances of form applicable to individual features and where no datum is used, involve the characteristics of flatness, straightness, circularity, and cylindricity (see Fig. 1).

These characteristics are used to specify form tolerances for single surface, element, or size features. No datum is involved, because the "relationship" of the feature is to a perfect counterpart of itself (i.e., flatness relates to a plane, circularity relates to a circle, etc.). If a datum were to be used in such applications, the controlled feature would also be its own datum. This, of course, would be redundant, confusing, and incorrect and, therefore, is not done.

Flatness

Flatness is the condition of a surface having all elements in one plane (see Fig. 1).

Flatness Tolerance. A flatness tolerance specifies a tolerance zone confined by two parallel planes within which the entire surface must lie.

Flatness Tolerance Application. Figure 17 shows how a flatness symbol is applied. The symbol is interpreted to read: "This surface shall be flat within .002 total tolerance zone over the entire surface." Note that the .002 tolerance zone is total variation. To be acceptable, the entire actual surface must fall within the parallel plane extremities of the .002 tolerance zone.

A flatness tolerance is a form control of all elements of a surface as it compares to a simulated perfect geometric counterpart of itself. The perfect geometric counterpart of a flat surface is a plane. The tolerance zone is established as a width or thickness zone relative to this plane as established from the actual part surface.

Note that Figure 17 shows that the extremities or high points of the surface determine one limit or plane of the tolerance zone, with the other limit or plane being established as .002 (the specified tolerance) parallel to it.

Since flatness tolerancing control is essentially a relationship of a feature to itself, no data references are required or proper. Also, note that since flatness is a form tolerance controlling surface elements only, it is not applicable to RFS, MMC, or LMC considerations.[3]

In the absence of a flatness tolerance specification, the size tolerance and method of manufacture of a part will exercise some control over its flatness. However, when a flatness tolerance is specified, as applicable to a single surface, the flatness tolerance zone must be contained within the size tolerance limits.[4] It cannot be additive to the size tolerance. Where necessary, the terms "must not be concave" or "must not be convex" may be added beneath the feature control frame.

Straightness

Straightness is a condition where an element of a surface or an axis is a straight line (see Fig. 1).

---

[3] Under special circumstances, where the desired control is applicable to the total thickness of sheet metal, square rods, and so on, the methods described for straightness on an RFS, MMC, or LMC basis may be used (tolerance applicable to size feature and dimension).

[4] Under special circumstances, a note specifically exempting the pertinent size dimension, such as "perfect form at MMC not required," may be specified.

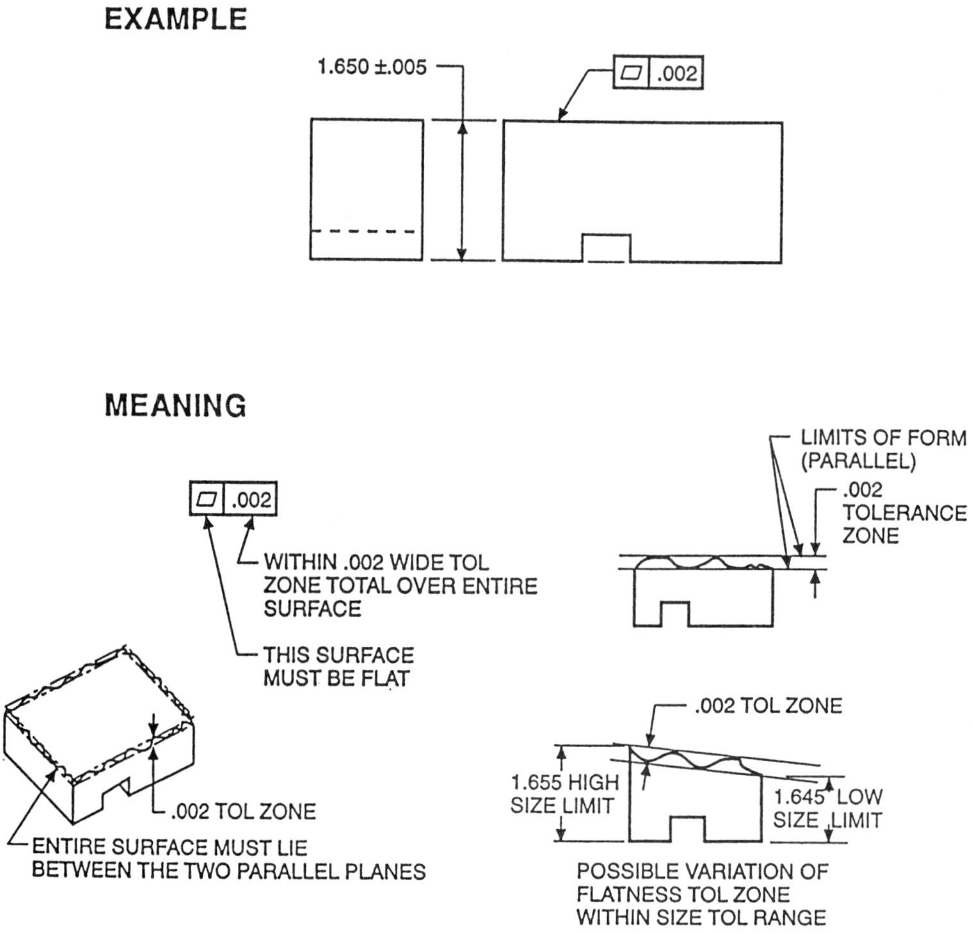

Figure 17 Examples of flatness specifications.

Straightness Tolerance. A straightness tolerance specifies a tolerance zone within which the considered element or axis must lie.

Straightness Tolerance Application. A straightness tolerance is applied in the view where the elements to be controlled are represented by a straight line.

Straightness Tolerance—Surface Element Control: Straightness tolerance is typically used as a form control of individual surface elements such as those on cylindrical or conical surfaces. Since surfaces of this kind are made up of an infinite number of longitudinal elements, a straightness requirement applies to the entire surface as controlled in single line elements in the direction specified.

Figure 18 illustrates straightness control of individual longitudinal surface elements on a cylindrical part. Note that the symbol is directed to the feature surface, not to the dimension lines. The straightness tolerance must be less than the size tolerance.

All circular elements of the surface must be within the specified size tolerance and the boundary of perfect form at MMC. Also, each longitudinal element of the

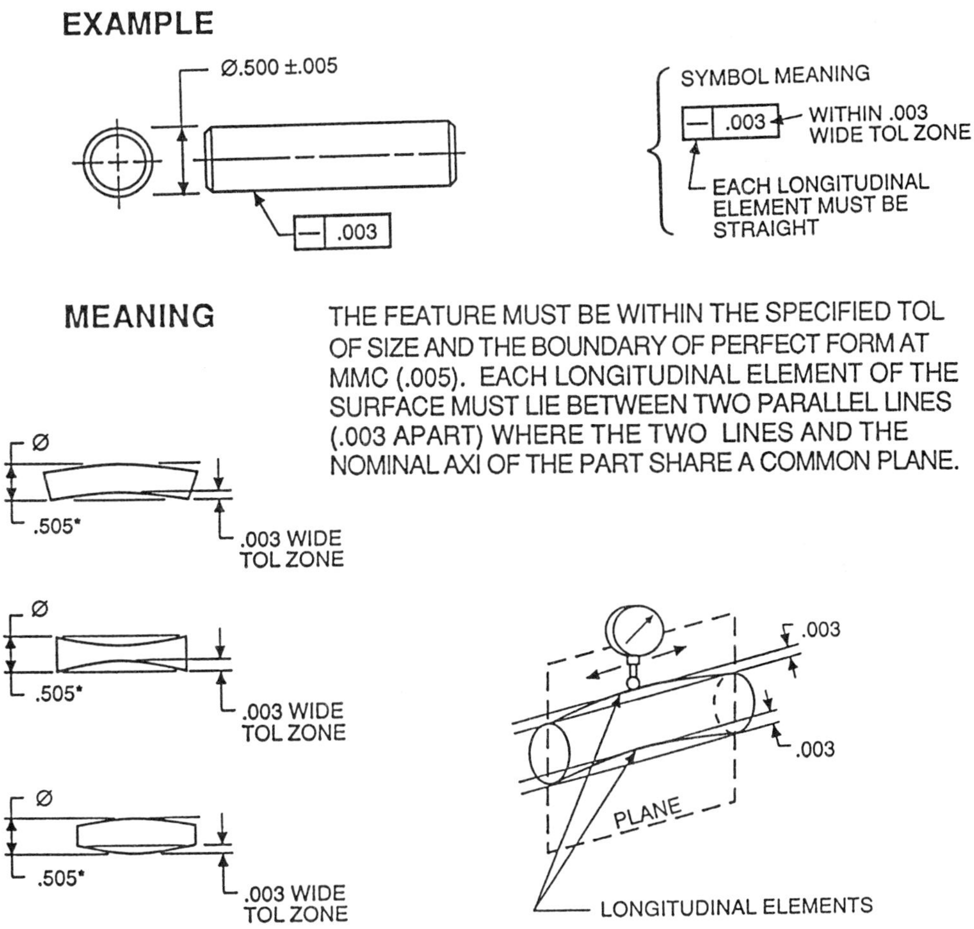

Figure 18 Straightness control of individual longitudinal surface elements on a cylindrical part.

surface must lie in a tolerance zone defined by two parallel lines spaced apart by the amount of the prescribed tolerance where the two lines and the nominal axis of the part share a common plane.

Note: Since surface element control is specified, the tolerance zone applies uniformly whether the part is of bowed, waisted, or barreled shape (see Fig. 18).

Straightness Tolerance RFS and MMC: Where function of a size feature permits a collective result of size and form variation, the RFS or MMC principles may be used. In this instance, where the appropriate symbology and specifications are used, the part is not confined to the perfect form at MMC boundary. All sectional elements of the surface are to be within the specified size tolerance, but the total part surface may exceed the perfect form at MMC boundary to the extent of the straightness tolerance. This principle may be applied to individual size features such as pins, shafts, bars, and so on, where the longitudinal elements are to be specified with a straightness tolerance independent of, or in addition to, the size tolerance (see Figs. 19 and 20).

Straightness—RFS: Where a cylindrical feature is to be controlled on an RFS basis as in the following figure, the feature control symbol must be located with the size dimension or attached to the dimension line, and the diameter symbol must precede the straightness tolerance (see Fig. 19).

Straightness—MMC: Where a cylindrical feature has a functional relationship with another feature, such as a pin or shaft and a hole, the control of straightness on an MMC basis may be desirable. If the pin or shaft, for example, is to fit into a hole of a given diameter, the collective effect of the pin size and its straightness error must be considered in relationship to the hole size minimum (i.e., their virtual conditions must be considered relative to one another).

EXAMPLE

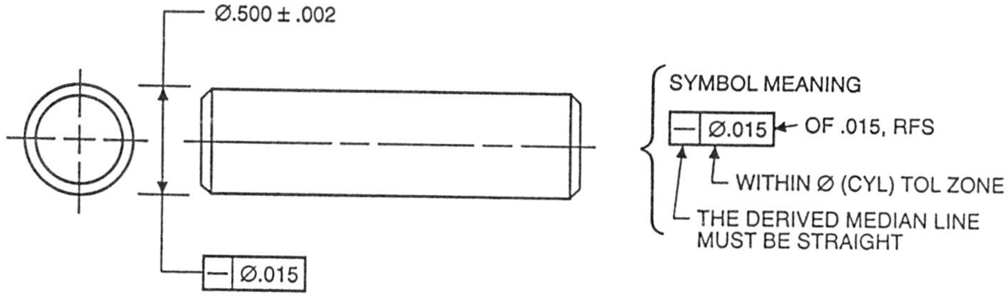

MEANING

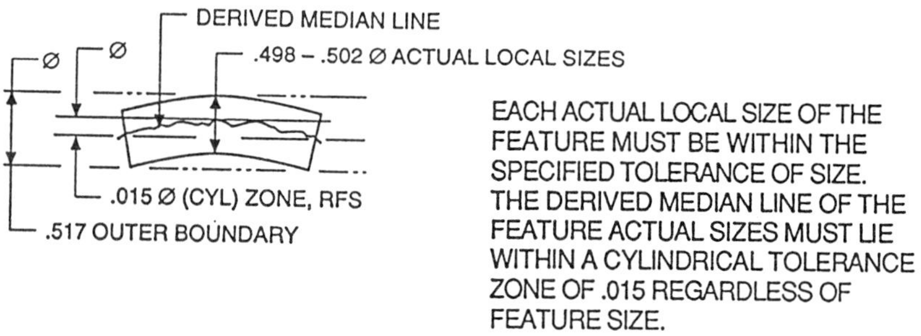

Figure 19 Straightness control of cylindrical features on an RFS basis. Author Advisory: In these cases, the straightness tolerance need not be greater than the size tolerance; however, it is a common situation. Where the part length may approach ten times the part diameter or more, maintaining the perfect form MMC boundary may be impractical. Therefore, this method could be a necessary alternative in the design considerations.

EXAMPLE

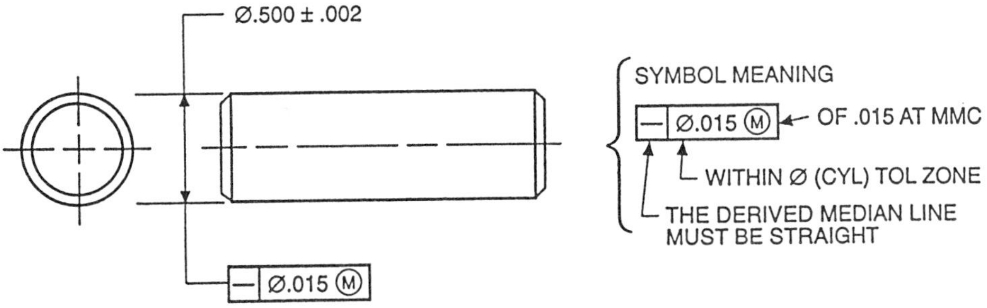

MEANING

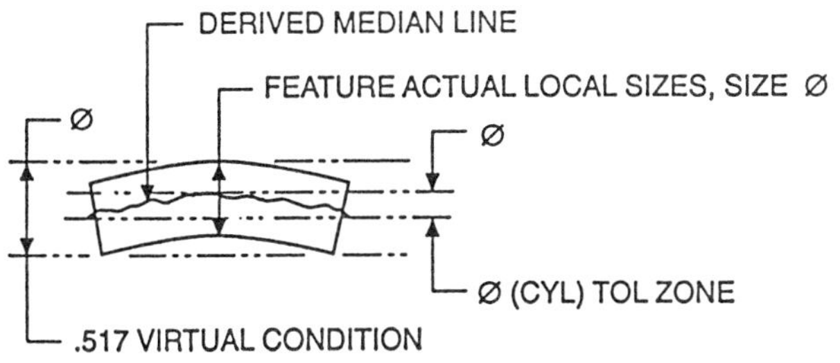

ACTUAL LOCAL SIZE Ø	Ø TOL ZONE IS
.502 MMC	.015
.501	.016
.500	.017
.499	.018
.498 LMC	.019

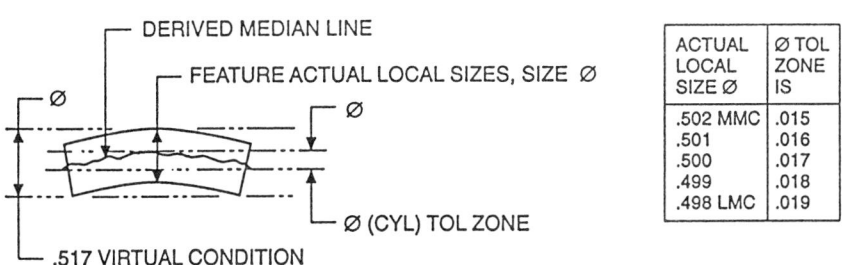

EACH ACTUAL LOCAL SIZE OF THE FEATURE MUST BE WITHIN THE SPECIFIED TOLERANCE OF SIZE. THE DERIVED MEDIAN LINE OF THE FEATURE MUST LIE WITHIN A CYLINDRICAL TOLERANCE ZONE OF .015 AT MMC. AS THE ACTUAL LOCAL SIZES OF THE FEATURE DEPART FROM MMC, AN INCREASE IN THE STRAIGHTNESS TOLERANCE IS ALLOWED WHICH IS EQUAL TO THE AMOUNT OF SUCH DEPARTURE.

Figure 20 Straightness control on an MMC basis.

By stating the requirements on an MMC basis, the allowable straightness tolerance may increase an amount equal to the actual local size departure from MMC. The feature control symbol must be located with the size dimension or be attached to the dimension line; the diameter symbol must precede the straightness tolerance; and the MMC symbol must be inserted following the tolerance. In this manner, maximum tolerance is achieved, part fit is guaranteed, and functional gaging techniques may be used (see Fig. 20).

Circularity

Circularity (see Fig. 1) is the condition on a surface of revolution where:

1. For a feature other than a sphere, all points of the surface intersected by any plane perpendicular to a common axis are equidistant from their axis.
2. For a sphere, all points of the surface intersected by any plane passing through a common center are equidistant from that center.

Circularity Tolerance. A circularity tolerance specifies a tolerance zone bounded by two concentric circles within which each circular element of the surface must lie and applies independently at any plane as described above.

Circularity Tolerance Application. Limits of size exercise control of circularity within the size tolerance. Often this provides adequate control. However, where necessary to further refine form control, circularity tolerancing can be used on any figure of revolution or circular cross section (Fig. 21). The circularity tolerance must be less than the size tolerance except for those parts subject to free state variation.

Cylindricity

Cylindricity is a condition of a surface of revolution in which all points of the surface are equidistant from a common axis (see Fig. 1).

Cylindricity Tolerance. A cylindricity tolerance specifies a tolerance zone bounded by two concentric cylinders within which the surface must lie.

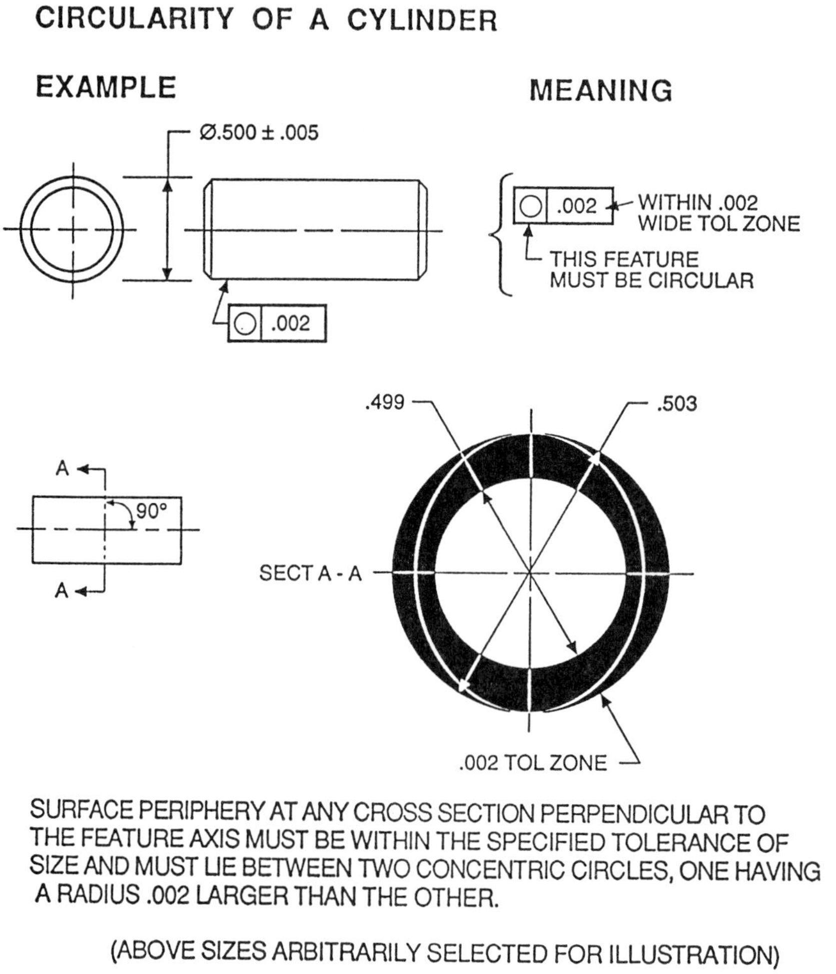

Figure 21 Roundness tolerancing can be used on any figure of evolution or circular cross section.

Cylindricity Tolerance Application. Limits of size exercise control of cylindricity within the size tolerance. This control is often adequate. However, where more refined form control is required, cylindricity tolerancing can be used. Note that in cylindricity, unlike circularity, the tolerance applies simultaneously to both circular and longitudinal elements of the entire surface (see Fig. 22).

Tolerances of Orientation, Related Features

Tolerances of form used on related features which require a datum involve the characteristics of perpendicularity, angularity, and parallelism (see Fig. 1).

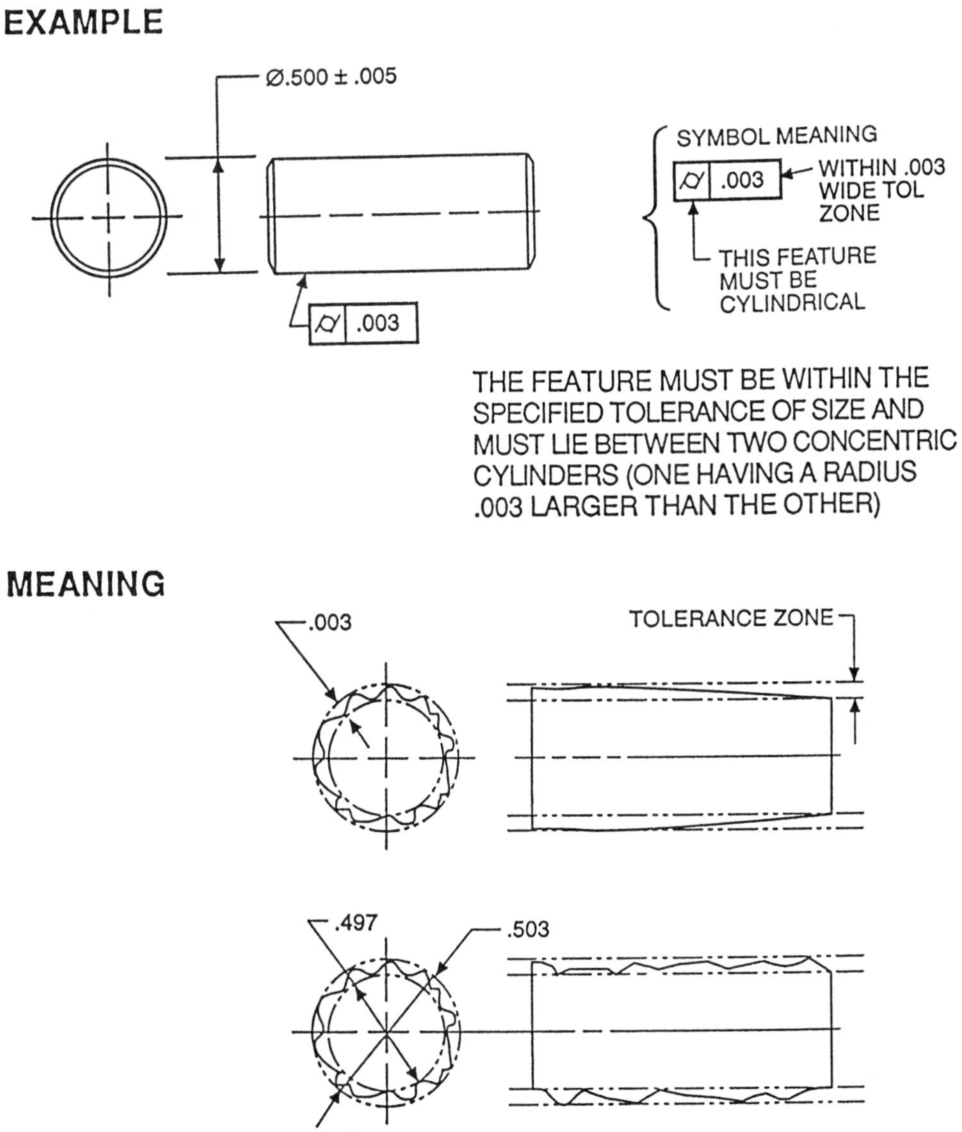
EXAMPLE
Ø.500 ± .005
.003
SYMBOL MEANING
.003
WITHIN .003 WIDE TOL ZONE
THIS FEATURE MUST BE CYLINDRICAL
THE FEATURE MUST BE WITHIN THE SPECIFIED TOLERANCE OF SIZE AND MUST LIE BETWEEN TWO CONCENTRIC CYLINDERS (ONE HAVING A RADIUS .003 LARGER THAN THE OTHER)
MEANING
.003
TOLERANCE ZONE
.497
.503

Figure 22 Application of cylindricity tolerancing.

These characteristics are used to describe orientation tolerances of single surface, element, or size features and are always related to a datum.

Tolerances of perpendicularity, angularity, and parallelism are also sometimes referred to as *attitude* tolerances. When applied to flat surfaces, they invoke a control of "flatness" to that surface as well to the extent of the form tolerance specified. When perpendicularity is applied to size features, rule 2 and RFS versus MMC or LMC considerations must be taken into account.

Perpendicularity

Perpendicularity is the condition of a surface, median plane, or axis at a right angle (90°) to a datum plane or axis.

Perpendicularity Tolerance. A perpendicularity tolerance specifies one of the following (as shown in Fig. 23):

1. A tolerance zone defined by two parallel planes perpendicular to a datum plane or axis within which:
   a. The surface of a feature must lie (see Fig. 23A)
   b. The median plane of a feature must lie (see Fig. 23B)
2. A tolerance zone defined by two parallel planes perpendicular to a datum axis within which the axis of the considered feature must lie (see Fig. 23C)
3. A cylindrical tolerance zone perpendicular to a datum plane within which the axis of the considered feature must lie (see Fig. 24D)
4. A tolerance zone defined by two parallel lines perpendicular to a datum plane or datum axis within which a line element of the surface must lie (see Fig. 25E).

Perpendicularity Tolerance Application. Figure 24 illustrates perpendicularity tolerance as applied to a surface. Figure 25 shows perpendicularity tolerance applied to a size feature. Note that the diameter symbol is used to indicate that a cylindrical tolerance zone is desired and that MMC has been specified.

Under "meaning," it can be noted that the tolerance of perpendicularity increases if the actual mating size of the feature departs from MMC. Functional

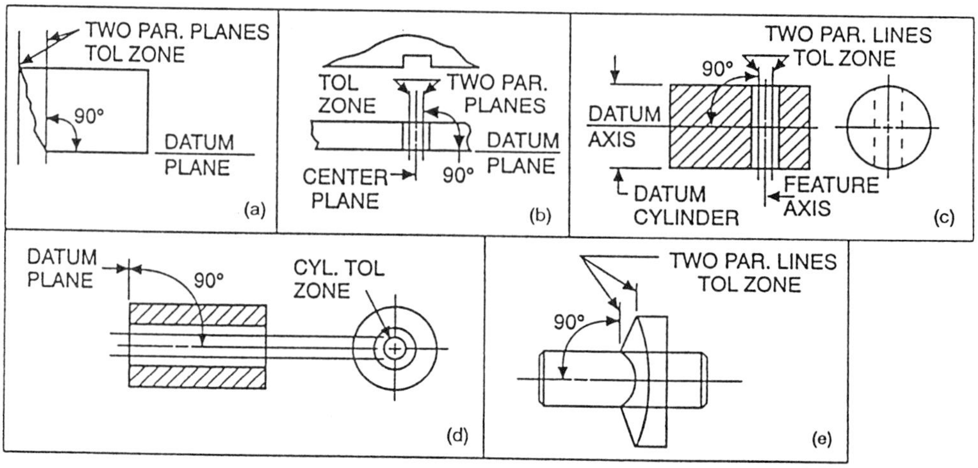

Figure 23 Perpendicularity tolerance.

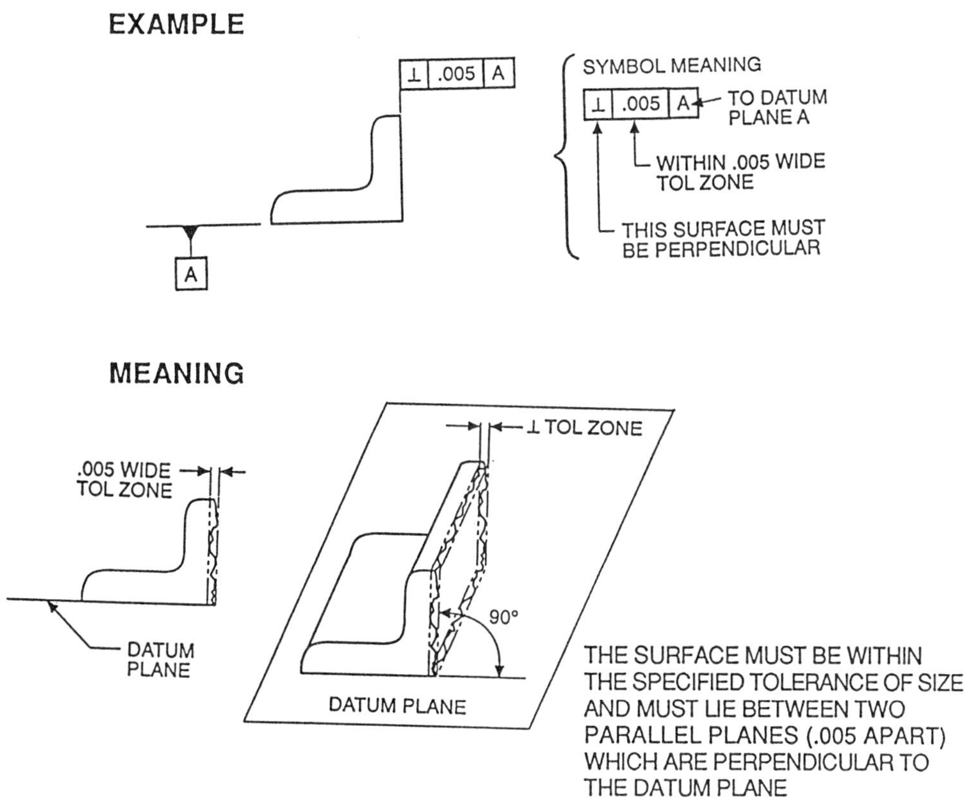

Figure 24 Perpendicularity tolerance as applied to a surface. Note that the perpendicularity tolerance applied to a plane surface controls flatness if a flatness tolerance is not specified (that is, the flatness will be at least as good as the perpendicularity).

gaging principles can be utilized as well in such an application and represents mating part assembly.

Angularity

Angularity is the condition of a surface, axis, or median plane which is at a specified angle (other than 90°) from a datum plane or axis (see Fig. 1).

Angularity Tolerance. Angularity tolerance is the distance between two parallel planes, inclined at the specified angle to a datum plane or axis, within which the toleranced surface, axis, or median plane must lie.

Angularity Tolerance Application. Figure 26 shows a part with a surface angular requirement. Note that the symbol meaning is: "This surface must be at 45° within a .005-wide tolerance zone in relation to datum plane A."

The meaning also shows how the tolerance zone is established. Note that the angular tolerance zone is at 45° basic (exact) from the datum plane A. To be acceptable, the entire angular surface must fall within this tolerance zone. All features of the part must separately be contained within their individual limits of part size.

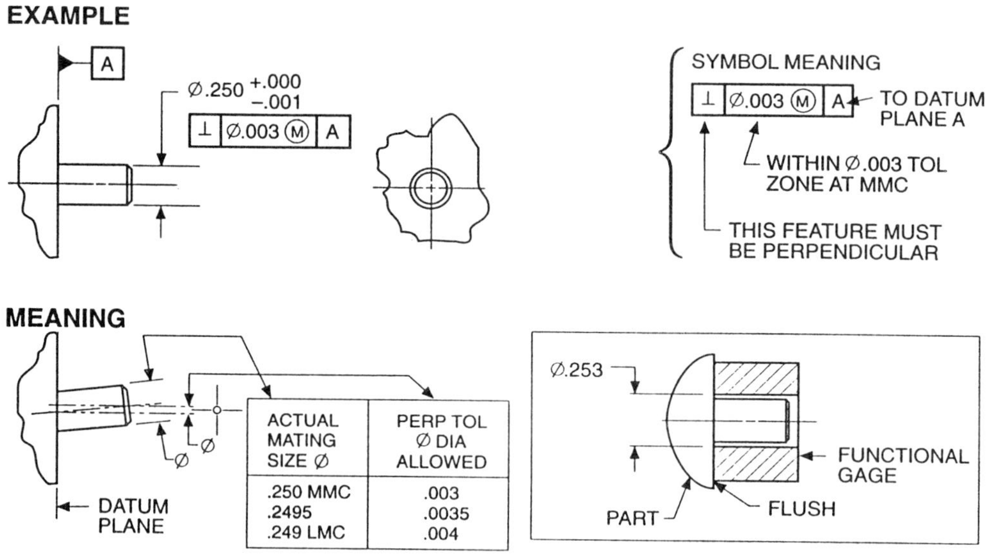

THE FEATURE MUST BE WITHIN THE SPECIFIED TOLERANCE OF LOCATION. WHEN THE FEATURE IS AT MMC .250, THE PERPENDICULARITY TOLERANCE IS .003 DIAMETER. AS THE FEATURE ACTUAL MATING ENVELOPE SIZE DEPARTS FROM MMC (GETS SMALLER), AN INCREASE IN TOLERANCE IS PERMITTED EQUAL TO THAT AMOUNT OF THAT DEPARTURE. VIRTUAL CONDITION IS Ø.253.

Figure 25 Perpendicularity tolerance as applied to a surface: Cylindrical feature at mmc, datum a plane. Note that the ∅ symbol is required to indicate a diameter (cylindrical) tolerance zone.

Parallelism

Parallelism is the condition of a surface or axis which is equidistant at all points from a datum plane or axis (see Fig. 1).

Parallelism Tolerance. A parallelism tolerance specifies:

1. A tolerance zone defined by two planes or lines parallel to a datum plane (or axis) within which the considered feature axis or surface must lie
2. A cylindrical tolerance zone parallel to a datum axis within which the axis of the considered feature must lie

Parallelism Tolerance Application. Note that in Figure 27, the bottom surface has been selected as the datum and the top surface is to be parallel to datum plane A within .002. This figure illustrates the tolerance zone and the manner in which the surface must fall within the tolerance zone to be acceptable. Note that the tolerance zone is established parallel to the datum plane A. Note also that the parallelism tolerance, when applied to a plane surface, controls flatness tolerance if a flatness tolerance is not specified (i.e., the implied flatness will be at least as good as the parallelism).

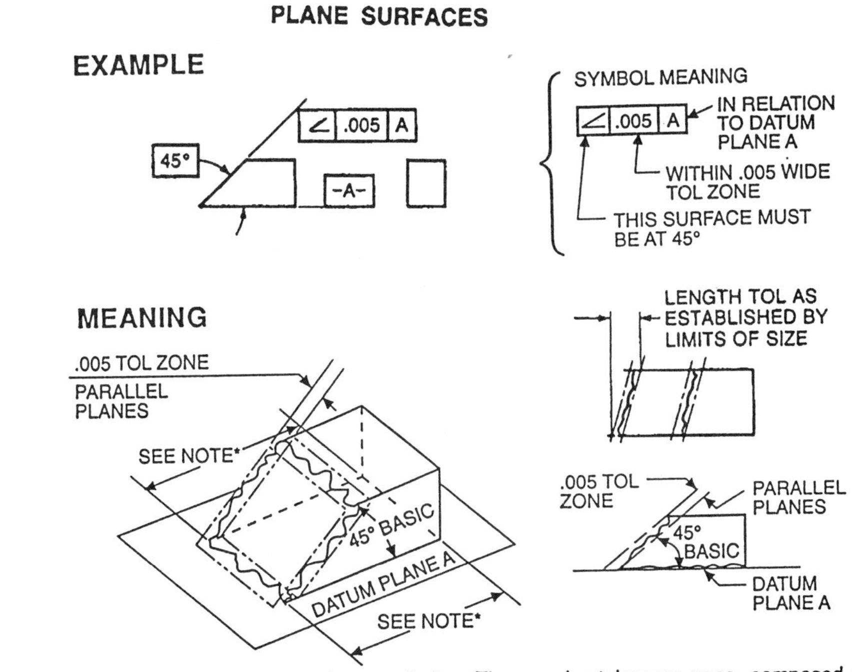

NOTE Part must be within the tolerance limits. The angular tolerance zone, composed of two parallel planes .005 apart, is 45° BASIC to the datum plane A. This tolerance one is established by contact of the outermost of the two planes with the extremities of the angular surface and with the outer plane parallel and inward at .005 distance. The entire surface must fall within this tolerance zone to be acceptable. Actually, this also controls the *flatness* of the surface to .005. Note that the angular surface extremities must be within both size and angular tolerance. See the examples below.

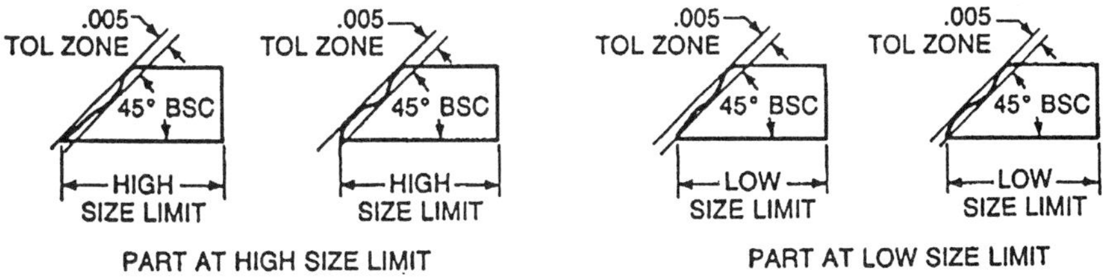

Figure 26 Plane surfaces: Illustration of a part with surface angular requirement.

Figure 28 shows parallelism tolerance applied to a size feature with the datum feature also a size feature. MMC is applied to the feature controlled with the datum feature at RFS (rule 2).

Under "meaning," it can be noted that the tolerance of parallelism increases if the feature actual mating size departs from MMC. The feature must also meet location tolerances. It should be noted that the parallelism tolerance controls only the feature orientation, not location.

Profile Tolerancing

Profile tolerancing is of two varieties and involves the characteristics of profile of a line and profile of a surface.

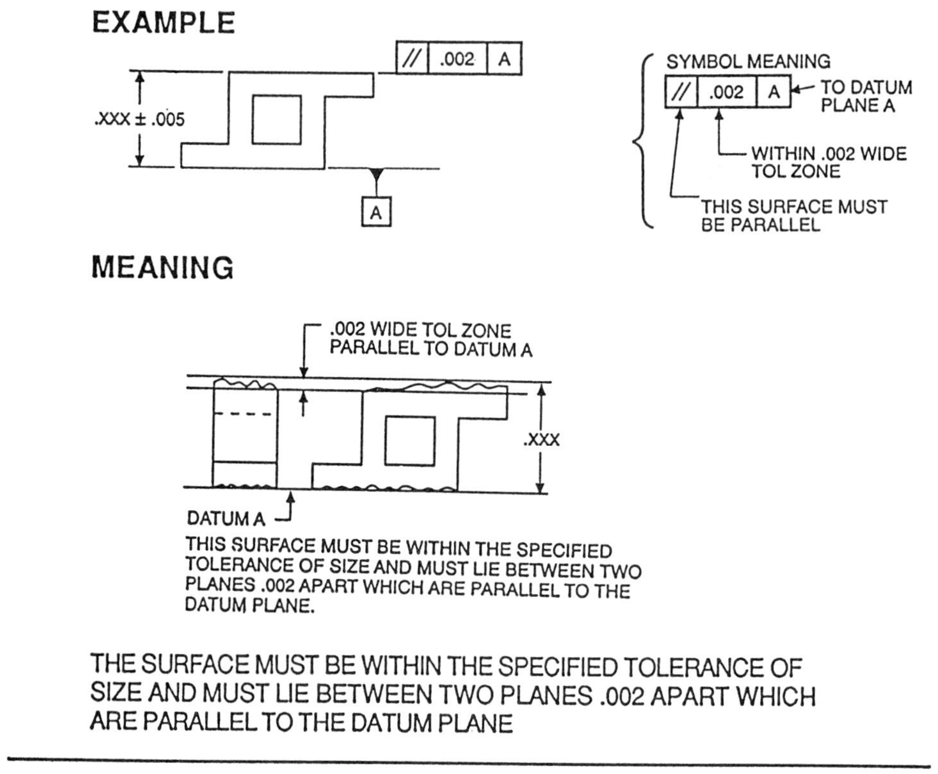

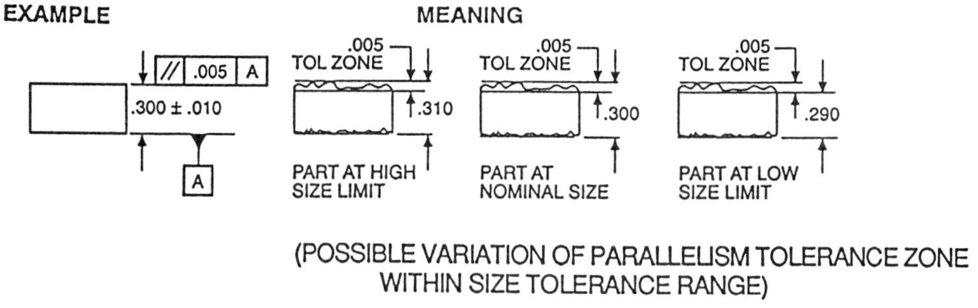

Figure 27 Surface to datum plane: Tolerance zone and the way the surface must fall within tolerance zone.

According to the design requirement, these characteristics may be applied to an individual feature, such as a single surface or element, or to related features, such as a single surface or element relative to a datum or datums.

Profile Tolerance. Profile tolerancing is a method used to specify a uniform amount of variation of a surface or line elements of a surface. A profile tolerance (either bilateral or unilateral) specifies a tolerance zone normal to the basic profile at all points of the profile, within which the specified part surface profile or line profile must lie.

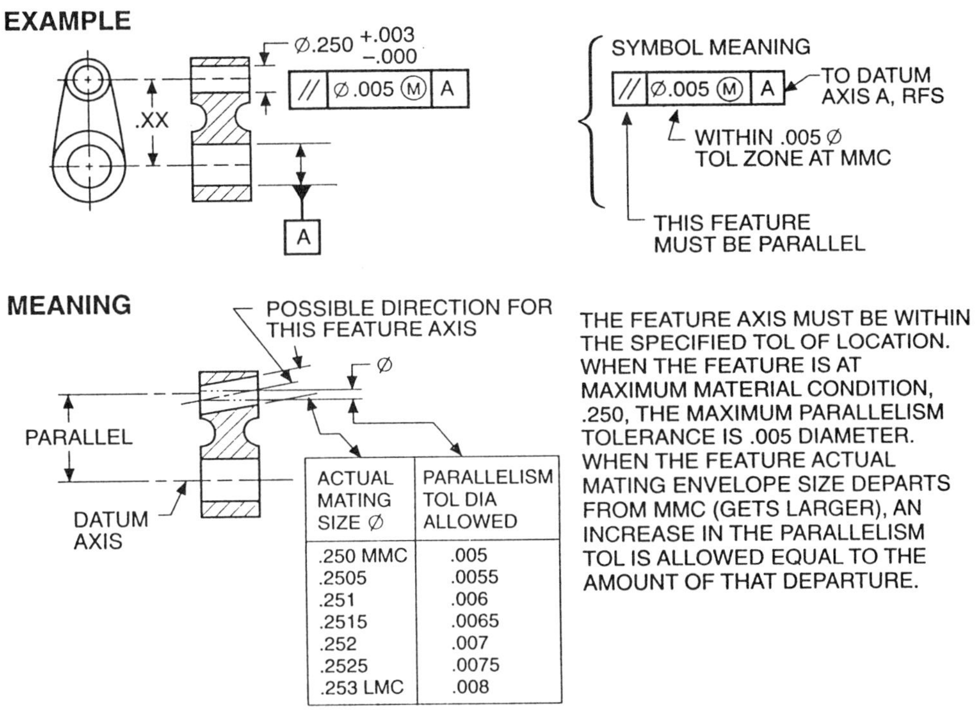

ACTUAL MATING SIZE Ø	PARALLELISM TOL DIA ALLOWED
.250 MMC	.005
.2505	.0055
.251	.006
.2515	.0065
.252	.007
.2525	.0075
.253 LMC	.008

Figure 28 Parallelism tolerance applied to a size feature with the datum feature also a size feature. Feature at MMC, Datum Feature at RFS. Note that the Ø symbol is required to indicate a diameter (cylindrical) tolerance zone.

Profile Tolerance Application. Profile tolerancing is an effective method for controlling lines, arcs, irregular surfaces, or other unusual part profiles. Profile tolerances are usually applied to surface features but may also be applied to a line (element on a feature surface). In either case, these requirements must be specified in association with the desired profile in a plane of projection (view) on the drawing as follows:

1. An appropriate view or section is drawn which shows the desired basic profile in true shape.
2. The profile is defined by basic dimensions. This dimensioning may be in the form of located radii and angles, or it may consist of coordinate dimensioning to points on the profile.
3. Depending on design requirements, the tolerance may be divided bilaterally to both sides of the true profile or applied unilaterally to either side of the true profile. Where an equally disposed bilateral

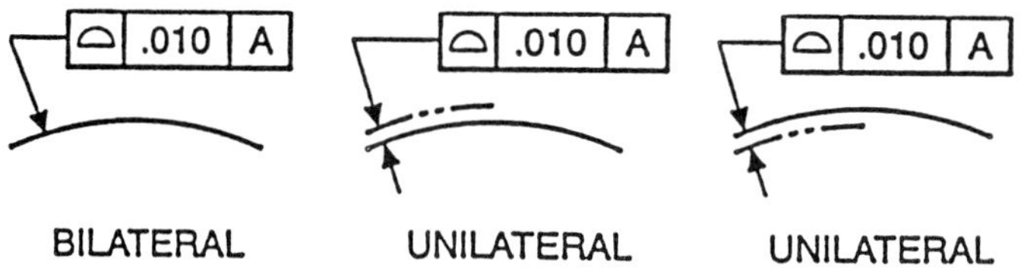

Figure 29 Profile tolerance application.

> tolerance is intended, it is only necessary to show the feature control symbol with a leader directed to the surface. For an unequally disposed or unilateral tolerance, phantom lines are drawn parallel to the true profile to indicate clearly the tolerance zone inside or outside the true profile. One end of a dimension line is extended to the feature control symbol.

See Figure 29 for details.

Two Types of Profile Tolerance. In practice, a profile tolerance may be applied either to an entire surface or to individual line element profiles taken at various cross sections through the part. The two types or methods of controlling profile are profile of a surface and profile of a line.

Profile of a Surface. The tolerance zone established by profile of a surface tolerance is a three-dimensional zone or total control across the entire length and width or circumference of the feature; it may be applied to parts having a constant cross section or to parts having a surface of revolution. Usually, the profile of a surface requires datum references (see Fig. 30a).

Profile of a Line. The tolerance zone established by the profile of a line tolerance is a two-dimensional zone extending along the length of the feature considered; it may apply to the profiles of parts having a varying cross section, such as a propeller, aircraft wing, nose cone, or to random cross sections of parts where it is not desired to control the entire surface as a single entity. The profile of a line may, or may not, require datum references (see Fig. 30b).

Figure 31 illustrates surface profile control and introduces a datum system.

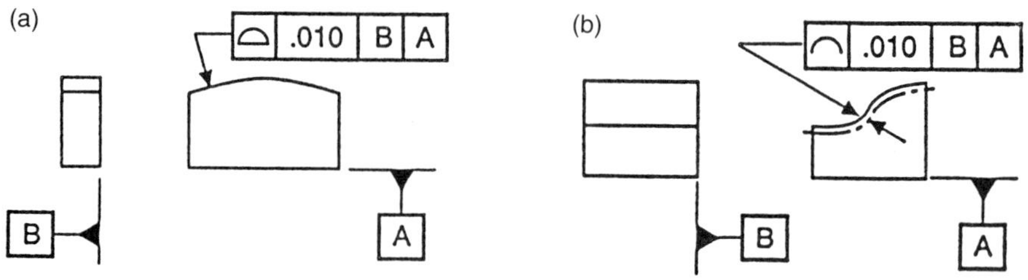

Figure 30 (a) Profile of a surface. (b) Profile of a line.

⌓ PROFILE OF A SURFACE

EXAMPLE

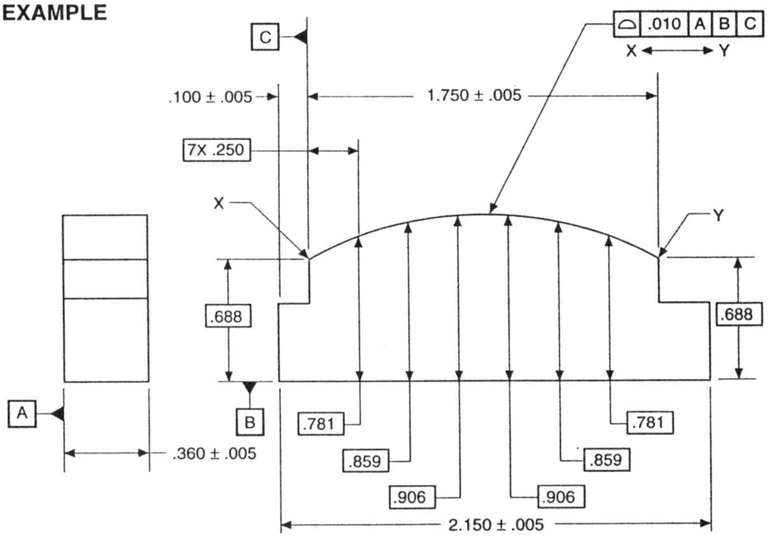

MEANING

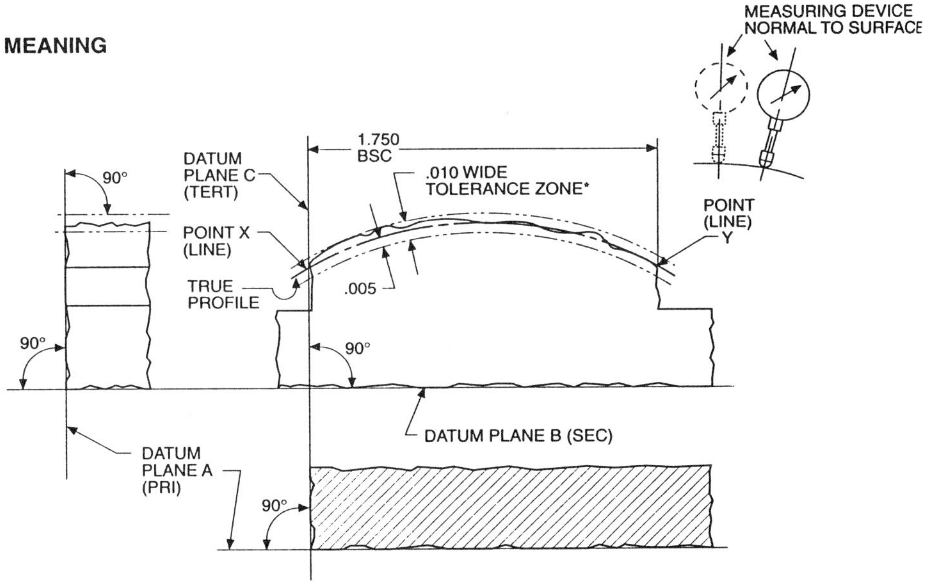

Figure 31 Illustration of surface profile control and introduction of a datum system. There is a .010 total wide zone (.005 each side of true profile). The surface between the two profile boundaries .010 apart, equally disposed about the true profile. The profile boundaries are perpendicular to datum plane A and positioned with respect to datum planes B and C.

Where line elements of a surface are to be specifically controlled or controlled as a refinement of size or surface profile control, the profile of a line characteristic may be used. Line profile control is applied in a manner similar to the application of surface profile.

Figure 32 illustrates line profile control as a refinement of size. As with surface profile, line profile must be shown in the drawing view in which it applies. The tolerance zone is established in the same way as the surface profile. However, its tolerance zone is disposed about each element of the surface. Therefore, the tolerance zone applies for the full length of each element of the surface (in the view in which it is shown), but only for the width or height at a cutting plane bisecting the element (see Fig. 29).

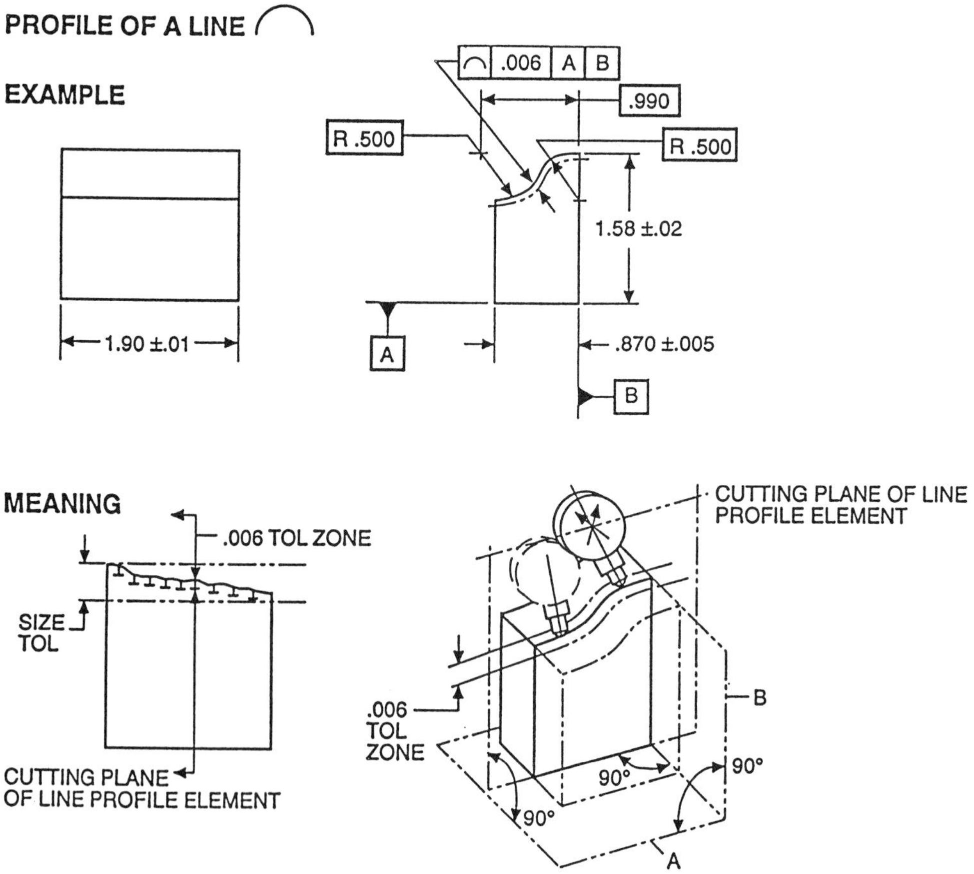

Figure 32 Line profile control as a refinement of size. Any profile line element of the surface (elements in the view in which the symbol is shown and element represented by the true profile) must be within the specified .006 tolerance zone, which is unilaterally disposed from the true profile. The profile tolerance zone must be contained within the size tolerance zone.

Runout

Runout is the composite deviation from the desired form and orientation of a part surface of revolution during full rotation (360°) on a datum axis (see Fig. 1).

Runout Tolerance

Runout tolerance states how far an actual surface or feature is permitted to deviate from the desired form and orientation implied by the drawing during full rotation (360°) of the part on a datum axis.

Runout Tolerance Application

Runout tolerancing is a method used to control the composite surface effect of one or more features of a part relative to a datum axis. Runout tolerance is applicable to rotating parts in which this composite surface control is based on the part function and design requirement. A runout tolerance always applies on an RFS basis; that is, size variation has no effect on the runout tolerance compliance.

Each feature considered must be within its individual runout tolerance when rotated 360° about the datum axis. The tolerance specified for a controlled surface is the total tolerance or full indicator movement (FIM) in terms of common inspection criteria. Former terms, full indicator reading (FIR), and total indicator reading (TIR) have the same meaning as FIM.

A runout tolerance is a relationship between surfaces or features: therefore, a datum (or datums) is required. Runout tolerance may be applied as "circular" or "total," based on the part functional requirements. Figures 33a and 33b illustrate "circular" versus "total" runout application on a similar part.

Figure 34 extends the principles of the previous examples. It is a shaft of multiple-diameters about a common datum axis C-D, with each feature, including the datum features, stating an individual runout tolerance. In addition to having both circular and total runout control, each datum feature has a cylindricity tolerance.

C. Coaxial Features: Selection of Proper Control

There are four methods of controlling interrelated coaxial features:

1. Runout tolerance (circular or total) (RFS)
2. Position tolerance (MMC or RFS)
3. Concentricity tolerance (RFS)
4. Profile tolerance (RFS)

Any of these methods will provide effective control. However, it is important to select the most appropriate one both to meet the design requirements and to provide the most economical manufacturing conditions.

Following are recommendations to assist in selecting the proper control. If the need is to control only circular cross-sectional elements in a composite relationship to a datum axis RFS (e.g., multiple diameters on a shaft), use:

EXAMPLE | ↗ | .005 | A – B | **CIRCULAR RUNOUT**

If the need is to control the total cylindrical or peripheral surface in composite relative to a datum axis RFS (e.g., multiple diameters on a shaft, bearing mounting diameters), use:

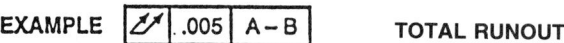

EXAMPLE | ⌰ | .005 | A – B | **TOTAL RUNOUT**

If the need is to control the total cylindrical or profile surface and its axis in a composite location relative to the datum axis on an MMC basis (e.g., on mating parts to assure interchangeability or assemble ability), use:

(IF MMC) EXAMPLE | ⌖ | Ø.005 Ⓜ | A Ⓜ | **POSITION**

(IF RFS) EXAMPLE | ⌖ | Ø.005 | A |

OR RFS DATUM | ⌖ | Ø.005 Ⓜ | A |

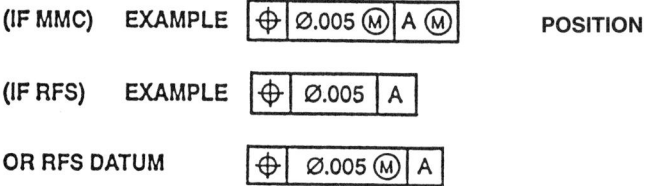

If the need is to control the axis of one or more features in composite relative to a datum axis RFS (e.g., to control the balance of a rotating part), use:

EXAMPLE | ◎ | Ø.005 | A – B | **CONCENTRICITY**

If the need is to control the total cylindrical or profile surface simultaneously with the size dimension(s) (using basic dimensions for both) relative to a datum axis, for example, precise fit, multidiameters, etc., use:

EXAMPLE | ⌓ | .005 | A | **PROFILE**

Tolerances of Location

Tolerances of location (see Fig. 1) state the persmissible variation in the specified location of a feature in relation to some other feature or datum. Tolerances of location refer to the geometric characteristics of position, concentricity, and symmetry.

Location tolerances involve features of size and relationships to center planes and axes. At least two features, one of which is a size feature, are required before location tolerancing is valid. Where function or interchangeability of mating part features is involved, the MMC principle may be introduced to great advantage. Perhaps the most widely used and best example of the application of this principle is position tolerancing.

The use of the position concept in conjunction with the maximum material condition concept provides some of the major advantages of the geometric tolerancing system.

Position

Position is a term used to describe the perfect (exact) location of a point, line, or plane of a feature in relation to a datum reference or other feature (see Fig. 1).

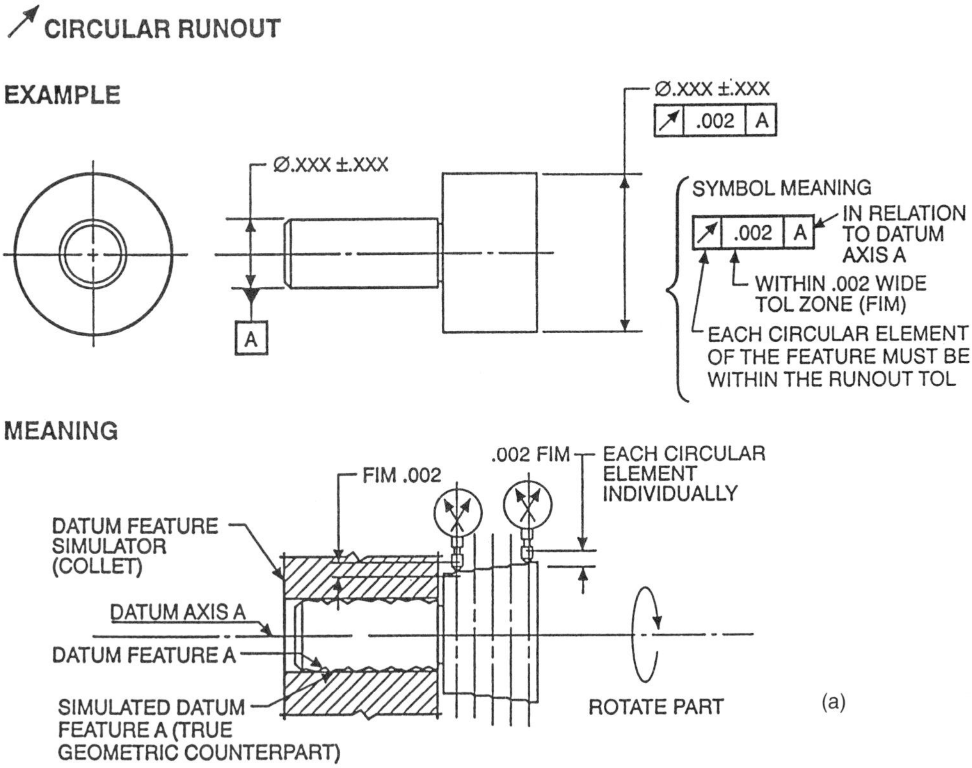

Figure 33 (a) Circular versus (b) total runout application on a similar part. In (b), the datum axis may be established by a single diameter (cylinder) of sufficient length, two diameters with sufficient axial separation, or a diameter and a face surface which is at right angles to it. Features selected as datums should, as much as possible, be functional to the part requirement (e.g., bearing mounted diameters, etc.).

Position Tolerance. A position tolerance is the total permissible variation in the location of a feature about its true (or exact) position. For cylindrical features (holes and bosses) the position tolerance is the diameter (cylinder) of the tolerance zone within which the axis of the feature must lie, with the center of the tolerance zone being at the exact position. For other features (slots, tabs, etc.), the position tolerance is the total width of the tolerance zone within which the center plane of the feature must lie, with the center plane of the zone being at the exact position.

Position Theory. The top illustration in Figure 35 shows a part with the hole pattern dimensioned and toleranced using the coordinate system. The bottom figure shows the same part dimensioned using the position system. Comparing the two approaches, the following differences are noted:

1. The derived tolerance zones for the hole centers result as square in the coordinate system and round in the position system.
2. The hole center location tolerance in the top illustration in Figure 35 is part of the coordinates (the 2.000 and 1.750 dimensions). In the bottom illustration, however, the location tolerance is associated with the hole

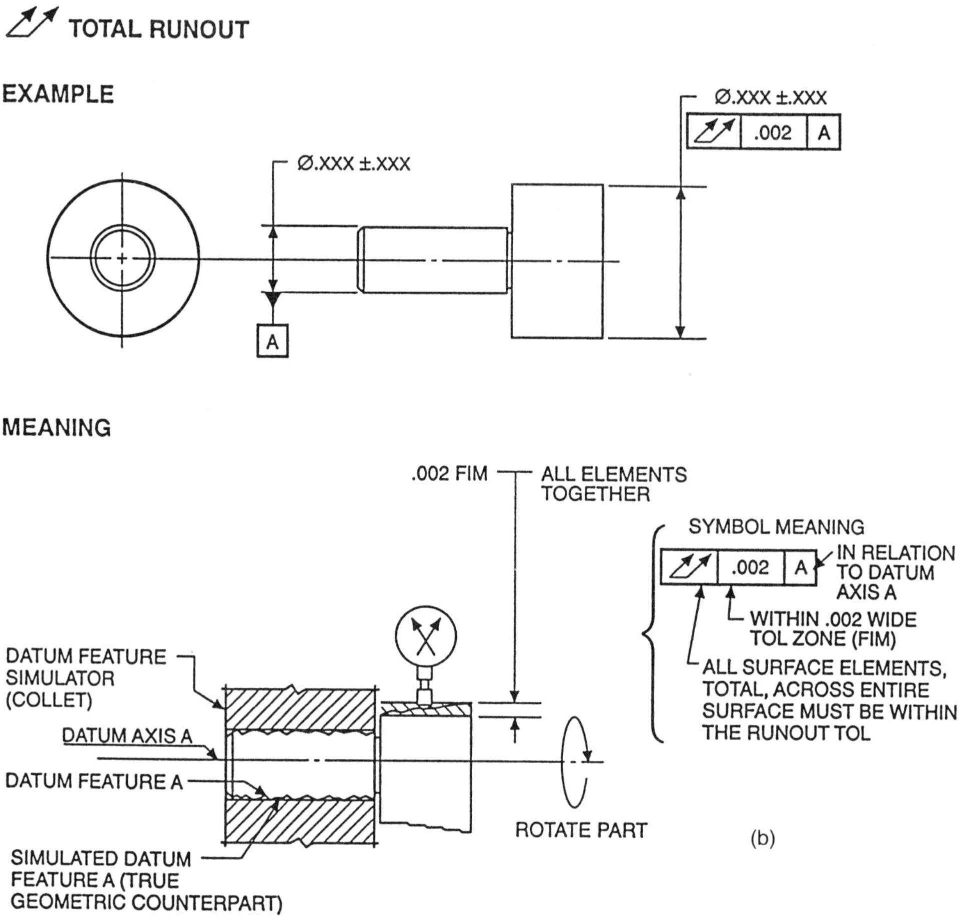

Figure 33 Continued

size dimension and is shown in the feature control symbol. The 2.000 and 1.750 coordinates are retained in the position application but are stated as basic or exact values.

For this comparison, the .005 square coordinate tolerance zone has been converted to a .007 position tolerance zone. The two tolerance zones are superimposed on each other in the enlargd detail.

The black dots represent possible inspected centers of this hole on eight separate piece parts. It is seen that if the coordinate zone is applied, only three of the eight parts are acceptable. However, with the position zone applied, six of the eight parts appear immediately acceptable.

The position diameter shaped zone can be justified by recognizing that the .007 diagonal is unlimited in orientation. Also, a cylindrical hole should normally have a cylindrical tolerance zone.

A closer analysis of the representative black dots and their position with respect to the desired exact location clearly illustrates the fallacies of the coordinate system when applied to a part such as that illustrated. The dot in the upper left

PART MOUNTED ON TWO FUNCTIONAL DIAMETERS (DATUMS)

EXAMPLE

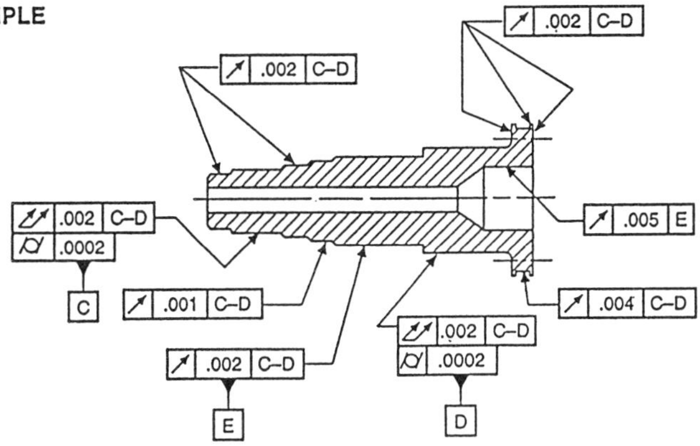

(SIZE DIMENSIONS & TOLERANCES NOT SHOWN FOR SIMPLICITY)

MEANING

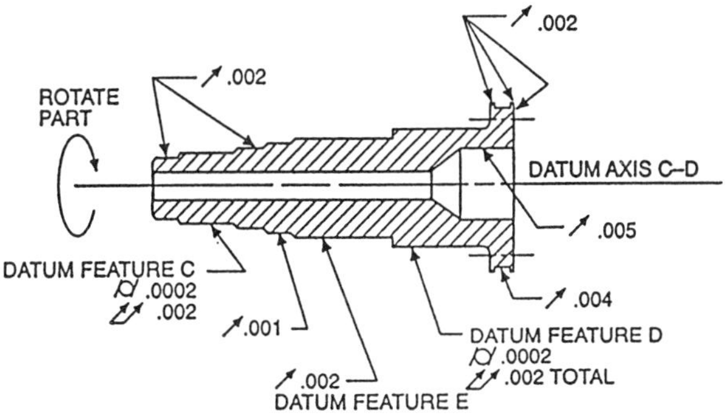

ALL VALUES FIM (TIR,FIR)

WHEN MOUNTED ON DATUMS C AND D, DESIGNATED SURFACES MUST BE WITHIN CIRCULAR RUNOUT (↗) TOLERANCE SPECIFIED. DATUMS C AND D MUST ALSO BE WITHIN TOTAL RUNOUT (⌰) TOLERANCE SPECIFIED AND CYLINDRICAL WITHIN .0002.

WHEN MOUNTED ON DATUM E, DESIGNATED SURFACE MUST BE WITHIN CIRCULAR RUNOUT (↗) TOLERANCE SPECIFIED.

Figure 34 Extension of previous examples on a shaft of multiple-diameters about a common axis C-D.

diagonal corner of the square zone and the dot on the left outside the square zone are in reality at nearly the same distance from the desired exact center. However, in terms of the square coordinate zone, the hole on the left is unacceptable by a wide margin, whereas the upper left hole is acceptable.

In normal calculation of position tolerances, the tolerance is derived, of course, from the design requirement, not from converted coordinates. The maximum material sizes of the features (hole and mating component) are used to determine this tolerance. Therefore, the .007 position tolerance of the example would be based on the MMC size of the hole (.247). As the hole size departs from MMC size, the position of the hole is permitted to shift off its "true position" beyond the original tolerance zone to the extent of that departure. The "bonus tolerance" of .013

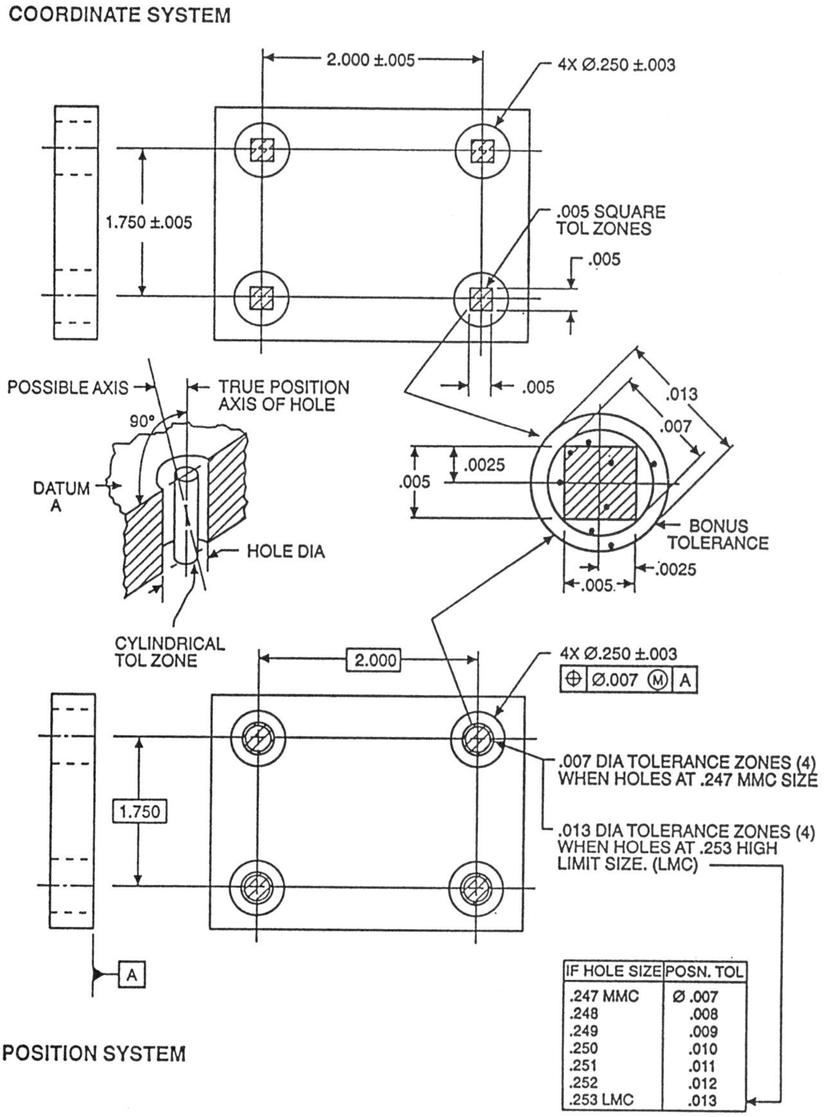

IF HOLE SIZE	POSN. TOL
.247 MMC	Ø .007
.248	.008
.249	.009
.250	.010
.251	.011
.252	.012
.253 LMC	.013

Figure 35 A part with hole pattern dimensioned and toleranced using coordinate system (top). Same part dimensioned using the geometric system (bottom).

illustrates the possible position tolerance should the hole be produced, for example, to its high limit size of .253. The tabulation in the lower part of the illustration shows the enlargement of the position tolerance zone as the hole size departs from MMC in production. Although only one hole has been considered in the explanation, the same reasoning applies to all the holes in the pattern relative to their respective "true positions."

Position tolerancing is ideally applied on mating parts in cases where fit, function, and interchangeability are the considerations. It provides greater production tolerances, ensures design requirements, and provides the advantages of functional inspection practices as desired.

Functional gaging techniques, familiar to a large segment of industry, are fundamentally based on the MMC position concept. It should be clearly understood, however, that functional gages are not mandatory in fulfilling MMC position requirements.

Some functional gaging principles are introduced in this section for the dual purpose of explaining the principles involved in positional tolerancing and for introducing the functional gaging technique as a valuable tool. A functional gage can often be considered as a representative mating part at its worst condition.

Position, although a locational tolerance, also includes form and orientation tolerance elements in composite. For example, as shown in Figure 35, perpendicularity is invoked as part of the control to the extent of the tolerance zone, for the depth of the hole. Further, the holes in the pattern are also parallel to one another within the positional tolerance.

Mating Parts, Floating Fastener. Position tolerancing techniques are most effective and appropriate in mating part situations. Figure 36, in addition to demonstrating the calculations required, also emphasizes the importance of decisions at the design stage to recognize and initiate the position principles.

The mating parts shown in Figure 36 are to be interchangeable. Thus the calculation of their position tolerances should be based on the two parts and their interface with the fastener in terms of MMC sizes (see Fig. 36).

The two parts are to be assembled with four screws. The holes in the two parts are to line up sufficiently to pass the four screws at assembly. Since the four screws ("fasteners") are separate components, they are considered to have some "float" with respect to one another. The colloquial term "floating fastener application" has been used popularly to describe this situation.

The calculations are shown in the upper right corner of Figure 36. Also note that, in this case, the same basic dimensions and position tolerances are used on both parts. They are, of course, separate parts and are on separate drawings. The calculations on the illustrated parts show the total permissible position tolerance of the holes on the two parts (i.e., .016).

As seen from Figure 36, part acceptance tolerances will increase as the hole actual mating sizes in the parts are actually produced and vary in size as a departure from MMC. From the .016 diameter tolerance calculated, the tolerance may increase to as much as .022, dependent on the actually produced hole size.

A possible functional gage is also shown in the illustration. The .190 gage pin diameters are determined by the MMC size of the hole, .206, minus the stated position tolerance of .016. In this example, the same functional gage can be used on both parts. Functional gages are, of course, not required with position application, but they do provide an effective method of evaluation where desired. Functional gaging may also be achieved by "soft gage" manipulation of coordinate measuring (CMM) data.

Mating Parts, Fixed Fastener. When one of two mating parts has "fixed" features, such as the threaded studs in Figure 37, the fixed fastener method is used in calculating position tolerances. The term *fixed fastener* is a colloquialism popularly used to describe this application. Both the term and the technique are applied to numerous other manufacturing situations, such as locating dowels, tapped holes, and so on.

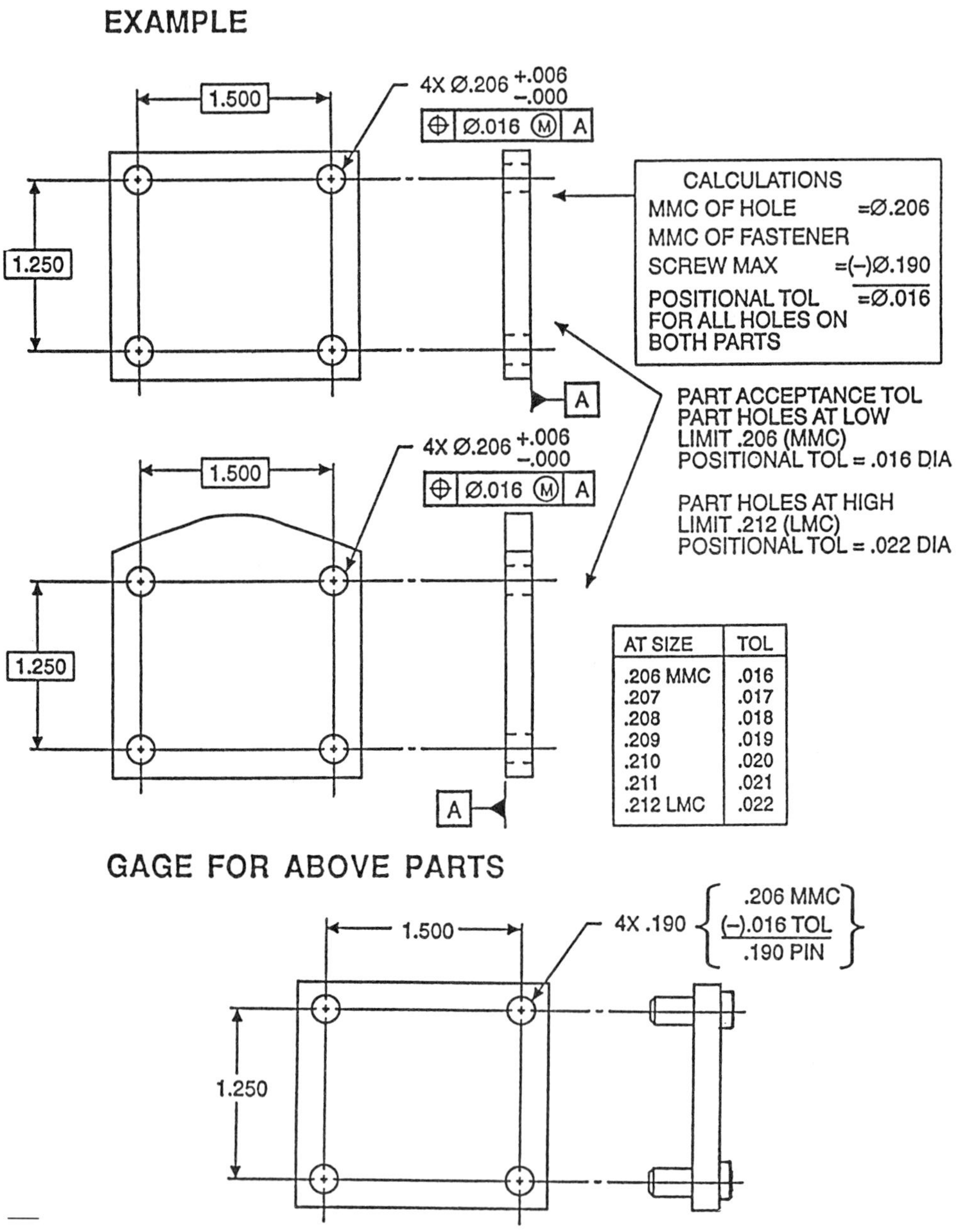

Figure 36 Specification using position tolerance for two mating parts and floating fasteners.

The advantages of the MMC principle as described for the floating fastener application also apply here. However, with a fixed fastener application, the difference between the MMC sizes of mating features must be divided between the two features, since the total position tolerance must be shared by the two mating features. In this example, the two mating features (actually four of each in each pattern) are the studs and the clearance holes. The studs must fit through the holes at assembly.

EXAMPLE

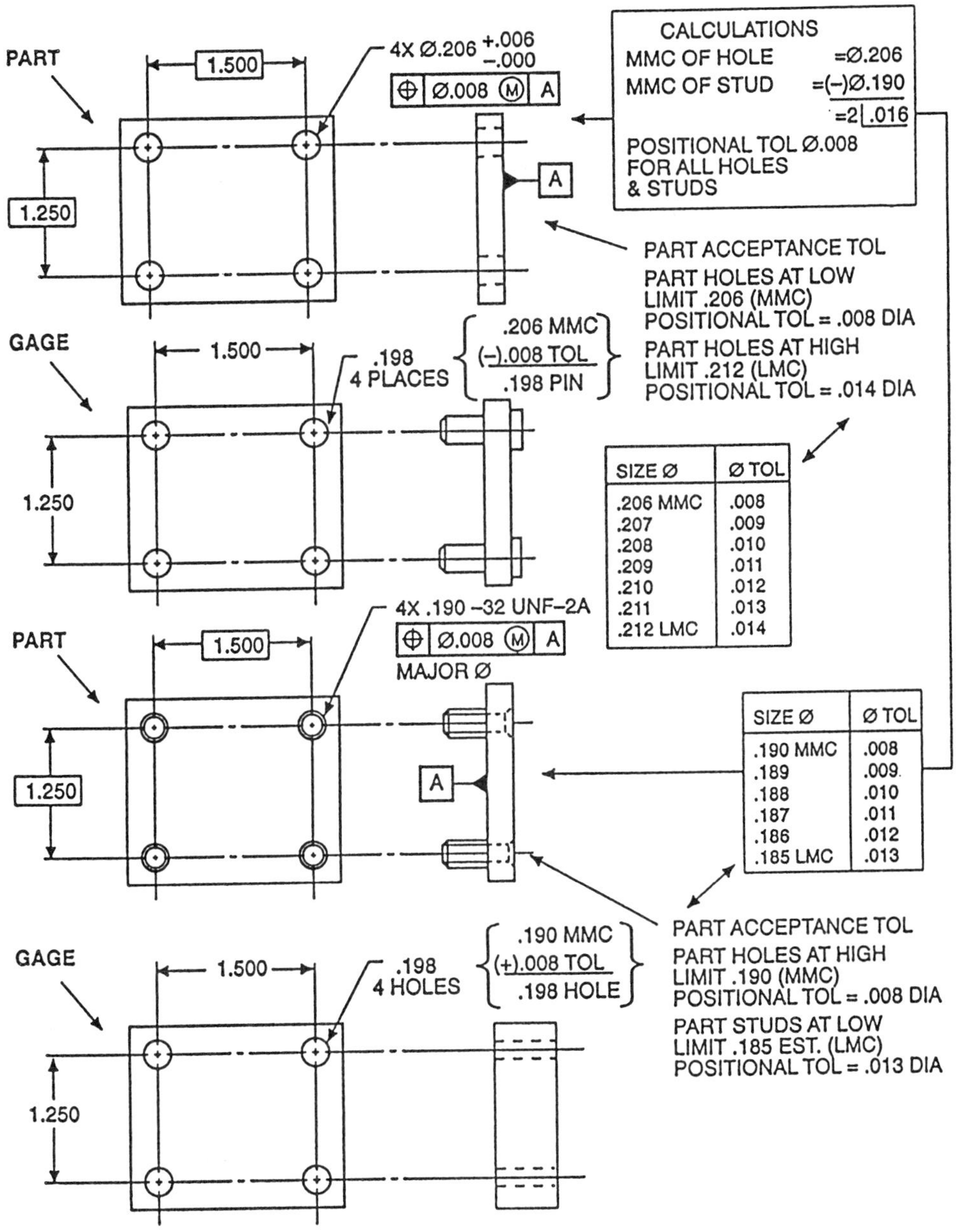

Figure 37 Relation to datums. Where part function is critical, specified datums are necessary.

Again, it is seen that the clearance of the mating features as they relate to each other at assembly determines the position tolerances. When one feature is to be assembled within another on the basis of the MMC sizes and "worst" condition of assembly, the clearance, or total tolerance, must be divided for assignment to each of the mating part features. In this case, the derived .016 was divided equally, with .008 diameter position tolerance assigned to each mating part feature (stud and hole). The total tolerance of .016 can be distributed to the two parts as desired, as long as the total is .016 (e.g., .010 + .006, .012 + .004). This decision is made at the design stage, however, and must be fixed on the drawing before release to production.

As the part features of both parts are produced, any departure in actual mating size from MMC will increase the calculated position by an amount equal to that departure. For example, the position tolerance of the upper part could possibly increase up to .014, and that of the lower part up to .013 dependent on the amount of departure from their MMC sizes. However, parts must actually be produced and sizes established before the amount of increase in tolerance can be determined.

Functional gages (shown below each part in the illustration) can be used for checking, and although their use is not a requirement, they provide a very effective method of evaluation if desired. Note that the functional gages resemble the mating parts; as a matter of fact, functional gages simulate mating parts at their worst condition. Functional gaging may also be achieved by "soft" gage manipulation of coordinate measuring machine (CMM) data.

Relation to Datums. Figure 38 shows a part where the datums are identified with the A, B, and C datum feature symbols and form tolerances are specified. Where part function, and thus the stated drawing requirements, are more critical, specified datums and greater geometric control are necessary.

In Figure 38, it was necessary to control the accuracy of the datum surfaces in their specific relationship to each other. To accomplish this, identification of the specific surfaces as datum references was required. Further, since the hole position pattern was critical in its orientation to the surfaces, datum identification was required for this purpose. With specification of the datum, precedence of the datum surfaces, is established.

Noncylindrical Features and Coaxial Features. The principles of positional tolerancing covered in the foregoing sections can also be applied to noncylindrical and coaxial features. Calculation of tolerances, use of MMC or RFS principles, datums, and so on, are similar.

Figures 39 and 40 are examples of a noncylindrical and coaxial feature application of position tolerancing.

Position Tolerance Verification. Often position tolerancing is misunderstood, or even avoided, by concern for its verification or inspection. That is, X and Y measuring equipment is envisioned as a possibility, which would appear illogical to the diametrical (cylindrical) tolerance zones of the common position tolerance requirement. Of course, this is not so, as many methods can be used in verifying position tolerance. The fact of the matter is that positional tolerancing opens up many *new* and *more valid* methods, such as functional gaging without losing any past methods.

Position tolerancing can be effectively verified with any of the coordinate measuring tools or machines presently used in open setup inspection, comparators,

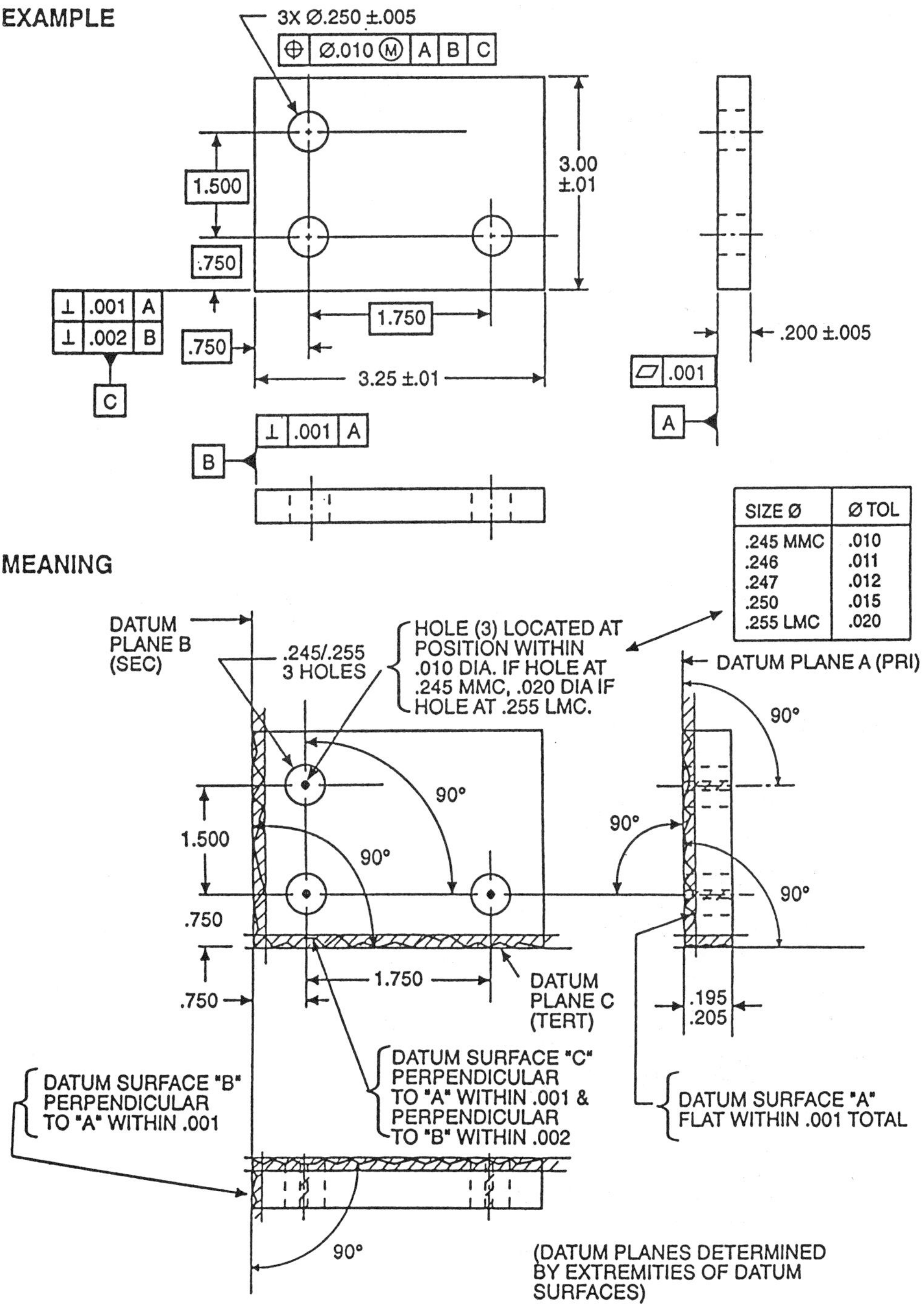

Figure 38 Example of position tolerances as related to datum targets (position tolerance—specified datums—form and orientation tolerances).

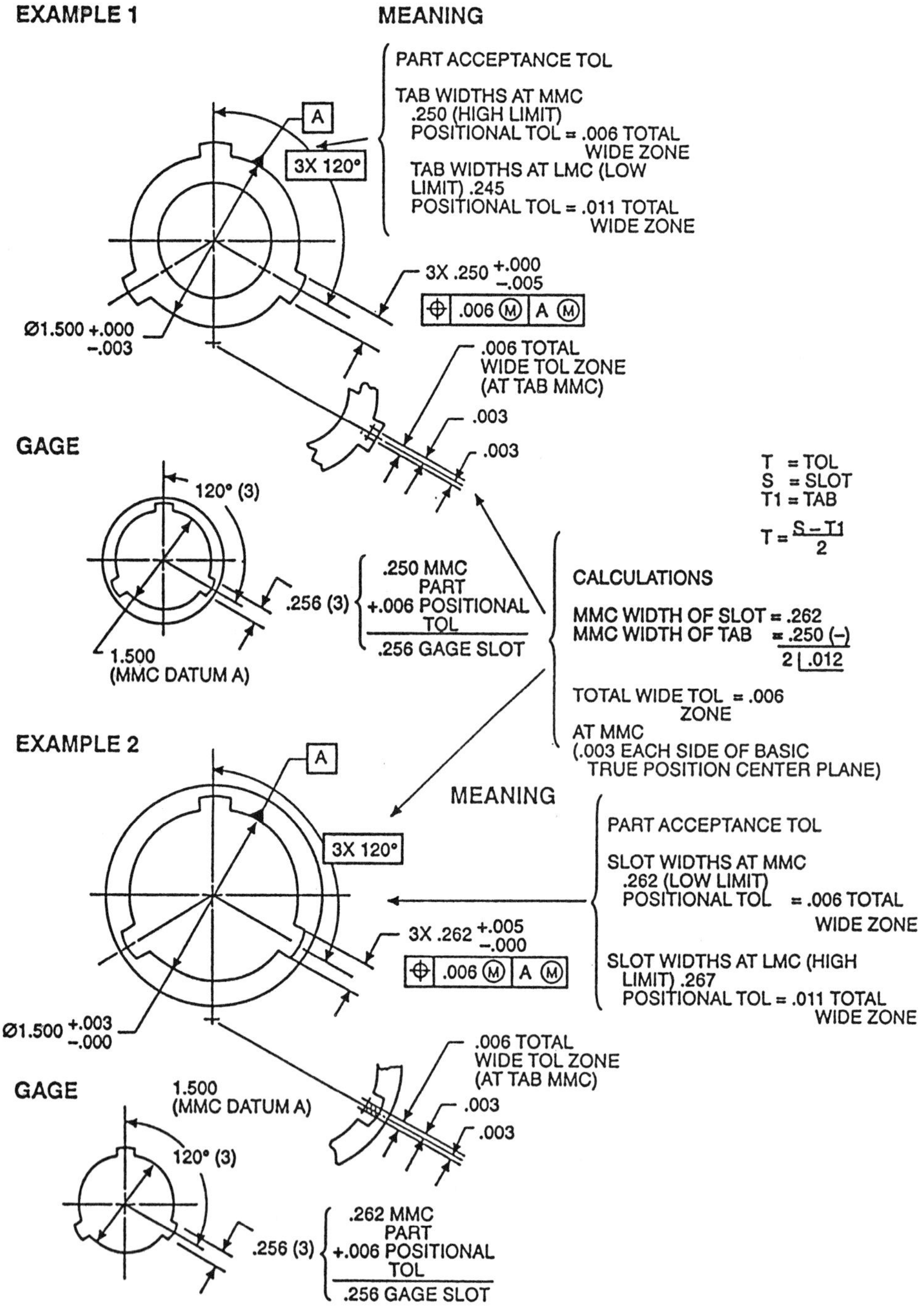

Figure 39 Example of positioning tolerance applied to noncylindrical part feature.

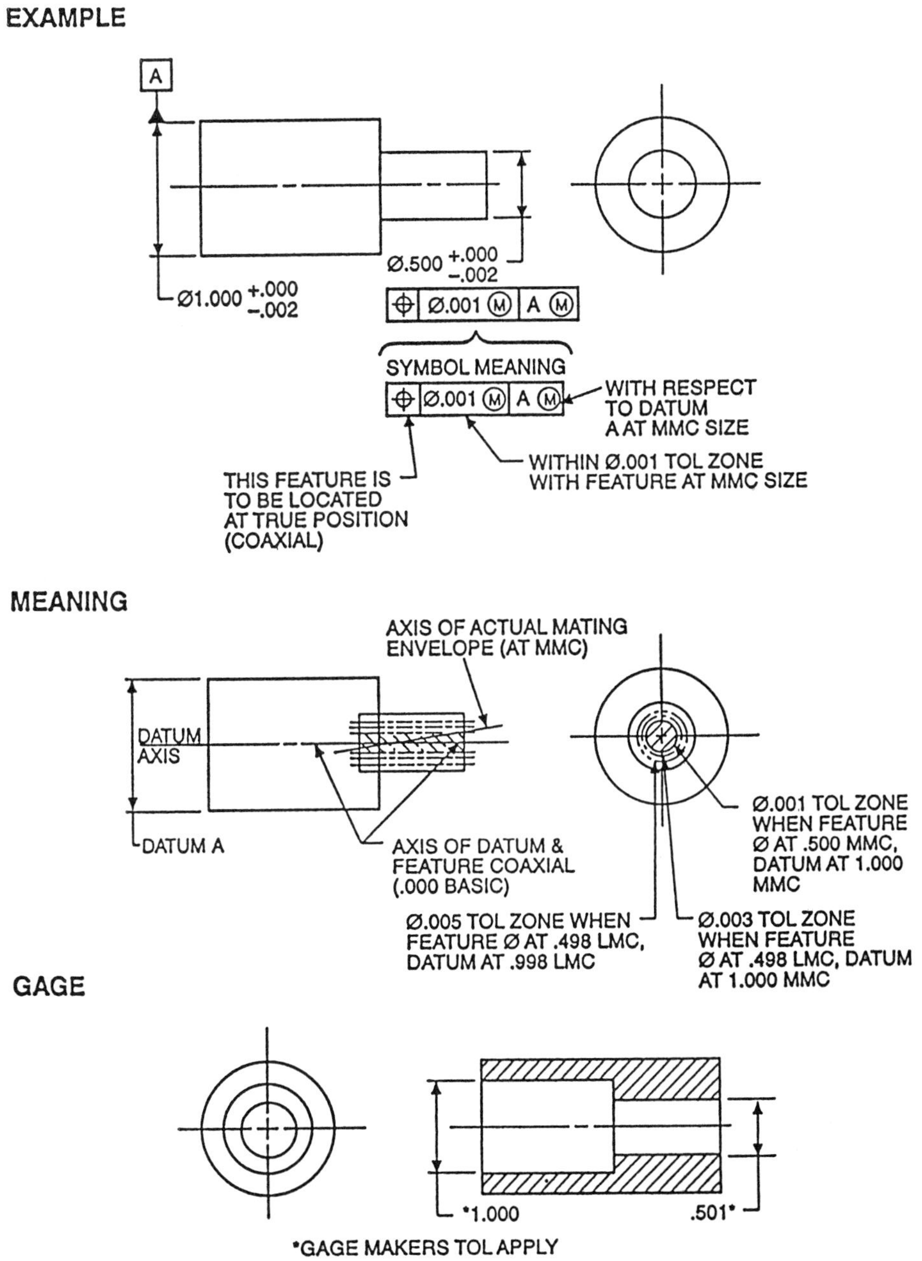

Figure 40 Example of position tolerance applied to coaxial features.

and functional gaging. Some functional gage applications appear earlier in this positional tolerancing section. Functional gaging is a recommended method where cost of the gage, production quantities, and so on, can be justified. It simulates part interface.

To illustrate one convenient method of inspecting a part with a coordinate measurement tool and making the translation to positional tolerance equivalents, Figure 41 is presented. The conversion illustrations in Figure 42 show representative coordinate results of an X and Y measurement process. By referring to the conversion table under the resulting differential value in X and Y, the translated positional value is derived. This, compared to the permissible position tolerance (and applying MMC principles when necessary), determines whether the hole is acceptable.

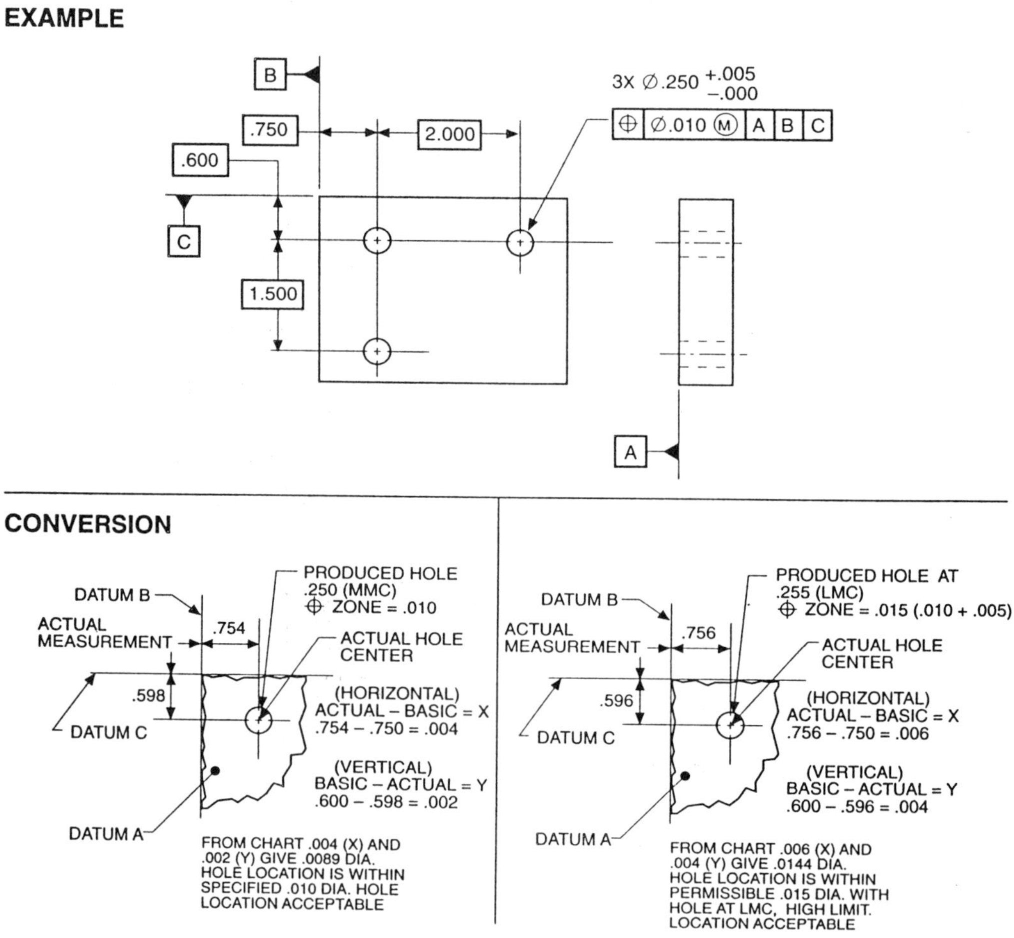

Figure 41 Illustration of one convenient method of inspecting a part with coordinate measuring tool.

CONVERSION OF COORDINATE MEASUREMENTS TO ⊕ POSITION LOCATION

Y																				
.020	.0400	.0402	.0404	.0408	.0412	.0418	.0424	.0431	.0439	.0447	.0456	.0466	.0477	.0488	.0500	.0512	.0525	.0538	.0552	.0566
.019	.0380	.0382	.0385	.0388	.0393	.0398	.0405	.0412	.0420	.0429	.0439	.0449	.0460	.0472	.0484	.0497	.0510	.0523	.0537	.0552
.018	.0360	.0362	.0365	.0369	.0374	.0379	.0386	.0394	.0403	.0412	.0422	.0433	.0444	.0456	.0469	.0482	.0495	.0509	.0523	.0538
.017	.0340	.0342	.0345	.0349	.0354	.0360	.0368	.0376	.0385	.0394	.0405	.0416	.0428	.0440	.0453	.0467	.0481	.0495	.0510	.0525
.016	.0321	.0322	.0325	.0300	.0335	.0342	.0349	.0358	.0367	.0377	.0388	.0400	.0412	.0425	.0439	.0452	.0467	.0482	.0497	.0512
.015	.0301	.0303	.0306	.0310	.0316	.0323	.0331	.0340	.0350	.0360	.0372	.0384	.0397	.0410	.0424	.0439	.0453	.0469	.0484	.0500
.014	.0281	.0283	.0286	.0291	.0297	.0305	.0313	.0322	.0333	.0344	.0356	.0369	.0382	.0396	.0410	.0425	.0440	.0456	.0472	.0488
.013	.0261	.0263	.0267	.0272	.0278	.0286	.0295	.0305	.0316	.0328	.0340	.0354	.0368	.0382	.0397	.0412	.0428	.0444	.0460	.0477
.012	.0241	.0243	.0247	.0253	.0260	.0268	.0278	.0288	.0300	.0312	.0325	.0339	.0354	.0369	.0384	.0400	.0416	.0433	.0449	.0466
.011	.0221	.0224	.0228	.0234	.0242	.0250	.0261	.0272	.0284	.0297	.0311	.0325	.0340	.0356	.0372	.0388	.0405	.0422	.0439	.0456
.010	.0201	.0204	.0209	.0215	.0224	.0233	.0244	.0256	.0269	.0283	.0297	.0312	.0328	.0344	.0360	.0377	.0394	.0412	.0429	.0447
.009	.0181	.0184	.0190	.0197	.0206	.0216	.0228	.0241	.0254	.0269	.0284	.0300	.0316	.0333	.0350	.0367	.0385	.0402	.0420	.0439
.008	.0161	.0165	.0171	.0179	.0189	.0200	.0213	.0226	.0241	.0256	.0272	.0288	.0305	.0322	.0340	.0358	.0376	.0394	.0412	.0431
.007	.0141	.0146	.0152	.0161	.0172	.0184	.0198	.0213	.0228	.0244	.0261	.0278	.0295	.0313	.0331	.0349	.0368	.0386	.0405	.0424
.006	.0122	.0126	.0134	.0144	.0156	.0170	.0184	.0200	.0216	.0233	.0250	.0268	.0286	.0305	.0323	.0342	.0360	.0379	.0398	.0418
.005	.0102	.0108	.0117	.0128	.0141	.0156	.0172	.0189	.0206	.0224	.0242	.0260	.0278	.0297	.0316	.0335	.0354	.0374	.0393	.0412
.004	.0082	.0089	.0100	.0113	.0128	.0144	.0161	.0179	.0197	.0215	.0234	.0253	.0272	.0291	.0310	.0330	.0349	.0369	.0388	.0408
.003	.0063	.0072	.0085	.0100	.0117	.0134	.0152	.0171	.0190	.0209	.0228	.0247	.0267	.0286	.0306	.0325	.0345	.0365	.0385	.0404
.002	.0045	.0056	.0072	.0089	.0108	.0126	.0146	.0165	.0184	.0204	.0224	.0243	.0263	.0283	.0303	.0322	.0342	.0362	.0382	.0402
.001	.0028	.0045	.0063	.0082	.0102	.0122	.0141	.0161	.0181	.0201	.0221	.0241	.0261	.0281	.0301	.0321	.0340	.0360	.0380	.0400
X	.001	.002	.003	.004	.005	.006	.007	.008	.009	.010	.011	.012	.013	.014	.015	.016	.017	.018	.019	.020

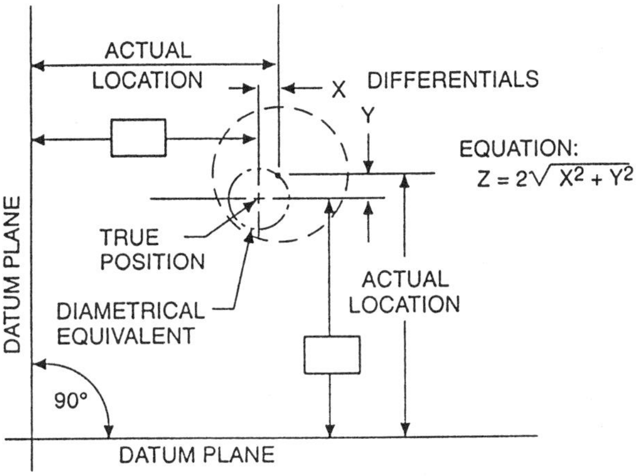

NOTE: The table above can be developed on your PC.

Use program $Z = 2\sqrt{X^2 + Y^2}$,

where Z = Diametrical Equivalent of X and Y differences in measurement.

Figure 42 Representative coordinate results of a X and Y measurement process.

Concentricity

Concentricity Tolerance. Concentricity is that condition where median points of all diametrically opposed elements of a figure of revolution (or correspondingly—located elements of two or more radially—disposed features) are congruent with the axis (or center point) of a datum feature (see Fig. 1).

Concentricity Tolerance Application. Concentricity is a type of location tolerancing. It always involves two or more basically coaxial features of size and controls the amount by which the axes of the features may fail to coincide. Concentricity tolerance, owing to its unique characteristic, is always used on an RFS basis. Where interrelated features are basically coaxial, we must first consider the possibility of using the more economical position or runout controls before considering concentricity.

The surface of a feature must be used to establish its axis. Therefore, all the irregularities or errors of form of a feature surface must be considered in establishing the axis. For instance, the surface may be bowed, out of round, and so on, in addition to being offset from its datum feature. This usually involves a complex inspection analysis of the entire surface, and therefore requires a more time-consuming and costly procedure.

Concentricity requirements are required less frequently than position or runout requirements. However, where concentricity is required, it provides effective control over the more unique applications of coaxial relationships. For example, concentricity might be applied to the coaxiality requirements of a tape-drive pulley or of a capstan on a computer mechanism or a motor generator rotor. Often where balance is required, the out-of-roundness or lobing effect (or possible other form errors) may be permissible, although it may exceed the conventional FIM requirement. Hence any basically symmetrical form of revolution (hexagons, cones, etc.) or consistently symmetrical variation of such a form could satisfy a concentricity tolerance where a runout requirement may not (Fig. 43).

Symmetry

Symmetry Tolerance. Symmetry is that condition where the median points of all opposed or correspondingly located elements of two or more feature surfaces are congruent with the axis or center plane of a datum feature (see Fig. 1).

Symmetry Tolerance Application. Like the two types of locational tolerance discussed previously, position and concentricity, symmetry deals with the location of actual features with respect to established center planes or axes. As its name implies, the purpose of any symmetry tolerance is to specify a symmetrical relationship for the toleranced feature, usually with the outside limits of the part for reasons of appearance, clearance, or fit to related or mating parts.

Figure 44 shows a part using symmetry tolerancing on the slot; the part width is established as the datum. The requirement is to relate the slot location to the outside width of the part. To simplify the explanation, size dimensions and tolerances of the slot and the datum width have been omitted.

Symmetry is a type of locational tolerancing. Where part features of a symmetrical shape are to be geometrically toleranced, it is recommended that MMC positional control be used instead of symmetry, if possible. Symmetry is, however, a valid characteristic and may be applied, if desired, on an RFS basis only (see Fig. 44).

Relation to Datum Targets. Where positional tolerancing relationships are to be related to specific points, lines, or areas as the functional requirement, or repeatability between manufacturing or inspection is of concern, datum targets may be used (Figs. 45 and 46).

CONCENTRICITY ◎

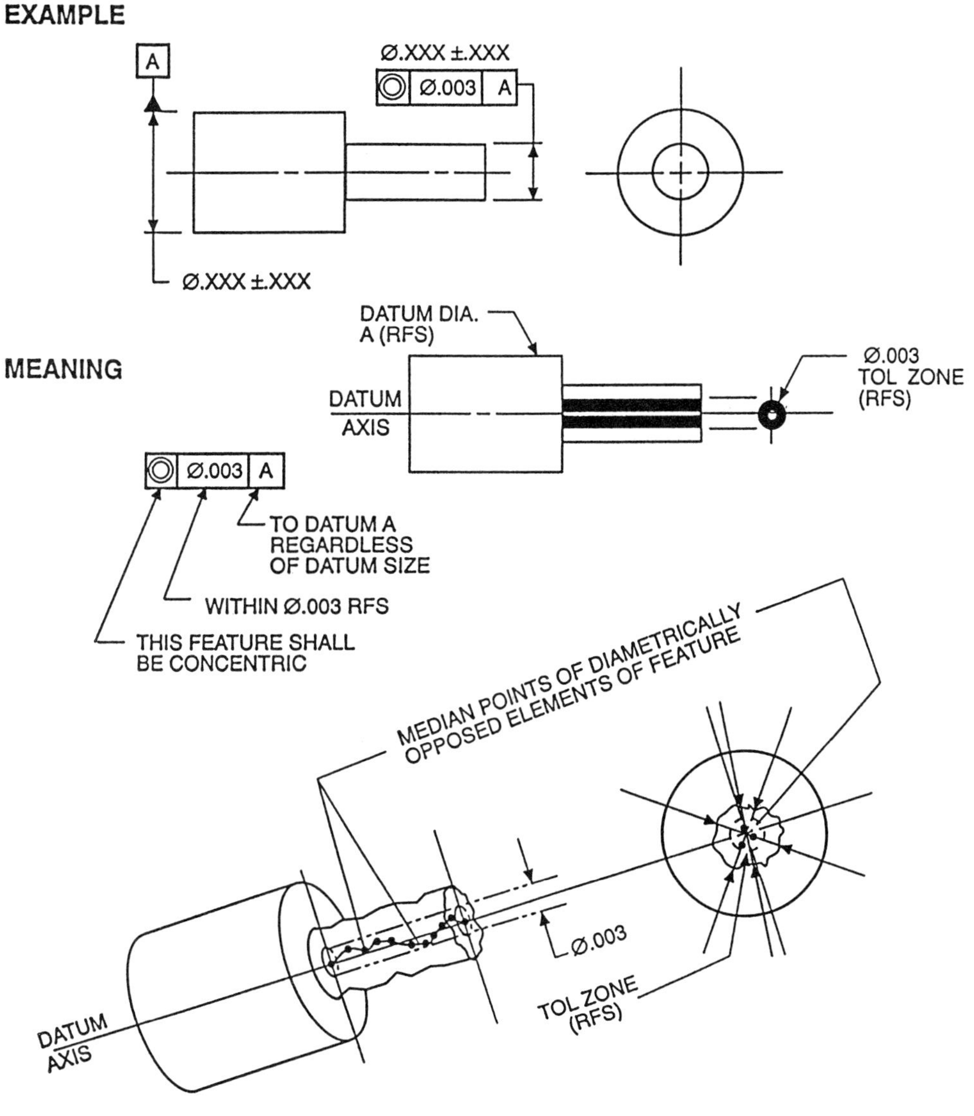

Figure 43 Any basically symmetrical form of revolution could satisfy a concentricity tolerance; a runout requirement may not.

D. Datum Targets

Where datum orientation is required on parts of irregular contour, such as castings, forgings, sheet metal, and so on, datum targets provide a valuable tool. Specified datum targets which serve as means of constructing special datum planes of orientation can be of three types: points, lines, or areas. Datum targets establish the

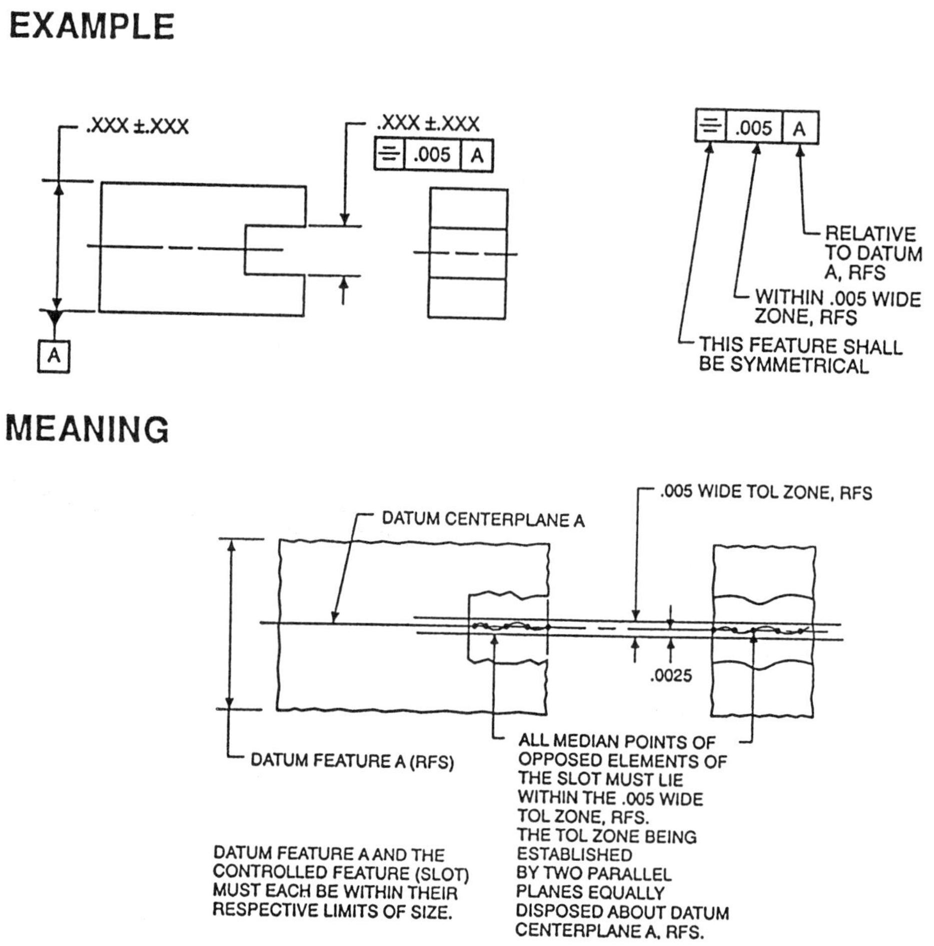

Figure 44 Positioning tolerances should be used in MMC applications.

necessary datum system framework and, in addition, ensure repeatable part location for manufacturing and inspection operations.

Datum targets are also used to indicate special, or more critical, design requirements where functional part feature relationships are to be indicated from specific points, lines, or areas on the part surface. Datum points and lines have been defined previously. A datum area is a datum established from a partial datum surface. On a drawing, a datum area is outlined with phantom lines and identified by diagonal slash lines. It may be of any shape.

The locations and/or sizes of datum points, lines, or areas are controlled by basic or untoleranced dimensions and imply exactness within standard tooling, gauging, or shop tolerances. Where necessary, toleranced locations or sizes may be used with datum target symbols.

Figures 45 and 46 illustrate the use of datum target symbols to establish datum planes and part orientation.

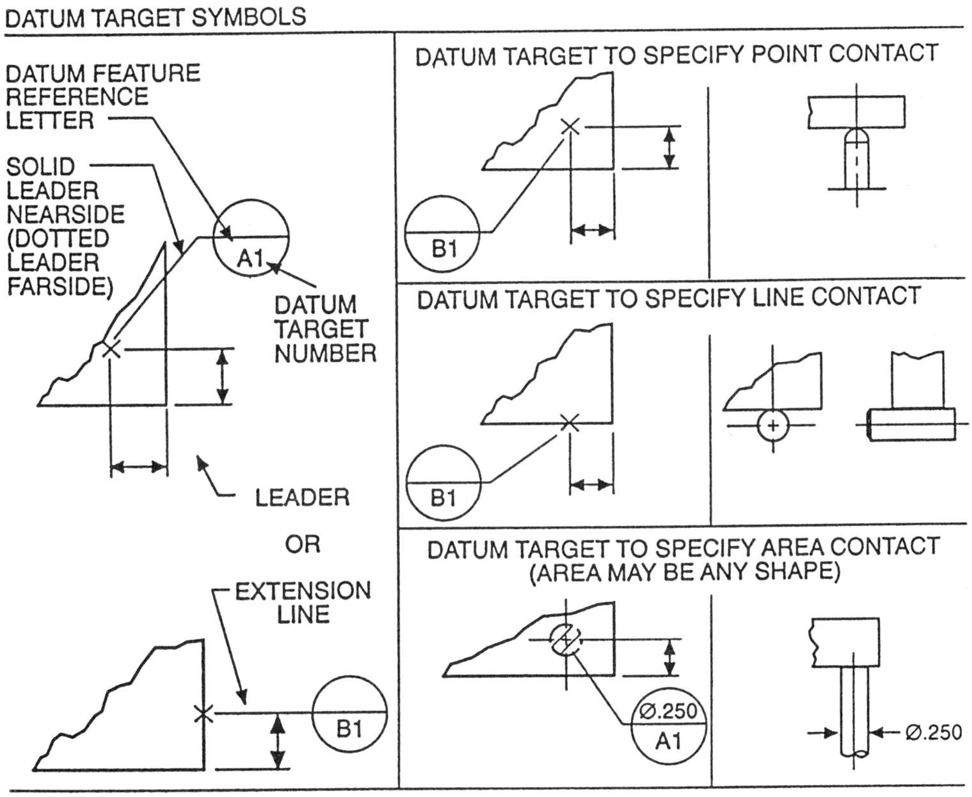

Figure 45 Datum target nomenclature.

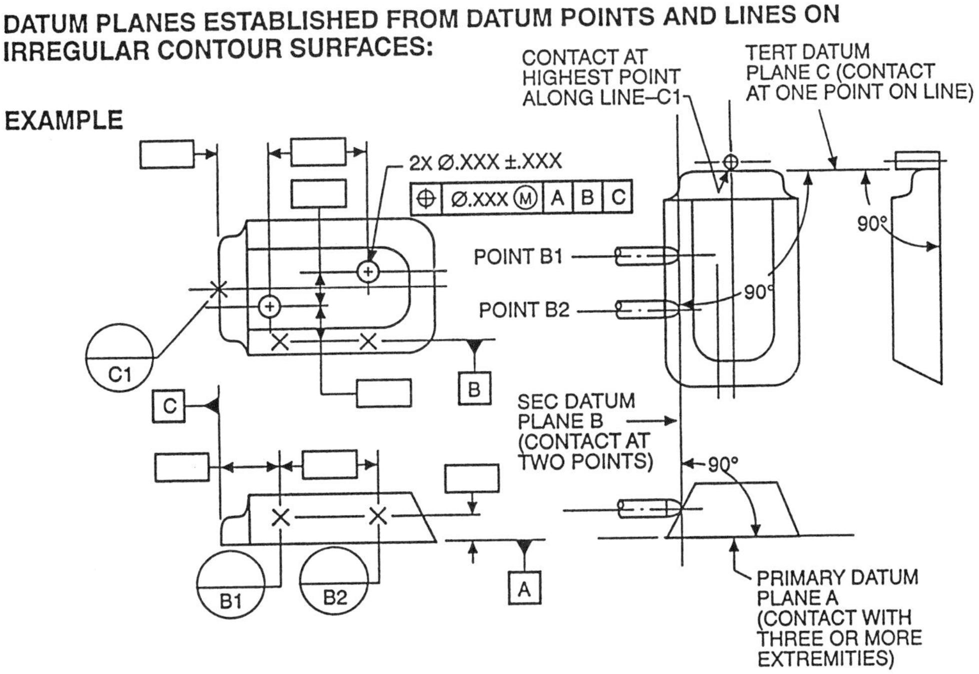

Figure 46 Datum target application.

ACKNOWLEDGMENTS

Acknowledgment is given to Addison-Wesley Publishing Company, Inc., Reading, Massachusetts; the American Society of Mechanical Engineers ASME); and the American National Standards Institute (ANSI), New York, for permission to extract certain material from the following texts and standards: L. W. Foster, *Geometrics II: The Application of Geometric Tolerancing Techniques* (Using the Customary Inch System), Revised 1986 Edition; L. W. Foster, *A Pocket Guide to Geometrics II: Dimensioning and Tolerancing* (Reference ANSI Y14.5M-1992); ANSI 14.5 M-1982 and ANSI/ASME Y14.5M-1994); and L. W. Foster, *Geometrics III: The Application of Geometric Tolerancing Techniques* (Using the Customary Inch System), © 1994. Addison-Wesley Publishing Company. Reprinted by permission of Addison-Wesley Longman, Inc. No further copying of this material is allowed without the prior written permission of Addison-Wesley Longman Inc.

BIBLIOGRAPHY

American National Standards Institute, *Dimensioning and Tolerancing*, ANSI Y14.5M-1982, ANSI/ASME Y14.5M-1994.

Foster, L. W., *A Pocket Guide to Geometrics II: Dimensioning and Tolerancing*, Addison-Wesley, Reading, MA, 1974, and later editions.

Foster, L. W., *Geometrics II: The Application of Geometric Tolerancing Techniques*, Addison-Wesley, Reading, MA, 1979, and later editions.

Foster, L. W., *Geometrics III: The Application of Geometric Tolerancing Techniques*, Addison-Wesley, Reading, MA, 1994.

FURTHER READING

National Tooling and Machining Association, Fort Washington, MD (latest editions), *Modern Geometric Dimensioning and Tolerancing*, *Answerbook for Geometric Dimensioning and Tolerancing (Inch)*.

Lowell W. Foster Associates, Inc., Minneapolis, MN, *Workbook and Answerbook for Geometric Dimensioning and Tolerancing (Metric or Inch) for Geo-Metrics II and Geo-Metrics III.*

27
Qualification Testing

ROBERT W. VINCENT*
Philip Crosby Associates, Inc.
Winter Park, Florida

I. INTRODUCTION

Product qualification testing is a defect-prevention program and is perhaps the last chance you have to assure that the new or newly redesigned product or service to be delivered to the customer is exactly like the requirements. Whether the "requirements" are in the form of drawings, specifications, a contract, or a law, they must appropriately reflect the kind of performance that the product or service should provide and include product safety and environmental considerations as well as such things as packaging, handling, use, and even misuse. Product qualification covers a review and test of all the requirements over the complete product or service cycle from the market survey through the end of the expected life of the product.

Although the bulk of the discussion of this chapter refers to products, services to customers can also be qualified. Qualifying a service against its requirements includes review and testing of policies, procedures, forms, and systems. Providing a trial number of proposed new insurance policy forms to insurance agents to use in completing new policies for customers is an example of a service qualification procedure.

II. FACTORS OF PRODUCT QUALIFICATION

Product qualification is a procedure that includes the following guidelines:

1. At least one unit of each type of product or service is subjected to specified tests and inspection to assure, at least as regards the test sample, that the product or service conforms to every requirement defined in its marketing and engineering specifications.

---

* Retired.

2. A product line is not made available to be sold to customers until one or more units have satisfactorily passed the qualification tests and until all related documentation is available and has been reviewed and judged acceptable.
3. Products should be requalified on a yearly basis or whenever major process, material design, or other changes have taken place.

III. DEFINITIONS

The following definitions apply to the product qualification program:

Product qualification: The inspection and testing of a finished unit of a product (or service) to verify that at least that test unit meets all of the marketing and engineering requirements that have been specified for the product or service.[1]

Requirements: The mechanical, physical, chemical, electrical, and functional characteristics that the product must possess or provide to meet specifications; that is, those requirements with which the product must comply to be acceptable.

Marketing specifications: The written requirements developed and approved by the manufacturing company's marketing department that reflect the known or anticipated wants and needs of the customers. Marketing specifications usually include pricing considerations.

Engineering specifications: The written requirements produced and developed by the manufacturing company's engineering department. The engineering specifications, in conjunction with the product's marketing specifications, reflect all the known and anticipated needs and wants of customers plus any requirements mandated by law. The combined specifications must include the requirements that the product must meet to comply with laws, regulations, societal and safety standards, life expectancy, and the like.[2] Engineering specifications generally include manufacturing cost considerations.

Qualification test specification: The written documents, produced by the engineering department and approved by the quality assurance department, that specify the required test and inspection procedures that must be used to evaluate the qualification samples(s) and the required performance that those samples must provide.

---

[1] Although even the simplest of products or services are candidates for formal qualification evaluations, the material in this chapter is primarily directed to the qualification of relatively complex consumer products such as snowthrowers and videocassette recorders. Products such as commercial aircraft and military equipment are mandatorily qualification tested.

[2] Life testing to verify the reliability and life expectancy of a product is an integral part of the product qualification program. Accelerated life testing is a specialized form of life testing and is performed early in the product design cycle to give preliminary assurance that the product will meet its reliability goals.

Qualification sample: A component, spare part, subassembly, or system built using the manufacturing processes that will be used to make regular production.

Process certification: The procedure, and the resulting documented report, that verifies that a manufacturing process has been evaluated and determined to be capable of meeting the requirements that have been specified for it. In some companies, process certification, or an analogous program, is required before manufacturing (operations) will accept responsibility for a newly developed manufacturing process from design engineering.

Test equipment certification: The procedure, and the resulting documented report, that verifies that a piece of test equipment or a measurement or test system has been evaluated and determined to be capable of the performance (e.g., accuracy, precision) that have been specified for it. Gage calibration and control and gage reproducibility and repeatability programs are associated programs.

Qualification test report: The summary report that presents to senior management the qualification test procedures and results that were obtained from the qualification testing.

IV. APPLICATION

Qualification (and requalification) testing evaluates both the design of the product and the processes used to make the product. All functional requirements within all environments specified in the engineering specification must be tested, examined, or inspected. This activity should be part of the original product plan.

The qualification procedure must include a formal inspection and acceptance of the samples against the engineering drawings prior to qualification testing. The production samples must be made by the intended manufacturing processes. Under certain conditions, samples can be manufactured using "soft" tooling if a requalification will occur using samples made with final hard tooling.

Because manufacturing processes change over time, it is necessary that product requalification be performed, typically, on a yearly basis or whenever any major change in design, materials, equipment, or other process condition occurs.

Initial qualification and any requalification test costs must be included in the budget and schedule of the development program. Gantt, PERT, or other program charts should be established to include costs and time as identifiable milestones.

V. POLICIES

To derive full benefit from a product qualification program and to avoid misunderstandings in its applications, there must be full agreement and commitment on the part of all levels of management regarding the company's quality policy. The quality policy must be in writing and understood and accepted by all managers.

Once the *quality* policy has been accepted by management, a *product qualification* policy should be issued, making it clear to all employees that management is serious about meeting the requirements. A sample qualification policy document is shown in Figure 1.

The policy states the company's position with respect to product qualification and specifically assigns the responsibility for applying the policy to the engineering and quality departments. This assignment is important for two reasons:

1. Unless both the required time and money for the qualification tests are budgeted, they will not occur.
2. In this example, the policy assigns the responsibility for budgeting the costs of the testing and performing the tests to the engineering department. The quality department is assigned the responsibility of monitoring or auditing the total operation and reporting the results to management. Regardless of which department actually performs which function in an actual setting, one group should perform the tests and the other should review the performance. Such an approach will ensure that the assigned responsibilities are performed in accordance with the written procedures.

In addition, a formal product qualification procedure should be created detailing the specific tasks that will be completed in any product qualification. A typical formal procedure is shown in Table 1.

VI. QUALIFICATION SCHEDULE

Once the qualification program has been accepted, the director of engineering must develop a qualification schedule as quickly as possible. The schedule must provide the planned dates for each key event in the test program. The schedule must be approved by the quality manager and should be reviewed at each design review meeting. Figure 2 is an example of such a schedule.

VII. REPORTING QUALIFICATION RESULTS

A. Qualification Test Log

Every event, nonstandard occurrence, and failure that occurs during a qualification test must be recorded in a qualification test log. Overall control of the test log is the responsibility of the quality program manager.

B. Interim and Final Reports

Qualification test reports are drafted to provide the general manager and immediate technical staff a clear understanding of the results of the qualification testing.

The reports must give the managers confidence that the tests have been carried out correctly and have been reported objectively. The reports must contain sufficient detail to carry out these purposes but must be, at the same time, as concise as possible. The report writer must not assume that the reader has detailed knowledge of the product, nor should understanding of the report depend on knowledge or reference to other documents such as the test specification. That is, the report must be self-contained. Special emphasis must be placed on the tests that have been

POLICY GUIDE	POLICY	Number
		Effective
	PRODUCT QUALIFICATION	Cancels
Affects SYSTEM	Signature	Dated
	DIRECTOR-QUALITY	Page 1 of 1

1.0 INTENT

1.1 The intent of this policy is to establish the requirements for the formal qualification of all products manufactured or purchased for sale by the company.

2.0 APPLICATION

2.1 Each new product together with its associated documentation such as operating manuals, repair manuals, and the like, shall be subjected to specified tests and inspections at specific stages in the product development cycle to assure that the product and its documentation conform to each requirement defined in the product's marketing and engineering specifications.

2.2 Once the product has been qualified, the need to requalify the product by repeating some or all of the tests and inspections on a yearly basis or as major changes are implemented.

2.3 New or substantially changed products shall not be made available for sale until qualification has been successfully completed.

3.0 RESPONSIBILITY

3.1 The qualification plan, performance, and evaluation is the joint responsibility of the engineering and quality assurance departments.

3.2 The engineering department is specifically responsible for generating product conformance specifications and for scheduling, budgeting, and performing the qualification tests.

3.3 The quality assurance department is specifically responsible for assuring that the tests and inspections are performed according to specification by auditing these inspections and tests, as well as reporting the results to management.

Figure 1 Qualification policy.

Table 1 Typical Qualification Procedure

STANDARD PROCEDURE No.
Subject: Product Qualification Procedure
Revision: Orig.; Dated 4-25

1.0 SCOPE

This standard procedure applies to product qualification projects.

2.0 RESPONSIBILITIES

The engineering and quality departments have the prime responsibility for product qualification. Personnel in both of these departments shall be assigned specific responsibilities for each qualification project.

2.1 The following assignments and actions will assure effective management of the program:

2.1.1 The director of engineering is responsible for establishing the qualification schedules.

2.1.2 The general manager and the product manager are responsible for approving the qualification schedules.

2.1.3 The director of engineering is responsible for the following actions:

2.1.3.1 Appointing an engineering project manager for each product to be qualified or requalified.

2.1.3.2 Including in the product development budget the costs required for qualification testing and, if necessary, requalification testing.

2.1.3.3 Charging test costs to the appropriate cost centers.

2.1.3.4 Making appropriate provision in the development program for test time and schedules.

2.1.3.5 Assigning qualification testing milestones such start of tests, completion of reports, etc., as key segments of the product development program.

2.1.3.6 Formally scheduling all qualification and requalification tests for the upcoming year.

2.1.4 The engineering project manager is responsible for the following activities:

2.1.4.1 Defining the exact configuration of the product to be tested.

2.1.4.2 Approving the qualification test schedule.

2.1.4.3 Ensuring that the product(s) to be tested is (are) made using final manufacturing methods, processes, and materials.

2.1.4.4 Ensuring that the required number of test samples are made and delivered according to the schedule.

2.1.4.5 Ensuring that sufficient personnel with sufficient expertise are available at the scheduled time.

2.1.4.6 Ensuring that necessary functional test equipment is calibrated and available in accordance with the test schedule.

2.1.4.7 Ensuring that the engineering test specification that defines the requirements that the product must meet has been written and approved by engineering, marketing, quality, and other functions and has been published according to schedule.

2.1.4.8 Ensuring that the qualification specification that defines the tests and inspections that are to be performed has been written and approved by engineering and quality and published according to schedule.

Table 1 Continued

STANDARD PROCEDURE No.
Subject: Product Qualification Procedure
Revision: Orig.; Dated 4-25

- 2.1.4.9 Ensuring that separate qualification documents have been prepared for spare parts, subassemblies, and packaging if required.
- 2.1.5 The director of quality assurance is responsible for the following activities:
  - 2.1.5.1 Designating a quality assurance project manager for each product that is to be qualified or requalified.
  - 2.1.5.2 Supplying personnel to assist or perform tests and inspections.
  - 2.1.5.3 Approve or disapprove qualification test reports.
- 2.1.6 The director of engineering is responsible for the correction of any product design deficiencies including those revealed during qualification testing.
- 2.1.7 The quality assurance project manager shall:
  - 2.1.7.1 Ensure that all aspects of the qualification (requalification) tests are performed correctly and according to schedule.
  - 2.1.7.2 Advise the general manager in the event that problems or delinquencies arise in the testing.
  - 2.1.7.3 Advise the controller of accrued costs of the qualification testing.
  - 2.1.7.4 Audit the measurable product qualification activities and report any unresolved deficiencies and omissions to the general manager.
  - 2.1.7.5 Provide the engineering project manager with assistance in drafting the qualification test specification, assuring that it is compatible with the engineering specification.
  - 2.1.7.6 Approve or disapprove the qualification test specification.
  - 2.1.7.7 Ensure that the qualification test log is correctly filled out and retained for future reference.
  - 2.1.7.8 Prepare and submit interim and final qualification test reports to the general manager.
  - 2.1.7.9 Issue a yearly summary of all product qualifications and requalifications.
  - 2.1.8.0 Audit corrective action and retesting that is necessary because of product failures and discrepancies during initial testing.

completed and whether the product passed or failed. Table 2 shows the items that should be included in a qualification test report. Interim and final qualification and requalification reports and all other referenced material should be appropriately filed as historical records for the life of the product.

VIII. MANAGEMENT TRAINING

So that the company can derive full benefit from the qualification test program, all management and supervisory personnel must understand the reasons for the program

Event	Planned start	Planned complete	Actual start	Actual complete	Person responsible
1. First draft engineering specification written, approved, and issued					
2. First draft qualification test specification written, approved, and issued					
3. Qualification test samples made, selected, and delivered					
4. Configuration control inspection					
5. Visual and mechanical inspection					
6. Product safety review and risk analysis (FMECA)					
7. Review of environmental considerations					
8. Functional test (standard conditions)					

Figure 2 Qualification schedule.

Event	Planned start	Planned complete	Actual start	Actual complete	Person responsible
9. Functional test (marginal conditions)					
10. Dry heat test					
11. XX-hr accelerated-life test					
12. Vibration					
13. Shock					
14. Altitude					
15. Salt spray					
16. Humidity					
17. Other environmental conditions					

Figure 2 Continued

and their role in it. Although the quality and engineering departments have prime responsibility for the qualification testing, all departments can contribute. Personnel from other departments can help in writing the test procedure, anticipating problems, and assuring that only qualified components are used to build the test units.

To help inform other managers of the concepts of product qualification, a training program can be established. Questions and answers such as those shown below can serve as the basis of such a training class.

A. Instructor's Guide

Q: Where do I get the money for product qualification testing?

A: Qualification testing should be a basic part of the product development process. The money required to perform the tests must be included when the product development program is started.

Q: We've instituted a formal qualification policy but I didn't budget any money. What do I do now?

A: This may have been a good argument before the product qualification policy was established. Now, the essential requirements of the qualification specification must be examined and at least the most critical portions must be done. If the money cannot be found in the product development program, perhaps it can be found in a

Table 2 Items to Be Included in a Qualification Test Report

1.	Title, identification of item being qualification tested, date of report, identification as interim or final report
2.	Approval signatures (unit's engineering program manager, quality director, and the director of engineering)
3.	Distribution (unit and group management)
4.	Summary (this must clearly state that the product is qualified or that it is not qualified and state the reason)
5.	Contents of the report
6.	Qualification test personnel (listing of personnel conducting the product qualification)
7.	Test location and date
8.	Specification (engineering, qualification, and others)
9.	Identification of samples tested and the configuration tested
10.	Assembly inspection and test status report of qualification samples (prior to qualification)
11.	List of certified test equipment used (range and accuracy and calibration date)
12.	Report on results of process certification at the time of qualification.
13.	Report on qualification of vendor parts, subassemblies, and components check
14.	Report on results of product safety and risk analysis (FMECA)
15.	Test log
16.	Corrective action (planned and accomplished)
17.	Repeat of failed tests (history)
18.	Summary of test results and conclusion

discretionary fund, or perhaps it can be expensed. It is crucial that every product development effort from here on in be funded for the qualification test work.

Q: I haven't got the people to do a product qualification program.

A: Again, this was a good excuse before product qualification became company policy. Product qualification is part of the development effort. It must have its share of scarce resources. It is not an off-again, on-again program.

Q: We don't have the necessary environmental test equipment to do product qualification testing.

A: Equipment can be leased or rented. There are also testing services that will provide certified tests for nominal fees.

Q: The product we would have to qualify is large and complex. Qualification testing of this equipment isn't practical.

A: Certainly, qualification testing of a large, complex system is different from that of a component. But the appropriate attitude is to determine how much can be done rather than saying that it can't be done. Some of the biggest problems with products in the field—and the biggest sources of customer dissatisfaction—could have been prevented by adequate qualification testing.

Methods can be developed for qualification testing of major subassemblies when the entire system is too large for environmental or other test chambers or equipment.

Q: Our newly drafted qualification procedure states that new processes must also be qualified. What does this mean?

A: The main emphasis of a qualification program is on "products (or services) for sale." If a new process is required to produce a new product, an appropriate qualification program will simultaneously qualify both the product and the process. If an existing process is used to produce a new product and later it is determined that a new process must be created for that same product, the cost of the qualification must be borne by the manufacturing or operations division as part of a cost-reduction or product-improvement program.

Q: We have too many new products to start qualification testing on all of them at the same time. How do we determine which ones to start first?

A: Determine which program(s) have the greatest real or potential impact and do those first. However, it may not be true that only a limited amount of testing and evaluation can be done if innovative effort is applied. Much more testing and evaluation might be possible than is at first apparent.

B. "Student Quiz"

Questions similar to those following could be used to determine the degree of understanding of the "students." The questions would usually be answered "true" or "false." Following each answer is discussion that expands on the material.

1. Qualification testing is carried out to reveal weaknesses in product design so that these can be eliminated before delivery to the customer. (Answer: False; and True)

The primary purpose of qualification evaluation is to prove that the product conforms to all of the requirements.

Product qualification must be considered to be a quality assurance procedure that is performed on products that the company is confident will meet all of the specified requirements. Nevertheless, it is true that qualification testing will usually reveal hidden weaknesses in the product. So, both answers are correct.

2. Qualification testing is solely the responsibility of the engineering department, because they are the only ones who understand the product—and they pay for the testing. (Answer: False)

Qualification evaluation must be set up as a joint function of the engineering and quality groups. There is no other practical way of getting the job done. The quality department needs the technical expertise of the engineering department, and the engineering department needs the quality department's ability to make sure that the programs happen, ensure that the programs are done in a disciplined way and properly reported on.

3. An important aspect of qualification testing is that it forces complete specification of the product. (Answer: True)

A company will often have subsystem specifications, partially complete system specifications, or obsolete specifications but not a well-documented, up-to-date system specification. To be able to complete a qualification test procedure requires that there be an engineering specification, maintenance manuals, customer literature, and other associated documentation.

During the review of these documents and the writing of the test procedure, the validity of the engineering specification is often questioned and changed.

4. It's OK to start shipping a new product line to customers even though the qualification testing isn't complete in order to meet promised delivery dates, provided that the quality manager approves. (Answer: False)

Many companies have shipped to customers product that did not pass final inspection or that did not complete—or perhaps even start—qualification testing. Even though it has been done, it is not permissible. It is an indication that management does not have the customer's interests in mind and is willing to risk the company's reputation.

Putting the quality manager in the position of responsibility for shipping questionable material is likewise an unacceptable situation. If there is not enough market to specify, develop, and qualify the product properly, management has probably made a mistake in developing it and they are making another mistake in forcing it to the market before everything is in place.

5. Qualification test reports are not written to the general manager because they contain too much technical material. (Answer: False)

The general manager wants to know if the product met all of the requirements. He or she has to be able to make a decision as to whether the product can be released. The qualification report should be written in such a way that the general manager can quickly understand it. Generally, this means not including reams of test data and mind-numbing technical jargon in the report. Raw data generally should not be included in the report, but a thorough analyses, summary, and recommendation should be. The data and the technical discussion can be included in an addendum.

6. Qualification testing must be conducted on products made using production tools and processes. (Answer: True)

There is risk involved in using handcrafted, model shop, or breadboarded products for qualification. Qualification testing is intended both to evaluate the product and to verify the production processes that will be used to make the product. Handcrafted or model shop made units don't test either the product or the process. Engineering testing on prototypes during development is technically called pretesting or prequalification. Formal product qualification can only be started when development is complete and all final specifications and drawings have been issued. Qualification test samples can be from pilot production but the process must be the final one.

7. A product that has passed all of its qualification tests should never give problems in the field. (Answer: False)

Especially in complex equipment, it is impossible to say that there will never be field failures even when qualification testing has been a success. Component life expectancies, for example, are usually given as averages and not absolute numbers. Thus some will fail earlier than the average and some later. Qualification tests are very valuable but, like any statistical evaluation or test program, should be used as one of a number of measures of product performance.

8. One product is enough for qualification testing. (Answer: It depends)

The statistical confidence of the number of unit products tested can be calculated if necessary. On very large or very expensive products, it may be logistically or economically impossible to test more than one unit, especially if the testing is considered destructive. In such cases, much sophisticated statistical analysis is required before test plans are finalized to be sure that greatest amount of information is obtained from the test. On smaller, less-expensive products, testing can intentionally be taken to product destruction or wearout to provide accurate measures of safety margins. It is often possible to break large systems down into smaller subassemblies, which can make testing easier and less costly.

FURTHER READING

Crosby, P., *Quality Is Free.* McGraw-Hill, New York, 1978.

Grenier, R. W., *Customer Satisfaction Through Total Quality Assurance,* Hitchcock, Wheaton IL, 1985.

28
Statistics Without Math

EUGENE W. ELLIS*
Consultant
Sanford, North Carolina

I. INTRODUCTION

To exist in today's competitive environment, industry must upgrade quality and decrease costs. Unfortunately, most companies—especially those in the United States—have the mistaken belief that improved quality must always be associated with increased costs. The basis of such a belief is the assumption that quality can only be improved by performing additional inspection after production to sort out all scrap and items that require rework. Without a doubt, such inspection, plus the rework and scrap involved, does create additional nonproductive costs in order to upgrade quality.

Many companies, however—especially those in Japan—have proven that it is possible to improve quality while *decreasing* costs. Such a program is achieved by discarding the old additional-inspection assumption and replacing it with a statistical process control system that "produces all things right the first time." Under such a "make it right" system, rework, scrap and additional inspection plus their costs are eliminated. The utopia of improving quality while decreasing costs can thus be achieved.

Unfortunately, however, very few successful process control programs exist in the United States. It is felt that this lack of success is due directly to the training that has been performed for most programs. In most cases, the training has been performed by statistical experts and the emphasis has been on the use of complex-sounding statistical/mathematical terms. Such terms simply "turn off" management and production workers, and thus the attempted process control system quickly fails. Experience shows that when the proposed program is presented using statistical

---

* Retired.

logic but not statistical terms, instead of being turned off, production workers are actually enthusiastic, because they can see that statistical process control is a most logical mode of operation.

Statistical application in industry actually involves three separate functions. These functions are statistical sampling, data analysis, and process control. Trained statisticians are needed for the data analysis and statistical sampling functions, but only statistical logic, without the need of statistical terms, is needed for process control. Separating process control logic from the data analysis and statistical sampling functions eliminates the need for the use of statistical terms. In addition, by using nomographs for most necessary mathematical calculations, a successful statistical process control program can exist without the need of complex-sounding statistical terms. Also, when a successful process control system does exist, the requirements for most data analysis and statistical sampling functions are eliminated.

Basic statistical logic indicates that in every process there are an infinite number of factors that can affect—to various degrees—the final product. Usually, only a few of these factors have any significant effect on the final product, and controls are established for these factors (e.g., operator's adjustments, speed of process). However, a great many other factors which are felt to have only an insignificant effect on the final product also exist and these are usually not controlled and often not even identified (e.g., meshing of gears of a machine in the process, variations in item placement within a fixture). Because these factors are not controlled, they have a random effect (no matter how small) from item to item in the process (Fig. 1).

II. PROCESS CENTER AND SPREAD

Although the noncontrolled factors individually have very little effect on the final product, there are so many of them that collectively they do prevent all items in the process from being produced identically. However, because the noncontrolled factors are operating independently and in a random manner (*plus* effect on some pieces, *minus* effect on others), they have a tendency to "average out," and thus most items are produced at or near the target point as established by the significant factors being controlled. But not all noncontrolled effects will average out, so some items will vary to different degrees from the target point.

This central tendency and spread caused by the noncontrolled factors can be visibly identified when a series of items produced by the process are measured and plotted as a histogram. A histogram plot represents the "population" of items being produced by the process, as shown in Figure 2. The target point and spread of the population will remain consistent provided that the controls on the significant factors are constant and no new noncontrolled factors are added to the process.

The recognition that each process produces a *population* of items rather than just a series of individual items is the basic statistical logic needed for a successful process control program. When the concept of a population is understood, it is logical that if the limits of the spread of the population are controlled to within requirements, the process will produce only conforming items.

Examination of the ideal population (Fig. 3) indicates that a successful process control system exists, because the target point of the population is held at or near nominal and the spread of the population is less than the width of the tolerance

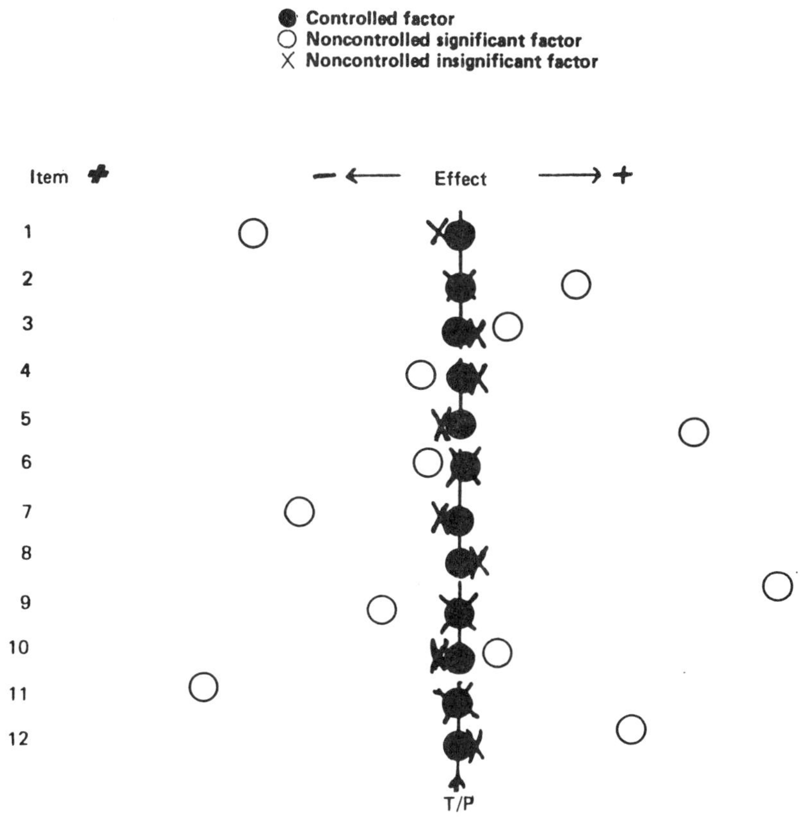

Figure 1 Factor effect on final product.

requirements. All other situations produce nonconforming items and costly inspection, rework, and scrap.

As the population's target point is the result of the controls established on selected significant factors, and shift in the target can only be due to changes in the established control points (Fig. 4). As the population's spread is the result of the random variations of the noncontrolled factors in the process, any spread problem can be caused only by significant changes in the process that have not been identified and are not under control.

It should be noted that often a noncontrolled spread factor can exist at a designated control point because the designated control point cannot be exactly established. A common example of this situation is an operation using a visual gage setting and adjustment at an approximate control point rather than at the exact control point established.

III. SPREAD AND TARGET WORKSHEET

For a successful process control system, the first and most important step is the identification of the population's target point and spread limits. The spread and

PRODUCTION MEASUREMENT RECORD
PART NO. 12345
DIMENSION 1.530-1.540
OPERATION NO.
GAGE DIVISIONS
TALLY
1 2 3 4 5 6 7 8 9 10 11 12 13 14 15 16 17 18 19 20 21 22 23
TOTAL
1.543
1.542
1.541
1.540
1.539
1.538
1.537
1.536
1.535
1.534
1.533
1.532
1.531
1.530
1
2
7
13
15
8
3
1
MAX.
MIN.

Figure 2 Histogram plot.

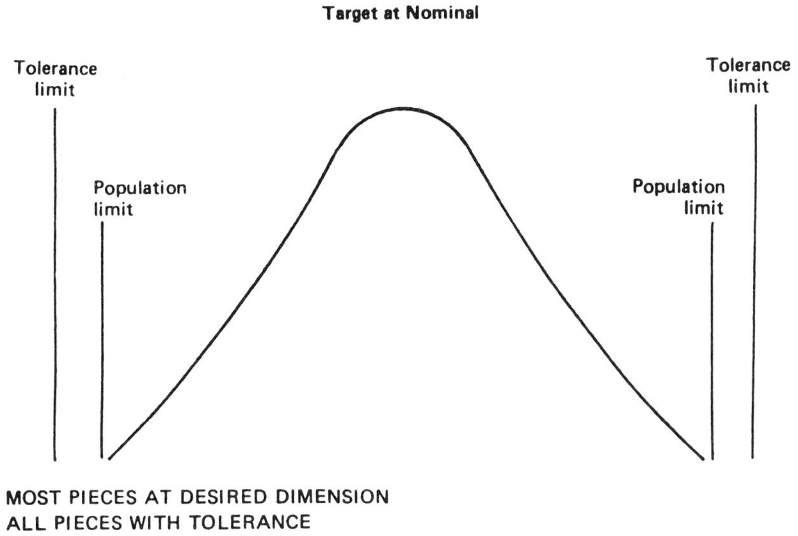

Figure 3 Ideal target at nominal.

target (S/T) worksheet is an excellent tool for such a determination (Fig. 5). Construction of the S/T worksheet is simple. Select a minimum of 25 representative items from the process and measure the same key feature on each part. The S/T worksheet is used as follows:

1. Select the cell boundaries/cell midpoint so that the spread of plotted data will cover 6–16 cells (keep the cell width consistent).
2. Record the data as a count of one each in the appropriate tally block.
3. Count the number of tallies in each block and record the total in the appropriate frequency block. Record the total of all tallies in sum block. Record twice the sum in the "2 × sum" block.
4. For each cell, add the accumulated frequency and the frequency of the cell above to the frequency of the appropriate cell and record the total as the accumulated frequency of the appropriate cell. (On the first tally cell at the top, the accumulated frequency will be equal to frequency, as the accumulated frequency and frequency of the cell above are both zero.) If no addition error is made, the last accumulated frequency figure will equal 2 × sum.
5. For each cell, divide the accumulated frequency by 2 × sum and multiply the result by 100. Record the answer in the "% over" column. (Note that 25-, 50-, or 100-piece samples make this calculation very simple.)
6. On the grid, match each cell's horizontal line with the "% over" vertical line that corresponds to the "% over" figure. Plot a circle on the grid for all "% over" amounts less than 10 or greater than 90. Plot X's for "% over" amounts 10 to 90.
7. Fit a straight line through the X points on the grid and draw a line across the grid. If the X points do not create a reasonable straight line, the data are not applicable for this analysis.

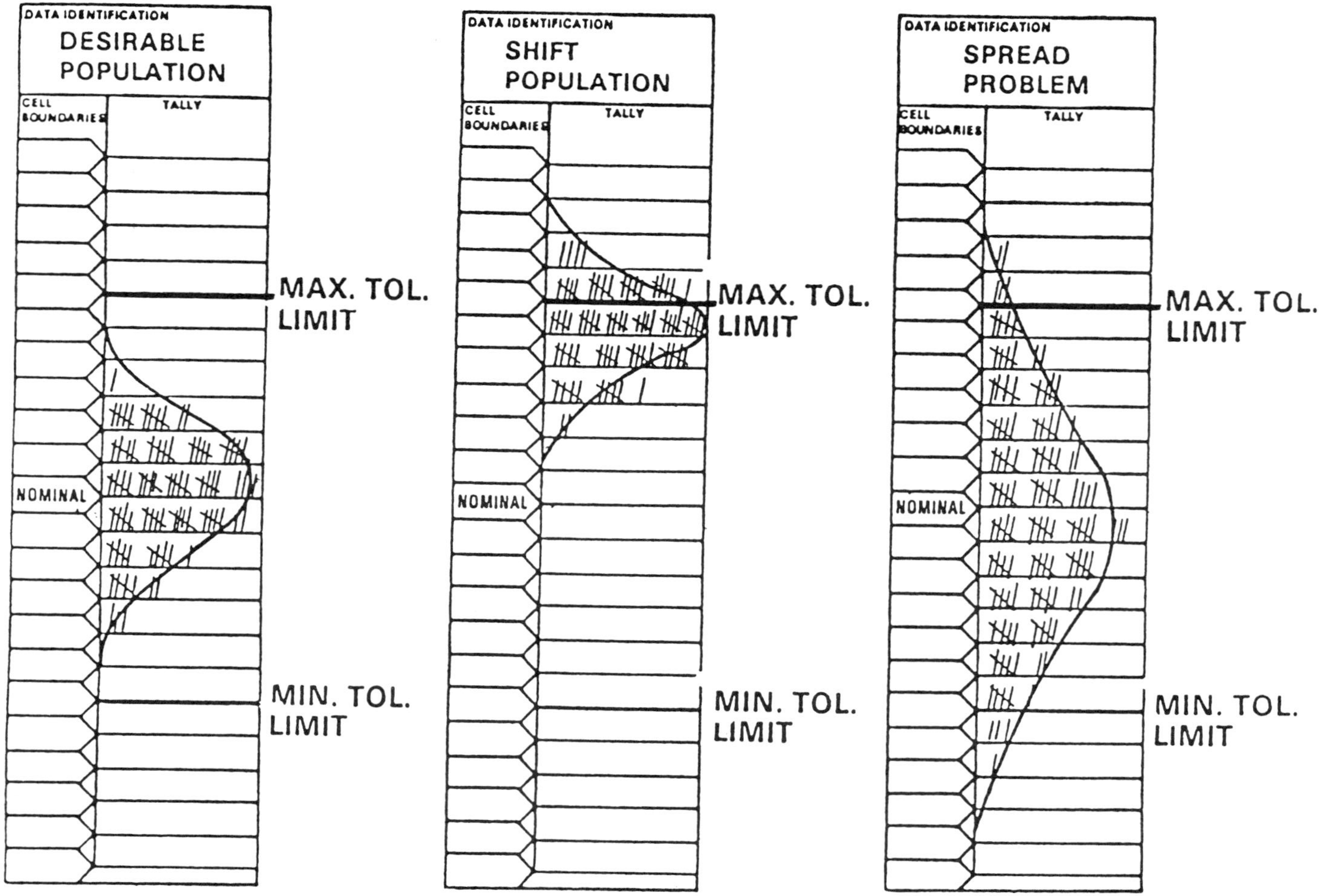

Figure 4 Shift and spread problems produce nonconforming items.

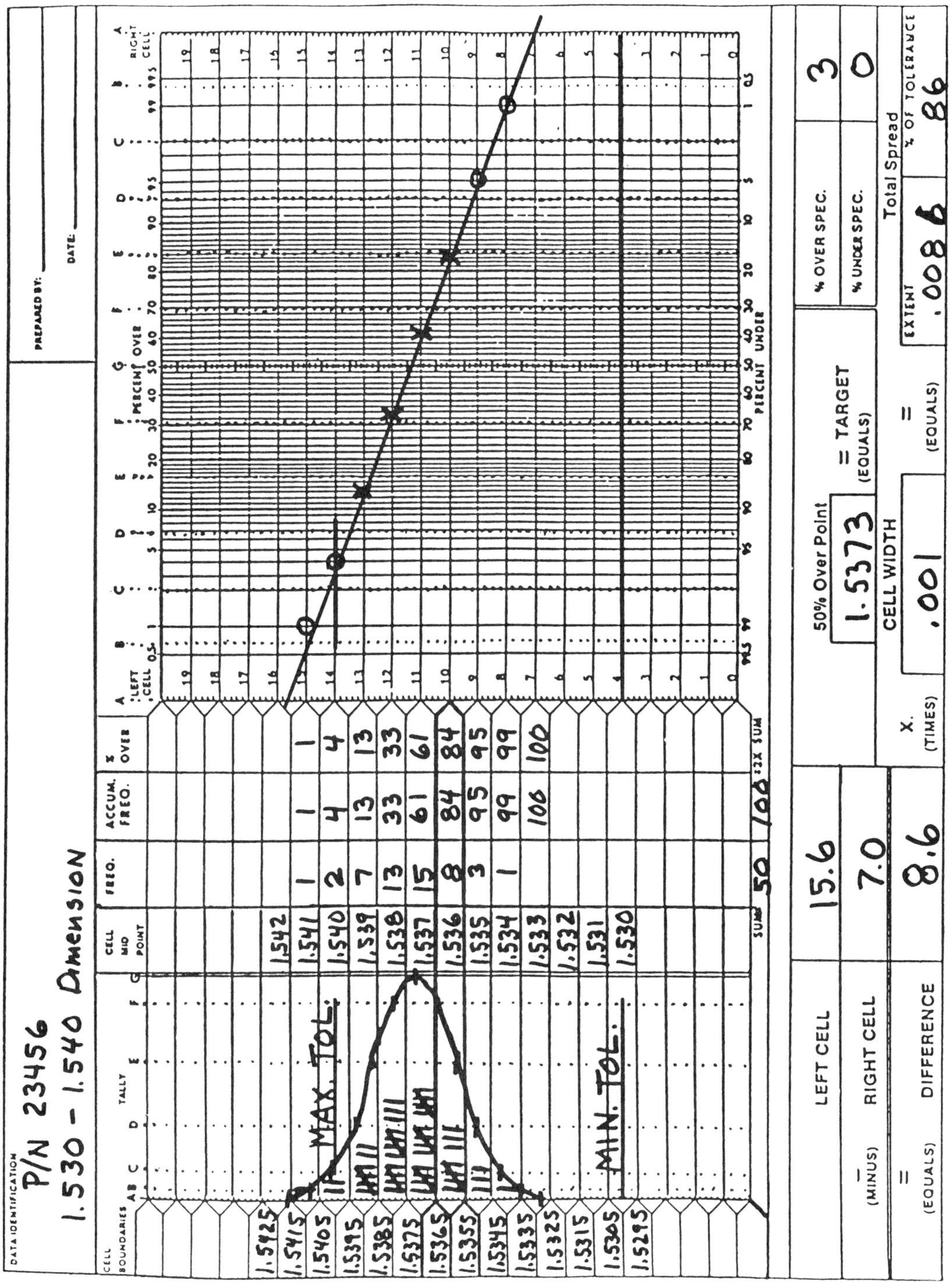

Figure 5 Spread and target worksheet.

8. The target of the population is the horizontal cell boundaries/cell midpoint point, which corresponds to the intersection of the drawn diagonal straight line and the "50% over" vertical line.
9. Determine the points where the diagonal straight line crosses the left cell and right cell vertical lines on the edge of the grid and record the points in the left cell and right cell blocks. Subtract the right cell point from the left cell point and record as "difference."
10. The total spread extent is the difference times the established cell width. The percent of tolerance is the extent divided by the total tolerance.
11. Determine the percent over or the percent under for any selected point by locating the point on the cell boundaries/cell midpoint scale and project horizontally to the fitted straight line and then vertically upward to the "% over" scale or vertically downward to the "% under" scale.
12. For a visual picture of the population, project horizontally from the intersection on the grid of the straight line and the leftmost vertical A line to the A line in the tally area and place a slash mark on the tally area line. Repeat for vertical lines B, C, D, E, F, and G, matching them with their corresponding tally lines. Draw a line connecting the slash marks for a picture of population. When issuing process reports such a picture is often very useful.

A fitted straight diagonal line across the grid indicates that the process is operating in a normal manner. If the X points on the grid form a smooth curve rather than a straight line, it is often an indication that one side of the population is being constrained. When such a situation exists, data transformation (the square of the square root of the basic data) will usually result in a straight line.

If tally points appear above or below where the fitted straight line passes the edges of the grid, it is an indication that such data are from a different population (different control points) than the population being studied. The existence of such points questions the consistency of the process.

Visual examination of the S/T worksheet often indicates unexpected variations in the process. For example, a normal process will always have approximately equally spread tails on both sides of the target point. If one of these tails is cut off, it is an indication that something not random is being done to the process, as tails will always exist in a normal process under control.

In a metal-working operation, a cutoff condition on the rework side of tolerance often indicates that the operator is taking time to inspect the work and is performing rework on those items that are outside tolerance. Such an operation creates nonproductive costs.

If the S/T worksheet indicates that the target point of the process is near nominal and the spread of the population is 80% or less than requirements, an "ideal" population exists. An ideal population will consistently produce all items to requirements provided that the target point is held near nominal and excessive spread does not suddenly happen.

It should be noted that personnel not completely familiar with statistical process control logic often fear that even when an ideal population exists, a "stray" item far outside the population limits might be produced. As such an item must be from a population with a different target point, and the designated controls establish the population's target point, such an item can be produced only when there has been a significant variation in an established control point. Such a variation is usually easily identified by the operator, and items produced under such a variation of control point should be removed from regular production.

IV. TARGET CONTROL WORKSHEET

However, even when an ideal population exists, it is necessary periodically to check the process to assure that neither the target point nor the spread of the process has changed significantly. The target control (T/C) sheet, shown in Figure 6, is an ideal tool for this purpose. The instructions for use of the T/C sheet are as follows:

1. In the blocks on the left-hand side of the chart, record, as indicated, maximum tolerance point, half maximum tolerance point, nominal tolerance point, half minimum tolerance point, and minimum tolerance point. The "green" band area of the sheet is between the half tolerance points and the "red" area is outside the maximum and minimum tolerance points.[1]
2. After setting up the process, measure and plot with circles the successive items produced.
3. Adjust the target point controls if two items within five are in the yellow band on the same side of the green band or if one item is in the red band.
4. Advance to the machine operation mode when five pieces in a row are in the green band.
5. Machine operation mode:
   a. Select a measurement frequency so that at least 20 items will be plotted prior to a necessary adjustment. (Be conservative at first and increase the frequency of checks as experience warrants.)
   b. Measure the item at the indicated frequency and if the measurement falls in the:
      (i) Green band—Do not adjust and continue frequency measurement.
      (ii) Yellow band—Measure the next piece. If this measurement is in:
         (a) Green band: do not adjust—continue frequency measurement.
         (b) Yellow band on the same side of nominal: Adjust.
         (c) Yellow band on the opposite side of nominal: Stop the process and correct the spread situation.
         (d) Red band: Adjust.
      (iii) Red band: Adjust.

The T/C sheet is not only an excellent method for process control but also furnishes documentary evidence that the process operated in control for the period indicated on the T/C worksheet. Such evidence is often very important in today's legal environment.

---

[1] Actual charts, which are available from a number of suppliers, are printed with red, green, and yellow zones. Some practitioners refer to them as stop and go charts or precontrol charts.

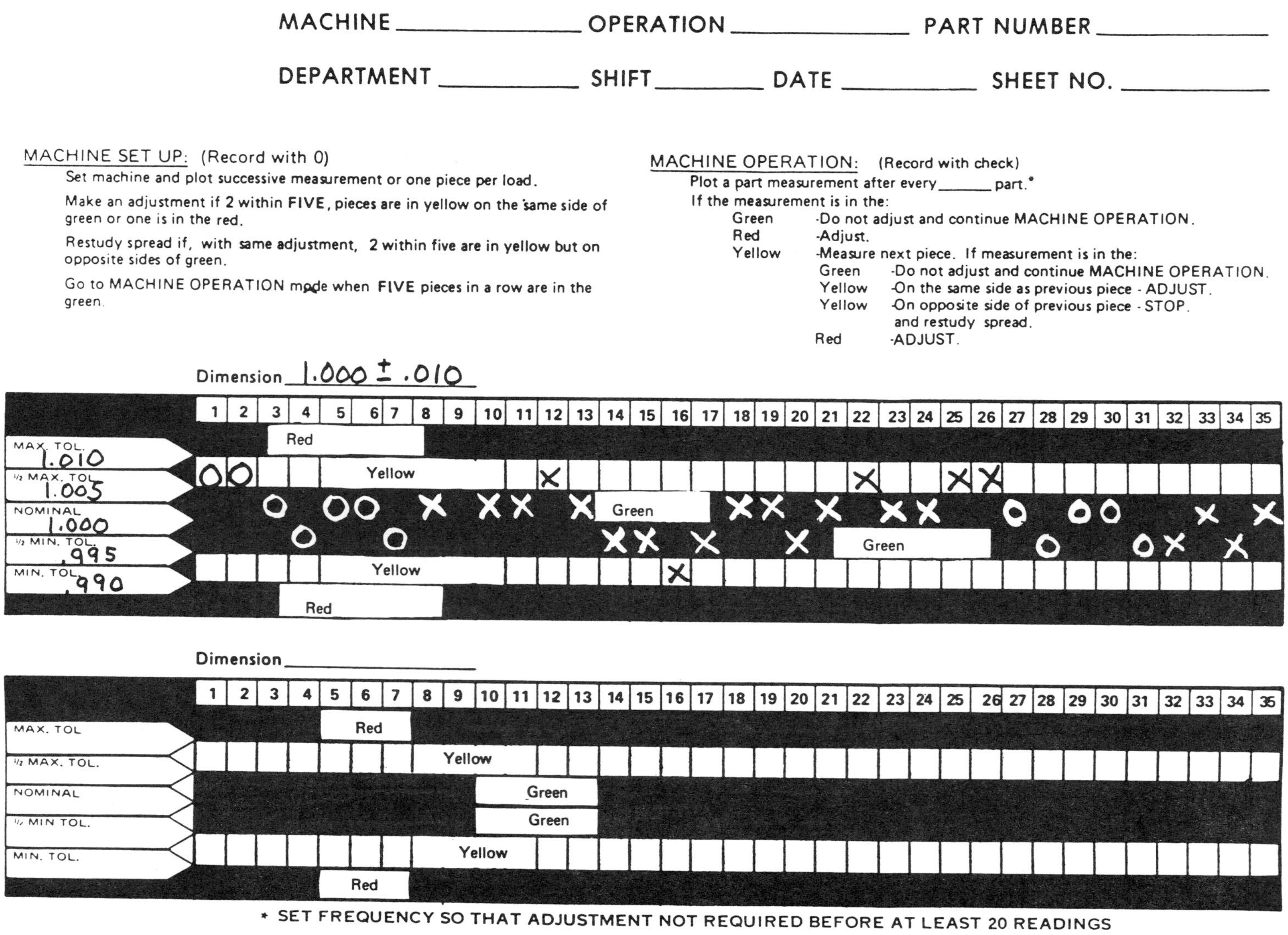
MACHINE ____________ OPERATION ____________ PART NUMBER ____________

DEPARTMENT __________ SHIFT ________ DATE __________ SHEET NO. __________

MACHINE SET UP: (Record with 0)

Set machine and plot successive measurement or one piece per load.

Make an adjustment if 2 within FIVE, pieces are in yellow on the same side of green or one is in the red.

Restudy spread if, with same adjustment, 2 within five are in yellow but on opposite sides of green.

Go to MACHINE OPERATION mode when FIVE pieces in a row are in the green.

MACHINE OPERATION: (Record with check)

Plot a part measurement after every ______ part.*

If the measurement is in the:

Green -Do not adjust and continue MACHINE OPERATION.

Red -Adjust.

Yellow -Measure next piece. If measurement is in the:

Green -Do not adjust and continue MACHINE OPERATION.

Yellow -On the same side as previous piece - ADJUST.

Yellow -On opposite side of previous piece - STOP. and restudy spread.

Red -ADJUST.

* SET FREQUENCY SO THAT ADJUSTMENT NOT REQUIRED BEFORE AT LEAST 20 READINGS

Figure 6 Target control sheet.

The target control sheet also eliminates many operator adjustments that are unnecessary and can cause nonconformance with the product. For example, if the operator does not recognize that the process produces a population with a spread that is not operator controlled, he or she can measure an item produced near tolerance and assume that a displaced target point exists. To correct the situation, the operator makes an adjustment to the process which can actually cause the population to shift and produce nonconforming items. Actually, the near-tolerance time could be part of the spread in a population correctly centered. A decision-making format to adjust or not to adjust is shown in Figure 7.

Unnecessary process adjustments can increase the spread of the process as well as causing nonconformances to exist. The target control sheet eliminates excessive adjustments and is also a control that the spread of the process has not increased significantly. This control is achieved by the requirement that the process has to be reexamined when two successive yellow band items, each on a different side of the green band area, are found.

When the S/T worksheet indicates that the spread of the population is greater than 80% of tolerance requirements, a spread problem exists. To correct a spread problem, additional factors not currently controlled must be identified and placed under control (Fig. 8). It should be noted that inexactness in setting current controls could be the factor needing better control (e.g., an operator setting a control approximately instead of exactly).

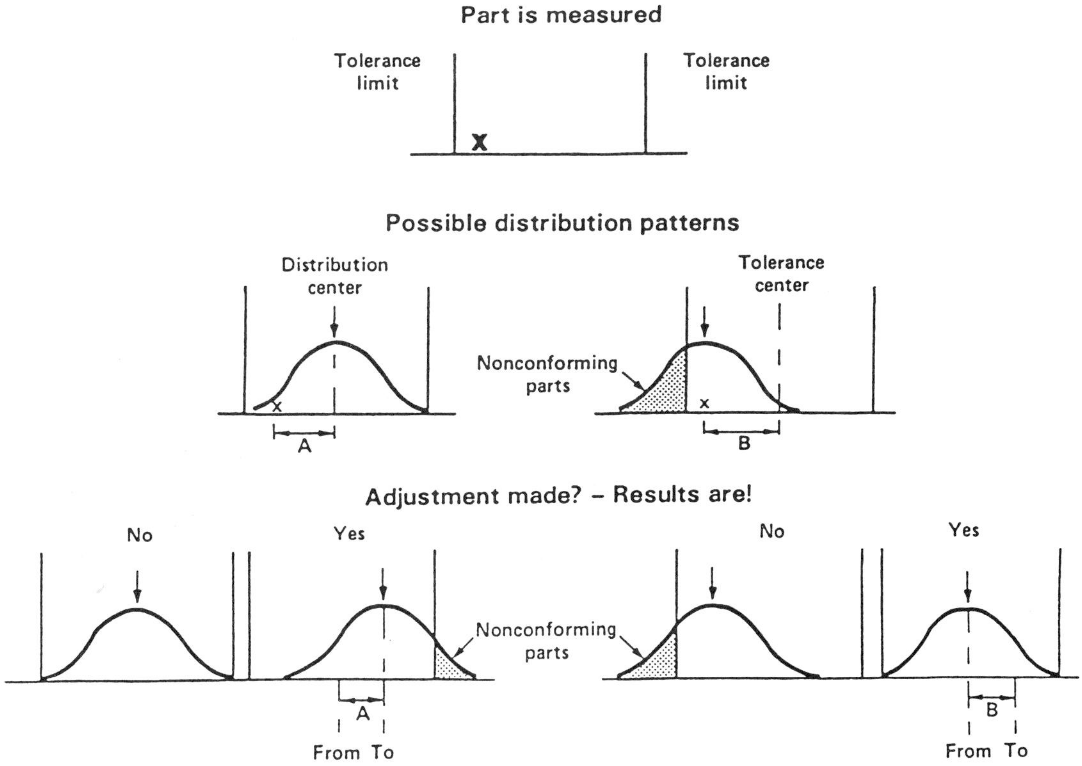

Figure 7 To adjust or not to adjust.

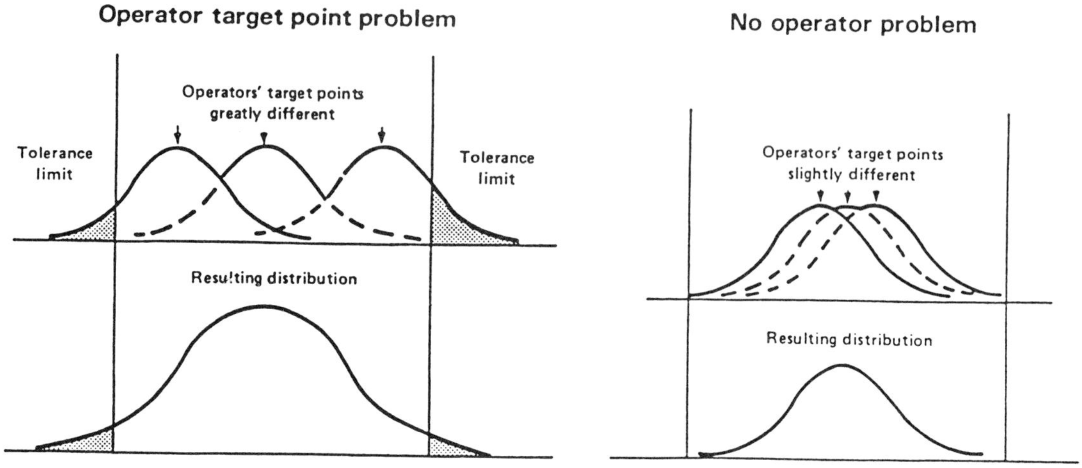

Figure 8 Identifying spread.

V. INHERENT SPREAD WORKSHEET

An examination of the process usually reveals several possible significant candidates that are not being controlled. The inherent spread worksheet (Fig. 9) is a tool to determine the effect of controlling an additional factor when it is not convenient to control the factor exactly for the series of at least 25 items needed for a S/T worksheet determination. Figure 9 is a three-piece worksheet.

Use for analysis the four-piece, three-piece, or two-piece inherent spread worksheet, depending on whether it is more convenient to control four, three, or two successive items at a time (the higher the number of items being controlled, the more precise are the results). The instructions for the inherent spread worksheets are as follows:

1. While holding the selected factor as nearly constant as possible, produce four, three, or two successive times.
2. Measure the items and record the measurements on the horizontal line 1 in the data area of the inherent spread worksheet selected.
3. Repeat steps 1 and 2 for 24 additional time intervals and record the data on horizontal lines 2 to 25. Note that the control on the factor selected can vary between the different sets of time-interval readings but must be constant during the items produced in a set.
4. Determine, for each set of readings, the difference between the largest and the smallest reading within the set and record result in appropriate range block.
5. Add and record the sum of the 25 range blocks.
6. Indicate this sum on the sum of ranges scale and draw a straight line to the maximum expected range index number that corresponds to the number of individual ranges involved.
7. Eliminate any individual range that is greater than the amount indicated by where the straight line crosses the maximum range scale. If a range is eliminated, recalculate a new sum of ranges.

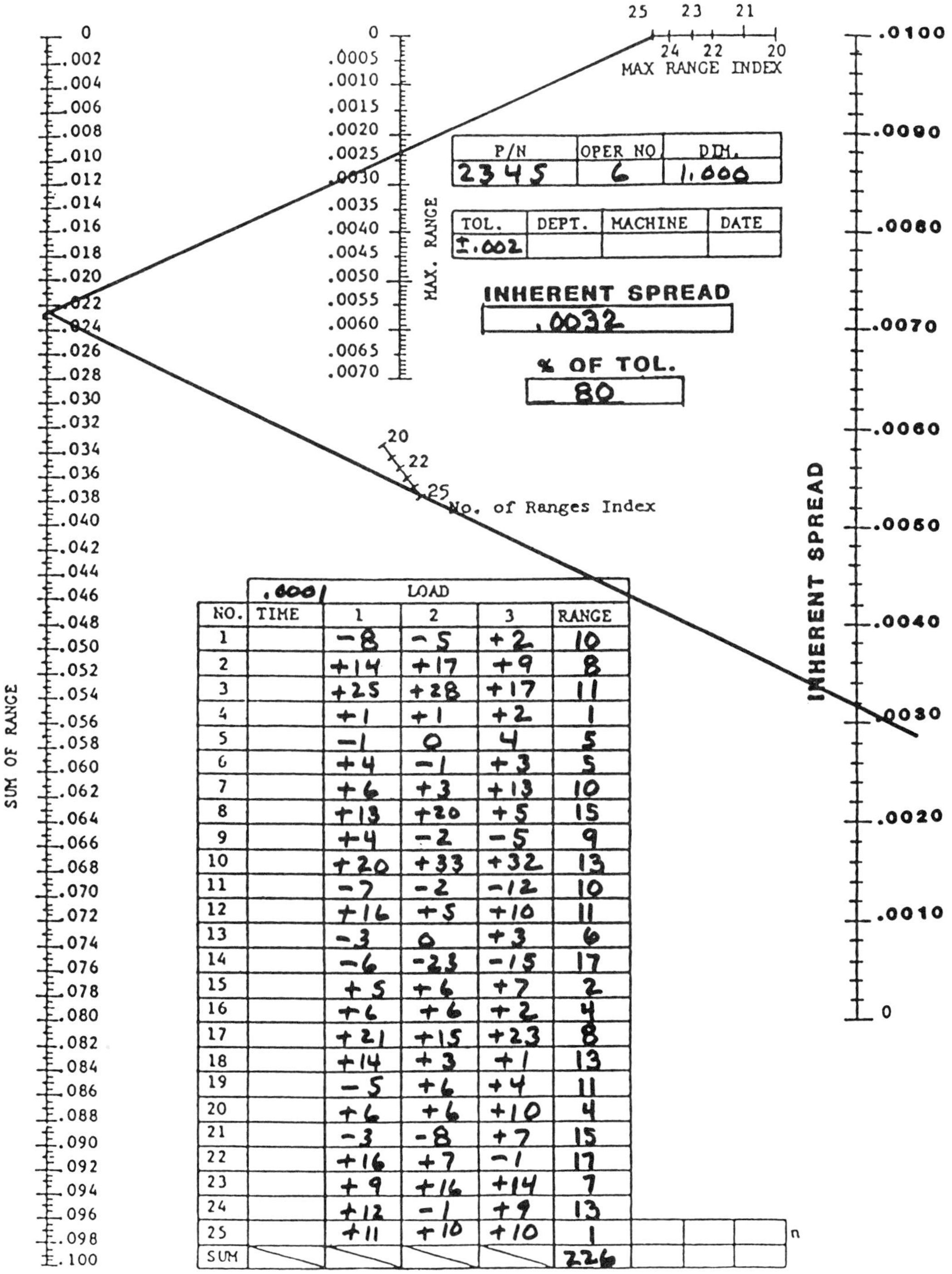

NO.	TIME	.0001 LOAD 1	2	3	RANGE
1		−8	−5	+2	10
2		+14	+17	+9	8
3		+25	+28	+17	11
4		+1	+1	+2	1
5		−1	0	4	5
6		+4	−1	+3	5
7		+6	+3	+13	10
8		+13	+20	+5	15
9		+4	−2	−5	9
10		+20	+33	+32	13
11		−7	−2	−12	10
12		+16	+5	+10	11
13		−3	0	+3	6
14		−6	−23	−15	17
15		+5	+6	+7	2
16		+6	+6	+2	4
17		+21	+15	+23	8
18		+14	+3	+1	13
19		−5	+6	+4	11
20		+6	+6	+10	4
21		−3	−8	+7	15
22		+16	+7	−1	17
23		+9	+16	+14	7
24		+12	−1	+9	13
25		+11	+10	+10	1
SUM					226

Figure 9 Inherent spread worksheet.

8. Indicate the final sum of ranges on the sum of ranges scale and draw a straight line to the number of ranges index point that corresponds to the number of ranges in the sum of ranges. Extend the straight line through the inherent spread scale. The reading on the inherent spread scale is the inherent spread of the process if the selected factor was held constant.

VI. THE FISHBONE DIAGRAM

If difficulty exists in identifying factors that might need control in the process, a "fishbone" diagram can be significant aid. To establish such a diagram:

1. Draw a long straight horizontal line to represent the process:

2. Establish a branch arrow to the horizontal line for each major contributor to the process and label each arrow.

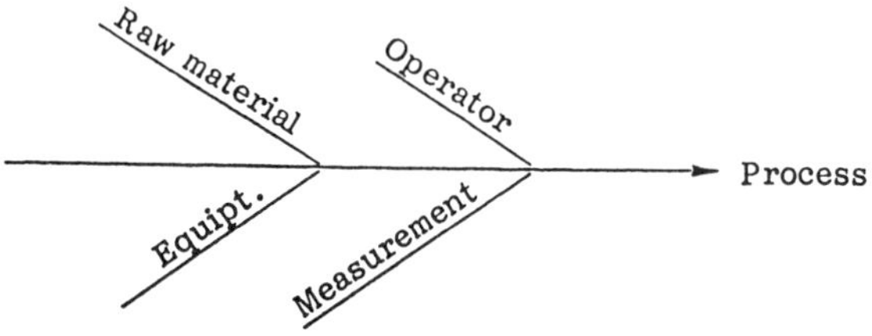

3. For each major contributor to the process branch arrow, establish branch arrows representing each significant factor of that function.

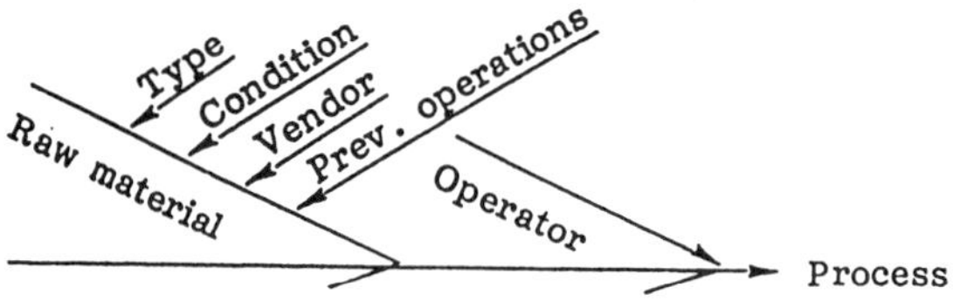

4. If necessary, establish smaller twigs to each significant factor's branch arrows, representing factors contributing to the significant factor involved.

The fishbone diagram creates a systematic analysis of all factors that contribute to a process and therefore can be very valuable in identifying factors that could have an important effect on the final product.

VII. SINGLE-ITEM WORKSHEET

In analyzing a process, a special situation exists when the process produces only one item per run. Using statistical logic, it is recognized that the process itself can produce a population of items with a center point and with a spread around the center point, and the single item produced must be part of that process population. Therefore, the single item itself could vary from its target point in relationship to the inherent spread of the process. The single-item worksheet is an excellent tool for determining the inherent spread of a single-item process when a series of similar single runs are produced by the same basic process (Fig. 10).

EQUIPMENT BRIDGEPORT OPERATION DIA. DATE________

PART NUMBER	REQUIREMENT	TARGET POINT	MEASUREMENT	DIFFERENCE
12345	1.0000 ±.002	1.0000	1.0002	+2
43217	1.5000 ±.003	1.5000	1.4996	−4
etc.	etc.	2.4000	2.4000	0
		.5000	.5009	+9
		3.2000	3.2001	+1
		1.0000	.9998	−2
		2.0000	1.9996	−4
		3.0000	3.0005	+5
		1.5000	1.5002	+2
		1.0000	1.0000	0
		.7500	.7503	+3
		.8750	.8754	+4
		1.0000	1.0002	+2
		2.0000	1.9998	−2
		1.0000	1.0001	+1
		1.5000	1.5004	+4
		3.5000	3.5003	+3
		.7500	.7502	+2
		1.0000	.9995	−5
		1.0000	1.0004	+4
		2.0000	2.0003	+3
		.8000	.8001	+1
		1.5000	1.5006	+6
		.9000	.8998	−2
		1.4000	1.4000	0

Figure 10 Single-item worksheet. Instructions: (1) For difference, subtract measurement from target point and record as plus or minus. All measurements must be result of first target point attempt. Do not record rework measurements. (2) Plot plus and minus differences on Spread & Target Worksheet. If individual readings are outside of indicated spread, eliminate such readings and replot. Final spread on Worksheet is inherent spread of involved operation.

1. For each of 25 different items produced one piece at a time on the basic process (e.g., different diameters on different pieces produced by the same basic Bridgeport Machine operation), record the part number, the requirement original target point, and a measurement after the first target point attempt.
2. Determine the difference between the target point and the resulting measurement. Record the difference as plus or minus.
3. Plot the 25 plus and minus differences on S/T worksheet (Fig. 11). If any individual readings are outside the indicated spread, eliminate these readings and replot.
4. The spread as indicated on the S/T worksheet is the inherent spread of the process. Expected variation of an individual items from its target point can be up to 50% of the indicated spread. (Example: If the S/T spread is 18 units, an individual piece to be produced by the process can vary up to ± 9 units from the target point selected.)

If after an inherent spread analysis it is determined that it is not economically feasible to decrease a process inherent spread to at least 80% of requirements (e.g., new expensive equipment would be needed), contact the design engineering personnel for possible relaxation of the tolerance so that the economical 80% spread requirement can be achieved (Fig. 12). If the tolerance width is increased to 125% of the minimum economically feasible spread, the 80% of tolerance requirement for the spread will be achieved. In almost all instances, the design engineers—if also trained in statistical logic—will agree to a reasonably extended tolerancing because they realize that such an extension will be to their advantage. (A successful process control program requires maximum cooperation between design engineering and manufacturing.)

In most industries a basic conflict usually exists between design engineering and manufacturing. The design engineer determines a desired requirement and then as a concession to manufacturing allows a tolerance around that desired requirement, although the desired requirement is always preferred. Manufacturing, on the other hand, usually considers all items within the tolerance limits equally good and therefore does not necessarily try to make all items at or near the desired requirement. This situation is especially true when the spread of the process is greater than the tolerance width because to prevent producing scrap items, manufacturing will target the process to the rework side of requirements (Fig. 13) (assuming each item can be inspected and those outside requirements can be reworked back into requirements). Such a rework technique has three serious drawbacks:

1. The 100% inspection and the rework operation are nonproductive costs.
2. Experience shows that isolated secondary rework operations often produce scrap which may or may not be identified.
3. Even if the inspection and rework operation is successful and no scrap items are produced, the targeting point near the rework side of tolerance produces the majority of conforming items near the extreme of tolerance rather than at the desired requirement (Fig. 13).

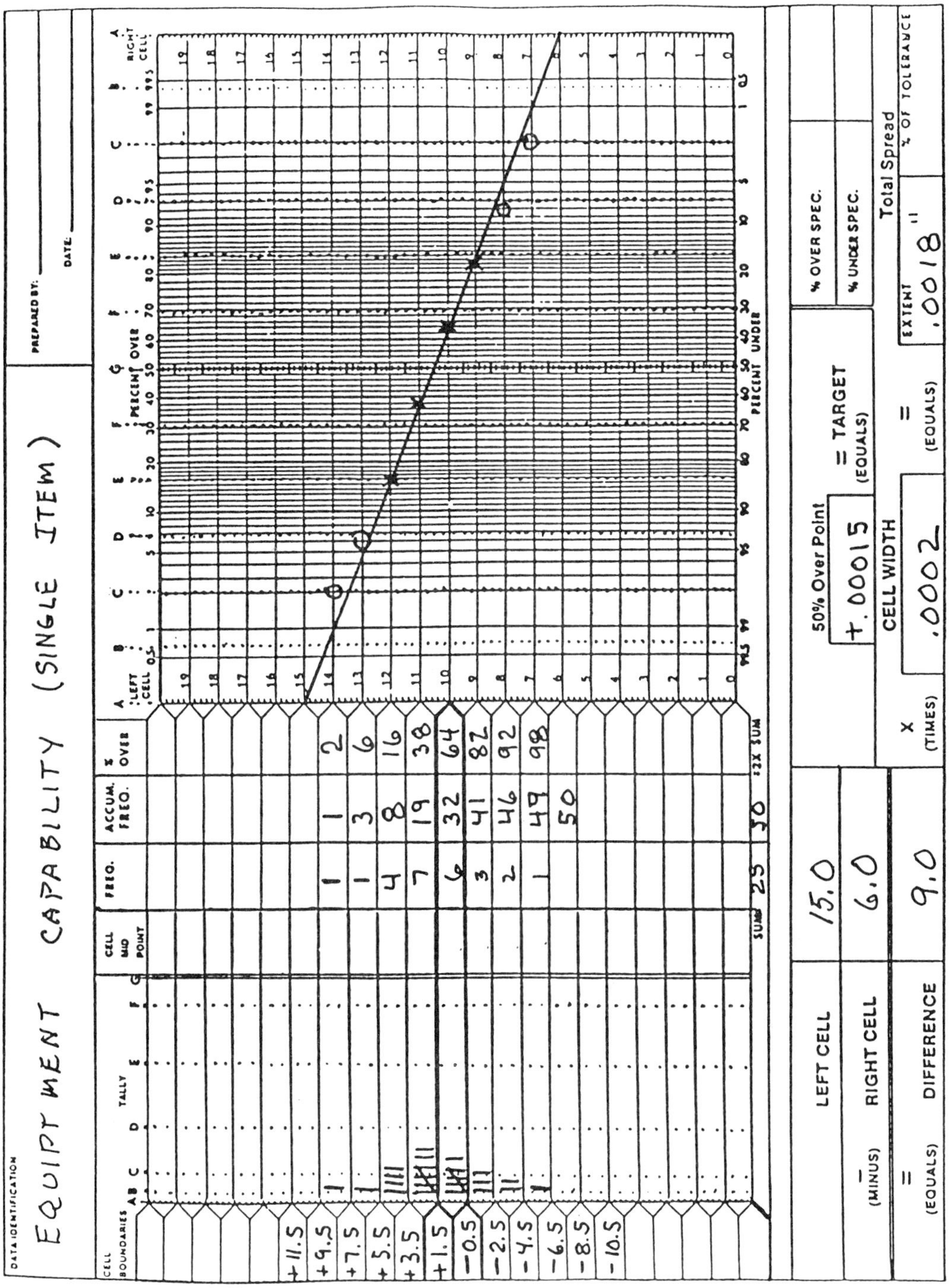

Figure 11 Spread and target worksheet.

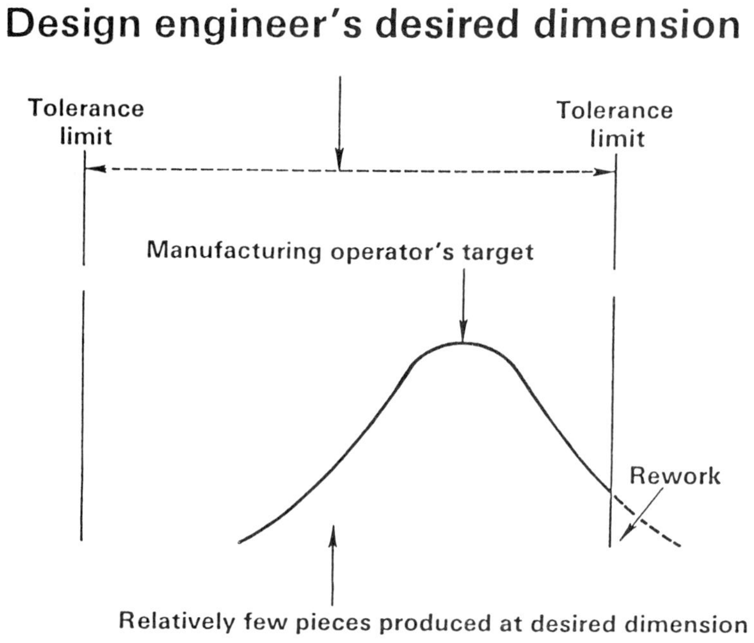

Figure 12 Target for rework.

Design engineers who understand statistical logic recognize the three drawbacks associated with too-narrow tolerances. In addition, they recognize that with compatible, sufficiently wide tolerances, only the few items in the tails of the process populations will be near the tolerance limits. The majority of the items will be at or near the desired requirement provided that manufacturing has successful process control of the operations (a population targeted at or near nominal and spread less than 80% of tolerance).

Because extending a tolerance is much more economical than making extensive changes to the process plus the fact that additional items will be produced near the desired requirement with only a few produced in the extended area, design engineers are often able to increase tolerances as necessary if assured that the items will be manufactured in a statistical process control environment (Fig. 14).

VIII. CONCLUSION

The steps for establishing a successful process control program follow.

1. Recognize that processes produce populations rather than individual items.
2. Establish a picture of the population.
3. If a shift problem exists, revise the established controls to correct the situation.
4. If a spread problem exists, select additional factors for control.
5. If it is not economically feasible to decrease spread to 80% or less of tolerance, ask the design engineers for an increased tolerance.

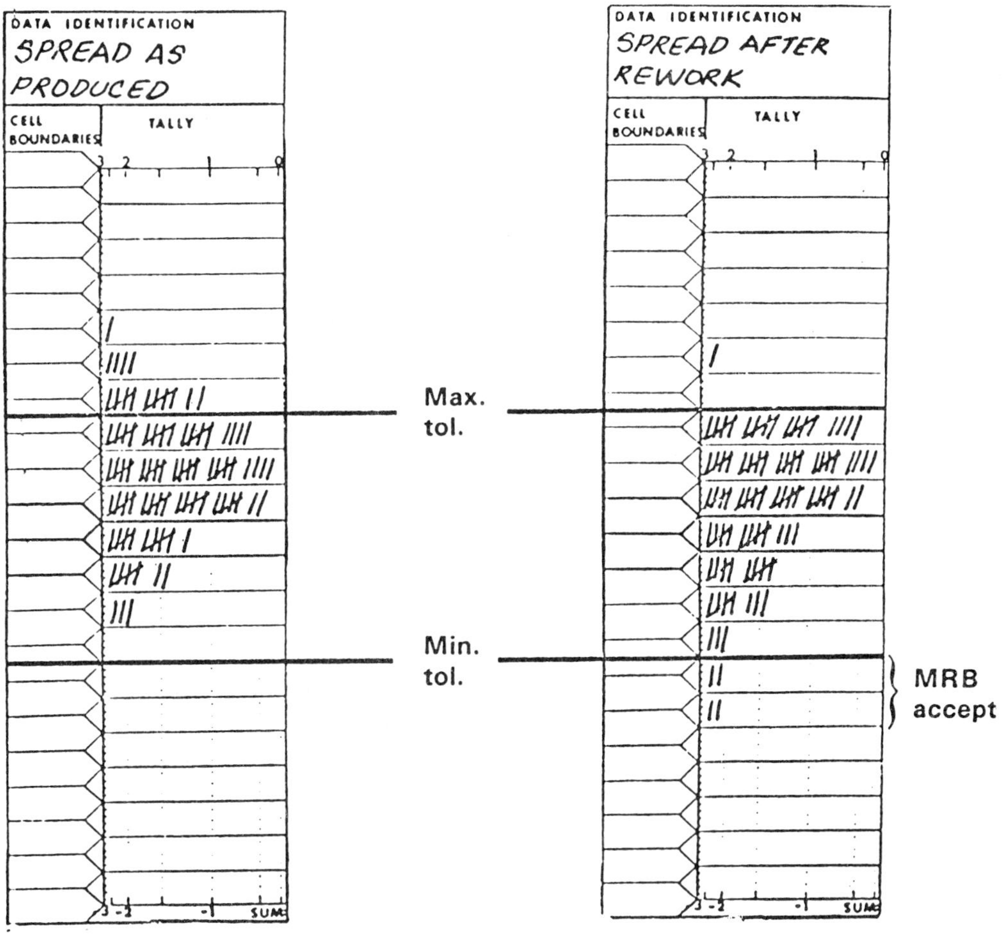

Figure 13 Population after rework.

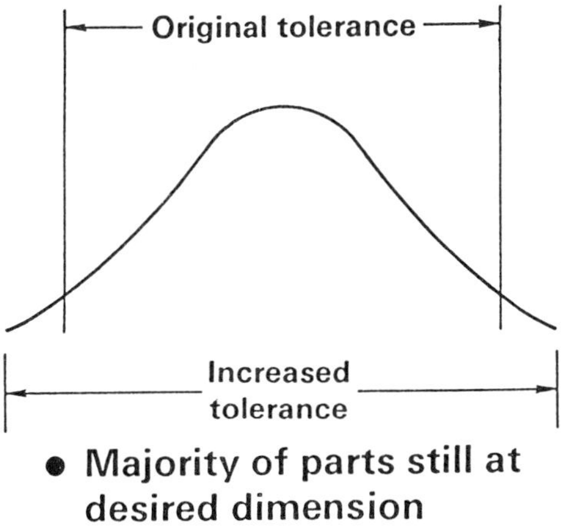

Figure 14 Population after increased tolerance.

6. When an ideal population exists, use a target control worksheet for assurance that only conforming items are being produced.

A successful process control operation produces only conforming items. Excessive nonproductive inspection costs, rework costs, and scrap costs are eliminated. Quality can be achieved while decreasing costs when statistical logic is used.

29

Pareto Charts and Histograms

JOHN JOURDAN HELDT
Free Lance Reliability Service
San Jose, California

I. INTRODUCTION

Charts of various kinds have long been used to present and explain the meaning of data. Pareto charts and histograms are two of the most useful of theses data presentation tools. Both charts are usually included in "the seven tools of quality." The "seven tools of quality" is a term used by some practitioners to refer to histograms, control charts, Pareto charts, flowcharts, checksheets, scatter diagrams, and fishbone charts.

Histograms and their close relatives, bar charts, are closely related to Pareto charts. In fact, Pareto charts are often called specialized histograms. Each, however, is used in a different way to convey information.

II. THE HISTOGRAM

The histogram uses a series of vertically oriented rectangles, called bars or cells, to depict the quantity of data contained either at a specific value or between two limits. (In general, the primary difference between histograms and bar charts is that bar charts are typically horizontally oriented.) A histogram that is commonly used in quality and probability tutorials shows the cumulative quantity of occurrences of each number that can turn up on repeated throws of two honest dies (or dice).

Developing the Histogram

Figure 1 is a histogram of weld strength data. The raw data used in the histogram are shown in Table 1. The difference in the amount of information contained in

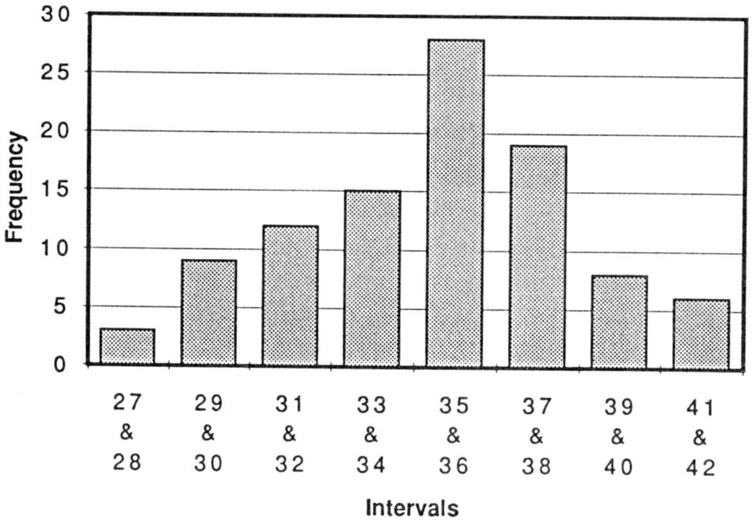

Figure 1 Histogram.

the histogram compared to the almost meaningless visual impact of the raw data in the table is immediately obvious.

The data shown in the table are *discrete* numbers (i.e., integers or whole numbers), which are relatively easy to set up. Many histograms are used to depict data that are defined as variables data or, more specifically, as continuous data (see Chapter 32). Histograms for variables data can be handled as easily as for discrete data.

To construct the histogram it is necessary to identify the numerically largest and smallest numbers in the raw data. These numbers (42 and 27) are shown in bold in Table 1. Guidelines for determining the number of intervals that would be used in creating a histogram are shown in Table 2. To calculate the "width" of each bar or "cell" of a histogram:

Table 1 Histogram Data

Terminal weld strength in lbs									
31	35	33	34	36	39	38	41	29	35
33	36	32	41	37	39	31	28	**42**	38
39	38	41	29	35	31	35	30	34	36
34	36	32	35	37	34	36	41	29	35
31	35	31	34	36	39	38	41	29	35
39	38	35	33	35	**27**	36	32	35	37
31	35	30	34	36	39	38	34	29	35
28	36	33	38	37	37	38	35	29	35
39	38	34	29	35	34	36	32	40	37
31	35	34	34	36	38	38	41	29	35

Table 2 Guidelines for Determining the Number of Intervals (Cells)

Number of data items	Number of intervals (cells)
50 or fewer	4–7
50–100	5–12
101–150	6–12
151–more	8 or more

1. Find the total number of integers represented in the data by subtracting the smallest number from the largest. In Table 1 the data limits are 42 and 27. If the data are continuous (i.e., variables data) the data would be rounded up and down. If, for example, the data limits in Table 1 were 41.68 and 27.32, they would be rounded to 42 and 27. The number of integers between 27 and 42 is $42 - 27 = 15$.

2. Choose the number of intervals or cells in which to distribute the data by referring to Table 2. Eight cells were chosen to set up this histogram.

3. Find the "data width" of each cell by dividing the number of integers from step one by the total number of proposed cells: $15/8 = 1.875$. It may be useful to round the cell width up or down to make the histogram more definitive.

4. Make a frequency tally, such as that shown in Figure 2, of the data.

5. Transfer the summary data from the frequency tally to a bar chart or histogram such as that shown in Figure 1. The cell heights, or magnitudes, in Figure 1 combine the tallies of each of two whole numbers (e.g., the tallies of the data numbers 27 and 28 are summed in bar number one, 29 and 30 in bar number two, and so on). When variables or continuous data are histogrammed, the "widths" of the bars must reflect the continuous nature of the data. The histogram bars might be labeled 27 to 29, 29.1 to 31, 31.1 to 33, and so forth.

III. THE PARETO CHART

The original discovery of the principle of what is now called the Pareto Chart has been attributed to an Italian economist named Vilfredo Pareto. Pareto, who later

Interval	Tally Sheet	Total
27 & 28	|||	3
29 & 30	𝍸 ||||	9
31 & 32	𝍸 𝍸 ||	12
33 & 34	𝍸 𝍸 𝍸	15
35 & 36	𝍸 𝍸 𝍸 𝍸 𝍸 |||	28
37 & 38	𝍸 𝍸 𝍸 ||||	19
39 & 40	𝍸 |||	8
41 & 42	𝍸 |	6

Figure 2 Weld strength frequency tally.

developed a logarithmic curve to fit the data, noted a "maldistribution of wealth" in that relatively few people became very wealthy while the majority of people remained poor. Later M. O. Lorenz developed a cumulative graph to illustrate Pareto's principle, and more recently Dr. Joseph Juran applied the principle of what he calls the "vital few and trivial many" causes of industrial problems such as scrap, defective products, late shipments, and the like. The intent of the modern Pareto chart is to graph data depicting problems and their causes, then to select the vital few—i.e., the most significant—causes of the problems for intense attention and correction. It has been accepted in industry that, as a general rule, 15 or 20% of the causes—the vital few—result in 80 to 90% of the problems.

A. Basic Pareto

The same data used in the development of the histogram were used to develop the Pareto chart in Figure 3 to illustrate the subtle but definite differences between the two techniques. In developing the histogram, the data cells were arranged in order of *ascending data values,* left to right. The Pareto, however, is developed by gathering groups of associated data together and arranging them in cells, or bars, of *decreasing magnitude,* left to right, as shown in Figure 3.

As in the histogram, the data tallies, or summaries, are transferred from the tally sheet (Figure 4) to the chart shown in Figure 3. A chart such as this, arranged in descending value, is typically called a *basic Pareto chart.* It is easy to see that the bulk of the data (71%) falls between the data values 31 and 38. The "causes" of the values between 31 and 38 are the "vital few" causes. Generally such causes are fewer than 20% of the total number of causes.

B. Stacked Pareto

Figure 5 shows a variation of the basic Pareto. Here, the value, or height, of each column is shown added to the sum of the previous columns in what is often called a "stacked" Pareto. By tying the upper corners of the cells together with straight lines as shown, an "ogive" curve effect takes place that often helps the analyst visualize the implications of the data.

Interval	Tally Sheet	Total
35 & 36	HHH HHH HHH HHH HHH III	28
37 & 38	HHH HHH HHH IIII	19
33 & 34	HHH HHH HHH	15
31 & 32	HHH HHH II	12
29 & 30	HHH IIII	9
39 & 40	HHH III	8
41 & 42	HHH I	6
27 & 28	III	3

Figure 3 Weld strength frequency arranged in Pareto chart format.

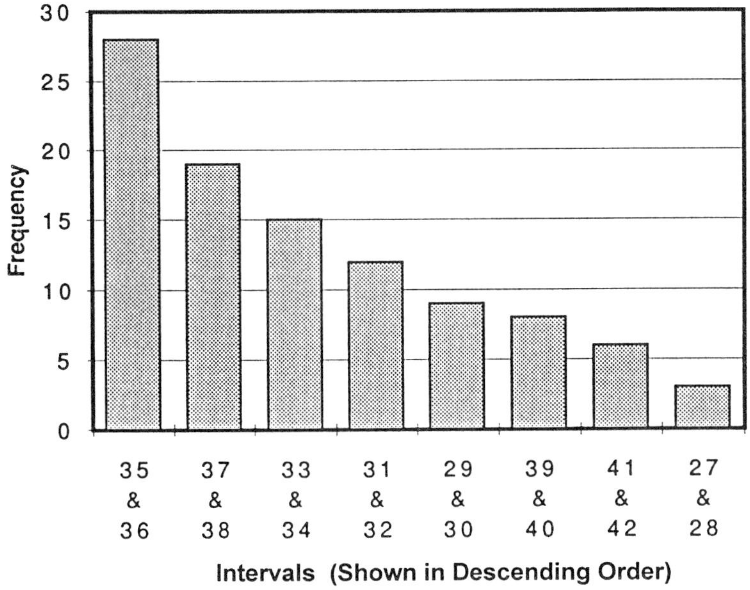

Figure 4 Basic Pareto chart.

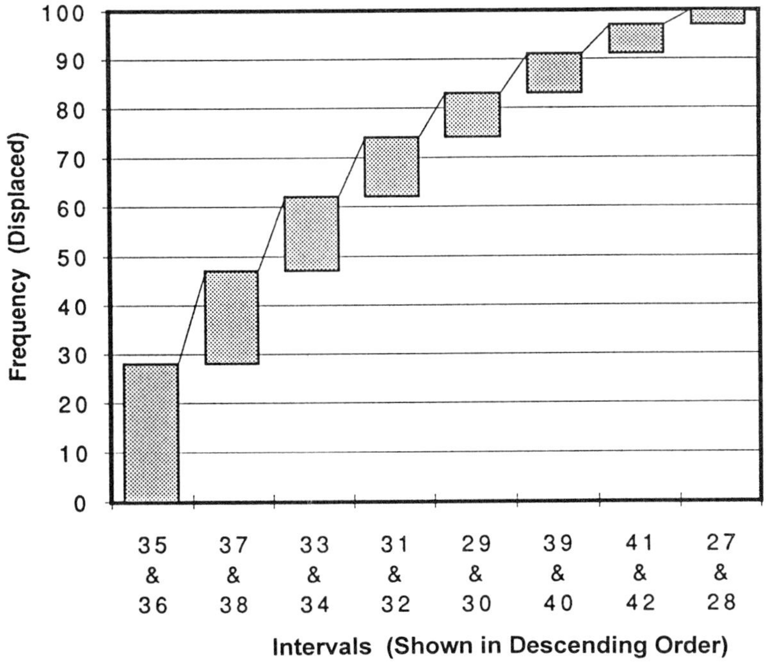

Figure 5 Stacked Pareto chart.

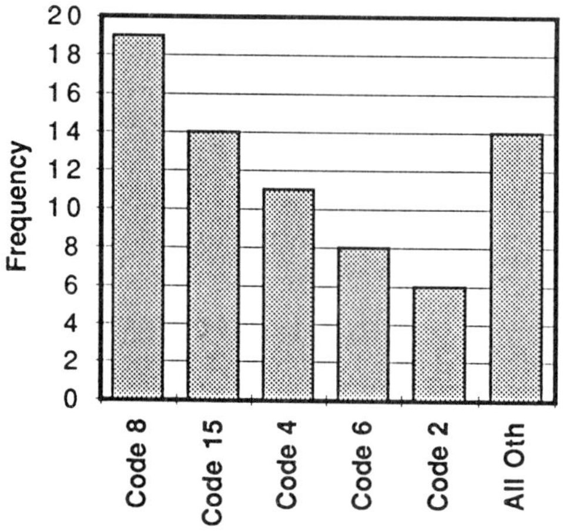

Figure 6 Pareto analysis by defect.

C. Pareto Analysis by Frequency of Defect

Figure 6 shows a Pareto analysis that describes the number of occurrences of certain defects in a finished complete assembly. The defect codes and numbers of occurrences are shown in Figures 7 and 8. Clearly, "mistuned", the defect indicated by code 8, is the first priority for correction. Since categories such as "all others" (miscellaneous, etc.) are usually indistinct, they generally are not strong candidates until everything else is done. Code 15, "transformer or filter," would be the second cause to correct in this example.

D. Pareto Analysis by Subassembly

Figure 9 shows a Pareto analysis based on the frequency of problems with subassemblies of the finished unit. The subassemblies are identified as mother board,

Code	Workmanship Defects	Code	Component Defects
1	Wiring Error	11	Semi-conductor
2	Short (actual/Latent)	12	Resistor
3	Wrong polarity	13	Capacitor
4	Missing component	14	Inductor
5	Damaged component	15	Transformer or Filter
6	Wrong value component	16	Switch or Relay
7	Improper Soldering		
8	Mistuned		
9	Strapping error		
10	All others		

Figure 7 Defect descriptions and code numbers.

ASSEMBLY	Defect Codes																
	1	2	3	4	5	6	7	8	9	10	11	12	13	14	15	16	Total
Moth Bd		5		11	5	3		11									35
Pwr Su							1	2				3			14		20
Mem Bd	2		2			4		3									11
All Oth		1				1	1	3									6
TOTALS	2	6	2	11	5	8	2	19	0	0	0	3	0	0	14	0	72

Figure 8 Pareto matrix (frequency of occurrence).

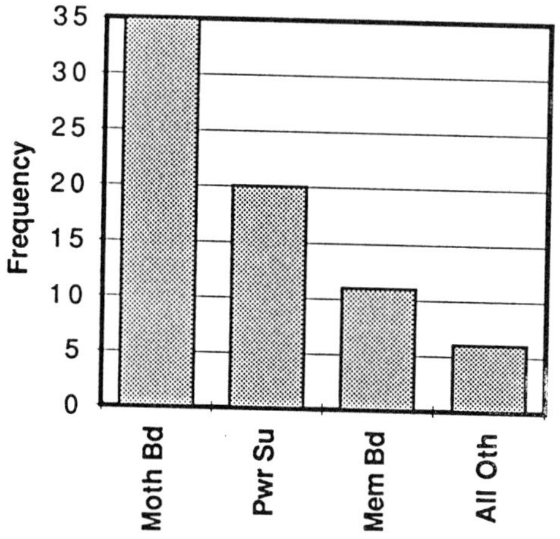

Figure 9 Pareto analysis by assembly.

power supply, memory board, and all others. Using this analysis the mother board is clearly the first target for improvement. Nearly half of all defects occurred in the mother board.

E. Pareto Analysis by Dollar Cost of Defects

Figure 10 shows a Pareto analysis of the dollar cost of defects by defect code. Dollar cost of correction of defects by subassembly are shown in Figure 11a and 11b. While defects 8 and 15 are still the prime targets for improvement, defect code number 1 (Figure 10) has replaced defect code number 4 (Figure 6) for third place.

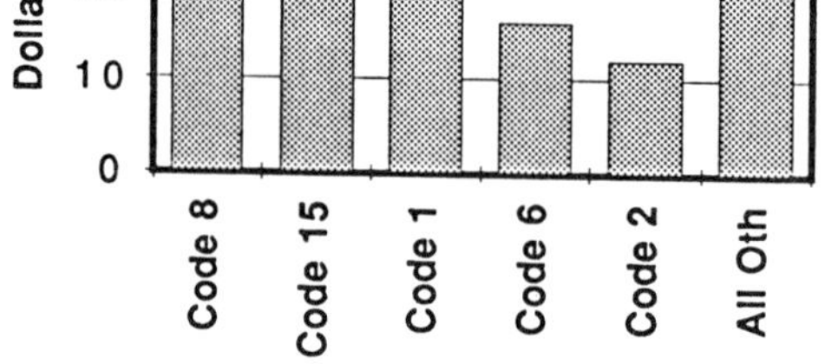

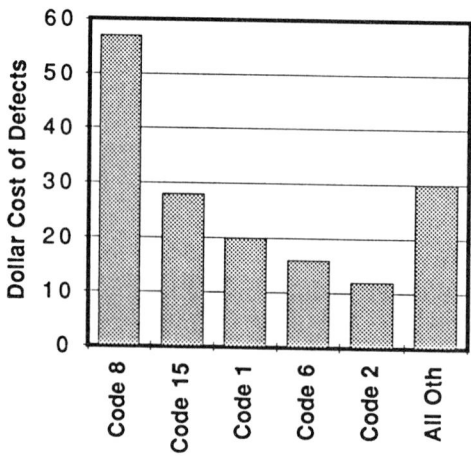

Figure 10 Pareto cost analysis by defect code.

Code	Work Defects	Cost
1	Wiring Error	$10
2	Short (Act/Lat)	$2
3	Wrong polarity	$3
4	Missing comp	$1
5	Damaged comp	$1
6	Wrong val comp	$2
7	Improper Solder	$1
8	Mistuned	$3
9	Strapping error	$1
10	All others	$5

Code	Comp Defects	Cost
11	Semi-conductor	$5
12	Resistor	$2
13	Capacitor	$2
14	Inductor	$2
15	X-former or Filter	$5
16	Switch or Relay	$4

Note: All cost factors are in Dollars.

(a)

Figure 11 (a) Defect descriptions and defect costs; (b) Pareto matrix by dollar value of defects.

ASSEMBLY	Defect Codes																
	1	2	3	4	5	6	7	8	9	10	11	12	13	14	15	16	Total
Moth Bd		$10		$11	$5	$6		$33									$65
Pwr Su							$1	$6				$6			$28		$41
Mem Bd	$20		$6			$8		$9									$43
All Oth		$2				$2	$1	$9									$14
TOTALS	$20	$12	$6	$11	$5	$16	$2	$57	$0	$0	$0	$6	$0	$0	$28	$0	$163

(b)

Figure 11 Continued

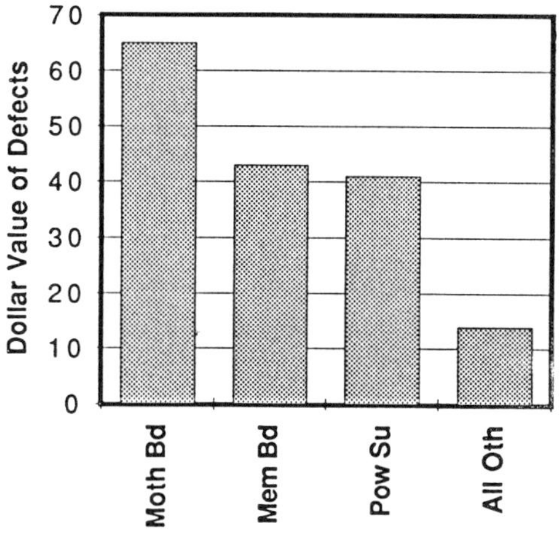

Figure 12 Pareto cost analysis by subassembly.

It is not unusual that the total cost of occurrence of one defect is quite a bit higher than the cost of occurrence of another defect that occurs more often. An expensive sealed unit that failed, a defective component that is difficult to replace, or a critical adjustment that is time consuming to make are usually very costly.

F. Pareto Analysis by Subassembly Repair Cost

Figure 12 shows a Pareto cost analysis by subassembly. Here, the mother board continues to be the primary focus of corrective action. However, the total defect cost for the memory boards has moved into second place slightly ahead of power supplies (compare Figure 9).

G. Pareto Analysis Summary

Since Pareto analysis is relatively easy to set up, it is often useful to compare the impact of occurrences, costs, effects on production, employee moral, and the like.

While Pareto analysis in this chapter has dwelt on defects and their results, the analysis is equally valuable for evaluating positive occurrences. Sales improvements resulting from promotional campaigns, productivity improvements resulting from bonus distributions, and many similar positive occurrences can be easily evaluated.

30
Process Capability Studies

MAE-GOODWIN TARVER*
Quest Associates Ltd.
Forest Park, Illinois

I. INTRODUCTION

A. Definitions

Process: The combination of people, machines, and materials used to manufacture or fabricate a product (1).

Capability: The capacity or ability of a process to reproduce product characteristics such as mass, dimension, color, etc., within specified requirements.

Process capability study: The systematic study of a process to determine its ability to meet specifications or tolerance limits under normal operating conditions (NOC).

Normal operating conditions (NOC): Conditions that occur when a plant or manufacturing process is operating under approved instructions, using specified raw materials, and trained, experienced operators; it includes different operators, different lots of raw materials or components, tool wear, or other factors that may be affected by time.

Machine capability study: The systematic study of a machine under somewhat controlled conditions to determine its inherent variation; machine adjustments are not made, usually a single operator is used, and a single lot of raw material or components is used (2). For this reason, machine capability studies are usually confined to equipment development shops.

Natural tolerance: The inherent or natural variation of a process when it is in a steady state (state of control); numerically, it is expressed as the arithmetic average of one measurable characteristic of the process ± 3 standard deviations or $\pm 3\sigma'$. A standard deviation is the root-mean-square variation around the arithmetic average.

---

* Retired.

B. Purpose of Capability Studies

To keep satisfied customers, producers must meet customer quality requirements. A very important requirement is that significant quality characteristics be held within specified tolerance limits. The major purpose of process capability studies, therefore, is to discover whether a process is in a steady state (state of statistical control), and if it is, whether the product will meet the customer's quality requirements. In other words, is the process "in-control" and "capable."

It is also important to note that process performance is not the same as process capability (3). Process capability means that only "natural" or inherent variation is affecting the process.

When a process is in a steady state, no extraneous factors are "rocking" the process; only "natural" process variation is present. When this state of control is encountered, the observed process variation is considered to be economically irreducible under the current operating conditions of the process.

C. Scope of the Method

Process capability studies have a wide range of application in industry. Problems in almost all industries will fall into one or more of the following categories:

- Quality: too many defective units being produced
- Cost: excessive scrap, rework, or repair, low yields, and so on
- Lack of information: process trends, effect of new materials, methods, tooling, and so on
- Engineering problems: design, specifications, quotations, and so on

The concept of analyzing process variation can be used to solve problems in all industries. The exact method used for this purpose, however, will be industry dependent. For example, high-speed, continuous operations are easily studied using the control chart system because groups of samples can be drawn from the process under normal operating conditions over a long period of time. Job-shop operations, on the other hand, run one product line for a short period of time and then change over to another product line. The usual control chart may not be appropriate for such operations. Special methods for analyzing process variation must also be used for the process industries because they produce relatively homogeneous liquids, slurries, and pulps. Multiple samples per time period may not be appropriate under these circumstances.

D. Questions Answered by Capability Studies

The information obtained from each study must answer the following questions (3,4):

- Was the process in a state of statistical control during the study?
- Does the process meet the specified tolerance? If not, can it do so by centering the process average on the nominal value?
- Is the process inherently capable of meeting the specified tolerance? If not, is it economically feasible to reduce the process spread?

All capability studies must be carefully planned to answer the foregoing questions. Data must be collected in a systematic manner to answer them; otherwise, invalid information will result.

E. Short-Run versus Long-Run Studies

There are two generally used methods for collecting information from continuous processes: (1) short-run capability studies, where a large number of consecutive product units are drawn; and (2) long-run studies, where samples are drawn from the process over a long period of time with a small number of product units appearing in each sample. The first method gives an instantaneous picture (or snapshot) of process variation and that will probably be the smallest variation which can be expected from the process. The second method gives a much more accurate picture of process variability since it exposes the process to factors that may disturb its stability and can result in increased process variation. Examples of each, together with their advantages and disadvantages, will be discussed in subsequent sections of this chapter.

Regardless of the type of capability study to be run, four steps are required (3):

1. Collecting data: determining what characteristic(s) to measure, how many sample units are needed, and the frequency of sampling
2. Plotting the data: using histograms or control charts
3. Interpreting data patterns: diagnosing any indicated quality problems
4. Isolating and removing assignable causes: uncovering the inherent process variation.

II. SHORT-RUN CAPABILITY STUDIES

A. Planning the Study

Draw consecutive sample units from the production stream whenever possible. If the sample size selected is too large to assure consecutive product units, draw them a few at a time over as short a time period as possible. The sampling frequency must be short enough to prevent "assignable causes" from enlarging the natural variation of the process under study. Some authors recommend tying a sample size to the production rate as follows (2,5):

1. If the process is running at 200 units per hour or more, sample 200 consecutive units.
2. If the process is running less than 200 units per hour, sample the equivalent of 1 hour's production but never less than 50 product units.

There is, however, a second way of selecting an appropriate sample size. For example, if you wish to have 95% assurance of including at least 95% of the possible product measurements within the maximum and minimum values found in the sample, Table 1 shows that at least 93 sample units are required. For this reason, 100 sample units were drawn from the process in the example shown for the short-run studies.

Table 1 Selection of Sample Size for Short-Run Capability Studies

X% of measurements within sample max and min	No. of samples required to include at least X% of measurements within sample extremes with this assurance (B)			
	99.7%	99%	95%	90%
99.7%	3100	2460	1757	1439
99	834	662	473	387
95	163	130	93	76
90	80	63	45	37
85	52	41	29	24

This table was calculated using the concepts given in Ref. 6 and the chi-square formula given in Ref. 7.

B. Example of a Short-Run Study

The short-run capability study will be illustrated using a blow-molding process for plastic bottles. The quality characteristic under discussion is the lip outside diameter (OD). A special type of closure will not fit properly if this dimension is too large or too small. The specification for this dimension had tentatively been set by the designer at 0.825 ± 0.014 in. One hundred consecutive bottles were sampled and the lip OD was measured on each of the samples. The "raw" data are shown in Figure 1. Table 2 illustrates a method for calculating the number of cells and the cell width for frequency distributions. Using these methods, a frequency distribution compiled from the raw data of Figure 1 is shown in Figure 2, and the histogram (together with the tentative specification limits) is shown in Figure 3. In Figure 3, a small part of the last step lies above the upper specification limit. Figure 4 shows the lip OD cumulative frequencies (from Fig. 2) plotted on normal probability paper. A straight line was eye-fitted to these plotted points between the 10 to 90% section of the graph. Because all of the points cluster closely about this line of best fit, for practical purposes the OD dimension data are considered to be normally distributed. (A more exact statistical method for this purpose is the chi-square goodness-of-fit test.)

Based on this assumption of normality, the long-run percentage of bottles expected to be outside the tentative specification limits for lip OD is calculated as follows, using information from Figures 1 and 2:

Step 1

$$K_H = \frac{\text{max. spec.} - \overline{X}}{S} = \frac{0.839 - 0.8254}{0.0049} = +2.78$$

$$K_L = \frac{\text{min. spec.} - \overline{X}}{S} = \frac{0.811 - 0.8254}{0.0049} = -2.94$$

QUALITY CHARACTERISTIC LIP OD (INCHES) PART NO. 00 – 000
PLANT NO. 1400 LINE NO. 1A DATE 4/9 NO. SAMPLES 100
COMMENTS MOLD #9 – LIP DROOP INSPECTOR Mary Doe
SAMPLE SIZE (N) 100 BOTTLES
MAX MEASUREMENT 0.838 MIN MEASUREMENT 0.814 in RANGE 0.024 in

1	0.822	0.823	0.820	0.824		
2	.818	* .838	.826	.824		
3	.820	.835	.816	.828		
4	.834	.826	.831	.816		
5	.819	.827	.826	.822		
6	.826	.825	.832	.833		
7	.816	.829	.819	.833		
8	.828	.824	.820	.817		
9	.833	.823	.827	.819		
10	* .814	.832	.830	.829		
11	.824	.822	.832	.827		
12	.823	.825	.821	.824		
13	.826	.825	.829	.833		
14	.823	.823	.827	.825		
15	.826	.829	.825	.826		
16	.823	.825	.828	.828		
17	.821	.817	.816	.819		
18	.825	.825	.825	.822		
19	.827	.824	.825	.827		
20	.827	.825	.820	.824		
21	.826	.822	.835	.834		
22	.818	.831	.824	.830		
23	.829	.822	.824	.830		
24	.826	.828	.831	.824		
25	.833	.820	.835	.823		

SPECIFICATIONS 0.825 ± 0.014 inch

Figure 1 Short-run capability study: process data sheet.

Step 2: Look up in Grant and Leavenworth (9) (or any other table of the cumulative areas under the normal curve) the equivalent areas (E.A.) for +2.78 and for −2.94 (9). You will get $K_H \approx 0.9973$ and $K_L \approx 0.0016$.

Step 3

$$\% \text{ Inside spec.} = 100(K_H \text{ E.A.} - K_L \text{ E.A.}) = 100(0.9973 - 0.0016) = 99.57\%$$

Step 4

$$\% \text{ Outside spec.} = 100 - \% \text{ inside} = 100 - 99.57 = 0.43\%$$

Table 2 Guidelines for Forming a Frequency Distribution—Variables Data

Guidelines	Example (from Fig. 1)
1. Calculate range of the data: $R = \max - \min$	1. $R = 0.838 - 0.814 = \underline{\underline{0.024}}$
2. Calculate the ideal number of cells [8]: $K = 1 + 3.322\ (\text{Log}_{10}N)$	2. $K = 1 + 3.322\ (\text{Log}_{10}\ 100) = 1 + 3.322\ (2)\ \underline{\underline{8}}$
3. Calculate the cell width, m: $m = R/K$	3. $m = 0.024/8 = \underline{\underline{0.003}}$
4. Decide upon cell boundaries of *first* cell (a) First cell midpoint is *near* the *min* measurement. (b) Locate cell boundaries to prevent a measurement from falling *exactly* upon them.	4. (a) min = $\underline{\underline{0.314}}$ (b) Lower cell limit = $0.814 - (0.003/2) = 0.814 - 0.0015 = \underline{\underline{0.8125}}$ Upper cell limit = $0.8125 + 0.003 = 0.8155$

In addition to the information above, a capability index (CI) can be calculated:

$$CI = \frac{6\sigma \text{ spread of process}}{\text{specification spread}}$$

$$CI = \frac{0.0296}{0.028} = 1.06$$

The interpretation of the capability index is as follows:

CI	Comments
0.75 or less	Excellent process; material substitutions can be made when necessary; quality control cost will be low.
Above 0.75 but less than 1.00	Process okay; be cautious about material substitutions; tight quality control required usually.
Above 1.00	Process not adequate for specification band; reduce variation or widen specifications.

The capability index in this example was 1.06; one of two decisions must be made before proceeding to a long-run capability study:

1. Work on the process to reduce its variation.
2. Widen the tentative specification limits.

The cost-effective solution, in this instance, was to widen the tentative specification limits from ±0.014 to ±0.018, a change of about 1.6σ. This was less costly than repairing molds or making new ones, since the specifications were tentative and were based only on designer experience.

NAME John Doe DATE 5/5

DISTRIBUTION Mold #9 - LIP OD

CELL NO.	CELL BOUNDARIES LOWER	UPPER	CELL MID-PT.	TALLY OF OBSERVATIONS (||||| = 5)					(1) d	(2) f	(3) Σfd	(4) Σfd^2	% REL. f	% CUM. f
1	0.8125	0.8155	0.814	|					−4	1	−4	16	1%	1%
2	.8155	.8185	.817	|||||	|||				−3	8	−24	72	8	9
3	.8185	.8215	.820	|||||	|||||	|			−2	11	−22	44	11	20
4	.8215	.8245	.823	|||||	|||||	|||||	|||||	|||	−1	23	−23	23	23	43
5	.8245	.8275	.826	|||||	|||||	|||||	|||||	||||| ||	0	27	0	0	27	70
6	.8275	.8305	.829	|||||	|||||	|||			1	13	13	13	13	83
7	.8305	.8335	.832	|||||	|||||	|			2	11	22	44	11	94
8	.8335	.8365	.835	|||||					3	5	15	45	5	99
9	.8365	.8395	.838	|					4	1	4	16	1	100
10														
11														
12														
13											−73			
14											+54			
15														
16														
17														
18														
19														
20														
										100 (Σf=N)	−19 (Σfd)	273 (Σfd^2)		

m = Cell Interval = 0.003

C = Assumed center = 0.826

$$\bar{X} = C + m\left(\frac{\Sigma fd}{N}\right) = 0.826 + 0.003\left(-19/100\right) = 0.826 - 0.00057 = 0.8254 \text{ in.}$$

$$S = m\sqrt{\frac{\Sigma fd^2 - \frac{(\Sigma fd)^2}{N}}{N-1}} = 0.003\sqrt{\frac{273 - \frac{(-19)^2}{100}}{100-1}} = 0.003\sqrt{\frac{273-3.61}{99}} = 0.003\times\sqrt{2.7211} =$$

$$0.003 \times 1.6496 = 0.0049$$

Natural Tolerance (Process Capability) = $\pm 3\sigma = \pm 3(0.0049) = \pm 0.0148$

Limits for Individual Units = $\bar{X} \pm 3\sigma = 0.8254 \pm 0.0148 = 0.811 - 0.840$ (Rounded)

Figure 2 Short-run capability study: frequency distribution of data.

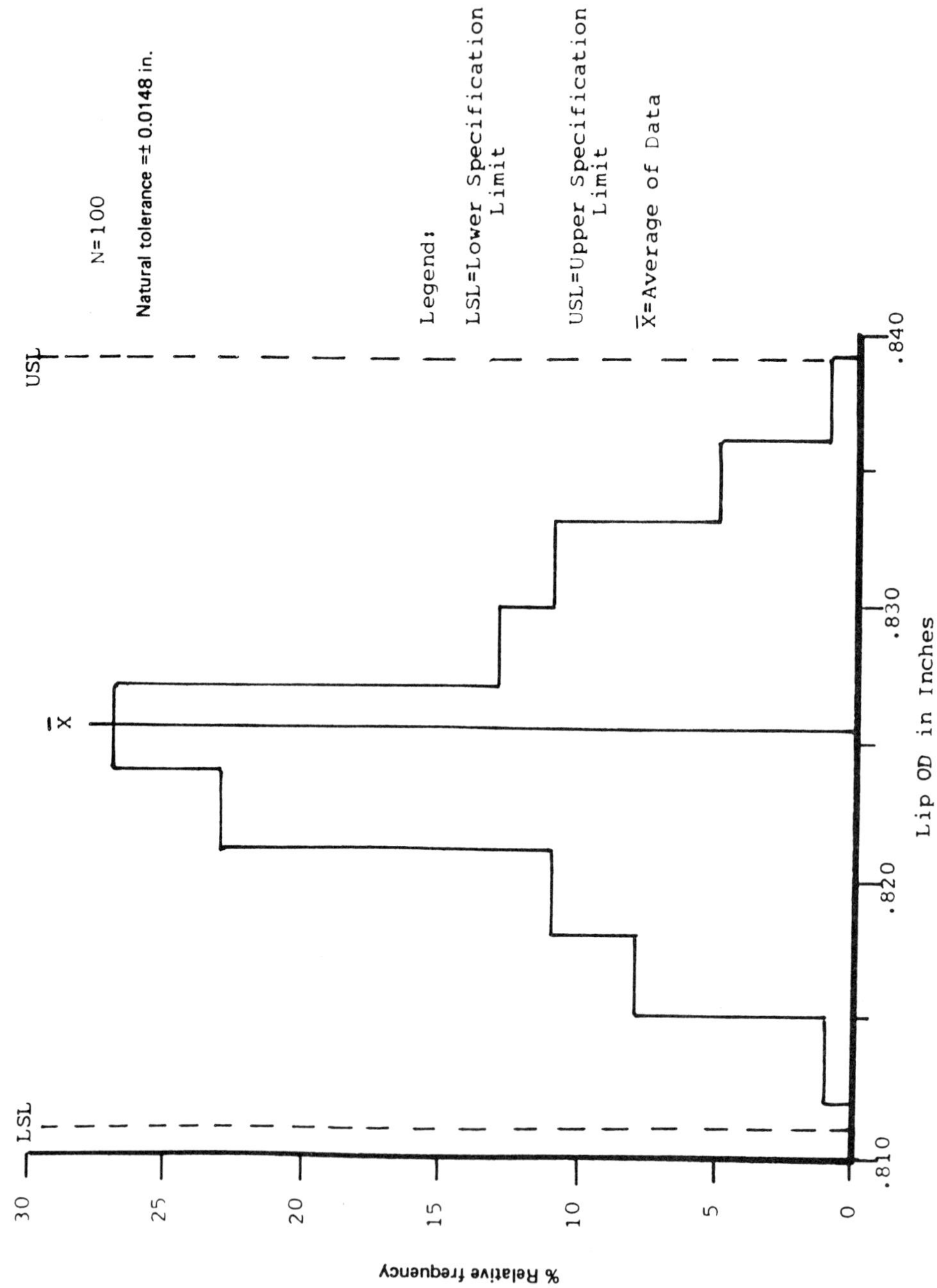

Figure 3 Histogram of lip OD data.

Lip O.D. in inches

Freq.	Accum Freq.	% Under	Mid Point
0	0	0.0	0.80500
0	0	0.0	0.80700
0	0	0.0	0.80900
0	0	0.0	0.81100
0	0	0.0	0.81300
5	5	2.5	0.81500
2	12	6.0	0.81700
11	25	12.5	0.81900
2	38	19.0	0.82100
23	63	31.5	0.82300
20	106	53.0	0.82500
7	133	66.5	0.82700
13	153	76.5	0.82900
3	169	84.5	0.83100
10	182	91.0	0.83300
3	195	97.5	0.83500
1	199	99.5	0.83700
0	200	100.0	0.83900
0	200	100.0	0.84100
0	200	100.0	0.84300
0	200	100.0	0.84500

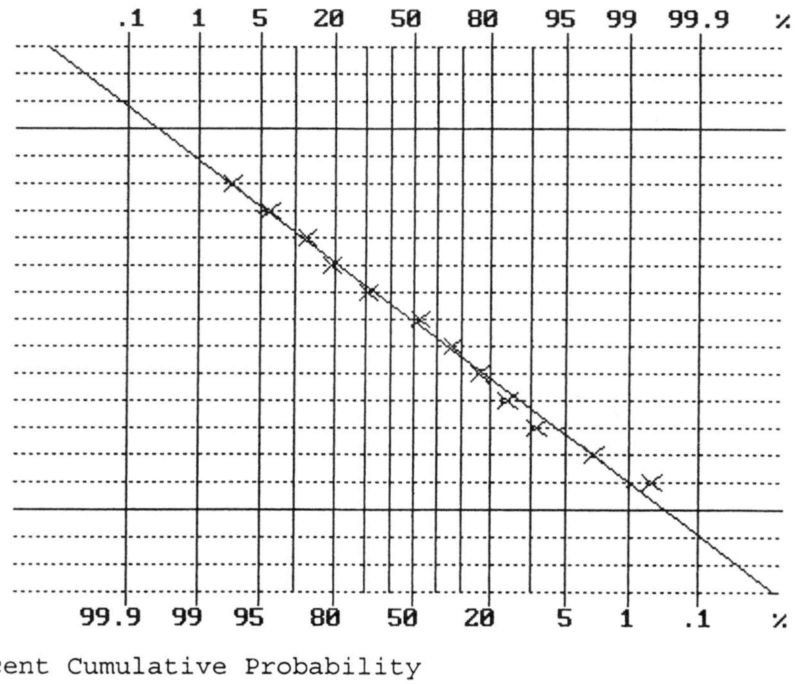

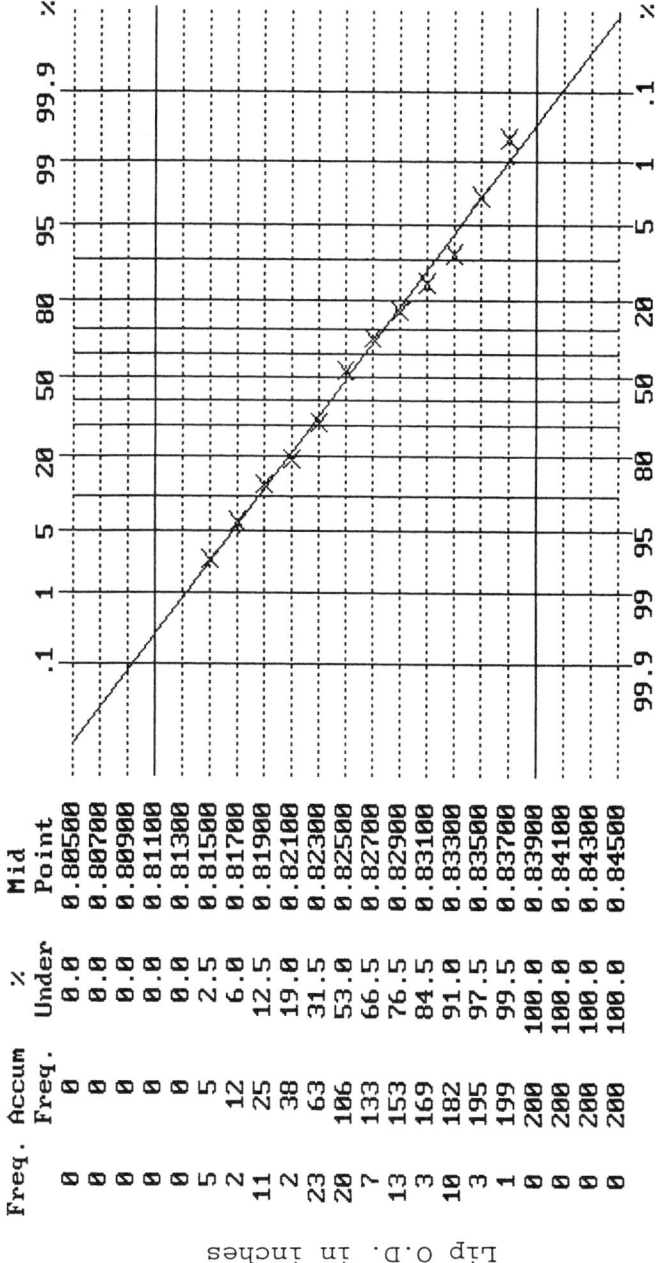

Figure 4 Probability plot. (Courtesy David C. Crosby, The Crosby Co.)

III. LONG-RUN CAPABILITY STUDIES

A. Planning the Study

This type of capability study must be run over a long enough period of time to include the effects of different operators, different lots of materials or components, and so on. The information resulting from a long-run capability study can be used for many purposes:

1. Establishing setup procedures
2. Appraising the adequacy of product specifications
3. Troubleshooting process problems
4. Improving the process by eliminating assignable causes of variation.

In planning long-run studies, the first questions that must be asked are: (1) Are the measurements reliable? and (2) How many sample units are required per subgroup? It is also assumed at this point that the measurements are variables data (dimensions, mass, etc.) and that process variation is much greater than that of the measuring process.

The number of samples per subgroup depends on the size of the shift from the process average that you wish to detect. The information shown in Table 3 is a guide for selecting subgroup size. A subgroup size of five sample units is often

Table 3 Selection of Subgroup Size for Long-Run Capability Studies—Variables Measurements

To detect this shift in either direction from the process average ($\overline{X}'$) (σ)	Use the following subgroup size
1.00	23
1.25	15
1.50	10
1.75	8
2.00	6
2.25	5
2.50	4
2.75	3
3.00	3
3.25	2

α = Risk of detecting a *false* difference above or below the process average = 0.025

β = Risk of overlooking a *true* difference above or below the process average = 0.10

$n = (K_{\alpha/2} + K_\beta)^2/(D)^2$

n = Subgroup size

$K_{\alpha/2}$ = Normal deviate associated with $\alpha/2$

K_β = Normal deviate associated with β

$D = (\overline{X} - \overline{X}')/S$

used for convenience, but its use is not obligatory. The cost of obtaining and inspecting each sample unit must also be considered. If these costs are very high, fewer samples per subgroup are used.

At least 20–25 subgroups are selected from the process. The frequency of sampling subgroups depends on how rapidly the process average shifts. A process whose average changes rapidly must be sampled more frequently than a process whose average changes slowly. The frequency of sampling is usually every hour or two when the process is a typical continuous process. Unless final inspection is considered to be part of the process, obtain the subgroups before this operation.

A "log" must be kept of all process adjustments and unusual occurrences, such as line downtime, material or supplier changes, and so on, during the time the study is run. These occurrences, although they may not seem important at the time, affect the interpretation of the study results.

Although it is possible to run process capability studies using attributes (fraction defective or numbers of defects) rather than variables (dimension, mass, etc.), this is not generally recommended because (3):

1. Studies based on attribute data cannot differentiate among out-of-control results caused by a shift in the process average, a change in the product specification, or excessive process variation causing more product to fall outside the limits (nonconforming product).
2. To achieve the same sensitivity as variable measurements, larger sample sizes must be used; hence sampling and inspection costs will be higher.

For process capability studies, when sampling from single-station processes, it is customary to draw consecutive sample units at each time period. However, sampling from processes containing multistation equipment, such as a multicavity plastic injection mold, is much more complex. Strictly speaking, each independent station is a separate process, but if 12, 24, 48, or more independent stations are present, it may not be cost effective to run that number of concurrent capability studies.

One way to solve this problem is to select a representative station and to study *only* this one. Practically, this may not be feasible since it is difficult in a high-speed process to identify the station from which a particular sample unit comes. If the product, as it is made, is identified with station numbers or letters, one station is easily sampled.

If the sample units cannot be easily identified by station, it is a generally accepted practice to sample the subgroup units at *random* from the output of the process. In this case the "inherent" process variation is a mixture of within-station and between-station variation and the two sources cannot be separated unless the numerical value of one of these sources has been calculated from past data. A second problem is that a sample size is often selected which is an "even" function of the number of stations. For example, if there are 12 independent stations, a subgroup size of 2, 3, 4, or 6 is not recommended, but one containing 5, 7, or 9 product units is recommended because these will discourage false cyclical data patterns. Just as in the short-run capability studies, great care must be taken to assure that normal operating conditions (NOC) are maintained throughout the capability study. Any departures from NOC must be noted in the log.

B. Example of a Long-Run Process Capability Study: Precision Industry

A staking operation was required in the process of riveting a metal tongue to a flat surface during the fabrication of Framis Widgets. The rivet head height after staking was a critical quality characteristic. If rivet heights were too low, the tongue would be detached when an 8-lb pull was applied; if the height were too high, the head would not be broad enough to retain the metal tongue when this force was applied.

A capability study was run to investigate the quality characteristics of this staking operation before customer complaints became excessive. Based on preproduction runs, the quality control specification for rivet head height was set at 74 ± 6 thousandths of an inch. In this capability study it was important to detect a shift of 3σ units in either direction from the specification nominal (or midpoint) with a high degree of assurance *if* a shift occurred. Referring to Table 3, this meant that three consecutive riveted assemblies must be sampled at each time period. Because considerable line downtime had occurred in the past, a sampling frequency of 30 minutes was specified for the study.

The data resulting from this capability study are shown in Figure 5 and the capability study control charts for subgroup averages and ranges are shown in Figure 6. The limits for these charts were calculated from the factors shown in Table 4 by the methods given in Grant and Leavenworth (9). In Figure 6, all the plotted points are within the range and the average chart control limits. This is the first condition for a controlled or steady-state process: the absence of excessive time-to-time variation when chart limits are calculated from capability study data (10). The second condition for a controlled process is that data points plotted on control charts must vary in a random manner about their centerlines. This means that plotted points must not show the presence of trends, cycling, or discontinuities (11). If the plotted points show any of these effects, assignable causes of variation are present in the process even though all the plotted points are inside control chart limits. If any one of these conditions for control is not met, assignable causes must be identified and removed before calculating the natural tolerance or inherent capability of the process, because a statement of process capability implies that the process is in a steady state.

A simple test for detecting nonrandom data patterns on control charts is the runs test. This test determines whether the distribution of plotted points above and below the control chart centerline follows a random pattern. The median value of individual measurements is frequently used for this purpose since it is that value (regardless of distribution shape) which divides the data exactly in half. If it is an average chart, we can safely use the control chart centerline for this purpose even though individual measurements are not normally distributed. (The central limit theorem is a great comfort to those who have little faith in the normality of industrial measurements.)

Runs above and below control chart centerlines are defined as one or more sequential points lying on one side of the line. For example, in Figure 6, the first two points plotted on the average chart constitute the first run, while the first point plotted on the range chart is its first run. If every two consecutive points are

DIMENSIONAL CONTROL RECORD

PRODUCT FRAMIS - WIDGET CODE 7857 PLANT # 75

OPERATION RIVET STAKING LINE NO. 7 MACHINE 7A

DATE 6/29 SHIFT 1 + 2 OPERATOR JDL + KR INSPECTOR JD + MS

SAMPLE IDENTIFICATION

MEAS. NO.	AM 7:30	7:55	8:30	9:10	9:25	10:00	10:30	11:05	*NOON 12:00	PM 12:45	1:15	1:35
1	72	70	75	70	74	72	70	74	73	75	72	72
2	70	72	70	74	72	73	70	74	73	70	74	74
3	72	73	75	70	74	73	72	75	70	70	76	70
4												
5												
TOTAL	214	215	220	214	220	218	212	223	216	215	222	216
AVG.	71.3	71.7	73.3	71.3	73.3	71.7	70.7	74.3	72.0	71.7	74.0	72.0
RANGE	2	3	5	4	2	1	2	1	3	5	4	4

MEAS. NO.	PM 2:00	2:25	2:55	3:30	3:55	4:30	5:00	5:25	* 6:00
1	74	75	72	74	74	73	72	72	72
2	72	70	73	77	74	75	70	76	73
3	76	70	73	74	75	75	72	75	74
4									
5									
TOTAL	222	215	218	225	223	223	214	223	219
AVG.	74.0	71.6	72.6	75.0	74.3	74.3	71.3	74.3	73.0
RANGE	4	5	1	3	1	2	2	4	2

COMMENTS * LINE DOWN - STUCK METAL

HT. OF RIVET HEAD (IN 1000 THS OF AN INCH)

Specification: Rivet Head Height after Staking : 74 ± 6 thousandths

See Table X for Control Chart Factors

GRAND AVG. = $\frac{\text{Sum of Totals}}{\text{Total No. Meas.}}$ = 4587/63 = $72.81 = \bar{\bar{X}}$

AVERAGE RANGE = $\frac{\text{Sum of Ranges}}{\text{Total No. Groups}}$ = 60/21 = $2.86 = \bar{R}$

Figure 5 Long-run capability study: rivet-staking dimensional control record.

connected by a straight line, it is very easy to count the total number of runs: count the number of times the connecting lines cross the centerline and add one to the total number of crossings.

In Figure 6, we count 9 runs for the range chart and 14 runs for the average chart. Entering Table 5 with 21 plotted points, we see that the limits of runs for

Subgroup Size (n) = 3

1. Grand Average ($\bar{X}$) = 72.81 thous.
2. Average Range ($\bar{R}$) = 2.86 thous.
3. Limits for Average Chart
   a. Upper Control Limit ($UCL_{\bar{X}}$)

   $\bar{X} + A_2\bar{R}$ = 72.81 + 1.02 (2.86) = 75.73

   b. Lower Control Limit ($UCL_{\bar{X}}$)

   $\bar{X} - A_2\bar{R}$ = 72.81 − 1.02 (2.86) = 69.89
4. Limits for Range Chart
   a. Upper Control Limit (UCL_R)

   $D_4\bar{R}$ = 2.57 (2.86) = 7.35

   b. Lower Control Limit (LCL_R)

   $D_3\bar{R}$ = 0 (2.86) = 0
5. Process Capability or Natural Tolerance:

   $\pm 3\sigma' = \pm 3 (\bar{R}/d_2)$ = ± 3 (2.86/1.693) = ± 5.07 thous.
6. Specification: 73 ± 6 thousandths

   % Outside of Specifications:

   $\frac{USL - \bar{X}}{\sigma'}$ = (79 − 72.81)/1.69 = + 3.66 ≈ 99.989 %

   $\frac{LSL - \bar{X}}{\sigma'}$ = (67 − 72.81)/1.69 = − 3.44 ≈ 0.003 %

   % Inside (By Subtraction) = 99.98⁺ %

   % Outside = 100 − % Inside = 0.02 %
7. Capability Index = 6 Spread of Process/6 Spread of Specification:

   CI = 10.14 thous. 12.0 thous. = 0.85

   Process is OK but requires watching

Figure 5 Continued

20 points are 6–15. Since the actual numbers for both charts are within these limits (the number of runs on the average chart misses the upper limit by one), there is no *strong* evidence that systematic process disturbances occurred during this capability study. To obtain additional information about the random distribution of points about control chart centerlines, we also examine the *length* of runs on either side of the line. An article by Frederick Mosteller, in 1941, gave the 95% limiting values for lengths of runs. A summary of this information is as follows (11).

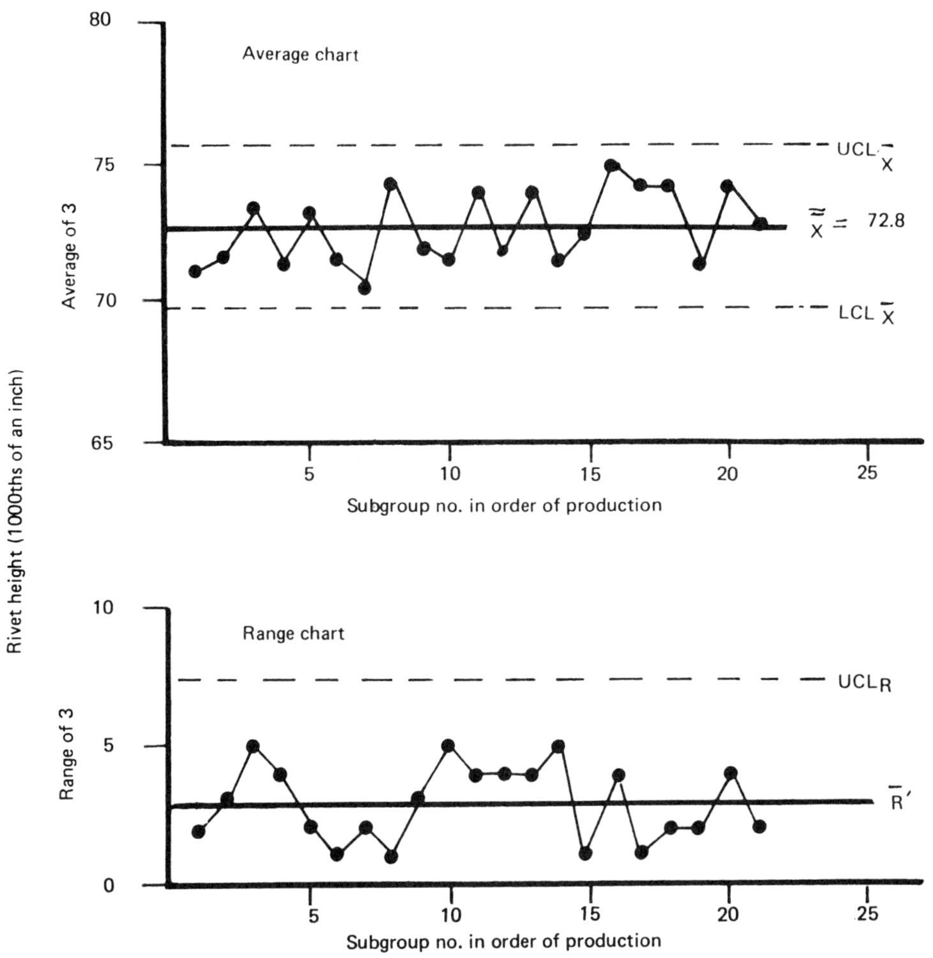

Figure 6 Control charts: final height of rivet head, subgroup size (n) = 3.

Number of plotted points	Maximum run length about centerline
10–14	5 successive points
15–19	6 successive points
20–29	7 successive points
30–35	8 successive points

In addition, the Western Electric (now, AT&T) SQC Handbook recommends the following criteria for judging out-of-control conditions for average charts (12):

1. If two out of three successive points fall at or outside the 2σ limits, assignable causes of variation are very likely to be present.
2. If four out of five successive points fall at or outside the 1σ limits, assignable causes of variation are likely to be present.

Table 4 Factors for X-Bar and R Charts

Chart for	Central Line	Chart Limits
Averages (X-bar)	$\overline{\overline{X}}$	$\overline{\overline{X}} \pm A_2 (\overline{R})$
Ranges (R)	$\overline{R}$	D_3R and $D_4\overline{R}$
Process Capability	–	$\pm 3\ \sigma' = \pm 3\ (\overline{R}/d_2)$

No. of Sample Units	Multipliers for chart limits and process capability			
	A_2	D_3	D_4	d_2
2	1.880	0	3.267	1.128
3	1.023	0	2.575	1.693
4	0.729	0	2.282	2.059
5	0.577	0	2.115	2.326
6	0.483	0	2.004	2.534
7	0.419	0.076	1.924	2.704
8	0.373	0.136	1.864	2.847
9	0.337	0.184	1.816	2.970
10	0.308	0.223	1.777	3.078

ASTM Manual on Quality Control of Materials, January, 1951, has very extensive tables of control chart factors.

For similar criteria for range charts, consult the Western Electric reference.

In our rivet height capability study (see Fig. 6), the longest run on the range chart contains six points (it misses its limit by one), while the longest run on the average chart is only three points. On the average chart, only two plotted points are at or beyond the 2σ limits (one point above and one point below the centerline). Again, there is no *strong* evidence of systematic process disturbances. We may therefore calculate the natural tolerance (process capability) of the rivet heights.

The natural tolerance of the rivet heights was ±5.1 thousandths (see the calculation sheet for Fig. 5), while the specification tolerance was ±6.0 thousandths. Comparing these two values, we obtain a capability index of 0.85, which is somewhat larger than the ideal ratio of 0.75. The significance of this difference is evaluated by plotting the capability study data against control chart limits calculated from the rivet-staking specifications (13). These plots are shown in Figure 7. Although the process average was below the specification nominal (72.8 versus 74 thousandths), none of the plotted points were outside the 3σ limits of the average and range charts. If the rivet-staking process average is adjusted to the 74-thousandths nominal value, the process is very capable of staying within the specification limits.

Table 5 Number and Length of Runs on Either Side of Control Chart Centerline (95% Confidence Limits)

	Excessive fluctuation	Insufficient fluctuation
No. of plotted points	If no. of runs > the no. given below	If no. of runs < the no. given below
10	9	2
11	9	3
12	10	3
13	11	3
14	12	3
15	12	4
16	13	4
17	13	5
18	14	5
19	15	5
20	15	6
25	18	8
30	21	10
35	24	12
40	27	14
45	30	16
50	33	18

This chart is based upon the binomial distribution: $P = .50$ and N = number of plotted points.
Too few runs (insufficient fluctuation) reflect trends, discontinuities, or long-term periodicity. Too many runs (excessive fluctuation) reflect "mixtures," overadjustment of the process, or short-term periodicity.

C. Example of a Long-Run Capability Study: Process Industry

Process industries have unique characteristics: (1) they modify raw materials by biological, chemical, or physical processes to make the final product; (2) samples taken during processing often have different chemical compositions and quality characteristics from those of the finished product; and (3) processes making the finished products are usually technically complex, requiring highly skilled plant personnel. Some examples of process industries are pulp and paper processes, pharmaceuticals, processing of raw materials to form resins and plastic materials, and food processing or packaging.

Capability studies of these processes sometimes require special types of control charts. When, for example, a critical quality characteristic is the percent solids in a homogeneous organic slurry, *one sample* of slurry is representative of the entire

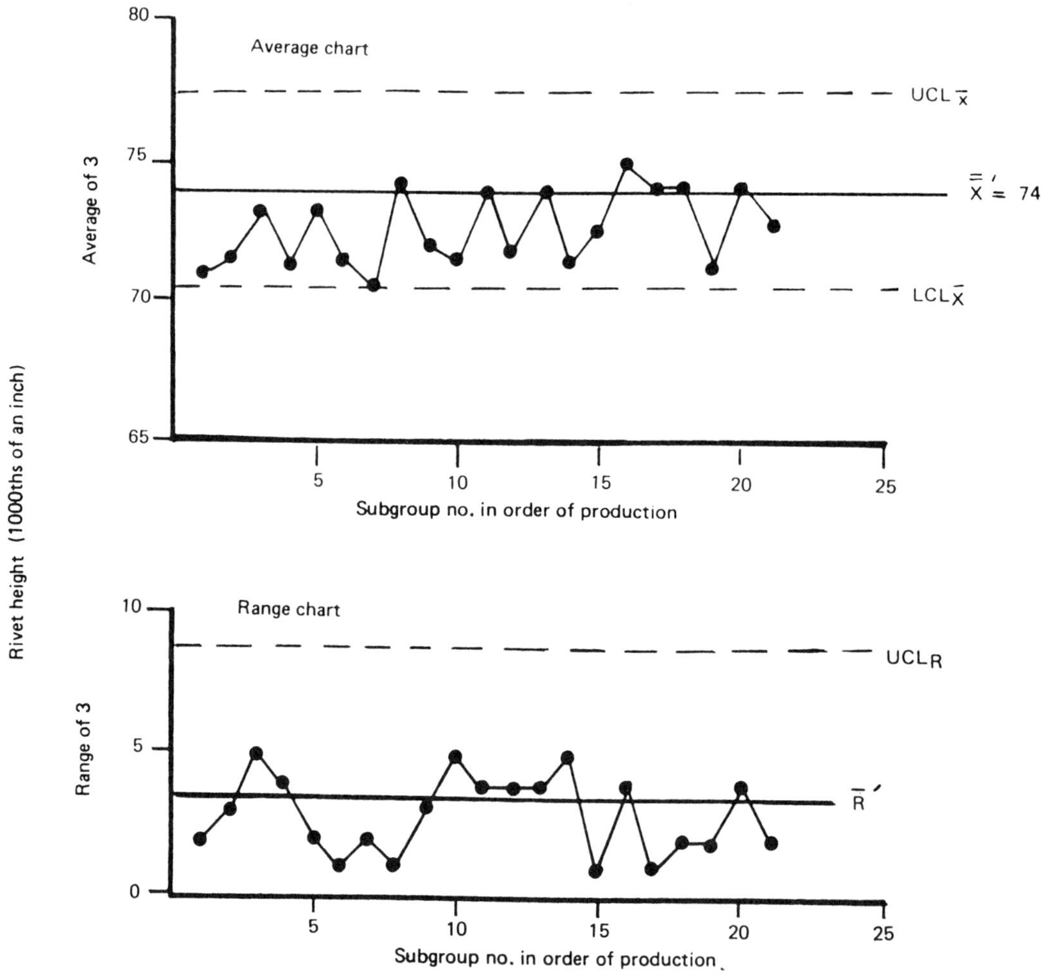

Figure 7 Data from Figure 6 plotted against control chart limits.

batch if it is drawn as the batch is continuously stirred in a large tank. Control charts for individual measurements are notoriously insensitive to small process disturbances; hence Cowden suggests using moving-average and moving-range charts for this type of process (14).

A second-type of process industry exists which combines both precision industry and process industry functions. When this combination occurs, capability studies are run in the usual way and the data are plotted using traditional control charts. An example of this process-precision combination is found in the food industry (e.g., capability studies of filling equipment) (Fig. 8) (15). Note that the range chart is not out of control, indicating that the filling equipment is inherently *capable* of filling within ±1.09 g. The average net weights, however, are very much out of control. A log kept during this study showed that the 3.3 hours of downtime could be "paretoized" (see Chapter 29) as follows:

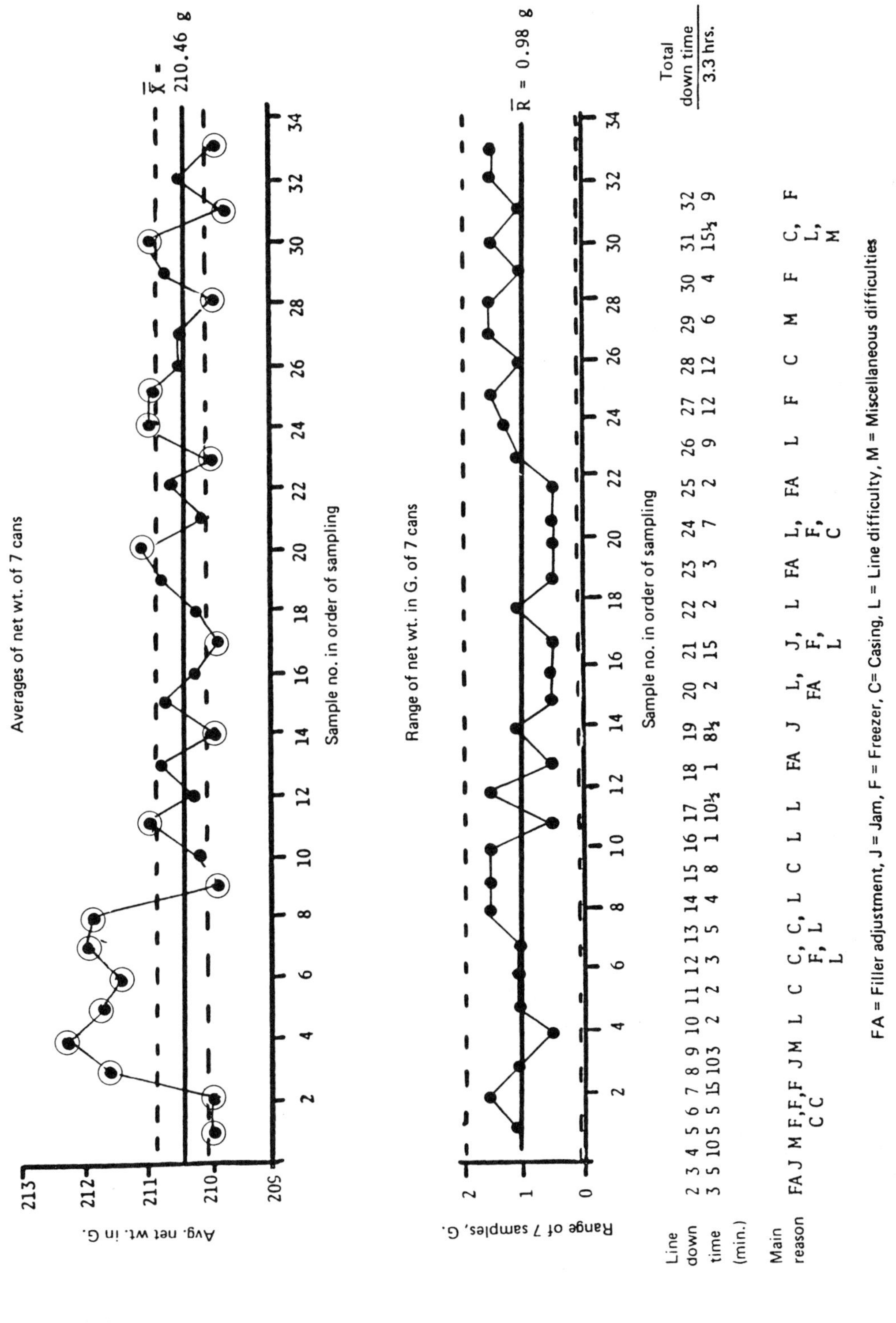

Figure 8 Filler control charts.

Unit operation	% of total downtime
Freezer	28.8
Time-feed problems	21.7
Caser	19.6
Miscellaneous	12.6
Seamer	12.2
Filler	5.1
Total	100

Obviously, the unit operation studied (filling equipment) was a minor cause of the quality problems in this process. To control filling operations effectively, other operations must be brought under control starting with the freezer.

IV. INTERPRETING CONTROL CHART PATTERNS IN CAPABILITY STUDIES (16)

A. Cycles

Cycles are short trends that repeat the pattern over a long enough period of time—processing variables come and go at somewhat regular intervals of time. The causes of cycles are as follows:

Average chart	Range chart
Temperature, humidity	Operator fatigue
Eccentricity	Maintenance
Differences in gages	Day shift/night shift
Worn threads	Tool wear

B. Gradual Change in Level

Some process elements originally affect only a few pieces, but the number affected increases with time. The process may settle down to a new level.

Average chart	Range chart
Gradual introduction of new materials, etc.	Better fixtures, better methods, greater skill of operators
Wear in tools, bearings, etc.	

C. Sudden Shift in Level

Shifts in process levels are characterized by a strong change in one direction—a number of points appear on only one side of the centerline.

Average chart	Range chart
New operator	New operator
New inspector	New equipment
New machine	Changes in material

D. Trends

Trends are characterized by a strong movement of data, up or down.

Average chart	Range chart
Tool wear	Increasing trends
Deterioration of stored solutions	Tool wear
Aging of materials	
Seasonal effects	
Changes in standards	

E. Freaks

Freaks are defined as a single unit or group of units that differ greatly from others.

Average chart	Range chart
Wrong settings	Accidental damage
Error in measuring	Incomplete operation
Incomplete operation	Breakdown of operations

F. Groups or Bunches

A group is a clustering together of similar measurements.

Average chart	Range chart
Measuring problems	Freaks in data
Change in calibration	Mixture of distributions

G. Instability

Process instability means that erratic data fluctuations occur which are too wide for control limits.

Average chart	Range chart
Overadjustment of machine	Untrained operators
Carelessness of operator	Mixture of material
Differences in test pieces	Machine needs repairing
	Assemblies off center

H. Mixtures

Average chart	Range chart
Affects this chart more frequently	Different lots of material
Is a combination of two different differences in gages levels—different lots of material	

REFERENCES

H. C. Charbonneau, G. L. Webster. *Industrial Quality Control.* Prentice-Hall, Englewood Cliffs, NJ, 1978.

Western Electric Co.*Statistical Quality Control Handbook,* 2nd ed. Mack Printing Co., Easton, PA (4th printing, 1970), Sec. II, Pts A and F.

The Bendix Corporation, Kansas City, MO. *Capability Study Handbook* (G-42), revised June 1968, prepared by SQC Dept. 416.

Shortcuts for machine capability studies. *Quality Management and Engineering*, February 1972.

J. W. Tukey. Non-parametric estimation II, Statistically equivalent blocks and tolerance regions—the continuous case. *Annals of Mathematical Statistics* 18:529–539, 1947.

H. Scheffé, J. W. Tukey. *Annals of Mathematical Statistics* 15:217, 1944.

H. Arkin, R. R. Colton. *Statistical Methods*, 5th ed. Barnes & Nobel Books, New York, 1970, p 3, footnote 2.

E. L. Grant, R. S. Leavenworth. *Statistical Quality Control*, 5th ed. McGraw-Hill, New York, 1972.

C. A. Bennett, N. L. Franklin. *Statistical Analysis in Chemistry and the Chemical Industry*, Wiley, New York, 1954, pp 663–677.

F. Mosteller. Note on application of runs to quality control charts. *Annals of Mathematical Statistics* V. 12:232, 1941.

Western Electric Company, Inc. *Statistical Quality Control Handbook*, 2nd ed. Mack Printing Co., Easton, PA, 1958, pp 25–27.

ASTM Manual on Quality Control of Materials. ASTM, Philadelphia, Jan. 1951, p. 72.

D. J. Cowden. *Statistical Methods in Quality Control*, Prentice-Hall, Englewood Cliffs, NJ, 1957, pp 342–345.

M.-G. Tarver. Systematic process analysis. *Quality Progress*, V. 7(9):28–32.

Western Electric SQC Handbook. Mack Printing Co., Easton, PA, 1956, pp 149–162.

31

Process Control

CHARLES A. CIANFRANI
Casmar Consulting Group
Green Lane, Pennsylvania

I. INTRODUCTION

Process control is a vital element of virtually every industrial enterprise, since profitability, productivity, and product quality depend very fundamentally on effective total control of the process producing a product. The most intense applications of process control are found in the process industries, such as chemicals, drugs, food, pulp and paper, primary metals (both ferrous and nonferrous), fabricated materials, glass, and textiles, and in petroleum refining, water and wastewater management, and generation and control of electric power. Typical processes requiring control include the conversion of crude oil into gasoline, wood pulp into paper, coal or oil into electric power, and iron ore into steel.

Processing can be performed in either a continuous manner (a continuous process) such as in petroleum refining or in a noncontinuous manner (a batch process) such as in heat treating. The types of properties that are typically controlled include temperature, pressure, flow, level, humidity, voltage, frequency, and analytical chemical properties such as pH and ionic concentration.

This chapter is directed at providing an understanding of the basic concepts of process control and how these concepts are applied in industrial environments. Such an understanding will enable effective communication with plant personnel in their own language on day-to-day quality issues and to be intelligently involved in the selection of new process control equipment, since such equipment has a significant effect on product quality. The approach used in this chapter is qualitative and intuitive and does not rely on complex formulas or abstract theory.

II. PROCESS CONTROL: A HISTORICAL PERSPECTIVE

Process control originated along empirical lines, based on intuition and experience. A good historical example of rudimentary but effective process control is the heat treatment of metals accomplished by a blacksmith, who watched flame color, material color, smoke, and time, to determine when the heat-treating task was completed. Since there were so many variables to watch, luck was a principal factor in such efforts.

As processes became larger in scale and more complex to meet volume and economic requirements, control was accomplished by using many people making individual judgments on a loosely coordinated basis (Fig. 1). In the 1950s the indication of process status and control was consolidated into control rooms (Fig. 2), where a number of plant operators could maintain overall surveillance of the process from one location. Also in this time period control actions began to be dictated by electronic controllers rather than by operators. Consequently, process control operations were performed in a more consistent and uniform fashion.

In the 1960s and early 1970s process control continued to become more sophisticated and more centralized due to the improved capability of indicating and controlling instrumentation (Fig. 3). More information was available about process phenomena and in a more usable form. Fewer operators were required and the results achieved became even more predictable. Also, process optimization (to increase throughput and lower costs) began to make a significant contribution to overall process operation. Although analog instruments and systems were predominant at this time, digital-based systems began to make their presence felt.

During the early 1970s the capabilities of digital-based process control systems began to make them more desirable than analog systems. Digital systems (Fig. 4) were able to process more information faster and make this information available to an operator in a more usable form than were equivalent analog systems.

The next major evolutionary step occurred in the mid 1970s when distributed process control systems made their debut (Fig. 5). In such systems the control

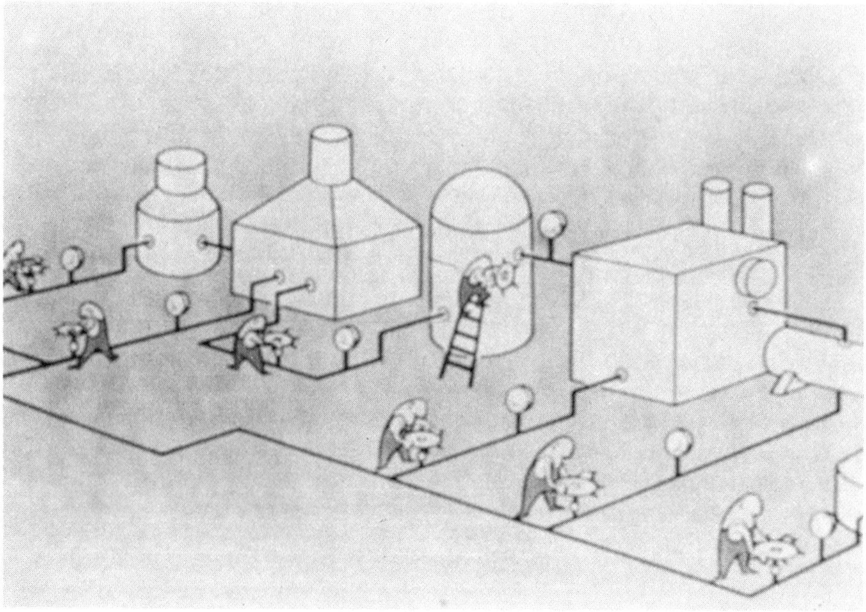

Figure 1 Early process control; many people, many individual judgments.

Figure 2 During the 1950s, process control moved into control rooms.

decisions were distributed into control rooms throughout the plant which were connected to a central control room where a computer and a chief operator could oversee the entire process. The distributed control approach combined the features, advantages, and benefits of all the previous approaches while eliminating most of the disadvantages.

Since the advent of distributed process control, advances have been made primarily in communications between field sensors and control rooms, in approaches used for interfacing the control system with operators (i.e., the human-machine interface), and in the basic speed and capability of the hardware elements of process control systems. Also, the power of the microprocessor and associated integrated-circuit chips has given rise to a plethora of process control system packages optimized for specific marketplace applications. All of the currently available sophistication, however, does not affect the validity of or the need to understand the basic concepts of process control that follow.

III. BASIC CONCEPTS OF PROCESS CONTROL

A. Simple Process

To illustrate the basics of process control, consider a simple process (Fig. 6) composed of a source of raw feed liquid (A) which flows into a tank at a variable rate and must be heated so that it leaves the tank at pipe (B) as hot feed at temperature T.

Figure 3 In the 1960s and 1970s, more sophisticated and improved instrumentation was developed.

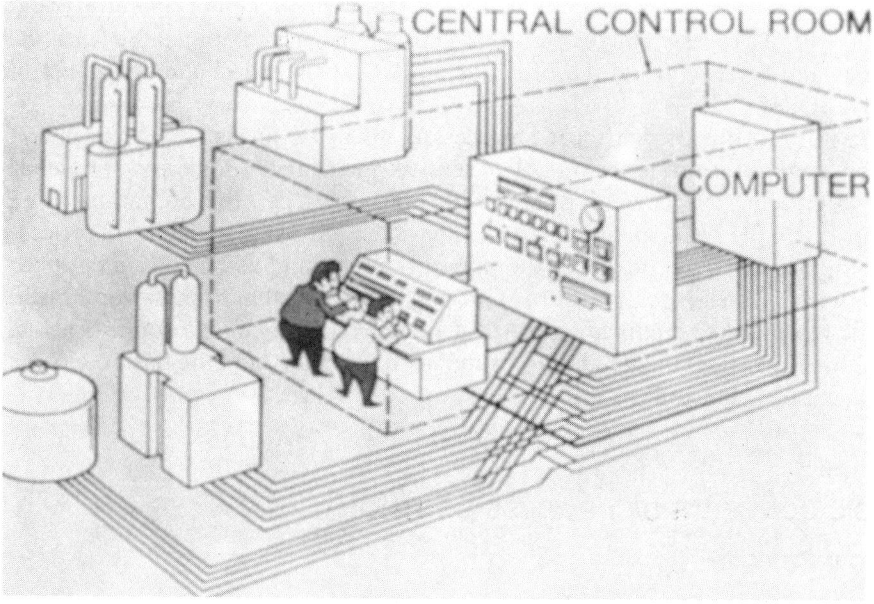

Figure 4 New digital systems in the early 1970s processed information faster, and data were more usable to the operator.

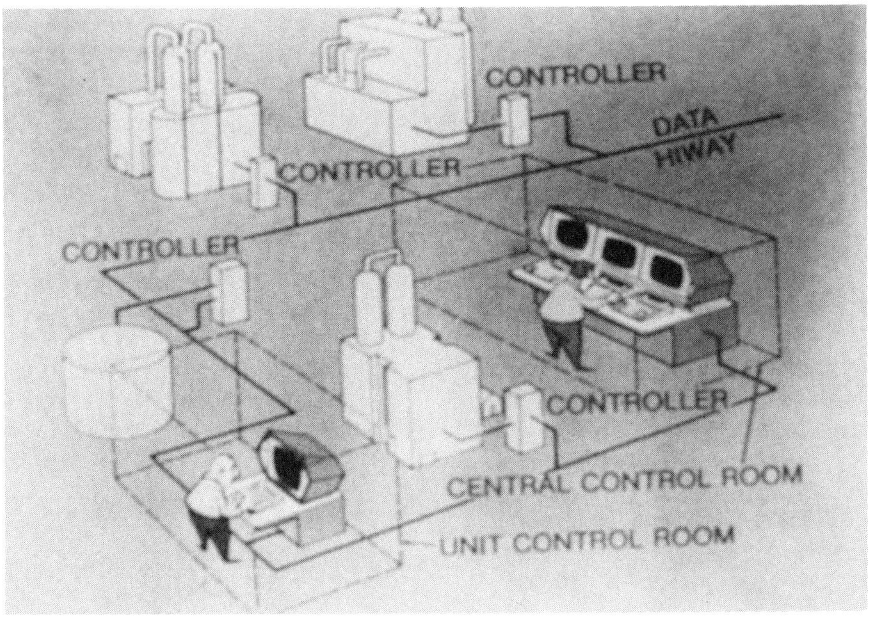

Figure 5 Distributed process control systems made their debut in the mid-1970s.

To heat the raw feed there is hot oil available (C) which flows through a heat exchanger.

The temperature of the feed liquid leaving the tank will be directly affected by the amount of hot oil passing through the heat exchanger. Therefore, the desired temperature, T, for the hot feed can be maintained by regulating the amount of hot oil, C, flowing through the heat exchanger. It is also desirable to assure that the level of feed in the tank does not exceed the top of the tank and that the tank does not become empty.

To decide how to regulate the hot oil flow, and hence the temperature of the raw feed, it is necessary to know the temperature of the raw feed liquid. This can be accomplished in many ways. Most simply, a mercury thermometer could be held in the tank and read directly. A more appropriate approach would be to sense the temperature of the feed with a suitable *temperature sensor* (such as a thermocouple) and to display the value of the temperature on a chart recorder. Similarly, a float and a float-level recorder could be used to indicate liquid level. Finally, assume that (1) there are valves on the hot line (V_C) and on the output pipe of the tank (V_B), (2) the valves can be either fully open or fully closed, (3) initially the level of the tank is constant, and (4) the main concern is maintaining control of the temperature T at some desired value, say 150°C, which will be called the *set point*.

On-Off Control

Now, how could an operator go about maintaining the hot feed output from the tank (B) at a specific temperature, T? The operator can fully open or fully close the two valves, V_C and V_B, thereby regulating the amount of hot oil passing through the heat exchanger and the amount of hot feed flowing out of the tank. This form

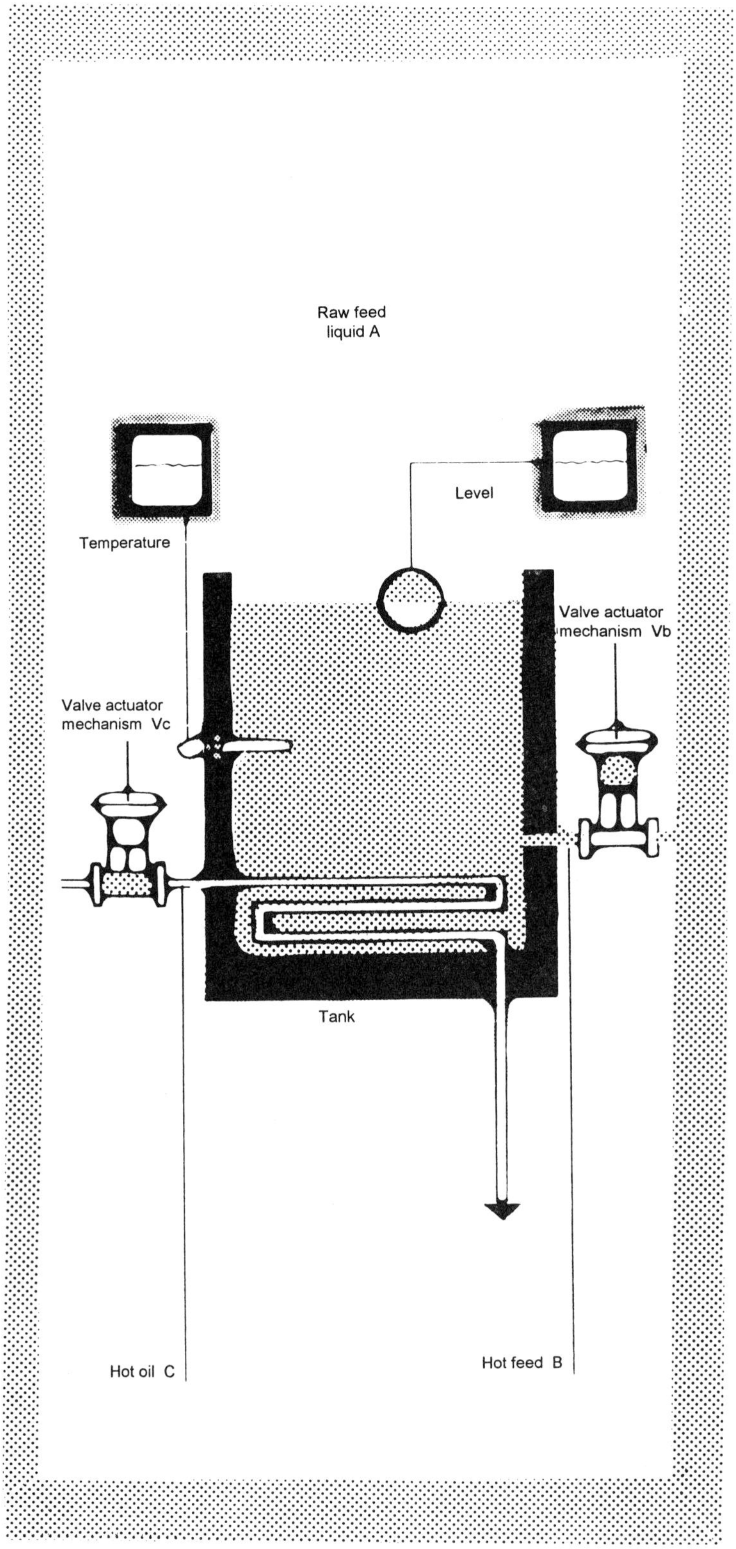

Figure 6 Basic process control in a simple process.

of control is called two-position or on-off control. The operator is comparing the indication of temperature on the chart recorder with a mental target of the set point and is opening valve V_C when the indicated temperature is below 150°C and closing V_C when the temperature is above 150°C.

The indication of raw feed liquid temperature on the chart recorder would show oscillation about the set point of 150°C. The primary reason for this is that the heating effect of the hot oil is either completely supplied or completely withheld from the liquid in the tank. The amount of the oscillation above and below the set point depends on the lags of the process and the attention of the operator to the difference (or *error*) between the set point of 150°C and the temperature of the tank. The tank temperature in this example is known as the *process variable*, a control term that is used to identify a quantity in the process that is changing.

Proportional Control

It is obvious that although on-off control will provide a degree of control over the process, something better is possible and desirable. If the process is viewed as a balance between energy in and energy out, it is reasonable to assume that smoother control of the output liquid will result if a steady flow of hot oil, C, is maintained, rather than alternating between full flow of hot oil and no flow of hot oil (i.e., on-off control). This control concept is analogous to taking a shower. It is more pleasant to have a steady stream of water at about 35°C than to alternate between water at 55°C and water at 10°C.

But what is the correct rate of hot oil flow? The correct rate of oil flow to hold the temperature of the output liquid at set point is obviously related to the rate of liquid flow into and out of the tank. Therefore, the on-off *control mode* must be modified. This can be accomplished by first establishing a steady flow value for hot oil that at average operating conditions tends to hold the process variable (the tank temperature) at the set point (i.e., 150°C). Once the flow value for the hot oil has been established, increases or decreases of the process variable from the set point (call this the *error*) can be used to cause corresponding increases or decreases in the hot oil flow.

The concept of taking corrective action in proportion to changes in the deviation of the set point from the process variable (i.e., changes in the error) is the concept of *proportional control*. To implement proportional control on a process, the control valves used must be of a type that can be positioned at any degree of flow from fully opened to fully closed. In addition, a mechanism to move the valve is required, such as an electric motor or a pneumatic valve positioner.

With such a system configuration (Fig. 7) the operator can manually make gradual (proportional) adjustments to the hot oil valve as the temperature of the tank deviates from the set-point temperature. Control action (i.e., moving the hot oil control valve) should be relatively infrequent since a steady flow of hot oil will be maintained that is in close balance to the average needs of the process. The degree of sensitivity of valve change to error is called the *proportional gain*. The term that describes the new valve position where the process variable equals the set point is *manual reset*.

Proportional control is certainly much more effective than on-off control (and in practice costs more to implement), but it does have a serious deficiency. If there

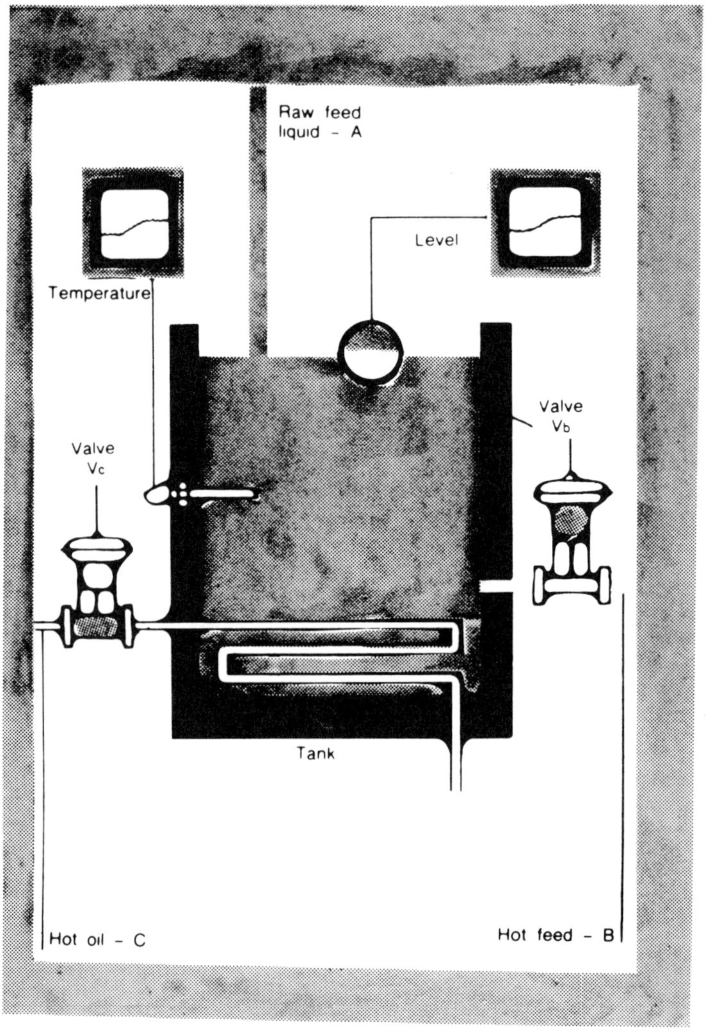

Figure 7 System configuration of proportional control on a process.

are frequent disturbances or process upsets, the process temperature hardly ever stays at set point. There is only so much proportional gain that can be applied before the process becomes unstable. Frequent and/or large changes in manual reset are not practical. Therefore, something still better than proportional control may be required.

Integral Control

If the manual reset adjustment mentioned above could be made automatically, the offset error due to load changes would be eliminated. One approach to automatic adjustment of reset is to move the valve at a speed proportional to the deviation of the process variable from the set point. In other words, process control would be enhanced if a piece of instrumentation (such as a controller) would, through electronics, automatically move the valve faster if the deviation from set point became larger, and conversely move the valve more slowly if the deviation became smaller. When there exists no deviation from setpoint, no valve motion would occur.

Figure 8 illustrates what happens to the process illustrated in Figure 6 when there is a change in the raw feed flow. Without any control, when the raw feed flow (A) increases, the temperature in the tank drops below the set point. With proportional control, the hot oil control valve opens and the process temperature rises to a level close to the set point but with a temperature offset. With automatic

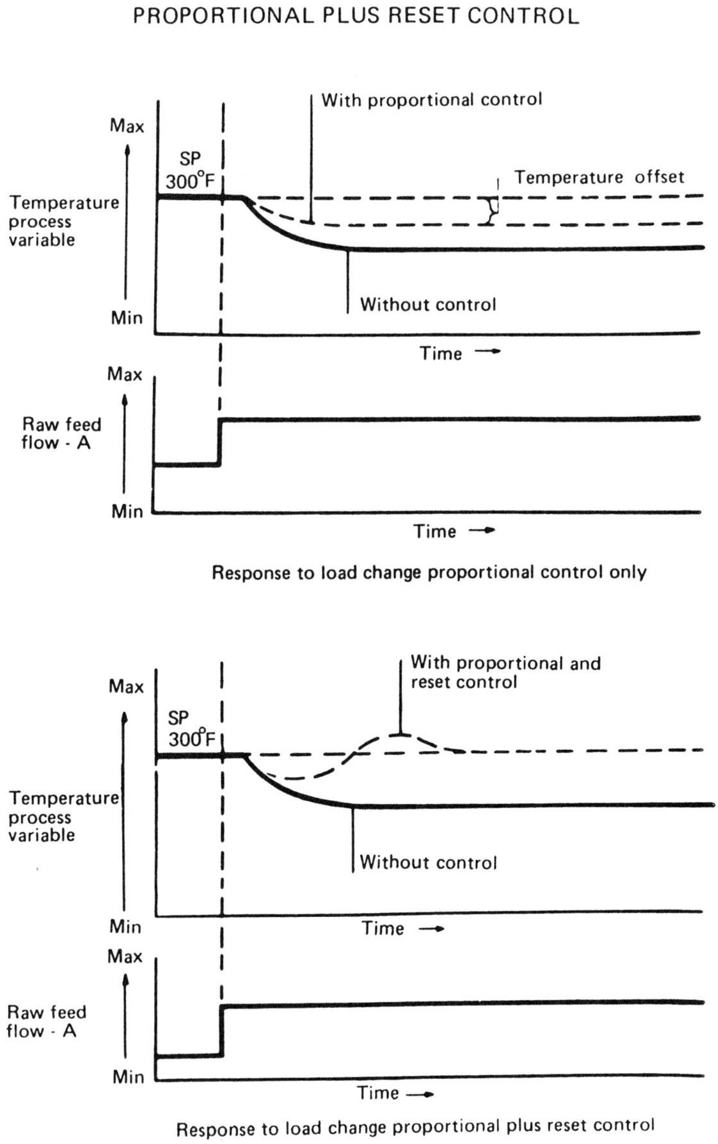

Figure 8 Process operation shown in Figure 6.

adjustment of the offset, the error is eliminated (after an initial overshoot of the set point).

The term used to describe the automatic adjustment of the reset is *integral control*. This name arises from the fact that the valve position is related to the integral of the error which has existed since time zero. When proportional and integral control are combined in the manner just described, the control form or control mode is called *two-mode* or *proportional-integral control*, commonly referred to as PI control.

Derivative Control

Finally, it seems reasonable to take one last step in addition to proportional control and integral control—taking control action based on the rate of change of an error signal. That is, the valve can be made to move proportionally in response to a changing deviation from the set point. This additional correction exists only when the error is *changing*. It disappears when the error stops changing, even though the size of the error may still be large. Such control action is called *derivative control*. Derivative control enhances process control because its contribution to control action is significant when the rate of change of the error signal is large and no contribution to control action is present when there is no rate of change of the error signal. Derivative control is illustrated in Figure 9. When derivative control is combined with proportional control and integral control, the control form is called *three-mode* or *PID control*.

A Few More Basic Concepts of Process Control

There are a few additional process control terms that should be considered in conjunction with those described earlier: open-loop control, closed-loop control,

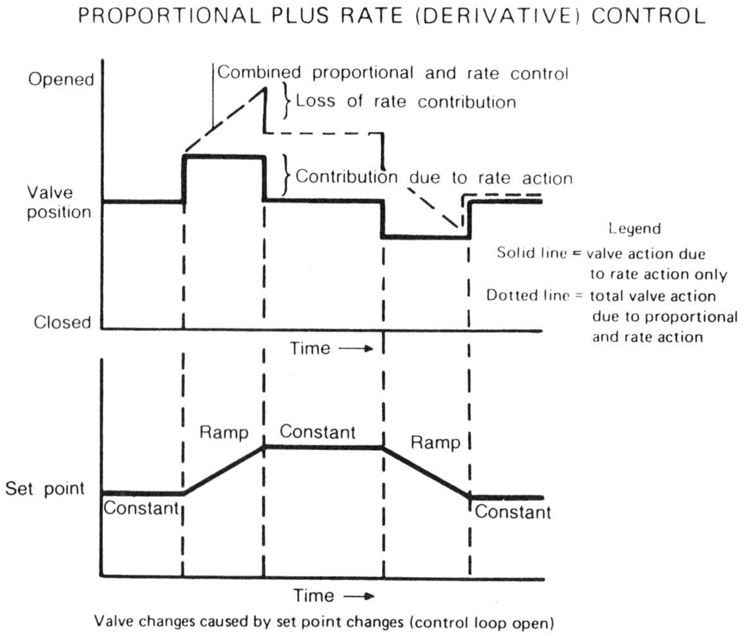

Figure 9 Schematic of derivative control.

and feedback. At times, control signals may be applied to a process based on information that is not directly obtained from the process variable; that is, the process is being controlled indirectly or inferentially. Under such conditions the process is considered to be operating under *open-loop control.* Another way of saying this is that the measured value is *not* compared directly with the set point. Experts in the process control field believe that many process control loops are operated in such an open-loop manner.

Conversely, a control loop is operating in a *closed-loop* manner when the results of the control manipulations are *compared* to the set point. The process by which the comparison against the set point takes place is called *feedback.* The comparison against set point is very important. Unless the results of the control manipulations are compared against an objective (i.e., the set point), continuous human interaction will be required. With comparison to the set point occurring continuously, corrections can automatically be made as required. Feedback is therefore a most fundamental and necessary part of an effective process control system.

B. Simple Process Control System

Now that the basic concepts of process control have been presented, it is appropriate to consider the nature and function of a typical process control system in an industrial environment. The key elements of a system to control a process are sensors, controllers, and final control elements. Sensors—such as temperature sensors (i.e., thermocouples) and level sensors—tell us what is happening with the process. Temperature and level sensors are only a few of the many kinds of sensors that can be used in a process control system. There are sensors for the detection of pressure, humidity, flow, pH, conductivity, turbitity, speed, and many other physical and chemical phenomena. The proper selection and application of sensors is a specialized art/science that is a critical element of any process control system. Controllers take inputs from sensors and determine what can be done to assure that the process behaves as expected. Controllers implement control actions, such as proportional, integral, and derivative control. Final control elements, such as valves, are used to implement the instructions of the controllers. In other words, a process control system includes *inputs* to controllers (from a thermocouple, for example), the controllers, and *outputs* from controllers to final control elements such as valves.

Knowing and understanding a few control forms and how a controller works however is not the same as understanding process control. There are additional factors to be considered. Processes have limits of controllability. Also, control loops can interact. Control of one variable at the most desired set point might preclude control of another variable at its optimum point. Also, proper tuning of control loops is certainly an art. It is easy to see—after the fact—why an integral action setting of x and a derivative action setting of y are appropriate. When one must actually establish the controller settings on line for a process, the "appropriate" settings are not nearly as obvious.

The basics of process control were illustrated earlier by means of a simple process that involved level, temperature, and flow control. The basics can be illustrated further by looking at a block diagram of a process and the minimum equipment necessary to control it.

Figure 10 illustrates a simple control loop where information is fed back from the output of the process in a way that effectively modifies the input to the process

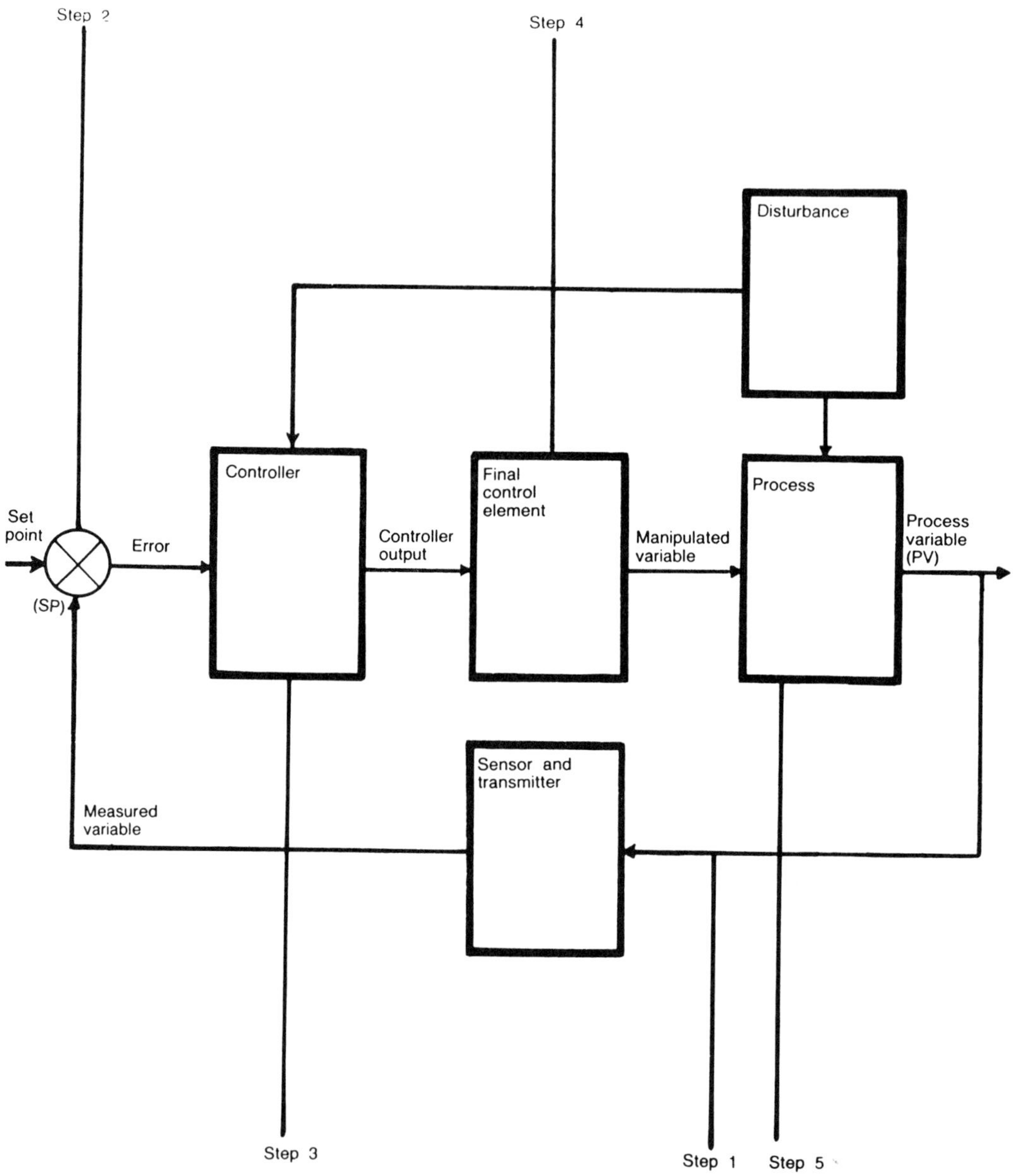

Figure 10 Simple control loop. Data fed back from the output modify the input to the process.

to maintain process operation at the desired condition. Five steps are essential to achieving good process control for such a control loop.

1. The process variable that best represents the desired condition of the final product must be measured. This depends on the specific product. Measurements such as temperature, pressure, and flow are commonly used.

2. The process variable is compared with the desired value for that variable (i.e., the set point) and an error signal is generated.
3. The error signal is applied to the controller.
4. The controller determines the signal that must be applied to the final control element, such as a valve.
5. The process is continually manipulated in response to the output of the controller so that the difference between the process variable and the set point is as close to zero as the sensitivity of the system permits.

Another way of saying this is that the process control system must effectively *measure* what is to be controlled, *compare* the measured value with a set point (i.e., the desired value), and *correct* for any differences between the measured value and the desired value. The process of measuring, comparing, and correcting is the foundation of process control.

IV. CONCLUSION

The objective of this chapter was to provide an understanding of the basic concepts of process control. It has covered the key elements of a process control system, including sensors, controllers, and final control devices, and has illustrated how these elements can be configured to control a process. The basic control approaches or modes of control—on-off control, proportional control, integral control, and derivative control—were described and an intuitive explanation of the nature of each control form was presented. Throughout the chapter, common control terms were defined and explained. Overall, a foundation was established that will permit the quality practitioner to understand and appreciate the nature of process control systems and their contribution to optimum process operation.

There is a considerable body of knowledge that goes beyond the basics in every area discussed above. Selection and application of sensors is an area of specialization to which some engineers devote a lifetime. Development of basic and advanced control strategies and techniques for optimizing processes is similarly a segment of technology that provides full-time employment for many engineers. Tuning of controllers is an artform developed through many hours of exposure to actual operating conditions, coupled with a sound understanding of the theoretical bases of process control. The determination of acceptable cost-performance trade-offs must balance at least initial instrumentation investment, ongoing maintenance costs, process throughput, product quality, and system reliability. Finally, understanding the hardware and software used to achieve process control is becoming more and more difficult as the microprocessor continues to spawn new and increasingly exotic offerings from process control instrumentation suppliers.

The challenge confronting those who must interact with the world of process control is to learn and understand the basics, and to go beyond the basics with personal initiative to acquire the knowledge necessary to be able to communicate with specialists. It is not an easy challenge, but it can be an extremely interesting, intellectually satisfying, and professionally rewarding experience.

As the 1990s draw to a close, workstations based on reduced instruction set computing (RISC), more powerful microprocessors, and peripheral devices of much greater speed and capacity continue to dramatically improve the capability of process control systems. In addition, the human-machine interface has become increasingly operator-friendly by the incorporation of "windows" technology and high-resolution monitors. Also, a wide variety of software is becoming available to provide better information faster to process operators. In even newer control systems, embedded software may incorporate artificial intelligence or fuzzy logic to enhance process control.

One of the most interesting recent developments is the evolution of communications among the elements of a process control system. For many years communication from sensors to controllers was via a 4-20 ma analog dc signal level. An alternate approach that has achieved acceptance is called the HART protocol. In the HART protocol, digital pulses are imposed on the traditional 4-20 ma signal.

An even more complicated approach is called "Fieldbus." In Fieldbus the 4-20 ma analog signals are completely replaced by digital signals.

By the end of the twentieth century, the communications aspects of process control will be more clearly defined and users will be able to capitalize on the new dimensions of functionality and capability.

FURTHER READING

An Evolutionary Look at Process Control/1, Honeywell, Inc., Fort Washington, PA.

Shinskey, F. G., *Process Control Systems*, 2nd ed., McGraw-Hill, New York.

Tucker, G. K., and D. M. Wills, *A Simplified Technique of Control System Engineering*, 3rd ed., Honeywell, Inc., Fort Washington, PA.

32

Attributes Inspection

JOHN JOURDAN HELDT
Free Lance Reliability Service
San Jose, California

I. INTRODUCTION

Attributes inspection can be described as "go" or "no go" inspection. The unit, part, or service is either good or bad. While we would like everything to always be perfect, it is obvious that there may be a tradeoff between the cost and value of perfection versus the cost and value of replacing a part at the use level.

When inspection is automatic—i.e., relegated to a programmed or computerized machine—"perfection" is the way to go. But when the cost of perfection is a substantial percentage of the cost of the product—as it may be if a lot of human inspection is applied—it may well be time to start looking for tradeoffs. For example, the cost of 100% human inspection of an item such as a self-tapping screw compared to evaluating the production lot using attributes inspection is high enough for a tradeoff investigation. Periodic inspection of production samples costing less than 1% of the product cost could be used to assure less than 2% bad parts. (Note that all factors need to be considered before looking for tradeoff values. For example, a bad screw could cause the stoppage of an automatic screw feeder, at a cost much higher than the 100% inspection process.) The screw producer could afford to furnish a few extra screws per hundred and split the difference with the customer.

II. COST OF CONTROL

When sequential sampling is evaluated in terms of attribute or go/no-go data (also known as pass/fail, present/absent, and conforming/nonconforming), one of the four kinds of go/no-go charts described in Table 1 could be used to chart the data in order to evaluate trends and assess the control of the process for management. The most important features of any attribute chart are the control limits. When all of

Table 1 Commonly Used Attribute Charts

p chart	Used to graph the percentage, or fraction, of defects or defective parts[a] that are contained in a sample. Although the samples are usually of constant size, the p chart—since it tabulates defects as proportions—is especially suited for data taken from samples of varying sizes.
np chart	Used to graph the number of defective[a] items in a sample. The samples must be of constant size to give meaning to trend analysis and control estimates.
c chart	Graphs the number of defects[a] per unit or the number that are contained in small samples of constant size.
d chart	Graphs the number of defects[a] in a single unit or the number that are contained in small samples that are not constant in size.

[a] While the "occurrences" are usually defects or defective items in quality control/assurance work, the techniques for analyzing attributes apply to any attribute-type characteristic. Red/not red, rain/no rain, absent/not absent, and car crash/no car crash are examples of attributes that can be charted and analyzed.

the data points on the chart are within the limits and there are no indications of data trends, the process is said to be in control. More important, an out-of-limits point indicates a high probability that an assignable cause of variation is present. In addition, a trend in an attribute chart's data points shows immediate proof—for better or worse—of any attempts to improve the product as a result of changes to the process.

In most organizations, even those not yet charting processes, go/no-go data is already accessible. Even where there is no formal inspection, attribute-type data is still available. Scrap is tabulated; attendance records are kept; defective materials are sorted. Often, this type of data is easy to arrange in control chart form. When more go/no-go data are needed, they are usually easy to obtain and can be tabulated inexpensively. Once defects and defectives have been defined, go/no-go visual inspection or pass/fail data using go/no-go tools is easy to initiate.

The attribute chart graphs the number of defects along the time line showing when the samples were taken. The *percent-defective chart* (i.e., the p chart) is shown in Figure 1 to illustrate go/no-go charts using constant sample sizes.

Sampling guidelines for the percent defective chart are: relatively large samples, taken frequently. A typical sample size is 100, although sample sizes may be larger or smaller. (One rule of thumb makes the ideal sample size one that contains an average of two defectives.) In Figure 1, the sample size, n, is equal to 200; i.e., each sample taken consists of 200 individual items. In the event that the sample size must be varied, control limits must be calculated using the sample size of each of the individual samples. This means that the numerical value of the control limits will change from sample to sample. (Sample size variance of 5 or 10% usually does not mandate recalculation of control limits.)

The calculations for the average number of defects and the upper and lower control limits for Figure 1a are shown as follows.

Calculations for Figure 1a:

$$\text{Average } (\overline{p}) = \frac{\sum_{i=1}^{n} p_i}{\sum_{i=1}^{n} n_i} = \frac{106}{200 \times 13} = 0.040769 = 4.1\%$$

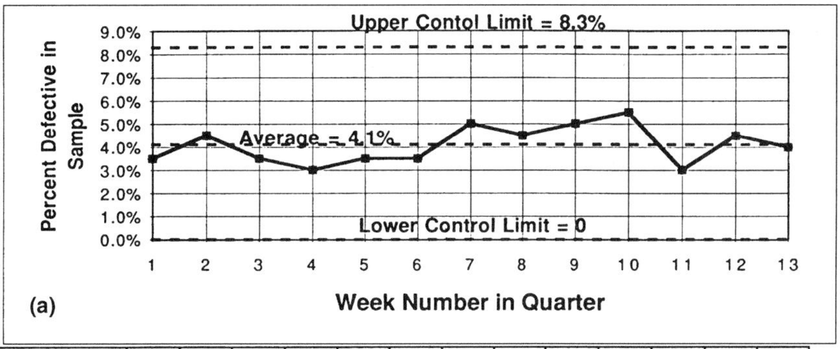

Average 4.1%

	1	2	3	4	5	6	7	8	9	10	11	12	13	Total
Percent Def.	3.5%	4.5%	3.5%	3.0%	3.5%	3.5%	5.0%	4.5%	5.0%	5.5%	3.0%	4.5%	4.0%	
5 Other	2		2	2	1	1	4	2	2	2		1	2	21
4 Finish					1	1	1		1	1		1	1	7
3 Crack		2	2	2	1	1			2	1		1	1	13
2 Color	4	3	1	2	2		1	3	4	2	4	4	2	32
1 Scratch	1	4	2		2	4	4	4	1	5	2	2	2	33
Number Def.	7	9	7	6	7	7	10	9	10	11	6	9	8	106
Number Insp.	200	200	200	200	200	200	200	200	200	200	200	200	200	2,600
(b)	1	2	3	4	5	6	7	8	9	10	11	12	13	

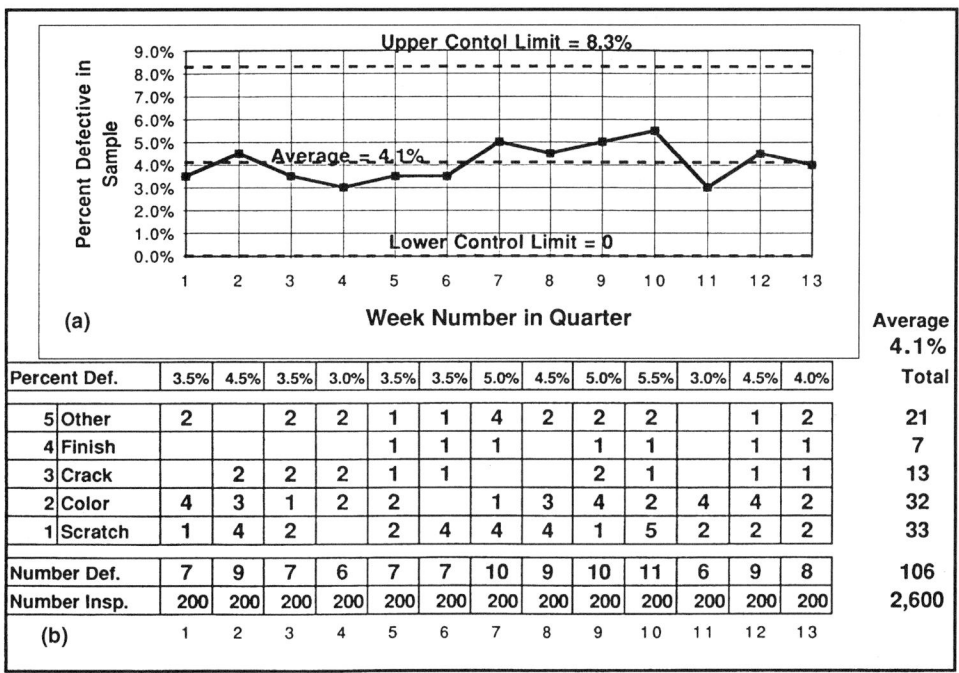

Figure 1 (a) Percent defective (p) chart; (b) attributes inspection data (number of defectives found in each sample).

The upper control limit is

$$UCL_p = \bar{p} + 3\sqrt{\frac{\bar{p} \times (100 - \bar{p})}{n}} = 4.1 + 3\sqrt{\frac{4.1 \times (100 - 4.1)}{200}}$$

$$UCL_p = 4.1 + 3\sqrt{\frac{393.19}{200}} = 4.1 + 3\sqrt{1.966} = 4.1 + 3(1.4)$$

$$UCL_p = 4.1 + 4.2 = 8.3\%$$

The lower control limit is

$$LCL_p = \bar{p} - 3\sqrt{\frac{\bar{p} \times (100 - \bar{p})}{n}} = 4.1 - 3\sqrt{\frac{4.1 \times (100 - 4.1)}{200}}$$

$$LCL_p = 4.1 - 3\sqrt{\frac{393.19}{200}} = 4.1 - 3\sqrt{1.966} = 4.1 - 3(1.4)$$

$$LCL_p = 4.1 - 4.2 = -0.1\% \quad (\text{use } LCL_p = 0)$$

These values, as they are shown on the p chart, provide the means of evaluation of the process. The process data are shown in Figure 1b.

When all the points are within the control limits, as they are in Figure 1a, it means the process is stable and the sampling continues. When one point is out of control, it is the signal to look for an assignable cause. [To signal out of control, the point must be outside the control limit (a point on the line does not signal out of control). The chance than an out-of-control point does not signal out of control is the same as the probability that a product falls outside the +3 sigma limit. When the lower control limit is zero, a point on the zero line usually signals that the process is getting better, but it is wise to look for an assignable cause when there are two or more zeros are plotted.] Most organizations also develop special evaluation rules to signal the presence of trends. The most universal rule indicating an improvement in the process is this: When 70% or more of the points on the chart are below the average, it is time to compute a new average and control limits as the first steps toward consolidation of the process improvement.

In Figure 2, a *number-defective chart*, commonly called the "np" chart, is used to graph the number of defective parts from each sample. [When the sample size (n) is multiplied by the fraction defective (p′), the result is simply the number of defectives in the sample, hence the name *number-defective* or np chart.] For this chart, the samples must be constant in size to give meaning to trend analysis and control computation.

Number-defective charts are the same as percent-defective charts except that all data is calculated and plotted as whole numbers. The formulas and calculations for the np chart are given as follows.

In the number-defective (np) chart, the average is calculated by dividing the total number of defective parts by the total number of samples (n) as shown below:

$$\text{Average } (n\bar{p}) = \frac{\sum_{i=1}^{n} p_i}{N} = \frac{106}{13} = 8.154$$

The control limits are calculated using the formulas below:

$$\text{UCL}_{np} = n\bar{p} + 3\sqrt{n\,(\bar{p}) \times \left(1 - \frac{n\bar{p}}{n}\right)}$$

$$\text{UCL}_{np} = 8.154 + 3\sqrt{8.154 \times \left(1 - \frac{8.154}{200}\right)}$$

$$\text{UCL}_{np} = 8.154 + 3\sqrt{8.154 \times 0.9592)} = 8.154 + 3\sqrt{7.82}$$

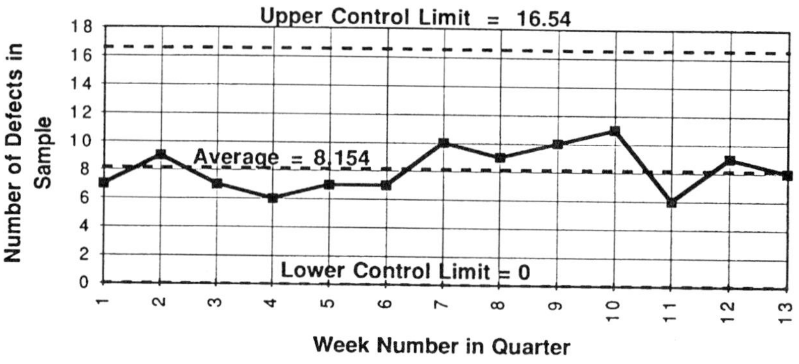

Figure 2 Number defective (np′) chart.

$$UCL_{np} = 8.154 + 3(2.796)$$

$$UCL_{np} = 8.154 + 8.390 = 16.54$$

$$LCL_{np} = n\bar{p} - 3\sqrt{n(\bar{p}) \times \left(1 - \frac{n\bar{p}}{n}\right)}$$

$$LCL_{np} = 8.154 - 3\sqrt{8.154 \times \left(1 - \frac{8.154}{200}\right)}$$

$$LCL_{np} = 8.154 - 3\sqrt{8.154 \times 0.9592)} = 8.152 - 3\sqrt{7.82}$$

$$LCL_{np} = 8.154 - 3(2.796)$$

$$LCL_{np} = 8.154 - 8.390 = -0.236 \quad (\text{use } LCL_{np} = 0)$$

For comparison, refer to the data in Figure 1a.

The techniques and use of the number of defects, or c, chart, and the number of defects per unit, or d, chart are essentially the same. The c chart tallies the number of defects—not defectives—in a sample. The d chart, as shown in Figure 3a, tallies the number of defects—or nonconformances—in a single item; for example, the number of problems in a new truck received by a dealer.

Charting the number of defects per truck might be the dealer's way of grading the manufacturer's manufacturing and shipping inspection operations. Obviously, when the chart indicates that the manufacturer's processes are out of control—or that the average number of defects per vehicle is too high—the dealer has an excellent backup for a request for larger "make-ready" payments.

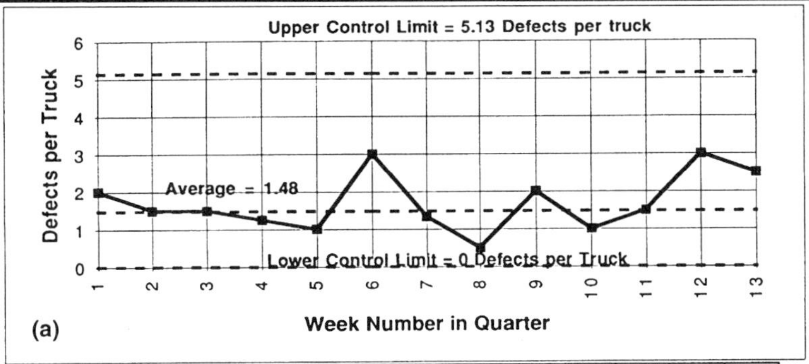

	1	2	3	4	5	6	7	8	9	10	11	12	13	
Defects/Truck	2.0	1.5	1.5	1.3	1.0	3.0	1.3	0.5	2.0	1.0	1.5	3.0	2.5	Average 1.48
														Totals
5 Body	1		1	2		1	2					1	2	10
4 Finish						1	1			1			1	4
3 Engine	1	2	1	2	1				2					9
2 Interior			1				1	2			3			7
1 Software		1		1		1						2	2	7
Number Def.	2	3	3	5	1	3	4	2	2	1	3	3	5	37
Number Insp.	1	2	2	4	1	1	3	4	1	1	2	1	2	25
(b)	1	2	3	4	5	6	7	8	9	10	11	12	13	

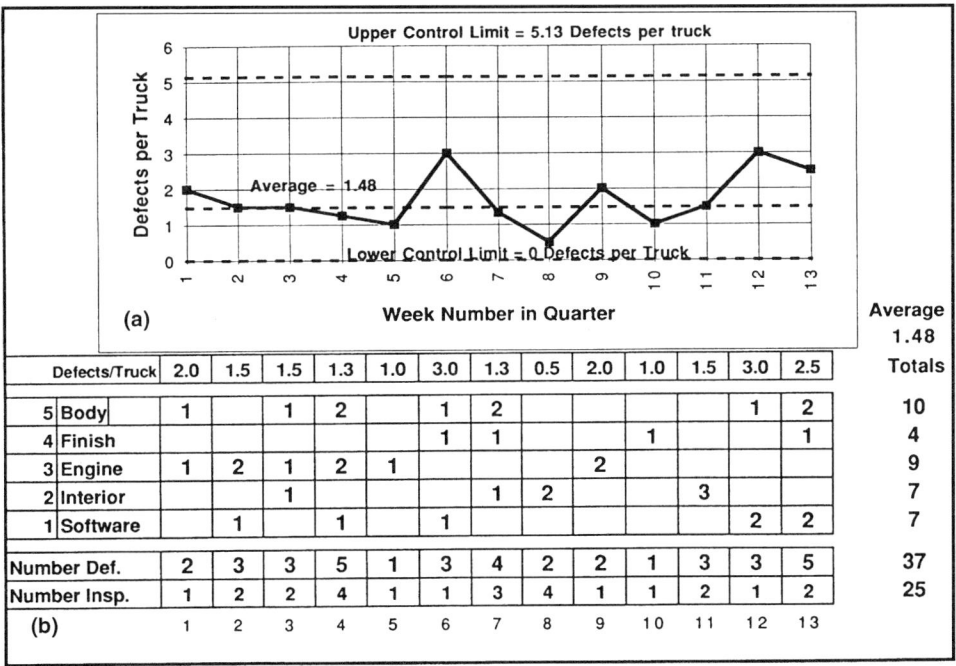

Figure 3 (a) Defects per unit (d) chart; (b) data for short.

The d chart can be used, for example, for monitoring attendance. The number of employees in a department provides a stable sample size. Excessive absenteeism is quickly noted, and an out-of-statistical-control condition can be immediately examined for cause.

The formulas and calculations for the d chart are as follows

$$\bar{d}_{(\text{defects/truck})} = \frac{\Sigma\text{defects}}{\Sigma\text{truck}}$$

$$\bar{d}_{(\text{no. absent})} = \frac{\Sigma \text{ total absences}}{\Sigma(\text{no. of days}) \times (\text{no. in the dept.})} (100)\%$$

In both cases, the control limits are

$$\text{UCL}_d = \bar{d} + 3\sqrt{\bar{d}}$$

$$\text{LCL}_d = \bar{d} - 3\sqrt{\bar{d}}$$

Truck example: When 37 defects have been found on 25 trucks, then

$$\bar{d} = \frac{37}{25} = 1.48 \text{ defects per truck}$$

$$\text{UCL}_d = 1.48 + 3\sqrt{1.48} = 1.48 + 3(1.216)$$

$$\text{UCL}_d = 1.48 + 3.648 = 5.13 \text{ defects per truck}$$

$$\text{LCL}_d = 1.48 - 3\sqrt{1.48} = 1.48 - 3(1.216)$$

$$\text{LCL}_d = 1.48 - 3.648 = 0 \text{ defects per truck}$$

III. SAMPLE INSPECTION

Sampling inspection, like attributes inspection, is based on discrete data. Discrete data are defined as those data that can have only specific whole number values. While statisticians discuss families with 2.3 children "on the average," a real family cannot have halves, quarters, or negative numbers of children.

Unlike the normal distribution, discrete distributions are seldom symmetrical about their average. The data points in discrete distributions tend to crowd near zero at one end, or "tail," of the distribution, then to skew off toward the other tail, as shown in Figure 4. (This distribution, like that in Fig. 7c, is described as skewed to the right.)

The most commonly used discrete distributions are the hypergeometric, the binomial, and the Poisson.

A. Hypergeometric Distribution

The *hypergeometric distribution* is considered to be the most accurate of the discrete distributions. It is sometimes called the "distribution for sampling, without replacement." The formula for the hypergeometric distribution follows:

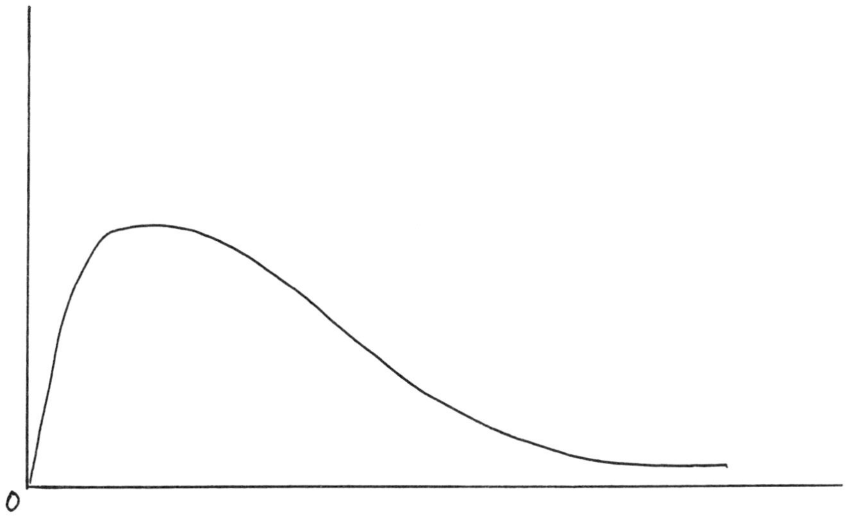

Figure 4 Typical discrete data curve.

$$P(d) = \frac{C_{(n-d)}^{(N-D)} \times C_d^D}{C_n^N}$$

where N = lot size; D = number of nonconforming items in the lot; n = sample size; and d = exact number of nonconformances in the sample for which the probability is to be calculated. In words, this formula says that the probability of finding exactly d nonconforming items (in a sample of size n) is equal to all of the possible combinations of conforming items $[C_{(n-d)}^{(N-D)}]$ multiplied by all of the possible combinations of nonconforming items $[C_d^D]$ divided by the total number of combinations that can be drawn from the lot in samples of size n $[C_n^N]$.

To summarize, the hypergeometric distribution:

- Is considered to be the most accurate of all the functions that deal with discrete distributions
- Is known as the statistical analysis method for sampling without replacement (of samples)
- Has been, in the past, often approximated by the binomial or Poisson distributions because the hypergeometric was considered to be too unwieldy
- Will likely be more often used since the math involved can now be done accurately and easily on a desktop computer.

B. Binomial Distribution

The *binomial distribution* formula is

$$P(d) = C_d^n \times q'^{(n-d)} \times p'^d = \frac{n!}{d! \times (n - d)!} \times q'^{(n-d)} \times p'^d$$

where p′ = fraction of nonconforming items in a lot or resulting from a process (the binomial p′ = the hypergeometric D/N) [Note that p′ + q′ = 1 (or q′ = 1 −

$p')]$; n = sample size; and d = exact number of nonconformances in the sample for which the probability is to be calculated.

Using the binomial distribution, probability errors get smaller as the lots grow larger.

Sometimes the binomial distribution is referred to as the "point binomial" or as the "Bernoulli distribution." The binomial is used most often for calculating the probabilities associated with continuing manufacturing processes (i.e., the lot sizes are very large and are added to on a regular basis), and the nonconformances (d) in a sample are determined on a periodic basis to estimate the fraction of defectives parts (p) that the process is producing. When the prime mark is applied, i.e., p', it indicates our calculations concerning the probabilities of finding d number of defects in a sample is based on the assumption that p' is fixed and constant.

In words, the binomial distribution states that the probability of finding exactly d nonconforming items in a sample of size n is the combination of n items taken d ways, multiplied by q' raised to the n minus d power, multiplied by p' raised to the d power.

To summarize, the binomial distribution

- Is not as accurate as the hypergeometric distribution, but is useful for establishing discrete probabilities in high production or high throughput areas.
- Has calculations that tend to be easy to perform since there are only three parameters to work with.
- Is sometimes called an analysis techniques used with *sampling with replacement*. The term "with replacement" leads to a little error when the lot sizes are small and the samples are not, in fact, replaced. Nonreplacement is typical in production inspection.
- Is sometimes called the Bernoulli distribution or the point binomial distribution.

C. The Poisson Law

The Poisson Law, which Simeon Denis Poisson developed in the distribution that bears his name, brought discrete probabilities into the realm of the common man. He developed this formula by taking the limit of the binomial equation as n approached infinity and as p' approached zero. He then defined c' as the *expected value*, which means that his formula has only two parameters.

c' = expected value ($c' = np'$, from binomial, or $c' = n \times D/N$, from hypergeometric)

c = number of nonconforming items in the sample for which the probability of occurrence can be calculated

The formula used to calculate probabilities according to the Poisson law is

$$P(c) = \frac{e^{-c'} \times c'^{c}}{c!}$$

Note: e is the constant which is the base for natural (or Napierian) logarithms (e = 2.7182818...).

In the past, powers of e (including negative powers) tables were used to develop a Poisson table. Now, a personal computer can be used to develop Poisson tables, which can be used directly for handbook level discrete quality probability uses.

In words, the Poisson formula says the probability of finding exactly c nonconformances (or, actually, any specified occurrence) in a sample of size n is c raised to the expected value's power, times the expected value raised to the number of nonconformances (or other specified occurrence), all divided by the factorial of the number of nonconformances (or occurrences). The expected value, called e(c′) or np′, is the sample size (n) times the true fraction defective (p′).

To summarize, the Poisson distribution

- Is the easiest for calculations because there are only two parameters
- Offers handbook confidence for discrete probability applications
- Loses accuracy rapidly when the sample size is small, say, less than 20 items (or occurrences), or when the number of nonconformances (or occurrences) exceeds 10% of the lot (or universe)
- Is based on the concept of expected value, i.e., np′. The concept of expected value makes understanding the math much easier.

The Poisson distribution is generally considered to be the most often used distribution in quality applications.

IV. APPLICATIONS OF POISSON'S LAW

Applications of Poisson's law are described in the following sections.

A. The Summation of Poisson Terms Table

The body of Table 2 shows the probability that c or fewer events (event numbers are across the top of the table) will occur when the expected value (np′) is the number shown in the left column of the table.

This Poisson table extends to 14 events (c) from np′ = 0.01 to np′ = 20. Within the limits of the spreadsheet software, the number of events can be extended as far as desired. The expected values can likewise be expanded.

Another name for this table is the Cumulative Poisson Table. Table details are summarized here:

1. The expected values (c′ or np′) are shown in the left-hand column. Some people speak of the expected value as the "average." Expected value is used here to indicate that the poisson distribution is not symmetrical about the expected value. (The term *expected value* is used for discrete distributions, and the term *average* is used for the normal distribution).
2. c numbers are shown across to top of the table.
3. To find the value of exactly c for any np′, the value in the c − 1 column must be subtracted from the value in the c column on that same line.

B. Poisson Distribution, when np′ = 3.0

Poisson's cumulative distribution shown in Figure 5a is used to develop a specific distribution by subtracting the previous probability from each successive probability

Table 2 Summation of Poisson Terms (The Probability of c or Fewer Occurrences of an Event That Has an Expected Value Equal to np′)

c --> np′	0	1	2	3	4	5	6	7	8	9	10	11	12	13	14
0.01	0.990														
0.03	0.970	1.000													
0.05	0.951	0.999													
0.10	0.905	0.995	1.000												
0.20	0.819	0.982	0.999												
0.30	0.741	0.963	0.996	1.000											
0.40	0.670	0.938	0.992	0.999											
0.50	0.607	0.910	0.986	0.998											
0.60	0.549	0.878	0.977	0.997	1.000										
0.70	0.497	0.844	0.966	0.994	0.999										
0.80	0.449	0.809	0.953	0.991	0.999	1.000									
0.90	0.407	0.772	0.937	0.987	0.998	1.000									
1.00	0.368	0.736	0.920	0.981	0.996	0.999									
1.25	0.287	0.645	0.868	0.962	0.991	0.998	1.000								
1.50	0.223	0.558	0.809	0.934	0.981	0.996	0.999								
1.75	0.174	0.478	0.744	0.899	0.967	0.991	0.998	1.000							
2.00	0.135	0.406	0.677	0.857	0.947	0.983	0.995	0.999	1.000						
2.25	0.105	0.343	0.609	0.809	0.922	0.973	0.992	0.998	0.999						
2.50	0.082	0.287	0.544	0.758	0.891	0.958	0.986	0.996	0.999	1.000					
2.75	0.064	0.240	0.481	0.703	0.855	0.939	0.978	0.993	0.998	0.999					
3.00	0.050	0.199	0.423	0.647	0.815	0.916	0.966	0.988	0.996	0.999	1.000				
3.25	0.039	0.165	0.370	0.591	0.772	0.889	0.952	0.982	0.994	0.998	0.999				
3.50	0.030	0.136	0.321	0.537	0.725	0.858	0.935	0.973	0.990	0.997	0.999	1.000			
3.75	0.024	0.112	0.277	0.484	0.678	0.823	0.914	0.962	0.985	0.995	0.998	0.999			
4.0	0.018	0.092	0.238	0.433	0.629	0.785	0.889	0.949	0.979	0.992	0.997	0.999	1.000		
4.5	0.011	0.061	0.174	0.342	0.532	0.703	0.831	0.913	0.960	0.983	0.993	0.998	0.999	1.000	
5.0	0.007	0.040	0.125	0.265	0.440	0.616	0.762	0.867	0.932	0.968	0.986	0.995	0.998	0.999	1.000
5.5	0.004	0.027	0.088	0.202	0.358	0.529	0.686	0.809	0.894	0.946	0.975	0.989	0.996	0.998	0.999
6.0	0.002	0.017	0.062	0.151	0.285	0.446	0.606	0.744	0.847	0.916	0.957	0.980	0.991	0.996	0.999
6.5	0.002	0.011	0.043	0.112	0.224	0.369	0.527	0.673	0.792	0.877	0.933	0.966	0.984	0.993	0.997
7.0	0.001	0.007	0.030	0.082	0.173	0.301	0.450	0.599	0.729	0.830	0.901	0.947	0.973	0.987	0.994
7.5	0.001	0.005	0.020	0.059	0.132	0.241	0.378	0.525	0.662	0.776	0.862	0.921	0.957	0.978	0.990
8.0		0.003	0.014	0.042	0.100	0.191	0.313	0.453	0.593	0.717	0.816	0.888	0.936	0.966	0.983
9.0		0.001	0.006	0.021	0.055	0.116	0.207	0.324	0.456	0.587	0.706	0.803	0.876	0.926	0.959
10.0			0.003	0.010	0.029	0.067	0.130	0.220	0.333	0.458	0.583	0.697	0.792	0.864	0.917
11.0			0.001	0.005	0.015	0.038	0.079	0.143	0.232	0.341	0.460	0.579	0.689	0.781	0.854
12.0			0.001	0.002	0.008	0.020	0.046	0.090	0.155	0.242	0.347	0.462	0.576	0.682	0.772
14.0					0.002	0.006	0.014	0.032	0.062	0.109	0.176	0.260	0.358	0.464	0.570
16.0						0.001	0.004	0.010	0.022	0.043	0.077	0.127	0.193	0.275	0.368
18.0							0.001	0.003	0.007	0.015	0.030	0.055	0.092	0.143	0.208
20.0								0.001	0.002	0.005	0.011	0.021	0.039	0.066	0.105

along the np′ = 3.0 line. The resultant histogram shown in Figure 5c shows the exact probability of occurrence for each of the event numbers.

C. Inspection Plan Curves for n = 100, c = 4, Single Sample Plan

Steps to construct the operating characteristic (OC) and average outgoing quality (AOQ) curves for an n = 100, c = 4 single sample plan are shown in Figure 6.

1. Select 4 or 5 probabilities of acceptance (P_a) from the column headed by c = 4 in Figure 6b.
2. Write the corresponding expected value next to the P_a, in the np′ column of Figure 6a.
3. Divide each value of np′ by 100 (n) and write the quotient in the column headed by p′ in Figure 6a.

c -->	0	1	2	3	4	5	6	7	8	9	10
np'											
0.01	0.990										
0.03	0.970	1.000									
0.05	0.951	0.999									
0.10	0.905	0.995	1.000								
0.20	0.819	0.982	0.999								
3.00	0.050	0.199	0.423	0.647	0.815	0.916	0.966	0.988	0.996	0.999	1.000

(a)

c	3.00 line from Table		Prev. Probability		Exact Prob.
0	0.050	=		=	5.0%
1	0.199	-	0.050	=	14.9%
2	0.423	-	0.199	=	22.4%
3	0.647	-	0.423	=	22.4%
4	0.815	-	0.647	=	16.8%
5	0.916	-	0.815	=	10.1%
6	0.966	-	0.916	=	5.0%
7	0.988	-	0.966	=	2.2%
8	0.996	-	0.988	=	0.8%
9	0.999	-	0.996	=	0.3%
10	1.000	-	0.999	=	0.1%

(b)

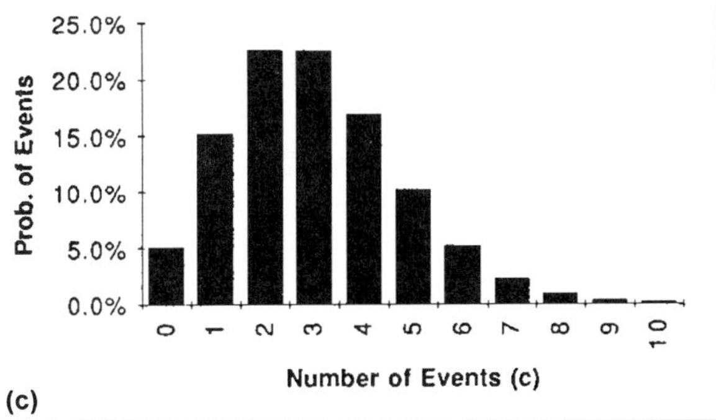

(c)

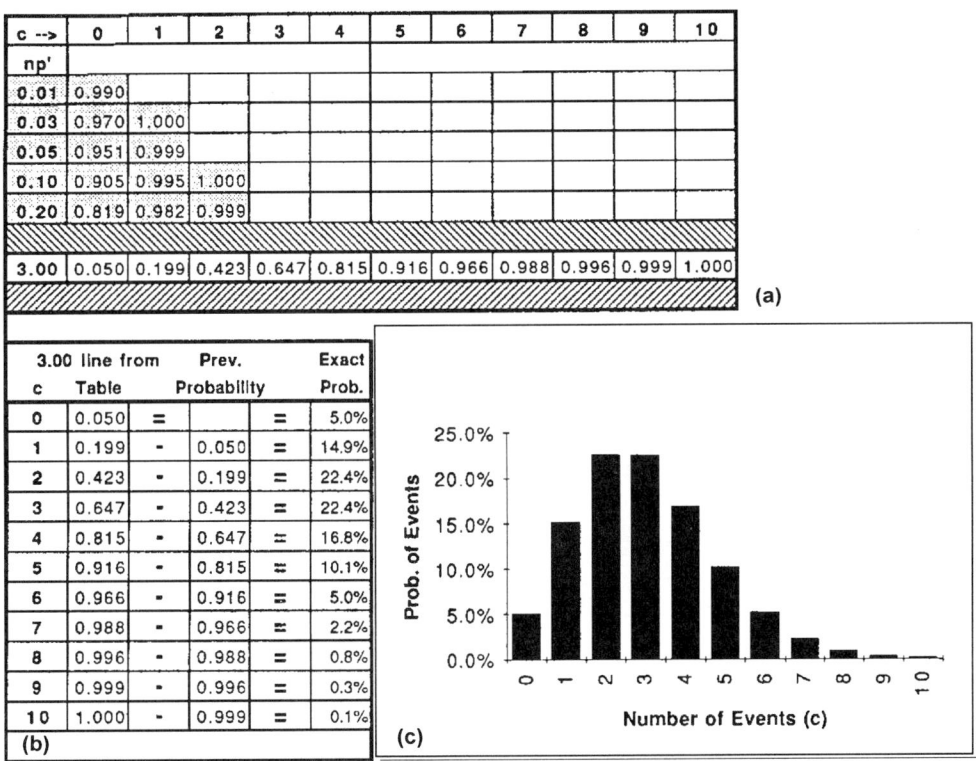

Figure 5 (a) Cumulative Poisson distribution (probability of c or fewer occurrences of an event that has an expected value equal to np′); (b) calculations; (c) distribution np′ = 3.0.

4. Multiply $P_a \cdot p'$ to get the average outgoing quality (AOQ).

5. Construct the OC curve by plotting each pair of values (P_a, p′). Sketch the actual curve by connecting the dots, as illustrated in Figure 6c. Construct the AOQ curve by plotting each pair of values ($P_a \cdot p'$, p′) and sketching the curve. [Note: In the example shown in Figure 6d, you will note the maximum value of the AOQ curve is identified as the average outgoing quality limit (AOQL). This AOQL is the real limit of liability of the sample plan, since on the average, the quality of the product going into the stock room is 2.54% only when the incoming population quality is 3.5%. The average outgoing quality reaches this maximum because all rejected lots are screened 100%. In this case, 3.5% is the point at which the 100% screened lots plus the lots that were accepted by the plan add up to the maximum overall lot percentage of defects.]

6. Looking at the OC curve, it is easy to see for a process that is constantly 4.5% defective that about half the lots presented will be accepted according to this plan (i.e., $P_a = 53\%$). This also means that half of the rejected lots will be screened and entered into stock with no defects included. [Actually, this is not totally accurate, since any defective parts found in screening will be replaced by parts which are also 4.5% defective. This means that 0.05 times 0.05 defective parts will be included in the *clean lots* being submitted to stock. But this fraction is so small, we can neglect it for this argument.] Half the lots will be accepted as presented and will

Data for the OC and AOQ Curves			
Prob of Ac	Expect Val	T Fract Def	AOQ
Pa	np'	p'	Paxp'
1.000		0.000	0.0000
0.947	2.00	0.020	0.0189
0.725	3.50	0.035	0.0254
0.532	4.50	0.045	0.0239
0.358	5.50	0.055	0.0197
0.100	8.00	0.080	0.0080
0.008	12.00	0.120	0.0009

(a)

c -->	0	1	2	3	4
np'					
0.01	0.990				
0.03	0.970	1.000			
0.05	0.951	0.999			
0.10	0.905	0.995	1.000		
0.20	0.819	0.982	0.999		
0.30	0.741	0.963	0.996	1.000	
0.40	0.670	0.938	0.992	0.999	
0.50	0.607	0.910	0.986	0.998	
0.60	0.549	0.878	0.977	0.997	1.000
0.70	0.497	0.844	0.966	0.994	0.999
0.80	0.449	0.809	0.953	0.991	0.999
0.90	0.407	0.772	0.937	0.987	0.998
1.00	0.368	0.736	0.920	0.981	0.996
1.25	0.287	0.645	0.868	0.962	0.991
1.50	0.223	0.558	0.809	0.934	0.981
1.75	0.174	0.478	0.744	0.899	0.967
2.00	0.135	0.406	0.677	0.857	0.947
2.25	0.105	0.343	0.609	0.809	0.922
2.50	0.082	0.287	0.544	0.758	0.891
2.75	0.064	0.240	0.481	0.703	0.855
3.00	0.050	0.199	0.423	0.647	0.815
3.25	0.039	0.165	0.370	0.591	0.772
3.50	0.030	0.136	0.321	0.537	0.725
3.75	0.024	0.112	0.277	0.484	0.678
4.0	0.018	0.092	0.238	0.433	0.629
4.5	0.011	0.061	0.174	0.342	0.532
5.0	0.007	0.040	0.125	0.265	0.440
5.5	0.004	0.027	0.088	0.202	0.358
6.0	0.002	0.017	0.062	0.151	0.285
6.5	0.002	0.011	0.043	0.112	0.224
7.0	0.001	0.007	0.030	0.082	0.173
7.5	0.001	0.005	0.020	0.059	0.132
8.0		0.003	0.014	0.042	0.100
9.0		0.001	0.006	0.021	0.055
10.0			0.003	0.010	0.029
11.0			0.001	0.005	0.015
12.0			0.001	0.002	0.008
14.0					0.002

(b)

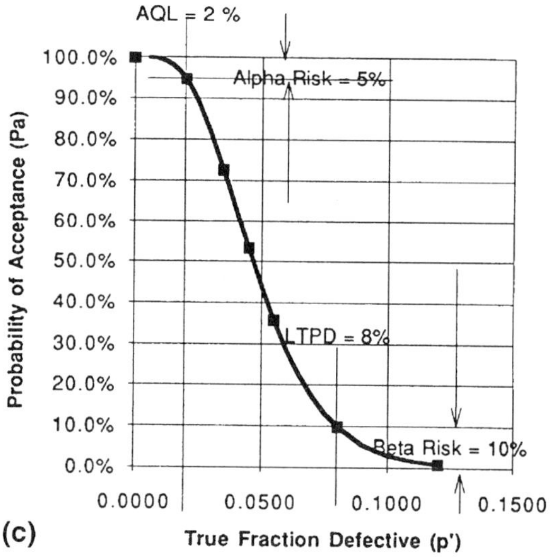

(c)

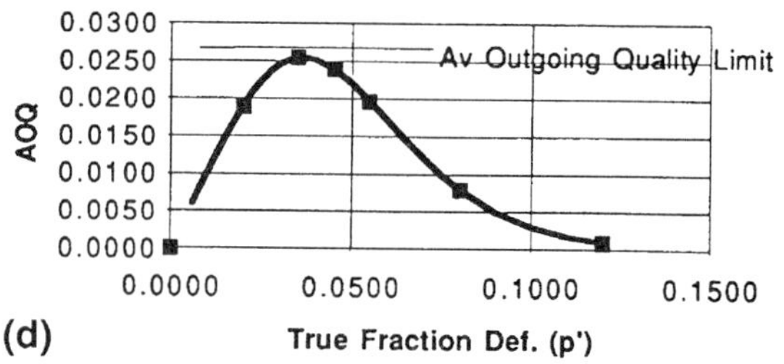

(d)

Figure 6 OC and AOQ curves for n = 100, c = 4: (a) data; (b) table of Poisson cumulative terms; (c) OC curve; (d) AOQ curve.

be entered into stock at 4.5% defective; the average outgoing quality (AOQ) for any fraction defective (p′) of incoming quality, will, in the long run, be entered into stock as the probability of acceptance (P_a) times the true fraction defective (p′) [In this case, for example, 0.53 (P_a) times 0.045 (p′) = 0.0239 (AOQL).] With this in mind, we can envision the highest point on the average outgoing quality graph ($P_a \cdot p'$ versus p') as the average outgoing quality limit (AOQL).

D. Sample Plan Parameters: AQL, LTPD, α and β Risk, and AOQL

The sample size (n) and the acceptance number (c) of any single sample plan define the operating characteristic (OC) curve and average outgoing quality (AOQ) curve for that sample plan. (Note: The definitions of the sample plan parameters are the same as those shown in Ref. 1.) There is only one OC curve and AOQ curve for any combination of n and c. There are four other entities common to OC and AOQ curves that we should know. These are described and defined here:

1. AQL, acceptable quality level. This is a nominal lot or process level considered satisfactory by the consumer (usually identified in terms of true fraction defective (p'), as, for example, AQL = 0.02, or this plan has a 2% AQL. As defined by MIL-STD-105E, "the AQL is a designated value of percent defective (or defects per 100 units) for which lots will be accepted most of the time by the sampling procedure being used" (2).

2. LTPD, lot tolerance proportion (or percent) defective. A nominal fraction defective to be accepted only a small portion of the time offered. This is usually identified in terms of true fraction defective (p') as LTPD = 0.08, or this plan has only a 10% probability of accepting product that is 8% (LTPD) defective.

3. Alpha (α) risk. This is also known as the producer's risk: the probability of rejecting a lot which is acceptable. It is usually identified as the percentage of the probability of acceptance (P_a) between 100% acceptance and the probability of rejecting lots when the true fraction defective of the lot equals the AQL (i.e., the probability for rejecting a good lot). To most people, the alpha risk is taken to be 5% (i.e., the probability of acceptance of a lot which meets the AQL is equal to 95%).

4. Beta (β) risk. This is also known as the consumer's risk: the probability of accepting a lot or a process when it is rejectable. The beta risk is usually identified as probability of acceptance (P_a) of a lot with quality as bad as the consumer is ever willing to accept. The P_a associated with the LTPD is 10%.

5. AOQL, average outgoing quality limit. This is the maximum of all of the AOQs for all of the ps for a plan. The AOQL is regarded by many to be the *true* limit of liability for the sample plan. Three things should be noted. First, the AOQL occurs only when the true fraction defective is such that $P_a \cdot p'$ maximizes. In other words, when the incoming quality is good, the average outgoing quality has to be good. When the incoming quality is bad, the AOQ will be good, because the worse the incoming quality, the higher the percentage of rejected lots (i.e., screened lots). Second, AOQL = AQL for tightened sampling plans according to MIL-STD-105(2). Third, AOQL's as described here are approximate:

$$\text{exact AOQL} = \text{AOQL}\left[1 - \frac{\text{sample size (n)}}{\text{lot or batch size (N)}}\right]$$

The multiplier, $[1 - (n/N)]$, takes into account the bad parts removed from the sample inspected of all of the accepted lots.

E. OC and AOQ Curve

The table in Figure 7 shows 21 blocks. These blocks can be used to make an estimate of the three points that define the AQL, the LTPD, and the AOQL. These three points are adequate to make fairly accurate sketches of OC and AOQ curves (these sketches become easier and easier to do with practice).

n =		c= 0	
Pa	np'	p'	Pa x p'
0.95	0.05		
0.37	1.00		
0.10	2.30		

n =		c= 1	
Pa	np'	p'	Pa x p'
0.95	0.36		
0.52	1.63		
0.10	3.89		

n =		c= 2	
Pa	np'	p'	Pa x p'
0.95	0.82		
0.60	2.30		
0.10	5.32		

n =		c= 3	
Pa	np'	p'	Pa x p'
0.95	1.37		
0.69	2.80		
0.10	6.68		

n =		c= 4	
Pa	np'	p'	Pa x p'
0.95	1.97		
0.71	3.60		
0.10	7.99		

n =		c= 5	
Pa	np'	p'	Pa x p'
0.95	2.61		
0.73	4.35		
0.10	10.53		

n =		c= 6	
Pa	np'	p'	Pa x p'
0.95	3.29		
0.75	5.10		
0.10	10.53		

n =		c= 7	
Pa	np'	p'	Pa x p'
0.95	3.98		
0.77	5.80		
0.10	11.77		

n =		c= 8	
Pa	np'	p'	Pa x p'
0.95	4.70		
0.79	6.50		
0.10	12.99		

n =		c= 9	
Pa	np'	p'	Pa x p'
0.95	5.43		
0.80	7.30		
0.10	14.21		

n =		c= 10	
Pa	np'	p'	Pa x p'
0.95	6.17		
0.81	8.10		
0.10	15.41		

n =		c= 11	
Pa	np'	p'	Pa x p'
0.95	6.92		
0.82	8.80		
0.10	16.60		

n =		c= 12	
Pa	np'	p'	Pa x p'
0.95	7.69		
0.83	9.60		
0.10	17.78		

n =		c= 13	
Pa	np'	p'	Pa x p'
0.95	8.46		
0.83	10.40		
0.10	18.96		

n =		c= 14	
Pa	np'	p'	Pa x p'
0.95	9.25		
0.85	11.10		
0.10	20.13		

n =		c= 15	
Pa	np'	p'	Pa x p'
0.95	10.04		
0.85	11.90		
0.10	21.29		

n =		c= 16	
Pa	np'	p'	Pa x p'
0.95	10.83		
0.85	12.80		
0.10	22.45		

n =		c= 17	
Pa	np'	p'	Pa x p'
0.95	11.63		
0.85	13.60		
0.10	23.61		

n =		c= 18	
Pa	np'	p'	Pa x p'
0.95	12.44		
0.86	14.40		
0.10	24.76		

n =		c= 19	
Pa	np'	p'	Pa x p'
0.95	13.25		
0.86	15.20		
0.10	25.90		

n =		c= 20	
Pa	np'	p'	Pa x p'
0.95	14.07		
0.87	15.90		
0.10	29.06		

Figure 7 AQL, LTPD, and AOQL approximator.

There are blocks provided for each value of acceptance numbers (c), from c = 0 to c = 20. For OC and AOQ curve estimates beyond c = 20, similar data using P_a = 95%, 80%, and 10% can be derived from a larger Poisson table.

Figures 7–9 illustrate uses for the OC and AOQ curve approximator.

n = 200		c= 3		
Pa	np'	p'	Pa x p'	Pa
0.95	1.37	0.0069	0.0065	95%
0.69	2.80	0.0140	0.0097	69%
0.10	6.68	0.0334	0.0033	10%

AQL = 0.69%

AOQL = 0.97%

LTPD = 3.34%

(a)

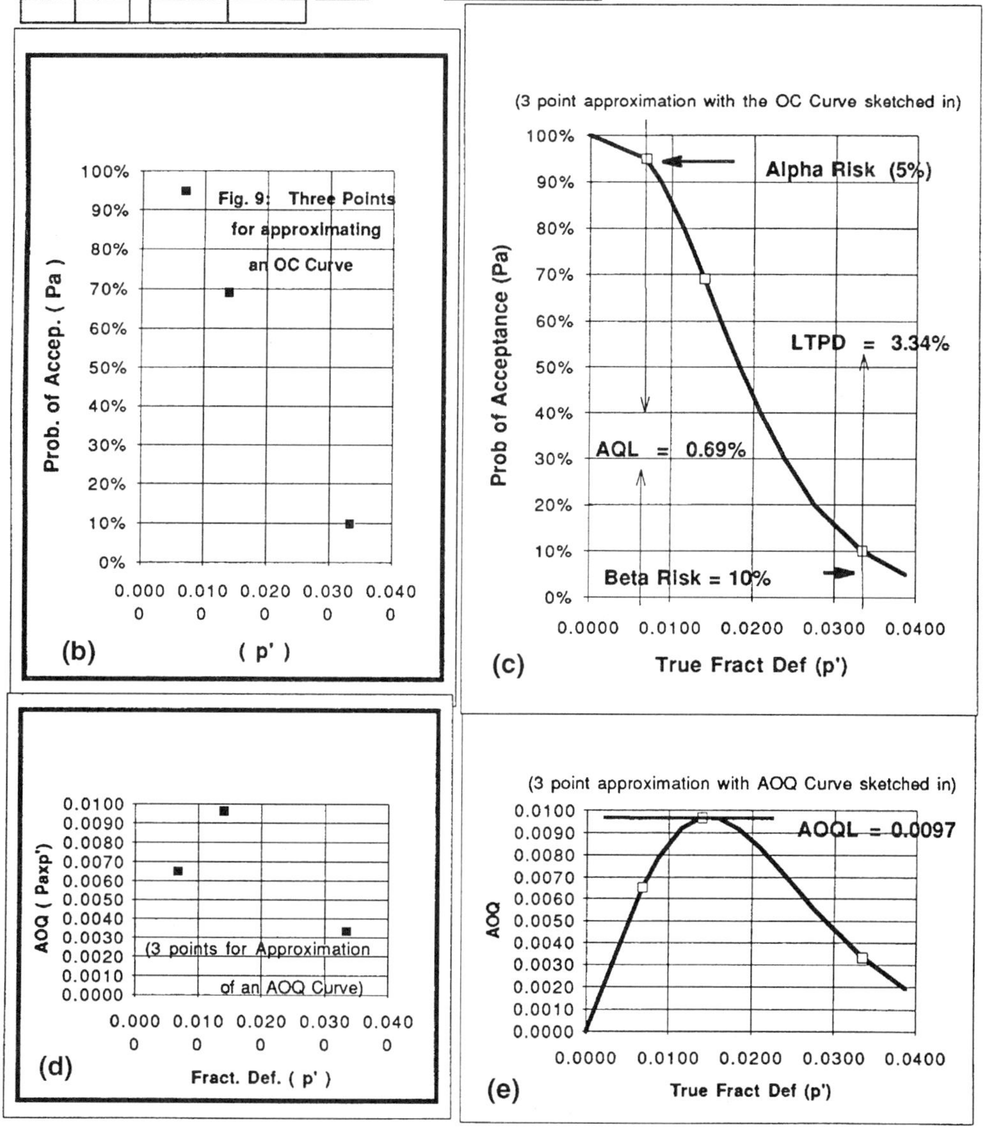

Figure 8 OC and AOQ curves for n = 200, c = 3: (a) data; (b) P_a versus p′; (c) OC curve; (d) AOQ versus p′; (e) AOQ curve.

(n = 69, c = 2: n = 328, c = 10: n = 578, c = 17)

n =	69		c =	2
Pa	np'		p'	Pa x p'
0.95	0.82		0.012	0.011
0.60	2.30		0.033	0.020
0.10	5.32		0.077	0.008

n =	328		c =	10
Pa	np'		p'	Pa x p'
0.95	6.17		0.019	0.018
0.81	8.10		0.025	0.020
0.10	15.41		0.047	0.005

n =	578		c =	17
Pa	np'		p'	Pa x p'
0.95	11.63		0.020	0.019
0.85	13.60		0.024	0.020
0.10	23.61		0.041	0.004

(a)

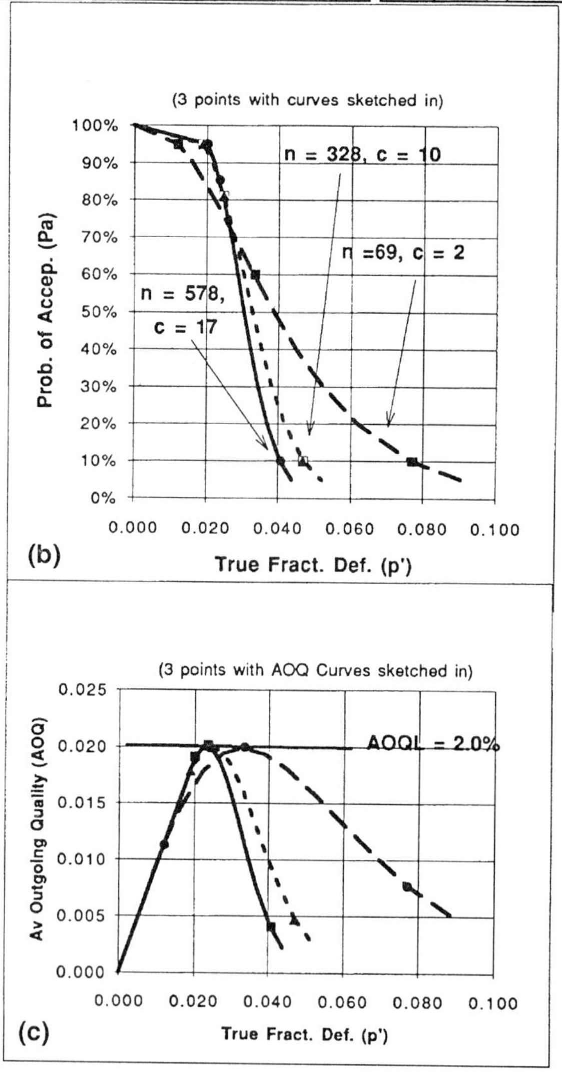

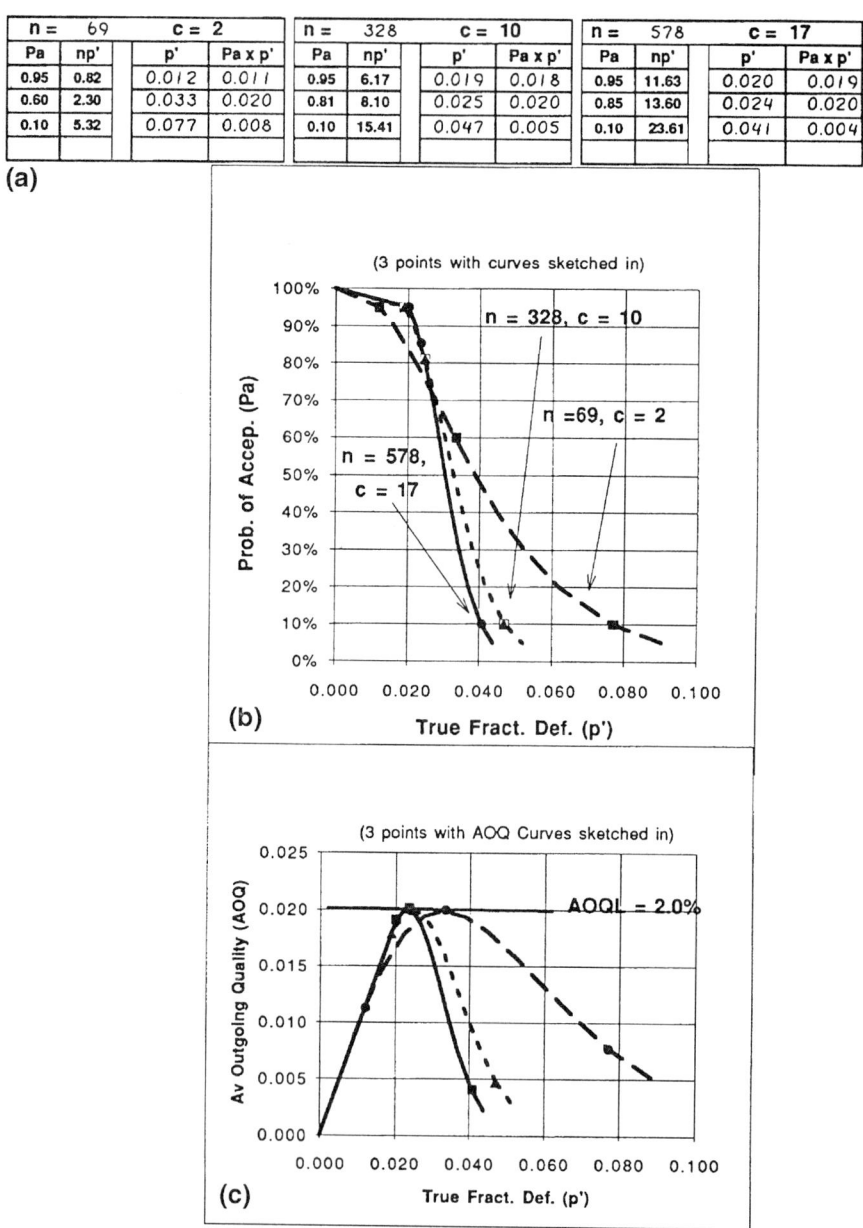

Figure 9 AOQL = 2% curves: (a) data; (b) probability of acceptance; (c) average outgoing quality.

F. n = 200, c = 3 OC, and AOQ Curves

The c = 3 block from Figure 7 is excerpted from the approximator for use in Figure 8. The following steps are used to find the points to be plotted.

Step 1. Write "200" in the "n = ____" area in the c = 3 block.
Step 2. Divide each np′ by 200 and record the value in the p′ column.
Step 3. Plot the points for the OC curve on the P_a vs. p′ graph.
Step 4. Multiply each p′ by its P_a and record this value in the $P_a \cdot p'$ column.
Step 5. Plot the points for the AOQ curve on the AOQ Curve ($P_a \cdot p'$ versus p′) graph.
Step 6. Sketch the OC curve and the AOQ curve by connecting the points.
Step 7. Evaluate sample plan as follows: AQL, LTPD, AOQL.

Note: The graphs of Figure 9 show the three points as plotted, with sketching the actual curve. At the right the curves are sketched in and the AQL, LTPD, and AOQL are identified.

G. AOQL = 2% Curves

The c = 2 block, the c = 10 block, and the c = 17 block were arbitrarily chosen from Figure 7 to illustrate the development of three sample plans (shown in Figure 9) that each have an AOQL = 2.0%. The approximator can be used to choose sample plans for any desired AOQL. In this case, it was an arbitrary choice to develop three sample plans, each with an AOQL of 0.02 (2%). The OC Curve approximator was used to determine that the three sample plans would have accept numbers (c) of 2, 10, and 17. The sample sizes (n) from the c = 2, c = 10, and c = 17 blocks were found by multiplying the middle P_a by its corresponding np′ and dividing this product by the desired AOQL, as shown here:

$$n \text{ (for } c = 2 \text{ plan)} = \frac{P_a \cdot np'}{\text{desired AOQL}} = \frac{0.60 \times 2.30}{0.02} = \frac{1.38}{0.02} = 69$$

$$n \text{ (for } c = 10 \text{ plan)} = \frac{P_a \cdot np'}{\text{desired AOQL}} = \frac{0.81 \times 8.10}{0.02} = \frac{6.561}{0.02} = 328$$

$$n \text{ (for } c = 17 \text{ plan)} = \frac{P_a \cdot np'}{\text{desired AOQL}} = \frac{0.85 \times 13.60}{0.02} = \frac{11.56}{0.02} = 578$$

The three points for each curve that were used to sketch the curves shown in Figure 9 were developed in the same way as the three points used in Figure 8. The points used to sketch the curves are shown on the finished curves instead of graphing them separately. The following are comparisons and comments:

1. The plan with the largest sample (578) appears to be the most stringent on the graph. This means the OC curve is closest to vertical, with AQL = 2% and LTPD = 4.1%.
2. The plan with the smallest sample (69) is the least stringent (LTPD = 7.7%).
3. All three plans, without regard to stringency, show AOQL = 2.0 on the AOQ graph, as predicted from the approximator.

H. LTPD = 10% Curves

The "c = 0" block, the "c = 7" block and the "c = 14" blocks were chosen from the approximator to illustrate the development of three sample plans (shown in

Figure 10) that each have an LTPD = 10%. To find the sample size (n) from the c = 0, c = 7, or the c = 14 blocks, divide the np′ at p_a = 10% by the desired LTPD (p′ = 0.10)

(n = 23, c = 0: n = 118, c = 7: n = 201,c = 14)

n =	23	c= 0		n =	118	c= 7		n =	201	c= 14	
Pa	np'	p'	Pa x p'	Pa	np'	p'	Pa x p'	Pa	np'	p'	Pa x p'
0.95	0.05	0.002	0.002	0.95	3.98	0.034	0.032	0.95	9.25	0.046	0.044
0.37	1.00	0.043	0.016	0.77	5.80	0.049	0.038	0.85	11.10	0.055	0.047
0.10	2.30	0.100	0.010	0.10	11.77	0.100	0.010	0.10	20.13	0.100	0.010

(a)

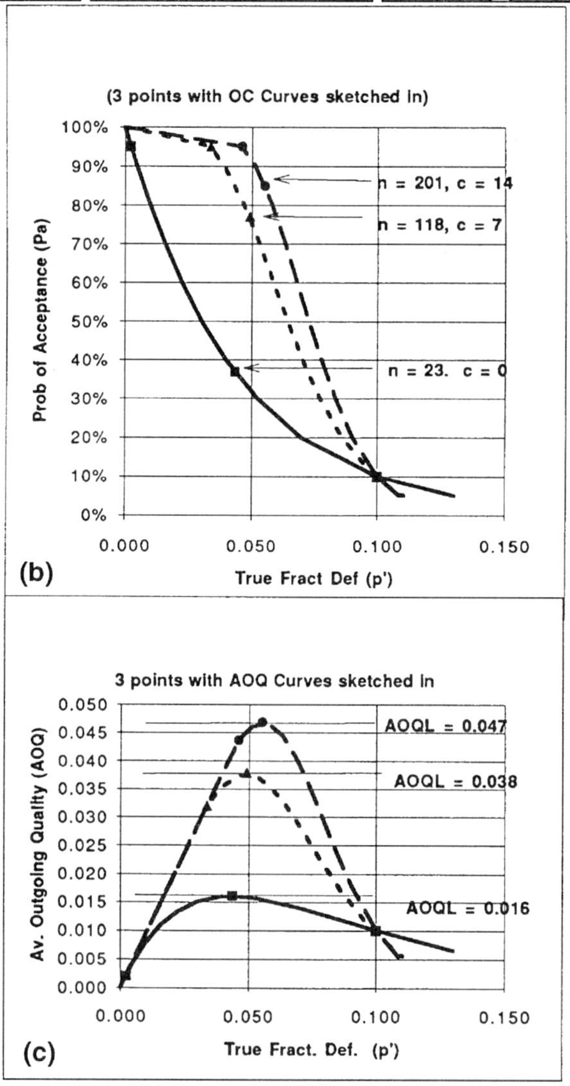

Figure 10 LTDP = 10% curves: (a) data; (b) probability of acceptance; (c) average outgoing quality.

$$n \text{ (for } c = 0 \text{ plan)} = \frac{np'}{\text{desired LTPD}} = \frac{2.30}{0.10} = 23$$

$$n \text{ (for } c = 7 \text{ plan)} = \frac{np'}{\text{desired LTPD}} = \frac{11.77}{0.10} = 118$$

$$n \text{ (for } c = 14 \text{ plan)} = \frac{np'}{\text{desired LTPD}} = \frac{20.13}{0.10} = 201$$

The following are comparisons and comments

1. The plan with the smallest sample (23) appears to be the most stringent on the graph. This may be too stringent since the OC curve shows this plan has only $P_a = 40\%$ when incoming lots are at $p' = 4.0$, which is very close to the AQL for either of the other sample plans. In other words, either of the other plans will accept $p' = 4.0$ for 90 or 95% of the time.
2. The plan with the largest sample (201) is the least stringent (AQL = 4.6%).
3. All three plans, without regard to stringency, show LTPD = 10.0 as predicted from the approximator.

V. DISCRETE DISTRIBUTION SUMMARY

In addition to the discrete applications shown here, the Poisson distribution can also be used to determine:

1. Operating characteristic curves for double and multiple sample plans (3).
2. MTBF values based on life test data and operating characteristic curves for life testing. (See Chapter 18 for more information on the development of MTBF values and life test OC curves.)

REFERENCES

1. Burr, I., *Statistical Quality Control Methods*, Chapter 9, Marcel Dekker, Inc., New York, 1976.
2. MIL-STD-105E, Paragraph 3.1.
3. Heldt, J. J., *a.k.a. Sam Poisson,* Hitchcock Publishing Co., Carol Stream, Ill., 1985.

FURTHER READING

Deming, W. E., *Out of the Crisis,* MIT Center for Advanced Engineering Study, Cambridge, MA, 1986.

Duncan, A. J., *Quality Control and Industrial Statistics,* 3rd Ed., Richard D. Irwin, Inc., Homewood, IL, 1965.

Grant, E. L. and Leavenworth, R. S., *Statistical Quality Control,* 6th Ed., McGraw-Hill, New York, 1988.

Small, B. B. (ed.), *Statistical Quality Control Handbook,* AT&T Technologies, Indianapolis, IN, 1985.

33
Variables Inspection

JOHN JOURDAN HELDT
Free Lance Reliability Service
San Jose, California

I. INTRODUCTION

Variables data[1] (also known as continuous data) are data that can take on any value between specified limits. Discrete data, on the other hand, can take on only exact values. The number of spots that can turn up on throws of two honest dice are representative of discrete data. Any nonfractional number between 2 and 12 can turn up. Three dots can turn up, but 3.5 cannot. A 7 can turn up but not a 7.26. As a general definition, discrete data are those numbers that are represented by what is called the set of "whole numbers": that is, 0, 1, 2, 3,..., but not 1.1 or 2.7.

As an example of variables data, when the output voltage of a 5-volt power supply is measured under variations in load, there will be variations within the measurements of output voltage. Using a basic voltmeter, the voltages measured might be between 4.9 and 5.1 volts. Using a more sophisticated measuring instrument, the measured voltages might be between 4.942 and 5.064 volts. As a more general definition, continuous data can take on any value between any two possible values between the specified limits. In the power supply example, if output voltages of 4.92 and 4.93 volts are possible, then it must be possible to have an output voltage of 4.925 volts.

[1] Examples of variables data are area of city building lots in square feet; diameter of copper wires in mils; ductility of aluminum in percent elongation in two inches; height of volleyball nets in feet; motor shaft end play in micro inches; net weight of cereal boxes in grams; pin diameters in centimeters; power supply output under load, in volts; signal to noise ratio in decibels; and tensile strength of 17-7 steel in thousands of pounds per square inch.

Variables data samples are used to make accurate estimates of the average (X-bar)[2] for the population represented by the sample. The variation within the total population can also be estimated based on the data variation within the sample. The standard deviation of the data is usually represented by the Greek lowercase letter sigma (σ). Sigma, or standard deviation, is the square root of the data variance. Variance is far less often used in quality control than is the standard deviation.

For more than 100 years, X-bar and σ tables of normal distribution data have been used to make estimates of the characteristics of populations. (Uses for the normal distribution in statistical quality control are well known and are delineated in every basic statistical text. This section will be devoted to statistical quality control and statistical process control, which is a modern outgrowth of the normal distribution.)

II. STATISTICAL QUALITY CONTROL

A. The X-Bar and R Chart

The X-bar and R chart, developed by Walter Shewhart in the 1920s, is a tool that is often used in evaluating the variation of processes. The Shewhart X-bar and R chart, also called a variables control chart, is considered the backbone of statistical quality control.

The data for a Shewhart X-bar and R chart are derived by making variables measurements on samples in what are termed *rational subgroups*. A subgroup is a sample—most often of four to five items—that is drawn from the process to be evaluated. A rational subgroup is a group of sample items drawn from the process in such a way that within-subgroup variation is minimized. Generally, this means that each subgroup contains items consecutively drawn from the process.

A basic concept of quality control is that a distribution of the averages (the X-bars) of subgroups drawn from any single distribution will be essentially normally distributed. As a result, the concepts of making statistically valid estimates from the normal distribution can be applied to the X-bar and R chart to evaluate industrial processes.

B. Making the X-Bar and R Control Chart

1. Determine which characteristics in the process are most critical and determine how to measure those you wish to evaluate.
2. Measure the selected characteristic periodically using a small subgroup. Usually, the subgroup will consist of five parts.[3] A total of 100 individual

---

[2] $\overline{X}$, is the symbol for the average. It is usually read as "X-bar." An overall or "grand" average, or average of the averages as used with the X-bar and R chart, is represented by the symbol $\overline{\overline{X}}$ (read as X-double-bar).

[3] Should the measurement taken be difficult (or expensive) to perform, you may wish to use four or fewer samples in your subgroup. Likewise, subgroups as large as 20 may be used for easy-to-measure characteristics. Subgroups of 10 or more are seldom used, but classic texts (e.g., Grant and Leavenworth, 1988) contain tables for subgroups of 20 or more.

measurements should be collected before the upper and lower control limits for the $\overline{X}$ and R chart can be determined. (Twenty subgroups of 5 units will be used for illustration in Figures 3, 4, and 5.) This means when 5 parts per subgroup are measured every hour, a total of 100 individuals will have been measured at the end of 20 hours of operation. Or, when 3 parts are measured every half hour, 17 hours of operation will be required before a total of 100 individual parts will have been measured.

3. As each subgroup is measured, the average and range of each subgroup must be calculated. To calculate the average of the subgroup, all of the measurements need be totaled and then divided by the number of measurements in the subgroup. To find the range of each subgroup, one simply subtracts the smallest measurement from the largest measurement in the subgroup, as long as the total number of measurements in the subgroup is two or more.
4. When a total of 100 individual parts have been measured and the range and average for each subgroup have been computed, the average of the ranges ($\overline{R}$), (called R-bar) and the average of the averages ($\overline{\overline{X}}$) (also called the grand average or "X-double-bar") may be calculated.
5. Using the formulas for the upper and lower control limits, $UCL_R = D_3(\overline{R})$ and $LCL_R = D_4(\overline{R})$, the control limits for the ranges may be reckoned.[4] The upper control limit for the ranges may be used to establish the scale for the range chart. With the control limits in place on the R chart, all of the ranges should be plotted as a function of the time that the measurements were made. Should any of the ranges be out of limits (i.e., *out of statistical control*), work should be discontinued while the process is evaluated for stability.[5] It is important to remember that when any ranges are out of control (one or more readings outside the control limits), any interpretation of trends of the averages is meaningless.
6. When the ranges are "in control," the control limits for the averages ($UCL_{\overline{X}}$ and $LCL_{\overline{X}}$) maybe be calculated using the formulas:

   $$UCL_{\overline{X}} = \overline{\overline{X}} + A_2\overline{R} \text{ and } LCL_{\overline{X}} = \overline{\overline{X}} - A_2\overline{R}$$

   These control limits may be used to set the range for the $\overline{X}$ chart. (Variable control charts for manual plotting of $\overline{X}$ and R are available from the American Society for Quality Control. ASQC charts were not used in Figures 3, 4, and 5. These charts were generated using software.)

   Once the control limits are installed, all of the averages—the $\overline{X}$s—should be plotted on the graph. One or more $\overline{X}$s outside of limits is cause to stop for reevaluation of the stability of the process.

---

[4] LCLR will always be 0 when the subgroup size is 6 or less, since $D_4 = 0$ for subgroups less than 7.

[5] When one or two out of control points can be suspected of being the result of the instabilities of a new process (i.e., not expected to recur), it is common practice to discard all data connected with these one or two points. The remaining data are used to recompute the control limits. When all points fall within the recomputed limits, the process may continue to run.

7. Once there are no points out of control (i.e., outside the control limit) on either the $\overline{X}$ or R chart, the process is considered to be stable and the continuation of $\overline{X}$s and Rs, of subgroups, falling within their respective control limits is accepted as statistical proof that the process is "in control." (When an item of data is plotted and falls on the control limit line, it is considered acceptable. Outside the control limit means just that.)

C. Statistical Control

A process is said to be in statistical control when all assignable causes for variation have been removed. In other words, all of the variations that were not due to random causes have been eliminated by removal of the cause or by controlling the condition.

The process is doing as well as it can. This means that with routine maintenance and constant vigilance against assignable causes, we may expect the process to continue to operate predictably. This also means that the process must be changed if it is to be made better (i.e., a new buffer added to a chemical process or a complete overhaul with new bearings and new ways for a precision manufacturing machine).

D. Statistical Quality Control

One of the most important aims of the Shewhart chart is to alert operators to any outside influences (assignable causes) that might disrupt the control of the process. X-bar and R charts are more than adequate to pinpoint variations other than normal process perturbations (chance causes).

However, controlling the process did not take into account the actual specification limits. Shewhart took it for granted that the manufacturing people would be using process equipment that could make product within the engineering specifications. Unfortunately, later practitioners were complacent, or uninformed, resulting in an occasional process that was in control but not completely within specification limits. Most of these errors were the result of an incomplete understanding of the differences between the distribution of the $\overline{X}$s (averages of the subgroups of individual measurements) and the distribution of the individual Xs (individual measurements). Often little care was taken to keep the process centered between the specifications: As long as the process was in control, it was deemed to be good enough. Genichi Taguchi was the first to show the value of keeping the average of the process as close as possible to the center point between the specification limits.

III. STATISTICAL PROCESS CONTROL

Modern statistical process control has been designed to assure that the centerline, or average, of any process stays as close as possible to the point that is half way between the upper and lower specifications for the process. At the same time, the

SPC process encourages narrowing of the limits of the process. (Today, it is popular to talk about *six sigma quality* control. Six sigma quality control means that 99.83% of all product made under this system will fall within half of the tolerance allowed. Another truism of six sigma quality control is the statement, "less than one part per million will fall outside the specification limits!")

SPC is an add on to statistical quality control. The control features designed by Shewhart for SQC are extrapolated to estimate the parameters of the process universe and these are used to enhance the centering and narrowing of the process. Factors and constants used in SPC are shown in Table 1.

Coding: A Statistical Aid

For a number of reasons, including ease of data entry and space limitations on the chart, most variables data are *coded*. Generally, there are only hints given in actual examples of control charts as to what the *zero equals* and *unit equals* blocks on the charts mean. The following is an example of the coding procedure as it is used in statistical work; the advantages and formulation are also demonstrated.

Advantages of Coding

1. Calculations, such as averages, control limits, and the like, are simplified. All of the coded numbers are expressed as simple integers.
2. These integers are easier to read because of the limited table areas allowed for individual measurements.
3. Coding and uncoding is simple and easy (see instructions below).
4. Coding simplifies X-bar and R charts for easier reading, understanding and interpretation.
5. With more practice, the coding process becomes even more helpful and useful.

Coding Procedure

To find the coded value of any number:

1. Subtract the "zero equals" value from the number.
2. Divide the remainder by the "unit of measure" value.

Decoding Procedure

To uncode any coded number (except standard deviations {σ}, ranges {R}, or other measures of variation).

1. Multiply the coded number by the "unit of measure" value.
2. Add the "Zero equals" value to this product.

To uncode standard deviations {σ}, ranges {R}, or other measures of variation:

1. Multiply the coded number by the "unit of measure" value. (explanation: The "zero equals" value is *subtracted out* of σ and R values in their calculations.)

Table 1 Factors and Constants

Factors for the Chart[a]		
$\overline{R}$	$= \Sigma R/k$	= Average of the ranges of the individual subgroups
$\overline{\overline{X}}$	$= \Sigma X/k$	= Average of the averages of the individual subgroups
A_2	= This factor is used to determine the upper and lower control limits for his $\overline{X}$ chart (A_2 depends on the size of the individual subgroup)	
$UCL_{\overline{X}}$	$= \overline{\overline{X}} + A_2\,\overline{R}$	= The upper control limit for the X-bar chart
$UCL_{\overline{X}}$	$= \overline{X} - A_2\,\overline{R}$	= The lower control limit for the X-bar chart
D_3	= This factor is used to determine the upper control limit for his R chart (D_3 depends on the size of the individual subgroup)	
D_4	= This factor is used to determine the upper control limit for his R chart (D_4^6 depends on the size of the individual subgroup)	
UCL_R	$= D_3\,(\overline{R})$	= The upper control limit for the R chart
LCL_R	$= D_4\,(\overline{R})$	= The lower control limit for the R chart
Factors for Capability Estimates: Statistical Process Control[b]		
d_2	= This factor is used to estimate the standard deviation for the universe (d_2 depends on the size of the individual subgroups used in the determination of the ranges)	
3σ	$= (3/d_2)(\overline{R})$	= Half the process capability (48.65% of all individual parts will fall between the average and $+3\sigma$ ADD)
UL_X	$= \overline{\overline{X}} + (3/d_2)(\overline{R})$	= The upper limit of the population of the individuals
LL_X	$= \overline{\overline{X}} - (3/d_2)(\overline{R})$	= The lower limit of the population of the individuals
6σ	$= (6/d_2)(\overline{R})$	= The process capability; sometimes called the machine capability or the process spread [98.3% of all individual parts will fall between $+3\sigma$ and -3σ (i.e., $UL_X - LL_X$)]
US	= Upper specification: upper limit of acceptable measurements for the total population	
LS	= Lower specification: lower limit of acceptable measurements for the total population	
US − LS	= The range of acceptable measurements for the total population	
Capability Estimates: Statistical Process Control[c]		
C_p	$= \text{Capability index} = \dfrac{\text{part tolerance}}{6\sigma} = \dfrac{US - LS}{(6/d_2)(\overline{R})}$	
C_{pk}	$= \text{The smaller of } \dfrac{US - \overline{\overline{X}}}{3\sigma} \text{ and } \dfrac{\overline{\overline{X}} - LS}{3\sigma}$	
CR	$= \text{Capability ratio} = \dfrac{6\sigma}{\text{part tolerance}} = \dfrac{(6/d_2)(\overline{R})}{US - LS}$	

[a] Data collectors are usually cautioned against being *too regular* in their time of collections. This prevents a *shop-wise* operator from manipulating the outcome by *selecting* the sets of parts to be measured.
[b] Normally, new variables control charts are not plotted until we are sure that all points will be within the control limits.
[c] $C_p = 2.0$ and a centered process is the aim of *six sigma quality*.

Coding Example

When "zero equals" = 1.495 inches and "unit of measurement" = 0.001 inches:

True value	Coded value	Coded parameters
Data		
1.497	2	
1.499	4	Coded average = 5.0
1.500	5	Coded standard deviation (σ) = 2.236
1.501	6	
1.503	8	

Uncoded average = 5.0 × (unit of measure) + (0 equals)
= 5.0 × 0.001 + 1.495 = 1.500″

Uncoded standard deviation = 2.236 × (unit of measure)
= 2.236 × 0.001 = 0.002236″

In essence, SPC data are coded by subtracting the "zero equals" value from each item of data. Then the "adjusted" data item is multiplied by the "unit equals" value. Finally, the "coded" value is entered into the data block.

Decoding of averages or other parameters of central measurement—such as modes and medians—is done by reversing the process; that is, multiply the coded value by the "unit equals" value and add the result to the "zero equals" value. *Note:* the number written in the "zero equals" block may be chosen arbitrarily. Obviously, judicious choice of this value can make data processing easier.

Ranges, standard deviations, or other parameters for variation are decoded simply by multiplying the value by the "unit equals" value.

All data must be decoded when the parameters are entered into the formulas shown in Table 1. This means that the capability index (C_p), product capability (C_{pk}), and any other SPC parameters will be shown in uncoded values.

IV. SPC EXAMPLES

Figures 1, 2, and 3 illustrate the steps of SPC.

A. A Process in Control But Not Capable

Figure 1 shows the data in coded form that are used to create the centerless grinding operation X-bar and R chart shown in Figure 2. The value *Zero = 1.450* was chosen so that there would be no negative numbers in the coded data. Negative numbers are difficult for some people to work with.

																		Zero = 1.450"		*Unit = 0.001"*
Part Name (Product) *YAV-115: Precision Shaft*							Operation (Process) *Centerless Grind*							Specification Limits *1.500 ± 0.015"*						
DATE	7-1	7-1	7-1	7-1	7-1	7-1	7-1	7-2	7-2	7-2	7-2	7-2	7-2	7-2	7-3	7-3	7-3	7-3	7-3	7-3
TIME	8A	9A	10A	11A	1P	2P	3P	8A	9A	10A	11A	1P	2P	3P	8A	9A	10A	11A	1P	2P
Sample 1	61	52	55	58	55	56	57	52	56	60	62	52	52	52	45	59	69	53	60	63
Sample 2	61	62	56	60	67	55	61	64	65	52	61	59	43	52	60	63	54	58	53	53
Sample 3	57	59	52	59	53	62	52	58	53	61	59	52	62	56	59	57	56	42	43	60
Sample 4	56	62	60	49	66	61	60	41	55	41	61	53	59	60	57	53	42	62	50	63
Sample 5	60	67	59	42	60	59	51	56	56	55	55	54	55	59	51	49	62	59	59	60
Sum	295	302	282	268	301	293	281	271	285	269	298	270	271	279	272	281	283	274	265	299
Average	59	60	56	54	60	59	56	54	57	54	60	54	54	56	54	56	57	55	53	60
Range	5	15	8	18	14	7	10	23	12	20	7	7	19	8	15	14	27	20	17	10
Notes																				

Figure 1 Variables control chart 1 (X-bar and range).

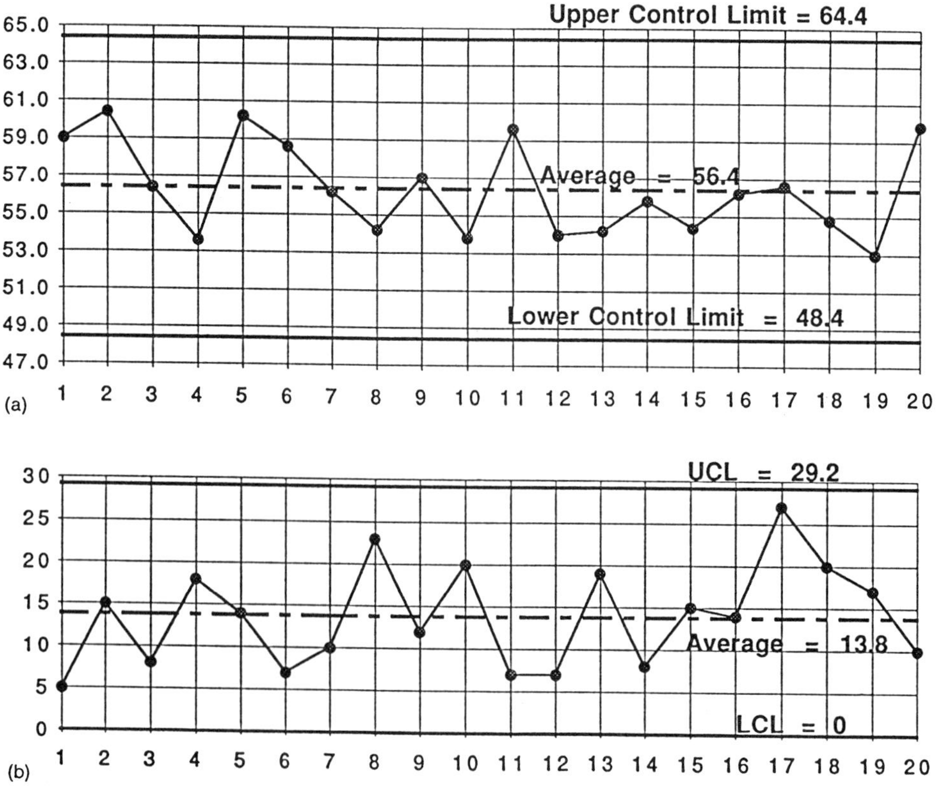

Figure 2 (a) X-bar chart; (b) range chart.

The increment of measurement (unit = 0.001) means that the first sample reading (61), uncoded, is 0.061″ added to 1.450″. In other words, this means that the first item of data represents a shaft that was measured as 1.511 inches (i.e., [61 × 0.001] + 1.450).

The second row of the title block shows: the part name "YAV-115: Precision Shaft", the name of the operation: "Centerless Grind" and the specification limits: "1.500 ± 0.015". Blocks for other information, such as "operator," "machine," or "assembly line," can be added to the title block as desired.

Data block: The row across the top of the data block shows the date on which the data were collected. The row below shows the approximate hour. The five lines titled "Samples 1 through 5" provide the blocks for entering the five measurements of each subgroup of five. Each of these numbers have been entered as "coded" values.

As an example, the actual measurements used in the first subgroup of data are listed here showing how the actual values are changed into the coded values:

Sample	Actual value	Subtract "0 ="	Diff.	Divide by "unit ="	Coded value
Sample 1	1.511″	−1.450	=0.061	÷0.001″	=61
Sample 2	1.511″	−1.450	=0.061	÷0.001″	=61
Sample 3	1.507″	−1.450	=0.057	÷0.001″	=57
Sample 4	1.506″	−1.450	=0.057	÷0.001″	=56
Sample 5	1.510″	−1.450	=0.060	÷0.001″	=60
				Sum	295
				Average	69
				Range	5

The "Sum" row in the data block shows the sum of all of the coded values for the intersecting subgroup column.

The "Average" row in the data block shows the average of all of the coded values for the intersecting subgroup column.

The "Range" row in the data block shows the range of all of the coded values for the intersecting subgroup column.

Each subgroup average and range is aligned directly below the data used to determine that average and range. Neither of these charts were plotted until after $\overline{\overline{X}}$, $\overline{R}$, and the control limits have been calculated. Determination of these values is shown in Table 2; $\overline{\overline{X}}$, $\overline{R}$, and the control limits are laid out in Figure 2.

Since all of the averages and ranges are well within the control limits, the limits for the individual values of the population are calculated in Table 3. Table 3 shows the method of estimating the process limits for individual items using the formula X-double-bar ±3 sigma′ (three sigma prime). Sigma prime is defined as the standard deviation of the *population* from which the sample is drawn. The numerical value of sigma prime is estimated by dividing the average range—R-bar—by the constant factor d_2. The estimated limits for individuals—*for the one parameter being evaluated*—is an upper limit (UL_x) of 1.524 in. +, and a lower

Table 2a Control Limit Calculations

$\overline{R}$	=	$\frac{\Sigma R}{k}$	=	$\frac{276}{20}$	=	13.8
$\overline{\overline{X}}$	=	$\frac{\Sigma \overline{X}}{k}$	=	$\frac{1127.8}{20}$	=	56.4
$A_2\overline{R}$	=	0.577×13.8			=	8.0
$UCL_{\overline{X}}$	=	$\overline{\overline{X}} + A_2\overline{R}$			=	64.4
$LCL_{\overline{X}}$	=	$\overline{\overline{X}} - A_2\overline{R}$			=	48.4
$UCL_{\overline{R}}$	=	$D_4\overline{R}$	=	2.114×13.8	=	29.2

Subgroups 1–20 are included.

Table 2b Factors for Control Limits

n	A_2	D_4	d_2	$3/d_2$
2	1.880	3.258	1.128	2.659
3	1.023	2.574	1.693	1.772
4	0.729	2.282	2.059	1.457
5	0.577	2.114	2.326	1.290
6	0.483	2.004	2.534	1.184

limit (LL_x) of 1.488 in. +. These values should be compared to the blueprint or drawing limits of 1.515 in. (US) and 1.485 in. (LS). If other features of the part are to be evaluated, a separate analysis must be performed for each one.

All of the measurements for individuals are background values used to make the process control analyses in Figure 2. Values shown are uncoded. In addition to the best estimate of the average (i.e., $\overline{\overline{X}}$) and the standard deviation (σ') for the population universe, the upper and lower specifications are also shown. C_p and C_{pk} are derived from these values.

Table 3 Limits for Individuals

$\overline{\overline{X}}$	=	$56.4 \times 0.001 + 1.450$		=	1.5064″
$\frac{3}{d_2}\overline{R}$	=	$1.290 \times 13.8\ (0.001)$		=	0.017802″
UL_X	=	$\overline{\overline{X}} + \frac{3}{d_2}\overline{R}$	=	=	1.524202″
LL_X	=	$\overline{\overline{X}} - \frac{3}{d_2}\overline{R}$	=	=	1.488598″
US	=		=	=	1.515″
LS	=		=	=	1.485″
US − LS	=		=	=	0.030″
$6\sigma = \frac{6}{d_2}\overline{R}$	=	2×0.017802		=	0.035604

Compare with specifications of tolerance limits.

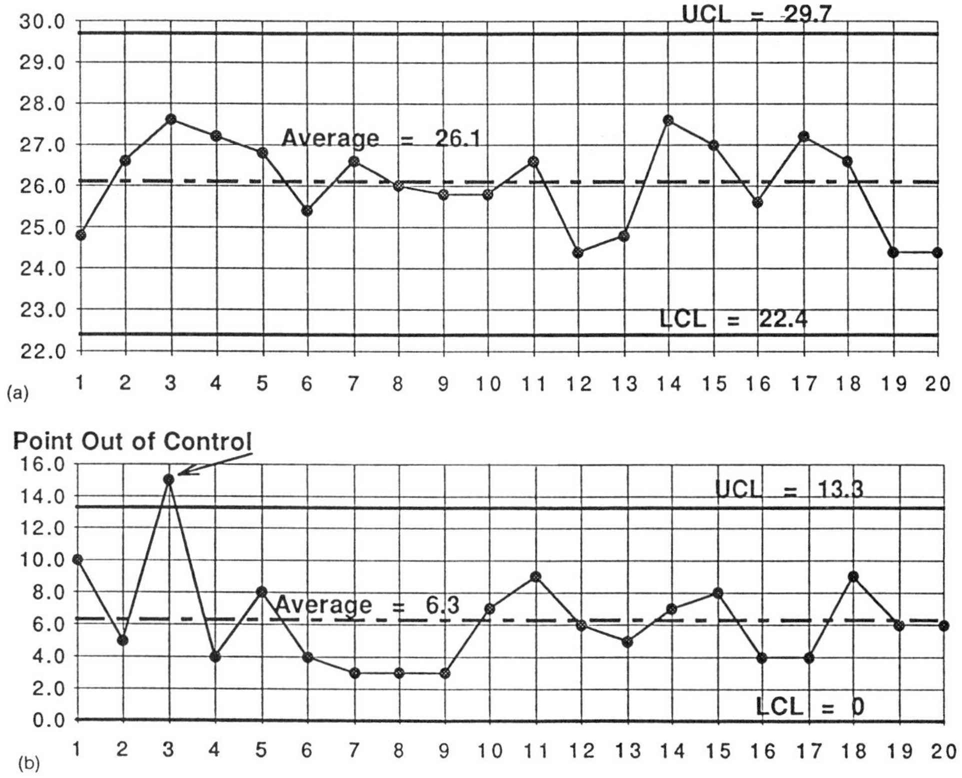

Figure 3 Centerless grinding: (a) X-bar and (b) range charts.

$$\text{Capability index (Cp)} = \frac{\text{part tolerance}}{6\sigma}$$

Cp is the value used to determine whether the process can meet the specification requirements. It is the indicator that shows how much of the process output can be fitted within the specification limits when the process population average exactly coincides with the center points between the upper and lower specification limits.

$$\text{Capability index (Cp)} = 0.8426$$

When Cp = 1, it means that for an exactly centered process, all parts between -3σ and $+3\sigma$ (i.e., 99.7%) will be within specification. When $C_p = 1.33$, it means that for an exactly centered process, all parts between -4σ and $+4\sigma$ (i.e., more than 99.97%) will be within specification. When $C_p = 0.666$, it means that for an exactly centered process, all parts between -2σ and $+2\sigma$ (i.e., 95.4%) will be within specification. $C_p = 0.8426$, and if there is an exactly centered process, all parts between -2.53σ and $+2.53\sigma$ (i.e., a little less than 98.9%) will be within specification. This means that under the best of circumstances, this process is barely capable, since this C_p predicts that at least 1% will always be out of limits.

$$\text{Process capability index } (C_{pk}) = \text{the smaller of: } \frac{US - \overline{\overline{X}}}{3\sigma}, \text{ or } \frac{\overline{\overline{X}} - LL}{3\sigma}$$

$$\text{For this example, } C_{pk} = \frac{\overline{\overline{X}} - LL}{3\sigma} = 1.202$$

This is not the C_{pk}, since it is not the smaller of the two values. Since this value is larger than 1.000, it means that almost all of the population will be larger than LL, the lower specification limit.

$$C_{pk} = \frac{US - \overline{\overline{X}}}{3\sigma} = 0.483$$

The smaller $C_{pk} = 0.483$ is indeed the C_{pk}. Interpreting the 0.483 value means that the upper specification will fall on the $+1.45\sigma$ value of the specification. $+1.45\sigma$ predicts 92.65% of all of the parts will fall below UL, the upper specification limit. It also means that 7.53% of the parts will be larger than UL.

When C_{pk} is less than 1.0, it means that the process is currently producing parts that will be over or under specification. Which C_{pk} formula gives the lower yield indicates which specification is violated.

In general, a process is "capable" when the average is close to the centerline between UL and LL, and $C_{pk} = 1.3$ or more. ($C_{pk} = 1.3$ means that the specification limits include all values of the total population to the $\pm 3\ \sigma$ limits.)

When the two C_{pk} values are unequal, they provide a comparison to show which way to shift the process average to bring optimum process performance.

The methods of calculating and evaluating the capability index (C_p) and the process capability index (C_{pk}) are described as follows.

Zero = 0 | Unit = 0.001"

Part Name (Product): *YAV-115: Motor Shaft* | Operation (Process): *End Play* | Specification Limits: *0.015"/0.045"*

DATE	7-1	7-1	7-1	7-1	7-1	7-1	7-1	7-2	7-2	7-2	7-2	7-2	7-2	7-2	7-3	7-3	7-3	7-3	7-3	7-3
TIME	8A	9A	10A	11A	1P	2P	3P	8A	9A	10A	11A	1P	2P	3P	8A	9A	10A	11A	1P	2P
Sample 1	21	28	31	27	32	26	25	26	25	30	32	23	27	29	27	27	27	32	23	23
Sample 2	20	28	35	27	24	24	27	25	26	23	28	25	24	27	30	26	27	28	25	25
Sample 3	30	23	20	29	25	28	27	28	25	25	25	27	26	23	24	23	29	25	27	27
Sample 4	27	26	24	25	25	24	28	25	28	26	23	26	22	29	31	25	25	23	26	26
Sample 5	26	28	28	28	28	25	26	26	25	25	25	21	25	30	23	27	28	25	21	21
Sum	124	133	138	136	134	127	133	130	129	129	133	122	124	138	135	128	136	133	122	122
Average	25	27	28	27	27	25	27	26	26	26	27	24	25	28	27	26	27	27	24	24
Range	10	5	15	4	8	4	3	3	3	7	9	6	5	7	8	4	4	9	6	6
Notes																				

Figure 4 Variables control chart 2 (X-bar and range).

Table 4a Control Limits for Figure 3

$\overline{R}$	=	$\frac{\Sigma R}{k}$	=	$\frac{126}{20}$	=	6.3
$\overline{\overline{X}}$	=	$\frac{\Sigma \overline{X}}{k}$	=	$\frac{521.2}{20}$	=	26.1
$A_2\overline{R}$	=	0.057×6.3			=	3.6
$UCL_{\overline{X}}$	=	$\overline{\overline{X}} + A_2\overline{R}$			=	29.7
$LCL_{\overline{X}}$	=	$\overline{\overline{X}} - A_2\overline{R}$			=	22.5
$UCL_{\overline{R}}$	=	$D_4\overline{R}$	=	2.114×6.3	=	13.3

Subgroups 1–20 are included.

Table 4b Factors for Control Limits

n	A_2	D_4	d_2	$3/d_2$
2	1.880	3.258	1.128	2.659
3	1.023	2.574	1.693	1.772
4	0.729	2.282	2.059	1.457
5	0.577	2.114	2.326	1.290
6	0.483	2.004	2.534	1.184

Note: When the variables control chart is being developed for a new production process and one or two points are "out of control," it is acceptable to eliminate all data for the offending point or points and refigure the control limits. Thus, Figure 4 uses all of the good data to refigure the variables control chart.

Process control analysis for centerless grinding job:

$$\text{Capability index, CI} = \frac{\text{part tolerance}}{6\sigma} = \frac{US - LS}{\frac{6}{d_2}\overline{R}}$$

When the capability index equals 1, it means that the process is barely capable. The centerline of the process (average) must coincide with the exact center between the upper and lower specifications for the 3 sigma points of the process to fall exactly on the specification's upper limit (UL) and lower limit (LL).

When the capability index is less than 1, the process can never fit between the upper and lower specifications unless the process is changed (improved).

A process capability index equal to 1.3 or more is considered to be the least acceptable for competition in today's market.

$$\frac{\text{Part tolerance}}{6\sigma} = \frac{US - LS}{\frac{6}{d_2}\overline{R}} = \frac{0.030}{0.035604} = 0.8426$$

$$C_{pk} = \text{the smaller of } \frac{US - \overline{\overline{X}}}{3\sigma} \text{ and } \frac{\overline{\overline{X}} - LL}{3\sigma}$$

$$\frac{US - \overline{\overline{X}}}{3\sigma} = \frac{1.515 - 1.5064}{0.017802} = 0.483$$

																Zero = 0			Unit = 0.001"
Part Name (Product) *YAV-115: Motor Shaft*						Operation (Process) *End Play*							Specification Limits *0.015"/0.045"*						
DATE	7-1	7-1	7-1	7-1	7-1	7-1	7-2	7-2	7-2	7-2	7-2	7-2	7-2	7-3	7-3	7-3	7-3	7-3	7-3
TIME	8A	9A	11A	1P	2P	3P	8A	9A	10A	11A	1P	2P	3P	8A	9A	10A	11A	1P	2P
Sample 1	21	28	27	32	26	25	26	25	30	32	23	27	29	27	27	27	32	23	23
Sample 2	20	28	27	24	24	27	25	26	23	28	25	24	27	30	26	27	28	25	25
Sample 3	30	23	29	25	28	27	28	25	25	25	27	26	23	24	23	29	25	27	27
Sample 4	27	26	25	25	24	28	25	28	26	23	26	22	29	31	25	25	23	26	26
Sample 5	26	28	28	28	25	26	26	25	25	25	21	25	30	23	27	28	25	21	21
Sum	124	133	136	134	127	133	130	129	129	133	122	124	138	135	128	136	133	122	122
Average	25	27	27	27	25	27	26	26	26	27	24	25	28	27	26	27	27	24	24
Range	10	5	4	8	4	3	3	3	7	9	6	5	7	8	4	4	9	6	6
Notes																			

Figure 5 Refigured control chart 2 (X-bar and range).

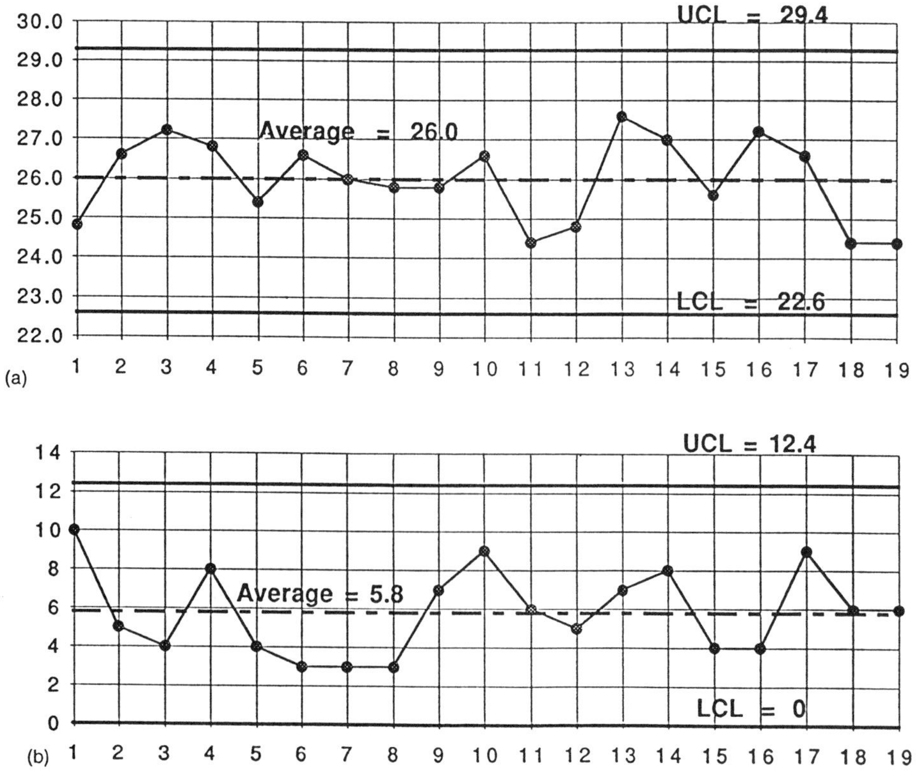

Figure 6 (a) Revised X-bar chart. (b) Revised range chart.

Table 5a Control Limits for Figure 6

$\overline{R}$	=	$\frac{\Sigma R}{k}$	=	$\frac{111}{19}$	=	5.8
$\overline{\overline{X}}$	=	$\frac{\Sigma \overline{X}}{k}$	=	$\frac{493.6}{19}$	=	26
$A_2\overline{R}$	=	0.577×5.8			=	3.4
$UCL_{\overline{X}}$	=	$\overline{\overline{X}} + A_2\overline{R}$			=	29.4
$LCL_{\overline{X}}$	=	$\overline{\overline{X}} - A_2\overline{R}$			=	22.6
$UCL_{\overline{R}}$	=	$D_4\overline{R}$	=	2.114×5.8	=	12.4

Subgroups 1–19 are included.

Table 5b Factors for Control Limits

n	A_2	D_4	d_2	$3/d_2$
2	1.880	3.258	1.128	2.659
3	1.023	2.574	1.693	1.772
4	0.729	2.282	2.059	1.457
5	0.577	2.114	2.326	1.290
6	0.483	2.004	2.534	1.184

$$\frac{\overline{\overline{X}} - LL}{3\sigma} = \frac{1.5064 - 1.485}{0.017802} = 1.202$$

Use of the C_{pk} index will indicate what kind of adjustment (if any) must be made to the process average to get the optimum yield and quality from that process. When the capability index is less than 1, at least one of the C_{pk} formulas will yield less than 1. When the capability index is greater than 1, and one of the C_{pk} formulas is less than 1, it indicates that the process average can be adjusted to improve product acceptance. In general, adjustment of the process should continue until the values of both "halves" of both capability index formulas are equal—thus indicating that the process is "centered"—and C_{pk} is 1.33 or higher. Such a process is generally considered to be "capable."

Since the upper estimated limit of 1.524 in. for individuals in the population as calculated in Table 3 is greater than the blueprint limit of 1.515 in., this process is considered to be not capable.

B. A Process Not in Control

The X-bar and R chart shown in Figure 3 was constructed using the data shown in Figure 4. Control limits for Figure 3 are shown in Table 4. Note that data point number 3 on the X-bar chart is above the control limit, thus this process is considered to be out of statistical control. No further analysis was done using these data.

C. Reviewing the Process Data

Figures 5 and 6 show the analysis required to reanalyze the data in Figures 3 and 4. In Figure 5, the third subgroup of samples (in Figs. 3 and 4) has been deleted.

Table 6 Limits for Individuals

$\overline{\overline{X}}$	=	$26 \times 0.001 + 0$			=	0.026″
$\frac{3}{d_2}\overline{R}$	=	$1.290 \times 5.8\ (0.001)$			=	0.0075″
UL_X	=	$\overline{\overline{X}} + \frac{3}{d_2}\overline{R}$	=		=	0.0335″
LL_X	=	$\overline{\overline{X}} - \frac{3}{d_2}\overline{R}$	=		=	0.0185″
US	=		=		=	0.045″
LS	=		=		=	0.015″
US − LS	=		=		=	0.030″
6σ	=	$\frac{6}{d_2}\overline{R}$	=	2×0.00748	=	0.015
C_{pk}	=	$\frac{US - \overline{\overline{X}}}{3\sigma}$	=	$\frac{0.045 - 0.026}{0.0075}$	=	2.533
or:						
		$\frac{\overline{\overline{X}} - LL}{3\sigma}$	=	$\frac{0.026 - 0.015}{0.0075}$	=	1.466

Compare with specifications of tolerance limits.
Note: $C_p = 2.0$ and a centered process is the aim of *six sigma quality*.

It should be noted that this is not an attempt to "fake" the analysis but a review of what the process would likely do if the assignable cause(s) of variation—such as affected the third subgroup of data—were removed. Having deleted the "offending" subgroup, the analysis continues with the remaining 19 subgroups.

Calculations for the revised control chart limits are shown in Table 5; Figure 6 is the resulting control chart. Note that the chart now shows statistical control. In doing initial process capability estimates, it is sometimes necessary to delete several subgroups of data. Capability calculations are shown in Table 6.

FURTHER READING

Deming, W. E., *Out of the Crisis,* MIT Center for Advanced Engineering Study, Cambridge, MA, 1986.

Duncan, A. J., *Quality Control and Industrial Statistics,* 3rd Ed., Richard D. Irwin, Inc., Homewood, IL, 1965.

Grant, E. L. and Leavenworth, R. S. *Statistical Quality Control,* 6th ed., McGraw-Hill, New York, 1988.

Shewhart, W. A., *Economic Control of Quality of Manufactured Product,* W. A. Shewhart, D. Van Nostrand C., New York, 1931, Reprinted by the American Society for Quality Control, Milwaukee, 1980.

Small, B. B. (ed.), *Statistical Quality Control Handbook,* AT&T Technologies, Indianapolis, IN, 1985.

34
Statistical Sampling

JOHN JOURDAN HELDT
Free Lance Reliability Service
San Jose, California

I. INTRODUCTION

There are many formal inspection sampling plans. And there are more than a few informal plans developed and used within individual companies.

Among the better known formal plans are Dodge-Romig; Shainin (or Hamilton Standard) Lot Plot; Military Standard 414 (MIL-STD-414); and a variety of continuous sampling plans (CSP) that are found in MIL-STD-1235. Among the typical informal plans are a variety of what are usually called C = 0, or C-Zero plans that are generally based on small sample sizes and acceptance on zero defective units.

With all of these formal and informal plans, however, it is likely that the best known and most often used plan is MIL-STD-105 (now in revision E).

II. MIL-STD-105

To many people, MIL-STD-105 is only a card or set of procedure sheets that describe *single sampling plans for normal inspection*. Actually, MIL-105 is much more. In addition to being a tried and true sampling system, the standard includes the following elements.

A. A Vendor Rating Plan

In addition to normal inspection plans, MIL-105 has sampling plans for *tightened inspection* and for *reduced inspection*. Section 4.7, *Switching Procedures*, shows how and when to go to tightened or reduced inspection and the conditions under which the switches are made. When used as a required outgoing inspection plan, this system rewards the good supplier by reducing the total amount of work involved in the outgoing inspection sampling system. It penalizes the poor supplier by imposing more stringent sampling. The average outgoing quality limit (AOQL) of the tightened sample plan is designed to be equal to the acceptable quality level (AQL). This means that, in the long run, product quality inspected using a tightened inspection plan will not exceed the limit of a satisfactory process average for percent defective. When used as an incoming inspection plan, the standard quickly shows which suppliers are performing well and which ones are not.

B. Evaluation of Sample Plans

The operating characteristic (OC) Curves shown in Tables X-A through X-S of MIL-STD-105 are provided for the evaluation of each of the sampling plans. The use of an OC curve permits the operator to evaluate the risks involved in sampling against the quality level requirements of the product.

C. A Total Work-Reduction Scheme

Equivalent double and multiple sampling plans are provided for every single sampling plan in the standard. Table IX of the standard compares the average sample size of double and multiple sampling plans to the analogous single sampling plan. Empirically, the use of double sampling plans appears to reduce the amount of sampling to about 60% of the analogous single plan.

Multiple-sampling, which in the worst case requires the evaluation of seven samples before decision, appears to bring the total work effort to less than 50% of the single plan.

Of course, setup time for an inspection job is the same whether the total sample is 10 or 100. But it remains that the use of the single sampling plan adds more than 25% to the workload, on the average. Double and multiple sampling plans do require somewhat more bookkeeping than single plans because of the additional samples and cumulative accept/reject numbers.

The fact remains that a good-quality product is accepted with much less total inspection by double or multiple sampling plans. Likewise, a poor-quality product is rejected more often at an early stage. It is only when the product quality is mediocre that the total amount of work, on the average, maximizes. Even so, the total number inspected (the ATI) is less.

D. Plans for Limiting the Risks Involved in Sampling

Average Outgoing Quality Limit (AOQL)

Many practitioners believe that, statistically, the limit of the risk involved in sampling inspection is the AOQL associated with the sampling plan. Indeed, MIL-STD-105 limits sampling risks involved in accepting product from poor suppliers by imposing a tightened inspection: a sampling system that has an AOQL that is equal to the plan's acceptable quality level (AQL).

Table V of MIL-105 provides for the determination of a specified AOQL for both normal and tightened inspection single sample plans.

Limiting Quality (LQ)

There are practitioners who believe that the risk in sampling can be reduced by limiting the defect rate that has a low probability of acceptance. Table VI-A of MIL-105 can be used to determine sample plans with specific defect rates for a 10% probability of acceptance. Table IV-B is for use in determining sample plans with specific defect rates for a 5% probability of acceptance.

Seven Inspection Levels in MIL-STD-105

Special inspection levels S-1, S-2, S-3, and S-4 are usually used when the inspection procedure is very expensive or very time consuming. General inspection levels I, II, and III are usually used for routine inspection. Generally, level II is selected for use. Level I is used when sampling for defects that are classified lower on the seriousness scale. Level III is used when inspecting for the most serious defects.

E. Selection of a Single Sample Plan

As an example of selecting a single sample inspection plan consider the situation in which the lot size is 12,000 pieces. The plan specified is for an AQL of 0.025%, and it is determined that general inspection level II will be used. Conditions for normal, tightened, and reduced inspection plans will be evaluated.

Normal Inspection Plan

Step 1. Go to Figure 1, which is a reproduction of Table I of MIL-STD-105 (*Sample size code letters*), to find the sample size code letter for the sample plan.

Lot or batch size	Special inspection levels				General inspection levels		
	S-1	S-2	S-3	S-4	I	II	III
2 to 8	A	A	A	A	A	A	B
9 to 15	A	A	A	A	A	B	C
16 to 25	A	A	B	B	B	C	D
26 to 50	A	B	B	C	C	D	E
51 to 90	B	B	C	C	C	E	F
91 to 150	B	B	C	D	D	F	G
151 to 280	B	C	D	E	E	G	H
281 to 500	B	C	D	E	F	H	J
501 to 1200	C	C	E	F	G	J	K
1201 to 3200	C	D	E	G	H	K	L
3201 to 10000	C	D	F	G	J	L	M
10001 to 35000	C	D	F	H	K	M	N
35001 to 150000	D	E	G	J	L	N	P
150001 to 500000	D	E	G	J	M	P	Q
500001 and over	D	E	H	K	N	Q	R

Code Letter

Figure 1 Table I: Sample size code letters.

Step 2. Go down the left-hand column "Lot or batch size" to the line "10001 to 35000," which includes the lot size of 12,000.

Step 3. Go across (to the right) to the column under the heading "General inspection levels II."

Step 4. Find the sample size code letter, which is "M."

Step 5. Turn to Figure 2, which is a reproduction of MIL-STD-105 Table II-A: *Single sampling plans for normal inspection.*

Step 6. Go down the left-hand column, "Sample size code letter," to line "M."

Step 7. Go across (to the right) to the column under the heading "0.025." Note that there is an arrow pointing downward through the "M" row. This arrow means that a sample size of 315 cannot be used for this plan. Instead, the next plan down, i.e., the plan at line "N" where sample size equals 500, must be used with an "accept" if zero defective items are found in the 500 piece sample, and a "reject" if one defective item is found in the sample.

The final inspection plan, then, is this: Draw 500 items, at random, from the 12,000-piece lot. Inspect the samples. If no defective items are found, accept the lot. If one or more defective items are found, reject the lot.

Tightened Inspection Plan

Follow the steps 1 through 4 in the preceding section.

Step 5. Turn to Figure 3, which is a reproduction of MIL-STD-105 Table II-B: *Single sampling plans for tightened inspection.*

Step 6. Go down the left-hand column, "Sample size code letter," to line "M."

Step 7. Go across (to the right) to the column with the heading "0.025." Note that there is an arrow pointing downward through both the "M" line and the "N" line. The sampling plan thus required is shown on the line labeled "P."

The final tightened inspection plan, then, is this: Draw 800 items, at random, from the lot of 12,000 pieces. Inspect the samples. If there are no defective items in the sample, accept the lot. If there are one or more defective items in the sample, reject the lot.

Reduced Inspection Plan

Follow steps 1 through 4 given previously.

Step 5. Turn to Figure 4, which is a reproduction of MIL-STD-105 Table II-C: *Single sampling plans for reduced inspection.*

Step 6. Go down the left-hand column, "Sample size code letter," to the line labeled "M."

Step 7. Go across (to the right) to the column labeled "0.025." Note that there is an arrow pointing downward through the "M" line. The sampling plan thus required is shown on the line labeled "N."

The final reduced inspection plan, then, is this: Draw 200 items, at random, from the lot of 12,000 pieces. Inspect the samples. If there are no defective items, accept the lot. If there are one or more defective items reject the lot.

Compare Specified Sample Sizes

Note the large difference in the number of sample items required to be inspected under each of the three plans. Reduced plans make a good supplier's life easier; tightened plans penalize a poor supplier.

Sample size code letter	Sample size	0.010 Ac Re	0.015 Ac Re	0.025 Ac Re	0.040 Ac Re	0.065 Ac Re	0.10 Ac Re	0.15 Ac Re	0.25 Ac Re	0.40 Ac Re	0.65 Ac Re	1.0 Ac Re	1.5 Ac Re	2.5 Ac Re	4.0 Ac Re	6.5 Ac Re	10 Ac Re	15 Ac Re	25 Ac Re	40 Ac Re	65 Ac Re	100 Ac Re	150 Ac Re
		Acceptable Quality Levels (normal inspection)																					
A	2	↓	↓	↓	↓	↓	↓	↓	↓	↓	↓	↓	↓	↓	↓	0 1	↓	↓	1 2	2 3	3 4	5 6	7 8
B	3	↓	↓	↓	↓	↓	↓	↓	↓	↓	↓	↓	↓	↓	0 1	↑	↓	1 2	2 3	3 4	5 6	7 8	10 11
C	5	↓	↓	↓	↓	↓	↓	↓	↓	↓	↓	↓	↓	0 1	↑	↓	1 2	2 3	3 4	5 6	7 8	10 11	14 15
D	8	↓	↓	↓	↓	↓	↓	↓	↓	↓	↓	↓	0 1	↑	↓	1 2	2 3	3 4	5 6	7 8	10 11	14 15	21 22
E	13	↓	↓	↓	↓	↓	↓	↓	↓	↓	↓	0 1	↑	↓	1 2	2 3	3 4	5 6	7 8	10 11	14 15	21 22	30 31
F	20	↓	↓	↓	↓	↓	↓	↓	↓	↓	0 1	↑	↓	1 2	2 3	3 4	5 6	7 8	10 11	14 15	21 22	↑	↑
G	32	↓	↓	↓	↓	↓	↓	↓	↓	0 1	↑	↓	1 2	2 3	3 4	5 6	7 8	10 11	14 15	21 22	↑	↑	↑
H	50	↓	↓	↓	↓	↓	↓	↓	0 1	↑	↓	1 2	2 3	3 4	5 6	7 8	10 11	14 15	21 22	↑	↑	↑	↑
J	80	↓	↓	↓	↓	↓	↓	0 1	↑	↓	1 2	2 3	3 4	5 6	7 8	10 11	14 15	21 22	↑	↑	↑	↑	↑
K	125	↓	↓	↓	↓	↓	0 1	↑	↓	1 2	2 3	3 4	5 6	7 8	10 11	14 15	21 22	↑	↑	↑	↑	↑	↑
L	200	↓	↓	↓	↓	0 1	↑	↓	1 2	2 3	3 4	5 6	7 8	10 11	14 15	21 22	↑	↑	↑	↑	↑	↑	↑
M	315	↓	↓	↓	0 1	↑	↓	1 2	2 3	3 4	5 6	7 8	10 11	14 15	21 22	↑	↑	↑	↑	↑	↑	↑	↑
(N)	500	↓	↓	0 1	↑	↓	1 2	2 3	3 4	5 6	7 8	10 11	14 15	21 22	↑	↑	↑	↑	↑	↑	↑	↑	↑
P	800	↓	(0 1)	↑	↓	1 2	2 3	3 4	5 6	7 8	10 11	14 15	21 22	↑	↑	↑	↑	↑	↑	↑	↑	↑	↑
Q	1250	0 1	↑	↓	1 2	2 3	3 4	5 6	7 8	10 11	14 15	21 22	↑	↑	↑	↑	↑	↑	↑	↑	↑	↑	↑
R	2000	↑	↑	1 2	2 3	3 4	5 6	7 8	10 11	14 15	21 22	↑	↑	↑	↑	↑	↑	↑	↑	↑	↑	↑	↑

▽ = Use first sampling plan below arrow. If sample size equals, or exceeds, lot or batch size, do 100 percent inspection.
△ = Use first sampling plan above arrow
Ac = Acceptance number
Re = Rejection number

SINGLE NORMAL

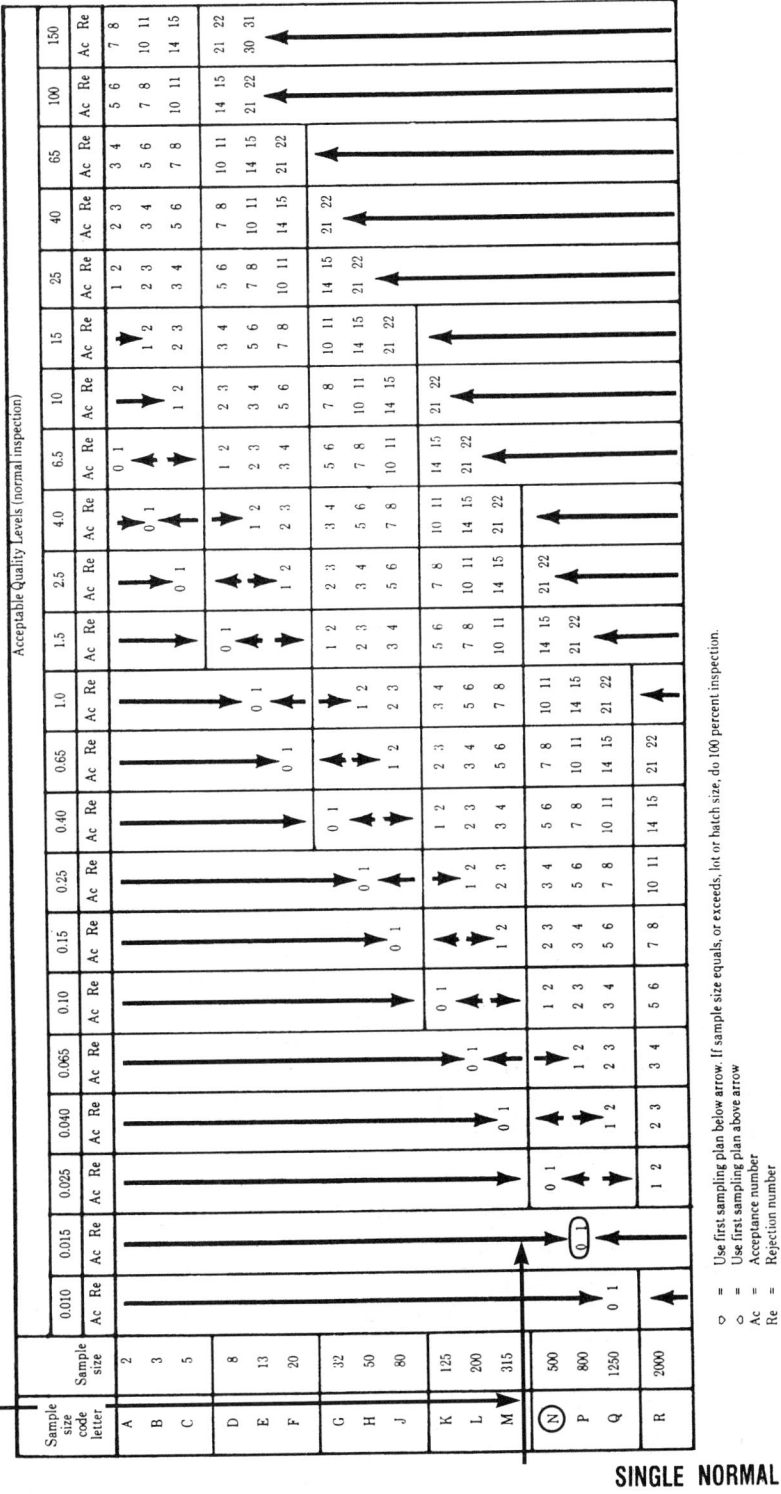

Figure 2 Table II-A: Single sampling plans for normal inspection (master table).

Sample size code letter	Sample size	0.010	0.015	0.025	0.040	0.065	0.10	0.15	0.25	0.40	0.65	1.0	1.5	2.5	4.0	6.5	10	15	25	40	65	100	150
		Ac Re	Ac Re	Ac Re	Ac Re	Ac Re	Ac Re	Ac Re	Ac Re	Ac Re	Ac Re	Ac Re	Ac Re	Ac Re	Ac Re	Ac Re	Ac Re	Ac Re	Ac Re	Ac Re	Ac Re	Ac Re	Ac Re
A	2	↓	↓	↓	↓	↓	↓	↓	↓	↓	↓	↓	↓	↓	↓	↓	↓	↓	↓	1 2	2 3	3 4	5 6
B	3	↓	↓	↓	↓	↓	↓	↓	↓	↓	↓	↓	↓	↓	↓	0 1	↓	↓	1 2	2 3	3 4	5 6	8 9
C	5	↓	↓	↓	↓	↓	↓	↓	↓	↓	↓	↓	↓	↓	0 1	↓	↓	1 2	2 3	3 4	5 6	8 9	12 13
D	8	↓	↓	↓	↓	↓	↓	↓	↓	↓	↓	↓	↓	0 1	↓	↓	1 2	2 3	3 4	5 6	8 9	12 13	18 19
E	13	↓	↓	↓	↓	↓	↓	↓	↓	↓	↓	↓	0 1	↓	↓	1 2	2 3	3 4	5 6	8 9	12 13	18 19	27 28
F	20	↓	↓	↓	↓	↓	↓	↓	↓	↓	↓	0 1	↓	↓	1 2	2 3	3 4	5 6	8 9	12 13	18 19	↑	↑
G	32	↓	↓	↓	↓	↓	↓	↓	↓	↓	0 1	↓	↓	1 2	2 3	3 4	5 6	8 9	12 13	18 19	↑	↑	↑
H	50	↓	↓	↓	↓	↓	↓	↓	↓	0 1	↓	↓	1 2	2 3	3 4	5 6	8 9	12 13	18 19	↑	↑	↑	↑
J	80	↓	↓	↓	↓	↓	↓	↓	0 1	↓	↓	1 2	2 3	3 4	5 6	8 9	12 13	18 19	↑	↑	↑	↑	↑
K	125	↓	↓	↓	↓	↓	↓	0 1	↓	↓	1 2	2 3	3 4	5 6	8 9	12 13	18 19	↑	↑	↑	↑	↑	↑
L	200	↓	↓	↓	↓	↓	0 1	↓	↓	1 2	2 3	3 4	5 6	8 9	12 13	18 19	↑	↑	↑	↑	↑	↑	↑
M	315	↓	↓	↓	↓	0 1	↓	↓	1 2	2 3	3 4	5 6	8 9	12 13	18 19	↑	↑	↑	↑	↑	↑	↑	↑
N	500	↓	↓	↓	0 1	↓	↓	1 2	2 3	3 4	5 6	8 9	12 13	18 19	↑	↑	↑	↑	↑	↑	↑	↑	↑
(P)	800	↓	↓	(0 1)	↓	↓	1 2	2 3	3 4	5 6	8 9	12 13	18 19	↑	↑	↑	↑	↑	↑	↑	↑	↑	↑
Q	1250	↓	0 1	↓	↓	1 2	2 3	3 4	5 6	8 9	12 13	18 19	↑	↑	↑	↑	↑	↑	↑	↑	↑	↑	↑
R	2000	0 1	↑	↓	1 2	2 3	3 4	5 6	8 9	12 13	18 19	↑	↑	↑	↑	↑	↑	↑	↑	↑	↑	↑	↑
S	3150		↑	1 2								↑	↑	↑	↑	↑	↑	↑	↑	↑	↑	↑	↑

Column heading: Acceptable Quality Levels (tightened inspection)

▽ = Use first sampling plan below arrow. If sample size equals or exceeds lot or batch size, do 100 percent inspection.
△ = Use first sampling plan above arrow.
Ac = Acceptance number
Re = Rejection number

SINGLE TIGHTENED

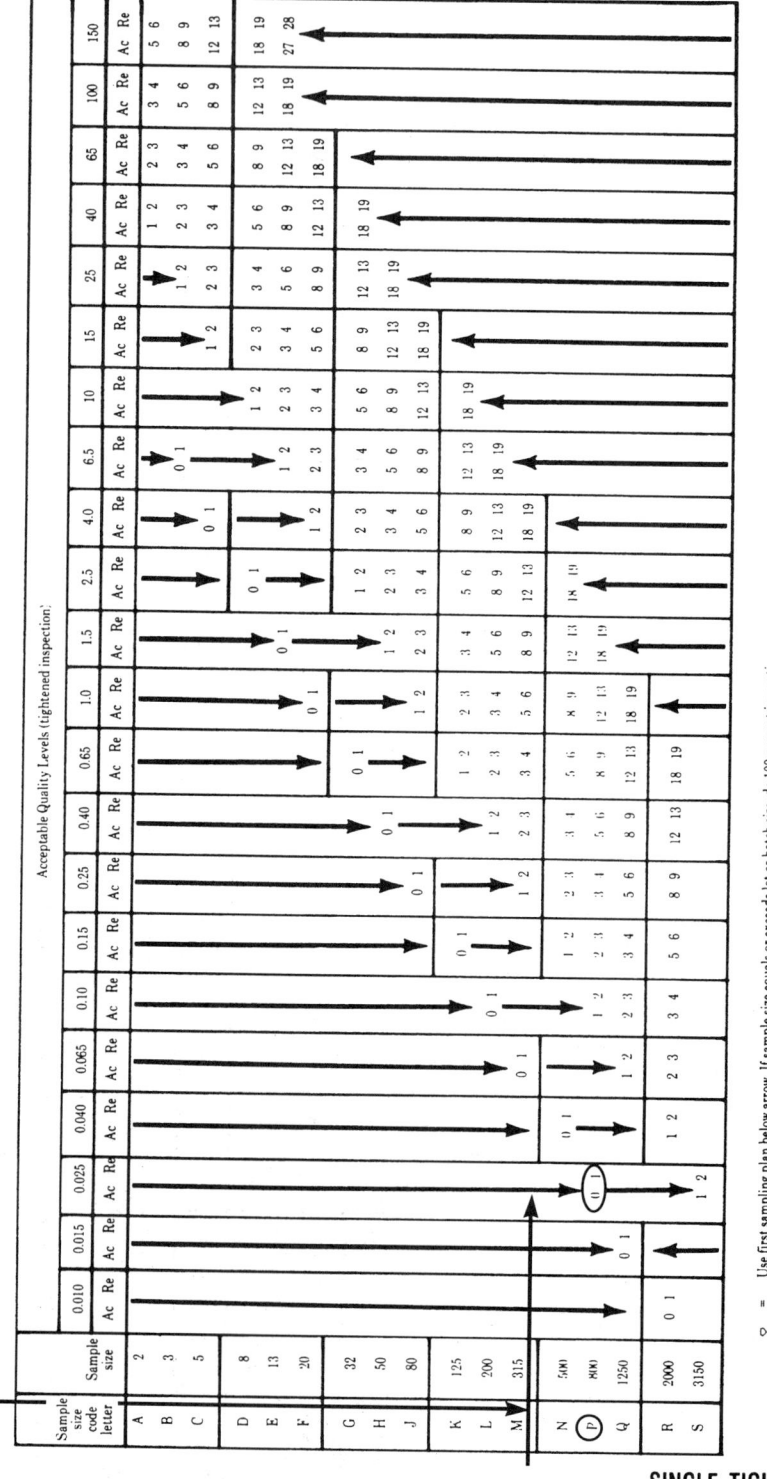

Figure 3 Table II-B: Single sampling plans for tightened inspection (master table).

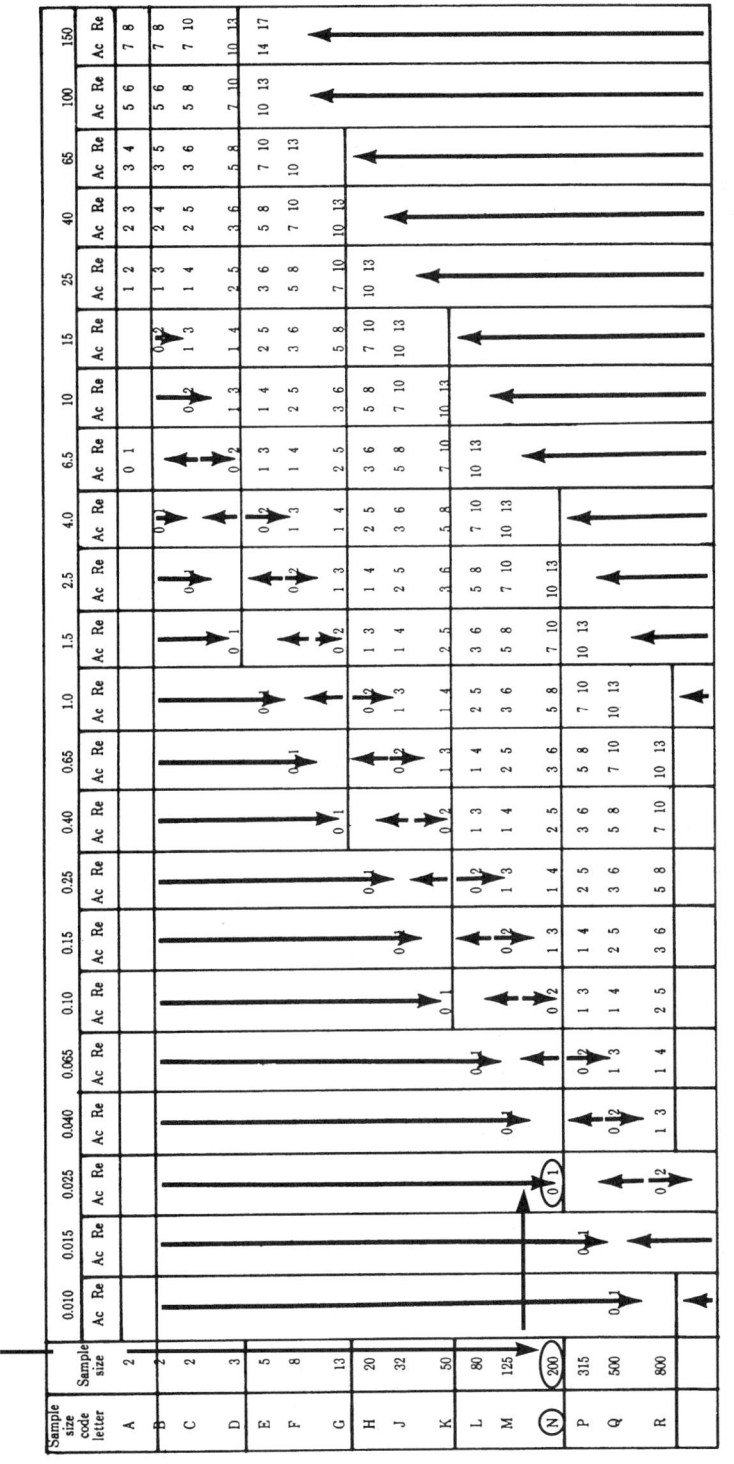

SINGLE REDUCED

Sample size code letter	Sample size	0.010	0.015	0.025	0.040	0.065	0.10	0.15	0.25	0.40	0.65	1.0	1.5	2.5	4.0	6.5	10	15	25	40	65	100	150
		Ac Re	Ac Re	Ac Re	Ac Re	Ac Re	Ac Re	Ac Re	Ac Re	Ac Re	Ac Re	Ac Re	Ac Re	Ac Re	Ac Re	Ac Re	Ac Re	Ac Re	Ac Re	Ac Re	Ac Re	Ac Re	Ac Re
A	2															0 1			1 2	2 3	3 4	5 6	7 8
B	2	↓	↓	↓	↓	↓	↓	↓	↓	↓	↓	↓	↓	↓	0 1	↑	↓	0 2	1 3	2 4	3 5	5 6	7 8
C	2	↓	↓	↓	↓	↓	↓	↓	↓	↓	↓	↓	↓	0 1	↑	↓	0 2	1 3	1 4	2 5	3 6	5 8	7 10
D	3	↓	↓	↓	↓	↓	↓	↓	↓	↓	↓	↓	0 1	↑	↓	0 2	1 3	1 4	2 5	3 6	5 8	7 10	10 13
E	5	↓	↓	↓	↓	↓	↓	↓	↓	↓	↓	0 1	↑	↓	0 2	1 3	1 4	2 5	3 6	5 8	7 10	10 13	14 17
F	8	↓	↓	↓	↓	↓	↓	↓	↓	↓	0 1	↑	↓	0 2	1 3	1 4	2 5	3 6	5 8	7 10	10 13	↑	↑
G	13	↓	↓	↓	↓	↓	↓	↓	↓	0 1	↑	↓	0 2	1 3	1 4	2 5	3 6	5 8	7 10	10 13	↑	↑	↑
H	20	↓	↓	↓	↓	↓	↓	↓	0 1	↑	↓	0 2	1 3	1 4	2 5	3 6	5 8	7 10	10 13	↑	↑	↑	↑
J	32	↓	↓	↓	↓	↓	↓	0 1	↑	↓	0 2	1 3	1 4	2 5	3 6	5 8	7 10	10 13	↑	↑	↑	↑	↑
K	50	↓	↓	↓	↓	↓	0 1	↑	↓	0 2	1 3	1 4	2 5	3 6	5 8	7 10	10 13	↑	↑	↑	↑	↑	↑
L	80	↓	↓	↓	↓	0 1	↑	↓	0 2	1 3	1 4	2 5	3 6	5 8	7 10	10 13	↑	↑	↑	↑	↑	↑	↑
M	125	↓	↓	↓	0 1	↑	↓	0 2	1 3	1 4	2 5	3 6	5 8	7 10	10 13	↑	↑	↑	↑	↑	↑	↑	↑
(N)	(200)	↓	↓	(0 1)	↑	↓	0 2	1 3	1 4	2 5	3 6	5 8	7 10	10 13	↑	↑	↑	↑	↑	↑	↑	↑	↑
P	315	↓	0 1	↑	↓	0 2	1 3	1 4	2 5	3 6	5 8	7 10	10 13	↑	↑	↑	↑	↑	↑	↑	↑	↑	↑
Q	500	0 1	↑	↓	0 2	1 3	1 4	2 5	3 6	5 8	7 10	10 13	↑	↑	↑	↑	↑	↑	↑	↑	↑	↑	↑
R	800		↑	0 2	1 3	1 4	2 5	3 6	5 8	7 10	10 13		↑	↑	↑	↑	↑	↑	↑	↑	↑	↑	↑
		↑	↑									↑	↑	↑	↑	↑	↑	↑	↑	↑	↑	↑	↑

↓ = Use first sampling plan below arrow. If sample size equals or exceeds lot or batch size, do 100 percent inspection.
↑ = Use first sampling plan above.
Ac = Acceptance number
Re = Rejection number
† = If the acceptance number has been exceeded, but the rejection number has not been reached, accept the lot, but reinstate normal inspection (see 10.1.4).

Figure 4 Table II-C: Single sampling plans for reduced inspection (master table).

F. Selection of a Double Sampling Plan

Determine the double sampling plan criteria for an inspection job where the lot size is 8,500 pieces, the specified AQL is 0.25 percent, and general inspection level II is to be used.

Step 1. Go to Figure 1 (*Sample size code letters*).

Step 2. Go down the column for lot or batch size to the line "3,201 to 10,000," which includes the lot size of 8,500 pieces.

Step 3. Go across (to the right) to the column under the heading "General inspection level II."

Step 4. Find the sample size code letter, which is "L."

Step 5. Turn to Figure 5, which is a reproduction of MIL-105 Table IIIA.

Step 6. Go down the left-hand column, "Sample size code letter," to the line labeled "L."

Step 7. Go across (to the right) to the column under the heading "0.25."

The double sampling plan then, is:

Sample size code letter "L"	Accept number	Reject number
First sample: 125	0	2
Second sample: 125	1	2

In other words, draw a sample, at random, of 125 items from the lot. Inspect the items. If there are no defective items, accept the lot. If there are two or more defective items, reject the lot. If there is one defective item draw another 125 items from the remainder of the lot. Inspect the items. If in the total of 250 items drawn from the lot there is a total of one or fewer defective items, accept the lot. If there are two or more defective items reject the lot. Note that the number of samples drawn and the number of defective items found are cumulative.

G. Selection of a Multiple Sampling Plan

Determine the multiple sampling plan criteria for an inspection job in which the lot size is 350 items, the AQL is 2.5%, and general inspection level II is to be used.

Step 1. Go to Figure 1.

Step 2. Go down the left-hand column, "Lot or batch size," to the line "281 to 500," which includes the lot size of 350 pieces.

Step 3. Go across (to the right) to the column under the heading "General inspection level II." Find the sample size code letter "H."

Step 4. Turn to Figure 6, which is a reproduction of MIL-105 Table IV-A: *Multiple sampling plans for normal inspection.*

Step 5. Go across (to the right) to the column under the heading "2.5."

Step 6. Find the inspection plan.

Sample size code letter	Smpl	Sample size	Cumulative sample size	Acceptable Quality Levels (normal inspection)																					(See 9.4 and 9.5)
				0.010	0.015	0.025	0.040	0.065	0.10	0.15	0.25	0.40	0.65	1.0	1.5	2.5	4.0	6.5	10	15	25	40	65	100	150
				Ac Re	Ac Re	Ac Re	Ac Re	Ac Re	Ac Re	Ac Re	Ac Re	Ac Re	Ac Re	Ac Re	Ac Re	Ac Re	Ac Re	Ac Re	Ac Re	Ac Re	Ac Re	Ac Re	Ac Re	Ac Re	Ac Re
A				↓	↓	↓	↓	↓	↓	↓	↓	↓	↓	↓	↓	↓	↓	•	↓	↓	•	•	•	•	•
B	1st	2	2	↓	↓	↓	↓	↓	↓	↓	↓	↓	↓	↓	↓	↓	•	↑	↓	0 2	0 3	1 4	2 5	3 7	5 9
	2nd	2	4																	1 2	3 4	4 5	6 7	8 9	12 13
C	1st	3	3	↓	↓	↓	↓	↓	↓	↓	↓	↓	↓	↓	↓	•	↑	↓	0 2	0 3	1 4	2 5	3 7	5 9	7 11
	2nd	3	6																1 2	3 4	4 5	6 7	8 9	12 13	18 19
D	1st	5	5	↓	↓	↓	↓	↓	↓	↓	↓	↓	↓	↓	•	↑	↓	0 2	0 3	1 4	2 5	3 7	5 9	7 11	11 16
	2nd	5	10															1 2	3 4	4 5	6 7	8 9	12 13	18 19	26 27
E	1st	8	8	↓	↓	↓	↓	↓	↓	↓	↓	↓	↓	•	↑	↓	0 2	0 3	1 4	2 5	3 7	5 9	7 11	11 16	17 22
	2nd	8	16														1 2	3 4	4 5	6 7	8 9	12 13	18 19	26 27	37 38
F	1st	13	13	↓	↓	↓	↓	↓	↓	↓	↓	↓	•	↑	↓	0 2	0 3	1 4	2 5	3 7	5 9	7 11	11 16	↑	↑
	2nd	13	26													1 2	3 4	4 5	6 7	8 9	12 13	18 19	26 27		
G	1st	20	20	↓	↓	↓	↓	↓	↓	↓	↓	•	↑	↓	0 2	0 3	1 4	2 5	3 7	5 9	7 11	11 16	↑	↑	↑
	2nd	20	40												1 2	3 4	4 5	6 7	8 9	12 13	18 19	26 27			
H	1st	32	32	↓	↓	↓	↓	↓	↓	↓	•	↑	↓	0 2	0 3	1 4	2 5	3 7	5 9	7 11	11 16	↑	↑	↑	↑
	2nd	32	64											1 2	3 4	4 5	6 7	8 9	12 13	18 19	26 27				
J	1st	50	50	↓	↓	↓	↓	↓	↓	•	↑	↓	0 2	0 3	1 4	2 5	3 7	5 9	7 11	11 16	↑	↑	↑	↑	↑
	2nd	50	100										1 2	3 4	4 5	6 7	8 9	12 13	18 19	26 27					
K	1st	80	80	↓	↓	↓	↓	↓	•	↑	↓	0 2	0 3	1 4	2 5	3 7	5 9	7 11	11 16	↑	↑	↑	↑	↑	↑
	2nd	80	160									1 2	3 4	4 5	6 7	8 9	12 13	18 19	26 27						
L	1st	125	125	↓	↓	↓	↓	•	↑	↓	0 2	0 3	1 4	2 5	3 7	5 9	7 11	11 16	↑	↑	↑	↑	↑	↑	↑
	2nd	125	250								1 2	3 4	4 5	6 7	8 9	12 13	18 19	26 27							
M	1st	200	200	↓	↓	↓	•	↑	↓	0 2	0 3	1 4	2 5	3 7	5 9	7 11	11 16	↑	↑	↑	↑	↑	↑	↑	↑
	2nd	200	400							1 2	3 4	4 5	6 7	8 9	12 13	18 19	26 27								
N	1st	315	315	↓	↓	•	↑	↓	0 2	0 3	1 4	2 5	3 7	5 9	7 11	11 16	↑	↑	↑	↑	↑	↑	↑	↑	↑
	2nd	315	630						1 2	3 4	4 5	6 7	8 9	12 13	18 19	26 27									
P	1st	500	500	↓	•	↑	↓	0 2	0 3	1 4	2 5	3 7	5 9	7 11	11 16	↑	↑	↑	↑	↑	↑	↑	↑	↑	↑
	2nd	500	1000					1 2	3 4	4 5	6 7	8 9	12 13	18 19	26 27										
Q	1st	800	800	•	↑	↓	0 2	0 3	1 4	2 5	3 7	5 9	7 11	11 16	↑	↑	↑	↑	↑	↑	↑	↑	↑	↑	↑
	2nd	800	1600				1 2	3 4	4 5	6 7	8 9	12 13	18 19	26 27											
R	1st	1250	1250	↑	↑	0 2	0 3	1 4	2 5	3 7	5 9	7 11	11 16	↑	↑	↑	↑	↑	↑	↑	↑	↑	↑	↑	↑
	2nd	1250	2500			1 2	3 4	4 5	6 7	8 9	12 13	18 19	26 27												

↓ = Use first sampling plan below arrow. If sample size equals or exceeds lot or batch size, do 100 percent inspection.
↑ = Use first sampling plan above arrow.
Ac = Acceptance number
Re = Rejection number
• = Use corresponding single sampling plan (or alternatively, use double sampling plan below, where available).

DOUBLE NORMAL

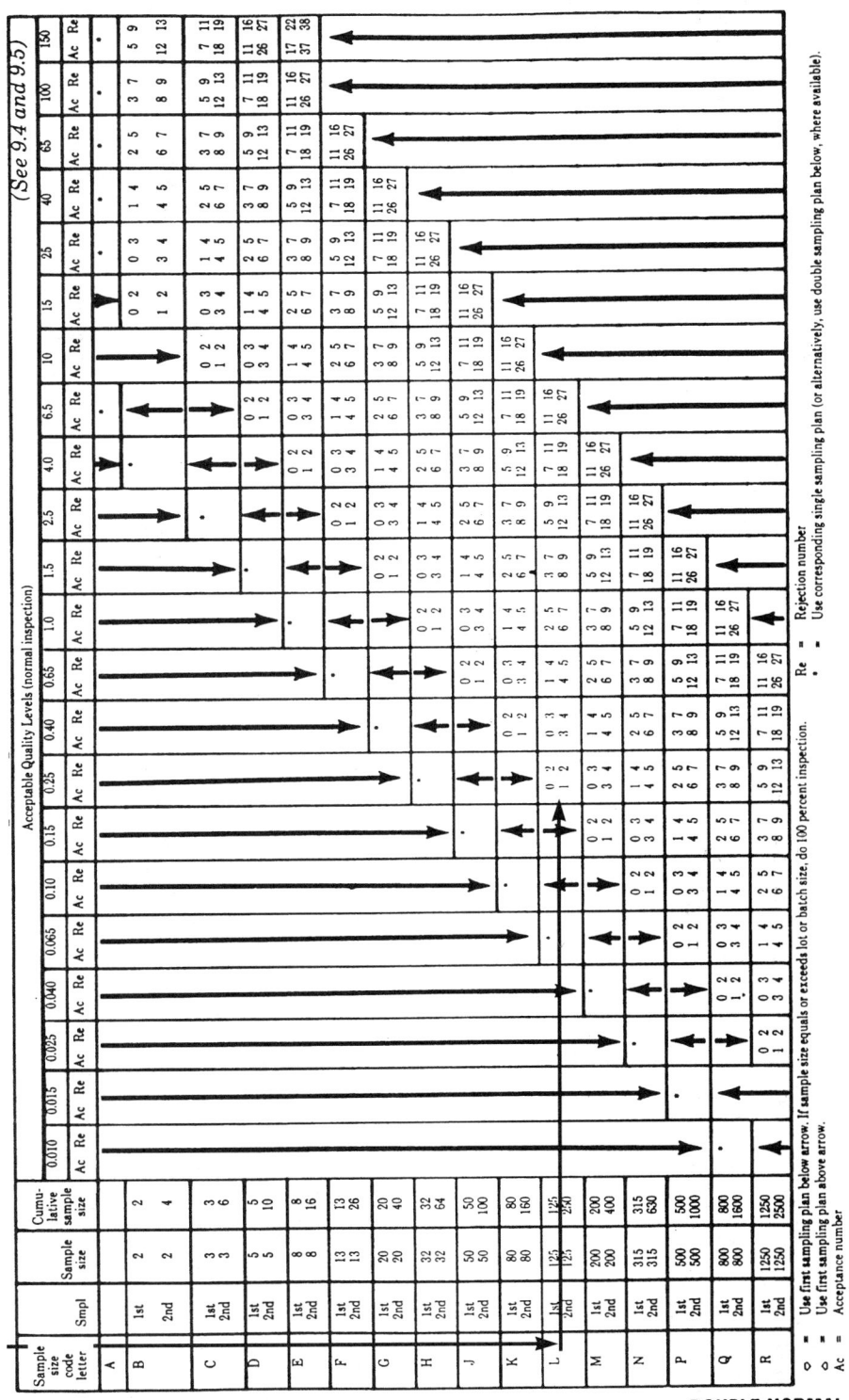

Figure 5 Table III-A: Double sampling plans for normal inspection (master table).

MULTIPLE NORMAL

Acceptable Quality Levels (normal inspection), AQL 0.010–0.65 (each cell: Ac Re)

Sample size code letter	Smpl	Sample size	Cumulative sample size	0.010	0.015	0.025	0.040	0.065	0.10	0.15	0.25	0.40	0.65
A				↓	↓	↓	↓	↓	↓	↓	↓	↓	↓
B				↓	↓	↓	↓	↓	↓	↓	↓	↓	↓
C				↓	↓	↓	↓	↓	↓	↓	↓	↓	↓
D	1st	2	2	↓	↓	↓	↓	↓	↓	↓	↓	↓	↓
	2nd	2	3										
	3rd	2	6										
	4th	2	8										
	5th	2	10										
	6th	2	12										
	7th	2	14										
E	1st	3	3	↓	↓	↓	↓	↓	↓	↓	↓	↓	↓
	2nd	3	6										
	3rd	3	9										
	4th	3	12										
	5th	3	15										
	6th	3	18										
	7th	3	21										
F	1st	5	5	↓	↓	↓	↓	↓	↓	↓	↓	↓	•
	2nd	5	10										
	3rd	5	15										
	4th	5	20										
	5th	5	25										
	6th	5	30										
	7th	5	35										
G	1st	8	8	↓	↓	↓	↓	↓	↓	↓	↓	•	↑
	2nd	8	16										
	3rd	8	24										
	4th	8	32										
	5th	8	40										
	6th	8	48										
	7th	8	56										
H	1st	13	13	↓	↓	↓	↓	↓	↓	↓	•	↑	↓
	2nd	13	26										
	3rd	13	39										
	4th	13	52										
	5th	13	63										
	6th	13	78										
	7th	13	91										
J	1st	20	20	↓	↓	↓	↓	↓	↓	•	↑	↓	# 2
	2nd	20	40										# 2
	3rd	20	60										0 2
	4th	20	80										0 3
	5th	20	100										1 3
	6th	20	120										1 3
	7th	20	140										2 3

Acceptable Quality Levels (normal inspection), AQL 1.0–1000 (each cell: Ac Re)

Sample size code letter	Smpl	Sample size	Cumulative sample size	1.0	1.5	2.5	4.0	6.5	10	15	25	40	65	100	150	250	400	650	1000
A				↓	↓	↓	↑	•	↓	↓	•	•	•	•	•	•	•	•	•
B				↓	↓	↓	•	↑	↓	‡	‡	‡	‡	‡	‡	‡	‡	‡	‡
C				↓	↓	•	↓	↓	‡	‡	‡	‡	‡	‡	‡	‡	‡	‡	↑
D	1st	2	2	↓	•	↑	↓	# 2	# 2	# 3	# 4	0 4	0 5	1 7	2 9	4 12	6 16	↑	↑
	2nd	2	3					# 2	0 3	0 3	1 5	1 6	3 8	4 10	7 14	11 19	17 27		
	3rd	2	6					0 2	0 3	1 4	2 6	3 8	6 10	8 13	13 19	19 27	29 39		
	4th	2	8					0 3	1 4	2 5	3 7	5 10	8 13	12 17	19 25	27 34	40 49		
	5th	2	10					1 3	2 4	3 6	5 8	7 11	11 15	17 20	25 29	36 40	53 58		
	6th	2	12					1 3	3 5	4 6	7 9	10 12	14 17	21 23	31 33	45 47	66 68		
	7th	2	14					2 3	4 5	6 7	9 10	13 14	18 19	25 26	37 38	53 54	77 78		
E	1st	3	3	•	↑	↓	# 2	# 2	# 3	# 4	0 4	0 5	1 7	2 9	4 12	6 16	↑	↑	↑
	2nd	3	6				# 2	0 3	0 3	1 5	1 6	3 8	4 10	7 14	11 19	17 27			
	3rd	3	9				0 2	0 3	1 4	2 6	3 8	6 10	8 13	13 19	19 27	29 39			
	4th	3	12				0 3	1 4	2 5	3 7	5 10	8 13	12 17	19 25	27 34	40 49			
	5th	3	15				1 3	2 4	3 6	5 8	7 11	11 15	17 20	25 29	36 40	53 58			
	6th	3	18				1 3	3 5	4 6	7 9	10 12	14 17	21 23	31 33	45 47	65 68			
	7th	3	21				2 3	4 5	6 7	9 10	13 14	18 19	25 26	37 38	53 54	77 78			
F	1st	5	5	↑	↓	# 2	# 2	# 3	# 4	0 4	0 5	1 7	2 9	↑	↑	↑	↑	↑	↑
	2nd	5	10			# 2	0 3	0 3	1 5	1 6	3 8	4 10	7 14						
	3rd	5	15			0 2	0 3	1 4	2 6	3 8	6 10	8 13	13 19						
	4th	5	20			0 3	1 4	2 5	3 7	5 10	8 13	12 17	19 25						
	5th	5	25			1 3	2 4	3 6	5 8	7 11	11 15	17 20	25 29						
	6th	5	30			1 3	3 5	4 6	7 9	10 12	14 17	21 23	31 33						
	7th	5	35			2 3	4 5	6 7	9 10	13 14	18 19	25 26	37 38						
G	1st	8	8	↓	# 2	# 2	# 3	# 4	0 4	0 5	1 7	2 9	↑	↑	↑	↑	↑	↑	↑
	2nd	8	16		# 2	0 3	0 3	1 5	1 6	3 8	4 10	7 14							
	3rd	8	24		0 2	0 3	1 4	2 6	3 8	6 10	8 13	13 19							
	4th	8	32		0 3	1 4	2 5	3 7	5 10	8 13	12 17	19 25							
	5th	8	40		1 3	2 4	3 6	5 8	7 11	11 15	17 20	25 29							
	6th	8	48		1 3	3 5	4 6	7 9	10 12	14 17	21 23	31 33							
	7th	8	56		2 3	4 5	6 7	9 10	13 14	18 19	25 26	37 38							
H	1st	13	13	# 2	# 2	# 3	# 4	0 4	0 5	1 7	2 9	↑	↑	↑	↑	↑	↑	↑	↑
	2nd	13	26	# 2	0 3	0 3	1 5	1 6	3 8	4 10	7 14								
	3rd	13	39	0 2	0 3	1 4	2 6	3 8	6 10	8 13	13 19								
	4th	13	52	0 3	1 4	2 5	3 7	5 10	8 13	12 17	19 25								
	5th	13	63	1 3	2 4	3 6	5 8	7 11	11 15	17 20	25 29								
	6th	13	78	1 3	3 5	4 6	7 9	10 12	14 17	21 23	31 33								
	7th	13	91	2 3	4 5	6 7	9 10	13 14	18 19	25 26	37 38								
J	1st	20	20	# 2	# 3	# 4	0 4	0 5	1 7	2 9	↑	↑	↑	↑	↑	↑	↑	↑	↑
	2nd	20	40	0 3	0 3	1 5	1 6	3 8	4 10	7 14									
	3rd	20	60	0 3	1 4	2 6	3 8	6 10	8 13	13 19									
	4th	20	80	1 4	2 5	3 7	5 10	8 13	12 17	19 25									
	5th	20	100	2 4	3 6	5 8	7 11	11 15	17 20	25 29									
	6th	20	120	3 5	4 6	7 9	10 12	14 17	21 23	31 33									
	7th	20	140	4 5	6 7	9 10	13 14	18 19	25 26	37 38									

↓ = Use first sampling plan below arrow. If sample size equals, or exceeds, lot or batch size, do 100 percent inspection.
↑ = Use first sampling plan above arrow (refer to preceding page, when necessary).
Ac = Acceptance number
Re = Rejection number
# = Acceptance not permitted at this sample size.
‡ = Use corresponding double sampling plan (or alternatively, use multiple sampling plan below, where available).
• = Use corresponding single sampling plan (or alternatively, use multiple sampling plan below, where available).

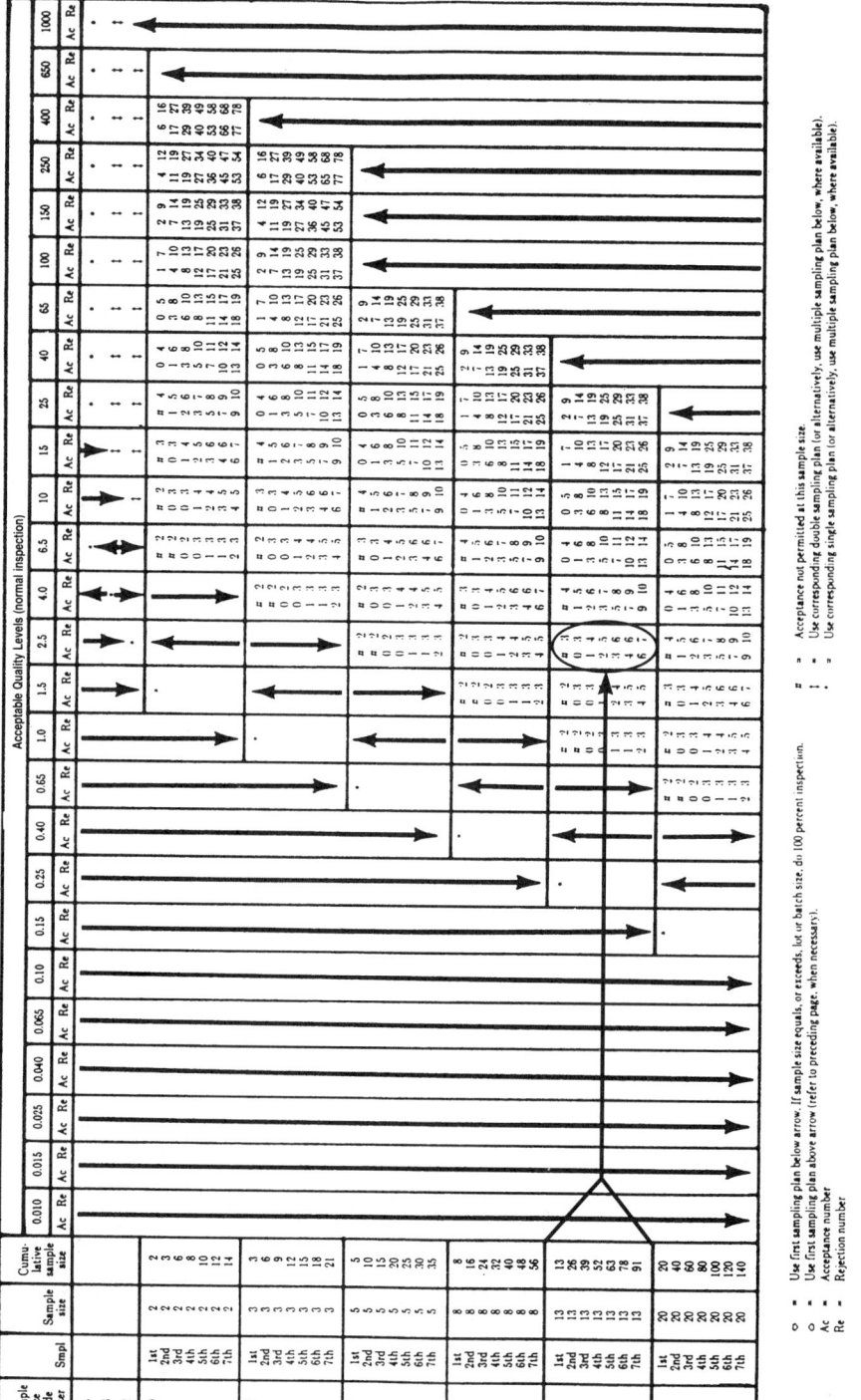

Figure 6 Table IV-A: Multiple sampling plans for normal inspection (master table).

Sample size code letter "H"	Sample size	Cumulative sample size	Accept number	Reject number
First sample	13	13	#	3
Second	13	26	0	3
Third	13	39	1	4
Fourth	13	52	2	5
Fifth	13	65	3	6
Sixth	13	78	4	6
Seventh	13	91	6	7

In other words, the multiple sampling plan is as follows. Take a sample of 13 items, at random, from the lot. Inspect the items. If there are 3 or more defective items, reject the lot. If there are fewer than 3 defective items, take another sample of 13 items (# indicates that acceptance at this level is not permitted). Inspect the second sample. If there are no defective items in the first two samples (total: 26 items inspected), accept the lot. If there is a total of three or more defective items in the 26 piece sample, reject the lot. If there are one or two defective items, draw a third sample of 13 items from the lot. Continue until the total number of defective items equals or exceeds the reject number or is equal to or less than the accept number for the total number of items inspected. A conclusion is forced at or before the seventh sample inspection. Note that the number of samples drawn and the number of defective items found are cumulative.

III. AN OVERVIEW OF SAMPLING INSPECTION

The concepts of sampling inspection—MIL-STD-105 and AQL in particular—have long been debated. Some practitioners believe that any sampling inspection scheme is basically wrong. W. Edwards Deming, for example, describes an "all or nothing" inspection strategy in his book *Out of the Crisis*. Other practitioners state that sampling inspection is wrong because all product should be made right in the first place.

Some practitioners believe that it is incorrect to use MIL-105 as a lot-by-lot inspection plan. They cite paragraph 4.2 of MIL-105: "The AQL is the maximum percent defective . . . that for purposes of sampling inspection, can be considered satisfactory as a *process average*." These practitioners believe that the use of MIL-STD-105 asks the question: Can I assume that the process from which I drew these samples has not deteriorated to the point where the average number of defective items exceeds the AQL? If no more than the acceptance number of defective items is discovered in the specified sample size, then the answer to the question is *yes*. In other words, is MIL-STD-105 a process monitoring tool rather than a lot inspection plan? Typically, these practitioners believe that lot-by-lot inspection plans should be based on the hypergeometric distribution or a similar statistical plan (see Chapter 32).

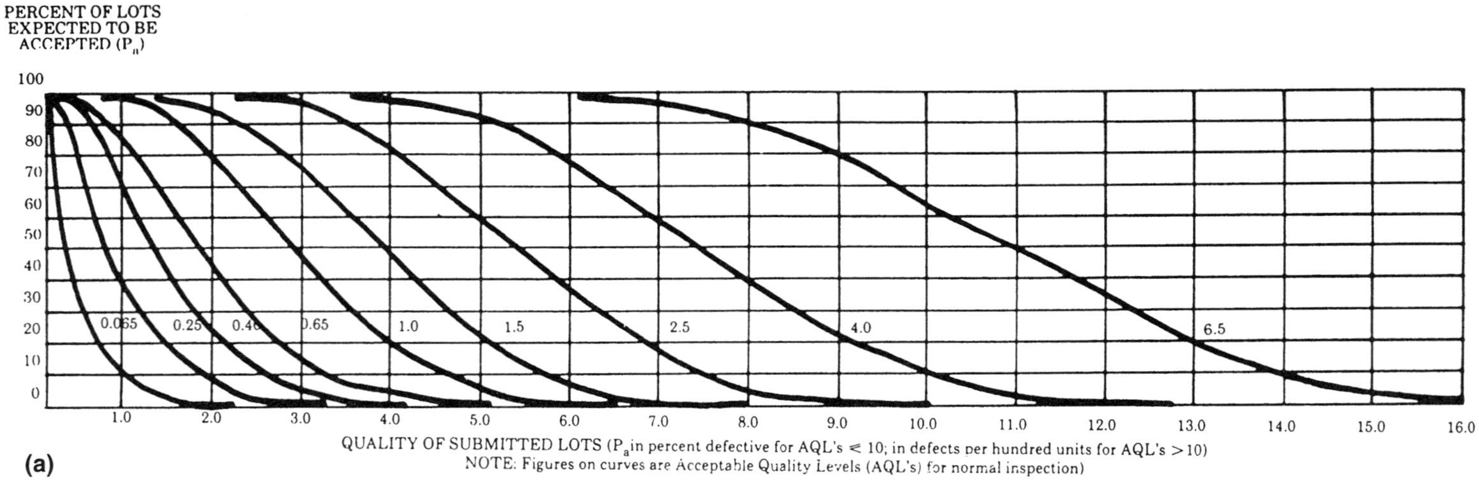

(a)

P_a	Acceptable Quality Levels (normal inspection)											
	0.065	0.25	0.40	0.65	1.0	1.5	><	2.5	><	4.0	><	6.5
	p (in percent defective or defects per hundred units											
99.0	0.0051	0.075	0.218	0.412	0.893	1.45	1.75	2.39	3.05	3.74	5.17	6.29
95.0	0.0256	0.178	0.409	0.683	1.31	1.99	2.35	3.09	3.85	4.62	6.22	7.45
90.0	0.0525	0.266	0.551	0.873	1.58	2.33	2.72	3.51	4.32	5.15	6.84	8.12
75.0	0.144	0.481	0.864	1.27	2.11	2.98	3.42	4.31	5.21	6.12	7.95	9.34
50.0	0.347	0.839	1.34	1.84	2.84	3.84	4.33	5.33	6.33	7.33	9.33	10.8
25.0	0.693	1.35	1.96	2.56	3.71	4.84	5.40	6.51	7.61	8.70	10.9	12.5
10.0	1.15	1.95	2.66	3.34	4.64	5.89	6.50	7.70	8.89	10.1	12.4	14.1
5.0	1.50	2.37	3.15	3.88	5.26	6.57	7.22	8.48	9.72	10.9	13.3	15.1
1.0	2.30	3.32	4.20	5.02	6.55	8.00	8.70	10.1	11.4	12.7	15.3	17.2
	0.10	0.40	0.65	1.0	1.5	><	2.5	><	4.0	><	6.5	><
	Acceptable Quality Levels (tightened inspection)											

Note: All values given in above table based on Poisson distribution as an approximation to the Binominal

(b)

Figure 7 Table X-L: Table for sample size code letter "L": (a) operating characteristic (OC) curves for single sampling plans (curves for double and multiple sampling are matched as closely as practicable); (b) tabulated values for OC curves for single sampling plans.

Still other practitioners, who have long used the standard as a lot-by-lot sampling and inspection plan, cite paragraph 4.3: "When a (purchaser) designates some specific value of AQL for a certain defect or group of defects, he indicates to the supplier that (the purchasers's) acceptance plan will accept the great majority {number not specified but generally accepted to be 95 out of 100} of the lots or batches that the supplier submits, provided that the process average level of percent defective . . . in these lots . . . be no greater than the designated value of AQL." The words "be no greater than" have led some practitioners to refer to AQL as "acceptable quality limit." This, however, is not a correct definition.

Perhaps the best way to look at MIL-STD-105 is that it is a tool—a tool that can be used in a number of ways. MIL-105 can be used successfully as a lot-by-lot inspection scheme. It can also be used as a process monitoring scheme.

IV. OPERATING CHARACTERISTICS AND PROBABILITY

As with most statistical procedures, sampling inspection relies heavily on the laws of probability. AQLs and AOQLs are measured "in the long run." It must be noted that AQL is an average: any given lot of product (or performance of services) may be better or worse than the AQL while the process average nevertheless meets the AQL.

Figure 7 shows OC curves and the data table for code letter "L" (single sampling plans). The Y-axis of the OC curve graph lists the probability of acceptance of a given lot of product that contains the percentage of defective units noted on each curve. The furthest right-hand curve is the OC curve for lots of product that are 6.5% defective (or, any occurrences that could take place 6.5% of the time). Note that while this sampling plan will (probably) accept every lot that is 6.0% defective or less, there is about a 5% chance that a lot of 6.5% defective—the plan AQL—will be rejected. To a production manager this would imply that inspection is rejecting good product. This 5% risk is called the alpha risk—the risk of rejecting product that is actually acceptable. Alpha risk is part of the cost of using sampling inspection techniques. On the other hand, this same inspection plan would accept production lots that are as bad as 14% defective about 10% of the time. This risk is called beta risk and is also a cost of using sampling inspection.

FURTHER READING

Deming, W. E., *Out of the Crisis*, MIT Center for Advanced Engineering Study, Cambridge, MA, 1986.

Duncan, A. J., *Quality Control and Industrial Statistics*, 3rd Ed., Richard D. Irwin, Inc., Homewood, IL, 1965.

Grant, E. L. and Leavenworth, R. S., *Statistical Quality Control*, 6th Ed., McGraw-Hill, New York, 1988.

Small, B. B. (ed.), *Statistical Quality Control Handbook*, AT&T Technologies, Indianapolis, IN, 1985.

35
Metrology

GEORGE O. RICE*
Rockwell International/Boeing North America
Anaheim, California

I. INTRODUCTION

The subject of calibration of measuring and test equipment is one that is usually not treated (or only treated in a limited way) in a handbook on quality control. The increasing emphasis on this vital aspect of measurement by various governmental agencies, the implications of measurement error on safety and product liability, and the increasing awareness of the economics of measurement in business have all contributed to the need for including the subject in this handbook. This chapter, then, is intended to introduce the concept of calibration into the quality control program through an overview of the field and to relate the concepts involved to the various tasks of the quality control practitioner.

Although this chapter discusses the concepts of metrology standards, calibration, and so on, in an industrial setting, the conclusions are equally applicable to nonindustrial institutions, large and small. As will be seen, it is a fundamental tenet of this chapter that calibration of measuring and test equipment is both necessary and cost-effective. As in the more typical quality control functions, the costs associated with calibration depend on the protection desired; seen as a preventative measure, calibration costs are offset by the better decisions made in the research laboratory or production floor.

In the United States, the present sense of the term *metrology* entered the industrial vocabulary in general use probably around the end of World War II, although it was in less frequent use for some time prior to that. Even today, measurement specialists (metrologists) must take time to explain that metrology is neither the science of weather forecasting (meteorology) nor the science of the properties of metal (metallurgy). Metrology is, of course, important to both of these sciences, as it is to all science and industry.

---

* Retired.

Literally, *metrology* means "science of measurement" and is derived from two Greek elements: metro, meaning "measure" or "measurement," and logy, meaning law or science. In its early use, and in its use in some areas of the world today, the word *metrology* has had a more restricted use than the very broad sense of its literal meaning, referring to measurements of the physical attributes of an artifact (e.g., length, mass, shape). However, in this chapter, as in most scientific and industrial institutions in the United States, the word *metrology* is used in its broadest sense, limited only by the context within which it occurs.

In the early years of its industrial use in the United States, *metrology* was normally used as the name of the organizational element responsible for the measurement references of the institution. Thus it was almost exclusively reserved as a name for the standards laboratory of academic, industrial, and governmental organizations other than the National Institute of Standards and Technology (NIST). Today, it is not unusual (although by no means typical) for the word *metrology* to encompass all elements of an organization's calibration and gage control system, excluding only the end use of the measurement device.

Metrology, then, in the organizational sense, is the collection of people, equipment, facilities, methods, and procedures assembled to assure correctness or adequacy of measurements. Many institutions also append such related responsibilities as preventative and remedial service and, in some cases, inventory management tasks, such as accountability and utilization analysis. In this chapter we limit the organizational meaning of the term to responsibilities that pertain to correctness or adequacy of measurements.

Lord Kelvin said: "[If] you can measure what you are speaking about and can express it in numbers, you know something about it; but when you cannot measure it, when you cannot express it in numbers, your knowledge is of a meager and unsatisfactory kind." All material objects and a good number of nonmaterial characteristics are measured in some way daily. People weigh themselves, the contents of canned goods are specified by volume or weight, gasoline is purchased by the gallon or liter, TV programming is based on the measurement of time, manufacturing plants measure parts that must be assembled or fitted together by other plants, aircraft and space vehicles must be assembled from thousands of parts which have been measured to a tolerance to assure that they will work together, and so on. Measurement in one form or another pervades all aspects of our lives. The existence of measurement capability at home, in the marketplace, in schools, in manufacturing, and in research is loosely connected within a country by a network referred to as the national measurement system. Informal measurements comprise the bulk of these and normally rely on judgment or experience as the means for validating (calibrating) their accuracy. Formal measurements, those in the marketplace and generally under the aegis of state or local weights and measures officials and those in industry, are by several orders of magnitude fewer in number but are of much more significance to the American economy (1).

To assure that measurements made at one place by one person yield (within a stated uncertainty) the same result as the same measurements made at other places by other persons, the national measurement system in the United States has evolved a hierarchy by means of which the uncertainty of measurement at each transfer is quantified. In regulated areas (i.e., principally weights and measures and industries under contract to or regulated by government) these conditions are usually well specified and adequately controlled. This hierarchical system provides a measurement capability which is highest at NIST from which the system flows outward and downward to various standardizing laboratories in industry and government, culminating in the marketplace (formally) or in the myriad of informal measurements made by most people daily (2).

The National Institute of Standards and Technology is this nation's custodian of the standards of measurement. Originally established by an act of Congress in 1901 as the National Bureau of Standards, the need for such a body had been seen by the framers of the Constitution. The two main campuses of the National Institute of Standards and Technology (NIST) are in Gaithersburg, Maryland, and Boulder, Colorado, where research in the phenomena of measurement and in the properties of materials and calibration of the reference standards submitted by laboratories from throughout the United States are carried out. The following is a generalization of the echelons of standards in the national measurement system:

National standards: including prototype and natural phenomena of SI (Système International, the worldwide organization of weight and measures standards) base units and reference and working standards for derived and other units

Metrology standards: reference standards of industrial or governmental laboratories

Calibration standards: working standards of industrial or governmental laboratories

Frequently, there are various levels within these echelons (3).

It is obvious that NIST cannot calibrate all of the standards of every institution in the United States that calibrates measuring equipment. The system that has evolved over the years, however, has proven to be workable and includes the following features: NIST calibrates directly or by means of measurement assurance programs the reference-level standards of those organizations requiring the highest level of accuracy. These organizations calibrate their own working-level standards and the reference or working-level standards of other laboratories. In turn, working-level standards are used to calibrate the measuring and test equipment used to measure or test products.

Calibration is the comparison of a measurement device or system of a known relationship to national standards with another device or system of an unknown relationship in order to estimate the uncertainty in the latter. It may include adjustment and/or repair to minimize the uncertainty. Calibration, which begins with the measurements made at NIST and culminates with those made in industrial calibration laboratories, is the process that assures consistency of measurements and provides the basis for interchangeability of parts and mass production.

The industrial standards laboratory is the custodian of a company's reference standards and provides a service similar to that which NIST performs for the nation except on a far smaller scale and usually on a scope that is considerably less broad. In some cases the industrial standards laboratory is able to make some measurements as accurately as those made by NIST. Reference standards maintained by the industrial metrology laboratories in the United States for calibration of working standards must have a known relationship to national standards. This may be accomplished by periodic calibrations by NIST or, in some cases, the maintenance of "natural phenomena" (e.g., atomic frequency standards or freezing-point temperature references). The measurements made in industrial standards laboratories are intended to support calibration of the companies' measuring and test equipment. This equipment is used in research and development, in production, in inspection, in testing, and in diagnosis. Its accuracy must be sufficient for the purpose intended and, consequently, the first principle in calibration work is that the measurement requirement (i.e., the intent of the measurement) must be known.

Measurements made in manufacturing have a variety of purposes or intents, although one, the production of an item that complies with its specifications, is often thought to be their sole purpose. In most cases the other purposes support this general purpose. Others, though, may support continuing research and development, quality assurance, or other not specifically production-related purposes. Quite often the measuring or test equipment's accuracy (or other capability) may be different for each purpose. That is, equipment that is used solely to determine that a product is working or has a certain general characteristic may require less accuracy than equipment used to determine that the product is working within established limits or that the characteristic is within a specified tolerance.

Some companies locate the group responsible for metrology and calibration within the organization responsible for the quality program; others do not. Organizational location of the calibration function is no more important to the quality program than the location of other functions necessary or vital to the success of that program. What is important is that there be awareness of the different responsibilities and communication on all matters of mutual interest and concern.

A meaningful and effective quality program depends on an adequate system of measuring and test equipment calibration and control. Measurements made in production with uncalibrated equipment or (perhaps worse) with equipment inadequately calibrated can lead to erroneous decisions—either in accepting non-conforming material or in rejecting conforming material. A close, if not organizational, relationship serves to minimize this potential for costly errors.

The remainder of this chapter explores the field of metrology: its concepts, its role in quality assurance, its role as a control system, and its management. In all these areas the intent is to provide an introduction of the field to the quality assurance practitioner and management, to allow meaningful communication and to provide an understanding of the metrology role within the company's quality control program.

II. CONCEPTS IN METROLOGY

A fundamental role of the metrology and calibration process is to assign accuracy or uncertainty statements to a measurement. *Accuracy* is the term used to describe the degree to which the measurement result reflects reality; accuracy is the statement made which defines how closely the measured value approximates the true value of a characteristic. Because the true value is unknowable, the accepted or specified value is normally used for this approximation. Most calibration work involves the determination or verification of the accuracy of a measuring instrument; most accuracy statements are given in percent, although higher-order accuracies are usually stated in parts per million (ppm).

Precision in measurement is the degree to which the result is repeatable under identical circumstances. In some respects, precision is a more important characteristic of a measurement than is accuracy. Measurements may be accurate and precise, accurate and imprecise, inaccurate and precise, and inaccurate and imprecise. The role of the metrology and calibration process is to quantify these measurement characteristics, to the extent required, by the design of experiments and careful consideration of all factors that could influence the results.

Error is an important consideration related to accuracy and is present in all measurements. The task of the metrologist, particularly in measurements that are at or near the state of the art, is to identify all contributions to measurement error and to quantify those contributions. There are basically two types of errors that may enter the measurement process: random and systematic. The latter is typified by fixed bias and may sometimes be removable. The former is probabilistic and must be estimated statistically. Both types of errors contribute to the total uncertainty of the measurement and to the accuracy assignable.

The entire metrology and calibration process, at least in the industrial sense, is intended to quantify measurement error contributions in respect to the national standards of measurement. This "traceability of measurements" has been a requirement for contractors doing business with the U.S. Department of Defense (DoD) for many years; it has received increased attention in the last few years because of two unrelated developments. These are the creation of federal requirements concerning quality assurance and calibration by agencies outside the DoD [e.g., Food and Drug Administration (FDA), National Research Council (FRC), Occupational Safety and Health Administration (OSHA)] and the widespread adoption of measurement assurance programs (MAPs) by NIST.

Traceability is a process intended to quantify a laboratory's measurement uncertainty in relationship to the national standards; it is based on analyses of error contributions present in each of the measurement transfers, from the calibration of the laboratory's reference standards by NIST, to the measurements made in the calibration transfer within the laboratory, and finally to the measurements made on a product. Evidence of traceability is normally required; it may be as simple as retention of certificates and reports of calibration or as complex as reproduction of the analyses demonstrating the uncertainties claimed for the measurements.

A laboratory that maintains its own reference standards (i.e., it relies on no laboratory other than NIST for calibration of its standards) must continuously monitor its own performance. Measurements on check standards, intercomparisons of standards and participation in measurement assurance programs sponsored by NIST are means to quantify laboratory error sources as well as to provide indications of the causes.

Measurement assurance is one of the more important concepts in the measurement field. Although the idea is not new, the application to the industrial metrology and calibration laboratory and ultimately to measurements made in factory operations has been given increased impetus in the last few years by the NIST and by the National Conference of Standards Laboratories.

A feature of MAPs (which undoubtedly gave rise to the feeling that MAPs affect only calibration laboratories) is that MAPs both provide traceability and quantify the participants' total uncertainty. Traditionally, calibrations by the NIST determine the accuracy and precision of the measuring instrument. MAPs, on the other hand, because the experiment involves measurements by participants in their own laboratories, are able to include not only the accuracy of the item, but also the contribution to error by the metrologist/technician, by the laboratory environment, and by the practices/procedures of the laboratory (4).

Measurement assurance, in addition to being a concept of importance to metrology and calibration laboratory managers, is one that should interest quality assurance personnel involved in testing and measurement. Most factory testing and measuring involves use of equipment whose accuracy has been determined through calibration. Little, if any, consideration is given to errors that may be contributed by the test operator, by his or her instructions or procedures, or by the environments in which the equipment is operated. Application of measurement assurance, including introduction of check fixtures or standards and control charts, particularly when state-of-the-art measurements are being made or where the measurements made are of critical importance to the production process, can serve to reduce errors.

Periodic recalibration of measuring and test equipment is accepted by most as necessary for measurement accuracy. A little more controversial is the question of determining the basis of the period of recalibration. There are a number of techniques in use to establish calibration intervals initially and to adjust the intervals thereafter. These methods include the same interval for all equipment in the user's inventory, the same interval for families of instruments [e.g., oscilloscopes, digital voltmeters (DVMs), gage blocks, etc.], and the same interval for a given manufacturer and model number. Adjustments of these initial intervals are then made for the entire inventory, for individual families, or for manufacturer and model numbers, respectively, based on analyses or history. A study conducted for NIST in connection with a review of government laboratory practices identifies these and other methods (5).

One method establishes the initial interval of an instrument based on the manufacturer model number. Adjustment of intervals is then made for the specific instrument based on its own performance and frequency of use. This method of interval adjustment is the only one known that considers the differences in production for the instrument and differences in application and environmental factors when the instrument is used (2).

The objective of adjusting the intervals between calibrations is to achieve a predetermined quality level. Quality level, sometimes referred to as reliability, is the fraction of instruments returned for calibration which are found to be within their specified tolerances. This information can also be used to estimate the fraction of instruments in use which are in tolerance (see Table 1).

The existence of instruments in use which are (or have a possibility of being) out of tolerance has given rise to considerable controversy over the years. At issue is the effect that use of out-of-tolerance measuring equipment has on the quality of products. On the one hand, logic suggests that an out-of-specification product is more likely to be accepted by an out-of-tolerance measuring instrument than by an in-tolerance measuring instrument. This position may be summarized by the phrase "Bad instruments buy bad products." The alternative view holds that the risk of accepting out-of-specification product (consumer or β risk) is probably less with out-of-tolerance instruments than with in-tolerance instruments. Or, even if it is not, the risk of rejecting in-specification product (producer or α risk) is so much greater than the β risk that producers will take action long before the consumer

Table 1 Tabulation of α and β Risks

	Accuracy ratio					
	10:1		4:1		2:1	
Test equipment condition in use	α	β	α	β	α	β
In tolerance	0.32	0.24	1.18	0.02	3.84	0.01
Out of tolerance by 25%	0.31	0.22	1.15	0.06	3.90	0.05
Out of tolerance by 50%	0.31	0.21	1.16	0.08	11.55	0.13
Out of tolerance by 100%	0.39	0.18	1.19	0.23	8.88	0.28
Out of tolerance by 200%	0.56	0.11	1.87	0.38	9.19	0.20
Out of tolerance by 300%	0.82	0.10	4.79	0.39	95.5	0

α risk = in-specification product rejected as percent of products produced; β risk = out-of-specification product accepted as percent of products produced.
Source: Ref. 6.

is affected. The controversy has led to several attempts by the DoD to revise its governing specification for contractors' calibration control systems. The goal is to require that data resulting from calibration of measuring equipment found to be out-of-tolerance be fed back to users. This "feedback" was to be used by quality assurance to determine the need for corrective action and to alert customers to potential out-of-specification products already delivered.

Calculation of α and β risks under several assumptions of process capability and measuring equipment accuracy indicates that measuring equipment that is out of tolerance has only minimal impact on product quality. For example, calculations with the following assumptions provide the risks displayed in Table 1 (6):

1. Process capability is $\pm 2\ \sigma$.
2. Measuring equipment has accuracy ratios of 10:1, 4:1, and 2:1.
3. Measuring equipment inaccuracy (except for 10:1 case) is subtracted from product specification tolerances to create acceptance limits (i.e., a guardband is formed).
4. Measuring equipment is found to be in tolerance, or 25, 50, 100, 200, and 300% out of tolerance.

III. MEASUREMENT AND QUALITY ASSURANCE

Calibration of measuring equipment acts as a reference for all quality control decisions; measurements made with equipment whose capabilities (accuracy, stability, precision, etc.) are unknown or unproven will frequently lead to decisions that are incorrect. In many cases these incorrect decisions do not manifest themselves immediately; some never do. The problem is that without a known reference the reasons for subsequently observed anomalies (i.e., the causes) may never be determined certainly.

The effect of calibration on the quality control program begins with the establishment of the product's specifications. These specifications come from research and development laboratory measurements as well as from the physics and mathematics of the design. In other words, measurements made in the research and development laboratory contribute in some way to the product specifications, which the quality control function must assure are being achieved in production.

Measurements made in receiving inspection, in fabrication or process control, in subassembly testing, and in systems testing are all made with the intent of verifying that some characteristic of the product conforms to its specifications. The effects of calibration on the economic well-being of the institution are more vividly seen here. Erroneous decisions (rejection of products that actually conform to the specifications or acceptance of products that actually do not) are costly: costly in terms of money, in terms of wasted resources, and in terms of reputation.

Because measurements form the basis for most quality assurance decisions, particularly those pertaining to acceptance of products, equipment selected to make product measurements must be adequate for the purpose. Adequacy is usually interpreted to mean that the measuring equipment has an accuracy that is somewhat better than the tolerance of the characteristic being measured. A rule of thumb, adopted by many organizations, is that the accuracy of the measuring instrument should be one-tenth the tolerance of the characteristic to be measured. Many other organizations have adopted a one-fourth rule. (In most cases this accuracy ratio will be stated in reverse. That is, the rule of thumb will require an accuracy ratio of 10:1 or 4:1.)

The use of accuracy ratios provides a simple means for the determination of measuring equipment adequacy. In many cases the accuracy-ratio rule of thumb is all that is necessary to assure that the measuring equipment is adequate; in some cases, particularly in the standards laboratory but also in other testing situations where measurement capability is at or near the state of the art, such simplified approaches are not adequate. More sophisticated analyses involving statistical techniques must be employed. The use of accuracy ratios recognizes that all measurement processes have errors. But if the ratios meet the rule of thumb (i.e., 10:1 or 4:1) these errors are sufficiently small so as to be ignored. Some organizations believe that "sufficiently small" is when the accuracy ratio is at least 10:1, whereas others believe that an error is still "sufficiently small" when the accuracy ratio reaches as low as 4:1.

There are, however, a number of measurement situations in almost any modern production operation where for various reasons, some technical and some economic, an accuracy ratio of 10:1 (or even 4:1) is not possible or desirable. In those cases the error in the measurement contributed by the measuring equipment is no longer "sufficiently small" as to be ignored; some means of compensating for the error must be provided. There are various statistical methods of compensation, but the most widely used (probably because it is the easiest) is a direct subtraction of the measuring equipment error from the specified product tolerance.

An example should clarify the concept. Consider a product with a specified requirement of one characteristic of 10 ± 1 V. This is a requirement for the product irrespective of how it is measured. If our rule for sufficiently small error is an accuracy ratio of at least 10:1, we must provide a measuring instrument capable of measuring 10 V to an accuracy of ± 0.1 V. If we had such an instrument, we would use it to measure the 10 V, and if the measured value were between 9 and 11 V (inclusive), we would say that the product met this requirement. Suppose, however, that the best measuring instrument we had could measure 10 V only to an accuracy of ±0.2 V. In this case our accuracy ratio is only 5:1 and, by our rule, the error of the measuring instrument is large enough that it cannot be safely ignored. Using the direct subtraction method, the instrument's error of ±0.2 V is subtracted from the product's specified tolerance of ±1 V to create a new set of acceptance limits: ±0.8 V. Now, when this instrument is used to measure the 10 V, we would say that the product met the requirement if the measured value were between 9.2 and 10.8 V (inclusive). This technique is called *guard banding*. A product with a true value between 9 and 9.2 V or between 10.8 and 11 V might be rejected by this method even though it would comply with the product's specification. This is considered prudent because an instrument with a ±0.2 V uncertainty without the guardband has a high probability of rejecting a product with a true value greater which is within the limits specified (10 ± 1 V).

As noted previously, calculation of the α and β risks associated with selected accuracy ratios and for various out-of-tolerance conditions of the measuring equipment discloses that the α risk is far higher than the β risk. Selection of higher accuracy ratios, then, is not usually made because of potential product jeopardy, but rather because of the costs associated with rejecting a product that is actually acceptable: a decision based on economics.

Two very important tasks for the calibration function are implied by this example. One is that measuring equipment capability (i.e., accuracy, precision, etc.) must be determined. This may be accomplished in two ways. First, the equipment is initially evaluated by the metrology function to determine its basic capability. (In many cases this step is not performed; basic capability is assumed to be what the instrument manufacturer claims.) Second, the instrument is calibrated periodically, thus updating the determination of the instrument's capability.

The second task, which is shared by the quality assurance function, is the determination of the calibration requirements. The calibration requirements are derived from the product's specified tolerance and are not necessarily the same as the instrument's capability. Knowledge of the calibration requirements, if they are less than the instrument's capability, allows the calibration organization to make calibration decisions based on economics. If the calibration requirements exceed the instrument's capability, the quality assurance organization may suggest a product tolerance change, may select an instrument with improved capability, or may establish calibration requirements based on a different accuracy ratio. These decisions are partly technical and partly economic.

IV. CALIBRATION CONTROL SYSTEM

A typical calibration program may involve all or most of the following tasks: (1) evaluation of equipment to determine its capability, (2) identification of calibration requirements, (3) selection of standards to perform calibration, (4) selection of methods/procedures to carry out the measurements necessary for the calibration, (5) establishment of the initial interval and the rules for adjusting the interval thereafter, (6) establishment of a recall system to assure instruments due for calibration are returned, (7) implementation of a labeling system to visually identify the instrument's due date, and (8) use of a quality assurance program to evaluate the calibration system (process control, audit, corrective action, etc.).

Selection of the standards, methods, and procedures to carry out the calibration includes the decision relating to where the calibration will be performed. Some instruments may require use of a laboratory's highest level of standards and thus calibration must be performed in the laboratory. Other instruments, however, may be calibrated in the using area by the transport of suitable standards to that area. Two methods are followed in this case. One, referred to as "in situ" calibration, requires external interface to the calibrated characteristics. When such interface exists, the instrument being calibrated remains in the specific location where it is used (e.g., in a rack-mounted configuration) and the calibration is performed. The second method, still performed in the using area, requires that the equipment be removed from its rack mounting to provide access to the measurement points used in calibration. Both of these methods have advantages over the more traditional method, which requires that the instrument be returned to a calibration laboratory. One advantage is that the calibration is performed with the environmental factors identical to those present when the instrument is used. Another advantage is that the instrument does not have to be transported, thus reducing the potential for damage or movement-induced changes in the instrument. A major advantage is that the measuring system which includes the instrument being calibrated is not "down" while the instrument is being transported to and from a laboratory. However, instruments that cannot be adjusted from the front panel or instruments requiring large or unique standards must be returned to the standards laboratory for calibration. In addition, instruments that during calibration in the using area are found to be out-of-tolerance must also be returned to the laboratory for repair.

The recall system must be designed to assure that the calibration organization and the using organization are both aware in advance that an instrument will be due for calibration. Depending on the number of instruments being controlled and their geographic location differences, the system may be as simple as a card file or as sophisticated as a fully automated data processing system. The more sophisticated the system, the more that can be expected from it beyond the basic purpose of providing recall notification (e.g., history of previous calibrations, interval assignments, labor standards or actual costs, parts replaced).

Labeling of instruments to display their calibration due dates visually is a companion feature to the recall system. Labels indicate (by dates, color codes, or similar symbols) the date the instrument is due for its next calibration. This visual identification may be used by the quality assurance organization to assure that the instrument is not used beyond its due date.

Intervals are established in a variety of ways, as discussed previously. Principal objectives of an interval adjustment program include minimizing the potential for out-of-tolerance instruments in using areas; minimizing the costs of calibration; and assuring the required accuracy of instrumentation. The effectiveness of the interval adjustment programs can be estimated by measuring the average interval and its trend and by measuring the quality level.

Quality-level goals that vary from about 75% to above 95% have been established by different organizations. The relationship between intervals and quality levels is complicated by such factors as age of equipment in the inventory (new items are added to inventories), the makeup of the inventory (mechanical instrumentation, electronic test equipment, fixtures, etc.), and the accuracy assignments of instruments in the inventory. However, the quality level is one indicator of the effectiveness of the interval adjustment program. When combined with other indicators, such as the average interval, the minimum and maximum observed intervals, and a corrective-action system that is triggered by low intervals, the quality level is a sound method for evaluation of the total calibration control system.

V. MANAGEMENT

Organizationally, the calibration function may report to almost any company element. Philosophically, particularly if the laboratory includes the metrology functions of standards development and maintenance, the calibration function is most in tune with engineering and quality assurance. In an organizational sense there has long been a close link between the company element responsible for quality control and the element responsible for equipment calibration. In many companies this link is positive and direct; the calibration group reports to the executive responsible for quality control. In others, the calibration group maintains close liaison with its quality control counterpart. Is there a correct organization approach? That is, is there one organizational structure that is most effective and, consequently, recommends itself for all companies to consider? The answer is no. Very effective programs have been implemented in companies with the calibration group reporting to quality control, to engineering, to plant maintenance, to manufacturing, and to the firm's chief executive officer. The key is not found in organizational location; rather, it is found in the firm's policy or philosophy.

Most effective calibration programs not only display a top-management philosophy which emphasizes the importance of measurement control and reflects a close relationship with quality control; they also adopt and implement strong quality control programs within the calibration process itself. The most significant development in this regard in the last decade has been the emergence of a control system referred to as *measurement assurance*. Although understood by many as a means for economically and most effectively transferring measurement traceability to the industrial standards and calibration laboratories, measurement assurance is more than this. Actually, measurement assurance is a control system for measurements made throughout the institution: research, development, manufacturing, and so on.

Calibration, then, is intimately related to quality control. Without calibration the quality control decisions pertaining to the product are at best questionable. Without quality control in calibration, performance is subject to the same deterioration that affects other operations without quality control: increased errors, increased costs, and so on.

Productivity is an important topic in most institutions today, particularly in the United States. Improving productivity is cited as a necessity to the control of inflation, to the improvement of the standard of living, to the improvement of the balance of payments, to the increase of employment, and several other characteristics. The calibration function can contribute to productivity improvement, and the single most important area that is subject to managed improvement of productivity of calibration functions is (as it is for most other organizations) in the labor content. Reduction of the labor involved in the various calibration tasks can measurably contribute to a company's productivity improvement. The most promising way to reduce labor is to use—or to increase the use of—computers and microprocessors. These may involve computers in data analyses or microprocessors in test automation. What is required is ingenuity coupled with a continuing concern for the adequacy of the measurement process. One study has shown that a particularly effective method to improve productivity is to identify and correct counterproductive practices (7).

The issues of importance to management of metrology and calibration laboratory include a capability in measurement adequate for the calibration tasks (a capability that includes equipment, personnel, facilities, and procedures), a budget sufficient to execute the responsibilities of calibration (including labor, expense, and equipment), and a well-defined mission statement for the calibration task. In many cases, the ability to forecast future measurement needs of the institution for periods in the mid to long term (i.e., 3 to 10 years) is an important asset of a metrology and calibration laboratory manager.

The capability of the laboratory to make the measurements required in carrying out its calibration tasks is a characteristic that must be assessed continuously. Coordination with the engineering and quality assurance organizations to provide the necessary capability when it is needed is an absolute requirement. This is of particular importance when the laboratory is operating at the state-of-the-art level or when new measurement requirements will push the laboratory toward the state of the art. Development of capabilities (including necessary coordination with the NIST) for state-of-the-art measurements may require several years; therefore, early planning is essential.

Budgets for the calibration tasks in some organizations are considered to be overhead, whereas in others they may be a combination of overhead and direct. Budgets should be based on an analysis of the tasks to be performed and should include labor budgets, expense budgets, and equipment (or capital) budgets. Budgeting methods (and the related budgeting difficulties) are normally unique to an institution. The principal requirements for a metrology and calibration laboratory manager in connection with budgeting are to understand his or her organization's budgeting system and to have a thorough grasp of his or her laboratory's tasks and their costs. Only in this way can the laboratory manager provide the forecasts essential to the approval of a budget that is adequate.

Control of the laboratory's costs (irrespective of the outcome of the budgeting process) is a measure of the manager's ability to balance the technical and economic needs of the calibration function. Control, of course, means meeting the established budgets for the tasks and identifying the tasks (and their costs) which change during the period of the budget to secure revisions to the budget (up or down). (It is just as important to identify budget decreases as budget increases for the financial well-being of the institution of which the calibration function is a part.)

Probably one of the most important steps that any laboratory manager can take in connection with assessing the adequacy of his or her measurement capabilities or with budgeting is the development of a well-defined mission statement. Such a statement should interrelate with and support the overall mission of the organization. Coupled with an objective evaluation of the laboratory's principal strengths and weaknesses, such a statement provides an excellent basis for organizational planning.

VI. CONCLUSION

Metrology and calibration are of significant importance to the quality assurance function whether or not they are organizationally a part of quality assurance. The principal role of metrology and calibration operations is to provide a measurement capability which is adequate for the institution. Whether the calibration function is organizationally within the quality assurance function or not, the need for continuing dialogue on matters of mutual interest cannot be overstated; a similar statement may be made in respect to the calibration function and engineering. In the final analysis, metrology is responsible for the integrity of measurement within our institutions, and measurements underlie the quality assurance program.

REFERENCES

1. R. C. Sangster, *Collected Executive Summaries: Studies of the National Measurement System, 1972–1975*, U.S. Department of Commerce, Washington, DC, August 1976.
2. G. O. Rice, "Measurement Systems and the Standards Laboratory," Workshop Conference on the Management of Laboratory Instruments, Cairo, Egypt, November 7–11, 1976. (Conference proceedings collected in a work titled *Management Systems for Laboratory Instrument Services*, Instrument Society of America, Research Triangle Park, NC, 1980.)
3. D. A. Mack, "Instrumentation Calibration," Workshop Conference on the Management of Laboratory Instruments, Cairo, Egypt, November 7–11, 1976. (Conference proceedings collected in a work titled *Management Systems for Laboratory Instrument Services*, Instrument Society of America, Research Triangle Park, NC, 1980.)
4. B. C. Belanger, "Measurement of Quality Control and the Use of NBS Measurement Assurance Program (MAP) Services," draft document (1980) to be published by U.S. Department of Commerce.
5. J. L. Vogt, *Optimizing Calibration Recall Intervals and Algorithms*, NBS Publication NBS-GCR-80-283, 1980.
6. R. F. Schumacher, "Feedback of Out-of-Tolerance Data," an internal study conducted by Rockwell International, Anaheim, CA, 1981.
7. R. M. Ranftl, *R&D Productivity: An Investigation of Ways to Improve Productivity in Technology-Based Organizations*, Hughes Aircraft Co., Culver City, CA, 1974, 1978.

FURTHER READING

ANSI/NCSL Z540-1-1994, *American National Standard for Calibration—Calibration Laboratories and Measuring and Test Equipment, General Requirements*, National Conference of Standards Laboratories, Boulder, CO, 1994.

ISO 12000-1, *Quality Assurance Requirements for Measuring Equipment*, International Organization for Standardization, Geneva, Switzerland, 1993. (Available from Global Engineering Documents, Englewood, CO.)

36
Dimensional Inspection Equipment

RONALD A. LAVOIE
Consultant
Providence, Rhode Island

I. INTRODUCTION

This chapter describes the dimensional inspection equipment needed for the various major inspection areas of a precision manufacturing operation. Although all the equipment listed would probably be needed in a large manufacturing operation, it will be apparent which items are vital even for a smaller-scale operation. The major inspection areas are considered to be incoming inspection, post-process manufacturing inspection, final or outgoing inspection, automatic and semiautomatic gaging, and the metrology laboratory.

The incoming inspection area inspects products from suppliers and perhaps the shop floor as well. It is an area that has to have a great deal of flexibility and be capable of measuring a broad spectrum of dimensional characteristics. Post-process gaging on the manufacturing floor may involve dedicated gages. Measurements are usually made by the machine operator shortly after a part is removed from the machine. Automatic and semiautomatic gages perform measurements either off- or on-line with little, if any, operator involvement. Finally, the metrology laboratory performs calibration functions on other gages as well as specialized dimensional analysis work.

II. INSPECTION EQUIPMENT

A. Surface Plate or Layout Inspection

Until the introduction of the coordinate measuring machine in the mid-1960s, the surface plate "layout" technique was the primary means for measuring complex parts and fixtures. Today, surface plates are generally limited to use in shops that do not have coordinate measuring machines or similar equipment, or when a few quick measurements are all that is needed. On rare occasions, a surface plate layout might be used for accuracy requirements that are greater than those possible with the coordinate measuring machine.

Typically, surface plate measuring techniques require highly skilled technicians and are 10–40 times slower than measurements made on a coordinate measuring machine.

A typical surface plate setup consists of the following elements:

Surface Plate

A granite surface plate can vary in size from as small as 1 ft^2 to massive plates 10 ft by 20 ft and 3 ft thick. These plates are usually of quartz granite. The top surface of this granite is lapped flat to tolerances on the order of 50 μin. This lapped surface establishes a reference plane from which parts can be accurately fixtured and measured.

Height Transfer Stand

A height transfer stand and test indicator are used to indicate a critical surface on a part and to transfer that setting to the height micrometer (Fig. 1). The reading on the height micrometer is the altitude from the surface plate to the critical surface on the part. Repeating this operation on another critical surface and subtracting the two readings will produce the distance between two surfaces on the part.

Height Standard

The height standard can be either mechanical or electronic. The mechanical version usually consists of a micrometer of 1-in. travel mounted on the top of a permanent stack of 1-in. steps usually 12–18 in high. This height micrometer establishes a very accurate measurement standard that is perpendicular to the surface plate. Movement of the micrometer lead screw allows the positioning of the 1-in. steps anywhere within the 1-in. travel of the micrometer in 0.0001-in. increments.

Electronic versions of the height micrometer include a precision linear transducer which produces a digital readout of 0.0005 or 0.0001 in resolution. These linear transducers are usually built into a height transfer stand, eliminating the need for a height micrometer. The accuracy of these systems varies substantially and is dependent on the mechanical rigidity and accuracy of the height transfer stand.

Gage Blocks

Gage blocks may also be stacked and placed vertically on the surface plate instead of on a height micrometer. This technique was popular before the advent of height micrometers. It is an extremely slow technique and is used only when the inherent high accuracy of gage blocks is required.

Test Indicators

There are mechanical and electronic test indicators. Mechanical test indicators are still popular and are used where transfer accuracies of 0.0001 in. are adequate. Electronic indicators include an electronic transducer with a lever-type contact and an amplifier. Multiple ranges and optimum viewing angle of the amplifier display

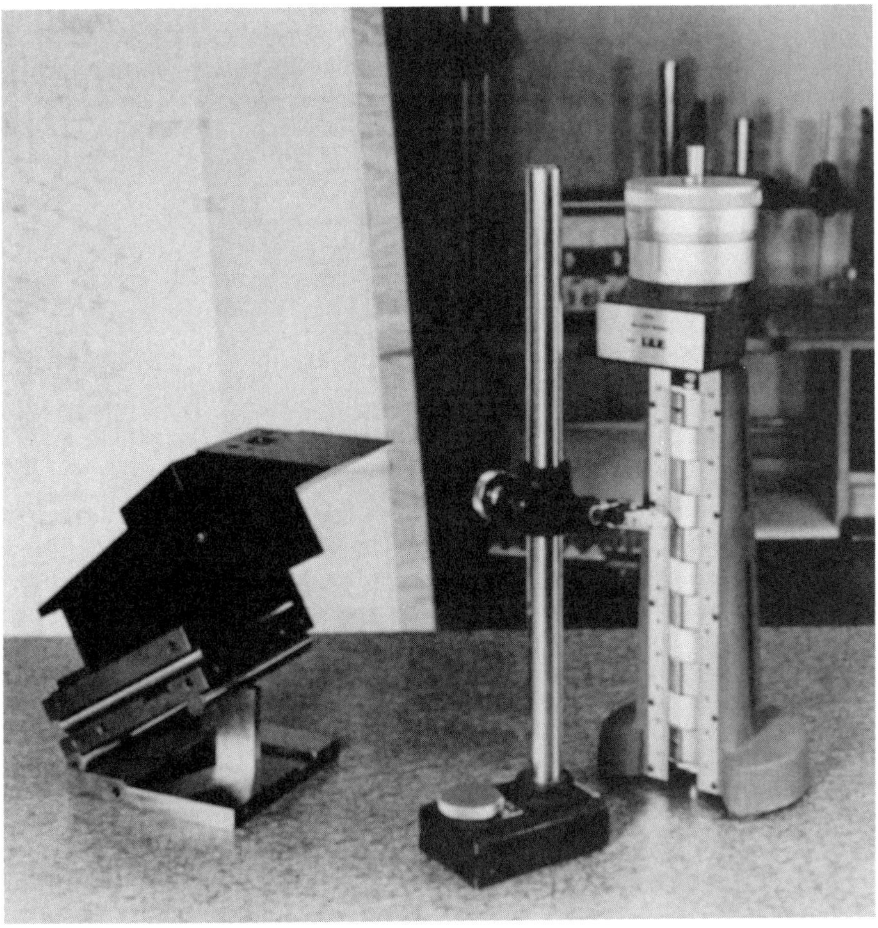

Figure 1 In a typical surface plate inspection operation, a height stand and test indicator are used to transfer a reading from the workpiece to a height standard. Proper orientation of the part is achieved by use of a sine plate.

are its primary features. However, electronic test indicators are approximately 10 times more expensive than the mechanical test indicator.

Sine Plates

Sine plates in conjunction with gage blocks create very precise angles; on the order of 15 arc seconds. They are used to measure angles using a surface plate. Simple and compound angle sine plates are available. The simple sine plate creates an angle in one direction. The compound sine plate permits creation of two angles oriented 90° to each other—a compound angle.

Fixturing

Surface plate inspection usually requires universal fixturing to facilitate accurate positioning of a variety of parts. Typical fixturing consists of magnetic blocks, V blocks, right-angle knees, parallels, and clamps.

B. Bench Inspection Station

Inspection stations usually incorporate inspection benches where measurements are made by hand. Surface plates are often a part of these benches. Measurements that can be easily made with hand gages are performed here, and should not be made on expensive equipment such as coordinate measuring machines and optical comparators.

Length Measuring Devices

Micrometers

At least a 0- to 1-in. depth micrometer and a multianvil micrometer are usually available at inspection stations. A central station may also have the less frequently used, larger, special-application micrometers (Fig. 2).

Calipers

Plain vernier calipers are the least expensive type available. They are used infrequently because they are difficult to read, resulting in reduced efficiency and errors of measurement. Dial calipers have more-or-less replaced the plain vernier caliper. They are easy to read, require less skill to operate, and produce fast, reliable readings. Electronic calipers with digital read-outs are also available; however, they are twice as expensive as dial calipers.

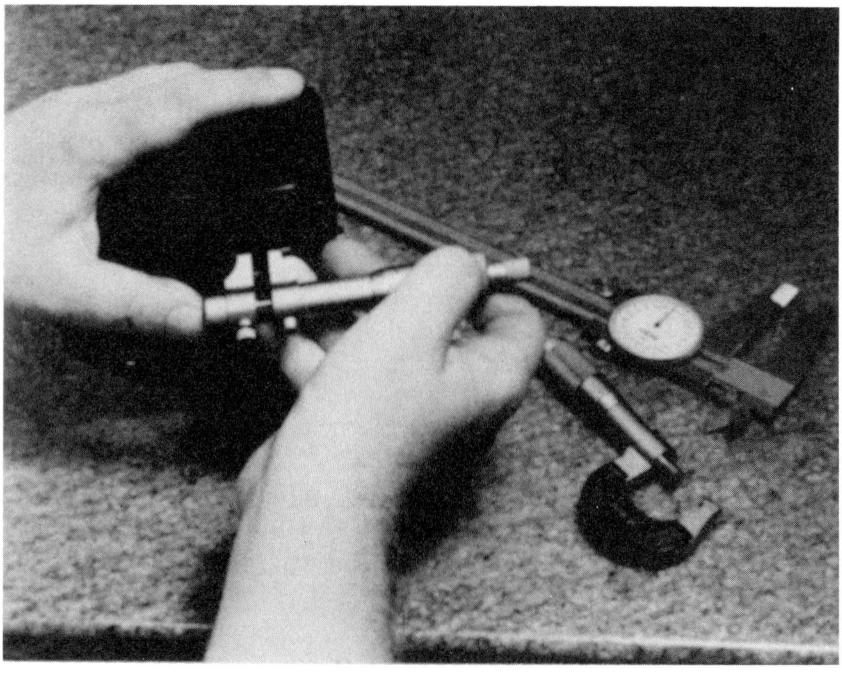

Figure 2 An inspector uses an inside micrometer to check a hole diameter. Other part dimensions will be checked using the dial calipers and micrometer.

Comparator Stand

A comparator stand and dial indicator afford rapid measurements of heights and diameters. The comparator stand is used for measuring large samples of the same dimension. With an electronic cartridge probe and amplifying system, it may be used for higher-accuracy requirements.

Bore Gaging

Pin Library

Pin gaging involves placing a series of pins at 0.001- or 0.0001-in. increments into a hole until one pin just fits and the next size does not. This is a coarse but effective means of measuring diameters. It is particularly useful in measuring the diameter of very small holes below the size practical for other types of bore-measuring devices. When using pin gages to measure holes, it is important to take into account the fact that a pin gage that is exactly the size of the hole will not enter the hole. For example, if a 0.150-in. pin will not enter a hole but a 0.1495-in. pin will, the hole may be anywhere from 0.1496 to 0.1500 in. in diameter. A company standard procedure for classifying the results of using pin gages will prevent arguments. One way is to classify the above example as 0.1495 go/0.1500 no-go. It is also important to note that if the hole is out of round, using pin gages to classify its diameter may be very inaccurate.

Dial Bore Gages

Dial bore gages have the advantage of being adjustable over a broad range of measuring sizes, typically from 0.5 to 13.0 in. They do not require special masters. They may be mastered with gage blocks or a micrometer. This setting is then transferred to the measured hole without loss of accuracy.

Inside Micrometer

The inside micrometer covers a broad range of measuring diameters. It is however, limited to holes at least 1 in. and larger. Accurate measurements with this device require careful manipulation of the micrometer to find the maximum diameter of the hole. It is only one-third as accurate as a dial bore gage.

Air and Electronic Plugs

Plugs are most often used in postprocess production situations. However, plugs are useful in incoming inspection for the measurement of popular hole sizes. Standardization of hole sizes by the engineering department can make it practical to utilize this type of bore gaging. These gages are desirable because of their simplicity of operation, but they are limited to a specific diameter and usually require master setting rings.

Telescoping Gages

Telescoping gages are spring-loaded cylinders that expand to fill the hole they are measuring. They are then locked into position and their overall dimension measured with a micrometer. They are difficult to operate, requiring considerable skill, and are limited to a 0.0005-in. accuracy.

Other Types

There are electronic and micrometer-type bore gages which do not require mastering. They read the inside diameter of a bore directly. Their high price has limited their use.

Miscellaneous

Simple measuring devices such as rules, protractors, and radius templates are commonly found in the bench inspection area.

C. Coordinate Measuring Machines

Since their introduction in the early 1960s, tens of thousands of coordinate measuring machines (CMMs) have been sold throughout the world. Today, practically every manufacturer of precision products has at least one CMM system. These machines measure most dimensions on complex parts 10–40 times faster than the former procedure of surface plate layout inspection. These systems are often used to inspect products produced by numerically controlled machining centers.

The more basic systems require only semiskilled operators, and their simplicity of operation, together with digital readouts, ensures reliable measurement results. They range in capacity from a 16-in. by 12-in. by 8-in. measuring cube to systems large enough to allow measurement of an auto body. Their accuracy potential is usually sufficient to satisfy the majority of production part inspection requirements.

Basic System

Basic coordinate measuring systems usually consist of air or ball-bearing ways which establish three axes: X, Y, and Z. The X axis establishes left-right motion; the Y axis, in-out motion; and the Z axis, vertical motion. Each of these axes has its own long-range linear transducer. The Z-axis shaft has a receiver at its lower end capable of accepting various measuring probes. A typical measurement involves placing a tapered probe into a hole and zeroing the three axes, then moving the tapered probe to a second hole. The digital display will indicate the distance between the holes.

Advanced Systems

Computer Assist

Many CMM systems have computers which are interfaced with the measuring system either as a stand-alone computer, a programmable calculator, or a microcomputer built within the digital display (Fig. 3). These computing systems speed up the operation of the coordinate measuring machine and can reduce inspection time by at least 50% over noncomputer systems. The computers perform such functions as the following:

1. Automatic axis alignment: This compensates for misalignment of the part datums to the axes of the coordinate measuring machine.
2. Coordinate conversion: The conversion of rectangular coordinates to polar coordinate readout.
3. Contour measurement: The measurement of contours "on the fly," by simply sweeping the probe across a contoured surface.
4. Centers of diameters and arcs: Determining the center position of arcs and diameters by placing the probe at three equally spaced positions on the diameter of consideration.

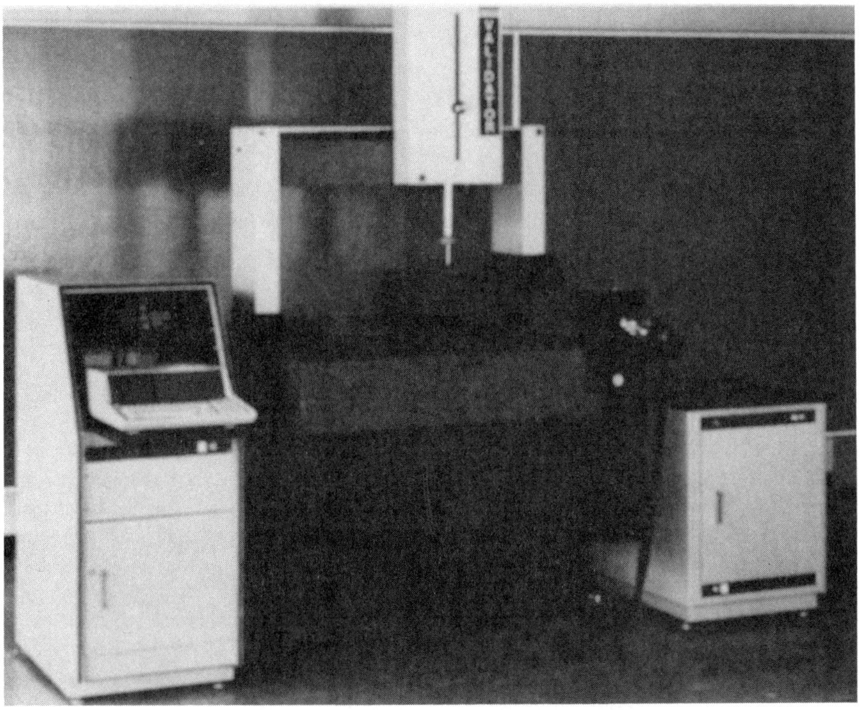

Figure 3 A direct computer-controlled coordinate measuring machine utilizing air bearings on all axes. (Courtesy of Brown & Sharpe Manufacturing Company.)

It is not uncommon for a computer package to incorporate 30 features similar to those just described. They increase the efficiency of coordinate measuring machine operation and minimize the probability of an operator error.

Servomotor Drive

Servomotor drive of the axes of the CMM in conjunction with computer control allows for automatic inspection of parts. The operator simply loads the part into the system, starts the operation, and the coordinate measuring machine performs all measurements automatically. These systems measure parts very rapidly without operator fatigue or involvement. However, they do require the programming of the system for each different type of part measured, and this is advantageous only if this programming time can be amortized over a substantial number of similar pieces.

Accessories

1. Touch probes: Touch probes contain a light-pressure transducer. When the tip of the probe contacts the surface to be measured, there is a slight deflection of the probe tip while the transducer responds. This movement of the probe tip is "zeroed out" during the setup program. The probe tip pressure against the surface to be measured is very light and therefore does not impart distorting forces to the coordinate measuring machine system, enhancing system accuracy.

2. Optical viewers: Optical viewing devices placed on the end of the Z axis allow establishing reference and measuring points by means of optical techniques. A microscope viewer is particularly useful on fragile surfaces that will not allow mechanical contact. However, operation of a microscope produces considerable operator fatigue.
3. TV viewers: TV viewing systems can be operated in conjunction with a computer to automate completely the optical detection of edges and center of holes. Generally, these systems are limited to X and Y axis measurement only. As imaging enhancing technology continues, these systems should continue to receive greater acceptance.

D. Optical Comparators

Optical comparators are used to project the profile or image of a precision part onto a large screen (Fig. 4). The viewing screen displays an image that is a precision magnification of the part. Most commonly, the part is magnified 10, 20, or 50 times

Figure 4 An optical comparator inspects the threaded portion of a splined shaft. The system features a 30-in. screen plus digital readout of the horizontal and vertical position of the part. The machine can be interfaced with instrumentation such as computers, printers, a vector positioner, and so on. (Courtesy of Jones & Lamson Metrology Systems, Waterbury Farrel Division of Textron, Inc.)

its original size. A part typically is measured or inspected on an optical comparator in one of two ways. A transparent overlay that contains two sets of lines that correspond to the maximum and minimum allowable dimensions (at the same magnification as the comparator is set for) of the part being measured can be fastened to the comparator screen. The part being measured is placed on the comparator table and its profile or shadow is superimposed on the overlay. If all points of the profile of the part being measured lie within the lines on the overlay the part is considered to be within tolerance. This technique is an "optical comparison" of the part with a standard. The second method involves a movable stage on which the part to be measured is placed. The location of the stage is controlled by mechanical or electronic micrometer thimbles. An edge, or a through hole, of the part is aligned with center crosshairs or a center bullseye on a specially etched screen that is a permanent part of the comparator. By turning the micrometer thimbles, the features of the part, such as edges and holes, can be aligned with the crosshairs to measure dimensions, radii, hole center to hole center, etc., with great accuracy.

The optical comparator is particularly ideal for the measurement of contours and characteristics too fragile to probe mechanically. Typical applications are the measurements of threads, plastic parts, and air foil contours.

Basic System

The optical comparator system consists of the following elements:

1. Collimated light source which projects a beam of light across the part to be measured
2. Staging fixture which permits placement of the part such that it interrupts the collimated beam of light
3. Lens system designed to amplify the shadow produced by the part: a precise amount, usually 10, 20, or 50 times the original part size
4. Frosted-glass viewing screen from 10 to 50 in. in diameter on which the amplified shadow is viewed.

Optical comparators are available either as bench or free-standing systems. Bench systems are usually limited to 10- to 14-in. screens and use a vertical light beam. Free-standing systems usually incorporate larger screens 30–50 in. in diameter with a horizontal light source.

Advanced Systems

Servomotor Drives

Large optical comparators with 30- or 50-in. screens often have a screen displaced to one side. This allows the operator to stand directly in front of the screen without having the light source and part staging area between him or her and the screen. Servomotor drives on the part staging fixture allow the operator to move the image up and down and left to right as well as adjusting focus without having to leave the viewing area. Linear transducers built into the fixturing also permit remote monitoring of the part movement.

Computer Assist

Computer features similar to those found in coordinate measuring machines have also been incorporated in optical comparator systems. As in the coordinate measuring machines, they improve operator efficiency and simplify system operation.

Edge Finder

The inclusion of an optical detector at the center of the screen eliminates the need to have an operator determine when a part "edge" has arrived at the screen center. This feature, in conjunction with servomotor drives and computer assist, completely automates optical comparators. The operator need only place the part onto the staging area and begin the measuring cycle. Each part measured this way requires considerable planning and programming. It is practical only when there are a sufficient number of parts to permit amortization of this initial programming effort.

E. Automatic Video Gages

Automatic video gaging systems employ a TV or other video camera, a light source, a part staging fixture, and a computer to automatically measure small (less than 5 in.) precision parts (Fig. 5). Any characteristic that can be measured on an optical comparator can be measured on one of these devices. They are compact, bench-mounted systems incorporating operator interactive programming and are usually employed in medium-volume production-type measurement.

III. POSTPROCESS GAGING

Single-Dimension Gaging

Single-dimension gaging is often referred to as dedicated gaging. Such gages are used mostly in high-volume operations and are used to measure the same characteristic repeatedly. The dedicated nature of these gages means simplicity of setup and operation.

Inside Diameter/Outside Diameter Gaging

Indicating Snap Gages

Indicating snap gages are a quick, reliable way to measure outside diameters. Parts may be measured in or out of the machine (Fig. 6). Snap gages are used to measure parts from 0 to 14 in. in diameter to an accuracy of 0.0001 in. Typically, this measuring range is covered by models that are adjustable over a range of 1 or 2 in.

Air Gaging

Air gaging systems employ the flow or back pressure of air to determine part size. A two-jet air plug consists of a body slightly smaller in diameter than the hole to be measured, with two air jets 180° apart (Fig. 7) The flow of air from these jets is dependent on the size of the hole the plug is measuring. Larger clearance produces a greater flow or less back pressure. This principle is used to produce a variety of air gages to measure such characteristics as hole diameter, out of round, outside diameters, dimensional relationships such as taper, parallelism, squareness, bend, twist, and center distance.

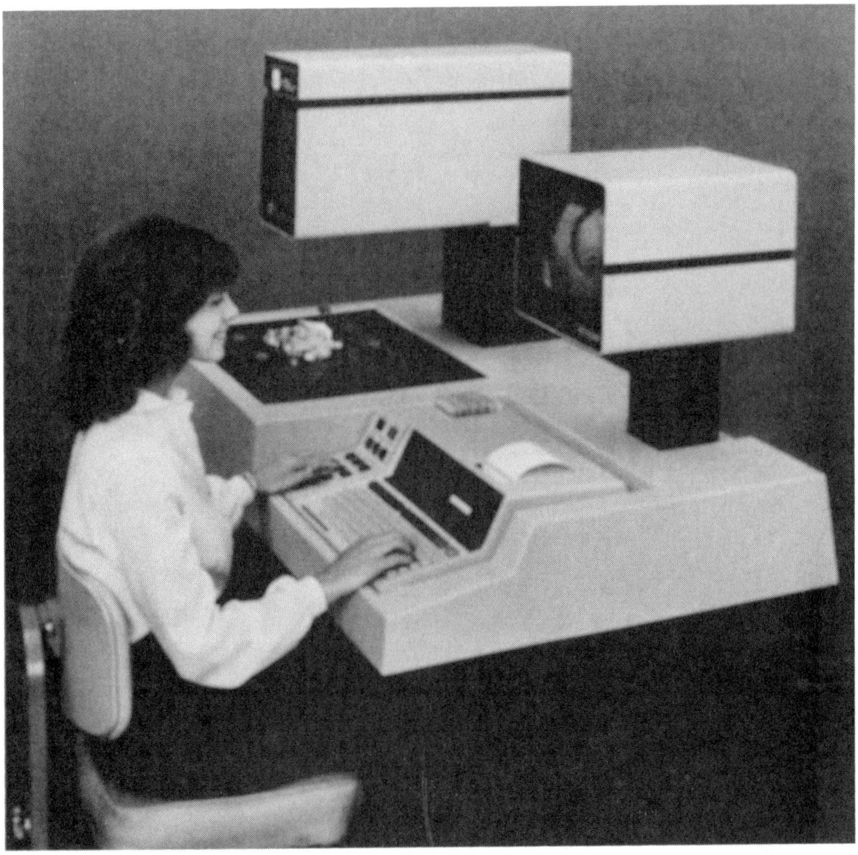

Figure 5 A new noncontact automatic gage with solid-state video imaging, a dedicated microprocessor control system, and computer analysis software. The gaging system is designed to provide accurate measurements of production-line parts, ranging from small electronic components to large cam shafts. (Courtesy of Optical Gaging Products, Inc.)

Air gages are dedicated to measuring a particular characteristic. However, they allow you to measure many parts faster, more conveniently, and accurately than any other gaging method (Fig. 8). Production workers do not require special training to use air gages. To check a hole, for instance, it is not necessary to develop skill in "rocking" the gage to find the true diameter. Merely insert the air plug in the hole and read the meter. It is as simple as that.

The noncontact characteristic of most air measuring units makes them particularly useful for checking soft, highly polished, thin-walled, or otherwise delicate material. Small gage heads and remote reading meters give air gages a distinct advantage in measuring multiple dimensions. Fixtures are smaller and remote meters permit placing contacts in positions that are inaccessible for other types of gages.

Air gages are readily adaptable to measuring parts in the machine. Their small gage heads make most dimensions accessible and the indicating meter can be located to make it clearly visible. A unique advantage is that the stream of air tends to

Figure 6 A snap gage measures an outside diameter of a part while still in the machine. (Courtesy of Federal Products Corporation.)

clean the measuring area from coolant or oil, providing accurate measuring without first cleaning the part.

Electronic Plugs

Electronic plugs are used for measuring the inside diameter of holes (Fig. 9). They operate in the same fashion as an air gage plug; however, they utilize mechanical contacts and an electronic transducer to measure size variations. Using only clean "electricity," the electronic plug is 50 times more energy efficient than air gaging, plus there is no risk of contamination and costly downtime due to "dirty" air.

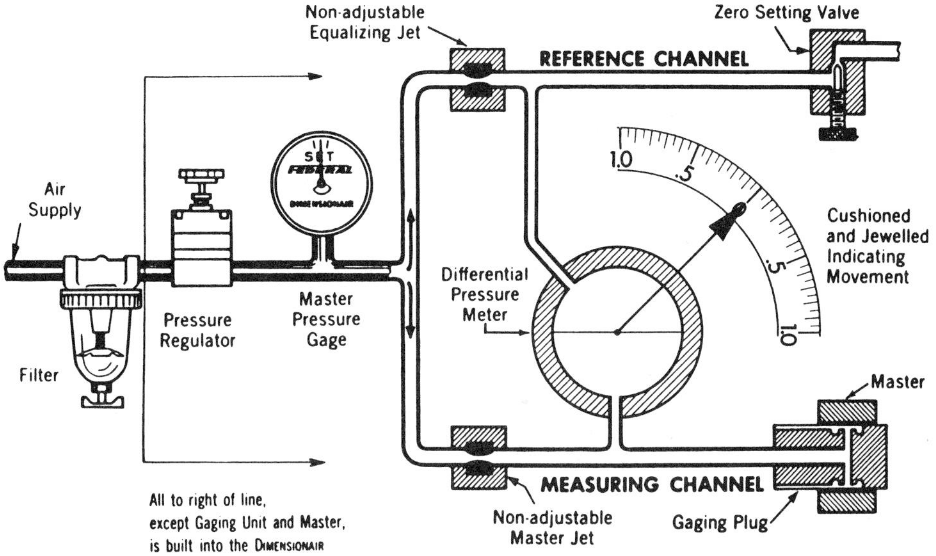

Figure 7 Basic components of a balanced air gaging system. With the balanced-type air system, the air from the supply line first passes through a regulator, then is divided into two channels. The air in one leg (the reference channel) escapes to the atmosphere through the adjustable zero restrictor, while the air in the opposite leg (the measuring channel) escapes to the atmosphere through the jets of the gage head. The two channels are bridged by an extremely precise indicating meter which responds immediately to any differential in air pressure between the two channels. This bridged system is similar to the familiar electrical Wheatstone bridge. (Courtesy of Federal Products Corporation.)

Electronic plugs are 20–40% more expensive than comparable air gaging systems. However, they do not require an air supply. Electronic plugs are not sensitive to surface finish variations. Air gaging cannot be used on surface finishes greater than 50 μin. or in parts made of porous materials.

Bench-Type Inside Diameter/Outside Diameter Comparators

These comparators are adjustable over a broad range of ID and OD sizes. Typically, 0.75–9.0 in. on IDs and 0.675–9.5 in. on ODs. Two- and three-point measurements for the detection of ovality and three-point out-of-round, respectively, are available with these devices. Measurements to 0.0001-in. accuracy are readily obtainable. The adjustable feature of these gages makes them particularly useful in applications where the dimensions checked change frequently.

Micrometers, Calipers, and Go/No-Go Gaging

Applications where only one or two parts are being made at a time do not lend themselves to the types of gages previously discussed. The direct-reading nature of micrometers and calipers are more suited for applications of this type.

A go/no-go ID gage consists of a minimum and a maximum size plug, representing the minimum and maximum tolerances of the ID. The operator should be able to introduce the minimum plug but not introduce the maximum plug. These gages have the advantage of being relatively inexpensive. However, they will determine only if the part is good or bad and will not indicate trends.

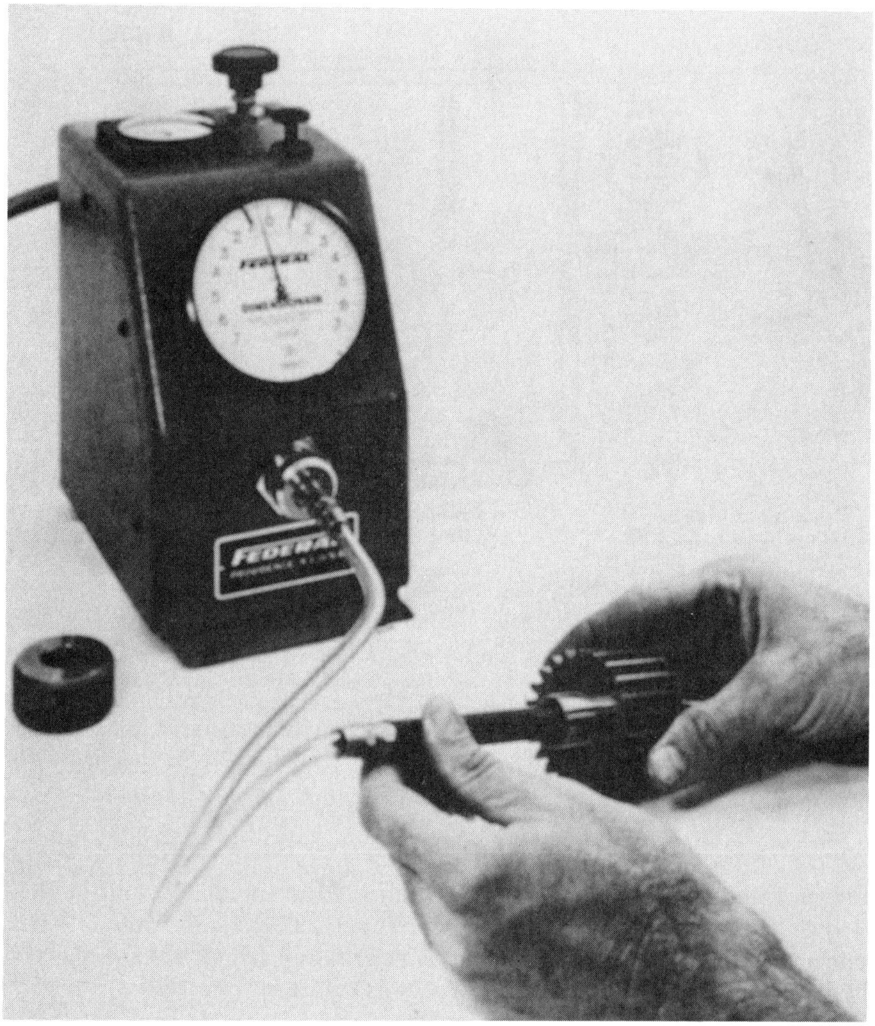

Figure 8 Ease of operation makes air gaging a fast accurate method for checking a particular characteristic. In this case (Dimensionair), an air plug is used to check the ID of a small gear. The master ring is shown next to the meter. (Courtesy of Federal Products Corporation.)

Depth Gages

Depth gages consist of a dial indicator, base, and a measuring probe which protrudes through the base. These instruments give accurate information (0.0001-in. resolution) at a glance, and are convenient for the inspection of slots, recesses, hole depths, keyways, and so on. They are available either as hand-held devices or bench-type gages. Interchangeable points of different lengths and mounts increase the gage versatility. Gages with measuring ranges of up to 3 in. are commonly available.

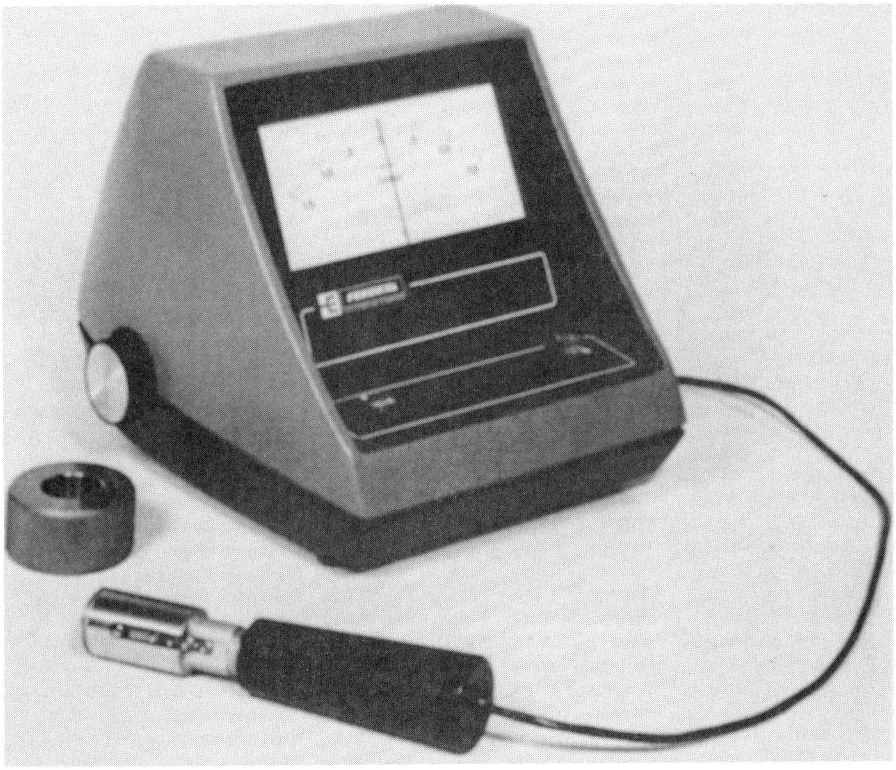

Figure 9 Fast, reliable bore measurement is possible with electronic plugs. (Courtesy of Federal Products Corporation.)

Fixture Gages

Fixture gages are specially designed gages for measuring one or more characteristics on a part. They are usually used in applications where general-purpose gaging will not do the job. The fixture consists of mechanical components that support and locate the part to be measured. These mechanical components orient the part such that indicating devices can measure various part characteristics.

Fixtures simplify measurements that would normally require specialized skills or be extremely difficult to perform. The operator simply places the part into the fixture and dial indicators or other types of electronic gaging display the condition of the part. Properly designed fixture gaging is an efficient, rugged gaging technique for high-production part inspection. They are expensive to design and build, and their use should be weighed against the cost of alternative, more universal measuring systems.

IV. AUTOMATIC GAGES

An automatic gage measures one or more characteristics on a part and makes size decisions automatically (Fig. 10). The part may be fed by hand, by means of a cable

Figure 10 Automatic gage for measuring large projectiles utilizes multisignal processing to classify parts as well as provide signals to machine operators when tolerances begin to approach reject area. In addition, the gage interfaces with a main computer to provide a continuously updated record of every measurement made and which machine produced the part measured. At the end of each gaging cycle the processor automatically verifies that the entire gage is up and running. (Courtesy of Federal Products Corporation.)

hoist and a conveyor, or out of a hopper. The gage may check 1, 2, or 50 dimensions. It may exercise corrective control or it may not. It may position, check, classify, segregate, record, memorize, store, transfer, mark, or dispose. It can do any or all of these things automatically, depending on need, but the gage is classified as an automatic because precise size determination—in millionths of an inch, if necessary—is made and communicated automatically. Traditionally, a gage is called semiautomatic if there is some manual involvement, but the gage still does the deciding for the operator.

An automatic gage can stand by itself as a final inspection device, and often there is no machine control involved. With increasing frequency, automatic gages are becoming part of the automated transfer line screening out bad parts and applying corrective action when part size shifts toward or beyond tolerance. Automatic gages perform rather complicated measurements, usually on high-precision parts that are impractical or impossible to obtain any other way.

A typical automatic gage consists of the following basic elements:

A part feed mechanism, which moves the part through the gaging system and disposes of the part at the end of the measuring cycle.

A programmable controller, which controls the processing of the part through the gaging system and its proper disposition.

The gaging station, or stations, where the part is oriented and measured via electronic or air transducers.

The instrumentation system that conditions the signals from the transducer into a meaningful dimension.

The incorporation of microprocessors in instrumentation systems has increased their capability. The more sophisticated systems contain all the electronic signal-processing capability needed for virtually any complex measurement application (Fig. 11). The only difference between applications is the instructions keyed into the system by the user through a built-in keyboard. Besides signal-processing capability, these systems are capable of automatic classification, tolerance limit setting, automatic zeroing to a master, and self-diagnostics.

Figure 11 The Multi-Signal Processor (MSP) system is a total, universal hardware package that can accommodate any number of gage applications, even if each is entirely different. The only difference between applications is the instruction set keyed into the MSP by the user through its built-in, user-interactive, instructive "prompting" routine. The MSP can use input signals in any combination that can be expressed as a mathematic equation, simple or complex, using "plus," "minus," "multiply," "divide," "square," "square root," and "absolute value." (Courtesy of Federal Products Corporation.)

V. METROLOGY LABORATORY

A. Calibration Functions

Calibration of high-accuracy standards such as master rings, gage blocks, and so on, is often performed by outside, independent laboratories. Only large companies have sufficient amounts of calibration to justify the purchase and maintenance of high-accuracy calibration equipment. The accuracy potential of laboratory equipment can be realized only in a controlled environment. Temperature, vibration, humidity, and cleanliness must be strictly controlled (Fig. 12).

Maximum accuracy potential can be obtained only by having: suitably calibrated masters; masters to be measured in good condition; high-quality, well-designed instrumentation in good working order; well-trained operators; and environment controls consistent with desired accuracy levels.

Master Ring and Disk Calibration

Master ring and disk calibration is performed on a horizontal master comparator (Fig. 13). It utilizes stacked gage blocks as a master setting device and compares

Figure 12 A precision measurement center is a precisely controlled environment for high-accuracy calibration of gage blocks, master rings, disks, and other precise measurement applications.

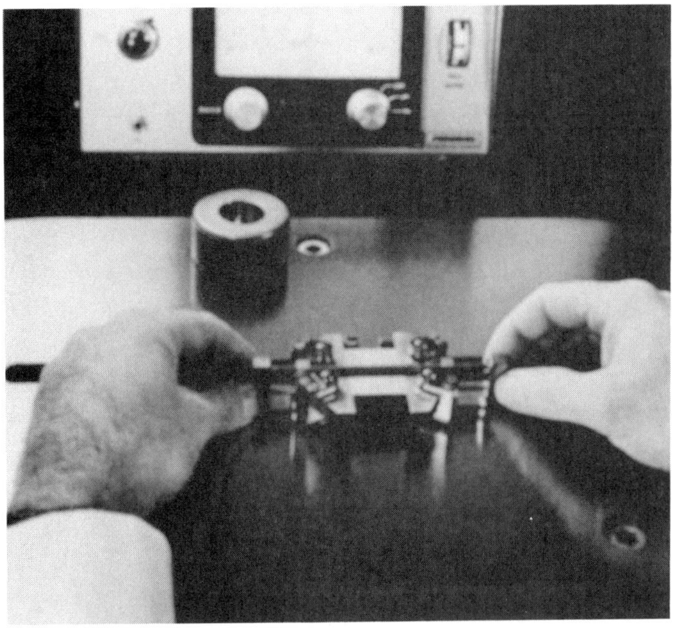

Figure 13 Gage blocks mounted in a nest are used to master a ring and disk comparator. (Courtesy of Federal Products Corporation.)

these master readings with that of the rings or disks to be calibrated. These systems have a measuring range of 0.040 to 13.250 in. and a resolution of 0.000001 in.

Gage Block Calibration

Comparison

The most popular method of calibrating gage blocks is to compare the test block against a master block of known error (Fig. 14). This method is most popular because it is the fastest method and produces adequate accuracy for most applications. A typical gage block comparator is adjustable to measure gage blocks from 0 to 4 in. The gage block comparator allows measurement of gage block error to 0.000001-in. resolution.

Interferometry

Grand master gage blocks are often calibrated by interferometry. This technique is recognized as being approximately twice as accurate as comparison checking of gage blocks; however, it is 10 times more time consuming.

Interferometric systems use light as their basic standard. Therefore, they do not require masters. Laboratories that use interferometric calibrators will occasionally calibrate an audit package of gage blocks. This package is then checked by the National Institute of Standards and Technology (NIST) to assure that the laboratory's measuring system is operating at a satisfactory level of accuracy.

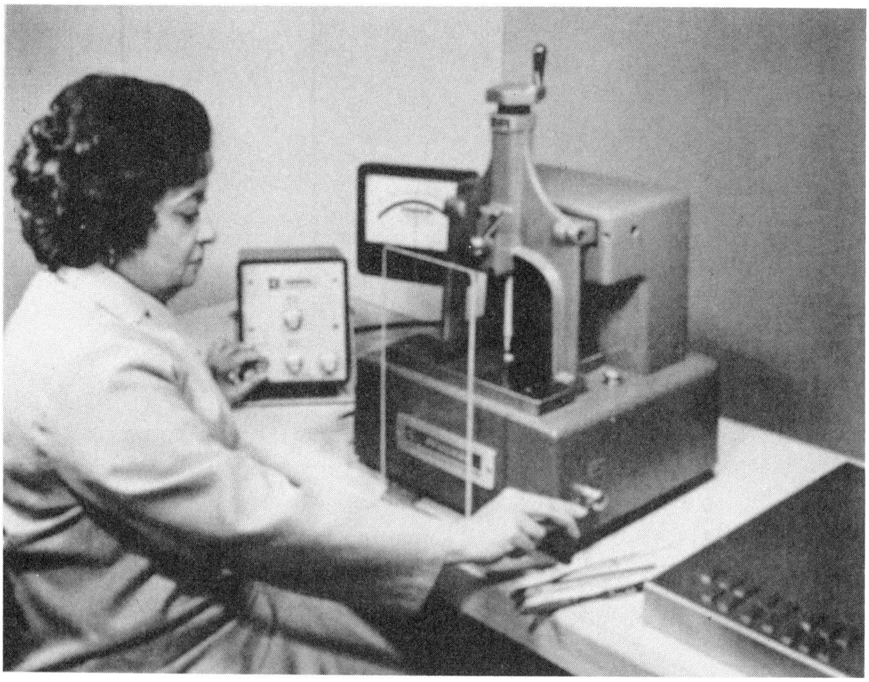

Figure 14 Gage block calibration takes place under tightest possible environmental controls. Above, a breath shield protects the measuring area from the warmth and humidity of the operator's breath. In the foreground gage blocks rest on a "soak plate" to achieve ambient temperature of the measuring environment. Utilizing a floating caliper design and true point-to-point measurement, this gage block comparator is guaranteed to calibrate within 0.000001 in. over its full range. (Courtesy of Federal Products Corporation.)

Threads and Plugs

Threads and plugs are usually checked in a comparator stand or a "supermicrometer" (Fig. 15). Threads are usually checked in a supermicrometer using the three-wire method. The three-wire method utilizes thread wires which are a specific size, depending on the pitch of the thread. Measuring "over the wires" at a specific load in the supermicrometer will produce a dimension proportional to the pitch diameter of the thread. The thread profile is usually checked in an optical comparator to ensure that the thread has not experienced excessive wear.

Plugs are also checked in a supermicrometer using hardened and flat cylindrical anvils at specific loads. The load and the diameter of the cylindrical anvil varies depending on the diameter of the plug. Threads and plugs are checked to a resolution of 0.000010 in.

Dial Indicators and Electronic Transducers

Dial indicators and transducers can be checked by mounting them in a comparator stand and displacing the contact with gage blocks. Special calibration devices are available in applications where the volume will justify their initial cost. The special calibration device is significantly more efficient than the gage block method (Fig. 16).

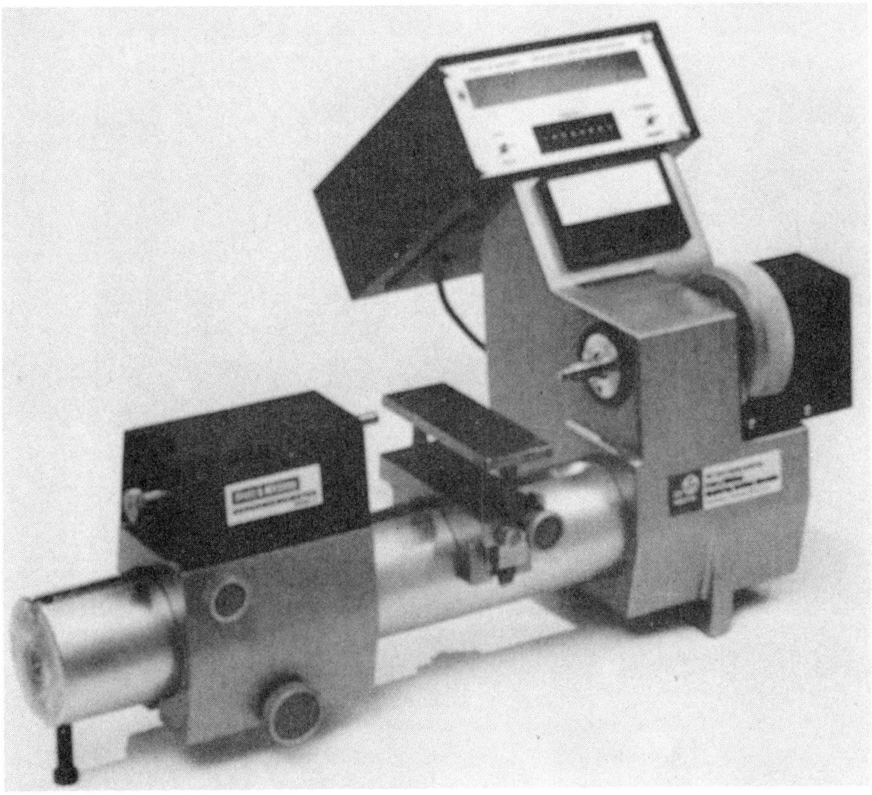

Figure 15 A supermicrometer can be used to check threads and plugs. (Courtesy of Pratt & Whitney Machine Tool Division.)

Surface Plate

Surface plates can be calibrated by auto collimation, laser interferometer, or electronic level. The auto collimation and laser interferometer involve the use of an accurate light beam reflected off a mirrored surface. The mirror is moved along the surface plate and variations in the topography of the surface plate cause this mirror angle to change and this angle is measured by the auto collimator or interferometer.

Auto collimation, the original method developed for the calibration of surface plates, has the liability of requiring two operators to perform the calibration. The laser interferometer system is often coupled to a programmable calculator so that the readings from the interferometer can be mathematically manipulated and then converted into a meaningful "map" of the surface plate.

The electronic level system consists of two electronic levels differentially coupled (Fig. 17). One level remains at a reference point on the surface plate and the other level is moved along a straight edge. The electronic level system offers simplicity of operation and relatively low cost compared to a laser interferometer. Levels can be interfaced to a programmable calculator to produce these same type of information as a laser interferometer system.

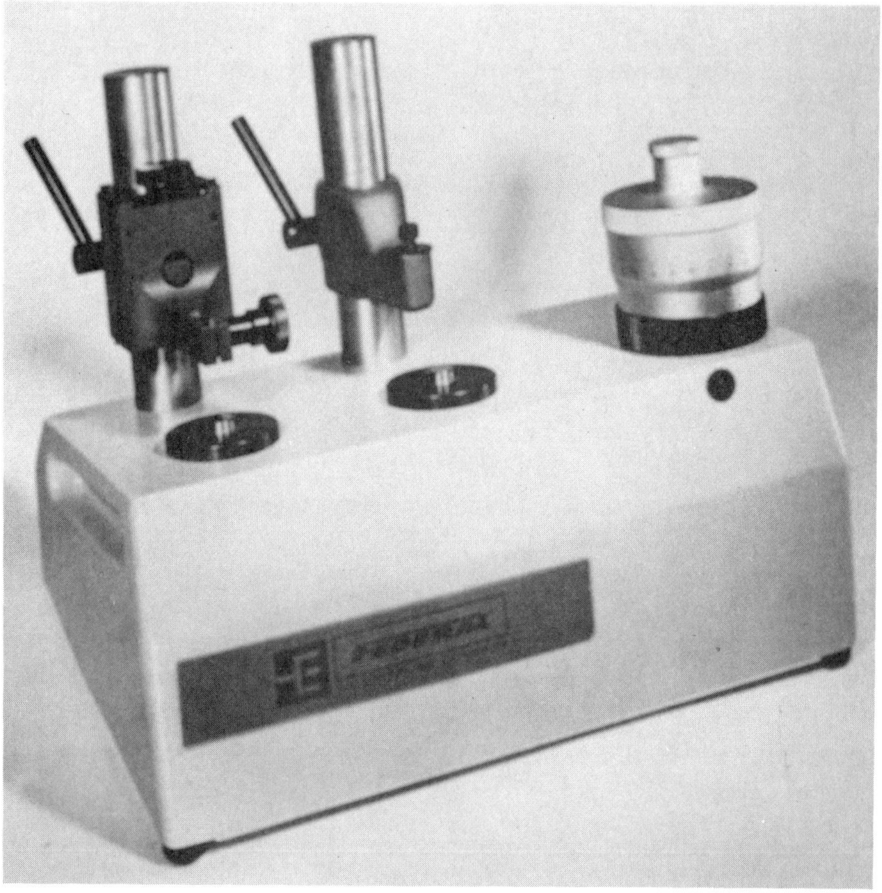

Figure 16 This universal calibrator allows precise monitoring of any type of gaging system: dial indicators, air and electronic probes. A unique micrometer and lever mechanism produces precise motion at either of two calibrating stations. One station for high-magnification gages, such as electronic systems, provides minimum graduations in increments of 0.000010 in. with a range of 0.100 in. For dial indicators or probes having ranges up to 0.500 in., the second station offers high-definition calibration with graduation value of 0.00050 in. Repeat is one-fifth graduation. (Courtesy of Federal Products Corporation.)

B. Dimensional Analysis

Calibration laboratories are often asked to perform other types of dimensional analysis to assist in the diagnosing of manufacturing process problems.

Surface Finish and Profile

A profiling system analyzes, computes, displays, and records linear profile and surface finish characteristics. A typical system consists of a probe, a precision linear drive, and a control center (Fig. 18). The control center usually includes a single- or dual-channel recorder, probe controls, digital displays, and space for special function modules.

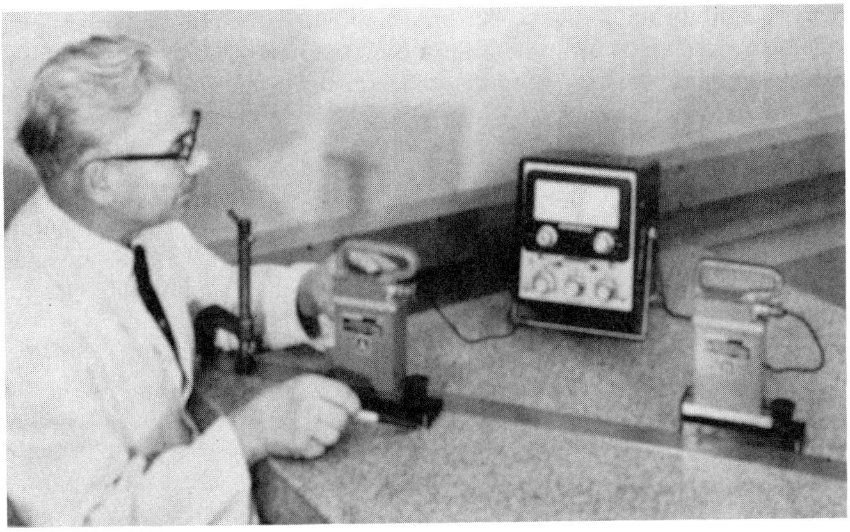

Figure 17 Electronic levels in surface plate calibration setup. (Courtesy of Federal Products Corporation.)

As the probe on the precision drive traverses the workpiece or test sample, the signal generated is conditioned, amplified, displayed, and recorded. These systems are capable of sensing displacements as small as 0.0000001 in. With this exceptional accuracy, surface profile measurement systems are suitable for a variety of laboratory and quality control tasks. They are especially useful for measuring

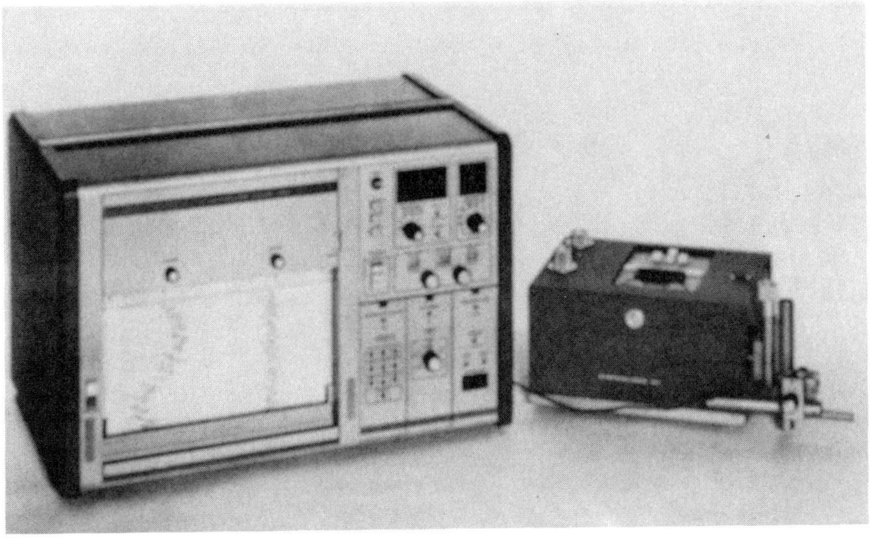

Figure 18 This surfanalyzer can be tailored to specific requirements by supplementing basic profile and roughness measurements with additional parameters such as roughness average, waviness, percent bearing area, and others. (Courtesy of Federal Products Corporation.)

deposits on thick- and thin-film microelectronic components. General applications include surface measurements of inside and outside diameters, gear teeth profiles, grooves and flats. They immediately identify such surface irregularities as bellmouth, runout, taper, and waviness.

Circular Geometry

Circular geometry gages are used to check roundness, concentricity, flatness, parallelism, and squareness of precision machined parts. Circular geometry systems consist of a precision spindle, one or more transducers, a signal conditioning system, and a data display and recording system (Fig. 19).

Circular geometry gages are capable of measuring to an accuracy level of a few microinches. Systems are available to measure parts as small as 0.04 in. inside diameter to as large as 60 in. in diameter and up to 5000 lb. Circular geometry systems make practical the early detection of geometrical errors. They also provide the prime means for fast, effective troubleshooting to correct the problems that produced them. As each corrective measure is taken, the geometry system can quickly show the progress being made, drastically reducing the time required to arrive at an effective solution. The savings in downtime, the reduction of scrap, and functional improvement in critical parts that result from the ability of geometric

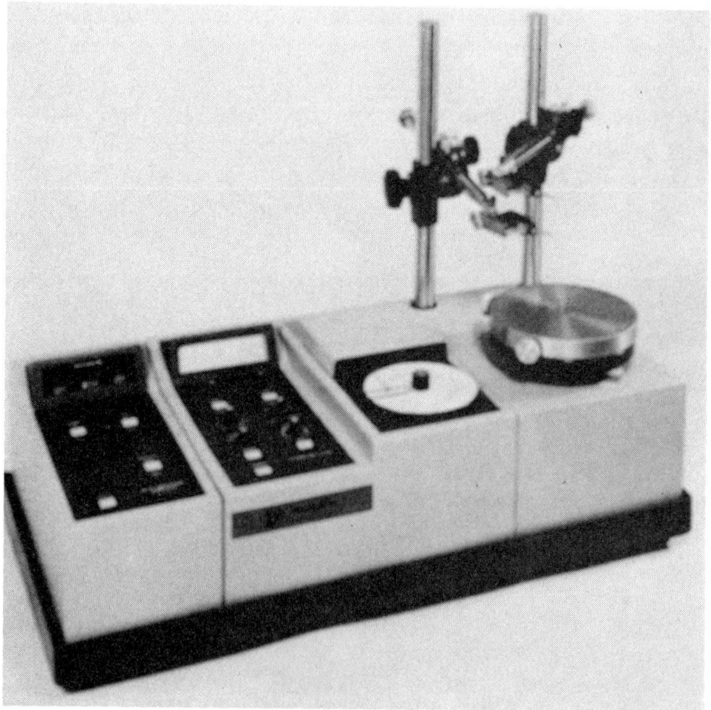

Figure 19 The geometry measurement system incorporates a rotary table supported by an air bearing spindle; an automatic centering computer that eliminates time-consuming mechanical centering of the workpiece; a digital readout of TIR from each gagehead as well as eccentricity and a polar chart recorder. (Courtesy of Federal Products Corporation.)

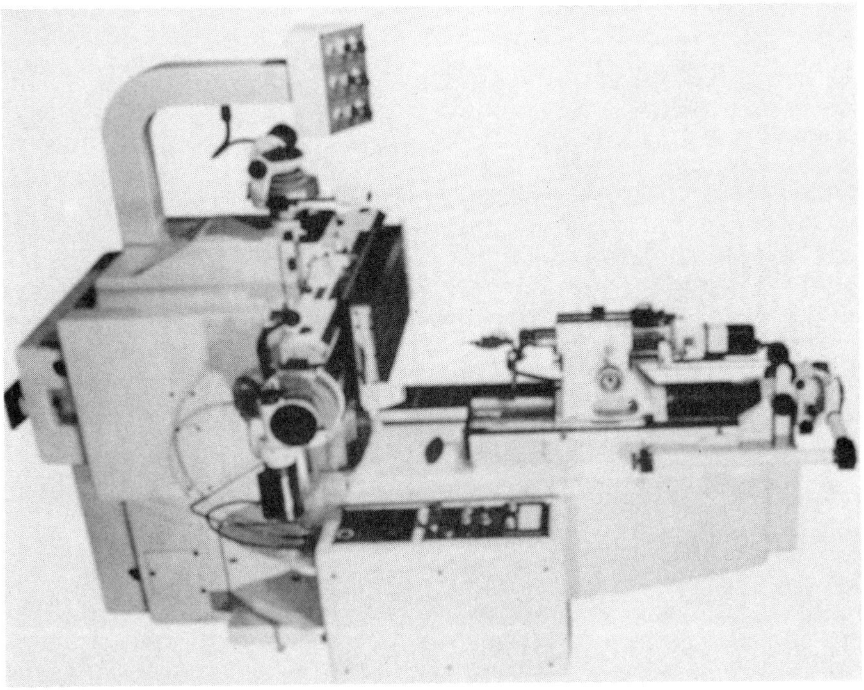

Figure 20 A universal measuring machine is a totally self-contained metrology center. Linear and geometric accuracies of the highest order are combined with a spindle that has a guaranteed trueness of rotation to 5 millionths of an inch (0.55 μm) TIR. With the use of the small-angle divider, angular determinations are made to an accuracy of ± 0.5 arc second. (Courtesy of Moore Special Tool Co., Inc.)

analysis bring the unseen into sharp, unmistakable focus and can easily amortize the cost of the system in a matter of days.

Measuring Machines

Measuring machines of the type used in laboratories are usually high-accuracy versions of coordinate measuring machines (Fig. 20). Their accuracies are typically 10 times greater than that expected from a coordinate measuring machine. They are useful for the analysis of complex profiles and characteristics which would not normally lend themselves to single-purpose gaging.

FURTHER READING

Busch, T., *Fundamentals of Dimensional Metrology*, Delmar, Alabany, NY.

Farago, F. T., *Handbook of Dimensional Measurement*, Industrial Press, New York.

Kennedy and Andrews, *Inspection and Gaging*, Industrial Press, New York.

Moore, W. R., *Foundations of Mechanical Accuracy*, Moore Special Tool Company, Bridgeport, CT.

The Society of Manufacturing Engineers, *Handbook of Industrial Metrology*, Prentice-Hall, Englewood Cliffs, NJ.

37
Microstructural Analysis

JAMES A. NELSON
Buehler Ltd.
Lake Bluff, Illinois

I. INTRODUCTION

The rapid development of Western industrial technology has occurred in relatively recent times. To a large degree, this is the result of a growing understanding of materials and their application to solve engineering problems. Early human progress toward this end was very slow because we had first to discover new materials and then learn by empirical means how to utilize them. Without an established material science, developments had to be based on trial and error and were therefore very slow.

Metals have long captivated human curiosity because they possess unique properties that could be altered thermally and mechanically to meet various needs. The first accidental discovery of metals is believed to have occurred about 8000 BC in Eurasia. It could have been a meteorite laying on the ground as depicted in Figure 1 or a bright nugget of gold found in a streambed. Metals were probably used first for jewelry, later as tools, utensils, and weapons.

Physical properties were the first method of characterizing metals. The enduring bright and lustrous appearance of gold was early appreciated by human beings, who discovered that it was harder than wood, but softer and less brittle than stone. When struck, it yielded rather than be fractured; when hammered repeatedly, it could be formed into thin sheets (leaf). By this tedious process, we learned to understand the physical properties of metals, natural alloys, and later our own contrived alloys. Even with this limited understanding of metals, based on physical properties, ancient civilization produced significant achievements, such as the lost wax casting process and the Damascus sword (1).

As history progressed, we learned that certain metals and alloys resisted weathering better than others, and that vegetable juices (acids) either darkened or brightened the polished surface of a particular metal or alloy. These natural acids

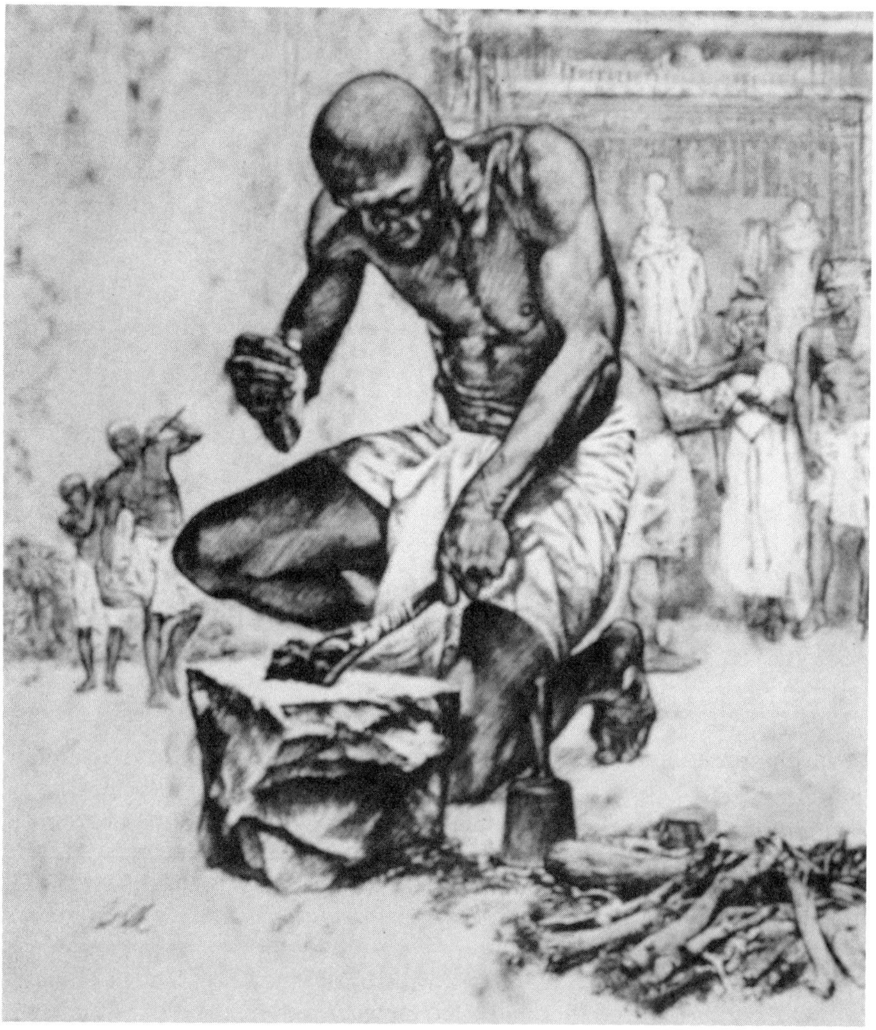

Figure 1 Metal from heaven: discovery of meteoric iron by the ancient Egyptians. (Courtesy of Basic Incorporated, from an original drawing by Paul Calle.)

were first used to decorate weapons and were an early application of primitive chemistry known as alchemy. Although alchemy was somewhat obsessed with the idea of producing gold from "base metals," it was nevertheless the forerunner of modern chemical analysis. As chemistry evolved, a more definite method of identifying materials was available that provided a more precise means of materials characterization.

Alchemists and philosophers had long speculated on the possibility of an internal structure in metals. This idea must have been inspired by orderly details they observed in wood and other materials. It was further encouraged by the observation of faint patterns on metallic surfaces, by the existance of treelike shapes (dendrites) in casting sinks, and by the faceted appearance of a coarse-grained metal fracture. Attempts to correlate the appearance of fractured surfaces to physical properties were common from AD 1500 to 1800. Founders of bronze bells, in the seventeenth century, believed a large grain fracture meant that more tin should be added; the color of a fracture was once used by iron makers to determine the metal's quality. These low-magnification observations were useful but not adequate to provide the information needed to understand the complex changes that occur in metallic alloys in various conditions of thermal and mechanical treatment. With the development of the scanning electron microscope (SEM), the study of fractures has again assumed significant proportions. The study of failure modes has been greatly aided by the scientific study of fractures.

The search for a discrete microstructure was pursued in both Great Britain and Europe. Most efforts to produce a satisfactory surface failed because the investigator did not adequately understand the characteristics of polished surfaces, or the abrasive processes that were used to achieve them. As a result, they succeeded, as C. S. Smith points out, in producing a burnished surface which did more to hide the microstructure than to reveal it.

It was not until Henry Sorby of Sheffield, England, applied careful preparation based on sound, proven principles that adequate surfaces were produced. Using simple but effective abrasive principles, which he had previously developed for rock analysis, Sorby achieved a polished surface which, when etched with a suitable acid, revealed clear microstructural detail. His success provided the basis for modern metallographic polishing and earned him the honorary title "Father of Metallography." He also adapted the transmitted light microscope to reflected light operation to produce the necessary bright-field illumination. Sorby also left various notes describing the microstructure he had observed. The names of some microstructures he observed [such as iron (ferrite), pearlite, and iron carbide] remain to this day.

II. WHAT IS MICROSTRUCTURAL ANALYSIS?

Microstructural analysis is the technique by which the internal microstructure of metal or alloy is made visible for analysis. The surface of a metallic part, such as the welded pieces shown in Figure 2a, cannot be directly examined to reveal the microstructure due to normal surface deformation and oxidation. To observe the internal microstructure, a part must be sectioned (cut) along a plane perpendicular to the area to be observed (Fig. 2a, dotted line). When the exposed and ground cut surfaces are etched (macroetched) with a solution consisting of equal parts of water and hydrochloric acid, coarse details such as the outline of the weld/parent-metal interface may be seen (Fig. 2b).

If the pieces are fully prepared, according to the sequence to be described later, fine microstructural details are revealed, as shown in Figure 2c. In this photomicrograph at 50× magnification, the principal microstructural features may be seen. They are the parent metal, heat-affected zone (HAZ), and the weld metal. Also visible is a void in the fusion line which, if large enough, could be considered a significant defect.

Microstructure consists of the details that are observed when we look at a polished and etched surface of a material. The details we observe are not the atomic structure, which is considerably finer detail than may be observed by light

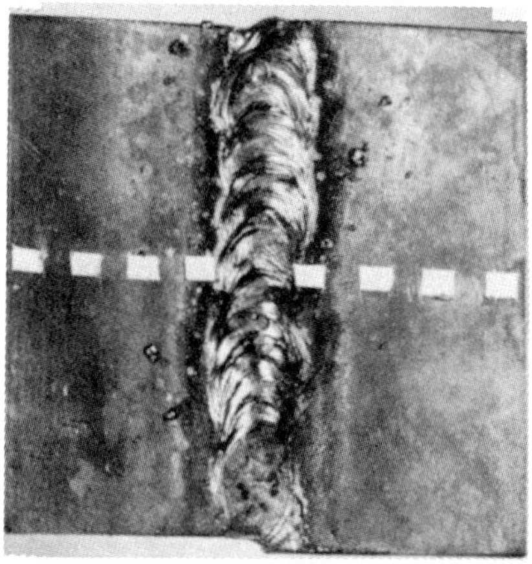

(a)

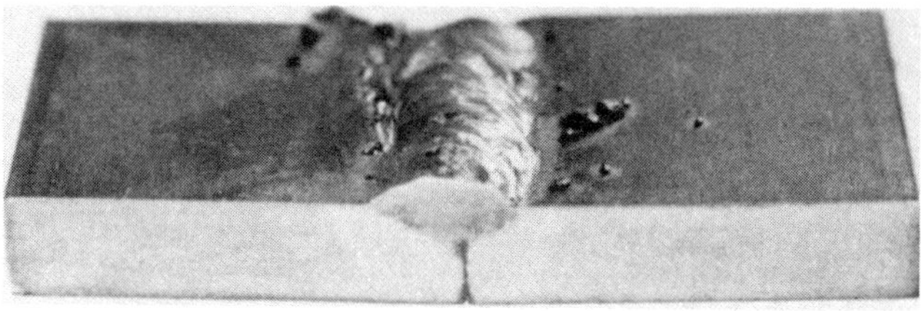

(b)

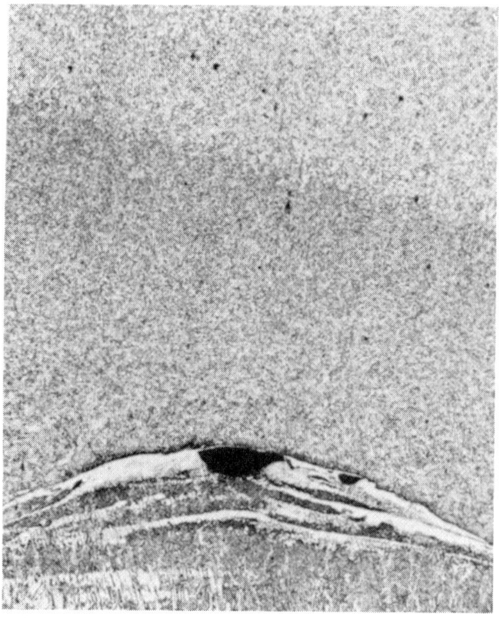

(c)

Figure 2 Metallographic sectioning of a weld: (a) surface appearance before sectioning (dashed line denotes plane of interest); (b) cross section showing plane of interest after grinding and macroetching; (c) polished and microetched view of weldment at 50× magnification.

microscopy. What we actually observe are the boundaries of crystal grains, various major alloy phases, precipitated intermetallic compounds, nonmetallic inclusions, and various inhomogeneities such as cracks and porosity.

The study of microstructure is significant because it enables us to observe the condition of a material, or to compare two or more materials that display different properties. Microstructure is the result of the chemistry plus the thermal and mechanical treatments to which the material has been subjected.

Figure 3 shows the dendritic microstructure of an as-cast alloy, which is characteristic of this condition. This microstructure is the result of an unequal distribution of chemical elements resulting from a particular rate of cooling. Figure 4 shows a nickel-based alloy which has been modified by thermal treatment to produce a

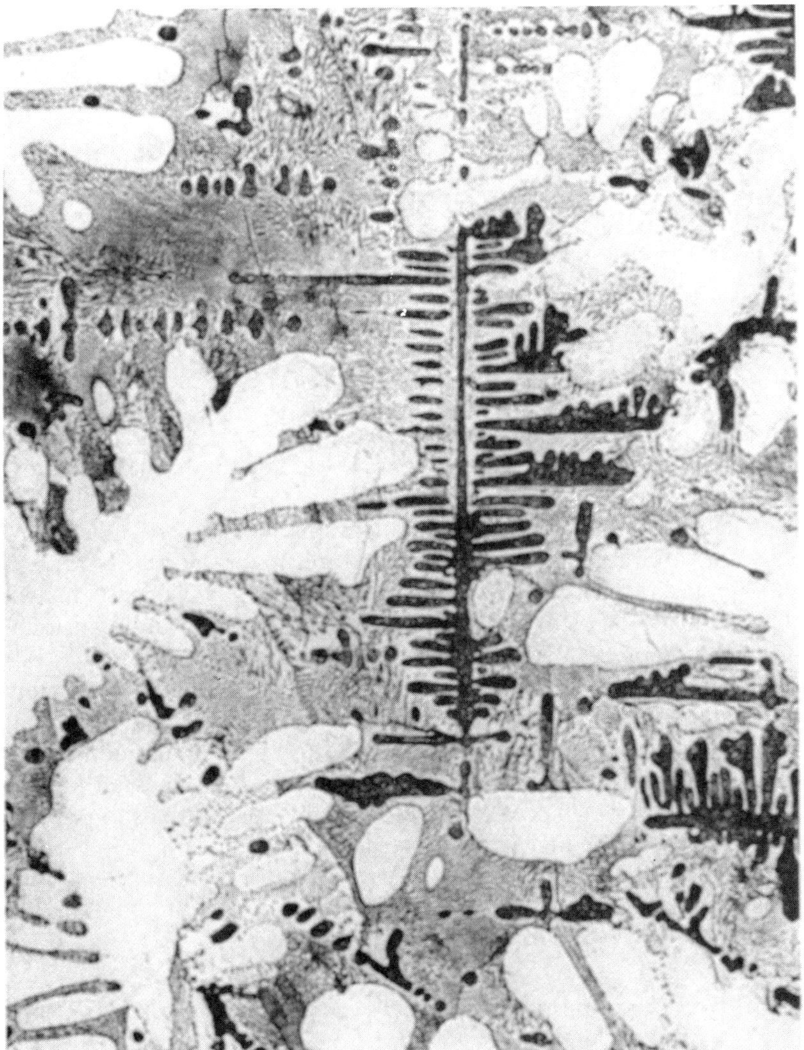

Figure 3 Dendrites in an as-cast alloy.

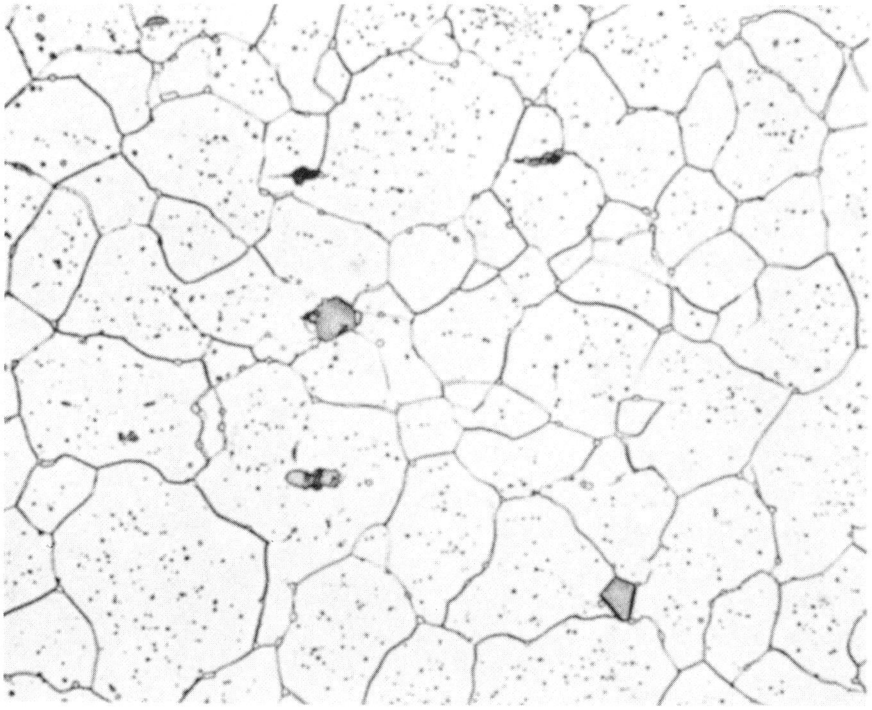

Figure 4 Crystalline microstructure of a nickel-based alloy. Magnification, 500×; etchant, 10% oxalic acid (electrolytic).

homogeneous crystalline microstructure. Metallography is able to reveal such differences.

It must be remembered that a polished metallographic sample provides a two-dimensional view of a three-dimensional microstructure. Figure 5a shows a SEM view of flake graphite in gray iron. To show this three-dimensional view, the matrix metal has been etched away, exposing the graphite flake. A normal two-dimensional light microscope examination of a similar alloy would appear like Figure 5b. Although this seems to be a severe limitation, the information obtained is extremely valuable because it still provides an accurate prediction of the three-dimensional condition. Microstructure alone cannot totally characterize a material, but must be interpreted along with chemical analysis and physical properties, plus all known thermal and mechanical history. The role of metallography, therefore, is to provide a tool that allows us to observe changes in microstructure which result from various processes. Using a reflected light microscope, the investigator may observe obvious defects and microstructural constituents, surface phenomenon, and nonmetallic inclusions.

III. THE PRINCIPLES OF METALLOGRAPHIC PREPARATION

We have already observed that clear, usable microstructures were not revealed until adequate preparation techniques were devised. In his notes, Sorby commented: "Anything approaching to a burnished surface is fatal to good results."

A burnished surface may be defined as: "To make shiney by rubbing." This technique is employed by silversmiths to hide surface blemishes and produce a bright lustrous surface. Although a good metallographic polish (one capable of revealing the true microstructure) will nearly always be smooth and lustrous, this superficial appearance is no guarantee of success. There is more to a well-prepared metallographic sample than meets cursory observation. A burnished surface may be bright and lustrous, but it also consists of a layer of deformed metal which is incapable of revealing the true microstructure. In fact, we may observe what appears to be the true microstructure but which, in reality, could be a pseudomicrostructure, sometimes referred to as an artifact. The microstructure that is observed is not characteristic of the sample material and therefore cannot be used as a basis for analysis.

Conversely, metallographic preparation by careful abrasive polishing produces a surface that is free from harmful deformation. The true microstructure may be observed and useful information is easily obtained. Various microstructural features may be distorted or obscured when incorrect preparation techniques are applied. For this reason, any preparation step that produces extensive mechanical deformation or fails to remove previously induced deformation is not acceptable. Similarly, any preparation technique that produces excessive surface heat is unacceptable.

Figure 6 shows cross-sectionally the various components of a deformation groove produced by a coarse abrasive grain or other source of deformation. The dark upper ray, or spike, represents the area of severest deformation consisting of crystalline grains that have been distorted, or fractured, beyond recognition. This layer is permanently distorted and must be completely removed, together with the area of less severe plastic deformation directly beneath. Further down, elastically deformed layers have been affected but have recovered their original microstructure. While elastic deformation may seem to have no consequence, it is possible that subtle microstructural changes have been produced which are not characteristic of the true microstructure. In certain metals and alloys, elastic deformation may be sufficient to produce deformation bands or twins. Additional background information on the nature of abrasive deformation has been published by L. E. Samuels (2) and G. Petzow (3).

To achieve a good metallographic polish by abrasive preparation, a systematic series of steps is required:

Sectioning
Planar grinding
Mounting
Fine grinding
Rough and final polishing

Each step must replace all abrasion damage produced by the previous steps with a lesser degree of abrasion until a satisfactory final polish is produced. In addition, each step must be performed with adequate lubrication and cooling to reduce the heat of friction. With rare exceptions, abrasive preparation for metallography is never performed without a coolant. In summary, sound metallographic preparation

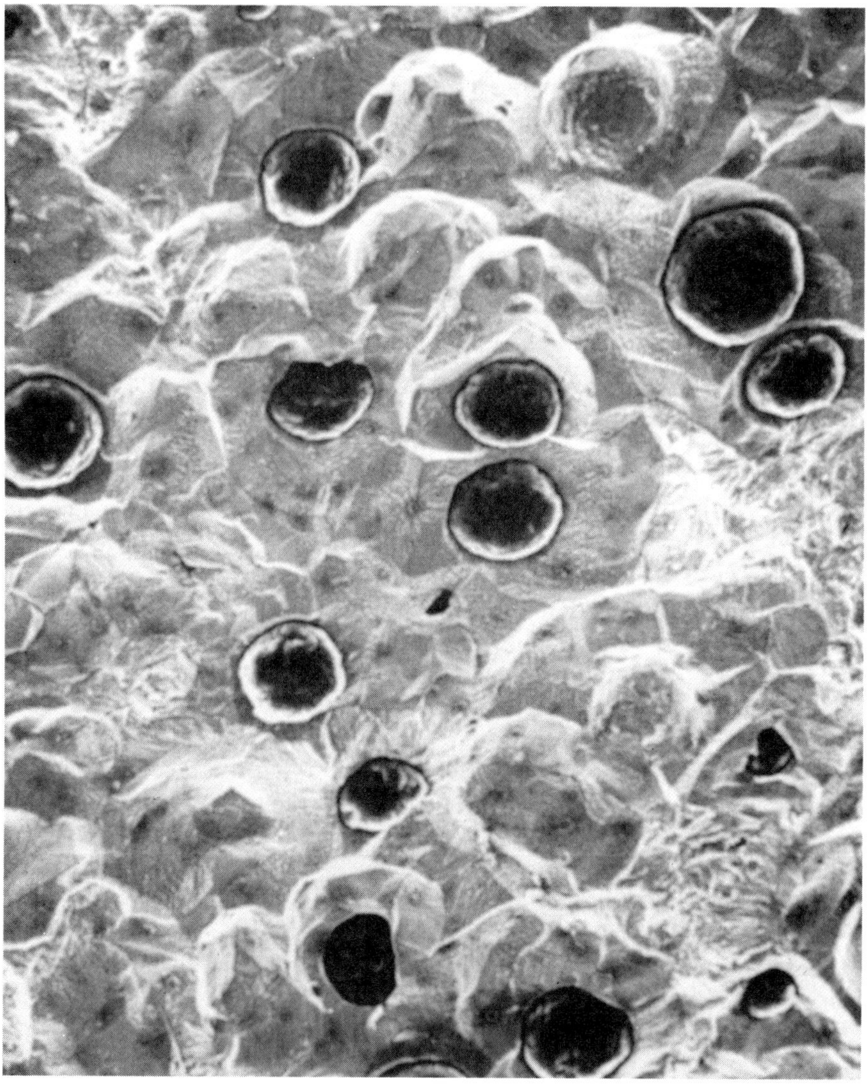

(a)

Figure 5 Comparison of ductile iron microstructures as revealed by (a) scanning electron microscopy and (b) light microscopy. (Courtesy of Foote Mineral Company.)

must produce a surface that is free of induced deformation, the surface must be flat, and the finish polish should be free of all scratches that could affect the analysis.

IV. THE PRACTICE OF ABRASIVE PREPARATION

A. Sectioning

To obtain a manageable-size sample, a portion is cut from the bulk material. Sectioning may be accomplished by various methods, including torch, band saw,

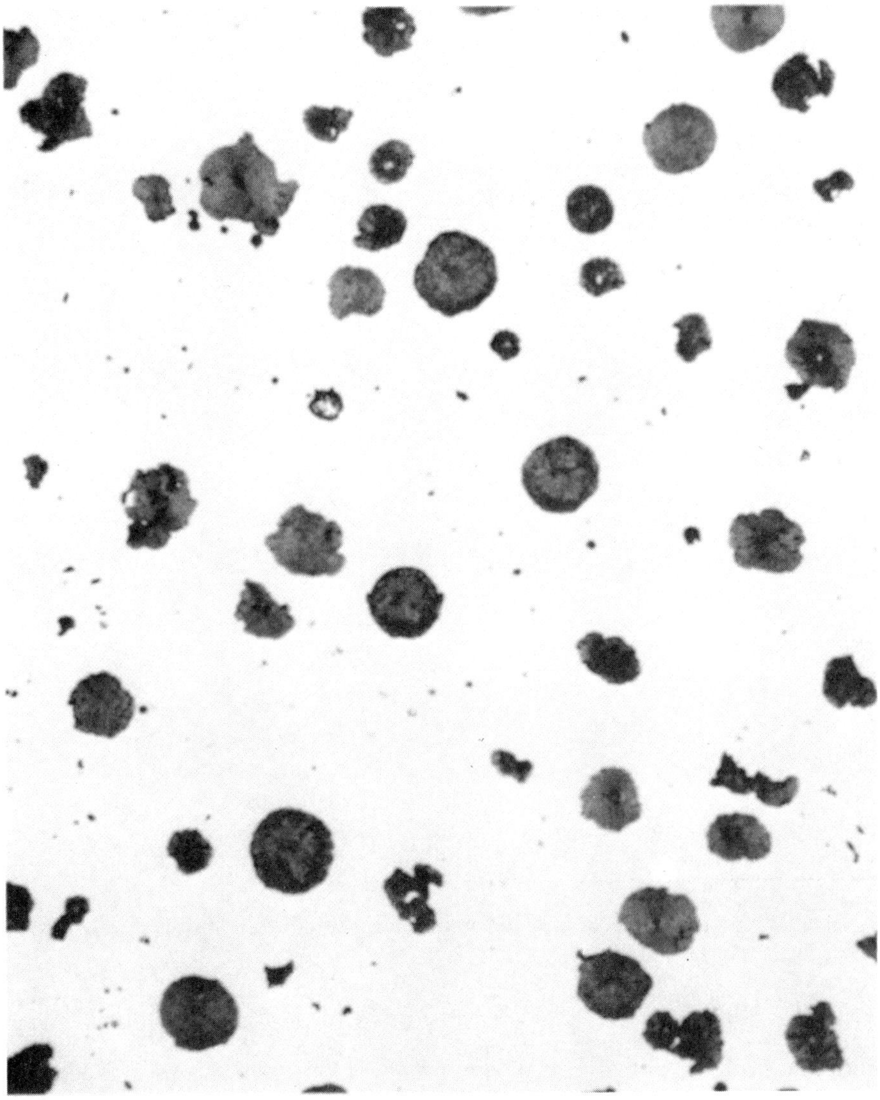

(b)

Figure 5 Continued

machining, or any other means of removing a portion of the material. Small parts may require only one cut, and very small parts are often mounted without any cutting. In this case, the desired plane of observation is reached by careful grinding. Regardless of the method used to obtain a sample for metallography, any gross mechanical deformation or heat effects must be kept clear of the area of concern, so that the microstructure will not be altered.

Abrasive Wheel Sectioning

Abrasive wheel sectioning, using a cutter similar to that shown in Figure 7, is the most rapid, economical, and gentle method of sectioning rigid materials. Metallographic

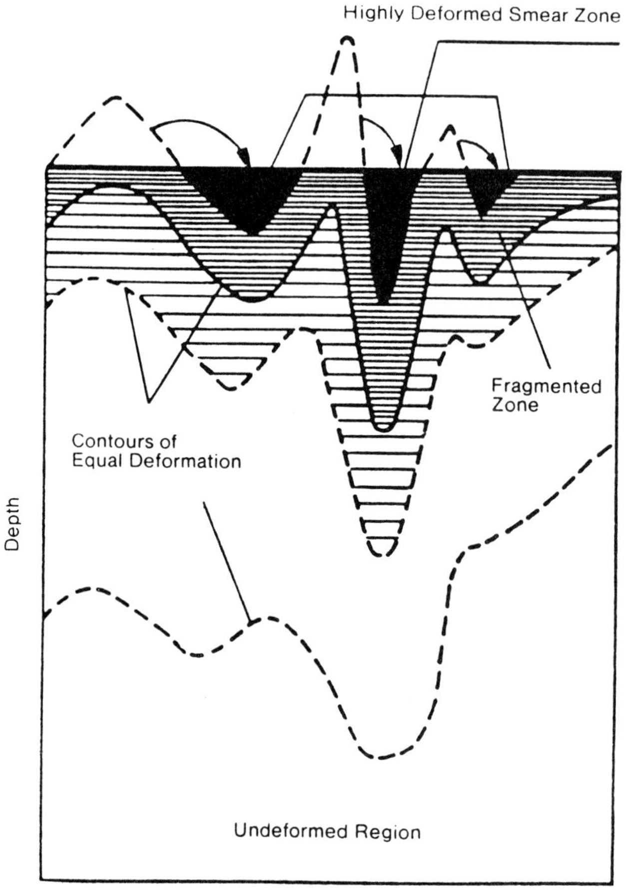

Figure 6 Schematic model of mechanical deformation resulting from abrasive preparation operations. (Courtesy of G. Petzow.)

cutters utilize a rotating abrasive disk consisting of abrasive grains bonded by a rubber or resin-rubber matrix. Effective cutting is achieved by an intentional breakdown of the matrix bond which allows fresh abrasive grains to be continually exposed. The sample material (workpiece) must be held firmly in a clamping device and, depending on the specific design, the abrasive wheel pivots to make contact with the sample or the sample is fed into the rotating wheel, which is in a fixed position. In either case, the sample must be contacted by the abrasive wheel so that cutting will proceed at a rate that does not produce excessive heat.

The coolant flow must be adequate to cool the sample and, ideally, should be aimed at the area of contact (the kerf) (4). Cutters having a fixed coolant nozzle provide a uniform application of coolant; that is, the volume of coolant is applied equally to both sides of the abrasive wheel. When the cutter has adjustable coolant nozzles, the operator must make sure they supply a uniform flow. Failure to do so will create unequal heat dissipation, causing the abrasive wheel to wear irregularly. This results in an angled contact surface (chisel-shaped) which will cause the abrasive

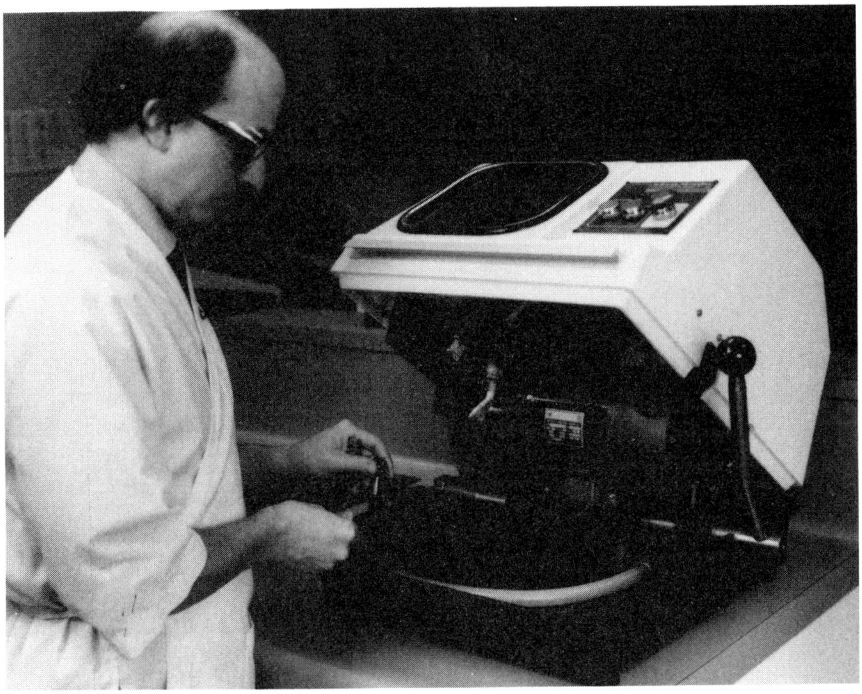

Figure 7 Abrasive cutter used to section rigid materials.

wheel to turn from the normal axis. The consequences of this condition are curved cuts and broken wheels.

Another prime requisite of good abrasive cutting is the wheel choice. There are various compositions available, but some simple guidelines are as follows:

Harder wheels for softer materials; softer wheels for harder samples
Alumina abrasive wheels for ferrous alloys and silicon carbide wheels for nonferrous metals and nonmetals
Diamond-rimmed wheels for cemented carbide and hard dense ceramics

Follow the manufacturer's recommendations and you will seldom go wrong.

Precision-Saw Sectioning

The sectioning of delicate or heat-sensitive materials presents a unique problem that cannot be solved by conventional abrasive cutters. Parts such as thin metal diaphragms, thin-walled tubing, electronic parts, printed circuit boards, and biomedical components are materials that must be handled with special care. For these applications, the precision saws provide an excellent solution (5).

The Isomet 1000 precision saw (shown in Fig. 8) utilizes a diamond-rimmed metal wafering blade which does not break down freely like the abrasive wheel described previously. Diamond or cubic boron nitride (CBM) abrasive grains are embedded in a solid powdered metal matrix attached to a metal disk. The wafering blade, which is in a fixed position, rotates at a relatively slow speed (50–300 rpm)

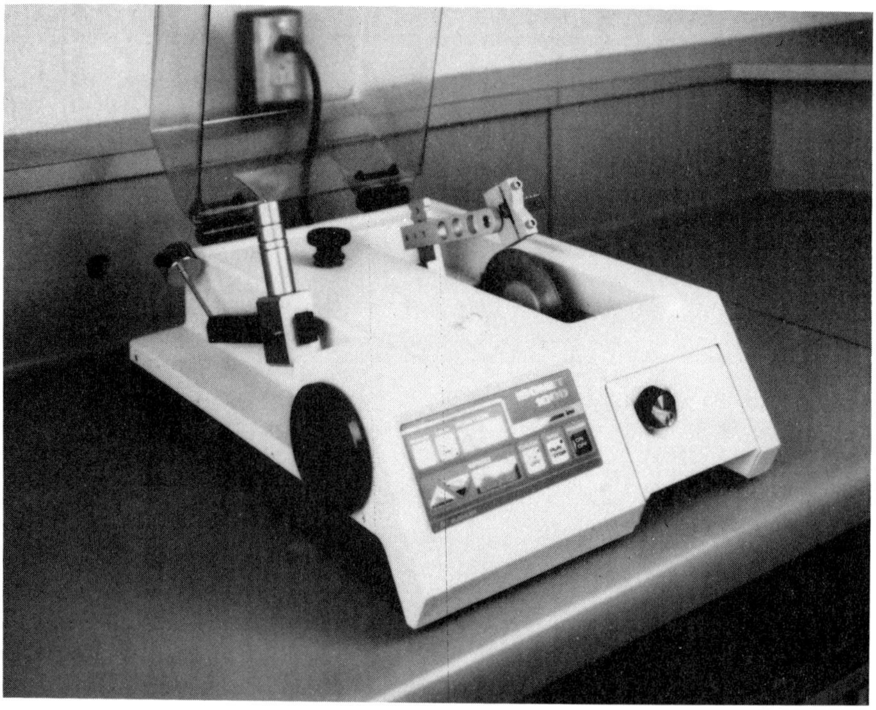

Figure 8 Use of a precision saw to section fragile, heat-sensitive printed wiring board assemblies.

and makes contact with the sample material, which is held by a pivoting specimen arm which feeds the sample into the wafering blade by gentle gravity pressure. With the table assembly the operator is able to locate the cutting plane precisely. As a result, cuts may be made at exact locations.

Although little perceptible heat is produced by the low-speed saw, use of a lubricant is nevertheless necessary to control heat that is concentrated at the point of contact between the abrasive and the workpiece. Lubricant is dragged by the blade to the sample, as the blade rotates through the lubricant contained in a tank at the base of the saw.

The unique features of the precision saw provide the conditions which are necessary to produce precision cuts without damage to the most delicate and heat-sensitive samples. Highly dissimilar materials, such as complex composites and extremely delicate composites such as solid-state electronic devices, may be sectioned at, or near, the plane of interest without inducing damage. The precision saw, and larger saws based on this concept, are therefore the answer to many sectioning problems which are beyond conventional cutting techniques.

B. Planar Grinding

Grinding with coarse abrasives (60–180 grit) is used primarily to correct poor surfaces or to obtain a particular plane for microscopic examination. Irregular surfaces may result from the use of torches or shop saws, and must be corrected before polishing proceeds. Rough grinding may also be used to remove heavy burrs, or burns, produced during previous cutting operations or heavy oxides resulting from heat treating that has been performed without a protective atmosphere.

Belt grinders are most commonly used because they are more convenient, economical, and utilize long-life, cloth-backed belts rather than less durable abrasive paper. Another advantage to belt grinding is the uniform material removal rate because the surface feet per minute (sfm) is identical at all contact positions. One disadvantage is the tendency of the belt to "whip," that is, to rise above the grinder table due to rotational centrifugal forces. As a result, the surface is not as flat as it is when using a disk grinder.

Disk grinders are used when extreme surface flatness and higher material removal rates are desired. Higher material removal rates are obtained because the effective sfm is significantly higher than is obtained with belt grinders. With the exception of special, high-performance composite abrasive disks, abrasive papers wear out more rapidly than belts that have a cloth backing.

Care must be observed during rough grinding to protect the sample from excessive exposure to heat and mechanical deformation. The operator must be protected from injury due to samples, or fragments, that might be hurled from the rapidly rotating belt or wheel. Some precautions to follow are:

Always wear safety glasses when grinding.
Do *not* use badly worn or torn abrasive disks or belts.
Avoid accidental contact, by any part of the body, with moving abrasive materials.
Use adequate coolant flow to reduce the surface heating of the sample.

C. Mounting

Metallographic sample preparation may be performed on unmounted samples if the edges are not important. However, they must be beveled to prevent snagging of the abrasive papers and cloths and prevent injury to the operator's fingers. To prevent these problems and provide a means of supporting highly irregular shaped samples, metallographic samples are normally encapsulated in a suitable plastic before further preparation. Mounting also minimizes rounding of sample edges and provides a safe, convenient means of holding the sample for both manual and automated preparation techniques.

Edge rounding is the tendency for polished edges of a sample to become a radius rather than be retained as a sharp edge. When this occurs, important microstructural details, such as surface layers, hardened cases, and corrosion products, may be distorted or unobservable. Figure 9 shows how edge rounding causes certain light rays to be reflected away from the rounded edge rather than return to the observer. If steps are taken to maintain edge flatness, nearly all light rays will be reflected back to the observer, thus revealing all the detail possible.

Resins used for mounting metallographic samples fall into two major categories: compression molding and castable resin molding. Compression molding is more economical and convenient to use and is the best choice for mounting samples that will withstand the heat and pressure required. Room-temperature curing resins are favored when delicate or heat-sensitive samples are encapsulated. The choice

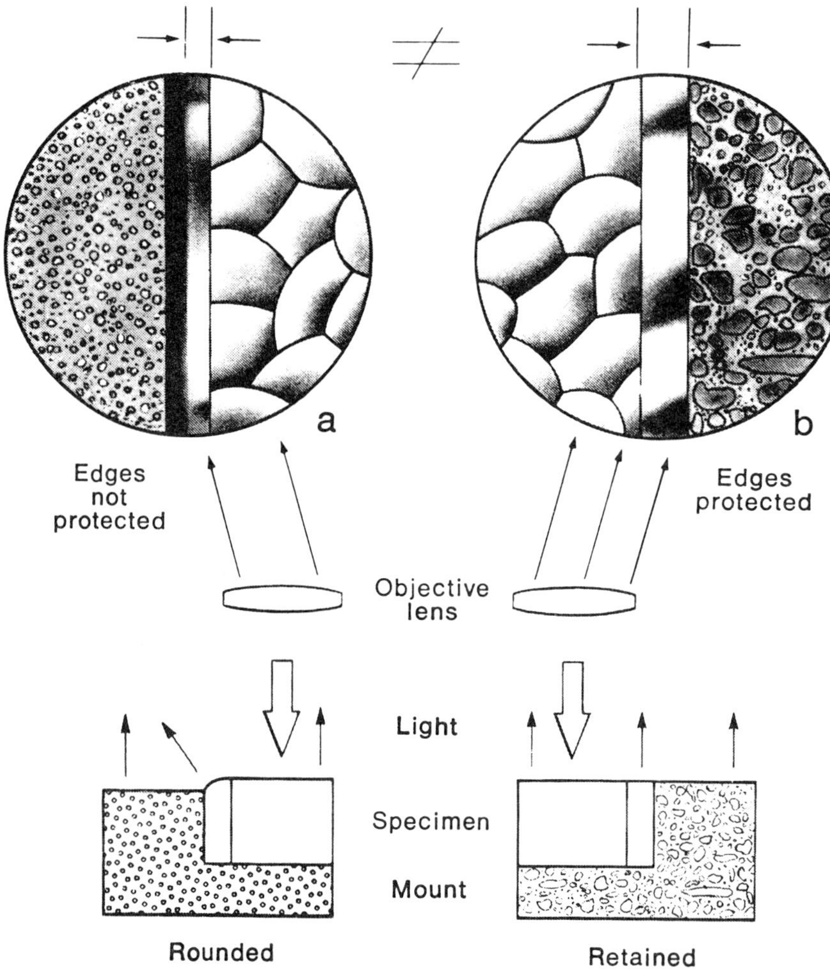

Figure 9 Effect of edge rounding on microstructural observations.

of mounting techniques and specific resin systems will depend on the sample material and how soon the completely cured mount is needed for further preparation.

In compression molding, the sample is placed on the elevated, movable ram located within the cavity of a mold cylinder. After the ram and sample have been lowered into the cavity, a powdered resin is poured in, or a prepressed resin slug is placed into the cavity. The stationary, upper ram is secured to the open top end of the cylinder and pressure is applied to the resin and sample. As a result of the pressure applied by the press (Fig. 10) and heat by an external heater, the resin melts and forms a hard, solid mass around the sample that may be removed after a suitable curing time (about 7–10 minutes).

Castable resins usually consist of two parts: a solid resin powder and liquid hardener, or a liquid resin and liquid catalyst. The sample is placed in a simple ring form on a smooth surface, each coated with a release agent that will prevent the sample from sticking to the mold and base. The mixed resin and hardener are

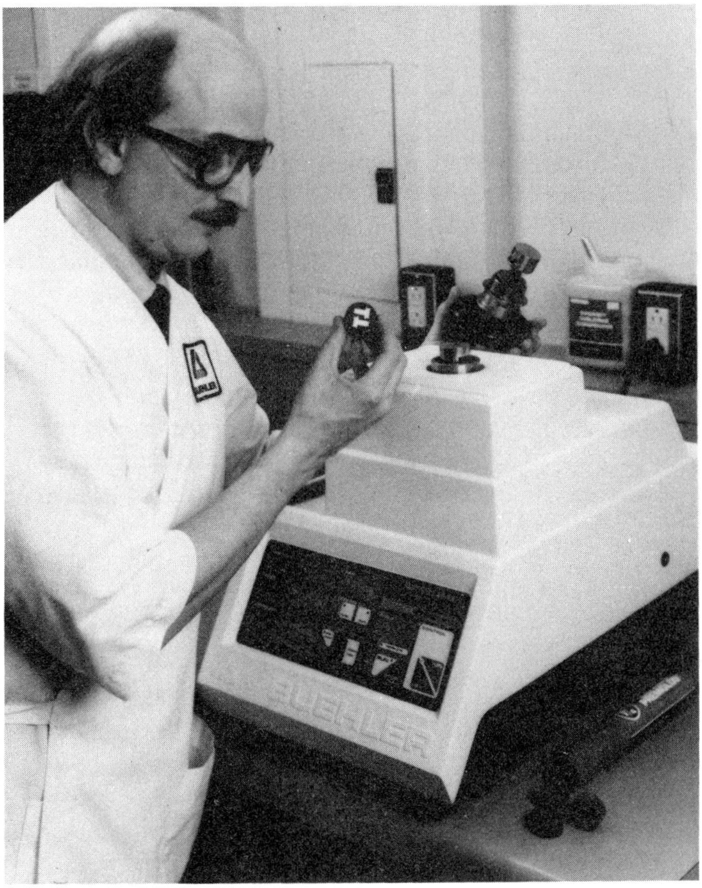

Figure 10 Using a mounting press to encapsulate a metallographic sample.

simply poured into the mold and allowed to cure without pressure or externally applied heat. In some cases, heat may evolve due to an exothermic reaction during curing. This may help accelerate the curing process, or it may become the source of defects that will reduce the value of the mount. It is also possible to use carefully applied external heat to accelerate the curing process.

Completely cured mounts usually have sharp edges and protruding resin "flash." If these edge conditions are not corrected, the mount may snag the rotating polishing cloths during the later stages of preparation. The operator may also experience discomfort from the sharp edge that is in contact with his or her fingers. Grinding is used to chamfer the mount edges so that these problems will be avoided. Another minor problem occurs in cured mounts when the sample protrudes from the face of the cured mount. This occurs due to the different expansion and contraction characteristics of the sample and the encapsulant. If the embedded sample and the surrounding mounting material are not on the same plane when fine grinding is begun, it will be difficult to achieve a common plane by fine grinding. A light application of coarse grinding (120–180 grit) is used to establish the common plane.

D. Fine Grinding

This is a series of grinding steps of decreasing coarseness which systematically reduces the level of abrasion so that subsequent cloth polishing will produce the desired final surface. Fine grinding is usually performed with 240-, 320-, 400-, and 600-grit abrasive papers in the form of stationary sheets or strips, or by using disks mounted on rotating polishing wheels.

Figure 11a illustrates the use of a manual fine grinder, a simple but efficient device. The various abrasive grits are positioned on separate platens so that the

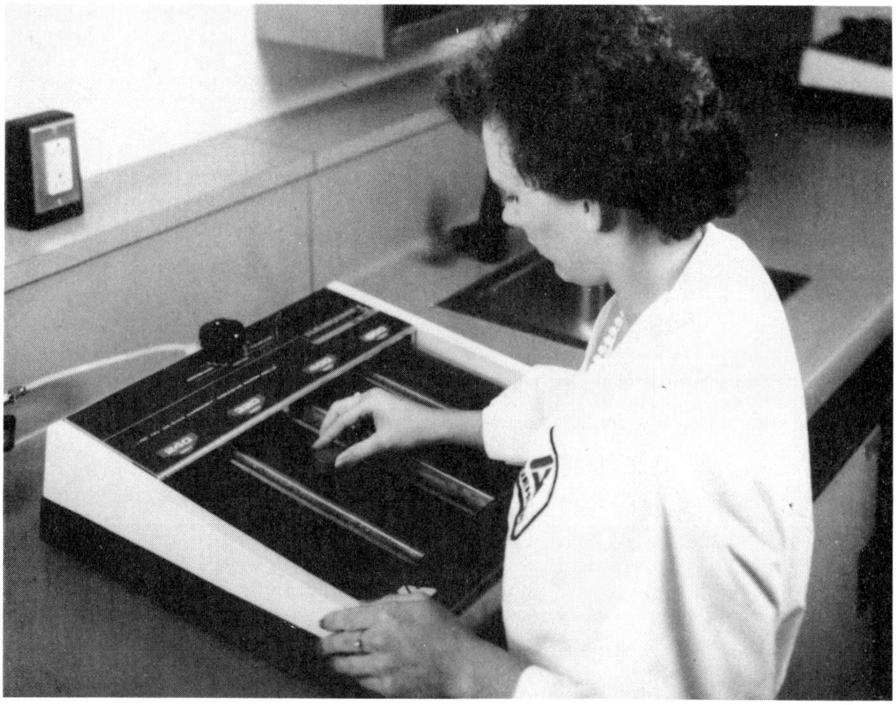

(a)

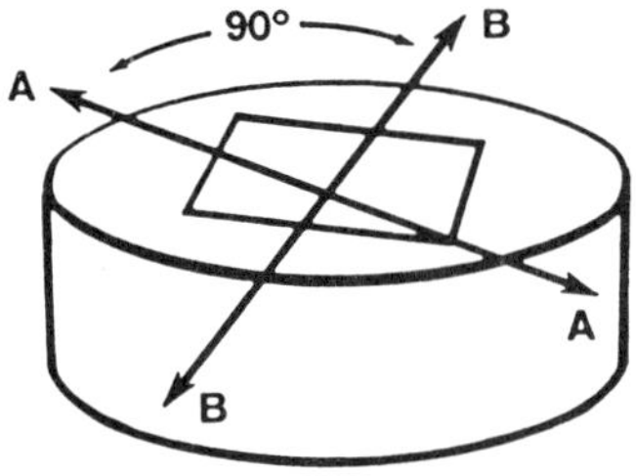

AA Direction of grinding in one grinding step
BB Direction of grinding in next step

(b)

Figure 11 Use of a manual fine grinder: (a) fine grinding performed on a roll grinder; (b) suggested directional motion for manual fine grinding.

operator may proceed to the next step with only a brief water rinse. To determine the duration of each step, the operator rotates the sample 90° (Fig. 11b) before the next step and is able to observe the progress by how completely the previous scratches have been removed. As in all metallographic sample cutting and grinding, lubrication (in this case water) is used to reduce friction at heat.

When fine grinding is performed on rotating wheels, less physical effort is required and an increase in sample production is usually achieved. Further increases in productivity may be realized by the use of semiautomatic devices which permit multiple samples to be prepared simultaneously. Larger and more modern high-volume sample preparation machines have also been developed which prepare up to 24 mounts at a time and utilize an advanced abrasive step which replaces the conventional multistep fine-grinding procedure. Fully automatic machines that polish samples without operator intervention are also available.

E. Rough and Final Polishing

Rough and final polishing complete the basic polishing sequence and are similar because they are both performed on cloth-covered rotating wheels. In the earlier history of metallographic sample preparation, both steps were performed as separate napped cloth steps, using a coarser and a finer alumina abrasive. With the combination it was difficult to remove completely the deformation and scratches produced by previous abrasive steps and still maintain flatness and retain all microstructural constituents. This was a particular problem when polishing samples with microstructures of vastly differing chemistry and hardness. With the introduction of diamond abrasives to metallography, rough polishing took on a new meaning, becoming a key step that was distinctily different from final polishing. Deformation and scratches produced by previous abrasive steps could be completely removed and the resultant surface had a quality that reduced final polishing to a more cosmetic role.

Rough polishing, as presently practiced, employs one or more steps using a napless polishing cloth which is charged with a diamond abrasive in the range 30- to 3-μm particle size (1 μm = 0.0004 in). A liquid extender is used to provide lubrication and to promote redistribution of the abrasive grain on the cloth. For most medium-hardness sample materials, such as non-heat-treated steels, a single step using 6-μm diamond abrasive on a napless cloth is sufficient. Additional steps are required as the hardness and complexity of the microstructure demands. There are no hard-and-fast rules that apply to such choices but, with experience, these decisions are more easily made.

Diamond is a necessity for rough polishing because it enables a metallographer to remove grinding deformation and scratches without producing edge rounding, microstructural relief, or pulled-out constituents. This is possible due to the extreme hardness of diamond and its durable cutting edges, which cut cleanly through hard and soft constituents. It is therefore desirable to spend more time in rough polishing, so that a minimum time is required for final polishing where polishing defects usually occur. Rough polishing is performed with high pressures and the sample should be rotated counter to the wheel rotation to avoid directional polishing effects, as illustrated in Figure 12.

Final polishing is usually performed on a slightly napped cloth which has been premoistened with distilled water and charged with a very fine (0.3–0.05 μm)

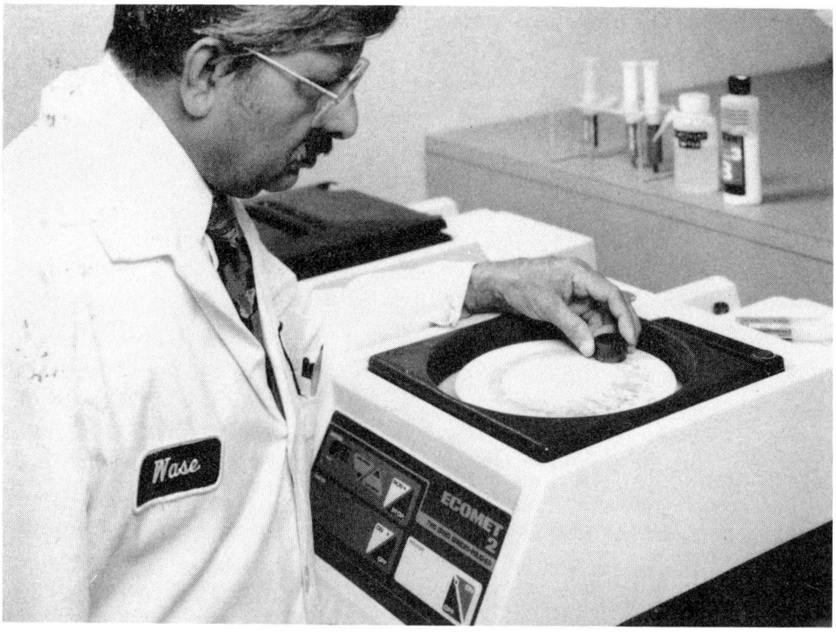

(a)

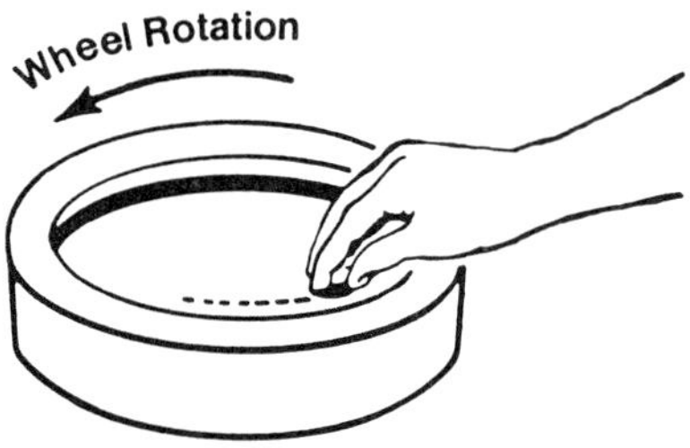

(b)

Figure 12 Use of a rotating wheel polisher/grinder to perform rough and final polishing: (a) rotating wheel polisher/grinder; (b) direction of polishing on a rotating wheel.

alumina abrasive slurry. Polishing times should be minimized to prevent edge rounding, relief, and pitting. The amount of moisture on the cloth is also a significant factor that affects the quality of final polishing. The cloth must be wet enough to prevent heating and deformation of the sample surface but dry enough to avoid the pitting of inclusions. An old rule of thumb said: "The cloth is correctly moistened when a sample lifted from cloth contact can be blown dry in 3 to 5 seconds with the operator's breath."

For more demanding applications, metallographers may choose to eliminate alumina and napped cloths entirely. In such cases, the final step would be 3-, 1-, or 0.25-μm diamond on a napless cloth. Although a very fine array of scratches will remain, their presence may be of less consequence than problems that may be created by excessive alumina polishing. This is often the case when the samples are intended for automatic image analysis, where microstructural relief and pitting would produce erroneous results.

Later developments in abrasives give the metallographer better control over the results and the ability to prepare difficult materials with superior results (6). Deagglomerated aluminas may be used to final-polish softer materials such as copper and aluminum alloys with fewer scratches and a less deformed surface. Conventional aluminas tend to leave excessive scratches owing to their natural tendency to agglomerate. These agglomerates act like coarser abrasive grains, producing scratches that are much larger than the actual abrasive grains should give. Deagglomerated aluminas consist of abrasive grains that are well separated and therefore produce a polished surface without coarse scratches.

Another help in solving final polishing problems is the colloidal silica polishing solution, which is a chemically active colloidal suspension of silica particles. This solution may be used alone or in addition to alumina slurry. Colloidal silica is used to produce brilliant, scratch-free surfaces on a wide variety of materials, including copper, aluminum, stainless steel, lead, and others which tend to deform readily and produce unclear microstructures.

F. Automatic Specimen Preparation

The previously described manual specimen preparation techniques are adequate for the preparation of a wide range of metal alloys and common polymers. However, if the required number of specimens per shift exceeds the limits of the equipment or the available workforce to complete, automation should be considered before hiring additional personnel. Some reasons are as follows.

Manual specimen preparation requires some skill and training to obtain acceptable results. In recent times, it has become more and more difficult to find trained metallographers who are capable of performing reliable specimen preparation. When a trained person is lost, another person must be trained, sometimes with less than satisfactory results. With automation, new hires are easier to train, because the grinding and polishing parameters are set and implemented by a machine such as shown in Figure 13.

With automation, output is significantly increased without any compromise of the quality of the specimen finish. In fact, automatically prepared specimens are nearly always better than those prepared manually. The reason is the better control of polishing parameters such as polishing pressure, wheel speed, and time.

Better control of the preparation parameters also makes it possible to prepare the more advanced materials such as ceramics, metal matrix composites, and thermally sprayed coatings that are very difficult to prepare using manual methods. These materials require higher pressures and longer preparation times and the use of advanced techniques such as power lapping to cope with their challenging properties such as high hardness and brittleness.

Finally, automation may eventually become a requirement to meet the newer tougher international quality standards that are impacting industry.

V. ANALYSIS OF POLISHED SAMPLES

A. Revealing the Microstructure

The as-polished surface produced by conventional abrasive techniques, other than the colloidal silica, are likely to have a very fine layer of deformation remaining.

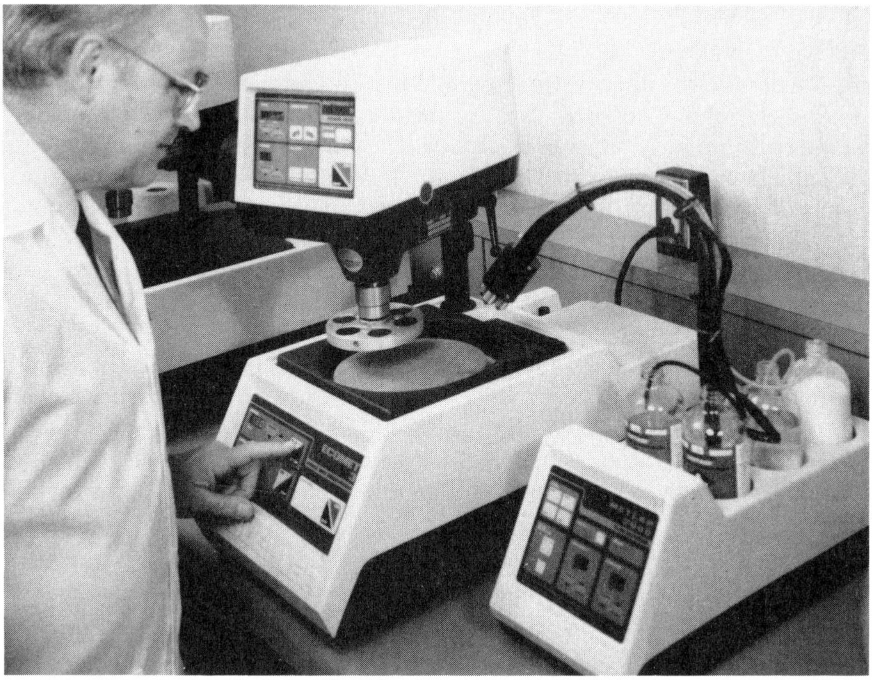

Figure 13 An automatic grinder/polisher used to prepare larger quantities of uniformly polished specimens for metallographic examination.

Although this may not obscure the microstructure, it could cause it to be less clearly defined, therefore more difficult to analyze. When residual scratches remain, or softer constituents such as graphite in cast iron are pitted or washed out, it is possible to restore the microstructure without a total repolish. One such technique is the etch-polish sequence, where the sample is lightly etched, then repolished on the final polishing wheel only enough to remove the etching effect.

This sequence may be repeated several times, if necessary, to remove residual deformation scratches and the effects of plucked constituents such as graphite in cast iron. Another technique, called "attack" polishing, utilizes an etchant which is applied directly to the final polishing wheel. The sample is simultaneously etched and polished and a better surface is produced because the remaining deformation is removed.

Metallographic etching is a separate post-polishing step that is used to reveal microstructural details which are not visible in the as-polished condition. Although nonmetallic inclusions, free graphite (cast irons) and numerous physical defects are visible without etching, far more information becomes available once the polished surface has been correctly etched. Although there are various etching techniques, the most common and useful type is chemical dissolution. This method employs a combination of solvent and chemical reagents, usually including one or more mineral acids which selectively attack various features of the microstructure, depending on their individual chemical and crystallographic properties. Specific features or constituents, such as grain boundaries, alloy phase boundaries, layer interfaces, and various precipitates, are delineated, depending on the specific etchant that is applied.

There are literally hundreds of etchants which have been devised by various investigators over the years, but a relatively few sample formulations are required to reveal the microstructures of most common alloys. Only proven, safe etching formulations should be selected and used according to printed directions and with accepted chemistry procedures. Etchants may be applied either by swabbing the sample surface, as shown in Figure 14, or by total immersion into the etching solution, as recommended by the information source.

Certain corrosion-resistant alloys, such as stainless steel and various nickel-based superalloys may be difficult to etch by chemical dissolution. In such cases it may be necessary to employ electrolytic etching where the sample is made the anode in an electrolytic cell. Etching occurs when a low-voltage DC current is applied, causing a more energetic attack of the sample surface.

B. Microscopes

Microscopes such as the one shown in Figure 15 are essential to reveal the microstructure to the investigator. Because the prepared metallographic sample has an opaque, polished surface, it is necessary to use a reflected light microscope rather than the transmitted light microscope used to view biological samples. Another

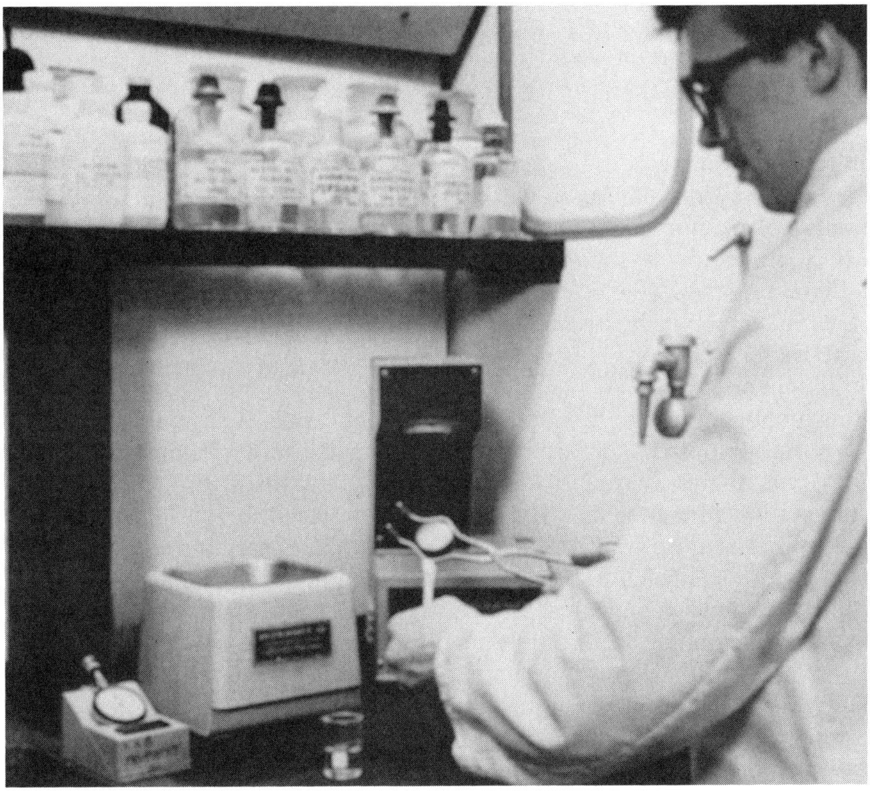

Figure 14 Etching a metallographic sample.

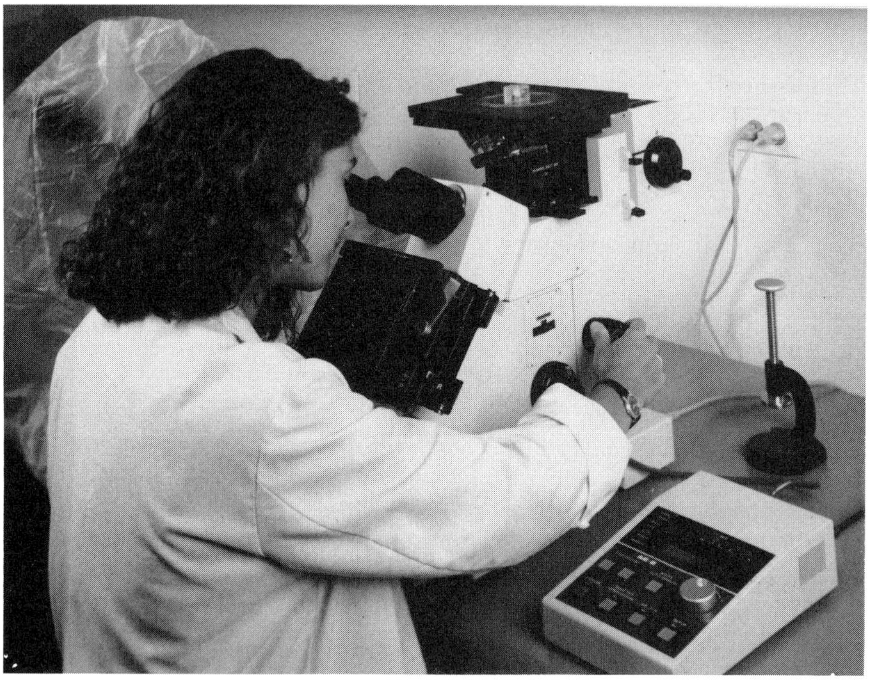

Figure 15 An inverted-stage metallograph used to examine, analyze, and photograph the microstructure of materials from polished specimens.

basic requirement of a metallurgical microscope is the use of vertical illumination. Stereomicroscopes, commonly used to observe unpolished and irregular surfaces, such as fractures, are not suitable for the examination of polished metallographic samples. In addition to the limited magnification range, they commonly employ oblique illumination, which gives a dark-field effect on polished samples. Darkfield illumination renders the microstructural details as bright lines in a dark background, rather than dark lines on a bright background which is characteristic of the desirable brightfield illumination.

Other types of illumination, such as polarized light and differential interference contrast, are useful adjuncts but are not the primary method of illumination used in metallographic analysis. Brightfield illumination is obtained through the use of vertical illumination where the light path strikes the polished surface at a perpendicular angle and therefore returns on the same path to the observer or film plane via a special mirror. The need for brightfield illumination was recognized early by Henry Sorby who stated in his notes in 1885: "With oblique illumination, a polished surface looks black, but with direct [vertical] illumination it looks bright and metallic."

In addition to the requirement for vertical illumination, a metallurgical microscope must have a sufficient magnification range to resolve various microstructures that will be encountered. If photomicrographs are required, an integral or accessory camera must be available. Many metallurgical microscopes utilize an inverted specimen stage; the sample is placed polished face down onto an aperture stage plate.

With this design, samples may be exchanged rapidly because the sample is self-leveling. When an erect stage microscope is used, a separate leveling operation must be performed.

VI. THE APPLICATION OF MICROSTRUCTURAL ANALYSIS

A correctly polished metallographic sample has no value unless the revealed microstructure is applied to solve practical problems. Figure 16 illustrates the wide range of information obtained from microstructural analysis. The actual utilization of microstructural information may be conveniently divided into three areas: materials characterization, process monitoring, and failure analysis.

A. Materials Characterization

Materials characterization was the goal of ancient philosophers and alchemists who wanted to understand the nature of matter. It was not until microstructural analysis became a practical reality that it was possible to develop new alloys by intention rather than by chance. When material properties became identified with specific microstructures, it was possible to evaluate visually the effects of chemical, thermal, and mechanical work variations. If, for example, a steel is rapidly quenched, it not only becomes hard but also has a characteristic needlelike microstructure. This visible condition therefore indicates that hardness has been achieved. Similarly, various other microstructures may be used as indicators of the success or failure of mechanical and thermal treatments.

Table 1 lists some of the common microstructures which are found in iron-carbon alloys, together with an indication of how they tend to affect the alloy properties. Because two, or more, and often as many as five of these microstructures may be present at one time, this table cannot be used as a practical guide because they are interacting. The effects of the microstructure are best understood and most useful when considered in the light of a knowledge of chemical composition and thermal history.

Various reference books have been compiled to provide a guide to identify materials using photomicrographs of typical microstructures. It is significant to note

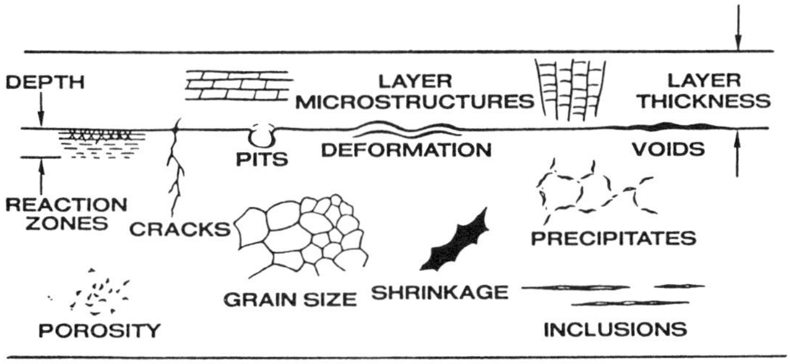

Figure 16 Microstructural details revealed by metallographic analysis.

Table 1 Effect of Microstructure of Ferrous Casting Alloy Properties

Microstructure	Identity	Impact
Austenite	Soft phase which forms first—usually transforms into other phases—seen only in certain alloys	Soft and ductile—low strength
Ferrite	Iron with elements in solid solution—soft matrix phase	Contributes ductility—little strength
Graphite	Free carbon in any size and shape	Improves machinability and damping properties—reduces shrinkage—reduces strength severely depending on shape
Cementite	Hard iron—carbon intermetallic phase	Imparts hardness, wear and corrosion resistance—severely reduces machinability
Pearlite	Laminar phase—alternate layers of ferrite and cementite	Contributes strength without brittleness
Martensite	Hard structure produced by specific thermal treatment	Hardest transformation structure—brittle unless tempered
Steadite	Iron-carbon-phosphorus eutectic—hard and brittle	Sometimes confused with ledeburite—aids fluidity in molten state—brittleness in solid state
Ledeburite	Massive eutectic phase composed of cementite and austenite (ledeburite transforms to cementite and pearlite on cooling)	Produces high hardness and wear resistance—virtually unmachinable

that these visual guides are nearly always organized according to compositional families and subdivided by thermal treatment histories.

The most effective application of microstructural analysis is to monitor the processes that are used to manufacture component parts rather than to inspect finished parts. It makes little sense to analyze the microstructure at final inspection only to discover that there are serious defects that originated in the starting material. One example is a circuit board manufacturer who failed to check incoming laminate, only to discover, after fusion, that the material was delaminated. Not only did this manufacturer lose money due to scrapped material, but their production schedule and reputation were jeopardized. To the printed circuit board industry, the primary value of metallography is qualitative analysis, primarily the search for defects.

In other industries and applications, there is an even greater need to analyze microstructure quantitatively. In the steel industry, for example, the analysis of nonmetallic inclusions has been and still is a critical measure of quality. Since the early days of steel making, it was known that the type, amount, and distribution of nonmetallic inclusions had a significant affect on its properties.

One of the earliest tools for inclusion analysis was a chart that was used to compare the inclusions observed in the microscope to the printed facsimiles displayed on the wall chart. There are many more applications for image analysis, each having its specific need for accurate quantitative data. However, instead of performing tedious, time-consuming, and not-too-accurate manual measurements, the analysis may be performed automatically without the variables introduced through operator bias or fatigue.

One such analyzer, shown in Figure 17, consists of a microscope equipped with a high-resolution CCTV camera, an automatically controlled specimen stage, and an autofocus program. The image information is processed by a computer that recognizes and counts summarization of the data. In addition to inclusion analysis, automatic image analyzers are capable of performing more than 40 types of field and feature analyses.

Concerned heat-treaters rely on microstructural analysis to provide evidence that a particular treatment has been successful. Qualitative visual analysis tells them if unwanted decarburization or retained austenite is present on treated parts. The addition of a measuring accessory to a microscope permits the accurate measurement of plated layers, total case depth, and depth of decarb. The previously described image analysis equipment can quickly and accurately calculate the percentage of retained austenite or carbides present.

Figure 18 shows a microhardness tester that is used to measure indentation hardness values on a polished specimen. It will accurately determine the effective case depth, a part of which has been hardened to a required minimum hardness such as Rockwell HRC 50. Figure 19 shows a series of Vickers microhardness indentations that traverse a hardened case and the adjacent core.

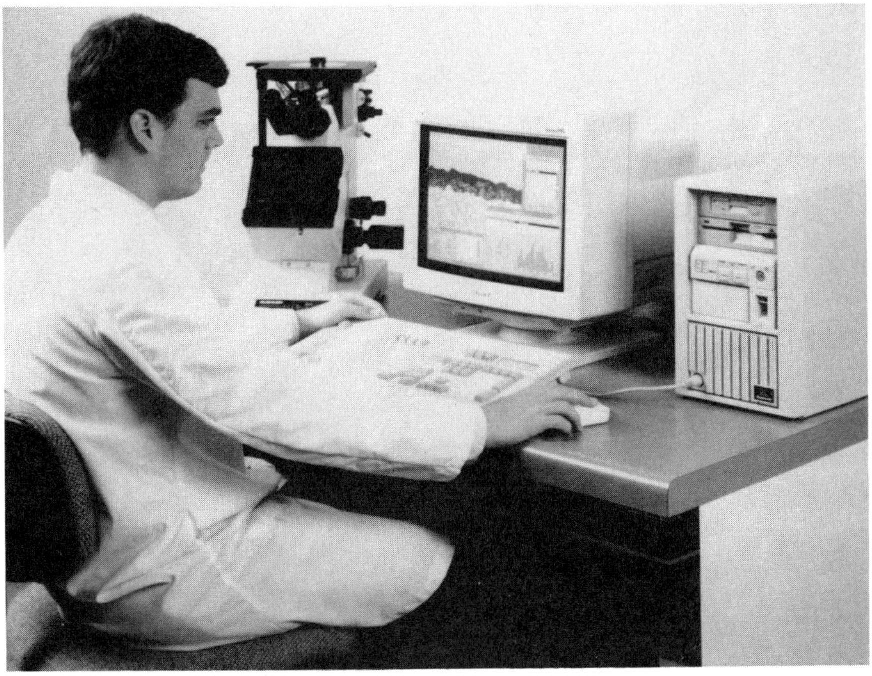

Figure 17 An automatic image analysis system used to perform rapid, accurate quantitative analysis.

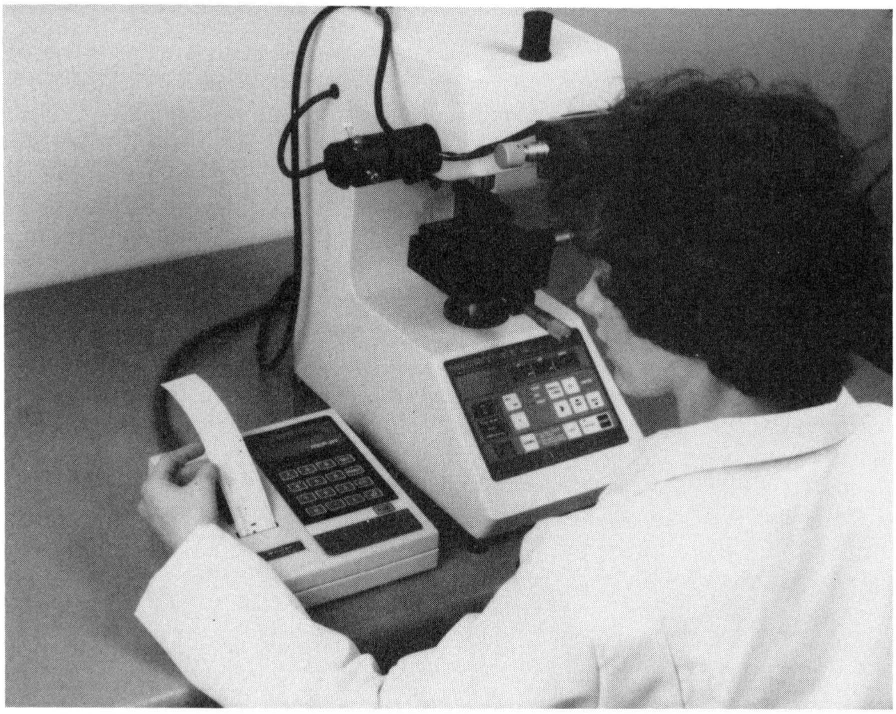

Figure 18 Microhardness tester with a data printer.

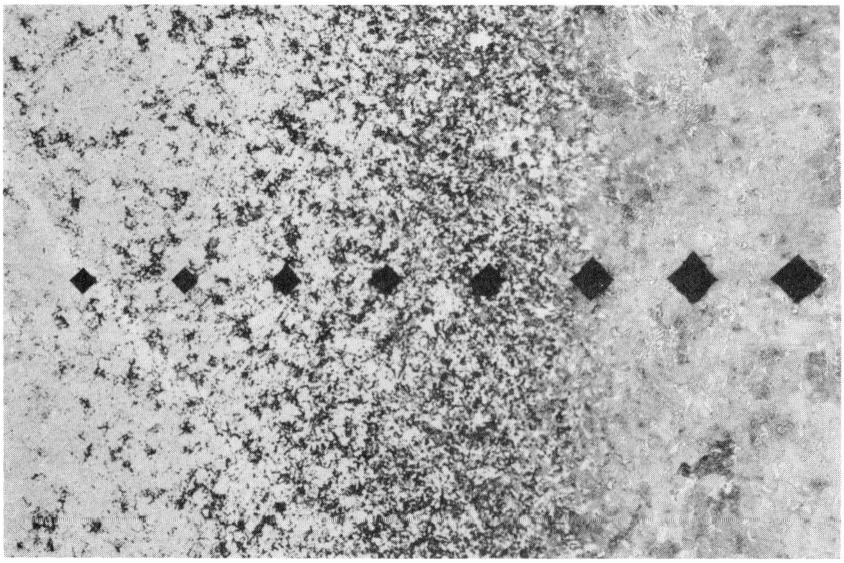

Figure 19 Vickers microhardness indentations across a hardened case (left) to softer core (right) Etchant: Picral. Magnification: ×100.

Fastener manufacturers are now required by the Fastener Quality Act PL-101-295 to perform various metallographic tests on specimens of various alloys used to produce fasteners for critical service applications. Figure 20 shows an etched view of the thread root of a fastener revealing an intergranular crack that would be sufficient cause for rejection in a grade marked fastener and some commercial fasteners as well.

B. Failure Analysis

Failure analysis is a valuable tool that has both negative and positive connotations. On the negative side, it is used to assign blame for the failure of a component, device, or assemblies in service. On the positive side, it is used to identify poor design by subjecting these parts to extreme operating conditions such as mechanical overload. When the part fails, it is analyzed to locate the weakest area that needs to be improved before placing it into service. Figure 21 is a cross section of a component lead that has been soldered into a circuit board throughhole. This photomicrograph clearly shows that, although the solder has wetted the hole pad, it has not adhered to the component lead itself. This condition would have been

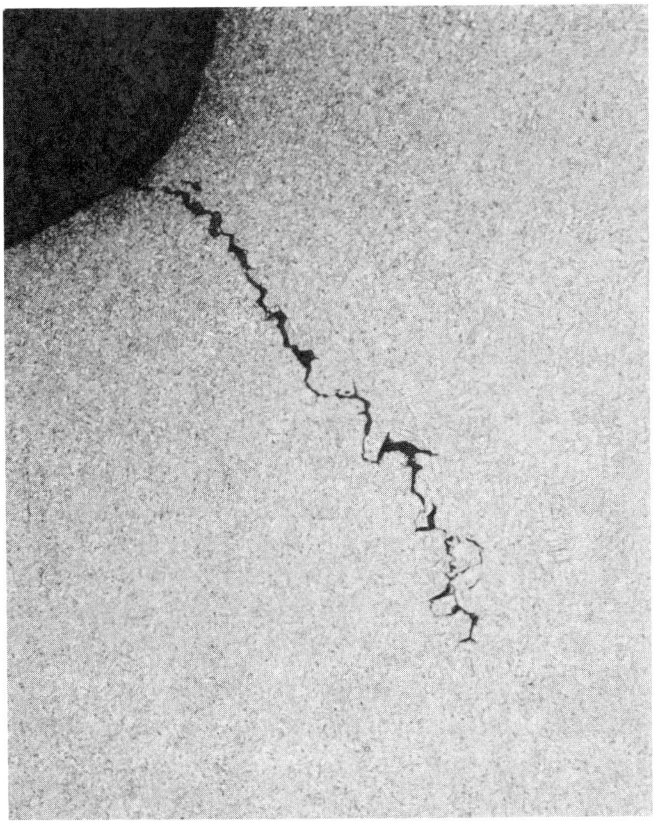

Figure 20 Intergranular crack in the thread root of a fastener. Etchant: picral. Magnification: ×100.

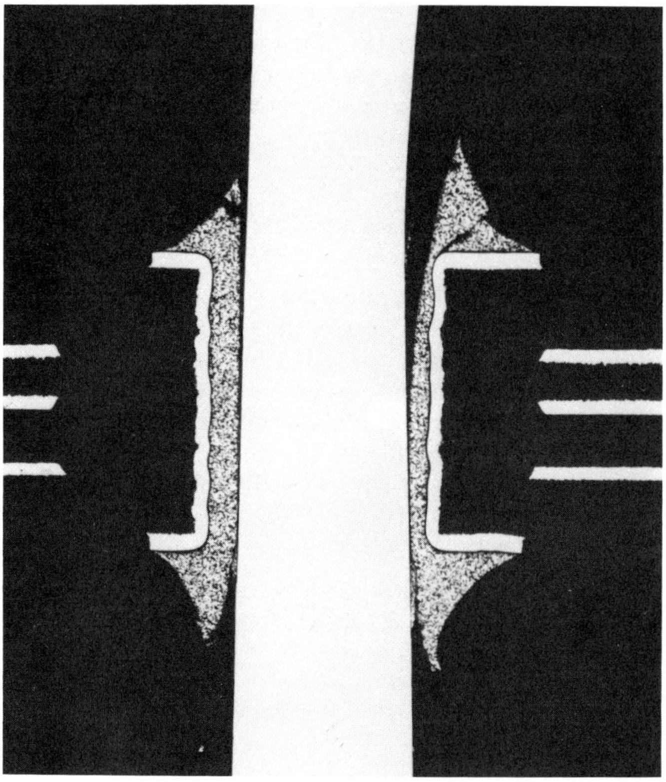

Figure 21 Nonwetting component lead in a PC board through-hole Etchant: None. Magnification: ×50.

detected by electrical testing, but it required cross sectioning to learn the full extent of the problem and what might have caused it. The most likely reason is contamination of the lead surface possibly due to improper storage that did not adequately protect the lead from a damaging environment.

Figure 22 is a cross section of a plated 70/30 brass decorative rivet such as is commonly used on casual clothing. The predominately intergranular and branched nature of the main and secondary cracks suggests stress corrosion or "season cracking" as it is sometimes called in copper-based alloys. This phenomenon usually occurs catastrophically when the brass part that is highly stressed from the forming operation comes in contact with an atmosphere containing ammonia. Close observation reveals that decorative plating penetrates even the secondary cracks, indicating that the cracks occurred before plating.

Heat-treaters who harden steel fasteners must maintain a furnace atmosphere that supplies sufficient carbon to produce a hardened case to withstand the wear that occurs during service installation. In through-hardening, the furnace atmosphere must be maintained at a high carbon content so that there will be no carbon lost to the atmosphere. If this carbon potential is too low, carbon will be removed from the surface and a softer microstructure will result. In an extreme case, if total decarburization occurs, extremely soft ferrite will result, as shown in Figure 23.

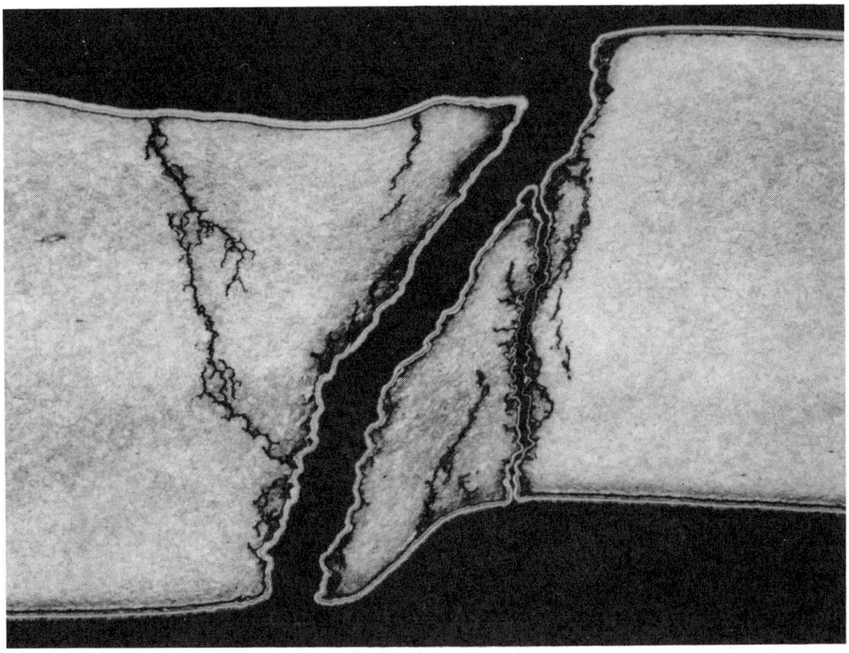

Figure 22 Stress corrosion cracking in a brass rivet. Etchant: NH_4OH — H_2O_2. Magnification: ×100.

Ferrite has little strength and resistance to corrosion, so its presence could lead to premature failure of a part. Visual examination will alert us to its presence but microhardness tests should be used to verify this identification.

VII. SUMMARY

Microstructural analysis, or metallography as some call it, has been practiced for more than 130 years, and its usefulness as a comprehensive method of qualitative and quantitative analysis has not slackened. Rather, those who practice it and those who produce the equipment and supplies to perform it have responded to the ever-changing needs of the materials scientist and various industries who need the information it provides.

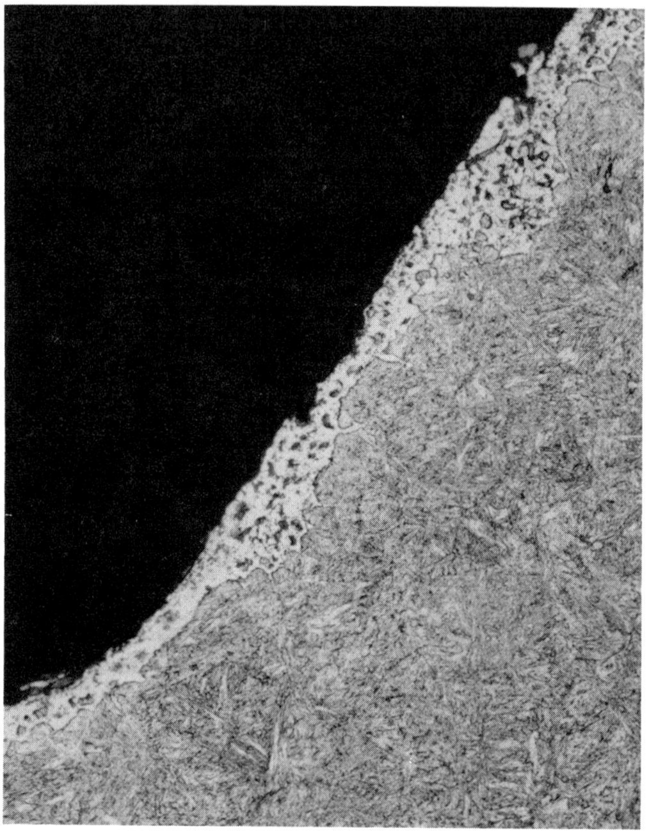

Figure 23 Decarburization at surface of the threads of a fastener. Etchant: 2% Nital. Magnification: ×400.

REFERENCES

1. C. S. Smith. *A History of Metallography*. University of Chicago Press, Chicago, 1960.
2. L. E. Samuels. *Metallographic Polishing by Mechanical Methods*, Vol. 3. Pitman, Melbourne, 1971, p 44.
3. G. Petzow. *Metallographic Etching*, Vol. 1, American Society for Metals. Metals Park, OH, 1978, pp 9–11.
4. J. A. Nelson and R. M. Westrich. Abrasive cutting in metallography. In *Metallographic Sample Preparation*, J. L. McCall and W. M. Mueller, eds. Plenum, New York, pp 41–54.
5. "The low speed saw." Metal Dig, 22: 1, 1983.
6. J. A. Nelson. New abrasives for metallography. In *Microstructural Science*, Vol. II. Proceedings of the 15th Annual Technical Meeting, Institute of Metallographic Sciences, 1983 pp 251–259.
7. G. F. Vander Voort. Inclusion measurement. In *Metallography Is a Quality Control Tool*. J. L. McCall and P. M. French, eds. Plenum, New York, 1980, pp 1–88.
8. J. C. Oppenheim. Introduction to basic quantitative image analysis. In *Microstructural Science*, G. Petzow, R. Paris, E. D. Albrecht, and J. L. McCall, eds. Elsevier, New York, 1981, pp 163–170.

9. J. A. Nelson. The final quality tool: failure analysis. Manufacturing Engineering, November 1982: 75–77.

FURTHER READING

American Society for Metals, *Metals Handbook, Eighth Edition*, Vol. 8, Metallographic Structures and Phase Diagrams, ASM, Metals Park, OH, 1973.

Petzow, G., *Metallographic Etching*, American Society for Metals, Metals Park, OH, 1982.

Samuels, L. E., *Metallographic Polishing by Mechanical Methods*, 3rd ed., American Society for Metals, Metals Park, OH, 1982.

Smith, C. S., *A History of Metallography*, University of Chicago Press, Chicago, 1960.

Van der Voort, G. F., *Metallography: Principles and Practices*, McGraw-Hill, New York, 1984.

38

Automatic Test Equipment

WILLIAM E. LAND
William Land Associates
Ione, California

I. INTRODUCTION

The purpose of automatic test equipment (ATE) has been to evaluate parts, components, subsystems, and complete systems. The commonly accepted definition of ATE has come to be understood as automatic electronic systems used in the manufacture and repair of electronic products. It should be understood that as technology advances, it becomes necessary to expand the definition of ATE to include any system operating under the control of computers, or microprocessors, which is used in the manufacture or repair of products and systems. In the years since the 1950s, indeed, we have seen the inclusions of diagnostic capability within the product or system itself, coming to be known as built in test equipment (BITE). These systems then include the automatic test equipment capability within themselves.

The first ATE systems were designed for the military services as an aid to maintenance of complex weapons systems. The low skill levels of new recruits, coupled with increasing difficulty in retaining trained, experienced technicians, led the U. S. Department of Defense to invest heavily in ATE. The test systems were intended to be the electronic equivalent of a field engineer as an aid in repairing missiles, radar systems, fighter planes, and other military weapons. ATE captured the skills of the best engineers and, through step-by-step programs, diagnosed faults and led the repair technician through a sophisticated maintenance procedure. As one example, which occurred in the late 1950s, a prime contractor was directed by the Naval Ordnance Department to build a system for shipboard use in the Talos Missile Program to determine the readiness of a missile prior to launch. The system, referred to as TATTE, was the first successful ATE intended for use in the Tactical Deployment of Weapons Systems.

Automatic test equipment for commercial purposes followed a different path. Semiconductor manufacturers, initially through in-house development, produced test equipment for use in their production lines. The objective of a selling price of pennies per unit dictated automated testing and sorting. Specialized test systems for each and every application proved very expensive. A merchant market developed in which manufacturers designed systems to cover as broad a spectrum of applications as possible. Most of the test equipment manufacturers were spin-offs from semiconductor groups, microwave electronic companies, and environmental test equipment manufacturers. Some of the early commercial automated test systems were spin-offs of complex weapon systems manufacturers.

The manufacturers of electronic equipment found the need for a third form of ATE, the subsystem tester. Circuit boards, power supplies, and wiring frames are the three primary subsystems used in electronic equipment. Even if the components of an electronic device are known to meet specifications, errors of both omission and commission can occur during the assembly process. Experience has proven that finding errors and problems is far easier and less expensive in the subsystem stage of manufacture than when a complex product is completely assembled.

It should be pointed out that automatic test equipments are now being developed and marketed as medical diagnostic equipment. Some examples are glucometers, used by diabetics to monitor blood sugar levels; automatic blood pressure testers, which are found in almost every drugstore; and systems which automatically control the temperature of premature babies. Most of this ATE requires little training for skilled operation, but there are complex systems which can only be operated by high-level medical technicians and physicians [e.g., ATE used to check and reset or reprogram pacemakers using radio frequency (RF) signals].

II. GENERATIONS OF ATE

Three generations of automatic test equipment have been developed so far. The first generation consisted of hardware only. The individual tests to be executed and the test programs were built into the equipment with minor programming changes accomplished by means of switches and patch panels or through numerical control entered from paper tape or punched cards. Because the systems were entirely hardware, they were generic machines intended for a preset set of tests on a given device or system. First-generation ATE tended to be slow, cumbersome, and expensive. The onrush of electronic development in the late 1950s and early 1960s made many first-generation machines obsolete as soon as they were built. Because of their hardware design, they could not be modified to meet new and more complex requirements.

The most efficient of these first-generation machines were controlled by punched paper tape. At worst, they require continual reprogramming. At best, they need a new copy of the punched tape for each new product. Equipment (lathes, mills, etc.) controlled by punched tape are called numerical controlled (NC) machines. Many NC machines are still in use today.

The late 1950s saw the introduction of small computers called minicomputers. These electronic processors had the ability to execute stored programs. (The programs are called software.) A small computer grafted onto existing hardware became the second generation of ATE. Languages and formats were developed so that test programs could be developed on the computer and stored in its memory or a peripheral storage device such as magnetic tape. Test results from the hardware were fed to the computer for storage and analysis. Second-generation ATE was characterized by the use of a computer as a programming and storage element.

The computer was often employed to control external power sources and measurement instruments in an effort to enhance system capability. However, it could not overcome the inherent lack of flexibility in the hardware. To change to a new device or system often required a major rebuilding of the system hardware.

During the 1960s, the third generation of ATE appeared. It was characterized by a concept that moved as much as possible of the complexity from hardware to software. The elements of the system power sources, stimulus generators, measurement instruments, and interconnection matrices are all operated under software control. Thus, within the capabilities of the individual subsections, the unit could test any device or system. A heavy burden was placed on the computer which had to execute a program for each test and which needed sufficient memory for long and complex test routines.

Minicomputer prices have dropped 10-fold for the processor and its peripherals and 1000-fold for memory in a decade, which is a trend that continues. With much of the hardware eliminated, ATE systems have become smaller, lighter, and less expensive. In the process, the move to software produced systems that are much easier to maintain, because the power of the computer was employed to aid the maintenance procedure, and because the amount of hardware to be maintained was significantly reduced.

The characteristics of the fourth generation of ATE are apparent. The limitation of the third generation is that a computer can execute only one instruction at a time. The next generation of ATE will use distributed processing. Many tasks, including testing, creating new tests, storing data, and so on, are ideally executed in parallel. Computer functions are implemented in micro form as single semiconductor chips called microprocessors (mPs). The cost of the microprocessors is so low that using multiple units in parallel is far more cost effective than the use of a single minicomputer. The development of software programs and operating systems to utilize the new miniature processors lags far behind the progress on hardware.

Once again, the manufacturers of military weapons systems have been at the forefront of this development. The only true fourth generation ATE at the present time is in use by the Defense Department in complex weapons systems. Fourth-generation ATE is currently in development for many commercial applications such as communications network managers/monitors which can diagnose a trouble and isolate it to one of several locations and there route the signals to allow continued operation (albeit with degraded capability) until the fault is *fixed* by field technician. This will no doubt be replaced sometime in the future for systems which can *heal themselves*.

III. ELECTRONIC ATE SYSTEM CLASSIFICATIONS

Automatic test equipment used in the electronics and communications industries are grouped into three broad categories by intended application: parts test, module test, and system test. They are also subdivided by the size and cost of the tester into two categories: small systems with price tags of $50,000 and under; as well as large mainframe systems, which typically cost from $75,000 to over $2,000,000.

Within the parts test family, there are various electronic component groups:

- *Passive components*: test resistors, capacitors, inductors, and transformers.
- *Discrete semiconductors*: include diodes, zeners, transistors, triacs, and optical devices.
- *Linear integrated circuits*: include operational amplifiers, voltage regulators, memory drivers, timers, audio circuits, data converters, and analog microprocessors.
- *Digital integrated circuits*: divided into three primary subdivisions: memory, standard circuits, and microprocessors. There are general-purpose digital test systems as well as ATE intended for a particular subdivision such as memory circuits.
- *Relay and power supply*: test subassemblies such as system components.

See Table 1, Applications for Automatic Test Systems, for the purpose and approach to the testing of the various electronic component groups.

The characteristics of the various component ATE vary widely, which makes designing a universal test system very difficult. Passive component test systems are typically limited to two or three terminal tests and can measure resistance, capacitance, and inductance. Discrete semiconductor testing requires a three-terminal tester with high-voltage and high-current capability, typically to 2000 V and 100 A. Precision power supplies, with both constant-current and constant-voltage capability, are needed for linear testing together with precision measurement circuitry and provision to connect to 128 terminals or more.

Digital testing requires only low voltage and moderate current but needs complex pattern generators, multiphase clocking capability, connection paths to 256 terminals, and the ability to measure very short time intervals accurately. To

Table 1 Applications for Automatic Test Systems

Testing classes	Purpose	Approach
Passive component	Accept/reject sorting	Electrical parameters tested
Semiconductor component	Accept/reject sorting	Electrical parameters tested
Bare board, wiring	Accept/reject	Ohmic short, open: leakage current
In-circuit board test	Troubleshooting repair	Ohmic or impedance testing to models or electrical isolation
Functional board	Accept/reject troubleshooting repair	Parametric testing with physical isolation comparison with known good board
Functional product	Accept/reject troubleshooting repair	Parametric testing at operating speed against design specification
Off-line maintenance	Accept/reject fault-isolate repair	Static, dynamic parametric testing against performance specification and failure mode information

support testing of linear/digital combination parts such as data converters and automotive engine controllers, linear test systems have been enhanced with slow-speed digital options. Some large digital test systems have had analog and even microwave test capability added. In general, however, full electronic component test capability requires at least four or five ATE systems.

Subassembly and board testing utilizes four types of ATE systems:

- *Bare board and wiring verification*: tests the printed conductors on circuit boards and back plane or card cage wiring. These testers measure resistance to locate open or shorted conductors.
- *In-circuit testing*: includes simple resistance testing to find shorts and opens and the capability of testing each component on the board to assure that it has been installed properly and is functional.
- *Functional test—comparison*: evaluates an unknown board by comparing its operating characteristics to known good boards.
- *Functional test—operating*: operates the board in a way that emulates its use in the end product and tests its functions.

Wiring verification and in-circuit testers are inexpensive themselves but often require elaborate fixtures. Contact to the board under test is via a "bed of nails" containing up to 3000 individual contacts. The board is held in place by mechanical or vacuum pressure. Although some beds of nails can service more than one type of board, often an individual fixture is needed for each board to be tested. Similarly, a complex set of cables and connectors is needed for wiring verification of back planes and card cages which often dwarf the test set.

There are specialized wiring verification testers manufactured which are intended only for cable assemblies and harnesses.

Functional testers evaluate a board from its connector, so complex bed of nails schemes are not needed. However, the board must be designed in such a way that all needed stimulus and measurement points are brought out to the connector; or a number of wires and clips are needed which connect to components on the board, a process that is time consuming and which can damage the board. Because of the limited number of connections to the board under test, the test operator is often provided with a hand probe which is used following instructions from the test system.

The current trend in high-volume electronic manufacturing is to use both in-circuit and functional testing. In-circuit testing is very effective at finding components inserted backward and shorts that could be difficult to locate during functional testing. Also, unskilled operators can be employed for in-circuit evaluation, as they just insert and remove boards. A functional test requires an operator with a high level of training and the tester itself is more expensive and complex. The functional test process moves much more quickly if circuit faults and defective components have been located and corrected.

To date, general-purpose ATE for final test at the systems level has been developed for the merchant market. Electronic products vary widely in final test requirements. A radio set is not similar to a large computer. Neither is tested in the same manner as for an electronic game or an automotive antiskid controller. System-level ATE is custom designed to test a particular item or product. The military, for example, will purchase an ATE system for the Phoenix missile and for the avionics systems in the F16 fighter aircraft. ATE for final tests resemble functional board testers with enhanced power capability and specialized stimulus and measurement instrumentation added. The availability of a wide variety of power supplies and test instruments that can be used as modules in a system has aided in the design and construction of ATE for final-test applications.

IV. SYSTEMS SIZE

Small ATE systems, nicknamed "benchtops," are popular in incoming-inspection applications because of their low cost. Many of the small systems currently on the market are first-generation designs, although newer versions using microprocessors are available for most applications. The microprocessor-based small systems usually cost two to three times as much as the first-generation benchtops. Their high initial cost can easily be justified by their lower cost of programming and reduced operator setup time.

Benchtop systems are generally constructed as an integral unit with control keyboard, display, and storage device built in. The original storage media were usually magnetic cards or magnetic tape in cartridge or cassette form. Miniature floppy disk drives are the rule rather than the exception.

Small systems are designed to test a particular product family. Usually, they can evaluate between 70% and 90% of devices within the family. Tests and devices that require special conditioning, unusual stimulus, high power, or a high pin count can usually be run. The only real disadvantage of benchtops was that they were slow; often many times slower than large mainframe systems. This disadvantage has been overcome with the advent of microprocessor or PC-controlled systems.

Mainframe ATE is characterized by use of a general-purpose computer and mass-storage device plus provision for a large number of accessories and options. It has the capability to test all the devices in a family, and it can be expanded and reconfigured as requirements change.

V. APPLICATIONS FOR ATE

Manufacturers of electronic components use ATE to test and classify their products. Semiconductor components are tested in wafer form to eliminate defective devices before the assembly process, during assembly to assure that damaged devices are eliminated, and at the end of the production cycle to determine final classification of each part before marking. Component manufacturers also employ ATE in the engineering laboratory to characterize new designs and in the quality assurance department.

The manufacturers of electronic equipment use ATE for incoming inspection of components, circuit boards, and power supplies. Although this inspection is the first step in the manufacturing process, the function is usually assigned as part of the quality assurance program. Once the circuit boards are assembled, they are tested on ATE for correct assembly (in circuit testing) or functionality or both. ATE is also employed to verify cables, harnesses, and back planes. Final tests may or may not use ATE. The current trend is to design a good deal of self-diagnostic capability into electronic products, which is a major aid in final tests. Military weapons systems almost always have ATE testing. Commercial practice has reached this level of sophistication for many applications. Many military systems have required built-in test equipment and makers of sophisticated commercial telecommunications systems, computers, facsimile machines, etc., have begun to adopt the built-in test equipment approach.

Incoming inspection and board test are often viewed as areas of trade off. Firms with strong quality assurance programs emphasize incoming inspection. It makes no sense to use bad components, and a certain portion of each batch is known statistically to be bad. Manual inspection methods generally preclude 100% inspection. Automatic test equipment makes full inspection a realizable goal. The rule of 10s applies, where if it costs $0.05 to find a defective part at incoming inspection, it will cost $0.50 to find the same bad part at the board level and $5 at the system level. In some organizations, there has been a tendency to skip component testing and to attempt to find all faults at the board level. Most quality engineers have learned that board testers are not the best component test systems. Their function is to spot manufacturing defects and to check board functions. The typical defects found during in-circuit board testing are shown in Table 2.

Field service has traditionally been very labor intensive. As the use of electronics has become part of a wide variety of products, a severe shortage of trained technicians has developed. Thus the situation that gave impetus to the development of ATE and BITE in the military is occurring in the commercial sector. To improve the productivity of service personnel, new classes of ATE are being developed for board test afield. In addition, regional service centers are being established which have board and system test facilities similar to those of the factory. A typical service operation today can lead the customer through BITE sequences and in many cases advise the customer of the exact action to be taken without the necessity of a service call.

VI. SYSTEM DESIGN

An automatic test system block diagram is shown in Figure 1. The processor is typically a general-purpose microcomputer which acts as a system manager. It also interprets program instructions, controls system signal flow and timing, allocates system resources, provides mathematical capability, stores test results, and provides human-machine interface via keyboard. In addition, depending on the design of the system, the computer may be active or passive in the execution of tests. It also can be used for automatic system calibration and for diagnostic maintenance aids.

Table 2 Typical Board-Test Results (%)

No defects found	67
Opens and shorts	16.5
Missing or wrong parts	6.6
Improper insertion	3.3
Defective components	3.3
Interactive functional failures	3.3

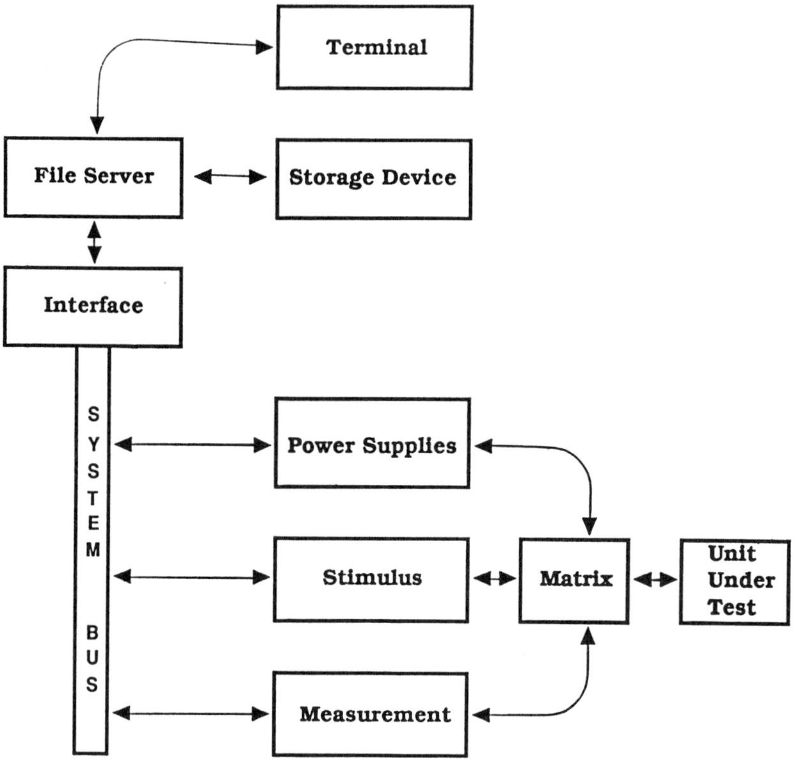

Figure 1 ATE block diagram.

The important computer characteristics are speed, the amount of memory it can address, and how arithmetic functions are handled. Speed is important because the computer must handle many tasks, a requirement that limits the use of microprocessors in large systems because they are relatively slow. A memory-limited system can be difficult to use, so the amount of memory—especially the *free* memory left once the operating program is loaded—is important. Test systems must perform multiple arithmetic calculations. Multiplication and division are slow when done in software, so computers that feature hardware implementation of multiplication and division are preferred.

The file server will have provision for one or more terminals. As a rule, one terminal per test head or operator is desirable plus one terminal for system control. The terminals are usually video display with a keyboard. In some instances, they may be microcomputers, themselves. Terminals communicate with the computer using the ASCII (American Standard Code for Information Interchange) code in a direct-current loop or using the American Standard RS-232 interface. A line printer is employed where large volumes of test results are to be printed. It typically uses a parallel computer interface designed specifically for a given printer-computer combination.

A number of mass storage devices are used in ATE systems. They include:

- *Hard disk*: a magnetic record that can store from 5 to 180 million bytes (1 byte is 8 bits) of information with fast access and high-speed data transfer.

- *Floppy disk*: a flexible magnetic record that can store from 128,000 to one million bytes of information. Access and transfer times are slower than with a hard disk. Floppy disks must be handled carefully, as they are easily damaged by dust and fingerprints.
- *Magnetic tape*: a tape employed in several forms, including reel to reel, cassette, and cartridge. The IBM-designed nine-channel tape format is often employed to back up disks in case they are damaged and to provide a transfer medium to separate computer systems.

The terminals and mass storage devices are called I/O (input/output) devices for the computer. They are interconnected via a high-speed parallel bus. The structure of the bus varies from manufacturer to manufacturer and for different computer models. Most ATE systems are controlled by extending the computer bus through an interface which provides isolation and noise reduction. The details of the bus vary, but a bus generally features high-speed parallel operation where parts of the tester are treated as I/O devices; or direct memory access to the computer can be provided where parts of the tester are treated as memory addresses.

In component and board testers where test times are measured in milliseconds, a high-speed bus is required. The standard for interconnection of test instruments, IEEE-STD-488, is a relatively slow bus structure. A series of instruments and power supplies can be interconnected to a computer via 488 to form a custom ATE system. Its operation would be sufficient for system final test or where individual test times are long. The 488 bus is also used to add and control extra instruments on component and board test systems where high-speed communication is not required.

Power supplies are a key element in automatic test systems. They vary widely in capability and accuracy. The power sources may be constant voltage, constant current, or both, called V/I sources. Most are programmable for voltage output and maximum current. The stability and forcing accuracy of the power sources are very important where precision measurements are to be made. Because long connecting cables are used which introduce both inductance and capacitance, power sources must be free of parasitic oscillations over a wide range of load conditions.

Power sources may be pulse types able to deliver high current for only short periods or units rated for continuous power. Discrete semiconductor test equipment usually employs pulse-type supplies, while linear IC and board testers have sources rated for continuous operation. V/I supplies can be employed as precision loads as well as sources by appropriate programming of voltage and current (positive voltage and negative current, for example). Precision power sources usually include measurement circuitry so that the current can be forced into a lead and so the voltage applied to the device or circuit under test can be checked.

Tests such as leakage current, breakdown voltage, and device gain can be performed using power supplies as the stimulus. Other sources of stimulus include audio, radio frequency (RF), pulse, and digital-pattern generators. Specialized stimulus devices are required for unique devices; the stereo-signal and croma generators needed in testing entertainment/linear integrated circuits are examples. Basic stimulus generators are designed into automatic testers as system resources programmed via software as required. Others are offered as options.

When a special stimulus is needed which is beyond the capability of the ATE systems, additional test equipment (controlled by the system) may be employed. Or conditioning, stimulus, and measurement circuits may be built for a particular device to be tested, called a test box, test package, or family board. Such units may be part of the mainframe or located at the device interface.

Basic DC and AC (audio frequency) measurement capability is standard in most automatic test systems. A radio frequency voltmeter is often an option, as is a counter for frequency, time, and period measurements. Where high resolution is needed, a counter with averaging capability is used.

System power, stimulus, and measurement resources are connected to the unit under test via a matrix. The size and complexity of the matrix varies widely, from two relays for polarity reversal in diode testers to several thousand switching paths for in-circuit board and cable testers. The important specifications for the matrix are its voltage and current capabilities and its frequency response. The matrix should have several leads per path to allow for remote voltage measurement. The matrix requires sufficient input paths for all system resources and accessories. The number of output paths needed is determined by the access points or pins at the unit under test.

At the output of the matrix is the interface to the unit under test. The interface may be a connector or adapter socket. It often contains bypass capacitors to prevent oscillation of the unit under test. If special test or conditioning circuits are needed, they are usually located in the interface.

As the use of automatic test systems has grown, so have the requirements for programming and for data analysis. Most commercial ATE manufacturers have developed a form of test system manager; a computer system with provision for large-capacity storage on disk or magnetic tape. Although details of the test system managers vary widely from manufacturer to manufacturer, from 4 to 16 test systems can be connected to the manager. It can store large libraries of test programs and transmit individual programs to the testers on command. The testers pass test data and lot results to the manager, which has programs for data reduction and for sophisticated statistical analysis.

Some managers have programs for daily and weekly summary reports. An important feature of test system managers is the ability to support user programming in one or more high-level languages. This allows users to write their own analysis and report programs.

VII. SOFTWARE

The most important element in any automatic test system is the computer software. It determines how the system operates and the ease with which new test programs can be created. The cost of program generation is substantial; often several times the cost of the system itself.

The operating system used with the computer may be an original design by the tester manufacturer or one of the systems popular in microcomputers. A custom-designed operating system is often more efficient and uses less computer memory space. However, it also usually lacks the flexibility of the general-purpose systems. Together with the operating system, the software for ATE includes a master operating program (MOP) or executive program. This software controls the tester, including such tasks as executing tests, communicating with terminals, data logging, and managing computer memory space. The executive programs are extensive and require a major portion of the computer memory.

Some ATE systems use a standard language, such as BASIC, FORTRAN, Pascal, or Atlas for program generation. Atlas has become a standard for military and aerospace system ATE. Component and board testers usually have a developed language roughly based on FORTRAN or BASIC, with enhancements and modifications to fit a particular requirement and set of hardware. There is no standardization of semiconductor-device tester languages. Each manufacturer and usually each individual test set has a specialized language. ATE also has sets of accessory software for system diagnostics, calibration, operation of optional equipment, and specialized testing such as high-reliability analysis. Computer controlled environmental test systems are usually programmed in BASIC or FORTRAN.

VIII. PROGRAMMING

Automatic test systems for discrete components and simple integrated circuits are usually preprogrammed by the manufacturer for all possible tests. Thus programming in the true sense of the word is not needed. The user builds test routines aided by the ATE system via prompting or menu display. In the prompting approach, when a particular test is requested the tester responds with a series of questions that request forcing functions and limits. The tester checks entries for appropriate responses and rejects errors. Menu entry is similar. At each step, the test system displays on a video screen all the possible alternatives available at a given point in a test routine. The user selects the desired step from the alternatives in the menu; then a new menu is displayed with a new set of alternatives.

Test systems for complex integrated circuits and circuit boards allow programming where one or more instructions are contained per line. In some systems, generalized tests can be created without forcing functions or limits. These generalized tests are called via software when needed. Forcing function information is inserted before the test executes. In this way, the programs for each device can be quite simple. Other testers require that all tests must be created for each device to be tested, which can be very laborious when many similar types of devices are to be tested. For in-circuit board testing, generalized software is available with the test systems where common components such as diodes, gates, flip-flops, operational amplifiers, and so on, are prewritten in modules. Thus a good deal of the programming task can be reduced to calling the appropriate software module. Cable and back plane testers usually have a self-programming feature where the pattern to be tested is "learned" from a known good unit.

Manufacturers offer prewritten software for many of the popular integrated circuits. They have also formed user groups where members contribute test routines and in return receive copies of programs from others in the group.

A test routine for a complex device or board can take a month or more to write. Because hundreds of instructions are involved, chances are high that there will be one or more mistakes in the finished program. Corrections of these mistakes requires a process called debugging. Modern ATE systems are equipped with hardware and software tools to aid in the debugging process. External access is provided to system resources, matrix connections, and control lines so that an oscilloscope or other test equipment can be employed to examine the dynamic performance of the system. Software aids include the ability to obtain a listing of the status of all system hardware at any point in a test routine. Such a list is valuable to show that all hardware was properly programmed at a given instant. An immediate execute mode of operation is also useful so that as changes are made in a routine, it can be executed to assure the desired result is achieved.

Computer-aided design techniques are increasingly being used to help in writing board- and system-test programs. A large-capacity computer can provide modeling and simulation techniques that dramatically reduce the time necessary to generate a complex program. Another trend is to use automatic test generation for semiconductor devices. In this approach, all tests are preprogrammed. The specifications for all devices to be tested are stored in a library. The library can be within the system, in a central manager, or in another computer. To create a test routine, the user specifies the devices to be tested and the sorting priority desired. An efficient test routine is built automatically using the data in the library.

IX. MAINTENANCE

Early ATE systems often totally neglected calibration and maintenance requirements. Later generations contain powerful software to aid in the maintenance process. The system as a whole is checked against an external or internal standard. Such diagnostic software provides an indication that the system is performing within specifications. If trouble is detected, specific diagnostics for a section of the unit can be run to pinpoint the problem to a particular circuit board or relay. Most ATE users do not attempt repair beyond changing boards and relays. Thus spare-boards kits are purchased with a test system. Manufacturers provide fast-turnaround board repair and a stock of spare boards at regional offices.

Modern ATE systems often contain both an operating and a maintenance manual within the computer. Special programs tell the maintenance technician what tools and test equipment are required to perform a particular task, provide notes about proper procedures, and detail step-by-step how to proceed. The sophisticated programs check the technician's work as he or she proceeds.

The calibration standards for the ATE system generally are removable as a unit, including the power supply, so that they can be sent to a laboratory for certification as a transfer standard. Traceability to the National Institute of Standards and Technology (NIST) is a requirement in some testing applications and a good procedure in all cases.

In older systems that use manual calibration, the need for recalibration is indicated by diagnostic software. Additional software programs set up individual sections of the system for calibration. A technician then proceeds to make whatever manual adjustments are required. An important feature of more current ATE is to use automatic calibration. In one approach, the system computer runs diagnostics continuously in the time spaces between test routines. Any errors that are found are corrected by inputs to digital-to-analog converters, which correct errors of scale or offset. The entire test system is recalibrated every few minutes. Another approach requires that the system be shut down once each shift. Extensive diagnostics are run and the error results found are stored in a software lookup table. When testing resumes, raw test results are automatically adjusted by the values stored in the lookup table and corrected to eliminate errors. Thus the system is self-correcting by means of software for error of calibration over a range of ±10%. Typically, the hardware contains no adjustments at all.

39
Testing Laboratories

RAY A. KLOTZ
World Class Quality Consulting Company
Escondido, California

JOHN JOURDAN HELDT
Free Lance Reliability Service
San Jose, California

I. INTRODUCTION

One estimate of the dollar value of testing laboratory services in 1993 listed total revenue of two billion dollars. There are, however, some five thousand test labs to share this revenue; thus the majority of these labs are small, independent, local, and specialized.

Table 1 lists many of the tests that U.S. test labs typically provide. Most labs, however, concentrate their services in one of the following specialties: (1) materials testing, (2) failure analysis, (3) environmental conditioning.

II. MATERIALS TESTING

Materials testing can generally be categorized as either destructive or nondestructive. As its name implies, destructive testing virtually always leaves the test specimen in a condition where it can no longer be used for its originally intended purpose. Firing a rifle bullet to measure its muzzle velocity is the classic example of a destructive test. X-raying is a classic example of a nondestructive materials test.

Tensile strength tests (see also Chapter 40) of metals are among the most common of destructive materials tests. Tables 2 and 3 list some properties of some common metals. Tensile strength is defined as the amount of stress applied to a material, typically measured in pounds (or kilograms) per square inch (per square centimeter), that just causes the material to fail. Tensile tests are often performed on "test coupons" that are dog-bone-shaped and have been cut out of the parent material. (At least one company, however, makes a specialized metal-cutting machine that produces "standard" tensile test coupons.) Material samples that are too

Table 1 Types of Tests Test Laboratories Provide

Acoustic testing and conditioning	Field and maintenance tests
Analytical tests	Film radiography
Biological testing	Filtered-particle tests
Chemical analysis and test	Fluoroscopy and x-ray image devices
Construction materials evaluation	Fundamental testing principles
Electrical testing and evaluation	High-voltage radiography
Electromagnetic interference (EMI) evaluation and test	Isotope-radiation sources
Engineering test	Liquid-penetrant test equipment
Environmental testing and conditioning	Liquid-penetrant test indications
Laser alignment	Liquid-penetrant test principles
Laser holographic systems	Magnetic-field test equipment
Laser measurement	Magnetic-field test principles
Lubrication evaluation and application testing	Magnetic-particle test equipment
Mechanical testing	Magnetic-particle test indications
Metallurgical testing	Magnetic-particle test principles
Metrology and calibration	Management and application of tests
Nondestructive test (NDT)	Natural frequency vibration tests
Nuclear testing and conditioning	Nondestructive testing laboratories
Optics/photometry/x-ray	Optical projectors and comparators
Plating, coatings, and finishes	Photoelastic-coating tests
Repair of rest gear	Radiation and particle physics
Stress analysis	Radiation detection and recording
Thermal testing and conditioning	Radiation protection
Vibration	Resistance strain-gage tests
Weld tests	Ultrasonic contact tests
Brittle-coating tests	Ultrasonic fields
Double-transducer ultrasonic tests	Ultrasonic immersion tests
Eddy current cylinder tests	Ultrasonic immersion test indications
Eddy current sphere and sheet tests	Ultrasonic resonance tests
Eddy current test automation	Ultrasonic test principles
Eddy current test equipment	Ultrasonic transducers
Eddy current test indicators	Vision and optics
Eddy current test principles	Visual inspection equipment
Eddy current tube tests	Xeroradiography
Electric current test principles	X-ray and isotope gaging
Electrified-particle test indications	X-ray control of weldments
Electrified-particle tests	X-ray diffraction and fluorescence
Electronic radiation sources	X-ray film processing
	X-ray interpretation

This listing is based on test categories shown in *Quality* Magazine's "1993 Buyer's Guide."

small in cross section to permit coupons—such as most electrical wire—are generally tested in the form that they will be used in actual service.

Hardness testing (covered in detail in Chapter 41) is most often classified as a nondestructive test. Because the indenter used in virtually all hardness tests leaves a perceptible flaw in the material surface, however, hardness testing of a bearing race, for example, would almost certainly be considered a destructive test.

Table 2 Mechanical Properties of Some Common Metals

Material	Tensile strength (psi)	Yield strength (psi)	Shear strength (psi)	Percentage elongation	Brinell hardness	Impact strength (ft-lbs)
Aluminum	13,000	5,000	9,500	45	23	—
Copper	32,000	10,000	22,000	45	40	29–113 IZOD
Gold	19,000	nil	—	45	25	—
Iron (alpha)	38,000	19,000	—	43–48	67	—
Lead	1,900	800	1,825	30	3.2–4.5	10.4 CHARPY
Silver	18,200	7,900	—	48	—	—

Wet and dry analytical tests are often used to verify the composition of materials. Spectrometers and spectroscopes—using x-ray diffraction, electrical sparks, and other methods of "exciting" the atomic structure of the material—and gas and liquid chromatography can provide extremely accurate analyses of material composition. Wet chemistry, in which classic chemical techniques are used, are routinely used to evaluate materials composition. Microstructural evaluation (see Chapter 37) has become a commonly used but powerful means of evaluating the crystalline structure of materials.

Impact strength tests are always considered to be destructive tests, because the test involves notching the material before it is hit with a calibrated hammer blow to measure the energy required to break the notched part or material.

III. FAILURE ANALYSIS

Although many companies, especially larger ones, often develop their own materials-evaluation labs, failure analysis, particularly in cases in which a product failed and caused injury or substantial property damage, tends to be the province of specialized independent laboratories. This specialization occurs because even relatively large companies could not afford the highly paid metallurgists and other experts for the (hopefully) occasional service they would provide. The failure analy-

Table 3 Physical Properties of Some Common Metals

Material	Coeff. of linear expansion (μin/°F)	Density (g/c^{3}m)	Melting point (degrees F)	Electrical resistivity (μ Ohm-cm)
Aluminum	13.3	2.71	1215	2.922
Copper	9.2	8.96	1981	1.673
Gold	7.9	19.32	1945.4	2.19
Iron (alpha)	6.5	7.87	2800	9.71
Lead	16.3	11.34	618	20.65
Silver	10.49	10.49	1760.9	1.59

sis laboratory, however, drawing on a wide range of companies, can generate sufficient bookings to support these experts.

An unfortunate but real example of failure analysis work is that which occurs when an airplane crashes. Usually all of the broken parts are brought into a building such as a hangar where the plane is reconstructed—as well as is possible.

A. Two Inspections

Typically, two inspections are made of all failed components. The first inspection is used to document the component as to its as-failed condition. This inspection may include pictures and descriptions of the damaged or broken component or assembly, including a thorough accounting of the steps taken in removing the failed elements of the assembly. When large assemblies, such as a crashed aircraft, are removed to the reconstruction site, or when components such as a twisted lever or an inoperative microcircuit are taken to the test laboratory, they are inspected a second time to document their as-received condition.

Once the as-received condition is documented, the actual failure analysis begins. When structural metal parts fail, there is generally one of four major types of failure that occurs. The presence of one (or more) of these modes provides vital clues to the kinds of forces that were applied and the length of time that the item was stressed to cause the failure.

B. Ductile Fracture

The ductile failure is so common it could be called the usual failure mode. This type of failure is due to the application of stress that exceeds the strength of the material. In most metals, the clue to a ductile failure is the obvious stretching (necking down) of the material before complete separation. The metal need not separate into two (or more) pieces, however, for a device to fail. Stress that exceeds the plastic limit, even if it doesn't exceed the tensile strength, of the material also typically causes device failure.

Ductile Failure Study

In an aircraft application, a mechanical linkage is used to change the electrical signals from an automatic terrain following computer (ATFC) into mechanical forces that are use to synchronize the airplane's flight for high-speed, low-level response to the terrain below.

On one flight, a pilot noted that the ATFC was a little slow in responding to the terrain below. Switching from the ATFC to manual, he gained altitude and returned to the base.

In the failure analysis that followed the flight, one of the load-bearing arms in the ATFC was found to be severely twisted. Material analysis showed that the arm had been fabricated from mild steel rather than the specified 304 stainless steel, which is much stronger.

All airplanes with ATFCs were immediately grounded until the material status of the arms was verified. In all, seven more mild steel arms were discovered.

C. Brittle Failure

The clue to brittle failure is the complete, or almost complete, lack of "necking down" of the material at the point of failure. Because there is almost no energy absorbed in a brittle failure, the result can be spectacular.

Brittle Failure Study

During a recent police action, troops were boarding a plane that was to take them back to the combat area after several days of rest and recuperation. As the plane was being readied for takeoff, a tank commander commented to an infantry officer, "Well, back to the same old stuff." The officer responded, "What same old stuff?" "Driveshafts that crystallize and shear off. Then we go out into the desert and try to replace them." "Don't you keep a log on the tanks?", asked the infantryman. "Yes, we do", was the tank commander's quick response. "We know that the average drive shaft crystallizes and fails after about 2100 hours of operation. I've had one fail at 1860 hours and one lasted for 2,711 hours of operation." After a moment, the infantry officer said, "Looks to me like you should pay attention to those logs. When you get to 1500 hours of operation, replace the shaft and make a fence post out of it."

After a period of time under stress and vibration, the metal of the drive shafts would fail, quite reliably, in a brittle failure mode.

D. Intergranular Failure

Intergranular failure, or separation, is due to a failure of a material's interstitular material. This occurs when the material between the crystals of a metal contains an embrittling component. A resulting failure looks much like a brittle fracture. Often very high-strength steels are subject to this type of failure because of intergranular corrosion that acts as a stress riser for rupture. Intergranular corrosion is often due to overheating the surface of a high-strength part during a grinding operation.

Intergranular Fracture Study

An aerospace manufacturer investigated a structural part of a landing gear that had failed. Intergranular failure was suspected. During the analysis, the part was chemically etched to reveal the material's grain structure. Following the etching, several white stripes several thousandths of an inch wide were noted. Further investigation revealed that the stripes were areas of the part in which the heat treatment had been affected by overheating during the finish grind operation. These deteriorated heat-treated areas provided the places in which stress corrosion would begin, ultimately affecting the part's strength.

E. Fatigue

Metal fatigue is a gradual fracturing of the material due to repeated and prolonged flexing, vibration, or similar stress. Unlike human muscular fatigue that can be corrected by rest, metal fatigue is permanently cumulative. The fracture mechanics of metal fatigue have been most intensely examined by the aircraft industry, which goes to great length to prevent such failures. Fatigue failure, however, also accounts for some 75% of all machinery failures.

Fatigue Failure Analysis

A test lab was contracted to test the safety feature of a new dual airbag configuration for a new model automobile. The plan was to drive five identical automobiles at different speeds into a concrete barrier using remote driving controls and crash dummies. To be evaluated was the relative passenger protection versus the speed of the crash.

When the test site was completed, the lab learned that it would be three weeks before the five new cars would be turned over to them. As much to test the site and the test equipment, the lab decided to test five cars that had been thoroughly used. They found five old cars with good bodies and ran them through the test procedure. The results were not as good as the test engineers had anticipated.

When the five new cars were eventually tested, the results indicated that the dual bag system offered fair protection to the front seat passengers up to 63 mph and excellent protection up to 56 mph. The older auto test results indicated fair protection to 52 mph and excellent protection up to 41 mph. (The speeds listed here are approximations for the purposes of this illustration. Actual figures are company private data. It is enough to know that there was a significant difference in safety versus speed even when the protection offered to the passenger was of the same sort and in the same condition.)

The difference in the test results was due to metal fatigue and corrosion that significantly reduced the crash-withstanding abilities of the older cars.

IV. ENVIRONMENTAL CONDITIONING

The electronics industry has made much use of environmental conditioning, or screening testing as it is sometimes called.[1] For example, entire lots of semiconductors[2] are artificially aged to speed the failure of marginal units. Since this testing involves variations of ambient temperature and vibration, it is sometimes referred to as "shake and bake" testing. The tests are generally designed to cause the same amount of stress as six months of normal operation; experience being that all premature failure[3] will occur during the first six months of normal operation.

---

[1] In addition, a visual inspection of the chip prior to encapsulation is believed to be able to remove 75% of the parts that will fail prematurely. There are some tests that are substituted for *pre-cap* inspection. (1) The Particle Impact Noise Detection (PIND) test vibrates each part in the lot. The pattern of the vibration is shown on an oscilloscope. Any loose material (particles) within the part will cause additional vibrations, which show up as noise on the top of the oscilloscope trace. (2) A sample from the lot is delidded and visually inspected. This will give an indication of lot quality at a confidence level dictated by the sample size.

[2] In this context, the term *semiconductor* includes all classes of diodes, transistors, microcircuits, and hybrid assemblies.

[3] Premature failures are caused by poor workmanship or substandard material. In reliability terms, premature failure is called "infant mortality."

Test labs use different screening programs for the various levels of semiconductor quality. As an example, the highest quality space program microcircuit costs about 100 times as much per unit as a commercial quality unit because of the testing and screening required. Commercial parts are usually minimally screened to a given level of acceptable quality (i.e., AQL); typically about 1.0%.

Many test labs have set up destructive physical analysis sections just for semiconductors. Failed units are decapped (i.e., opened) to determine the failure mode and the cause of the failure. Often, the chip is redesigned on the basis of the findings. Sometimes a more rugged part is specified to bring the circuit to a more robust configuration. In other cases, the circuitry may be changed to compensate for component weaknesses. Increasingly, independent test labs are being used to support design improvement by analyzing failures to determine the root cause of the failure. [FMEA (Failure Modes Effect and Analysis) is sometimes used. For more information, see MIL-STD-785 Reliability Program for Systems and Equipment Development and Production, and MIL-STD-1543 Reliability Program Requirements for Space and Launch Vehicles.]

Nuclear Hardening

A number of independent labs have established nuclear testing and nuclear hardening capabilities. Nuclear testing is performed to demonstrate that cables, semiconductors, or circuits are able to perform as specified when they are subjected to prescribed amounts of nuclear radiation for a specified period of time. These tests may be performed under load conditions wherein the inputs are applied to the unit or component and the outputs are monitored by telemetry from a safe distance. Radiation may be applied to unpowered units for the prescribed time, after which the devices are rapidly cooled and kept frozen until operational testing is performed. (Freezing the parts keeps them in the radiated state until testing can demonstrate degradation, if any, that has occurred.)

Nuclear hardening is similar to nuclear testing except that hardening is used as a part of the design program so that the units can be redesigned if they fail to perform under specified radiation.

FURTHER READING

Allen, Dell K., *Metallurgical Theory and Practices*, Homewood, IL: American Technical Publishers.

40

Tensile Testing

JOSEPH J. CIEPLAK
Acco Industries, Inc.
Bridgeport, Connecticut

I. INTRODUCTION

The basic principles of tension and compression testing have changed hardly at all since 1865, when the Riehle brothers of Philadelphia designed and manufactured the first tensile testing machine (Fig. 1). A known load is applied to a sample of the material being tested, and the deformation of the material under that load is measured. Throughout the test, the amount of deformation, is measured to determine the relative relationship of the deformation to the load.

A thicker sample of the same material requires a larger breaking load than that required by a smaller sample. It is therefore necessary to normalize the load and deformation into parameters that do not depend on sample size (Fig. 2). To do this, we convert load into stress by dividing the load applied by the area over which the load acts. The resulting units are called stress, and, in the English system, are expressed as pounds per square. In most cases, the load is normalized by testing specimens of specific length, width, and thickness.

To normalize deformation, we determine the change in length of the sample under a specified load and divide that result by its original length. The resulting parameter is called strain, and in the English system is indicated as inches per inch. Both of these definitions are properly referred to as engineering stress and engineering strain, as they do not take into account the necking down which occurs as a sample is stretched toward breaking.

II. LOAD-DEFORMATION CURVE

The normal result of a tension or compression test is a plot of load versus deformation on graph paper. This plot is called a stress-strain curve, or more accurately a

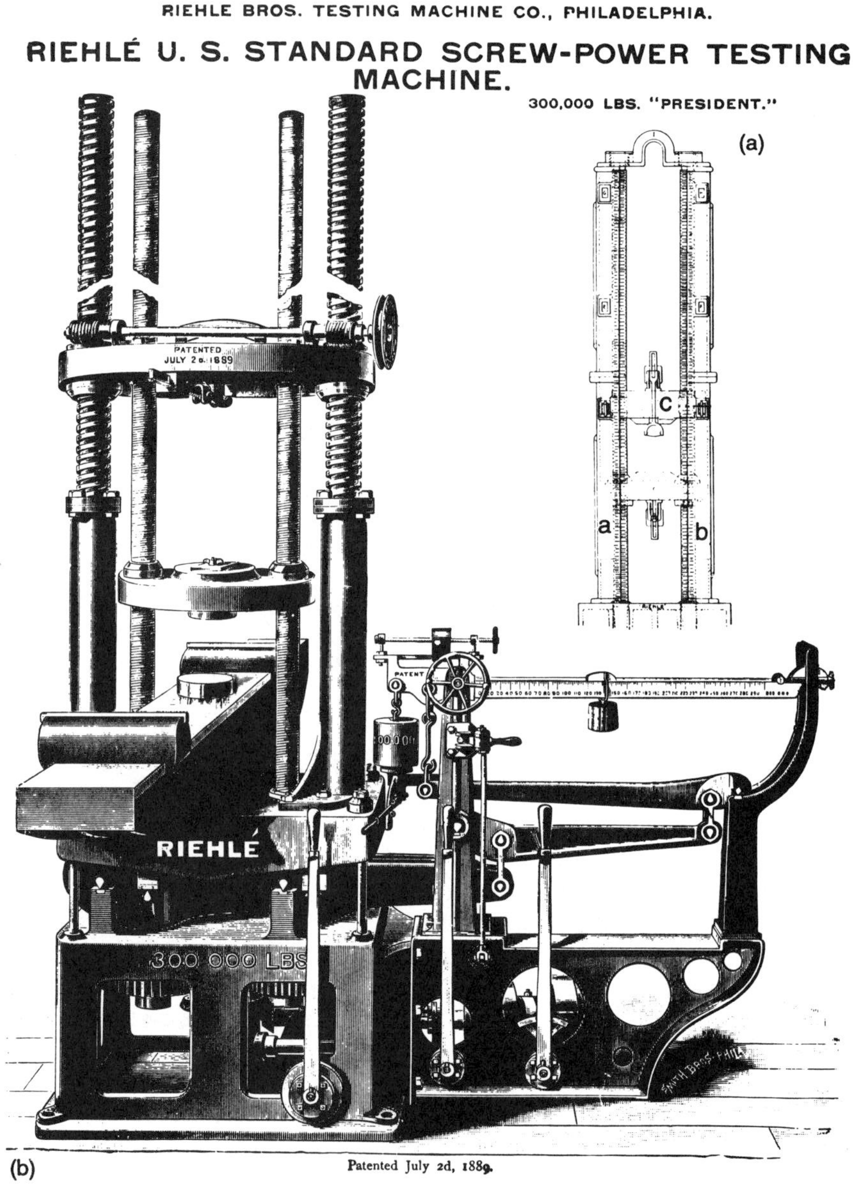

Figure 1 The Riehle Brothers' original tensile test machine: Sketch (a) shows columns A, B, and the upper grip head C.

load-deformation curve, and it is this graph which gives us the basic information used to calculate all our results.

In a typical stress-strain curve for metals, load is plotted on the left-hand vertical axis, and elongation is plotted on the horizontal bottom axis. A typical load-deformation curve is shown in Figure 3. The first portion of the curve between the zero point and point A is the elastic portion. In this region there is a direct and uniform relationship between the amount of load applied and the resulting elongation, represented by a straight line. These results are indicative of most

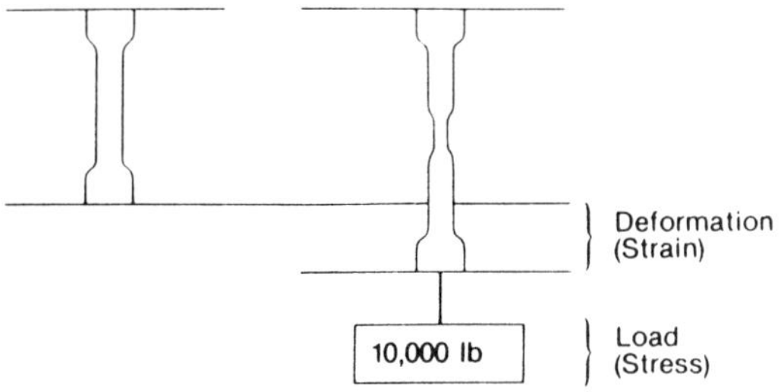

Figure 2 A load applied to a sample is called stress. The resulting deformation of the sample is called strain. Engineering stress and strain do not include the effects of sample neck-down.

metals. Other materials, such as plastic or rubber, do not normally exhibit a straight-line relationship anywhere on the curve. Up until point A, which is called the proportional limit, the test could be aborted and the sample would return to its original length with no permanent change having resulted from the test. Once the test has gone through the proportional limit, the material generally has undergone permanent deformation and no longer will be able to return to its original condition.

Points B and C are graphically determined yield points which represent the stress value at which the material being tested exhibits a specific limiting permanent deformation. This yield stress is generally regarded as the stress beyond which the material cannot be used in normal circumstances. Farther along the graph, point D represents the highest load or the highest stress seen by the sample during the testing process. This point is called the ultimate strength. All loading done after this point will be at a lower level as the sample "necks-down" (i.e., its cross-sectional

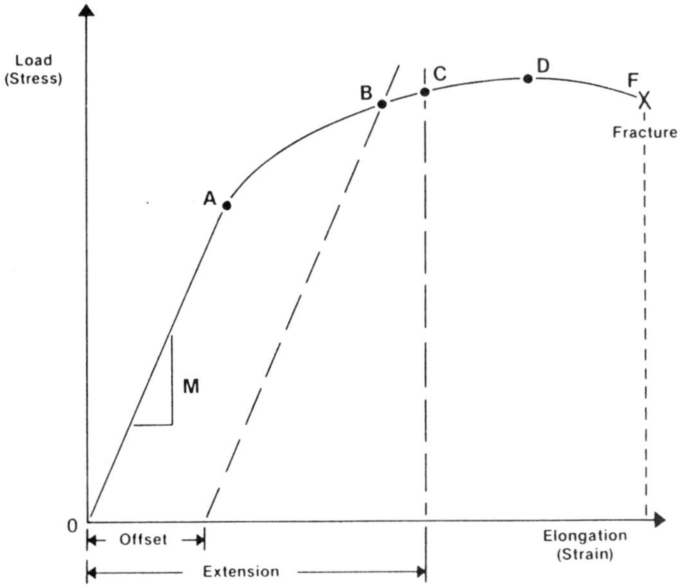

Figure 3 A typical load-deformation (stress/strain) curve.

area decreases as the elongation begins to increase rapidly). Point F is the fracture point at which the sample breaks and the test is concluded.

Once the graph has been drawn by the recording device connected to the tensile tester, the operator must graphically obtain the desired results. The slope of the line, up to point A, represents the completely elastic relationship of stress to the resulting strain. For metals, this is called Young's modulus and is obtained by dividing a specific change in stress by the corresponding change in strain. The Young's modulus of many metals can be expressed in millions of pounds per square inch.

Points B and C represent two different methods of determining the yield strength of the material. The choice of method depends on the material being tested and its normal characteristics. In the offset method, a particular strain offset is chosen, often 0.2% of the gage length, and a parallel line is then drawn 0.2% to the right of the zero starting point until it intersects the stress-strain curve. This point, B, is then considered the yield strength by the *offset method.* Similarly, the yield strength by the extension-under-load method is determined by moving the predetermined distance to the right of the zero point, often a 0.5% elongation, and then drawing a perpendicular line upward until it intersects with the stress-strain curve. This point, C, is then considered the yield point by the *extension-under-load method.* In both cases, the particular offset or extension being used is an integral part of the testing results and must be recorded in order for the results to be reproducible by other laboratories.

If we were testing other materials, we might be looking for different results. For instance, Young's modulus would be meaningless for rubber material, as there would be no straight-line portion of the curve. If we were testing a plastic material, we would be concerned with the tensile strength at the breaking point and the percent elongation at particular points during the loading process and at the breakpoint. Another calculation sometimes necessary for plastic testing is tensile energy absorbed (TEA). This result is obtained by integrating the area under the curve from the zero point to the fracture point. Although the tensile test itself is basically a simple test, the analysis of the results becomes the important task.

III. APPLICATIONS

Although a simple test, the tensile test provides information which can be used for many purposes. For example, in those cases where material is being made as part of a continuous process, specimens may be taken at specific intervals from the production line and tested on a universal testing machine (UTM) to determine the properties of the material being produced. If the test results show incorrect properties or a lack of homogeneity in the material, the laboratory can feed this information back to the manufacturing people to have the process adjusted to bring it to within the appropriate specifications.

Some products, such as steel, reinforced fiberglass, and wood composites, are sold with a guarantee that they conform to certain strength specifications. In these cases a tensile machine is used to qualify the products sold, and a copy of the results obtained from the test is often supplied with the material as proof of conformance.

Occasionally, it becomes necessary to verify certain materials to determine their mechanical properties before their use in a manufacturing process. In these circumstances, samples of an unknown or questionable material are tested in a tensile machine to ensure that they meet specifications.

Design engineers often choose a material to be used in a finished product by considering its structural characteristics, its formability, or its fatigue-resisting characteristics. In these cases, the design engineering group will compile information concerning tensile results and will probably specify tests to be run on the material they are considering for use in a critical application. In the never-ending search for stronger, lighter-weight, and less costly material, the universal testing machine is used routinely to compile research information for analysis by material scientists.

IV. UNIVERSAL TESTING MACHINE COMPONENTS

In its simplest form, a universal testing machine system (Fig. 4) includes a load frame where the test is actually performed. The load frame must, of course, be rugged enough for the application. Some means of control over the load frame is necessary. This control can be as simple as a hand wheel on a valve or as complex as a computer to control the loading and unloading process and the rates at which these are done. Generally, a recorder is used to record permanently the results of the test.

Grips or some other accessory device are used to interface between the sample being tested and the load frame itself. The action and use of the grips is often one of the most critical and least understood parts of the test. Many people who have been in the tensile testing industry for long periods of time consider the use and choice of grips to be an art, patiently and painstakingly learned through experience.

We also need a load-measuring device, such as a load cell, a pressure transducer, or a mechanical beam, to provide information concerning the load applied. Additionally, in most instances, we require information concerning the amount of elongation or deformation the sample has undergone. For these cases we use a

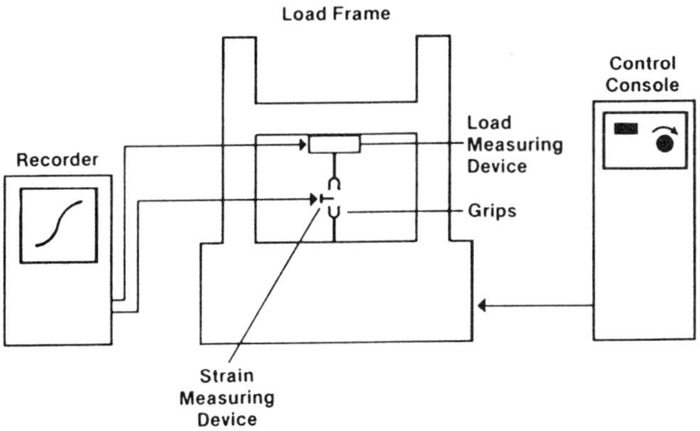

Figure 4 Block diagram of a universal testing machine. Either tension or compression can be applied to the sample.

strain measuring device such as an electronic extensometer, a crosshead motion transmitter, a deflectometer, or even an infrared extensometer.

V. FACTORS IN SELECTING A UTM

A great variety of factors determines the appropriate choice of a universal testing machine, including type of testing, load capacity, testing speeds, maximum elongation, data compilation, operator skill, location, and cost. These factors include:

1. Type of testing: Foremost among the questions that arise is whether the machine will be used for a single purpose, such as repeated testing of 0.505-in. test bars, or whether the machine will be used where versatility is required, such as in a professional testing lab, where testing may be performed on anything from aluminum foil and paper to concrete beams and reinforcing bars.

2. Load capacity: All the specimens to be tested must be reviewed and a determination made as to the highest load capacity required. In many cases it is possible to reduce the dimensions of the specimen if for some reason a higher-capacity machine is not available. Universal testing machines are generally purchased with the maximum capacity of the load frame as the determining factor, and with ranging modules or auxiliary load cells to provide testing capacities at intermediate positions below the maximum capacity of the load frame. For instance, a 60,000-lb load frame can easily be provided with full-scale ranges of 60,000, 30,000, 12,000, and 6000 lb. The use of auxiliary load cells would provide even lower-capacity full-scale ranges.

3. Testing speeds: Rubber generally requires testing at 20 in./min, while most metals require testing at much slower speeds, such as 0.2 in./min. You must determine beforehand the testing speed requirements of your samples and use a testing machine with the appropriate speed ranges.

4. Maximum elongation: Metals that elongate only a few percentage points do not require large travel of the crosshead. A 6-in. hydraulic ram provides sufficient movement. However, materials such as plastic and rubber may elongate 600 or 700% and require large crosshead travel. In these cases, a screw-powered load frame is required. Many labs also have special testing conditions that must be considered. For example, a wider-than-normal load frame may be necessary for those applications in which furnaces or ovens will be used in the test procedures. Some compression tests require the ability to apply loads offset from the centerline of the specimen. The testing machine being considered must have the ability to operate correctly under offset loading conditions.

5. Data compilation: If no permanent copy is required, a simple dial gage or digital display is all that is necessary. If permanent records must be kept, either a recording device or a printing device is necessary to supply copies of these results. If data are to be fed to a computer terminal, sophisticated data acquisition equipment is required. In situations where process control is being performed by the tensile test, it is necessary that high production rates be maintained. In these cases it would be wise to use a single-purpose machine set up specifically for the material being tested. The loss of versatility is more than made up for the increase in efficiency. If, however, we are discussing a research lab situation, the speed at which testing

is performed is not nearly as critical as the ability to perform a great variety of test functions.

6. Operator skill: The degree of machine control and data acquisition capability of the testing system affect the operator's skill requirements. In general, the more the machine does, the higher the skill level of the operator must be.

7. Location: The physical size of the tester must be accommodated by the space available. This is especially important for testers that are unusually high or which require a pit to be located beneath the load frame. Second, unusually high- or low-temperature conditions and environmental conditions, such as humidity, dust, and oil, must be considered, especially as the equipment becomes more sophisticated and uses more electronic components. Third, the electrical power available to the machine is also critical, again, especially as the tester becomes more sophisticated. Testing systems that use microprocessor control generally require a clean, isolated source of power and no fluctuations in either voltage or current. Induction motors, for example, will cause erroneous results and possible loss of memory to nearby electronic equipment. A power failure, even one lasting only seconds, will abort the test being performed and will probably cause loss of memory and in some computer systems, possibly loss of software.

8. Cost: The initial price of the equipment is only one cost element to be considered. Maintenance cost, personnel cost, and the cost of lost opportunity are other very real costs.

VI. CONTROLS

Controls for universal testing machines can either be fixed to the load frame or mounted on a separate console. The load-indicating module displays the load being applied at any point in time. An electronic module adjusts the signal coming from the load cell so that it is properly zeroed under start-test conditions. Another module controls the operating functions of the machine. A sweep dial indicates the speed of the ram. Other controls and pushbuttons turn the machine on or off, choose tension or compression modes, and adjust the speed. The signal conditioning module takes the signal from the LVDT extensometer, demodulates it, and sends it to an X-Y recorder.

Depending on the type of testing being done, it may be necessary for the universal testing machine to have some type of machine control ability. For example, the tester may be required to hold a particular load or a particular position or a particular strain. Additionally, you may want to perform a test that cycles between various loads, positions, or strain conditions. It may also be necessary to control the rate of load application or the rate of strain application.

This type of machine control is available from most microprocessor systems or by specific modules that can be added to the tester's console. Microprocessor-controlled equipment can simplify many machine control functions which formerly were time consuming and complicated to run.

The advent of computer and microprocessor control systems now makes it possible to eliminate much of the tedious deskwork and calculations that are usually necessary to transcribe the raw data into usable information. Without computer assistance, it is not uncommon for an operator to take 5 min to perform the test and another 10–15 min to perform the calculations. Many tensile testing systems are now available with data acquisition capabilities to perform a variety of calculations automatically to arrive at final results. Additionally, some systems contain enough memory to hold data from prior tests and intermix it with data from the current test in progress in order to determine average, mean, and standard deviation characteristics. Yield load, yield strength, modulus, ultimate strength, percent elongation, break load, break stress, creep, relaxation, energy under the curve, and mean and standard deviation are all examples of data acquisition requirements that can now be performed routinely and efficiently by microprocessor- and computer-based systems.

Microprocessor-controlled data acquisition systems can be either preprogrammed or user programmable. Either version allows the operator to control the load frame, measure load and strain, analyze the results obtained, compute appropriate parameters, output the information to a CRT screen, a recorder, or a peripheral computer device, and record the information. Preprogrammed units (Fig. 5) requires less operator skill and are generally less expensive to purchase, but they may not have the versatility of programmable units.

Preprogrammed units will use PROM (programmable read-only memory) chips and other microprocessor components to store the program and secure it from inadvertent loss or change, but PROM chips also keep the user from modifying the program at will. Programming entries and testing parameters are entered into the computer's memory through in the front keyboard panel, and the test is then performed according to these choices.

Programmable control systems generally require either a programmer to program the control and data acquisition functions, which would be stored on disk or magnetic tape, or purchase of manufacturer-supplied software to run the appropriate tests. Programmable controls offer more power and versatility to the user but generally require a high skill level, a larger investment, and a more controlled environment than do preprogrammed controllers.

While the test is running, the computer system will display the resulting data on the CRT screen. Upon completion of the test, the CRT screen (and printer and recorder if supplied) will display the finished data in the form requested.

VII. COMPUTER LIMITATIONS

A word of caution: As powerful as computer control is, it is no better than the information it receives and the programs it follows. Every computer-controlled program must follow an algorithm established by the programmer. This algorithm approximates as closely as possible the answers that would be obtained by a skilled operator doing manual calculations. The strength of the computer is that it does the computations in the same way every time; the answer is always the same given the same information.

However, there are certain limitations. For example, if we ask the computer to calculate the Young's modulus, it must take a particular stress value and divide it by a particular strain value over the straight-line portion of the graph, the region between zero and A shown in Figure 3. An operator would look at that line and by eye, be able to tell within reason where the straight-line portion lies.

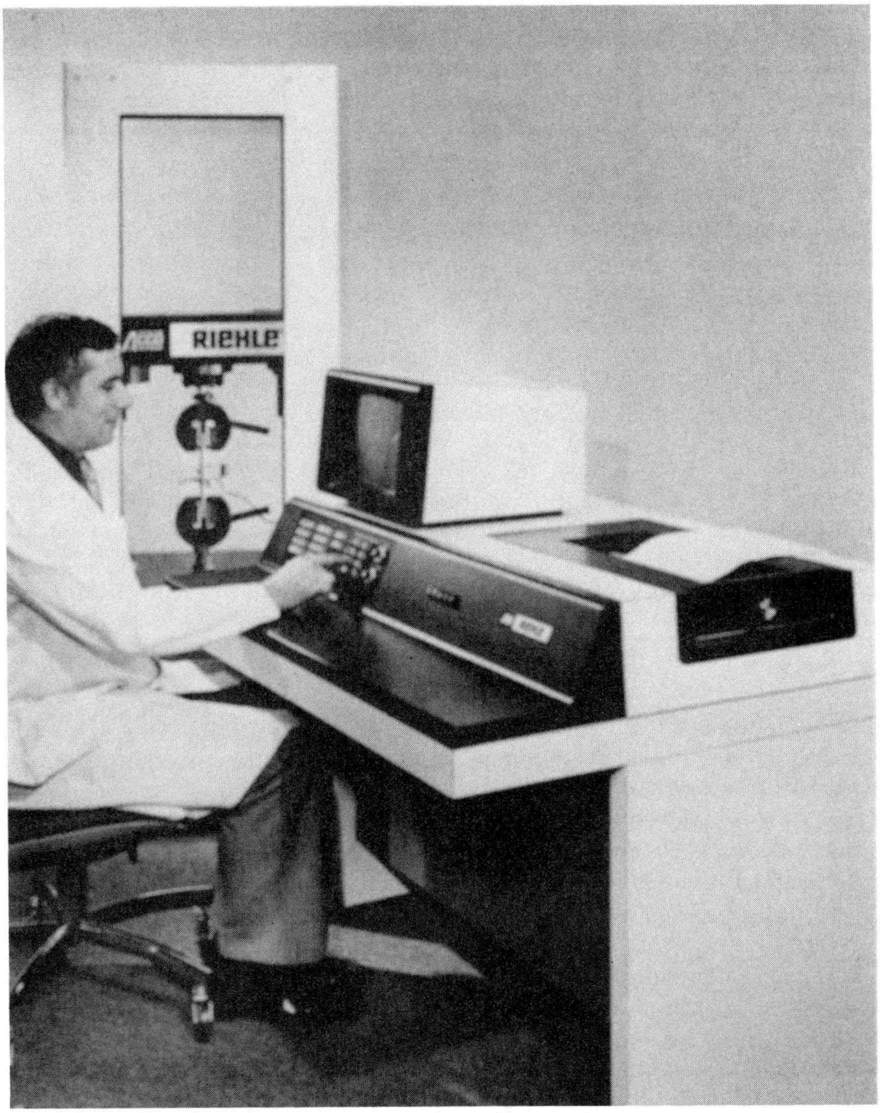

Figure 5. A modern, computer-controlled universal testing machine.

The computer must use other methods to determine the straight-line portion of the graph. We can program the computer to compare each value that it sees with a prior value and then determine whether the line is still straight, but we have to tell it how far off the line can be before it is no longer straight. If we make our parameters too tight, the computer will read each bump or glitch in the graph as the end of the straight-line portion and will make an erroneous determination of Young's modulus. If we make our algorithm parameters too loose, the computer may make an erroneous late decision on Young's modulus.

Often, the algorithm tells the computer to begin taking data at a certain percent of the load range, to end taking data at a higher percent of the load range, and to consider all data in between those two points as a straight line. This method works very well provided that the data are linear between those two positions. Essentially, the choice of the correct program is dependent on the characteristics of the material being tested, and it is not always self-evident in what ways the program characteristics and the material characteristics interact.

VIII. ACCESSORIES

Auxiliary load cells provide full-scale testing at loads below the capacity of the original tester. Many recording devices, such as an X-Y recorder, are available for plotting stress-strain curves. Strip-chart recorders plot stress or strain versus time. Strain and elongation measuring devices include electronic extensometers, deflectometers, crosshead motion transmitters, infrared extensometers, and optical encoders. Extensometers, for example, use knife edges to establish a fixed gage length in the unstressed state. As the load is applied, the bottom knife edge follows the deformation of the sample and mechanically transmits the deformation to a coil, which translates this mechanical movement into an electrical voltage. The electrical voltage is sent to a signal demodulator, where the signal is conditioned for plotting on a recorder.

Grips are used to hold the sample during the testing procedure. It is important that the grips attach themselves firmly to the sample and that no movement in the grips themselves occurs during the critical part of the testing process. It is also important to prevent the grip inserts from breaking or otherwise deforming the sample so as to cause premature failure and therefore erroneous data.

For heavy load applications, suspended grips are not normally acceptable because the loads are too great and will spring the housings. Open-front grip housings are quite rugged, as they must withstand high loads during the testing process. A rack-and-pinion arrangement is often used to open and close the wedge grips, which are located in the center of the open-front housings. The wedge grips are interchangeable depending on the size and type of sample being tested. Flat grips are used for flat samples, and vee grips are used for round samples. Some grip housings also provide hydraulic means of opening and closing the grips.

Closed grip housings require inserting grips through the top or bottom of the crosshead when they are changed or repositioned. This type of housing is the most cumbersome to use but is able to withstand the greatest loads.

41

Hardness Testing

ANTHONY DeBELLIS
Acco Industries, Inc.
Bridgeport, Connecticut

I. INTRODUCTION

The most commonly used indentation hardness tests are the Brinell, Rockwell, Rockwell superficial, and Knoop and Vickers microhardness tests. Briefly, the Brinell test is used for testing forgings and castings, especially cast iron. The Rockwell test is used for testing ferrous and nonferrous materials, hardened and tempered steel, case-hardened steel, sheet materials in the heavier gages, and cemented carbides. The Rockwell superficial test is also used for ferrous and nonferrous materials where lighter loads are required, such as testing thin case-hardened surfaces, as well as nitrided cases, decarburized surfaces, and sheet material in thin gages. Microhardness tests (Knoop and Vickers) are used for very small and thin parts as well as for case depth determinations. A description of these test methods follows.

II. BRINELL TEST

The Brinell method consists of indenting a test specimen with a steel ball (usually 10 mm) with a load that is usually 3000 kg but reduced to 500 kg for soft metals. The full load is applied for 10 to 15 seconds in the case of iron and steel, and for at least 30 seconds in the case of other metals. After the load is removed, the diameter of the indentation is measured. The Brinell hardness number is calculated by dividing the load applied by the surface area of the indentation:

$$\text{Brinell hardness number (HB)} = \frac{P}{\pi(D/2)\left[D - \sqrt{D^2 - d^2}\right]}$$

where

P = load (kg)
D = diameter of ball (mm)
d = diameter of indentation (mm)

The indentation diameter is the average of two readings made at right angles to each other, and the Brinell hardness number is determined by referring to tables similar to Table 1.

Hardened steel cannot be tested with a hardened steel ball by the Brinell method because the ball will flatten during penetration and a permanent deformation will take place. However, a hardened steel ball is satisfactory in testing softer metals. A significant error will be introduced in the Brinell number of values over 450 when high-grade hardened steel balls are used.

Carbide balls are recommended for Brinell testing of materials up to 630. Because of the difference in elastic properties between the steel and carbide balls, the type of ball used should be specifically reported where Brinell hardness values exceed 200. Brinell hardness numbers determined with a steel ball are designated HBS [e.g., 450 HBS (450 Hardness, Brinell Test, Steel Ball)] and with a carbide ball HBW [e.g., 450 HBW (450 Hardness, Brinell Test, Carbide Ball)].

Table 1 Brinnell Hardness Numbers

Dia. of Indentation	500 kg Load	1500 kg Load	3000 kg Load	Dia. of Indentation	500 kg Load	1500 kg Load	3000 kg Load	Dia. of Indentation	500 kg Load	1500 kg Load	3000 kg Load
2.00	158	473	945	3.50	50.3	151	302	5.00	23.8	71.5	143
2.05	150	450	899	3.55	48.9	147	293	5.05	23.3	70.0	140
2.10	143	428	856	3.60	47.5	143	285	5.10	22.8	68.5	137
2.15	136	409	817	3.65	46.1	139	277	5.15	22.3	67.0	134
2.20	130	390	780	3.70	44.9	135	269	5.20	21.8	65.5	131
2.25	124	373	745	3.75	43.6	131	262	5.25	21.4	64.0	128
2.30	119	356	712	3.80	42.4	128	255	5.30	20.9	63.0	126
2.35	114	341	682	3.85	41.3	124	248	5.35	20.5	61.5	123
2.40	109	327	653	3.90	40.2	121	241	5.40	20.1	60.5	121
2.45	104	314	627	3.95	39.1	118	235	5.45	19.7	59.0	118
2.50	100	301	601	4.00	38.1	115	229	5.50	19.3	58.0	116
2.55	96.3	289	578	4.05	37.1	112	223	5.55	18.9	57.0	114
2.60	92.6	278	555	4.10	36.2	109	217	5.60	18.6	55.5	111
2.65	89.0	267	534	4.15	35.3	106	212	5.65	18.2	54.5	109
2.70	85.7	257	514	4.20	34.4	104	207	5.70	17.8	53.5	107
2.75	82.6	248	495	4.25	33.6	101	201	5.75	17.5	52.5	105
2.80	79.6	239	477	4.30	32.8	98.5	197	5.80	17.2	51.5	103
2.85	76.8	231	461	4.35	32.0	96.0	192	5.85	16.8	50.5	101
2.90	74.1	222	444	4.40	31.2	93.5	187	5.90	16.5	49.6	99.2
2.95	71.5	215	429	4.45	30.5	91.5	183	5.95	16.2	48.7	97.3
3.00	69.1	208	415	4.50	29.8	89.5	179	6.00	15.9	47.8	95.5
3.05	66.8	201	401	4.55	29.1	87.0	174	6.05	15.6	46.9	93.7
3.10	64.6	194	388	4.60	28.4	85.0	170	6.10	15.3	46.0	92.0
3.15	62.5	188	375	4.65	27.8	83.5	167	6.15	15.1	45.2	90.3
3.20	60.5	182	363	4.70	27.1	81.5	163	6.20	14.8	44.4	88.7
3.25	58.6	176	352	4.75	26.5	79.5	159	6.25	14.5	43.6	87.1
3.30	56.8	171	341	4.80	25.9	78.0	156	6.30	14.2	42.8	85.5
3.35	55.1	166	331	4.85	25.4	76.0	152	6.35	14.0	42.0	84.0
3.40	53.4	161	321	4.90	24.8	74.5	149	6.40	13.7	41.3	82.5
3.45	51.8	156	311	4.95	24.3	73.0	146	6.45	13.5	40.5	81.0

A. Test Surface Preparation

The surface on which the Brinell indentation is to be made must be filed, ground, machined, or polished with emery paper (3/0 emery paper is suitable) so that the indentation diameter is clearly enough defined to permit its measurement. There should be no interference from tool marks. The surface should be representative of the material and not decarburized, case-hardened or otherwise superficially hardened to any extent.

B. Indentation Measurement

The diameter of the indentation is measured using a special microscope such as a toolmakers microscope to the nearest 0.01 mm (0.0004 in.). The error in reading the diameter should not exceed 0.01 mm, to keep the error in the Brinell number less than 1%. A stage micrometer is usually provided with the microscope and should be used frequently to check its calibration.

Brinell indentations may exhibit different surface characteristics. When some metals are tested there is a ridge around the impression extending above the original surface of the test piece; at other times the edge of the impression is below the original surface. In some cases there is no difference whatever. The first phenomenon is called a "ridging" type of impression and the second a "sinking" type. Cold-worked alloys generally have the former, and annealed metals the latter type of impression.

The definition of the Brinell number relates it to the surface area of the indentation. To determine surface area, it is necessary to measure the diameter of the indentation, assuming that this is the diameter of the indentation with which the ball was in actual contact. But in view of ridging- and sinking type impressions, there is a question as to the exact part of the visible indentation with which actual contact was made. In the case of ridging-type impressions the diameter of the indentation is greater than the true value, whereas with sinking-type impressions, the reverse is true. No way is known of making absolutely certain that the correct diameter is measured, and the judgment and experience of the operator introduce a personal factor into the test.

In some materials the brink of the indentation is poorly defined, especially when hardened steels—even with polished surfaces—are tested; the use of carbide balls produces a more distinct indentation.

Brinell indentations made on some materials are far from round; those on materials which have been subjected to considerable rolling or other cold working are elliptical in shape, whereas those on heat-treated steels are quite round. For indentations that are not circular, an average value of the Brinell number may be obtained by measuring the diameter in four directions approximately 45° apart.

C. Spacing of Indentations

For accurate results, indentations should not be made too close to the edge of a piece. Lack of sufficient supporting material on one side will cause the resulting indentation to be large and unsymmetrical. The error in Brinell number is negligible if the distance from the center of the indentation to any edge is not less than 2.5 times the diameter of the indentation.

Indentations cannot be made too close to one another. Under such conditions, the material may be cold-worked by the first indentation, or there may not be sufficient supporting material for the second indentation. The latter condition would produce too large an indentation, whereas the former may produce too small an indentation. The distance between centers of adjacent indentations should be at least three times the diameter of the indentation in order to have the error in the Brinell number of the order of less than 1%.

D. Selecting the Load

The standard loads for Brinell testing are 3000, 1500, and 500 kg. The load should be selected to keep the ratio of the diameter of the indentation to the diameter of the ball (d/D) greater than 0.24 and less than 0.60. When a ratio is less than 0.24, the resulting indentation is so small that errors in determining the diameter become a large proportion of the total diameter. Further, the test loss sensitivity and small differences in hardness values are not differentiated. For a ratio greater than 0.60, the test becomes supersensitive.

To meet these requirements for a given load the hardness must fall within the following ranges:

3000 kg: BHN 96–600
1500 kg: BHN 48–300
500 kg: BHN 16–100

E. General Precautions

When indentations are made on a curved surface with the 10-mm-diameter ball, the radius of the test specimen should not be less than 1 in.

Indentations should not be made within 2 1/2 times the diameter of indentation from each other or from the edge of the specimen.

The load should be applied at right angles (perpendicular) to the surface of the specimen and within 2°.

The thickness of the piece being tested should be such that no bulge or marking showing the effect of the load appears on the side of the test piece opposite the indentation.

The surface finish on which the indentation is to be made should be such that the indentation diameter is clearly defined

III. ROCKWELL TEST

The Rockwell test is the most commonly used of all indentation hardness tests. The test can be made within 5 to 10 sec depending on the size and hardness of the specimen. The hardness number is indicated directly on either a dial gage or digital readout.

A. Principle of Test

As shown in Figure 1, the Rockwell test consists of measuring the additional depth to which a steel ball or Brale diamond penetrator is forced by a heavy (major) load beyond the depth of a previously applied light (minor) load.

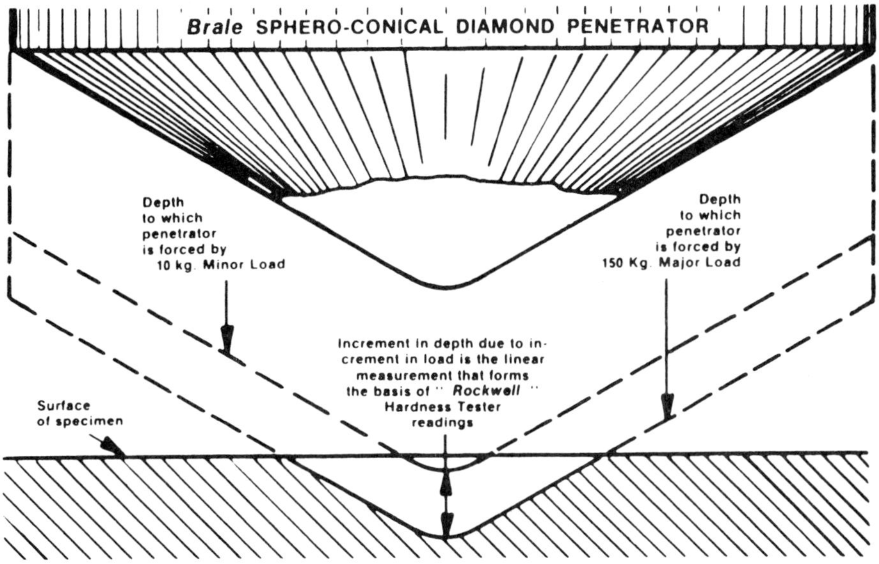

Figure 1 Spheroconical diamond penetrator.

The minor load is applied first and a reference or set position is established on the dial gage of the Rockwell tester. Then the major load is applied without moving the piece being tested, the major load is removed, and the Rockwell hardness number is automatically indicated on a dial gauge or digital readout.

The Brale diamond penetrator is used for testing material such as hardened steels and cemented carbides. The steel ball penetrators, available with 1/16-, 1/8-, 1/4-, and 1/2-in diameter, are used when testing materials such as steel, copper alloys, aluminum, and plastics, to name a few.

Rockwell hardness testing falls into two categories: regular Rockwell and superficial Rockwell. Regular Rockwell testing most often uses HRC (Hardness, Rockwell, "C" scale) and HRB (Hardness, Rockwell, "B" scale) ranges. Superficial testing often uses HR30N (Hardness, Rockwell, "30N" scale) and HR30T (Hardness, Rockwell, "30T" scale) ranges (see Tables 2 and 3). The dial gauge divisions on all Rockwell testers are such that high Rockwell hardness numbers represent hard materials and low numbers represent soft material.

Regular Rockwell Testing

In regular Rockwell testing the minor load is 10 kg and the major load can be either 60, 100, or 150 kg. No Rockwell hardness number is given by a number alone. It must always be prefixed by a letter signifying the value of the major load and type of penetrator. A letter has been assigned for every possible combination of load and penetrator, as shown in Table 2.

One Rockwell number represents a penetration of 0.002 mm (0.000080 in.). A reading of 60 HRC, for example, indicates penetration from minor to major load of $(100 - 60) \times 0.002$ mm $= 0.080$ mm or 0.0032 in. A reading of 80 HRB indicates a penetration of $(130 - 80) \times 0.002 = 0.100$ mm or 0.004 in.

Table 2 Regular Tester

Scale symbol	Penetrator	Major load in kilograms
B	1/16-in. ball	100
C	Brale	150
A	Brale	60
D	Brale	100
E	1/8-in. ball	100
F	1/16-in. ball	60
G	1/16-in. ball	150
H	1/8-in. ball	60
K	1/8-in. ball	150
L	1/4-in. ball	60
M	1/4-in. ball	100
P	1/4-in. ball	150
R	1/2-in. ball	60
S	1/2-in. ball	100
V	1/2-in. ball	150

Superficial Rockwell Testing

In superficial Rockwell testing the minor load is 3 kg and the major load can be either 15, 30, or 45 kg. As in regular testing, a Rockwell superficial hardness number must always be prefixed by the major load and a letter for the type of penetrator. A scale designation has been assigned for every possible combination of load and penetrator, as shown in Table 3.

One Rockwell superficial number represents a penetration of 0.001 mm or 0.000040 in. A reading of 80 HR30N, for example, indicates penetration from minor to major load of $(100 - 80) \times 0.001 = 0.020$ mm or 0.0008 in.

Table 3 Superficial Tester

Scale symbol	Penetrator	Major load in kilograms
15N	N Brale	15
30N	N Brale	30
45N	N Brale	45
15T	1/16-in. ball	15
30T	1/16-in. ball	30
45T	1/16-in. ball	45
15W	1/8-in. ball	15
30W	1/8-in. ball	30
45W	1/8-in. ball	45
15X	1/4-in. ball	15
30X	1/4-in. ball	30
45X	1/4-in. ball	45
15Y	1/2-in. ball	15
30Y	1/2-in. ball	30
45Y	1/2-in. ball	45

B. Scale Selection

In many instances Rockwell hardness tolerances are specified or are indicated on drawings. At times, however, you must select the Rockwell scale for a given test specimen or part. Knowledge of the factors governing the choice of the proper Rockwell scale is valuable in this situation, as the choice is not only between the regular hardness tester and superficial hardness tester, with three different major loads for each, but also between the Brale diamond penetrator and the 1/16-, 1/8-, 1/4-, and 1/2-in. diameter steel ball penetrators or a combination of 30 different scales.

In the event that no specification exists or there is doubt about the suitability of a specified scale, an analysis should be made of the controlling factors important in the selection of the proper scale (e.g., type of material and thickness of specimen).

C. Type of Material

ASTM Standard E18 lists all regular Rockwell scales and typical materials for which these scales are applicable. This list (reprinted in Table 4) provides an excellent starting point. Table 4 includes only the regular Rockwell scales; however, this information can be a helpful guide even when one of the superficial scales may be required. For example, note that the C, A, and D scales—all with the diamond penetrator—are used on hard materials such as steel and tungsten carbide. Any material in this hardness category would be tested with the diamond penetrator. The choice to be made is whether the C, A, D, 45N, 30N, or 15N scale is applicable. In any event, the possible scales have been reduced to six. The next step is to find the scale—whether it be regular or superficial—that will guarantee accuracy, sensitivity, and repeatability.

Table 4 Typical Rockwell Scale Applications

Scale symbol	Typical application of scales
B	Copper alloys, soft steels, aluminum alloys, malleable iron, etc.
C	Steel, hard cast irons, pearlitic malleable iron, titanium, deep casehardened steel and other materials harder than B 100
A	Cemented carbides, thin steel and shallow case-hardened steel
D	Thin steel and medium casehardened steel and pearlitic malleable iron
E	Cast iron, aluminum and magnesium alloys, bearing metals
F	Annealed copper alloys, thin soft sheet metals
G	Phosphor bronze, beryllium copper, malleable irons, Upper limit G 92 to avoid possible flattening of ball
H	Aluminum, zinc lead
K, L, M, P, R, S, V	Bearing metals and other very soft or thin materials. Use smallest ball and heaviest load that do not give anvil effect

D. Thickness of Specimen

The material immediately surrounding a Rockwell test is cold-worked. The extent of the area cold-worked depends on the type of material and any previous work hardening done to it. The depth of material affected has been found to be on the order of 10 times the depth of the indentation. Therefore unless the thickness of the material being tested is at least approximately 10 times the depth of the indentation, an accurate Rockwell test cannot be expected. This "minimum thickness" ratio of 10:1 should be regarded only as an approximation.

The depth of penetration for any Rockwell test can be calculated, but in actual practice this is not necessary, as "minimum thickness" charts are available. These minimum thickness values (Table 5) do follow the 10:1 ratio in some ranges, but are actually based on experimentation on varying thicknesses of low-carbon steels and on hardened and tempered strip steel.

Table 5 Minimum Specimen Thickness

Any greater thickness & hardness can be safely tested on indicated scale	*Rockwell* Superficial Hardness Tester			*Rockwell* Hardness Tester		
	15 N	30 N	45 N	A	D	C
	15 Kg.	30 Kg.	45 Kg.	60 Kg.	100 Kg.	150 Kg.
Thickness in inches	Diamond "N" "Brale" Penetrator			Diamond "Brale" Penetrator		
.006	92	–	–		–	–
.008	90	–	–	–	–	–
.010	88	–	–	–	–	–
.012	83	82	77	–		–
.014	76	80	74	–		–
.016	68	74	72	86	–	–
.018	X	66	68	84	–	–
.020	X	57	63	82	77	–
.022	X	47	58	78	75	69
.024	X	X	51	76	72	67
.026	X	X	37	71	68	65
.028	X	X	20	67	63	62
.030	X	X	X	60	58	57
.032	X	X	X	X	51	52
.034	X	X	X	X	43	45
.036	X	X	X	X	X	37
.038	X	X	X	X	X	28
.040	X	X	X	X	X	20

Any greater thickness & hardness can be safely tested on indicated scale	*Rockwell* Superficial Hardness Tester			*Rockwell* Hardness Tester		
	15T	30T	45T	F	B	G
	15 Kg	30 Kg.	45 Kg.	60 Kg.	100 Kg.	150 Kg
Thickness in inches	1/16" Ball Penetrator			1/16" Ball Penetrator		
.005	93	–	–	–	–	–
.010	90	87	–	–	–	
.015	78	77	77	–	–	–
.020	X	58	62	100	–	
.025	X	X	26	92	92	90
.030	X	X	X	67	68	69
.035	X	X	X	X	44	46
.040	X	X	X	X	20	22

A typical example of the use of these tables should be helpful. Consider a requirement to check the hardness of a strip of steel 0.014 in. thick, of approximate hardness 63 HRC. According to Table 5, the material must be approximately 0.028 in. for an accurate Rockwell C scale test. Therefore, this specimen should not be tested on the C scale. A handy tabulation to make at this point is the approximate converted hardness on the other Rockwell scales equivalent to 63 HRC. These values, taken from Wilson Conversion Chart No. 52, are 73 HRD, 83 HRA, 70 HR45N, 80 HR30N, and 91 HR15N. Referring once again to Table 4, for hardened 0.014-in. material, there are only three Rockwell scales to choose from: 45N, 30N, and 15N. The 45N scale is not suitable, as the material should be at least 74 HR45N. On the 30W scale, 0.014-in. material must be at least 80 HR30N—the material in question is 80 HR30N. On the 15N scale, the material must be at least 76 HR15N. This material is 91.5 HR15N. Therefore, either the 30N or the 15N scale may be used. After all limiting factors have been eliminated, and a choice exists between two or more scales, the scale applying the heavier load should be used. The heavier load will produce a larger indentation, covering a greater portion of the material, and a Rockwell hardness number more representative of the material as a whole will be obtained. In addition, the heavier the load, the greater the sensitivity of the scale.

The foregoing approach would also apply in determining which scale should be used to measure the hardness of the case for a case of known approximate depth and hardness. Minimum-thickness charts as well as the 10:1 ratio serve only as guides. After determining the Rockwell scale, based on minimum thickness values, an actual test should be made and the underside directly beneath the area of test examined to determine if the material was disturbed or if a bulge exists. If so, the material was not sufficiently thick for the applied load, resulting in a condition known as the "anvil effect," and the Rockwell scale applying the next lighter load should be used. On softer materials, the high stress concentration due to insufficient thickness will result in flow of the material.

When either anvil effect or flow exists, the Rockwell hardness number obtained may not be a true value. The use of several specimens piled one on top of the other is not recommended. The slippage between the contact surfaces of the several specimens makes a true value impossible to obtain.

E. Spacing of Indentation

An indentation hardness test cold-works the surrounding material, and if another indentation is placed within this cold-worked area, the Rockwell hardness test will be affected. Usually, the readings will be higher than on the virgin material.

If the indentation is placed too close to the edge of a specimen, the material will yield and the Rockwell hardness number decreased accordingly. Experience has shown that the distance from the center of the indentation to the edge of the specimen must be at least 2 1/2 diameters to assure an accurate test.

The distance from center to center of indentations must be at least three diameters. Usually, the softer the material, the more critical the spacing, but if a distance of three diameters is maintained, the indentations will be far enough apart for most materials.

F. Support for Test Piece

An important requirement of the Rockwell test is that the surface being tested be normal (at right angles) to the penetrator and that the piece being tested not move or slip as the major load is applied. The depth of indentation is measured by the movement of the plunger rod holding the penetrator; therefore, any slipping or moving of the piece will be followed by the plunger rod and the motion transferred to the dial gage, causing an error. As one point of hardness represents a depth of only 0.000080 in., a movement of only 0.001 in. could cause an error of over 10 Rockwell numbers. The support itself must be of sufficient rigidity to prevent its permanent deformation in use.

Sheet metal, small pieces, or pieces that do not have flat undersurfaces are tested on an anvil having a small, elevated, flat bearing surface. Pieces that are not flat should have the convex side on the bearing surface.

IV. MICROHARDNESS TESTING

The term *microhardness* usually refers to indentation tests made with loads up to 1000 g. The indenter is either the Vickers or the Knoop diamond indenter. The size of the indentation, which is extremely shallow, is such that it must be precisely determined using a measuring microscope with good resolving power. It is very important that the surface being tested be lapped flat and be free from scratches. For tests with loads of 100 g or lighter, a metallographic finish is necessary (see Chapter 37).

The microhardness test is generally used for testing the following:

Small precision parts
Surface layers
Thin materials and small-diameter wires
Exploration of small areas
Hardness of constituents
Hardness near the edge of cutting tools
Dental materials

A. Knoop Scale

The Knoop hardness number is applied load divided by the unrecovered projected area of the indentation. The Knoop indenter (Fig. 2) is a diamond ground to pyramidal form that produces a diamond-shaped indentation having an approximate ratio between long and short diagonals of 7 : 1. The pyramid shape has an included longitudinal angle of 172°30′ and included transverse angle of 130°0′. The depth of indentation is about one-thirtieth of its length. The Knoop hardness number is the load divided by the projected area. Both hard and brittle materials may be tested with the Knoop indenter.

The Knoop hardness number (HK) is the ratio of the load applied to the indenter, P (kgf) to the projected area A (mm^2). By formula,

$$HK = \frac{P}{A} = \frac{P}{C\ell^2}$$

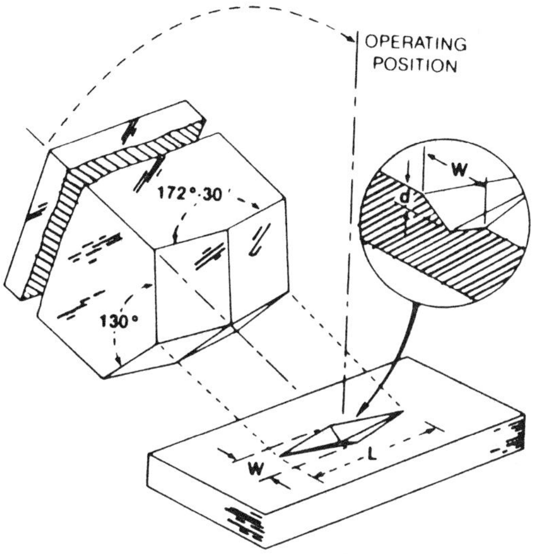

Figure 2 Knoop indenter.

where

P = applied load (kg)
A = unrecovered projected area of indentation (mm^2)
ℓ = measured length of long diagonal (mm)
C = 0.07028 = constant of indenter, relating projected area of the indentation to the square of the length of the long diagonal

The Knoop hardness number calculated from this equation can be found in ASTM Standard 384.

B. Vickers Scale

The Vickers hardness number is the applied load divided by the surface area of the indentation. The Vickers indenter is a diamond ground in the form of a square-based pyramid with an angle of 136° between faces (Fig. 3).

With the Vickers indenter, the depth of indentation is about one-seventh of the diagonal length. For certain types of investigation, there are advantages to such a shape. The diamond pyramid hardness (DPH) number [HV (Hardness, Vickers)] is the ratio of the load applied to the indenter, P (kgf), to the surface area of the indentation (mm^2), or

$$HV = \frac{2P \sin(\theta/2)}{d^2}$$

where

P = applied load (kg)
d = mean diagonal of the indentation (mm)
θ = angle between opposite faces of the diamond = 136°

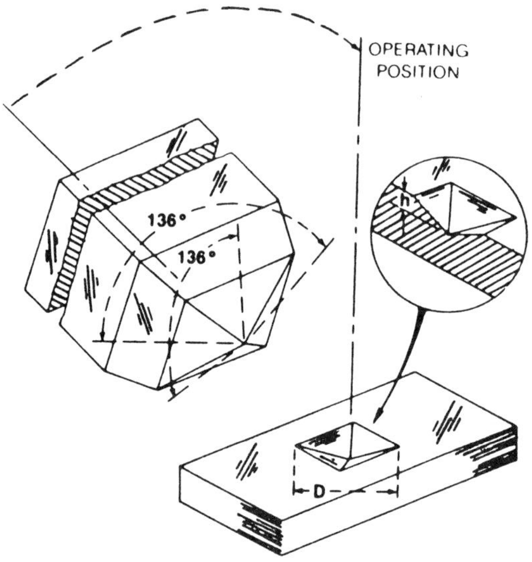

Figure 3 Diamond pyramid indenter.

The Vickers hardness number calculated from this equation can be found in ASTM Standard E384.

C. Knoop versus Vickers

Figure 4 is a comparison of indentation made with Knoop and Vickers under loads of 3000, 1000, 500, and 100 g on steel of approximate hardness 550 HV (1000-g load). For a given load, the Vickers indenter penetrates about twice as far into the specimen as the Knoop indenter, and the diagonal will be about one-third of the length of the Knoop indentation. Thus the Vickers test is less sensitive to surface

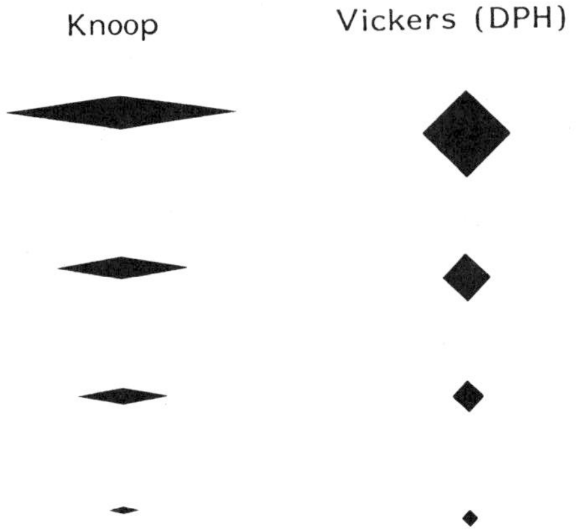

Figure 4 Indentation comparison. Knoop and diamond pyramid indentations under loads of 3000, 1000, 500, and 100 g. (Not to actual size.)

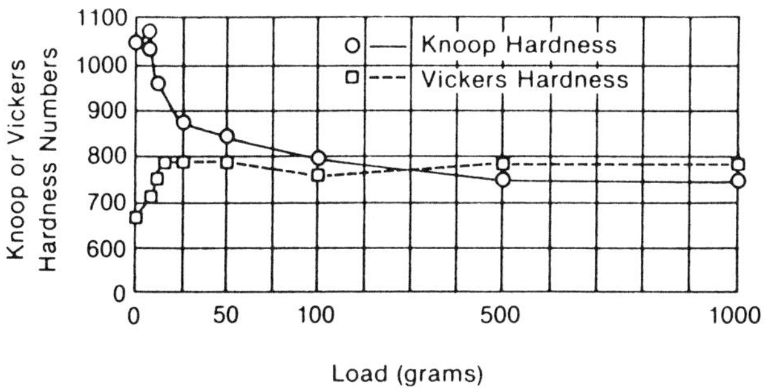

Figure 5 Hardness values as function of test load.

conditions than the Knoop test, and for equal loads the Vickers indentation, because of its shorter length, is more sensitive to errors in measuring the indentation.

D. Surface Preparation

To permit accurate measurement of the length of the Knoop indentation or diagonals of the Vickers indentations, the indentations must be clearly defined. A sharp indentation is in fact the criterion for surface preparation, and as a rule the lighter the test load, the higher the degree of surface finish required.

In many instances the piece to be tested for microhardness will also be used for metallographic examination, in which case mounting, polishing, and even etching are justified. More often than not, however, mounting is not necessary and only a simple polishing is required.

Many standard fixtures are available for supporting the majority of parts normally tested for microhardness. At the start of any testing work a piece should be placed on one of these fixtures and a test made. If the indentation is clearly

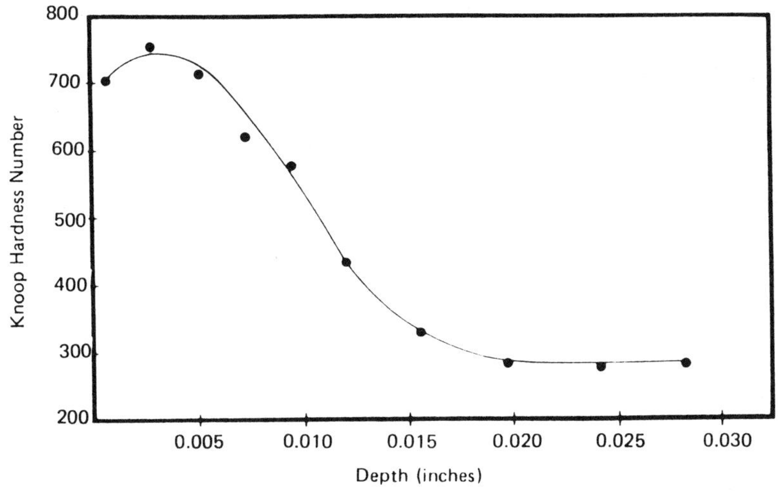

Figure 6 Knoop hardness gradient through carburized case with 1-kg load.

defined, no polishing is required. If the indentation is not clearly defined, the test surface should be polished with varying grades of emery paper until the indentation is sharp. If even 4/0 emery paper is not successful, polishing with diamond paste and a felt bob is recommended. When testing at loads below 100 g it may be necessary to go to a metallographic finish (refer to ASTM Designation E3).

E. Optical Equipment

Optical equipment used in microhardness testers for measuring the indentation must focus both ends of the indentation at the same time as well as be rigid and free from vibration. Lighting also plays an important role. Complete specifications of measurement, including the mode of illumination, are necessary in microhardness testing techniques. Polarized light, for example, results in definitely larger measurements than does unpolarized light. Apparently, this is caused by the reversal of the defraction pattern; that is, the indentation appears brighter than the background. In recording data, the magnification should be reported.

Dry objectives having the highest resolving power available are generally used, but oil immersion objectives may be necessary under some conditions. For dry lenses, with the highest numerical aperture that can be used, an accuracy of $\pm 0.5\ \mu m$ is possible. Also, the same observer can compare differences between two specimens to $\pm 0.2\ \mu m$ accuracy. It should, however, be added that considerable experience and care are necessary to obtain this accuracy.

F. Hardness Number Depends on Load

Prior to the advent of the microhardness tester, it had been assumed that the Vickers indenter (as well as other indenters giving geometrically similar indentations) produced a hardness number which was independent of the indenting load. Speaking very generally, this can be accepted for loads of approximately 1000 g and up. However, microhardness testing, when performed with loads of less than 500 g with the Knoop and 100 g with the Vickers indenter, is a function of the magnitude of the test load. In most instances, microhardness values, particularly Knoop values, decrease with increasing load, as Figure 5 illustrates.

Some observers have noticed, however, an initial increase in microhardness values with increasing load. This is followed by a range in which the hardness becomes independent of the load or decreases continuously to a constant value. This effect occurs with a wide range of materials, from those as soft as copper to fully hardened martensitic steel.

The apparent increase in hardness with decrease in load (in properly prepared surfaces, of course) is caused primarily by an error in the determination of the size of the indentation and in the elastic recovery of the indentation. As the size of the indentation decreases, a change in hardness will result. It may be related to the stress-strain curve of the material and the relationship between the size of the indentation and the constituents of the material.

From a practical standpoint, however, load dependence does not have to be a problem. The choice of load depends on the size and depth of indentation that are considered to be most desirable. Generally, the indentation is made as large as practicable to obtain the greatest accuracy possible. As long as a single load is

Figure 7 Photomicrograph of hardness gradient through carburized case.

used throughout a study, the load dependence takes on less significance. Using different loads in any particular investigation alters the hardness numbers as measured. The lighter the load, the more significant the change. As a general rule, any comparison of Knoop hardness numbers with loads under a value of 500 g and Vickers hardness numbers under a value of 100 g is incorrect unless the load dependence of the hardness is clearly pointed out. Indicate the load used when listing Knoop or Vickers numbers.

G. Case Depth Determinations

The microhardness test is extremely useful in determining "effective" case depth. Effective case depth is the distance normal to the surface of the hardened case where the hardness is a specified value. This hardness is usually 50 HRC. The microhardness test, usually the Knoop test, is the most accurate and repeatable method of determining the effective case depth by means of a hardness traverse. As illustrated in Figure 6, the hardness gradient is determined from a small distance from the surface inward on a cross-sectioned and polished specimen. 50 HRC (Hardness, Rockwell, "C" scale) converts to 542 HK (Hardness, Knoop), and therefore, for the example illustrated, the effective case depth is approximately 0.009 in (Fig. 7).

42

Food Process Control

BEN A. MURRAY
Consultant
Englewood, Florida

I. INTRODUCTION

The food industry has some unique problems. Every person is a food customer, and each one considers him or herself an expert. This belief is shared by the management of many food companies, and it is up to the quality manager to put food quality control on an objective, measurable basis.

Another problem within the food industry is that it is sometimes considered good to have some variation in the product, thus giving the food a homemade appearance. This philosophy, however, can lead to nonuniformity in all parts of the quality control system and must be recognized. If variations are desired, those variations must be carefully controlled within appropriate limits.

Quality is like water—both always flow downhill unless steps are taken to prevent the flow. In general, the professional food quality manager is like the pump that must always be lifting the standards and systems back to the proper level. Food quality managers must measure, report, and, when correction is necessary, insist on corrective action.

II. QUALITY ORGANIZATION

Food processing plants vary tremendously in quality assurance and quality control systems. Many small food processors have only one technically trained person and this person must organize and manage all technical activities—from new product development, to quality control, to dealing with the federal government on regulatory matters. In large food processing companies, specialized quality groups have been organized and practice detailed quality assurance and quality control methods.

It is important to organize the quality system in a food processing plan so that the measurement activities of the quality control group are:

1. Conducted properly and accurately
2. Reported to the proper management levels
3. Used to produce proper corrective action

If quality control reports to production management, there may be a tendency to ignore signs of upcoming problems and fail to take corrective action. Although some production supervisors are certainly quality minded, an organization that puts quality under production will eventually find conflicts of interest. It is important—and must be stressed over and over—that food quality control must not report to production. In most modern food processing companies, quality control reports directly to the plant manager.

A modern organization requires that standards and specifications have been set for all materials and finished products. It is necessary that reports on compliance to standards be written and circulated promptly. Problems that are not written down are not corrected. Whenever possible, statistical quality control (SQC) techniques should be implemented to monitor production line operations.

III. PEOPLE IN FOOD PROCESSING QUALITY

There are many extremely competent people in food quality control who have little formal training. However, if a food processing firm is organizing for the future, it will find that it is very desirable to hire people with training in at least one of the biological or chemical sciences. Many different disciplines must be involved in food processing quality control. These include chemistry, microbiology, entomology, and statistics. A certain familiarity with all of these is necessary so that proper measurement techniques can be used to prevent problems. Although employees without formal training can be very valuable, it is wise to build on their skills using formal technical training to develop a competent food quality control system. Probably the best background is a college degree in food technology.

IV. FOOD LABORATORIES

Food laboratories deal not only with chemical and microbiological testing, but also with an area called organoleptical testing. *Organoleptical* refers to responses of the senses, such as taste, smell, and texture.

Many of the chemical testing topics are covered in detail in manuals such as the *Official Methods of Analysis of the Association of Official Analytical Chemists.* Several container companies, trade associations and food departments of large universities have also developed excellent texts.

It is important that organoleptic testing be carefully controlled to remove personal bias from the tests. Most of these tests should be conducted on a *blind* basis; that is, the control sample should not be identified. If the control sample and several test samples are tested at the same time and the persons doing the evaluating are not able to distinguish between them, the tester can be certain that there are no outstanding differences. Sometimes, rather extreme methods must be taken so that the control samples cannot be identified. Different colored lights, removal of all labels and packaging, addition of foreign colors, and so on may need to be used.

Frequently, it is desirable to set up check laboratories within a food quality control system. The most important testing will be done at the plant level, as close to the time of production as possible. The goal is to detect unacceptable raw materials or processes before they are used in production. The same is true of finished product testing. If an unacceptable finished product is detected soon after production, the processing line can be stopped or altered so that a minimum amount of defective material is produced. Some of this testing can be done right on the processing line. The remainder is usually done in the plant's laboratory.

It is possible that testing techniques can drift away from the standard methods, thus it is important to occasionally run companion samples in another laboratory and then compare the results with the in-plant test results. Such tests are not performed as a way of accepting or rejecting product but are an audit of the performance of the system. If the companion shows that the test results are not in agreement, an immediate investigation should take place to determine the reason.

V. SPECIFICATIONS AND SYSTEMS

Exact specifications must be drawn up for raw materials, sanitation, in-process controls, the finished product, and other areas. Frequently, raw materials do not have precise specifications. It is only fair to suppliers—and purchasers—that specifications are committed to writing and agreed upon. In effect, the agreement becomes a contract. The specifications should include testing methods and permissible variations from target values. Provision for the rejection of raw materials that do not meet specifications must be included in the agreement.

Specifications for finished products should also be documented and agreed upon by the food processors customers. If the specification cannot be defined in chemical or microbiological terms, then color photographs, drawings, physical measurements and the like can be used to describe what the finished product is expected to look like. Shelf life and required storage conditions must also be included in the finished-product specification.

VI. INGREDIENT TESTING

Food processing plants vary from those that actually collect raw materials from the farmer's fields to those that assemble ingredients already partially processed or partially assembled, perhaps elsewhere. However, the same principles apply.

A. Field Crops

In the case of crops in the field (e.g., fruit and vegetables), testing starts with the maturity and quality of the raw materials while they are still growing. Peas, for instance, are carefully tested all during the final stages of ripening until they are harvested at what producers like to call "precisely the proper moment" to insure top quality and yields. If weather or other reasons prevent harvesting at the proper time, whole fields may be skipped and the peas harvested later for use dried or as seed stock.

B. Purchased Materials

In the case of raw materials or ingredients that have been purchased, such as flour, corn syrup, and the like, it is important that the food processor test upon receipt but before acceptance. At this point the specification becomes important. Before unloading, inspection should be made of the delivery vehicle to be certain that the ingredients are not contaminated with insects, rodents, or other foreign materials or odors. If the delivery appears to be contaminated it should be rejected before it is unloaded. If the delivery appears satisfactory, it is unloaded but should be kept in a holding areas until samples are examined by receiving inspection or a quality control laboratory. Incoming inspection is one of the most critical areas of food processing. It is possible to make a bad product from good raw materials, but it is generally not possible to make good products from bad raw materials.

After raw materials are tested and released for production, control tags should be applied so that the progress of the ingredients can be tracked throughout the manufacturing process. It must be possible to exactly locate ingredients and their point of use. Occasionally, representatives of regulatory agencies will arrive and ask for records showing usage of certain lots of raw materials. If records are not available, the entire plant's production may be embargoed. It is possible that a finished product may turn out to be defective; the quality control department must know exactly what ingredients were used.

VII. IN-PROCESS CONTROLS

Critical process control points must be determined and carefully documented for each part of the processing line. If problems occur, the problem point can be identified and corrective action taken. The goal, of course, is to detect problems before significant amounts of product are made.

For example, on a frozen-food processing line the temperature of the gravy as it enters the pot pie or dinner must be carefully controlled and recorded. If the temperature is too high, off flavors can result. If the temperature is not high enough, bacteria may survive and spoil the product.

Weights of individual components of frozen and other foods are also critical since these are prescribed by federal regulations.

VIII. FINISHED FOOD TESTING

Finished product testing includes chemical and microbiological tests as well as organoleptic tests.

A. Chemical and Microbiological Tests

At varying periods of time, products that have been labeled as containing certain quantities of nutrients must be tested to verify that the label statements are correct.

These are laboratory tests and must be carefully recorded and the records retained for several years to prove label compliance.

In addition, chemical tests should be performed to insure that:

1. All ingredients have been added
2. Proper processing has taken place
3. The product does not contain foreign materials

Microbiological testing must be included for frozen foods, and incubation and testing of canned goods must be made to assure that the cans have been sterilized and the product is safe for shipment.

B. Organoleptic Tests

Food processors will frequently compare their products with control samples or competitor's products that have been purchased "in the field." Organoleptic tests including taste, color, and texture are compared. The age of competitor's samples should be approximately the same as the products to be compared for the tests to be valid. Frequently it is not possible to know the age of products in grocery stores, but it is important to be sure that a fresh product is not compared to competitor's products that may be months old.

Shelf-life testing is done so that products may be removed from the field before they become unsatisfactory to the consumer. Usually it is possible to design accelerated shelf-life tests so that in a reasonably short time it is possible to determine if the product will stay usable for the expected shelf-life times—perhaps years in the case of some canned or bottled products.

The finished product itself is the company's last chance to be sure that the product will satisfy the consumer, will satisfy all regulatory agencies, and has been made according to approved procedures and formulas. It is important that all test documentation be retained to verify the quality level of the product, including weights, at the time the product was manufactured and shipped to the field.

IX. DISTRIBUTION QUALITY

Various processing methods produce products of different shelf lives. These shelf lives range from two days in the case of fresh bread, to four or five years in the case of certain canned goods.

A. Storage Conditions

Some food products are stored dry; some frozen; some canned; some in plastic pouches; and some, such as specialty breads, on the shelf at room temperature. It is important that the food processor understand the way that product quality might change during storage in the distribution chain.

In dry-storage warehouses in the South, for example, temperatures of 125°F may well be common in the summer. In northern areas, in the winter, temperatures are often well below freezing. Freezing can drastically change the texture of some canned products. Also, the storage of some products near others that have strong odors can completely change the acceptability of the finished product.

B. Field Testing

As a function both of stock turnover and conditions of storage, every product must be observed and tested in the field (i.e., from the store shelves). This is to make certain that the products will not deteriorate to the point where they will be unsatisfactory or even unsafe for the consumer prior to the "use before" date.

X. REGULATORY ASPECTS

Food products are carefully controlled not only by the Federal Government but by state and local regulatory agencies as well. Of course, safety is paramount. But proper weights, pricing, and the like are also watched with increased frequency in the marketplace.

In 1994, metric labeling became mandatory. Nutritional labeling is under constant study and is also mandatory. Quality control of food products requires that laws and regulations be carefully followed. Prevention is the operative word, because, in addition to the possibility of seizure and loss of finished product, food processors may be criminally liable for shipping misbranded, misprocessed, or contaminated products.

After the required laws and regulations have been identified, it is important to test the product to assure compliance and then document the tests. It is a key step for the processor to be able to prove that, while the product was in the plant, the processor took all reasonable precautions to manufacture, test, and report on the quality of the product.

Many food processing companies use consultants or rely on trade associations such as the Food Processors Association to keep up with the constantly changing food laws and with the occasional conflict between local, state, and federal laws.

XI. QUALITY IMPROVEMENT

During the last few years, the concepts of *quality improvement* and *total quality* have been used in some food processing plants to involve all of the plant personnel in quality assurance. The concept has often turned out to be a very efficient system for preventing problems that could affect quality. Additionally, the concept has proved to be a useful way of reducing costs through quality cost reduction (see Chapter 5).

Basically, quality improvement involves organizing the individual manufacturing groups into teams. The teams concentrate on the most costly and repetitive problems until they are solved. The overall thrust of the program is to prevent errors from happening rather than catching them after they do.

The individual worker's ideas and recommendations are used as raw material for the team's activities. Each team's progress is charted so that all workers are kept up to date on what is happening.

XII. GOOD MANUFACTURING PRACTICE

A crucial part of food processing is that of *good manufacturing practice* (GMP), or, more recently, *current* good manufacturing practice (cGMP). Not unlike the ISO 9000 series, GMP requires that proper manufacturing and operating procedures

Follow Procedure QQ7303 in evaluating areas and assigning ratings.

Use A = Acceptable and N = Not acceptable to rate areas

Inspector/Audit Leader Initials ________

Area	Date, Time, and Rating			
1. Storage				
(a) Floors				
(b) Walls and ceilings				
(c) Lights and fixtures				
2. Processing				
(a) Washing equipment				
(b) Peeling equipment				
(c) Slicing equipment				
(d) Cooking equipment				
(e) Conveyors				
(f) Drains, gutters, etc.				
(g) Floors, walls, etc.				
(h) Waste disposal				
(i) Lighting				
(j) Hoods and ventilation				
3. Packing				
(a) Fillers				
(b) Conveyors				
(c) Tanks, pipes				
(d) Floors, walls				

Figure 1 Typical food manufacturing plant inspection/audit checklist.

be approved, documented, followed, and audited. Audit records must be kept and used to guide corrective action when it is required.

Unlike ISO 9000, however, GMP or cGMP is not optional. Initially published in the U.S. Federal Register as Part 128 of the Code of Federal Regulations (CFR) and now codified as Part 110, cGMP regulations fall under the provisions of the federal Food, Drug and Cosmetics Act (FDCA) and are reviewed and enforced by the U.S. Food and Drug Administration (FDA). The FDA uses GMP regulations to control the risk of filth, microbiological or chemical poisoning, and other contaminants during manufacture. In effect, GMP and cGMP cover every aspect of food materials, production, people, facilities, and procedures.

One of the more powerful tools of cGMP is the plant inspection. A well-run plant relies on its own inspection conducted by quality assurance and other staff members to find problems and initiate corrective action. Figure 1 shows a portion of a typical inspection/audit checklist. In addition to detailed directions for conducting the audit, the procedure referred to at the beginning of the checklist would contain descriptions of conditions that would result in "not acceptable" ratings. These would require prompt corrective action. Formal inspections conducted by municipal, state, or FDA officials are routinely performed to assure that conditions meet the specified requirements.

FURTHER READING

Allan, D. V., and Murray, B. A., "Identifying and Removing Causes of Error in a Food Manufacturing Plant," *Food Technology*, 1972.

Allan, D. V., and Murray, B. A., "Identifying and Removing Causes of Error in a Food Manufacturing Plant," *Food Technology*, 1973.

The Almanac of the Canning, Freezing, Preserving Industries, Edward E. Judge & Sons, Inc., Westminster, MD, 1993.

American Society for Quality Control, *Food Processing Industry Quality System Guidelines*, Food, Drug, and Cosmetic Division, Food Industries Quality Guidelines Committee, ASQC, Milwaukee, WI, 1986

Amerine, M., Pangborn, R., and Roessier, E. B., *Principles of Sensory Evaluation of Food*, Academic Press, New York, 1965.

A Complete Course in Canning, The Canning Trade, Inc., 1975.

Crosby, P. B., *The Art of Getting Your Own Sweet Way*, McGraw-Hill, New York, 1979.

Crosby, P. B., *Quality is Free*, McGraw-Hill, New York, 1979.

GMP-GSP Guideline Rules for Food Plant Employees, L. J. Bianco Publishers, Northbrook, IL, 1992.

Gould, W. A., *Current Good Manufacturing Practice: Food Plant Sanitation*, CTI Publications, Baltimore, MD, 1994.

Kramer and Twigg, *Fundamentals of Quality Control for the Food Industry*, Vols. 1 and 2, AVI, Westport, CT 1970.

Laboratory Manual for Food Canners and Processors, Vols. 1 and 2, National Canners Association, AVI, Westport, CT, 1968.

Murray, B. A., "Balanced Quality Control," Food Industries Quality Control European Symposium, Madrid, 1973.

Murray, B. A., *Quality For Real Workbook*, Wilhelm Press, Englewood, FL, 1992.

Official Methods of Analysis of the Association of Official Analytical Chemists, 15th ed., 1990.

Puri, S. C., Ennis, D., Mullen, K., *Statistical Quality Control for Food and Agricultural Scientists*, G. K. Hall, Boston, 1979.

Appendix: Military Standard 105D

EDITOR'S NOTE

A number of long-used military standards, including MIL-STD-105, are no longer being required on new contracts by the U.S. military services. That these standards, which include MIL-I-45208 and MIL-Q-9858, are "no longer supported" by the U.S. government does not at all diminish their value as industrial programs. The American National Standards Institute (ANSI) and the American Society for Quality Control (ASQC) have, for example, issued a document numbered ANSI/ASQC-Z-1.4, 1993. Initial review indicates that this document represents little, if any, change from the basic elements of MIL-STD-105.

The following Appendix is MIL-STD-105-D. MIL-STD-105-D has been replaced in general usage by MIL-STD-105-E. We have included the previous revision as a reference and as a service to readers who may not have ready access to MIL standards or who are not able to obtain clear copies.

A general review indicates that the MIL-STD-105-E revisions are negligible. While most readers will be able to use the version in this book in the same way as they would use the E revision, readers are cautioned to verify inspection plans that are based on the following version.

MIL-STD-105D
29 April 1963

SUPERSEDING
MIL-STD-105C
18 July 1961

MILITARY STANDARD

SAMPLING PROCEDURES AND TABLES FOR INSPECTION BY ATTRIBUTES

MIL-STD-105D
29 APRIL 1963

DEPARTMENT OF DEFENSE
Washington 25, D.C.

SAMPLING PROCEDURES AND TABLES FOR INSPECTION BY ATTRIBUTES

MIL-STD-105D

29 APRIL 1963

1. This standard has been approved by the Department of Defense and is mandatory for use by the Departments of the Army, the Navy, the Air Force and the Defense Supply Agency. This revision supersedes MIL-STD-105C, dated 18 July 1961.

2. This publication provides sampling procedures and reference tables for use in planning and conducting inspection by attributes. This publication was developed by a working group representing the military services of Canada, the United Kingdom and the United States of America with the assistance and cooperation of American and European organizations for quality control. The international designation of this document is ABC-STD-105. When revision or cancellation of this standard is proposed, the departmental custodians will inform their respective Departmental Standardization Office so that appropriate action may be taken respecting the international agreement concerned.

3. The U.S. Army Munitions Command is designated as preparing activity for this standard. Recommended corrections, additions, or deletions should be addressed to the Commanding Officer, U.S. Army CBR Engineering Office, Attn: SMUCE-ED-S, Army Chemical Center, Maryland.

CONTENTS

TABLES

SAMPLING PROCEDURES AND TABLES FOR INSPECTION BY ATTRIBUTES

1. SCOPE

1.1 PURPOSE. This publication establishes sampling plans and procedures for inspection by attributes. When specified by the responsible authority, this publication shall be referenced in the specification, contract, inspection instructions, or other documents and the provisions set forth herein shall govern. The "responsible authority" shall be designated in one of the above documents.

1.2 APPLICATION. Sampling plans designated in this publication are applicable, but not limited, to inspection of the following:

a. End items.

b. Components and raw materials.

c. Operations.

d. Materials in process.

e. Supplies in storage.

g. Data or records.

h. Administrative procedures.

These plans are intended primarily to be used for a continuing series of lots or batches. The plans may also be used for the inspection of isolated lots or batches, but, in this latter case, the user is cautioned to consult the operating characteristic curves to find a plan which will yield the desired protection (see 11.6).

1.3 INSPECTION. Inspection is the process of measuring, examining, testing, or otherwise comparing the unit of product (see 1.5) with the requirements.

1.4 INSPECTION BY ATTRIBUTES. Inspection by attributes is inspection whereby either the unit of product is classified simply as defective or nondefective, or the number of defects in the unit of product is counted, with respect to a given requirement or set of requirements.

1.5 UNIT OF PRODUCT. The unit of product is the thing inspected in order to determine its classification as defective or nondefective or to count the number of defects. It may be a single article, a pair, a set, a length, an area, an operation, a volume, a component of an end product, or the end product itself. The unit of product may or may not be the same as the unit of purchase, supply, production, or shipment.

2. CLASSIFICATION OF DEFECTS AND DEFECTIVES

2.1 METHOD OF CLASSIFYING DEFECTS. A classification of defects is the enumeration of possible defects of the unit of product classified according to their seriousness. A defect is any nonconformance of the unit of product with specified requirements. Defects will normally be grouped into one or more of the following classes; however, defects may be grouped into other classes, or into subclasses within these classes.

2.1.1 CRITICAL DEFECT. A critical defect is a defect that judgment and experience indicate is likely to result in hazardous or unsafe conditions for individuals using, maintaining, or depending upon the product; or a defect that judgment and experience indicate is likely to prevent performance of the tactical function of a major end item such as a ship, aircraft, tank, missile or space vehicle. NOTE: For a special provision relating to critical defects, see 6.3.

2.1.2 MAJOR DEFECT. A major defect is a defect, other than critical, that is likely to result in failure, or to reduce materially the usability of the unit of product for its intended purpose.

2.1.3 MINOR DEFECT. A minor defect is a defect that is not likely to reduce materially the usability of the unit of product for its intended purpose, or is a departure from established standards having little bearing on the effective use or operation of the unit.

2.2 METHOD OF CLASSIFYING DEFECTIVES. A defective is a unit of product which contains one or more defects. Defectives will usually be classified as follows:

2.2.1 CRITICAL DEFECTIVE. A critical defective contains one or more critical defects and may also contain major and or minor defects. NOTE: For a special provision relating to critical defectives, see 6.3.

2.2.2 MAJOR DEFECTIVE. A major defective contains one or more major defects, and may also contain minor defects but contains no critical defect.

2.2.3 MINOR DEFECTIVE. A minor defective contains one or more minor defects but contains no critical or major defect.

3. PERCENT DEFECTIVE AND DEFECTS PER HUNDRED UNITS

3.1 EXPRESSION OF NONCONFORMANCE. The extent of nonconformance of product shall be expressed either in terms of percent defective or in terms of defects per hundred units.

3.2 PERCENT DEFECTIVE. The percent defective of any given quantity of units of product is one hundred times the number of defective units of product contained therein divided by the total number of units of product, i.e.:

$$\text{Percent defective} = \frac{\text{Number of defects}}{\text{Number of units inspected}} \times 100$$

3.3 DEFECTS PER HUNDRED UNITS. The number of defects per hundred units of any given quantity of units of product is one hundred times the number of defects contained therein (one or more defects being possible in any unit of product) divided by the total number of units of product, i.e.:

$$\text{Defects per hundred units} = \frac{\text{Number of defectives}}{\text{Number of units inspected}} \times 100$$

4. ACCEPTABLE QUALITY LEVEL (AQL)

4.1 USE. The AQL, together with the Sample Size Code Letter, is used for indexing the sampling plans provided herein.

4.2 DEFINITION. The AQL is the maximum percent defective (or the maximum number of defects per hundred units) that, for purposes of sampling inspection, can be considered satisfactory as a process average (see 11.2).

4.3 NOTE ON THE MEANING OF AQL. When a consumer designates some specific value of AQL for a certain defect or group of defects, he indicates to the supplier that his (the consumer's) acceptance sampling plan will accept the great majority of the lots or batches that the supplier submits, provided the process average level of percent defective (or defects per hundred units) in these lots or batches be no greater than the designated value of AQL. Thus, the AQL is a designated value of percent defective (or defects per hundred units) that the consumer indicates will be accepted most of the time by the acceptance sampling procedure to be used. The sampling plans provided herein are so arranged that the probability of acceptance at the designated AQL value depends upon the sample size, being generally higher for large samples than for small ones, for a given AQL. The AQL alone does not describe the protection to the consumer for individual lots or batches but more directly relates to what might be expected from a series of lots or batches, provided the steps indicated in this publication are taken. It is necessary to refer to the operating characteristic curve of the plan, to determine what protection the consumer will have.

4.4 LIMITATION. The designation of an AQL shall not imply that the supplier has the right to supply knowingly any defective unit of product.

4.5 SPECIFYING AQLs. The AQL to be used will be designated in the contract or by the responsible authority. Different AQLs may be designated for groups of defects considered collectively, or for individual defects. An AQL for a group of defects may be designated in addition to AQLs for individual defects, or subgroups, within that group. AQL values of 10.0 or less may be expressed either in percent defective or in defects per hundred units; those over 10.0 shall be expressed in defects per hundred units only.

4.6 PREFERRED AQLs. The values of AQL given in these tables are known as preferred AQLs. If, for any product, an AQL be designated other than a preferred AQL, these tables are not applicable.

5. SUBMISSION OF PRODUCT

5.1 LOT OR BATCH. The term lot or batch shall mean "inspection lot" or "inspection batch," i.e., a collection of units of product from which a sample is to be drawn and inspected to determine conformance with the acceptability criteria, and may differ from a collection of units designated as a lot or batch for other purposes (e.g., production, shipment, etc.).

5.2 FORMATION OF LOTS OR BATCHES. The product shall be assembled into identifiable lots, sublots, batches, or in such other manner as may be prescribed (see 5.4). Each lot or batch shall, as far as is practicable, consist of

5. SUBMISSION OF PRODUCT (continued)

units of product of a single type, grade, class, size, and composition, manufactured under essentially the same conditions, and at essentially the same time.

5.3 LOT OR BATCH SIZE. The lot or batch size is the number of units of product in a lot or batch.

5.4 PRESENTATION OF LOTS OR BATCHES. The formation of the lots or batches, lot or batch size, and the manner in which each lot or batch is to be presented and identified by the supplier shall be designated or approved by the responsible authority. As necessary, the supplier shall provide adequate and suitable storage space for each lot or batch, equipment needed for proper identification and presentation, and personnel for all handling of product required for drawing of samples.

6. ACCEPTANCE AND REJECTION

6.1 ACCEPTABILITY OF LOTS OR BATCHES. Acceptability of a lot or batch will be determined by the use of a sampling plan or plans associated with the designated AQL or AQLs.

6.2 DEFECTIVE UNITS. The right is reserved to reject any unit of product found defective during inspection whether that unit of product forms part of a sample or not, and whether the lot or batch as a whole is accepted or rejected. Rejected units may be repaired or corrected and resubmitted for inspection with the approval of, and in the manner specified by, the responsible authority.

6.3 SPECIAL RESERVATION FOR CRITICAL DEFECTS. The supplier may be required at the discretion of the responsible authority to inspect every unit of the lot or batch for critical defects. The right is reserved to inspect every unit submitted by the supplier for critical defects, and to reject the lot or batch immediately, when a critical defect is found. The right is reserved also to sample, for critical defects, every lot or batch submitted by the supplier and to reject any lot or batch if a sample drawn therefrom is found to contain one or more critical defects.

6.4 RESUBMITTED LOTS OR BATCHES. Lots or batches found unacceptable shall be resubmitted for reinspection only after all units are re-examined or retested and all defective units are removed or defects corrected. The responsible authority shall determine whether normal or tightened inspection shall be used, and whether reinspection shall include all types or classes of defects or for the particular types or classes of defects which caused initial rejection.

7. DRAWING OF SAMPLES

7.1 SAMPLE. A sample consists of one or more units of product drawn from a lot or batch, the units of the sample being selected at random without regard to their quality. The number of units of product in the sample is the sample size.

7.2 REPRESENTATIVE SAMPLING. When appropriate, the number of units in the sample shall be selected in proportion to the size of sublots or subbatches, or parts of the lot or batch, identified by some rational criterion.

7. DRAWING OF SAMPLES (continued)

When representative sampling is used, the units from each part of the lot or batch shall be selected at random.

7.3 TIME OF SAMPLING. Samples may be drawn after all the units comprising the lot or batch have been assembled, or samples may be drawn during assembly of the lot or batch.

7.4 DOUBLE OR MULTIPLE SAMPLING. When double or multiple sampling is to be used, each sample shall be selected over the entire lot or batch.

8. NORMAL, TIGHTENED AND REDUCED INSPECTION

8.1 INITIATION OF INSPECTION. Normal inspection will be used at the start of inspection unless otherwise directed by the responsible authority.

8.2 CONTINUATION OF INSPECTION. Normal, tightened or reduced inspection shall continue unchanged for each class of defects or defectives on successive lots or batches except where the switching procedures given below require change. The switching procedures given below require a change. The switching procedures shall be applied to each class of defects or defectives independently.

8.3 SWITCHING PROCEDURES.

8.3.1 NORMAL TO TIGHTENED. When normal inspection is in effect, tightened inspection shall be instituted when 2 out of 5 consecutive lots or batches have been rejected on original inspection (i.e., ignoring resubmitted lots or batches for this procedure).

8.3.2 TIGHTENED TO NORMAL. When tightened inspection is in effect, normal inspection shall be instituted when 5 consecutive lots or batches have been considered acceptable on original inspection.

8.3.3 NORMAL TO REDUCED. When normal inspection is in effect, reduced inspection shall be instituted providing that all of the following conditions are satisfied:

a. The preceding 10 lots or batches (or more, as indicated by the note to Table VIII) have been on normal inspection and none has been rejected on original inspection; and

b. The total number of defectives (or defects) in the samples from the preceding 10 lots or batches (or such other number as was used for condition "a" above) is equal to or less than the applicable number given in Table VIII. If double or multiple sampling is in use, all samples inspected should be included, not "first" samples only; and

c. Production is at a steady rate; and

d. Reduced inspection is considered desirable by the responsible authority.

8.3.4 REDUCED TO NORMAL. When reduced inspection is in effect, normal inspection shall be instituted if any of the following occur on original inspection:

a. A lot or batch is rejected; or

b. A lot or batch is considered acceptable under the procedures of 10.1.4; or

c. Production becomes irregular or delayed; or

d. Other conditions warrant that normal inspection shall be instituted.

8.4 DISCONTINUATION OF INSPECTION. In the event that 10 consecutive lots or batches remain on tightened inspection (or such other number as may be designated by the responsible authority), inspection under the provisions of this document should be discontinued pending action to improve the quality of submitted material.

9. SAMPLING PLANS

9.1 SAMPLING PLAN. A sampling plan indicates the number of units of product from each lot or batch which are to be inspected (sample size or series of sample sizes) and the criteria for determining the acceptability of the lot or batch (acceptance and rejection numbers).

9.2 INSPECTION LEVEL. The inspection level determines the relationship between the lot or batch size and the sample size. The inspection level to be used for any particular requirement will be prescribed by the responsible authority. Three inspection levels: I, II, and III, are given in Table I for general use. Unless otherwise specified, Inspection Level II will be used. However, Inspection Level I may be specified when less discrimination is needed, or Level III may be specified for greater discrimination. Four additional special levels: S-1, S-2, S-3 and S-4, are given in the same table and may be used where relatively small sample sizes are necessary and large sampling risks can or must be tolerated.

NOTE: In the designation of inspection levels S-1 to S-4, care must be exercised to avoid AQLs inconsistent with these inspection levels.

9.3 CODE LETTERS. Sample sizes are designated by code letters. Table I shall be used to find the applicable code letter for the particular lot or batch size and the prescribed inspection level.

9.4 OBTAINING SAMPLING PLAN. The AQL and the code letter shall be used to obtain the sampling plan from Tables II, III or IV. When no sampling plan is available for a given combination of AQL and code letter, the tables direct the user to a different letter. The sample size to be used is given by the new code letter not by the original letter. If this procedure leads to different sample sizes for different classes of defects, the code letter corresponding to the largest sample size derived may be used for all classes of defects when designated or approved by the responsible authority. As an alternative to a single sampling plan with an acceptance number of 0, the plan with an acceptance number of 1 with its correspondingly larger sample size for a designated AQL (where available), may be used when designated or approved by the responsible authority.

9.5 TYPES OF SAMPLING PLANS. Three types of sampling plans: Single, Double and Multiple, are given in Tables II, III and IV, respectively. When several types of plans are available for a given AQL and code letter, any one may be used. A decision as to type of plan, either single, double, or multiple, when available for a given AQL and code letter, will usually be based upon the comparison between the administrative difficulty and the average sample sizes of the available plans. The average sample size of multiple plans is less than for double (except in the case corresponding to single acceptance number 1) and both of these are always less than a single sample size. Usually the administrative difficulty for single sampling and the cost per unit of the sample are less than for double or multiple.

10. DETERMINATION OF ACCEPTABILITY

10.1 PERCENT DEFECTIVE INSPECTION. To determine acceptability of a lot or batch under percent defective inspection, the applicable sampling plan shall be used in accordance with 10.1.1, 10.1.2, 10.1.3, 10.1.4, and 10.1.5.

10.1.1 SINGLE SAMPLING PLAN. The number of sample units inspected shall be equal to the sample size given by the plan. If the number of defectives found in the sample is equal to or less than the acceptance number, the lot or batch shall be considered acceptable. If the number of defectives is equal to or greater than the rejection number, the lot or batch shall be rejected.

10.1.2 DOUBLE SAMPLING PLAN. The number of sample units inspected shall be equal to the first sample size given by the plan. If the number of defectives found in the first sample is equal to or less than the first acceptance number, the lot or batch shall be considered acceptable. If the number of defectives found in the first sample is equal to or greater than the first rejection number, the lot or batch shall be rejected. If the number of defectives found in the first sample is between the first acceptance and rejection numbers, a second sample of the size given by the plan shall be inspected. The number of defectives found in the first and second samples shall be accumulated. If the cumulative number of defectives is equal to or less than the second acceptance number, the lot or batch shall be considered acceptable. If the cumulative number of defectives is equal to or greater than the second rejection number, the lot or batch shall be rejected.

10.1.3 MULTIPLE SAMPLE PLAN. Under multiple sampling, the procedure shall be similar to that specified in 10.1.2, except that the number of successive samples required to reach a decision may be more than two.

10.1.4 SPECIAL PROCEDURE FOR REDUCED INSPECTION. Under reduced inspection, the sampling procedure may terminate without either acceptance or rejection criteria having been met. In these circumstances, the lot or batch will be considered acceptable, but normal inspection will be reinstated starting with the next lot or batch (see 8.3.4 (b)).

10.2 DEFECTS PER HUNDRED UNITS INSPECTION. To determine the acceptability of a lot or batch under Defects per Hundred Units inspection, the procedure specified for Percent Defective inspection above shall be used, except that the word "defects" shall be substituted for "defectives."

11. SUPPLEMENTARY INFORMATION

11.1 OPERATING CHARACTERISTIC CURVES. The operating characteristic curves for normal inspection, shown in Table X (pages 30-62), indicate the percentage of lots or batches which may be expected to be accepted under the various sampling plans for a given process quality. The curves shown are for single sampling; curves for double and multiple sampling are matched as closely as practicable. The O.C. curves shown for AQLs greater than 10.0 are based on the Poisson distribution and are applicable for defects per hundred units inspection; those for AQLs of 10.0 or less and sample sizes of 80 or less are based on the binomial distribution and are applicable for percent defective inspection; those for

AQLs of 10.0 or less and sample sizes larger than 80 are based on the Poisson distribution and are applicable either for defects per hundred units inspection, or for percent defective inspection (the Poisson distribution being an adequate approximation to the binomial distribution under these conditions). Tabulated values, corresponding to selected values of probabilities of acceptance (P_a, in percent) are given for each of the curves shown, and, in addition, for tightened inspection, and for defects per hundred units for AQLs of 10.0 or less and sample sizes of 80 or less.

11.2 PROCESS AVERAGE. The process average is the average percent defective or average number of defects per hundred units (whichever is applicable) of product submitted by the supplier for original inspection. Original inspection is the first inspection of a particular quantity of product as distinguished from the inspection of product which has been resubmitted after prior rejection.

11.3 AVERAGE OUTGOING QUALITY (AOQ). The AOQ is the average quality of outgoing product including all accepted lots or batches, plus all rejected lots or batches after the rejected lots or batches have been effectively 100 percent inspected and all defectives replaced by nondefectives.

11.4 AVERAGE OUTGOING QUALITY LIMIT (AOQL). The AOQL is the maximum of the AOQs for all possible incoming qualities for a given acceptance sampling plan. AOQL values are given in Table V-A for each of the single sampling plans for normal inspection and in Table V-B for each of the single sampling plans for tightened inspection.

11.5 AVERAGE SAMPLE SIZE CURVES. Average sample size curves for double and multiple sampling are in Table IX. These show the average sample sizes which may be expected to occur under the various sampling plans for a given process quality. The curves assume no curtailment of inspection and are approximate to the extent that they are based upon the Poisson distribution, and that the sample sizes for double and multiple sampling are assumed to be 0.631n and 0.25n respectively, where n is the equivalent single sample size.

11.6 LIMITING QUALITY PROTECTION. The sampling plans and associated procedures given in this publication were designed for use where the units of product are produced in a continuing series of lots or batches over a period of time. However, if the lot or batch is of an isolated nature, it is desirable to limit the selection of sampling plans to those, associated with a designated AQL value, that provide not less than a specified limiting quality protection. Sampling plans for this purpose can be selected by choosing a Limiting Quality (LQ) and a consumer's risk to be associated with it. Tables VI and VII give values of LQ for the commonly used consumer's risks of 10 percent and 5 percent respectively. If a different value of consumer's risk is required, the O.C. curves and their tabulated values may be used. The concept of LQ may also be useful in specifying the AQL and Inspection Levels for a series of lots or batches, thus fixing minimum sample size where there is some reason for avoiding (with more than a given consumer's risk) more than a limiting proportion of defectives (or defects) in any single lot or batch.

TABLE I — Sample size code letters *(See 9.2 and 9.3)*

Lot or batch size			Special inspection levels				General inspection levels		
			S-1	S-2	S-3	S-4	I	II	III
2	to	8	A	A	A	A	A	A	B
9	to	15	A	A	A	A	A	B	C
16	to	25	A	A	B	B	B	C	D
26	to	50	A	B	B	C	C	D	E
51	to	90	B	B	C	C	C	E	F
91	to	150	B	B	C	D	D	F	G
151	to	280	B	C	D	E	E	G	H
281	to	500	B	C	D	E	F	H	J
501	to	1200	C	C	E	F	G	J	K
1201	to	3200	C	D	E	G	H	K	L
3201	to	10000	C	D	F	G	J	L	M
10001	to	35000	C	D	F	H	K	M	N
35001	to	150000	D	E	G	J	L	N	P
150001	to	500000	D	E	G	J	M	P	Q
500001	and	over	D	E	H	K	N	Q	R

Code Letter

TABLE II-A — *Single sampling plans for normal inspection (Master table)*

(See 9.4 and 9.5)

Sample size code letter	Sample size	Acceptable Quality Levels (normal inspection)																					
		0.010	0.015	0.025	0.040	0.065	0.10	0.15	0.25	0.40	0.65	1.0	1.5	2.5	4.0	6.5	10	15	25	40	65	100	150
		Ac Re	Ac Re	Ac Re	Ac Re	Ac Re	Ac Re	Ac Re	Ac Re	Ac Re	Ac Re	Ac Re	Ac Re	Ac Re	Ac Re	Ac Re	Ac Re	Ac Re	Ac Re	Ac Re	Ac Re	Ac Re	Ac Re
A	2	↓	↓	↓	↓	↓	↓	↓	↓	↓	↓	↓	↓	↓	↓	0 1	↓	↓	1 2	2 3	3 4	5 6	7 8
B	3	↓	↓	↓	↓	↓	↓	↓	↓	↓	↓	↓	↓	↓	0 1	↑	↓	1 2	2 3	3 4	5 6	7 8	10 11
C	5	↓	↓	↓	↓	↓	↓	↓	↓	↓	↓	↓	↓	0 1	↑	↓	1 2	2 3	3 4	5 6	7 8	10 11	14 15
D	8	↓	↓	↓	↓	↓	↓	↓	↓	↓	↓	↓	0 1	↑	↓	1 2	2 3	3 4	5 6	7 8	10 11	14 15	21 22
E	13	↓	↓	↓	↓	↓	↓	↓	↓	↓	↓	0 1	↑	↓	1 2	2 3	3 4	5 6	7 8	10 11	14 15	21 22	30 31
F	20	↓	↓	↓	↓	↓	↓	↓	↓	↓	0 1	↑	↓	1 2	2 3	3 4	5 6	7 8	10 11	14 15	21 22	↑	↑
G	32	↓	↓	↓	↓	↓	↓	↓	↓	0 1	↑	↓	1 2	2 3	3 4	5 6	7 8	10 11	14 15	21 22	↑	↑	↑
H	50	↓	↓	↓	↓	↓	↓	↓	0 1	↑	↓	1 2	2 3	3 4	5 6	7 8	10 11	14 15	21 22	↑	↑	↑	↑
J	80	↓	↓	↓	↓	↓	↓	0 1	↑	↓	1 2	2 3	3 4	5 6	7 8	10 11	14 15	21 22	↑	↑	↑	↑	↑
K	125	↓	↓	↓	↓	↓	0 1	↑	↓	1 2	2 3	3 4	5 6	7 8	10 11	14 15	21 22	↑	↑	↑	↑	↑	↑
L	200	↓	↓	↓	↓	0 1	↑	↓	1 2	2 3	3 4	5 6	7 8	10 11	14 15	21 22	↑	↑	↑	↑	↑	↑	↑
M	315	↓	↓	↓	0 1	↑	↓	1 2	2 3	3 4	5 6	7 8	10 11	14 15	21 22	↑	↑	↑	↑	↑	↑	↑	↑
N	500	↓	↓	0 1	↑	↓	1 2	2 3	3 4	5 6	7 8	10 11	14 15	21 22	↑	↑	↑	↑	↑	↑	↑	↑	↑
P	800	↓	0 1	↑	↓	1 2	2 3	3 4	5 6	7 8	10 11	14 15	21 22	↑	↑	↑	↑	↑	↑	↑	↑	↑	↑
Q	1250	0 1	↑	↓	1 2	2 3	3 4	5 6	7 8	10 11	14 15	21 22	↑	↑	↑	↑	↑	↑	↑	↑	↑	↑	↑
R	2000	↑	↑	1 2	2 3	3 4	5 6	7 8	10 11	14 15	21 22	↑	↑	↑	↑	↑	↑	↑	↑	↑	↑	↑	↑

↓ = Use first sampling plan below arrow. If sample size equals, or exceeds, lot or batch size, do 100 percent inspection.
↑ = Use first sampling plan above arrow
Ac = Acceptance number
Re = Rejection number

TABLE II-B — *Single sampling plans for tightened inspection (Master table)*

(See 9.4 and 9.5)

Sample size code letter	Sample size	0.010	0.015	0.025	0.040	0.065	0.10	0.15	0.25	0.40	0.65	1.0	1.5	2.5	4.0	6.5	10	15	25	40	65	100	150
		Ac Re	Ac Re	Ac Re	Ac Re	Ac Re	Ac Re	Ac Re	Ac Re	Ac Re	Ac Re	Ac Re	Ac Re	Ac Re	Ac Re	Ac Re	Ac Re	Ac Re	Ac Re	Ac Re	Ac Re	Ac Re	Ac Re
A	2	↓	↓	↓	↓	↓	↓	↓	↓	↓	↓	↓	↓	↓	↓	↓	↓	↓	↓	1 2	2 3	3 4	5 6
B	3	↓	↓	↓	↓	↓	↓	↓	↓	↓	↓	↓	↓	↓	↓	0 1	↓	↓	1 2	2 3	3 4	5 6	8 9
C	5	↓	↓	↓	↓	↓	↓	↓	↓	↓	↓	↓	↓	↓	0 1	↓	↓	1 2	2 3	3 4	5 6	8 9	12 13
D	8	↓	↓	↓	↓	↓	↓	↓	↓	↓	↓	↓	↓	0 1	↓	↓	1 2	2 3	3 4	5 6	8 9	12 13	18 19
E	13	↓	↓	↓	↓	↓	↓	↓	↓	↓	↓	↓	0 1	↓	↓	1 2	2 3	3 4	5 6	8 9	12 13	18 19	27 28
F	20	↓	↓	↓	↓	↓	↓	↓	↓	↓	↓	0 1	↓	↓	1 2	2 3	3 4	5 6	8 9	12 13	18 19	↑	↑
G	32	↓	↓	↓	↓	↓	↓	↓	↓	↓	0 1	↓	↓	1 2	2 3	3 4	5 6	8 9	12 13	18 19	↑	↑	↑
H	50	↓	↓	↓	↓	↓	↓	↓	↓	0 1	↓	↓	1 2	2 3	3 4	5 6	8 9	12 13	18 19	↑	↑	↑	↑
J	80	↓	↓	↓	↓	↓	↓	↓	0 1	↓	↓	1 2	2 3	3 4	5 6	8 9	12 13	18 19	↑	↑	↑	↑	↑
K	125	↓	↓	↓	↓	↓	↓	0 1	↓	↓	1 2	2 3	3 4	5 6	8 9	12 13	18 19	↑	↑	↑	↑	↑	↑
L	200	↓	↓	↓	↓	↓	0 1	↓	↓	1 2	2 3	3 4	5 6	8 9	12 13	18 19	↑	↑	↑	↑	↑	↑	↑
M	315	↓	↓	↓	↓	0 1	↓	↓	1 2	2 3	3 4	5 6	8 9	12 13	18 19	↑	↑	↑	↑	↑	↑	↑	↑
N	500	↓	↓	↓	0 1	↓	↓	1 2	2 3	3 4	5 6	8 9	12 13	18 19	↑	↑	↑	↑	↑	↑	↑	↑	↑
P	800	↓	↓	0 1	↓	↓	1 2	2 3	3 4	5 6	8 9	12 13	18 19	↑	↑	↑	↑	↑	↑	↑	↑	↑	↑
Q	1250	↓	0 1	↓	↓	1 2	2 3	3 4	5 6	8 9	12 13	18 19	↑	↑	↑	↑	↑	↑	↑	↑	↑	↑	↑
R	2000	0 1	↑	↓	1 2	2 3	3 4	5 6	8 9	12 13	18 19	↑	↑	↑	↑	↑	↑	↑	↑	↑	↑	↑	↑
S	3150		↑	1 2								↑	↑	↑	↑	↑	↑	↑	↑	↑	↑	↑	↑

Column headings 0.010 to 150: Acceptable Quality Levels (tightened inspection)

↓ = Use first sampling plan below arrow. If sample size equals or exceeds lot or batch size, do 100 percent inspection.
↑ = Use first sampling plan above arrow.
Ac = Acceptance number
Re = Rejection number

SINGLE TIGHTENED

TABLE II-C — *Single sampling plans for reduced inspection (Master table)*

(See 9.4 and 9.5)

Acceptable Quality Levels (reduced inspection)†

Sample size code letter	Sample size	0.010	0.015	0.025	0.040	0.065	0.10	0.15	0.25	0.40	0.65	1.0	1.5	2.5	4.0	6.5	10	15	25	40	65	100	150
		Ac Re	Ac Re	Ac Re	Ac Re	Ac Re	Ac Re	Ac Re	Ac Re	Ac Re	Ac Re	Ac Re	Ac Re	Ac Re	Ac Re	Ac Re	Ac Re	Ac Re	Ac Re	Ac Re	Ac Re	Ac Re	Ac Re
A	2															0 1			1 2	2 3	3 4	5 6	7 8
B	2	↓	↓	↓	↓	↓	↓	↓	↓	↓	↓	↓	↓	↓	0 1	↑	↓	0 2	1 3	2 4	3 5	5 6	7 8
C	2	↓	↓	↓	↓	↓	↓	↓	↓	↓	↓	↓	↓	0 1	↑	↓	0 2	1 3	1 4	2 5	3 6	5 8	7 10
D	3	↓	↓	↓	↓	↓	↓	↓	↓	↓	↓	↓	0 1	↑	↓	0 2	1 3	1 4	2 5	3 6	5 8	7 10	10 13
E	5	↓	↓	↓	↓	↓	↓	↓	↓	↓	↓	0 1	↑	↓	0 2	1 3	1 4	2 5	3 6	5 8	7 10	10 13	14 17
F	8	↓	↓	↓	↓	↓	↓	↓	↓	↓	0 1	↑	↓	0 2	1 3	1 4	2 5	3 6	5 8	7 10	10 13	↑	↑
G	13	↓	↓	↓	↓	↓	↓	↓	↓	0 1	↑	↓	0 2	1 3	1 4	2 5	3 6	5 8	7 10	10 13	↑	↑	↑
H	20	↓	↓	↓	↓	↓	↓	↓	0 1	↑	↓	0 2	1 3	1 4	2 5	3 6	5 8	7 10	10 13	↑	↑	↑	↑
J	32	↓	↓	↓	↓	↓	↓	0 1	↑	↓	0 2	1 3	1 4	2 5	3 6	5 8	7 10	10 13	↑	↑	↑	↑	↑
K	50	↓	↓	↓	↓	↓	0 1	↑	↓	0 2	1 3	1 4	2 5	3 6	5 8	7 10	10 13	↑	↑	↑	↑	↑	↑
L	80	↓	↓	↓	↓	0 1	↑	↓	0 2	1 3	1 4	2 5	3 6	5 8	7 10	10 13	↑	↑	↑	↑	↑	↑	↑
M	125	↓	↓	↓	0 1	↑	↓	0 2	1 3	1 4	2 5	3 6	5 8	7 10	10 13	↑	↑	↑	↑	↑	↑	↑	↑
N	200	↓	↓	0 1	↑	↓	0 2	1 3	1 4	2 5	3 6	5 8	7 10	10 13	↑	↑	↑	↑	↑	↑	↑	↑	↑
P	315	↓	0 1	↑	↓	0 2	1 3	1 4	2 5	3 6	5 8	7 10	10 13	↑	↑	↑	↑	↑	↑	↑	↑	↑	↑
Q	500	0 1	↑	↓	0 2	1 3	1 4	2 5	3 6	5 8	7 10	10 13	↑	↑	↑	↑	↑	↑	↑	↑	↑	↑	↑
R	800	↑	↑	0 2	1 3	1 4	2 5	3 6	5 8	7 10	10 13	↑	↑	↑	↑	↑	↑	↑	↑	↑	↑	↑	↑

↓ = Use first sampling plan below arrow. If sample size equals or exceeds lot or batch size, do 100 percent inspection.
↑ = Use first sampling plan above.
Ac = Acceptance number
Re = Rejection number
† = If the acceptance number has been exceeded, but the rejection number has not been reached, accept the lot, but reinstate normal inspection (see 10.1.4).

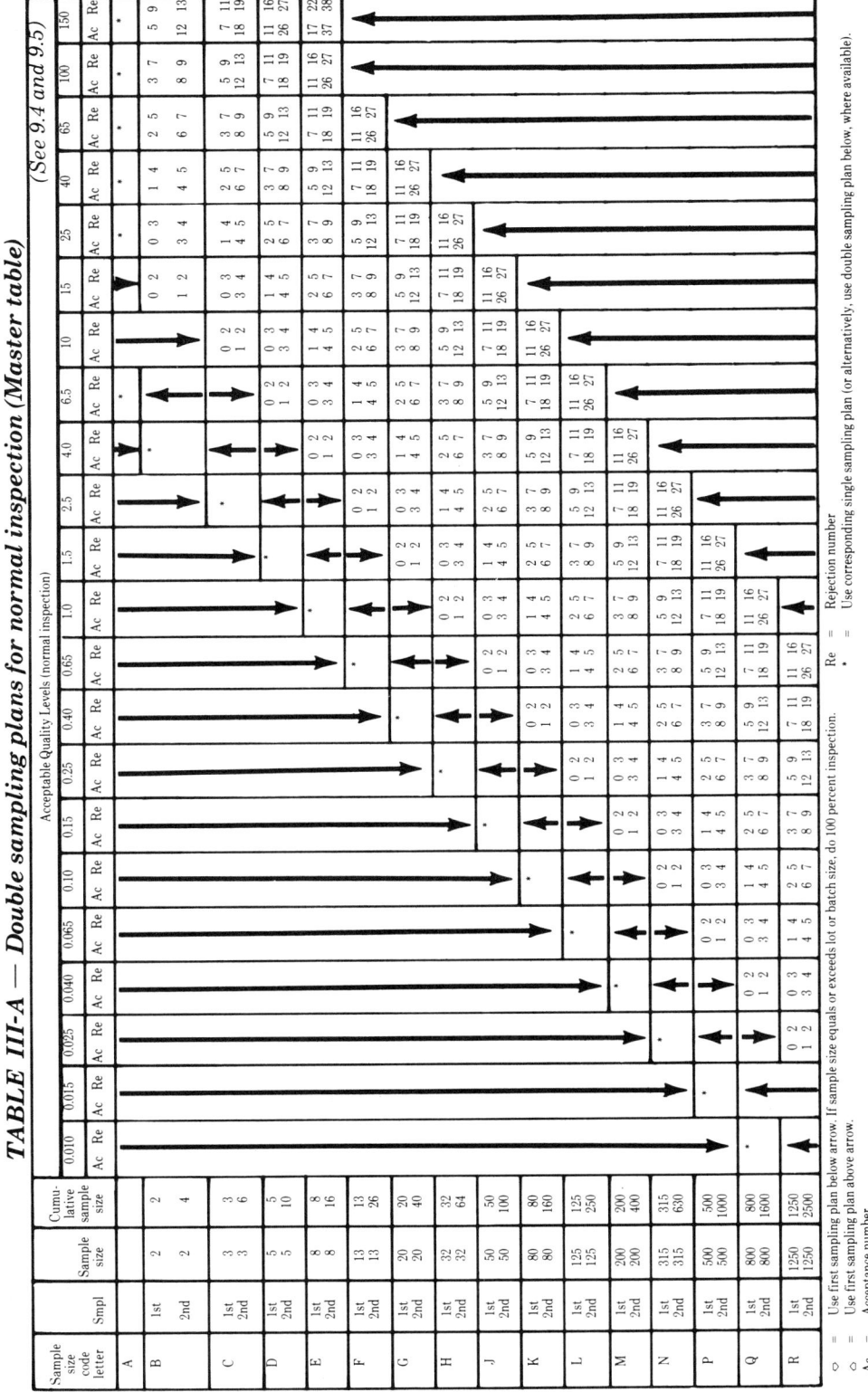

TABLE III-A — Double sampling plans for normal inspection (Master table)

(See 9.4 and 9.5)

Sample size code letter	Smpl	Sample size	Cumulative sample size	0.010	0.015	0.025	0.040	0.065	0.10	0.15	0.25	0.40	0.65	1.0	1.5	2.5	4.0	6.5	10	15	25	40	65	100	150
				Ac Re	Ac Re	Ac Re	Ac Re	Ac Re	Ac Re	Ac Re	Ac Re	Ac Re	Ac Re	Ac Re	Ac Re	Ac Re	Ac Re	Ac Re	Ac Re	Ac Re	Ac Re	Ac Re	Ac Re	Ac Re	Ac Re
A				↓	↓	↓	↓	↓	↓	↓	↓	↓	↓	↓	↓	↓	↓	*	↓	↓	*	*	*	*	*
B	1st	2	2	↓	↓	↓	↓	↓	↓	↓	↓	↓	↓	↓	↓	↓	*	↑	↓	0 2	0 3	1 4	2 5	3 7	5 9
	2nd	2	4																	1 2	3 4	4 5	6 7	8 9	12 13
C	1st	3	3	↓	↓	↓	↓	↓	↓	↓	↓	↓	↓	↓	↓	*	↑	↓	0 2	0 3	1 4	2 5	3 7	5 9	7 11
	2nd	3	6																1 2	3 4	4 5	6 7	8 9	12 13	18 19
D	1st	5	5	↓	↓	↓	↓	↓	↓	↓	↓	↓	↓	↓	*	↑	↓	0 2	0 3	1 4	2 5	3 7	5 9	7 11	11 16
	2nd	5	10															1 2	3 4	4 5	6 7	8 9	12 13	18 19	26 27
E	1st	8	8	↓	↓	↓	↓	↓	↓	↓	↓	↓	↓	*	↑	↓	0 2	0 3	1 4	2 5	3 7	5 9	7 11	11 16	17 22
	2nd	8	16														1 2	3 4	4 5	6 7	8 9	12 13	18 19	26 27	37 38
F	1st	13	13	↓	↓	↓	↓	↓	↓	↓	↓	↓	*	↑	↓	0 2	0 3	1 4	2 5	3 7	5 9	7 11	11 16	↑	↑
	2nd	13	26													1 2	3 4	4 5	6 7	8 9	12 13	18 19	26 27		
G	1st	20	20	↓	↓	↓	↓	↓	↓	↓	↓	*	↑	↓	0 2	0 3	1 4	2 5	3 7	5 9	7 11	11 16	↑	↑	↑
	2nd	20	40												1 2	3 4	4 5	6 7	8 9	12 13	18 19	26 27			
H	1st	32	32	↓	↓	↓	↓	↓	↓	↓	*	↑	↓	0 2	0 3	1 4	2 5	3 7	5 9	7 11	11 16	↑	↑	↑	↑
	2nd	32	64											1 2	3 4	4 5	6 7	8 9	12 13	18 19	26 27				
J	1st	50	50	↓	↓	↓	↓	↓	↓	*	↑	↓	0 2	0 3	1 4	2 5	3 7	5 9	7 11	11 16	↑	↑	↑	↑	↑
	2nd	50	100										1 2	3 4	4 5	6 7	8 9	12 13	18 19	26 27					
K	1st	80	80	↓	↓	↓	↓	↓	*	↑	↓	0 2	0 3	1 4	2 5	3 7	5 9	7 11	11 16	↑	↑	↑	↑	↑	↑
	2nd	80	160									1 2	3 4	4 5	6 7	8 9	12 13	18 19	26 27						
L	1st	125	125	↓	↓	↓	↓	*	↑	↓	0 2	0 3	1 4	2 5	3 7	5 9	7 11	11 16	↑	↑	↑	↑	↑	↑	↑
	2nd	125	250								1 2	3 4	4 5	6 7	8 9	12 13	18 19	26 27							
M	1st	200	200	↓	↓	↓	*	↑	↓	0 2	0 3	1 4	2 5	3 7	5 9	7 11	11 16	↑	↑	↑	↑	↑	↑	↑	↑
	2nd	200	400							1 2	3 4	4 5	6 7	8 9	12 13	18 19	26 27								
N	1st	315	315	↓	↓	*	↑	↓	0 2	0 3	1 4	2 5	3 7	5 9	7 11	11 16	↑	↑	↑	↑	↑	↑	↑	↑	↑
	2nd	315	630						1 2	3 4	4 5	6 7	8 9	12 13	18 19	26 27									
P	1st	500	500	↓	*	↑	↓	0 2	0 3	1 4	2 5	3 7	5 9	7 11	11 16	↑	↑	↑	↑	↑	↑	↑	↑	↑	↑
	2nd	500	1000					1 2	3 4	4 5	6 7	8 9	12 13	18 19	26 27										
Q	1st	800	800	*	↑	↓	0 2	0 3	1 4	2 5	3 7	5 9	7 11	11 16	↑	↑	↑	↑	↑	↑	↑	↑	↑	↑	↑
	2nd	800	1600				1 2	3 4	4 5	6 7	8 9	12 13	18 19	26 27											
R	1st	1250	1250	↑	↑	0 2	0 3	1 4	2 5	3 7	5 9	7 11	11 16	↑	↑	↑	↑	↑	↑	↑	↑	↑	↑	↑	↑
	2nd	1250	2500			1 2	3 4	4 5	6 7	8 9	12 13	18 19	26 27												

↓ = Use first sampling plan below arrow. If sample size equals or exceeds lot or batch size, do 100 percent inspection.
↑ = Use first sampling plan above arrow.
Ac = Acceptance number
Re = Rejection number
* = Use corresponding single sampling plan (or alternatively, use double sampling plan below, where available).

DOUBLE NORMAL

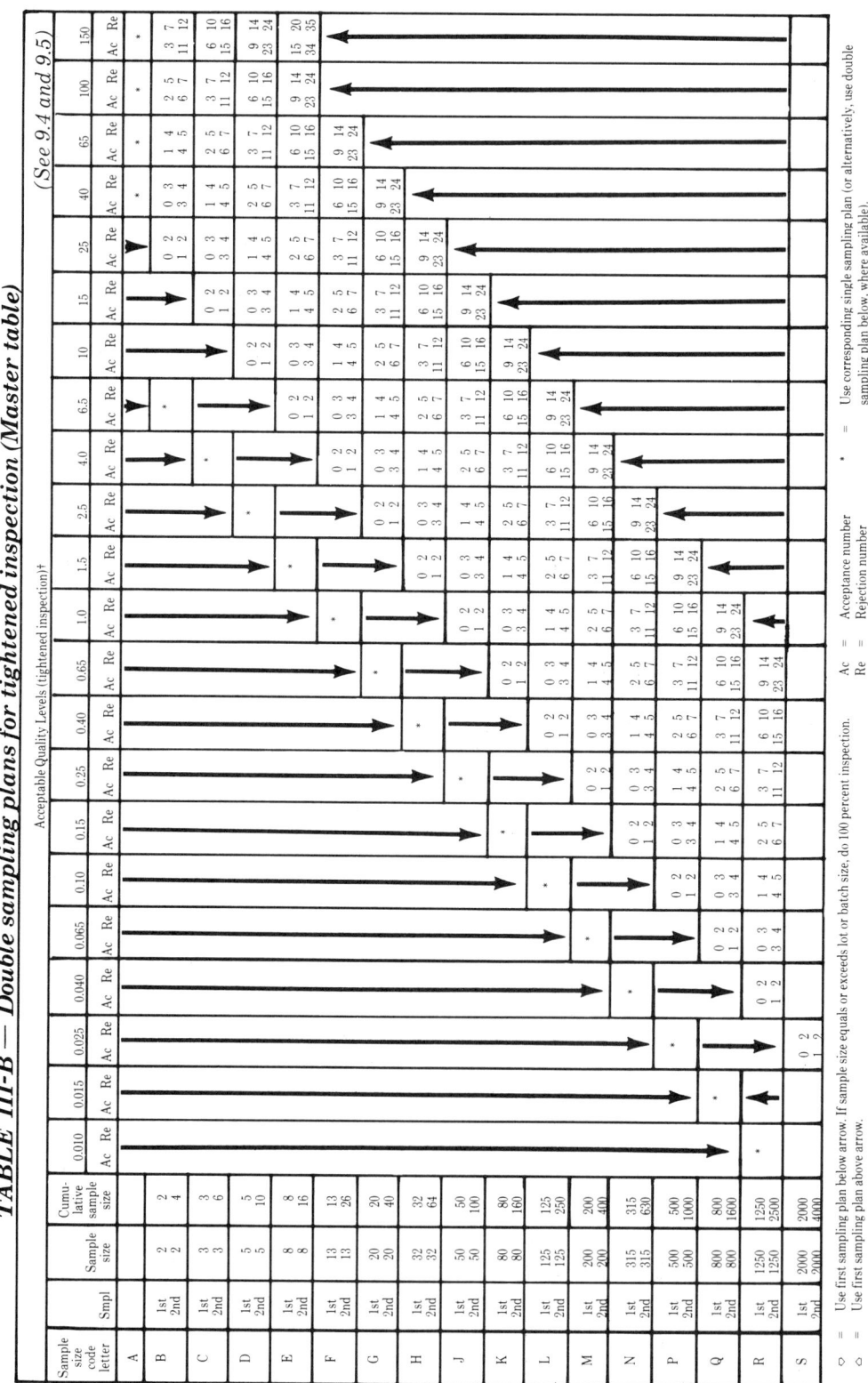

TABLE III-B — Double sampling plans for tightened inspection (Master table)

Acceptable Quality Levels (tightened inspection)† (See 9.4 and 9.5)

Sample size code letter	Smpl	Sample size	Cumu-lative sample size	0.010	0.015	0.025	0.040	0.065	0.10	0.15	0.25	0.40	0.65	1.0	1.5	2.5	4.0	6.5	10	15	25	40	65	100	150
				Ac Re	Ac Re	Ac Re	Ac Re	Ac Re	Ac Re	Ac Re	Ac Re	Ac Re	Ac Re	Ac Re	Ac Re	Ac Re	Ac Re	Ac Re	Ac Re	Ac Re	Ac Re	Ac Re	Ac Re	Ac Re	Ac Re
A				↓	↓	↓	↓	↓	↓	↓	↓	↓	↓	↓	↓	↓	↓	↓	↓	↓	↓	*	*	*	*
B	1st 2nd	2 2	2 4	↓	↓	↓	↓	↓	↓	↓	↓	↓	↓	↓	↓	↓	↓	*	↓	↓	0 2 1 2	0 3 3 4	1 4 4 5	2 5 6 7	3 7 11 12
C	1st 2nd	3 3	3 6	↓	↓	↓	↓	↓	↓	↓	↓	↓	↓	↓	↓	↓	*	↓	↓	0 2 1 2	0 3 3 4	1 4 4 5	2 5 6 7	3 7 11 12	6 10 15 16
D	1st 2nd	5 5	5 10	↓	↓	↓	↓	↓	↓	↓	↓	↓	↓	↓	↓	*	↓	↓	0 2 1 2	0 3 3 4	1 4 4 5	2 5 6 7	3 7 11 12	6 10 15 16	9 14 23 24
E	1st 2nd	8 8	8 16	↓	↓	↓	↓	↓	↓	↓	↓	↓	↓	↓	*	↓	↓	0 2 1 2	0 3 3 4	1 4 4 5	2 5 6 7	3 7 11 12	6 10 15 16	9 14 23 24	15 20 34 35
F	1st 2nd	13 13	13 26	↓	↓	↓	↓	↓	↓	↓	↓	↓	↓	*	↓	↓	0 2 1 2	0 3 3 4	1 4 4 5	2 5 6 7	3 7 11 12	6 10 15 16	9 14 23 24	↑	↑
G	1st 2nd	20 20	20 40	↓	↓	↓	↓	↓	↓	↓	↓	↓	*	↓	↓	0 2 1 2	0 3 3 4	1 4 4 5	2 5 6 7	3 7 11 12	6 10 15 16	9 14 23 24	↑	↑	↑
H	1st 2nd	32 32	32 64	↓	↓	↓	↓	↓	↓	↓	↓	*	↓	↓	0 2 1 2	0 3 3 4	1 4 4 5	2 5 6 7	3 7 11 12	6 10 15 16	9 14 23 24	↑	↑	↑	↑
J	1st 2nd	50 50	50 100	↓	↓	↓	↓	↓	↓	↓	*	↓	↓	0 2 1 2	0 3 3 4	1 4 4 5	2 5 6 7	3 7 11 12	6 10 15 16	9 14 23 24	↑	↑	↑	↑	↑
K	1st 2nd	80 80	80 160	↓	↓	↓	↓	↓	↓	*	↓	↓	0 2 1 2	0 3 3 4	1 4 4 5	2 5 6 7	3 7 11 12	6 10 15 16	9 14 23 24	↑	↑	↑	↑	↑	↑
L	1st 2nd	125 125	125 250	↓	↓	↓	↓	↓	*	↓	↓	0 2 1 2	0 3 3 4	1 4 4 5	2 5 6 7	3 7 11 12	6 10 15 16	9 14 23 24	↑	↑	↑	↑	↑	↑	↑
M	1st 2nd	200 200	200 400	↓	↓	↓	↓	*	↓	↓	0 2 1 2	0 3 3 4	1 4 4 5	2 5 6 7	3 7 11 12	6 10 15 16	9 14 23 24	↑	↑	↑	↑	↑	↑	↑	↑
N	1st 2nd	315 315	315 630	↓	↓	↓	*	↓	↓	0 2 1 2	0 3 3 4	1 4 4 5	2 5 6 7	3 7 11 12	6 10 15 16	9 14 23 24	↑	↑	↑	↑	↑	↑	↑	↑	↑
P	1st 2nd	500 500	500 1000	↓	↓	*	↓	↓	0 2 1 2	0 3 3 4	1 4 4 5	2 5 6 7	3 7 11 12	6 10 15 16	9 14 23 24	↑	↑	↑	↑	↑	↑	↑	↑	↑	↑
Q	1st 2nd	800 800	800 1600	↓	*	↓	↓	0 2 1 2	0 3 3 4	1 4 4 5	2 5 6 7	3 7 11 12	6 10 15 16	9 14 23 24	↑	↑	↑	↑	↑	↑	↑	↑	↑	↑	↑
R	1st 2nd	1250 1250	1250 2500	*	↑	↓	0 2 1 2	0 3 3 4	1 4 4 5	2 5 6 7	3 7 11 12	6 10 15 16	9 14 23 24	↑	↑	↑	↑	↑	↑	↑	↑	↑	↑	↑	↑
S	1st 2nd	2000 2000	2000 4000			0 2 1 2																			

▽ = Use first sampling plan below arrow. If sample size equals or exceeds lot or batch size, do 100 percent inspection.
△ = Use first sampling plan above arrow.

Ac = Acceptance number
Re = Rejection number

* = Use corresponding single sampling plan (or alternatively, use double sampling plan below, where available).

DOUBLE TIGHTENED

TABLE III-C — Double sampling plans for reduced inspection (Master table)

Acceptable Quality Levels (reduced inspection)† *(See 9.4 and 9.5)*

Sample size code letter	Smpl	Sample size	Cumulative sample size	0.010 Ac Re	0.015 Ac Re	0.025 Ac Re	0.040 Ac Re	0.065 Ac Re	0.10 Ac Re	0.15 Ac Re	0.25 Ac Re	0.40 Ac Re	0.65 Ac Re	1.0 Ac Re	1.5 Ac Re	2.5 Ac Re	4.0 Ac Re	6.5 Ac Re	10 Ac Re	15 Ac Re	25 Ac Re	40 Ac Re	65 Ac Re	100 Ac Re	150 Ac Re
A				↓	↓	↓	↓	↓	↓	↓	↓	↓	↓	↓	↓	↓	↓	*	↓	↓	*	*	*	*	*
B				↓	↓	↓	↓	↓	↓	↓	↓	↓	↓	↓	↓	↓	*	↑	↓	*	*	*	*	*	*
C				↓	↓	↓	↓	↓	↓	↓	↓	↓	↓	↓	↓	*	↑	↓	*	*	*	*	*	*	*
D	1st	2	2	↓	↓	↓	↓	↓	↓	↓	↓	↓	↓	↓	*	↑	↓	0 2	0 3	0 4	0 4	1 5	2 7	3 8	5 10
	2nd	2	4												*			0 2	0 4	1 5	3 6	4 7	6 9	8 12	12 16
E	1st	3	3	↓	↓	↓	↓	↓	↓	↓	↓	↓	↓	*	↑	↓	0 2	0 3	0 4	0 4	1 5	2 7	3 8	5 10	7 12
	2nd	3	6														0 2	0 4	1 5	3 6	4 7	6 9	8 12	12 16	18 22
F	1st	5	5	↓	↓	↓	↓	↓	↓	↓	↓	↓	*	↑	↓	0 2	0 3	0 4	0 4	1 5	2 7	3 8	5 10	↑	↑
	2nd	5	10													0 2	0 4	1 5	3 6	4 7	6 9	8 12	12 16		
G	1st	8	8	↓	↓	↓	↓	↓	↓	↓	↓	*	↑	↓	0 2	0 3	0 4	0 4	1 5	2 7	3 8	5 10	↑	↑	↑
	2nd	8	16												0 2	0 4	1 5	3 6	4 7	6 9	8 12	12 16			
H	1st	13	13	↓	↓	↓	↓	↓	↓	↓	*	↑	↓	0 2	0 3	0 4	0 4	1 5	2 7	3 8	5 10	↑	↑	↑	↑
	2nd	13	26											0 2	0 4	1 5	3 6	4 7	6 9	8 12	12 16				
J	1st	20	20	↓	↓	↓	↓	↓	↓	*	↑	↓	0 2	0 3	0 4	0 4	1 5	2 7	3 8	5 10	↑	↑	↑	↑	↑
	2nd	20	40										0 2	0 4	1 5	3 6	4 7	6 9	8 12	12 16					
K	1st	32	32	↓	↓	↓	↓	↓	*	↑	↓	0 2	0 3	0 4	0 4	1 5	2 7	3 8	5 10	↑	↑	↑	↑	↑	↑
	2nd	32	64									0 2	0 4	1 5	3 6	4 7	6 9	8 12	12 16						
L	1st	50	50	↓	↓	↓	↓	*	↑	↓	0 2	0 3	0 4	0 4	1 5	2 7	3 8	5 10	↑	↑	↑	↑	↑	↑	↑
	2nd	50	100								0 2	0 4	1 5	3 6	4 7	6 9	8 12	12 16							
M	1st	80	80	↓	↓	↓	*	↑	↓	0 2	0 3	0 4	0 4	1 5	2 7	3 8	5 10	↑	↑	↑	↑	↑	↑	↑	↑
	2nd	80	160							0 2	0 4	1 5	3 6	4 7	6 9	8 12	12 16								
N	1st	125	125	↓	↓	*	↑	↓	0 2	0 3	0 4	0 4	1 5	2 7	3 8	5 10	↑	↑	↑	↑	↑	↑	↑	↑	↑
	2nd	125	250						0 2	0 4	1 5	3 6	4 7	6 9	8 12	12 16									
P	1st	200	200	↓	*	↑	↓	0 2	0 3	0 4	0 4	1 5	2 7	3 8	5 10	↑	↑	↑	↑	↑	↑	↑	↑	↑	↑
	2nd	200	400					0 2	0 4	1 5	3 6	4 7	6 9	8 12	12 16										
Q	1st	315	315	*	↑	↓	0 2	0 3	0 4	0 4	1 5	2 7	3 8	5 10	↑	↑	↑	↑	↑	↑	↑	↑	↑	↑	↑
	2nd	315	630				0 2	0 4	1 5	3 6	4 7	6 9	8 12	12 16											
R	1st	500	500	↑	↑	0 2	0 3	0 4	0 4	1 5	2 7	3 8	5 10	↑	↑	↑	↑	↑	↑	↑	↑	↑	↑	↑	↑
	2nd	500	1000			0 2	0 4	1 5	3 6	4 7	6 9	8 12	12 16												

↓ = Use first sampling plan below arrow. If sample size equals or exceeds lot or batch size, do 100 percent inspection.
↑ = Use first sampling plan above arrow.
Ac = Acceptance number
Re = Rejection number
* = Use corresponding single sampling plan (or alternatively, use double sampling plan below, where available).
† = If, after the second sample, the acceptance number has been exceeded, but the rejection number has not been reached, accept the lot, but reinstate normal normal inspection (see 10.1.4).

DOUBLE REDUCED

TABLE IV-A — *Multiple sampling plans for normal inspection (Master table)*

(See 9.4 and 9.5)

Sample size code letter	Smpl	Sample size	Cumulative sample size	Acceptable Quality Levels (reduced inspection)																									
				0.010	0.015	0.025	0.040	0.065	0.10	0.15	0.25	0.40	0.65	1.0	1.5	2.5	4.0	6.5	10	15	25	40	65	100	150	250	400	650	1000
				Ac Re	Ac Re	Ac Re	Ac Re	Ac Re	Ac Re	Ac Re	Ac Re	Ac Re	Ac Re	Ac Re	Ac Re	Ac Re	Ac Re	Ac Re	Ac Re	Ac Re	Ac Re	Ac Re	Ac Re	Ac Re	Ac Re	Ac Re	Ac Re	Ac Re	Ac Re
A				↓	↓	↓	↓	↓	↓	↓	↓	↓	↓	↓	↓	↓	↑	•	↓	↓	•	•	•	•	•	•	•	•	•
B																	•	↑		‡	‡	‡	‡	‡	‡	‡	‡	‡	‡
C																•	↓	↓	‡	‡	‡	‡	‡	‡	‡	‡	‡	‡	↑
D	1st	2	2	↓	↓	↓	↓	↓	↓	↓	↓	↓	↓	↓	•	↑	↓	# 2	# 2	# 3	# 4	0 4	0 5	1 7	2 9	4 12	6 16	↑	↑
	2nd	2	3															# 2	0 3	0 3	1 5	1 6	3 8	4 10	7 14	11 19	17 27		
	3rd	2	6															0 2	0 3	1 4	2 6	3 8	6 10	8 13	13 19	19 27	29 39		
	4th	2	8															0 3	1 4	2 5	3 7	5 10	8 13	12 17	19 25	27 34	40 49		
	5th	2	10															1 3	2 4	3 6	5 8	7 11	11 15	17 20	25 29	36 40	53 58		
	6th	2	12															1 3	3 5	4 6	7 9	10 12	14 17	21 23	31 33	45 47	66 68		
	7th	2	14															2 3	4 5	6 7	9 10	13 14	18 19	25 26	37 38	53 54	77 78		
E	1st	3	3	↓	↓	↓	↓	↓	↓	↓	↓	↓	↓	•	↑	↓	# 2	# 2	# 3	# 4	0 4	0 5	1 7	2 9	4 12	6 16	↑	↑	↑
	2nd	3	6														# 2	0 3	0 3	1 5	1 6	3 8	4 10	7 14	11 19	17 27			
	3rd	3	9														0 2	0 3	1 4	2 6	3 8	6 10	8 13	13 19	19 27	29 39			
	4th	3	12														0 3	1 4	2 5	3 7	5 10	8 13	12 17	19 25	27 34	40 49			
	5th	3	15														1 3	2 4	3 6	5 8	7 11	11 15	17 20	25 29	36 40	53 58			
	6th	3	18														1 3	3 5	4 6	7 9	10 12	14 17	21 23	31 33	45 47	65 68			
	7th	3	21														2 3	4 5	6 7	9 10	13 14	18 19	25 26	37 38	53 54	77 78			
F	1st	5	5	↓	↓	↓	↓	↓	↓	↓	↓	↓	•	↑	↓	# 2	# 2	# 3	# 4	0 4	0 5	1 7	2 9	↑	↑	↑	↑	↑	↑
	2nd	5	10													# 2	0 3	0 3	1 5	1 6	3 8	4 10	7 14						
	3rd	5	15													0 2	0 3	1 4	2 6	3 8	6 10	8 13	13 19						
	4th	5	20													0 3	1 4	2 5	3 7	5 10	8 13	12 17	19 25						
	5th	5	25													1 3	2 4	3 6	5 8	7 11	11 15	17 20	25 29						
	6th	5	30													1 3	3 5	4 6	7 9	10 12	14 17	21 23	31 33						
	7th	5	35													2 3	4 5	6 7	9 10	13 14	18 19	25 26	37 38						
G	1st	8	8	↓	↓	↓	↓	↓	↓	↓	↓	•	↑	↓	# 2	# 2	# 3	# 4	0 4	0 5	1 7	2 9	↑	↑	↑	↑	↑	↑	↑
	2nd	8	16												# 2	0 3	0 3	1 5	1 6	3 8	4 10	7 14							
	3rd	8	24												0 2	0 3	1 4	2 6	3 8	6 10	8 13	13 19							
	4th	8	32												0 3	1 4	2 5	3 7	5 10	8 13	12 17	19 25							
	5th	8	40												1 3	2 4	3 6	5 8	7 11	11 15	17 20	25 29							
	6th	8	48												1 3	3 5	4 6	7 9	10 12	14 17	21 23	31 33							
	7th	8	56												2 3	4 5	6 7	9 10	13 14	18 19	25 26	37 38							
H	1st	13	13	↓	↓	↓	↓	↓	↓	↓	•	↑	↓	# 2	# 2	# 3	# 4	0 4	0 5	1 7	2 9	↑	↑	↑	↑	↑	↑	↑	↑
	2nd	13	26											# 2	0 3	0 3	1 5	1 6	3 8	4 10	7 14								
	3rd	13	39											0 2	0 3	1 4	2 6	3 8	6 10	8 13	13 19								
	4th	13	52											0 3	1 4	2 5	3 7	5 10	8 13	12 17	19 25								
	5th	13	63											1 3	2 4	3 6	5 8	7 11	11 15	17 20	25 29								
	6th	13	78											1 3	3 5	4 6	7 9	10 12	14 17	21 23	31 33								
	7th	13	91											2 3	4 5	6 7	9 10	13 14	18 19	25 26	37 38								
J	1st	20	20	↓	↓	↓	↓	↓	↓	•	↑	↓	# 2	# 2	# 3	# 4	0 4	0 5	1 7	2 9	↑	↑	↑	↑	↑	↑	↑	↑	↑
	2nd	20	40										# 2	0 3	0 3	1 5	1 6	3 8	4 10	7 14									
	3rd	20	60										0 2	0 3	1 4	2 6	3 8	6 10	8 13	13 19									
	4th	20	80										0 3	1 4	2 5	3 7	5 10	8 13	12 17	19 25									
	5th	20	100										1 3	2 4	3 6	5 8	7 11	11 15	17 20	25 29									
	6th	20	120										1 3	3 5	4 6	7 9	10 12	14 17	21 23	31 33									
	7th	20	140										2 3	4 5	6 7	9 10	13 14	18 19	25 26	37 38									

↓ = Use first sampling plan below arrow. If sample size equals, or exceeds, lot or batch size, do 100 percent inspection.
↑ = Use first sampling plan above arrow (refer to preceding page, when necessary).
Ac = Acceptance number
Re = Rejection number
# = Acceptance not permitted at this sample size.
‡ = Use corresponding double sampling plan (or alternatively, use multiple sampling plan below, where available).
• = Use corresponding single sampling plan (or alternatively, use multiple sampling plan below, where available).

TABLE IV-A — Multiple sampling plans for normal inspection (Master table) (Continued)

(See 9.4 and 9.5)

Acceptable Quality Levels (reduced inspection); each AQL cell gives Ac Re.

Sample size code letter	Smpl	Sample size	Cumulative sample size	0.010	0.015	0.025	0.040	0.065	0.10	0.15	0.25	0.40	0.65	1.0	1.5	2.5	4.0	6.5	10	15	25	40	65	100	150	250	400	650	1000
K	1st	32	32	↓	↓	↓	↓	↓	•	↑	↓	# 2	# 2	# 3	# 4	0 4	0 5	1 7	2 9	↑	↑	↑	↑	↑	↑	↑	↑	↑	↑
	2nd	32	64									# 2	0 3	0 3	1 5	1 6	3 8	4 10	7 14										
	3rd	32	96									0 2	0 3	1 4	2 6	3 8	6 10	8 13	13 19										
	4th	32	138									0 3	1 4	2 5	3 7	5 10	8 13	12 17	19 25										
	5th	32	160									1 3	2 4	3 6	5 8	7 11	11 15	17 20	25 29										
	6th	32	192									1 3	3 5	4 6	7 9	10 12	14 17	21 23	31 33										
	7th	32	224									2 3	4 5	6 7	9 10	13 14	18 19	25 26	37 38										
L	1st	50	50	↓	↓	↓	↓	•	↑	↓	# 2	# 2	# 3	# 4	0 4	0 5	1 7	2 9	↑	↑	↑	↑	↑	↑	↑	↑	↑	↑	↑
	2nd	50	100								# 2	0 3	0 3	1 5	1 6	3 8	4 10	7 14											
	3rd	50	150								0 2	0 3	1 4	2 6	3 8	6 10	8 13	13 19											
	4th	50	200								0 3	1 4	2 5	3 7	5 10	8 13	12 17	19 25											
	5th	50	250								1 3	2 4	3 6	5 8	7 11	11 15	17 20	25 29											
	6th	50	300								1 3	3 5	4 6	7 9	10 12	14 17	21 23	31 33											
	7th	50	350								2 3	4 5	6 7	9 10	13 14	18 19	25 26	37 38											
M	1st	80	80	↓	↓	↓	•	↑	↓	# 2	# 2	# 3	# 4	0 4	0 5	1 7	2 9	↑	↑	↑	↑	↑	↑	↑	↑	↑	↑	↑	↑
	2nd	80	160							# 2	0 3	0 3	1 5	1 6	3 8	4 10	7 14												
	3rd	80	240							0 2	0 3	1 4	2 6	3 8	6 10	8 13	13 19												
	4th	80	320							0 3	1 4	2 5	3 7	5 10	8 13	12 17	19 25												
	5th	80	400							1 3	2 4	3 6	5 8	7 11	11 15	17 20	25 29												
	6th	80	480							1 3	3 5	4 6	7 9	10 12	14 17	21 23	31 33												
	7th	80	560							2 3	4 5	6 7	9 10	13 14	18 19	25 26	37 38												
N	1st	125	125	↓	↓	•	↑	↓	# 2	# 2	# 3	# 4	0 4	0 5	1 7	2 9	↑	↑	↑	↑	↑	↑	↑	↑	↑	↑	↑	↑	↑
	2nd	125	250						# 2	0 3	0 3	1 5	1 6	3 8	4 10	7 14													
	3rd	125	375						0 2	0 3	1 4	2 6	3 8	6 10	8 13	13 19													
	4th	125	500						0 3	1 4	2 5	3 7	5 10	8 13	12 17	19 25													
	5th	125	625						1 3	2 4	3 6	5 8	7 11	11 15	17 20	25 29													
	6th	125	750						1 3	3 5	4 6	7 9	10 12	14 17	21 23	31 33													
	7th	125	875						2 3	4 5	6 7	9 10	13 14	18 19	25 26	37 38													
P	1st	200	200	↓	•	↑	↓	# 2	# 2	# 3	# 4	0 4	0 5	1 7	2 9	↑	↑	↑	↑	↑	↑	↑	↑	↑	↑	↑	↑	↑	↑
	2nd	200	400					# 2	0 3	0 3	1 5	1 6	3 8	4 10	7 14														
	3rd	200	600					0 2	0 3	1 4	2 6	3 8	6 10	8 13	13 19														
	4th	200	800					0 3	1 4	2 5	3 7	5 10	8 13	12 17	19 25														
	5th	200	1000					1 3	2 4	3 6	5 8	7 11	11 15	17 20	25 29														
	6th	200	1200					1 3	3 5	4 6	7 9	10 12	14 17	21 23	31 33														
	7th	200	1400					2 3	4 5	6 7	9 10	13 14	18 19	25 26	37 38														
Q	1st	315	315	•	↑	↓	# 2	# 2	# 3	# 4	0 4	0 5	1 7	2 9	↑	↑	↑	↑	↑	↑	↑	↑	↑	↑	↑	↑	↑	↑	↑
	2nd	315	630				# 2	0 3	0 3	1 5	1 6	3 8	4 10	7 14															
	3rd	315	945				0 2	0 3	1 4	2 6	3 8	6 10	8 13	13 19															
	4th	315	1260				0 3	1 4	2 5	3 7	5 10	8 13	12 17	19 25															
	5th	315	1575				1 3	2 4	3 6	5 8	7 11	11 15	17 20	25 29															
	6th	315	1890				1 3	3 5	4 6	7 9	10 12	14 17	21 23	31 33															
	7th	315	2205				2 3	4 5	6 7	9 10	13 14	18 19	25 26	37 38															
R	1st	500	500	↑	↑	# 2	# 2	# 3	# 4	0 4	0 5	1 7	2 9	↑	↑	↑	↑	↑	↑	↑	↑	↑	↑	↑	↑	↑	↑	↑	↑
	2nd	500	1000			# 2	0 3	0 3	1 5	1 6	3 8	4 10	7 14																
	3rd	500	1500			0 2	0 3	1 4	2 6	3 8	6 10	8 13	13 19																
	4th	500	2000			0 3	1 4	2 5	3 7	5 10	8 13	12 17	19 25																
	5th	500	2500			1 3	2 4	3 6	5 8	7 11	11 15	17 20	25 29																
	6th	500	3000			1 3	3 5	4 6	7 9	10 12	14 17	21 23	31 33																
	7th	500	3500			2 3	4 5	6 7	9 10	13 14	18 19	25 26	37 38																

↓ = Use first sampling plan below arrow. If sample size equals, or exceeds, lot or batch size, do 100 percent inspection.
↑ = Use first sampling plan above arrow (refer to preceding page, when necessary).
Ac = Acceptance number
Re = Rejection number
# = Acceptance not permitted at this sample size.
• = Use corresponding single sampling plan (or alternatively, use multiple sampling plan below, where available).

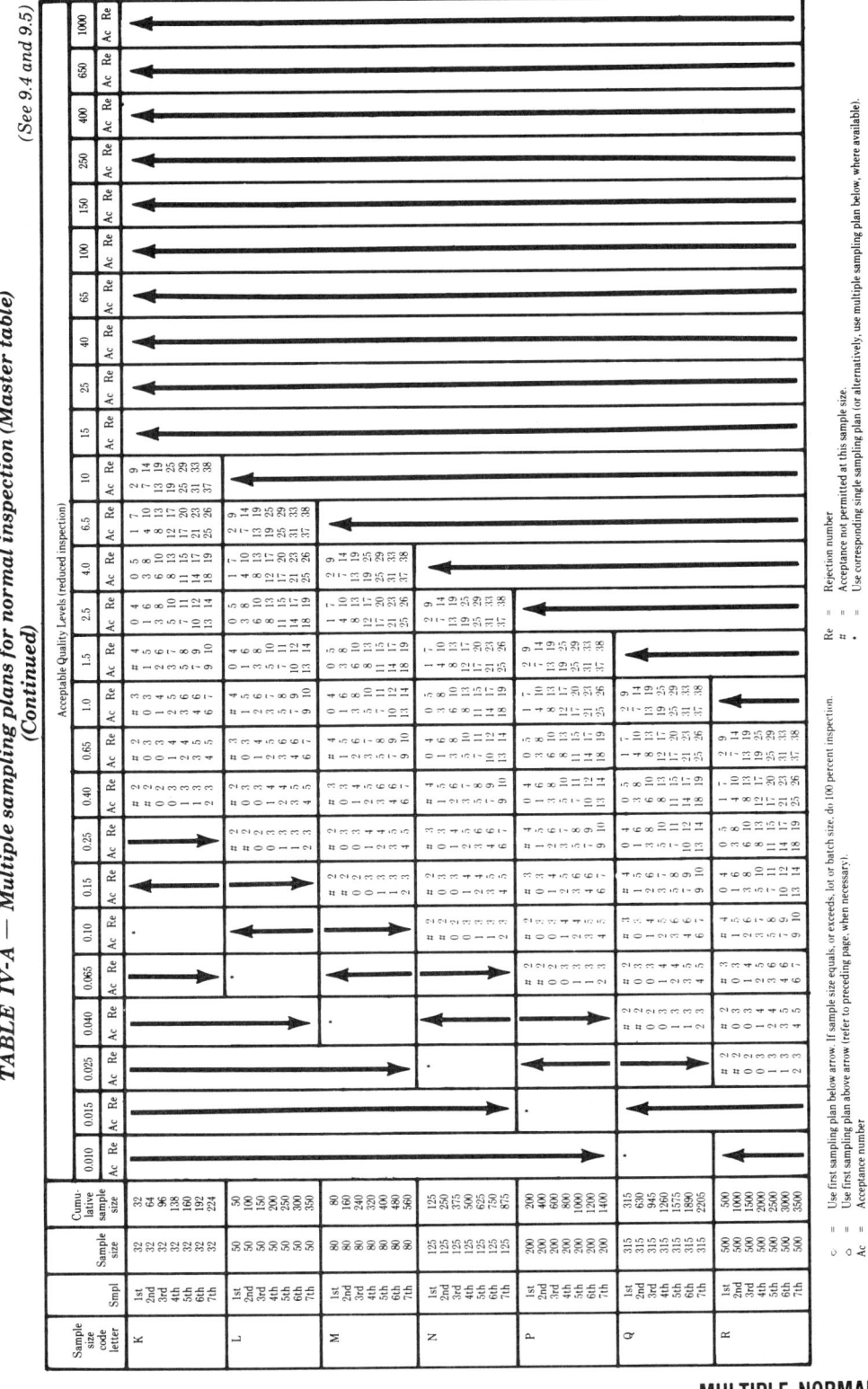

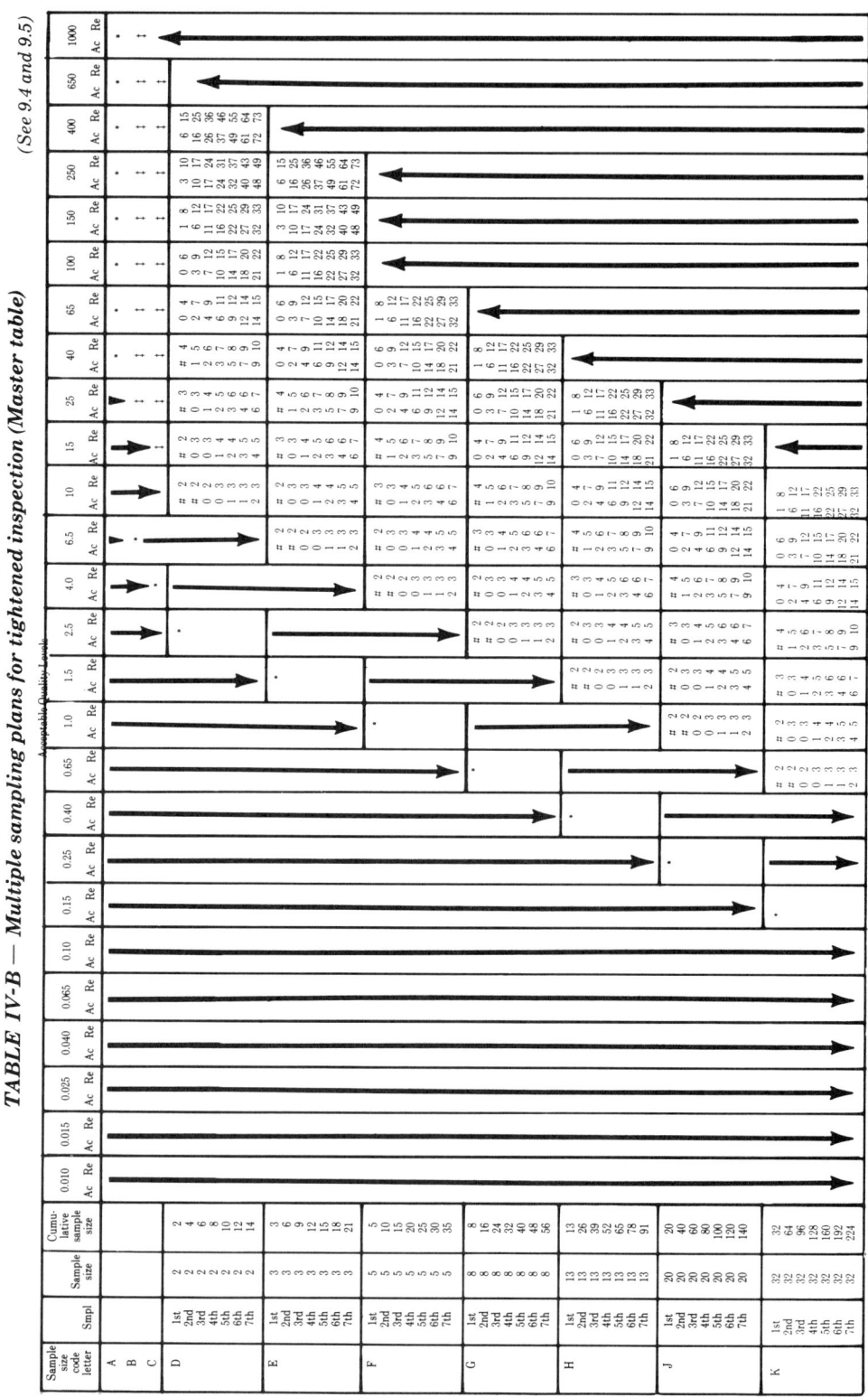

TABLE IV-B — Multiple sampling plans for tightened inspection (Master table)

(See 9.4 and 9.5)

Acceptable Quality Levels

Sample size code letter	Smpl	Sample size	Cumulative sample size	0.010	0.015	0.025	0.040	0.065	0.10	0.15	0.25	0.40	0.65	1.0	1.5	2.5	4.0	6.5	10	15	25	40	65	100	150	250	400	650	1000
				Ac Re	Ac Re	Ac Re	Ac Re	Ac Re	Ac Re	Ac Re	Ac Re	Ac Re	Ac Re	Ac Re	Ac Re	Ac Re	Ac Re	Ac Re	Ac Re	Ac Re	Ac Re	Ac Re	Ac Re	Ac Re	Ac Re	Ac Re	Ac Re	Ac Re	Ac Re
A				↓	↓	↓	↓	↓	↓	↓	↓	↓	↓	↓	↓	↓	↓	↓	↓	↓	↓	*	*	*	*	*	*	*	*
B				↓	↓	↓	↓	↓	↓	↓	↓	↓	↓	↓	↓	↓	↓	*	↓	↓	‡	‡	‡	‡	‡	‡	‡	‡	‡
C				↓	↓	↓	↓	↓	↓	↓	↓	↓	↓	↓	↓	↓	*	↓	↓	‡	‡	‡	‡	‡	‡	‡	‡	‡	↑
D	1st	2	2	↓	↓	↓	↓	↓	↓	↓	↓	↓	↓	↓	↓	*	↓	↓	# 2	# 2	# 3	# 4	0 4	0 6	1 8	3 10	6 15	↑	↑
	2nd	2	4	↓	↓	↓	↓	↓	↓	↓	↓	↓	↓	↓	↓		↓	↓	# 2	0 3	0 3	1 5	2 7	3 9	6 12	10 17	16 25	↑	↑
	3rd	2	6	↓	↓	↓	↓	↓	↓	↓	↓	↓	↓	↓	↓		↓	↓	0 2	0 3	1 4	2 6	4 9	7 12	11 17	17 24	26 36	↑	↑
	4th	2	8	↓	↓	↓	↓	↓	↓	↓	↓	↓	↓	↓	↓		↓	↓	0 3	1 4	2 5	3 7	6 11	10 15	16 22	24 31	37 46	↑	↑
	5th	2	10	↓	↓	↓	↓	↓	↓	↓	↓	↓	↓	↓	↓		↓	↓	1 3	2 4	3 6	5 8	9 12	14 17	22 25	32 37	49 55	↑	↑
	6th	2	12	↓	↓	↓	↓	↓	↓	↓	↓	↓	↓	↓	↓		↓	↓	1 3	3 5	4 6	7 9	12 14	18 20	27 29	40 43	61 64	↑	↑
	7th	2	14	↓	↓	↓	↓	↓	↓	↓	↓	↓	↓	↓	↓		↓	↓	2 3	4 5	6 7	9 10	14 15	21 22	32 33	48 49	72 73	↑	↑
E	1st	3	3	↓	↓	↓	↓	↓	↓	↓	↓	↓	↓	↓	*	↓	↓	# 2	# 2	# 3	# 4	0 4	0 6	1 8	3 10	6 15	↑	↑	↑
	2nd	3	6	↓	↓	↓	↓	↓	↓	↓	↓	↓	↓	↓		↓	↓	# 2	0 3	0 3	1 5	2 7	3 9	6 12	10 17	16 25	↑	↑	↑
	3rd	3	9	↓	↓	↓	↓	↓	↓	↓	↓	↓	↓	↓		↓	↓	0 2	0 3	1 4	2 6	4 9	7 12	11 17	17 24	26 36	↑	↑	↑
	4th	3	12	↓	↓	↓	↓	↓	↓	↓	↓	↓	↓	↓		↓	↓	0 3	1 4	2 5	3 7	6 11	10 15	16 22	24 31	37 46	↑	↑	↑
	5th	3	15	↓	↓	↓	↓	↓	↓	↓	↓	↓	↓	↓		↓	↓	1 3	2 4	3 6	5 8	9 12	14 17	22 25	32 37	49 55	↑	↑	↑
	6th	3	18	↓	↓	↓	↓	↓	↓	↓	↓	↓	↓	↓		↓	↓	1 3	3 5	4 6	7 9	12 14	18 20	27 29	40 43	61 64	↑	↑	↑
	7th	3	21	↓	↓	↓	↓	↓	↓	↓	↓	↓	↓	↓		↓	↓	2 3	4 5	6 7	9 10	14 15	21 22	32 33	48 49	72 73	↑	↑	↑
F	1st	5	5	↓	↓	↓	↓	↓	↓	↓	↓	↓	↓	*	↓	↓	# 2	# 2	# 3	# 4	0 4	0 6	1 8	↑	↑	↑	↑	↑	↑
	2nd	5	10	↓	↓	↓	↓	↓	↓	↓	↓	↓	↓		↓	↓	# 2	0 3	0 3	1 5	2 7	3 9	6 12	↑	↑	↑	↑	↑	↑
	3rd	5	15	↓	↓	↓	↓	↓	↓	↓	↓	↓	↓		↓	↓	0 2	0 3	1 4	2 6	4 9	7 12	11 17	↑	↑	↑	↑	↑	↑
	4th	5	20	↓	↓	↓	↓	↓	↓	↓	↓	↓	↓		↓	↓	0 3	1 4	2 5	3 7	6 11	10 15	16 22	↑	↑	↑	↑	↑	↑
	5th	5	25	↓	↓	↓	↓	↓	↓	↓	↓	↓	↓		↓	↓	1 3	2 4	3 6	5 8	9 12	14 17	22 25	↑	↑	↑	↑	↑	↑
	6th	5	30	↓	↓	↓	↓	↓	↓	↓	↓	↓	↓		↓	↓	1 3	3 5	4 6	7 9	12 14	18 20	27 29	↑	↑	↑	↑	↑	↑
	7th	5	35	↓	↓	↓	↓	↓	↓	↓	↓	↓	↓		↓	↓	2 3	4 5	6 7	9 10	14 15	21 22	32 33	↑	↑	↑	↑	↑	↑
G	1st	8	8	↓	↓	↓	↓	↓	↓	↓	↓	↓	*	↓	↓	# 2	# 2	# 3	# 4	0 4	0 6	1 8	↑	↑	↑	↑	↑	↑	↑
	2nd	8	16	↓	↓	↓	↓	↓	↓	↓	↓	↓		↓	↓	# 2	0 3	0 3	1 5	2 7	3 9	6 12	↑	↑	↑	↑	↑	↑	↑
	3rd	8	24	↓	↓	↓	↓	↓	↓	↓	↓	↓		↓	↓	0 2	0 3	1 4	2 6	4 9	7 12	11 17	↑	↑	↑	↑	↑	↑	↑
	4th	8	32	↓	↓	↓	↓	↓	↓	↓	↓	↓		↓	↓	0 3	1 4	2 5	3 7	6 11	10 15	16 22	↑	↑	↑	↑	↑	↑	↑
	5th	8	40	↓	↓	↓	↓	↓	↓	↓	↓	↓		↓	↓	1 3	2 4	3 6	5 8	9 12	14 17	22 25	↑	↑	↑	↑	↑	↑	↑
	6th	8	48	↓	↓	↓	↓	↓	↓	↓	↓	↓		↓	↓	1 3	3 5	4 6	7 9	12 14	18 20	27 29	↑	↑	↑	↑	↑	↑	↑
	7th	8	56	↓	↓	↓	↓	↓	↓	↓	↓	↓		↓	↓	2 3	4 5	6 7	9 10	14 15	21 22	32 33	↑	↑	↑	↑	↑	↑	↑
H	1st	13	13	↓	↓	↓	↓	↓	↓	↓	↓	*	↓	↓	# 2	# 2	# 3	# 4	0 4	0 6	1 8	↑	↑	↑	↑	↑	↑	↑	↑
	2nd	13	26	↓	↓	↓	↓	↓	↓	↓	↓		↓	↓	# 2	0 3	0 3	1 5	2 7	3 9	6 12	↑	↑	↑	↑	↑	↑	↑	↑
	3rd	13	39	↓	↓	↓	↓	↓	↓	↓	↓		↓	↓	0 2	0 3	1 4	2 6	4 9	7 12	11 17	↑	↑	↑	↑	↑	↑	↑	↑
	4th	13	52	↓	↓	↓	↓	↓	↓	↓	↓		↓	↓	0 3	1 4	2 5	3 7	6 11	10 15	16 22	↑	↑	↑	↑	↑	↑	↑	↑
	5th	13	65	↓	↓	↓	↓	↓	↓	↓	↓		↓	↓	1 3	2 4	3 6	5 8	9 12	14 17	22 25	↑	↑	↑	↑	↑	↑	↑	↑
	6th	13	78	↓	↓	↓	↓	↓	↓	↓	↓		↓	↓	1 3	3 5	4 6	7 9	12 14	18 20	27 29	↑	↑	↑	↑	↑	↑	↑	↑
	7th	13	91	↓	↓	↓	↓	↓	↓	↓	↓		↓	↓	2 3	4 5	6 7	9 10	14 15	21 22	32 33	↑	↑	↑	↑	↑	↑	↑	↑
J	1st	20	20	↓	↓	↓	↓	↓	↓	↓	*	↓	↓	# 2	# 2	# 3	# 4	0 4	0 6	1 8	↑	↑	↑	↑	↑	↑	↑	↑	↑
	2nd	20	40	↓	↓	↓	↓	↓	↓	↓		↓	↓	# 2	0 3	0 3	1 5	2 7	3 9	6 12	↑	↑	↑	↑	↑	↑	↑	↑	↑
	3rd	20	60	↓	↓	↓	↓	↓	↓	↓		↓	↓	0 2	0 3	1 4	2 6	4 9	7 12	11 17	↑	↑	↑	↑	↑	↑	↑	↑	↑
	4th	20	80	↓	↓	↓	↓	↓	↓	↓		↓	↓	0 3	1 4	2 5	3 7	6 11	10 15	16 22	↑	↑	↑	↑	↑	↑	↑	↑	↑
	5th	20	100	↓	↓	↓	↓	↓	↓	↓		↓	↓	1 3	2 4	3 6	5 8	9 12	14 17	22 25	↑	↑	↑	↑	↑	↑	↑	↑	↑
	6th	20	120	↓	↓	↓	↓	↓	↓	↓		↓	↓	1 3	3 5	4 6	7 9	12 14	18 20	27 29	↑	↑	↑	↑	↑	↑	↑	↑	↑
	7th	20	140	↓	↓	↓	↓	↓	↓	↓		↓	↓	2 3	4 5	6 7	9 10	14 15	21 22	32 33	↑	↑	↑	↑	↑	↑	↑	↑	↑
K	1st	32	32	↓	↓	↓	↓	↓	↓	*	↓	↓	# 2	# 2	# 3	# 4	0 4	0 6	1 8	↑	↑	↑	↑	↑	↑	↑	↑	↑	↑
	2nd	32	64	↓	↓	↓	↓	↓	↓		↓	↓	# 2	0 3	0 3	1 5	2 7	3 9	6 12	↑	↑	↑	↑	↑	↑	↑	↑	↑	↑
	3rd	32	96	↓	↓	↓	↓	↓	↓		↓	↓	0 2	0 3	1 4	2 6	4 9	7 12	11 17	↑	↑	↑	↑	↑	↑	↑	↑	↑	↑
	4th	32	128	↓	↓	↓	↓	↓	↓		↓	↓	0 3	1 4	2 5	3 7	6 11	10 15	16 22	↑	↑	↑	↑	↑	↑	↑	↑	↑	↑
	5th	32	160	↓	↓	↓	↓	↓	↓		↓	↓	1 3	2 4	3 6	5 8	9 12	14 17	22 25	↑	↑	↑	↑	↑	↑	↑	↑	↑	↑
	6th	32	192	↓	↓	↓	↓	↓	↓		↓	↓	1 3	3 5	4 6	7 9	12 14	18 20	27 29	↑	↑	↑	↑	↑	↑	↑	↑	↑	↑
	7th	32	224	↓	↓	↓	↓	↓	↓		↓	↓	2 3	4 5	6 7	9 10	14 15	21 22	32 33	↑	↑	↑	↑	↑	↑	↑	↑	↑	↑

MULTIPLE TIGHTENED

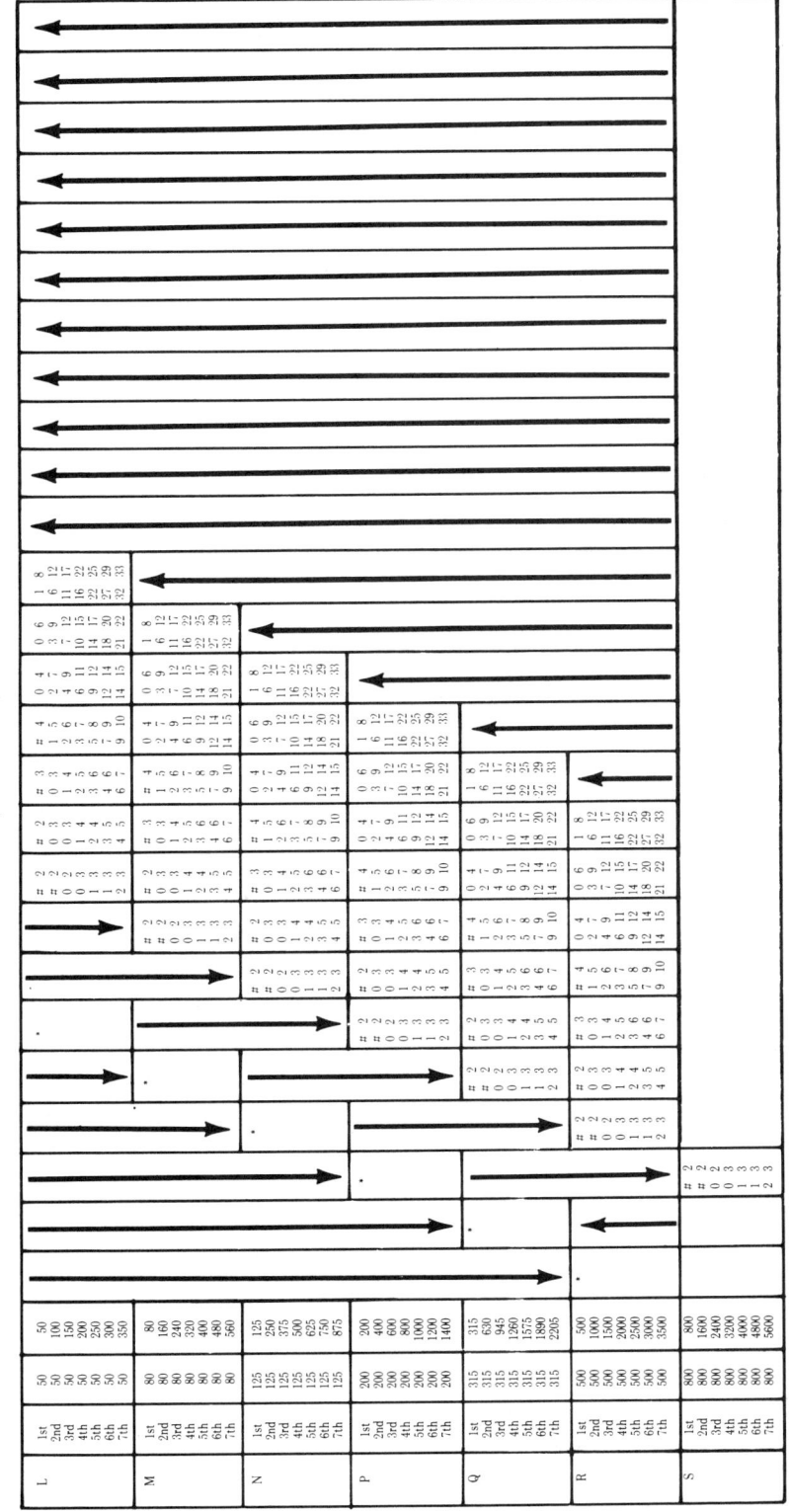

TABLE IV-B — *Multiple sampling plans for tightened inspection (Master table)* *(Continued)*

L	1st 2nd 3rd 4th 5th 6th 7th	50 50 50 50 50 50 50	50 100 150 200 250 300 350	↓	↓	↓	↓	↓	•	↓	↓	# 2 # 2 0 2 0 3 1 3 1 3 2 3	# 2 0 3 0 3 1 4 2 4 3 5 4 5	# 3 0 3 1 4 2 5 3 6 4 6 6 7	# 4 1 5 2 6 3 7 5 8 7 9 9 10	0 4 2 7 4 9 6 11 9 12 12 14 14 15	0 6 3 9 7 12 10 15 14 17 18 20 21 22	1 8 6 12 11 17 16 22 22 25 27 29 32 33	↑	↑	↑	↑	↑	↑	↑	↑	↑	↑	↑
M	1st 2nd 3rd 4th 5th 6th 7th	80 80 80 80 80 80 80	80 160 240 320 400 480 560	↓	↓	↓	↓	•	↓	↓	# 2 # 2 0 2 0 3 1 3 1 3 2 3	# 2 0 3 0 3 1 4 2 4 3 5 4 5	# 3 0 3 1 4 2 5 3 6 4 6 6 7	# 4 1 5 2 6 3 7 5 8 7 9 9 10	0 4 2 7 4 9 6 11 9 12 12 14 14 15	0 6 3 9 7 12 10 15 14 17 18 20 21 22	1 8 6 12 11 17 16 22 22 25 27 29 32 33	↑	↑	↑	↑	↑	↑	↑	↑	↑	↑	↑	↑
N	1st 2nd 3rd 4th 5th 6th 7th	125 125 125 125 125 125 125	125 250 375 500 625 750 875	↓	↓	↓	•	↓	↓	# 2 # 2 0 2 0 3 1 3 1 3 2 3	# 2 0 3 0 3 1 4 2 4 3 5 4 5	# 3 0 3 1 4 2 5 3 6 4 6 6 7	# 4 1 5 2 6 3 7 5 8 7 9 9 10	0 4 2 7 4 9 6 11 9 12 12 14 14 15	0 6 3 9 7 12 10 15 14 17 18 20 21 22	1 8 6 12 11 17 16 22 22 25 27 29 32 33	↑	↑	↑	↑	↑	↑	↑	↑	↑	↑	↑	↑	↑
P	1st 2nd 3rd 4th 5th 6th 7th	200 200 200 200 200 200 200	200 400 600 800 1000 1200 1400	↓	↓	•	↓	↓	# 2 # 2 0 2 0 3 1 3 1 3 2 3	# 2 0 3 0 3 1 4 2 4 3 5 4 5	# 3 0 3 1 4 2 5 3 6 4 6 6 7	# 4 1 5 2 6 3 7 5 8 7 9 9 10	0 4 2 7 4 9 6 11 9 12 12 14 14 15	0 6 3 9 7 12 10 15 14 17 18 20 21 22	1 8 6 12 11 17 16 22 22 25 27 29 32 33	↑	↑	↑	↑	↑	↑	↑	↑	↑	↑	↑	↑	↑	↑
Q	1st 2nd 3rd 4th 5th 6th 7th	315 315 315 315 315 315 315	315 630 945 1260 1575 1890 2205	↓	•	↓	↓	# 2 # 2 0 2 0 3 1 3 1 3 2 3	# 2 0 3 0 3 1 4 2 4 3 5 4 5	# 3 0 3 1 4 2 5 3 6 4 6 6 7	# 4 1 5 2 6 3 7 5 8 7 9 9 10	0 4 2 7 4 9 6 11 9 12 12 14 14 15	0 6 3 9 7 12 10 15 14 17 18 20 21 22	1 8 6 12 11 17 16 22 22 25 27 29 32 33	↑	↑	↑	↑	↑	↑	↑	↑	↑	↑	↑	↑	↑	↑	↑
R	1st 2nd 3rd 4th 5th 6th 7th	500 500 500 500 500 500 500	500 1000 1500 2000 2500 3000 3500	•	↑	↓	# 2 # 2 0 2 0 3 1 3 1 3 2 3	# 2 0 3 0 3 1 4 2 4 3 5 4 5	# 3 0 3 1 4 2 5 3 6 4 6 6 7	# 4 1 5 2 6 3 7 5 8 7 9 9 10	0 4 2 7 4 9 6 11 9 12 12 14 14 15	0 6 3 9 7 12 10 15 14 17 18 20 21 22	1 8 6 12 11 17 16 22 22 25 27 29 32 33	↑	↑	↑	↑	↑	↑	↑	↑	↑	↑	↑	↑	↑	↑	↑	↑
S	1st 2nd 3rd 4th 5th 6th 7th	800 800 800 800 800 800 800	800 1600 2400 3200 4000 4800 5600			# 2 # 2 0 2 0 3 1 3 1 3 2 3																							

▽ = Use first sampling plan below arrow. If sample size equals, or exceeds, lot or batch size, do 100 percent inspection.
△ = Use first sampling plan above arrow (refer to preceding page, when necessary).
Ac = Acceptance number

Re = Rejection number
# = Acceptance not permitted at this sample size.
↕ = Use corresponding double sampling plan (or alternatively, use multiple sampling plan below, where available).
• = Use corresponding single sampling plan (or alternatively, use multiple sampling plan below, where available).

TABLE IV-C — Multiple sampling plans for reduced inspection (Master table)

(See 9.4 and 9.5)

Sample size code letter	Smpl	Sample size	Cumulative sample size	Acceptable Quality Levels (reduced inspection)†																									
				0.010	0.015	0.025	0.040	0.065	0.10	0.15	0.25	0.40	0.65	1.0	1.5	2.5	4.0	6.5	10	15	25	40	65	100	150	250	400	650	1000
				Ac Re	Ac Re	Ac Re	Ac Re	Ac Re	Ac Re	Ac Re	Ac Re	Ac Re	Ac Re	Ac Re	Ac Re	Ac Re	Ac Re	Ac Re	Ac Re	Ac Re	Ac Re	Ac Re	Ac Re	Ac Re	Ac Re	Ac Re	Ac Re	Ac Re	Ac Re
A				↓	↓	↓	↓	↓	↓	↓	↓	↓	↓	↓	↓	↓	↓	•	↓	↓	•	•	•	•	•	•	•	•	•
B				↓	↓	↓	↓	↓	↓	↓	↓	↓	↓	↓	↓	↓	•	↑	↓	•	•	•	•	•	•	•	•	•	•
C				↓	↓	↓	↓	↓	↓	↓	↓	↓	↓	↓	↓	•	↑	↓	•	•	•	•	•	•	•	•	•	•	↑
D				↓	↓	↓	↓	↓	↓	↓	↓	↓	↓	↓	•	↑	↓	‡	‡	‡	‡	‡	‡	‡	‡	‡	‡	↑	↑
E				↓	↓	↓	↓	↓	↓	↓	↓	↓	↓	•	↑	↓	‡	‡	‡	‡	‡	‡	‡	‡	‡	‡	↑	↑	↑
F	1st	2	2	↓	↓	↓	↓	↓	↓	↓	↓	↓	•	↑	↓	# 2	# 2	# 3	# 3	# 4	# 4	0 5	0 6	↑	↑	↑	↑	↑	↑
	2nd	2	4													# 2	# 3	# 3	0 4	0 5	1 6	1 7	3 9						
	3rd	2	4													0 2	0 3	0 4	0 5	1 6	2 8	3 9	6 12						
	4th	2	6													0 3	0 4	0 5	1 6	2 7	3 10	5 12	8 15						
	5th	2	10													0 3	0 4	1 6	2 7	3 8	5 11	7 13	11 17						
	6th	2	12													0 3	1 5	1 6	3 7	4 9	7 12	10 15	14 20						
	7th	2	14													1 3	1 5	2 7	4 8	6 10	9 14	13 17	18 22						
G	1st	3	3	↓	↓	↓	↓	↓	↓	↓	↓	•	↑	↓	# 2	# 2	# 3	# 3	# 4	# 4	0 5	0 6	↑	↑	↑	↑	↑	↑	↑
	2nd	3	6												# 2	# 3	# 3	0 4	0 5	1 6	1 7	3 9							
	3rd	3	9												0 2	0 3	0 4	0 5	1 6	2 8	3 9	6 12							
	4th	3	12												0 3	0 4	0 5	1 6	2 7	3 10	5 12	8 15							
	5th	3	15												0 3	0 4	1 6	2 7	3 8	5 11	7 13	11 17							
	6th	3	18												0 3	1 5	1 6	3 7	4 9	7 12	10 15	14 20							
	7th	3	21												1 3	1 5	2 7	4 8	6 10	9 14	13 17	18 22							
H	1st	3	5	↓	↓	↓	↓	↓	↓	↓	•	↑	↓	# 2	# 2	# 3	# 3	# 4	# 4	0 5	0 6	↑	↑	↑	↑	↑	↑	↑	↑
	2nd	3	10											# 2	# 3	# 3	0 4	0 5	1 6	1 7	3 9								
	3rd	3	15											0 2	0 3	0 4	0 5	1 6	2 8	3 9	6 12								
	4th	3	20											0 3	0 4	0 5	1 6	2 7	3 10	5 12	8 15								
	5th	3	25											0 3	0 4	1 6	2 7	3 8	5 11	7 13	11 17								
	6th	3	30											0 3	1 5	1 6	3 7	4 9	7 12	10 15	14 20								
	7th	3	35											1 3	1 5	2 7	4 8	6 10	9 14	13 17	18 22								
J	1st	8	8	↓	↓	↓	↓	↓	↓	•	↑	↓	# 2	# 2	# 3	# 3	# 4	# 4	0 5	0 6	↑	↑	↑	↑	↑	↑	↑	↑	↑
	2nd	8	16										# 2	# 3	# 3	0 4	0 5	1 6	1 7	3 9									
	3rd	8	24										0 2	0 3	0 4	0 5	1 6	2 8	3 9	6 12									
	4th	8	32										0 3	0 4	0 5	1 6	2 7	3 10	5 12	8 15									
	5th	8	40										0 3	0 4	1 6	2 7	3 8	5 11	7 13	11 17									
	6th	8	48										0 3	1 5	1 6	3 7	4 9	7 12	10 15	14 20									
	7th	8	56										1 3	1 5	2 7	4 8	6 10	9 14	13 17	18 22									
K	1st	13	13	↓	↓	↓	↓	↓	•	↑	↓	# 2	# 2	# 3	# 3	# 4	# 4	0 5	0 6	↑	↑	↑	↑	↑	↑	↑	↑	↑	↑
	2nd	13	26									# 2	# 3	# 3	0 4	0 5	1 6	1 7	3 9										
	3rd	13	39									0 2	0 3	0 4	0 5	1 6	2 8	3 9	6 12										
	4th	13	52									0 3	0 4	0 5	1 6	2 7	3 10	5 12	8 15										
	5th	13	65									0 3	0 4	1 6	2 7	3 8	5 11	7 13	11 17										
	6th	13	78									0 3	1 5	1 6	3 7	4 9	7 12	10 15	14 20										
	7th	13	91									1 3	1 5	2 7	4 8	6 10	9 14	13 17	18 22										

↓ = Use first sampling plan below arrow. If sample size equals, or exceeds, lot or batch size, do 100 percent inspection.
↑ = Use first sampling plan above arrow (refer to preceding page, when necessary).
Ac = Acceptance number
Re = Rejection number
# = Acceptance not permitted at this sample size.
‡ = Use corresponding double sampling plan (or alternatively, use multiple sampling plan below, where available).
• = Use corresponding single sampling plan (or alternatively, use multiple sampling plan below, where available).
† = If, after the final sample, the acceptance number has been exceeded, but the rejection number has not been reached, accept the lot, but reinstate normal inspection (see 10.1.4).

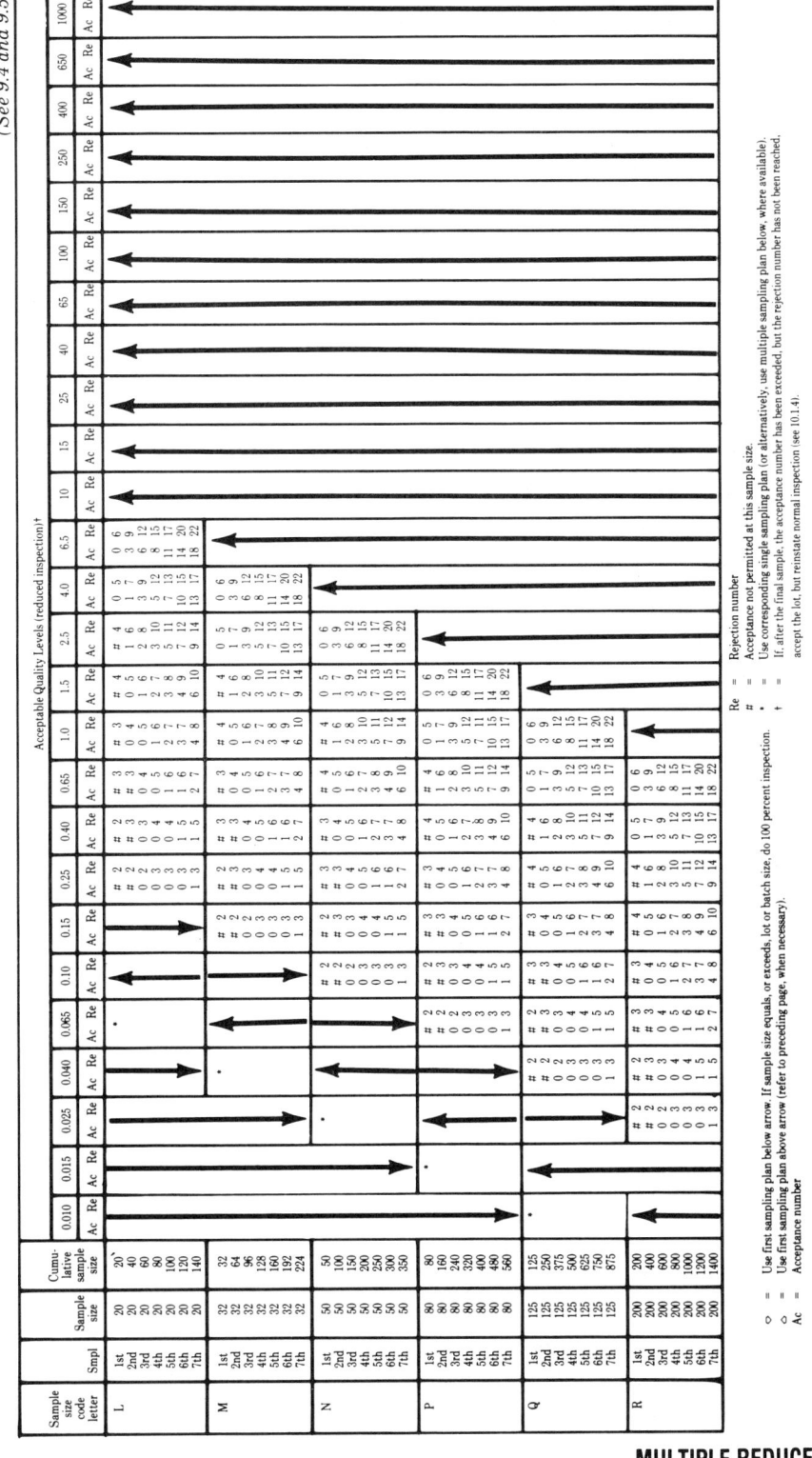

TABLE IV-C — Multiple sampling plans for reduced inspection (Master table) (Continued)

(See 9.4 and 9.5)

Sample size code letter	Smpl	Sample size	Cumulative sample size	Acceptable Quality Levels (reduced inspection)†																																								
				0.010		0.015		0.025		0.040		0.065		0.10		0.15		0.25		0.40		0.65		1.0		1.5		2.5		4.0		6.5		10	15	25	40	65	100	150	250	400	650	1000
				Ac	Re	Ac	Re	Ac	Re	Ac	Re	Ac	Re	Ac	Re	Ac	Re	Ac	Re	Ac	Re	Ac	Re	Ac	Re	Ac	Re	Ac	Re	Ac	Re	Ac	Re	Ac Re	Ac Re	Ac Re	Ac Re	Ac Re	Ac Re	Ac Re	Ac Re	Ac Re	Ac Re	Ac Re
L	1st	20	20`	↓		↓		↓		↓		•		↑		↓		#	2	#	2	#	3	#	3	#	4	#	4	0	5	0	6	↑	↑	↑	↑	↑	↑	↑	↑	↑	↑	↑
	2nd	20	40															#	2	#	3	#	3	0	4	0	5	1	6	1	7	3	9											
	3rd	20	60															0	2	0	3	0	4	0	5	1	6	2	8	3	9	6	12											
	4th	20	80															0	3	0	4	0	5	1	6	2	7	3	10	5	12	8	15											
	5th	20	100															0	3	0	4	1	6	2	7	3	8	5	11	7	13	11	17											
	6th	20	120															0	3	1	5	1	6	3	7	4	9	7	12	10	15	14	20											
	7th	20	140															1	3	1	5	2	7	4	8	6	10	9	14	13	17	18	22											
M	1st	32	32							•		↑		↓		#	2	#	2	#	3	#	3	#	4	#	4	0	5	0	6	↑												
	2nd	32	64													#	2	#	3	#	3	0	4	0	5	1	6	1	7	3	9													
	3rd	32	96													0	2	0	3	0	4	0	5	1	6	2	8	3	9	6	12													
	4th	32	128													0	3	0	4	0	5	1	6	2	7	3	10	5	12	8	15													
	5th	32	160													0	3	0	4	1	6	2	7	3	8	5	11	7	13	11	17													
	6th	32	192													0	3	1	5	1	6	3	7	4	9	7	12	10	15	14	20													
	7th	32	224													1	3	1	5	2	7	4	8	6	10	9	14	13	17	18	22													
N	1st	50	50					•		↑		↓		#	2	#	2	#	3	#	3	#	4	#	4	0	5	0	6	↑														
	2nd	50	100											#	2	#	3	#	3	0	4	0	5	1	6	1	7	3	9															
	3rd	50	150											0	2	0	3	0	4	0	5	1	6	2	8	3	9	6	12															
	4th	50	200											0	3	0	4	0	5	1	6	2	7	3	10	5	12	8	15															
	5th	50	250											0	3	0	4	1	6	2	7	3	8	5	11	7	13	11	17															
	6th	50	300											0	3	1	5	1	6	3	7	4	9	7	12	10	15	14	20															
	7th	50	350											1	3	1	5	2	7	4	8	6	10	9	14	13	17	18	22															
P	1st	80	80			•		↑		↓		#	2	#	2	#	3	#	3	#	4	#	4	0	5	0	6	↑																
	2nd	80	160									#	2	#	3	#	3	0	4	0	5	1	6	1	7	3	9																	
	3rd	80	240									0	2	0	3	0	4	0	5	1	6	2	8	3	9	6	12																	
	4th	80	320									0	3	0	4	0	5	1	6	2	7	3	10	5	12	8	15																	
	5th	80	400									0	3	0	4	1	6	2	7	3	8	5	11	7	11	11	17																	
	6th	80	480									0	3	1	5	1	6	3	7	4	9	7	12	10	15	14	20																	
	7th	80	560									1	3	1	5	2	7	4	8	6	10	9	14	13	17	18	22																	
Q	1st	125	125	•		↑		↓		#	2	#	2	#	3	#	3	#	4	#	4	0	5	0	6	↑																		
	2nd	125	250							#	2	#	3	#	3	0	4	0	5	1	6	1	7	3	9																			
	3rd	125	375							0	2	0	3	0	4	0	5	1	6	2	8	3	9	6	12																			
	4th	125	500							0	3	0	4	0	5	1	6	2	7	3	10	5	12	8	15																			
	5th	125	625							0	3	0	4	1	6	2	7	3	8	5	11	7	13	11	17																			
	6th	125	750							0	3	1	5	1	6	3	7	4	9	7	12	10	15	14	20																			
	7th	125	875							1	3	1	5	2	7	4	8	6	10	9	14	13	17	18	22																			
R	1st	200	200	↑				#	2	#	2	#	3	#	3	#	4	#	4	0	5	0	6	↑																				
	2nd	200	400					#	2	#	3	#	3	0	4	0	5	1	6	1	7	3	9																					
	3rd	200	600					0	2	0	3	0	4	0	5	1	6	2	8	3	9	6	12																					
	4th	200	800					0	3	0	4	0	5	1	6	2	7	3	10	5	12	8	15																					
	5th	200	1000					0	3	0	4	1	6	2	7	3	8	5	11	7	13	11	17																					
	6th	200	1200					0	3	1	5	1	6	3	7	4	9	7	12	10	15	14	20																					
	7th	200	1400					1	3	1	5	2	7	4	8	6	10	9	14	13	17	18	22																					

↓ = Use first sampling plan below arrow. If sample size equals, or exceeds, lot or batch size, do 100 percent inspection.
↑ = Use first sampling plan above arrow (refer to preceding page, when necessary).
Ac = Acceptance number

Re = Rejection number
# = Acceptance not permitted at this sample size.
• = Use corresponding single sampling plan (or alternatively, use multiple sampling plan below, where available).
† = If, after the final sample, the acceptance number has been exceeded, but the rejection number has not been reached, accept the lot, but reinstate normal inspection (see 10.1.4).

AOQL NORMAL

TABLE V-A — Average Outgoing Quality Limit Factors for Normal Inspection (Single sampling)

(See 11.4)

		Acceptable Quality Level																					
Code letter	Sample size	0.010	0.015	0.025	0.040	0.065	0.10	0.15	0.25	0.40	0.65	1.0	1.5	2.5	4.0	6.5	10	15	25	40	65	100	150
A	2															18			42	69	97	160	220
B	3														12			28	46	65	110	150	220
C	5													7.4			17	27	39	63	90	130	190
D	8												4.6			11	17	24	40	56	82	120	180
E	13											2.8			6.5	11	15	24	34	50	72	110	170
F	20										1.8			4.2	6.9	9.7	16	22	33	47	73		
G	32									1.2			2.6	4.3	6.1	9.9	14	21	29	46			
H	50								0.74			1.7	2.7	3.9	6.3	9.0	13	19	29				
J	80							0.46			1.1	1.7	2.4	4.0	5.6	8.2	12	18					
K	125						0.29			0.67	1.1	1.6	2.5	3.6	5.2	7.5	12						
L	200					0.18			0.42	0.69	0.97	1.6	2.2	3.3	4.7	7.3							
M	315				0.12			0.27	0.44	0.62	1.00	1.4	2.1	3.0	4.7								
N	500			0.074			0.17	0.27	0.39	0.63	0.90	1.3	1.9	2.9									
P	800		0.046			0.11	0.17	0.24	0.40	0.56	0.82	1.2	1.8										
Q	1250	0.029			0.067	0.11	0.16	0.25	0.36	0.52	0.75	1.2											
R	2000			0.042	0.069	0.097	0.16	0.22	0.33	0.47	0.73												

Note: For the exact AOQL, the above values must be multiplied by $\left(1 = \frac{\text{Sample size}}{\text{Lot or Batch size}}\right)$ (see 11.4)

TABLE V-B — *Average Outgoing Quality Limit Factors for Tightened Inspection (Single sampling)*

Code letter	Sample size	Acceptable Quality Level																					
		0.010	0.015	0.025	0.040	0.065	0.10	0.15	0.25	0.40	0.65	1.0	1.5	2.5	4.0	6.5	10	15	25	40	65	100	150
A	2																			42	69	97	160
B	3															12			28	46	65	110	170
C	5														7.4			17	27	39	63	100	160
D	8													4.6			11	17	24	40	64	99	160
E	13												2.8			6.5	11	15	24	40	61	95	150
F	20											1.8			4.2	6.9	9.7	16	26	40	62		
G	32										1.2			2.6	4.3	6.1	9.9	16	25	39			
H	50									0.74			1.7	2.7	3.9	6.3	10	16	25				
J	80								0.46			1.1	1.7	2.4	4.0	6.4	9.9	16					
K	125							0.29			0.67	1.1	1.6	2.5	4.1	6.4	9.9						
L	200						0.18			0.42	0.69	0.97	1.6	2.6	4.0	6.2							
M	315					0.12			0.27	0.44	0.62	1.0	1.6	2.5	3.9								
N	500				0.074			0.17	0.27	0.39	0.63	1.0	1.6	2.5									
P	800			0.046			0.11	0.17	0.24	0.40	0.64	0.99	1.6										
Q	1250		0.029			0.067	0.11	0.16	0.25	0.41	0.64	0.99											
R	2000	0.018			0.042	0.069	0.097	0.16	0.26	0.40	0.62												
S	3150			0.027																			

Note: For the exact AOQL, the above values must be multiplied by $\left(1 - \frac{\text{Sample size}}{\text{Lot or Batch size}}\right)$ (see 11.4)

AOQL TIGHTENED

TABLE VI-A — *Limiting Quality (in percent defective) for which* P_a = *10 Percent*
(for Normal Inspection, Single sampling)

(See 11.6)

Code letter	Sample size	Acceptable Quality Level															
		0.010	0.015	0.025	0.040	0.065	0.10	0.15	0.25	0.40	0.65	1.0	1.5	2.5	4.0	6.5	10
A	2															68	
B	3														54		
C	5													37			58
D	8												25			41	54
E	13											16			27	36	44
F	20										11			18	25	30	42
G	32									6.9			12	16	20	27	34
H	50								4.5			7.6	10	13	18	22	29
J	80							2.8			4.8	6.5	8.2	11	14	19	24
K	125						1.8			3.1	4.3	5.4	7.4	9.4	12	16	23
L	200					1.2			2.0	2.7	3.3	4.6	5.9	7.7	10	14	
M	315				0.73			1.2	1.7	2.1	2.9	3.7	4.9	6.4	9.0		
N	500			0.46			0.78	1.1	1.3	1.9	2.4	3.1	4.0	5.6			
P	800		0.29			0.49	0.67	0.84	1.2	1.5	1.9	2.5	3.5				
Q	1250	0.18			0.31	0.43	0.53	0.74	0.94	1.2	1.6	2.3					
R	2000			0.20	0.27	0.33	0.46	0.59	0.77	1.0	1.4						

TABLE VI-B — Limiting Quality (in defects per hundred units) for which $P_a = 10$ Percent (for Normal Inspection, Single sampling)

(See 11.6)

Code letter	Sample size	0.010	0.015	0.025	0.040	0.065	0.10	0.15	0.25	0.40	0.65	1.0	1.5	2.5	4.0	6.5	10	15	25	40	65	100	150
		Acceptable Quality Level																					
A	2															120			200	270	330	460	590
B	3														77			130	180	220	310	390	510
C	5													46			78	110	130	190	240	310	400
D	8												29			49	67	84	120	150	190	250	350
E	13											18			30	41	51	71	91	120	160	220	300
F	20										12			20	27	33	46	59	77	100	140		
G	32									7.2			12	17	21	29	37	48	63	88			
H	50								4.6			7.8	11	13	19	24	31	40	56				
J	80							2.9			4.9	6.7	8.4	12	15	19	25	35					
K	125						1.8			3.1	4.3	5.4	7.4	9.4	12	16	23						
L	200					1.2			2.0	2.7	3.3	4.6	5.9	7.7	10	14							
M	315				0.73			1.2	1.7	2.1	2.9	3.7	4.9	6.4	9.0								
N	500			0.46			0.78	1.1	1.3	1.9	2.4	3.1	4.0	5.6									
P	800		0.29			0.49	0.67	0.84	1.2	1.5	1.9	2.5	3.5										
Q	1250	0.18			0.31	0.43	0.53	0.74	0.94	1.2	1.6	2.3											
R	2000			0.20	0.27	0.33	0.46	0.59	0.77	1.0	1.4												

LQ (DEFECTS) 10.0%

LQ (DEFECTIVES) 5%

TABLE VII-A — Limiting Quality (in percent defective) for which $P_a = 5$ Percent (for Normal Inspection, Single sampling)

(See 11.6)

Code letter	Sample size	Acceptable Quality Level 0.010	0.015	0.025	0.040	0.065	0.10	0.15	0.25	0.40	0.65	1.0	1.5	2.5	4.0	6.5	10
A	2															78	
B	3														63		
C	5													45			66
D	8												31			47	60
E	13											21			32	41	50
F	20										14			22	28	34	46
G	32									8.9			14	18	23	30	37
H	50								5.8			9.1	12	15	20	25	32
J	80							3.7			5.8	7.7	9.4	13	16	20	26
K	125						2.4			3.8	5.0	6.2	8.4	11	14	18	24
L	200					1.5			2.4	3.2	3.9	5.3	6.6	8.5	11	15	
M	315				0.95			1.5	2.0	2.5	3.3	4.2	5.4	7.0	9.6		
N	500			0.60			0.95	1.3	1.6	2.1	2.6	3.4	4.4	6.1			
P	800		0.38			0.59	0.79	0.97	1.3	1.6	2.1	2.7	3.8				
Q	1250	0.24			0.38	0.50	0.62	0.84	1.1	1.4	1.8	2.4					
R	2000			0.24	0.32	0.39	0.53	0.66	0.85	1.1	1.5						

TABLE VII-B — Limiting Quality (in defects per hundred units) for which P_a = 5 Percent (for Normal Inspection, Single sampling)

(See 11.6)

Code letter	Sample size	Acceptable Quality Level																					
		0.010	0.015	0.025	0.040	0.065	0.10	0.15	0.25	0.40	0.65	1.0	1.5	2.5	4.0	6.5	10	15	25	40	65	100	150
A	2															150			240	320	390	530	660
B	3														100			160	210	260	350	440	570
C	5													60			95	130	160	210	260	340	440
D	8												38			59	79	97	130	160	210	270	380
E	13											23			37	48	60	81	100	130	170	230	310
F	20										15			24	32	39	53	66	85	110	150		
G	32									9.4			15	20	24	33	41	53	68	95			
H	50								6.0			9.5	13	16	21	26	34	44	61				
J	80							3.8			5.9	7.9	9.7	13	16	21	27	38					
K	125						2.4			3.8	5.0	6.2	8.4	11	14	18	24						
L	200					1.5			2.4	3.2	3.9	5.3	6.6	8.5	11	15							
M	315				0.95			1.5	2.0	2.5	3.3	4.2	5.4	7.0	9.6								
N	500			0.60			0.95	1.3	1.6	2.1	2.6	3.4	4.4	6.1									
P	800		0.38			0.59	0.79	0.97	1.3	1.6	2.1	2.7	3.8										
Q	1250	0.24			0.38	0.50	0.62	0.84	1.1	1.4	1.8	2.4											
R	2000			0.24	0.32	0.39	0.53	0.66	0.85	1.1	1.5												

TABLE VIII — Limit Numbers for Reduced Inspection

(See 8.3.3)

Number of sample units from last 10 lots or batches	Acceptable Quality Level																				
	0.010	0.015	0.025	0.040	0.065	0.10	0.15	0.25	0.40	0.65	1.0	1.5	2.5	4.0	6.5	10	15	25	40	65	100	150
20-29	*	*	*	*	*	*	*	*	*	*	*	*	*	*	*	0	0	2	4	8	14	22
30-49	*	*	*	*	*	*	*	*	*	*	*	*	*	*	0	0	1	3	7	13	22	36
50-79	*	*	*	*	*	*	*	*	*	*	*	*	*	0	0	2	3	7	14	25	40	63
80-129	*	*	*	*	*	*	*	*	*	*	*	*	0	0	2	4	7	14	24	42	68	*105*
130-199	*	*	*	*	*	*	*	*	*	*	*	0	0	2	4	7	13	25	42	72	115	177
200-319	*	*	*	*	*	*	*	*	*	*	0	0	2	4	8	14	22	40	68	115	181	277
320-499	*	*	*	*	*	*	*	*	*	0	0	1	4	8	14	24	39	68	113	189		
500-799	*	*	*	*	*	*	*	*	0	0	2	3	7	14	25	40	63	110	181			
800-1249	*	*	*	*	*	*	*	0	0	2	4	7	14	24	42	68	105	181				
1250-1999	*	*	*	*	*	*	0	0	2	4	7	13	24	40	69	110	169					
2000-3149	*	*	*	*	*	0	0	2	4	8	14	22	40	68	115	181						
3150-1999	*	*	*	*	0	0	1	4	8	14	24	38	67	111	186							
5000-7999	*	*	*	0	0	2	3	7	14	25	40	63	110	181								
8000-12499	*	*	0	0	2	4	7	14	24	42	68	105	181									
12500-19999	*	0	0	2	4	7	13	24	40	69	110	169										
20000-31499	0	0	2	4	8	14	22	40	68	115	181											
31500-49999	0	1	4	8	14	24	38	67	111	186												
500000 & Over	2	3	7	14	25	40	63	110	181	301												

*Denotes that the number of sample units from the last ten lots or batches is not sufficient for reduced inspection for this AQL. In this instance more than ten lots or batches may be used for the calculation, provided that the lots or batches used are the most recent ones in sequence, that they have all been on normal inspection, and that none has been rejected while on original inspection.

TABLE IX — Average sample size curves for double and multiple sampling (normal and tightened inspection)

(See 11.5)

Average Sample Size

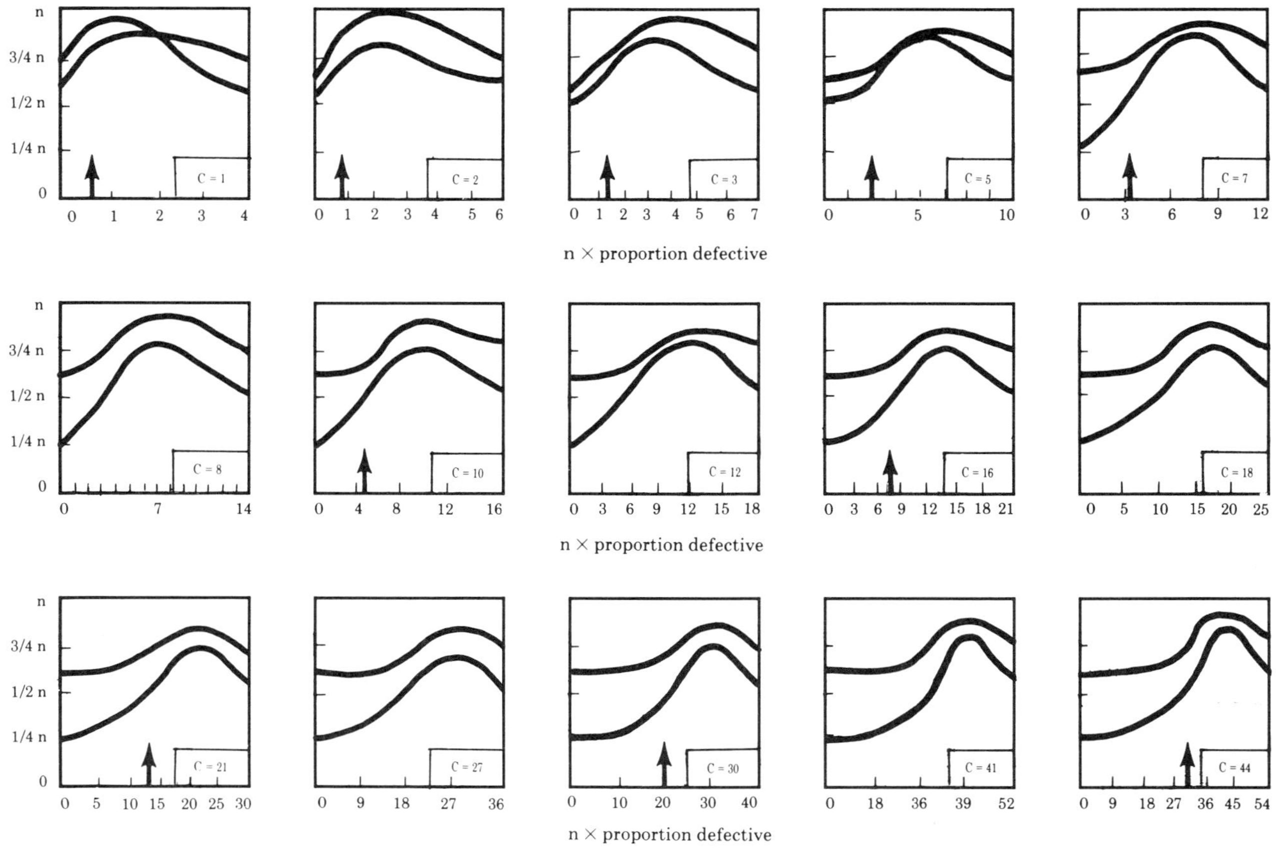

n = Equivalent single sample size   c = Single sample acceptance number   ↑ = AQL for normal inspection

AVERAGE SAMPLE SIZE

TABLE X-A — Tables for sample size code letter: A

CHART A — OPERATING CHARACTERISTIC CURVES FOR SINGLE SAMPLING PLANS

(Curves for double and multiple sampling are matched as closely as practicable)

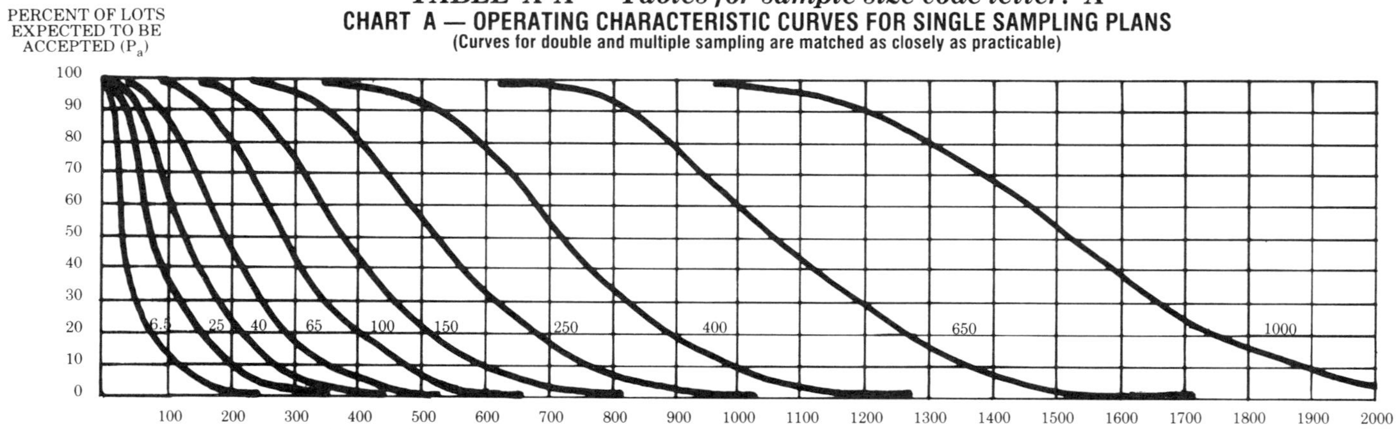

QUALITY OF SUBMITTED LOT (P_a in percent defective for AQL's ≤ 10; in defects per hundred units for AQL's > 10)
NOTE: Figures on curves are Acceptable Quality Levels (AQL's) for normal inspection.

TABLE X-A-1 — TABULATED VALUES FOR OPERATING CHARACTERISTIC CURVES FOR SINGLE SAMPLING PLANS

P_a	Acceptable Quality Levels (normal inspection)														
	6.5	6.5	25	40	65	100	150	X	250	X	400	X	650	X	1000
	p (in percent defective)	p (in defects per hundred units)													
99.0	0.501	0.51	7.45	21.8	41.2	89.2	145	175	239	305	374	517	629	859	977
95.0	2.53	2.56	17.8	40.9	68.3	131	199	235	308	385	462	622	745	995	1122
90.0	5.13	5.25	26.6	55.1	87.3	158	233	272	351	432	515	684	812	1073	1206
75.0	13.4	14.4	48.1	86.8	127	211	298	342	431	521	612	795	934	1314	1354
50.0	29.3	34.7	83.9	134	184	284	383	433	533	633	733	933	1083	1383	1533
25.0	50.0	69.3	135	196	256	371	484	540	651	761	870	1087	1248	1568	1728
10.0	68.4	115	195	266	334	464	589	650	770	889	1006	1238	1409	1748	1916
5.0	77.6	150	237	315	388	526	657	722	848	972	1094	1334	1512	1862	2035
1.0	90.0	230	332	420	502	655	800	870	1007	1141	1272	1529	1718	2088	2270
	X	X	40	65	100	150	X	250	X	400	X	650	X	1000	X
	Acceptable Quality Levels (tightened inspection)														

Note: Binominal distribution used for percent defective computations; Poisson for defects per hundred units.

TABLE X-A-2 — SAMPLING PLANS FOR SAMPLE SIZE CODE LETTER: A

Type of sampling plan	Cumu-lative sample size	Acceptable Quality Levels (normal inspection)																		Cumu-lative sample size
		Less than 6.5	6.5	><	10	15	25	40	65	100	150	><	250	><	400	><	650	><	1000	
		Ac Re	Ac Re	Ac Re	Ac Re	Ac Re	Ac Re	Ac Re	Ac Re	Ac Re	Ac Re	Ac Re	Ac Re	Ac Re	Ac Re	Ac Re	Ac Re	Ac Re	Ac Re	
Single	2	▽	0 1	Use Letter D	Use Letter C	Use Letter B	1 2	2 3	3 4	5 6	7 8	8 9	10 11	12 13	14 15	18 19	21 22	27 28	30 31	2
Double		▽	*				(*)	(*)	(*)	(*)	(*)	(*)	(*)	(*)	(*)	(*)	(*)	(*)	(*)	
Multiple		▽					•	•	•	•	•	•	•	•	•	•	•	•	•	
		Less than 10	><	10	15	25	40	65	100	150	><	250	><	400	><	650	><	1000	><	
		Acceptable Quality Levels (tightened inspection)																		

A

▽ = Use next subsequent sample size code letter for which acceptance and rejection numbers are available.

Ac = Acceptance number.

Re = Rejection number.

● = Use single sampling plan above (or alternatively use letter D).

(*) = Use single sampling (or alternatively use letter B).

TABLE X-B — Tables for sample size code letter: B

CHART B — OPERATING CHARACTERISTIC CURVES FOR SINGLE SAMPLING PLANS

(Curves for double and multiple sampling are matched as closely as practicable)

PERCENT OF LOTS EXPECTED TO BE ACCEPTED (P_a)

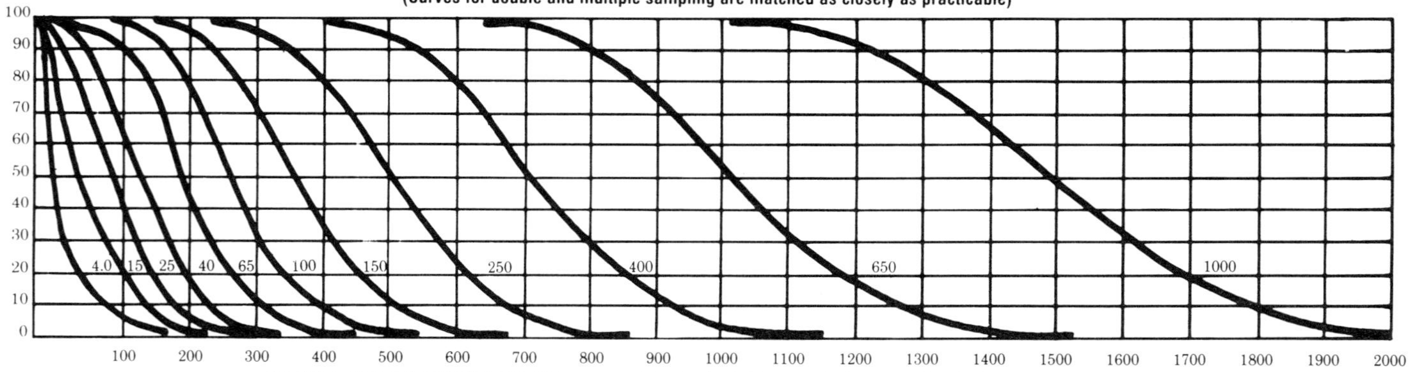

QUALITY OF SUBMITTED LOTS (P_a in percent defective for AQL's $\leq$ 10; in defects per hundred units for AQL's $>$10)

Note: Figures on curves are Acceptable Quality Levels (AQL's) for normal inspection.

TABLE X-B-1 — TABULATED VALUES FOR OPERATING CHARACTERISTIC CURVES FOR SINGLE SAMPLING PLANS

P_a		Acceptable Quality Levels (normal inspection)															
	4.0	4.0	15	25	40	65	100	X	150	X	250	X	400	X	650	X	1000
	p (in percent defective)	p (in defects per hundred units)															
99.0	0.33	0.34	4.97	14.5	27.4	59.5	96.9	117	159	203	249	345	419	573	651	947	1029
95.0	1.70	1.71	11.8	27.3	45.5	87.1	133	157	206	256	308	415	496	663	748	1065	1152
90.0	3.45	3.50	17.7	36.7	58.2	105	155	181	234	288	343	456	541	716	804	1131	1222
75.0	9.14	9.60	32.0	57.6	84.5	141	199	228	287	347	408	530	623	809	903	1249	1344
50.0	20.6	23.1	55.9	89.1	122	189	256	289	356	422	489	622	722	922	1022	1389	1489
25.0	37.0	46.2	89.8	131	170	247	323	360	434	507	580	724	832	1046	1152	1539	1644
10.0	53.6	76.8	130	177	233	309	392	433	514	593	671	825	939	1165	1277	1683	1793
5.0	63.2	99.9	158	210	258	350	438	481	565	648	730	890	1008	1241	1356	1773	1886
1.0	78.4	154	221	280	335	437	533	580	672	761	848	1019	1145	1392	1513	1951	2069
	6.5	6.5	25	40	65	100	X	150	X	250	X	400	X	650	X	1000	X
	Acceptable Quality Levels (tightened inspection)																

Notes: Binominal distribution used for percent defective computation; Poisson for defects per hundred units.

TABLE X-B-2 — SAMPLING PLANS FOR SAMPLE SIZE CODE LETTER: B

Type of sampling plan	Cumulative sample size	Acceptable Quality Levels (normal inspection)																				Cumulative sample size
		Less than 4.0	4.0	6.5	×	10	15	25	40	65	100	×	150	×	250	×	400	×	650	×	1000	
		Ac Re	Ac Re	Ac Re	Ac Re	Ac Re	Ac Re	Ac Re	Ac Re	Ac Re	Ac Re	Ac Re	Ac Re	Ac Re	Ac Re	Ac Re	Ac Re	Ac Re	Ac Re	Ac Re	Ac Re	
Single	3	▽	0 1	Use Letter A	Use Letter D	Use Letter C	1 2	2 3	3 4	5 6	7 8	8 9	10 11	12 13	14 15	18 19	21 22	27 28	30 31	41 42	44 45	3
Double	2	▽	*				0 2	0 3	1 4	2 5	3 7	3 7	5 9	6 10	7 11	9 14	11 16	15 20	17 22	23 29	25 31	2
	4						1 2	3 4	4 5	6 7	8 9	11 12	12 13	15 16	18 19	23 24	26 27	34 35	37 38	52 53	56 57	4
Multiple		▽	*				‡	‡	‡	‡	‡	‡	‡	‡	‡	‡	‡	‡	‡	‡	‡	
		Less than 6.5	6.5	×	10	15	25	40	65	100	×	150	×	250	×	400	×	650	×	1000	×	
			Acceptable Quality Levels (tightened inspection)																			

▽ = Use next subsequent sample size code letter for which acceptance and rejection numbers are available.
Ac = Acceptance number.
Re = Rejection number.
* = Use single sampling plan above (or alternatively use letter E).
‡ = Use double sampling plan above (or alternatively use letter D).

B

TABLE X-C — Tables for sample size code letter: C

CHART C — OPERATING CHARACTERISTIC CURVES FOR SINGLE SAMPLING PLANS

(Curves for double and multiple sampling are matched as closely as practicable)

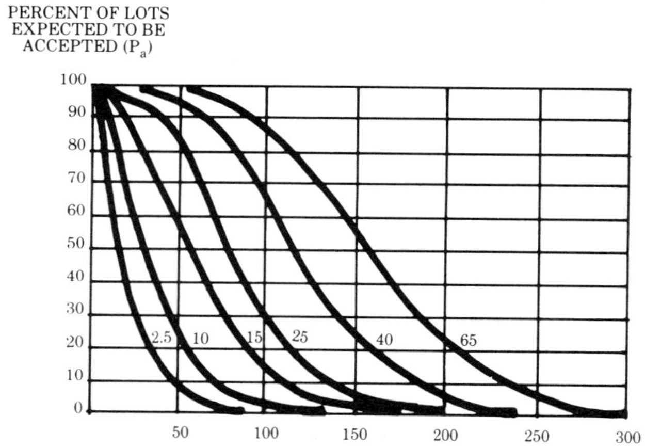

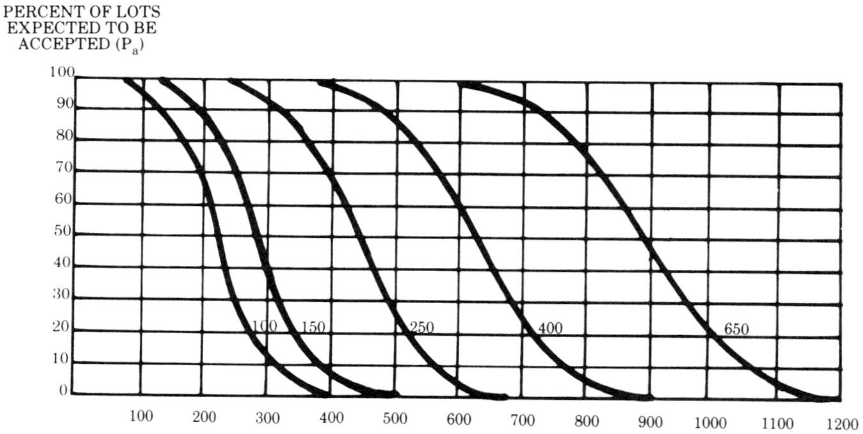

QUALITY OF SUBMITTED LOTS (P_a in percent defective for AQL's $\leqslant$ 10: in defects per hundred units for AQL's $>$10)

NOTE: Figures on curves are Acceptable Quality Levels (AQL's) for normal inspection

TABLE X-C-1 — TABULATED VALUES FOR OPERATING CHARACTERISTIC CURVES FOR SINGLE SAMPLING PLANS

P_a	Acceptable Quality Levels (normal inspection)																	
	2.5	10	2.5	10	15	25	40	65	×	100	×	150	×	250	×	400	×	650
	p (in percent defective)		p (in defects per hundred units)															
99.0	0.20	3.28	0.20	2.89	8.72	16.5	35.7	58.1	70.1	95.4	122	150	207	251	344	391	568	618
95.0	1.02	7.63	1.03	7.10	16.4	27.3	52.3	79.6	93.9	123	154	185	249	298	398	449	639	691
90.0	2.09	11.2	2.10	10.6	22.0	34.9	63.0	93.1	109	140	173	206	273	325	429	482	679	733
75.0	5.59	19.4	5.76	19.2	34.5	50.7	84.4	119	137	172	208	245	318	374	485	542	749	806
50.0	12.9	31.4	13.9	33.6	53.5	73.4	113	153	173	213	253	293	373	433	553	613	833	893
25.0	24.2	45.4	27.7	53.9	78.4	102	148	194	216	260	304	348	435	499	627	691	923	987
10.0	36.9	58.4	46.1	77.8	106	134	186	235	260	308	356	403	495	564	699	766	1010	1076
5.0	45.1	65.8	59.9	94.9	126	155	210	263	289	339	389	438	534	605	745	814	1064	1131
1.0	60.2	77.8	92.1	133	168	201	262	320	348	403	456	509	612	687	835	908	1171	1241
	4.0	×	4.0	15	25	40	65	×	100	×	150	×	250	×	400	×	650	×
	Acceptable Quality Levels (tightened inspection)																	

Note: Binominal distribution used for percent defective computation; Poisson for defects per hundred units.

TABLE X-C-2 — SAMPLING PLANS FOR SAMPLE SIZE CODE LETTER: C

Type of sampling plan	Cumulative sample size	Less than 2.5	2.5	4.0	><	6.5	10	15	25	40	65	><	100	><	150	><	250	><	400	><
		Ac Re	Ac Re	Ac Re	Ac Re	Ac Re	Ac Re	Ac Re	Ac Re	Ac Re	Ac Re	Ac Re	Ac Re	Ac Re	Ac Re	Ac Re	Ac Re	Ac Re	Ac Re	Ac Re
Single	5	▽	0 1	Use Letter B	Use Letter E	Use Letter D	1 2	2 3	3 4	5 6	7 8	8 9	10 11	12 13	14 15	18 19	21 22	27 28	30 31	41 42
Double	3	▽	*				0 2	0 3	1 4	2 5	3 7	3 7	5 9	6 10	7 11	9 14	11 16	15 20	17 22	23 29
	6						1 2	3 4	4 5	6 7	8 9	11 12	12 13	15 16	18 19	23 24	26 27	34 35	37 38	52 53
Multiple		▽	*				‡	‡	‡	‡	‡	‡	‡	‡	‡	‡	‡	‡	‡	‡
Acceptable Quality Levels (tightened inspection)		Less than 4.0	4.0	><	6.5	10	15	25	40	65	><	100	><	150	><	250	><	400	><	650

Column headers 2.5 through the last >< are Acceptable Quality Levels (normal inspection).

▽ = Use next subsequent sample size code letter for which acceptance and rejection numbers are available.
Ac = Acceptance number.
Re = Rejection number.
* = Use single sampling plan above (or alternatively use letter F).
‡ = Use double sampling plan above (or alternatively use letter D).

C

TABLE X-D — Tables for sample size code letter: D

CHART D — OPERATING CHARACTERISTIC CURVES FOR SINGLE SAMPLING PLANS

(Curves for double and multiple sampling are matched as closely as practicable)

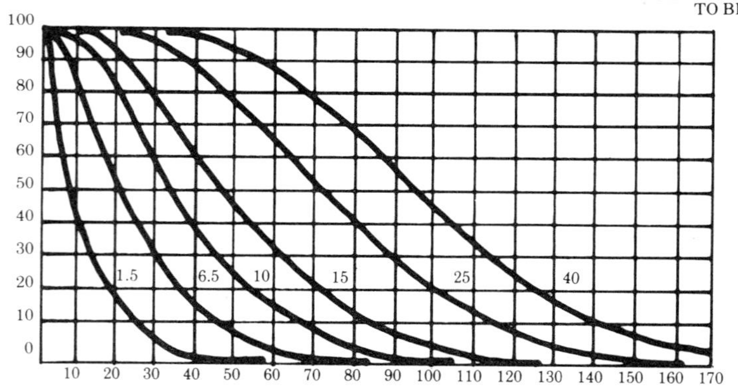

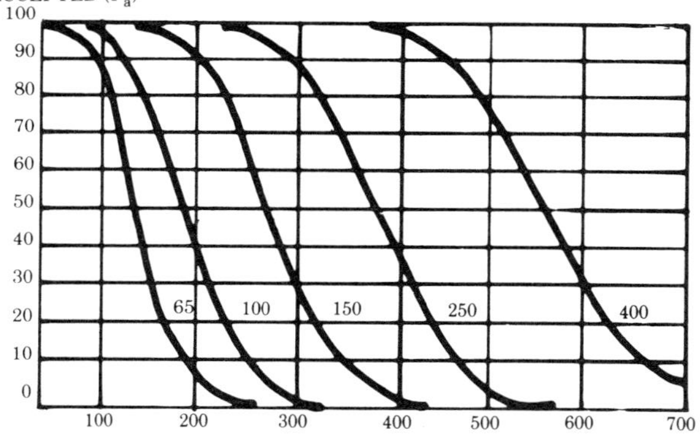

QUALITY OF SUBMITTED LOTS (P_a is percent defective for AQL's ≤ 10; in defects per hundred units for AQL's >10)

NOTE: Figures on curves are Acceptable Quality Levels (AQL's) for normal inspection.

TABLE X-D-1 — TABULATED VALUES FOR OPERATING CHARACTERISTIC CURVES FOR SINGLE SAMPLING PLANS

P_a	Acceptable Quality Levels (normal inspection)																		
	1.5	6.5	10	1.5	6.5	10	15	25	40	X	65	X	100	X	150	X	250	X	400
	p(in percent defective)			p (in defects per hundred units)															
99.0	0.13	2.00	6.00	0.13	1.86	5.45	10.3	22.3	36.3	43.8	59.6	76.2	93.5	129	157	215	244	355	386
95.0	0.64	2.64	11.1	0.64	4.44	10.2	17.1	32.7	49.8	58.7	77.1	96.1	116	156	186	249	281	399	432
90.0	1.31	6.88	14.7	1.31	6.65	13.8	21.8	39.4	58.2	67.9	87.8	108	129	171	203	268	301	424	458
75.0	3.53	12.1	22.1	3.60	12.0	21.6	31.7	52.7	74.5	85.5	108	130	153	199	234	303	339	468	504
50.0	8.30	20.1	32.1	8.66	21.0	33.4	45.9	70.9	95.9	108	133	158	183	233	271	346	383	521	558
25.0	15.9	30.3	43.3	17.3	33.7	49.0	63.9	92.8	121	135	163	190	218	272	312	392	432	577	617
10.0	25.0	40.6	53.9	28.8	48.6	66.5	83.5	116	147	162	193	222	252	309	352	437	478	631	672
5.0	31.2	47.1	59.9	37.5	59.3	78.7	96.9	131	164	180	212	243	274	334	378	465	509	665	707
1.0	43.8	58.8	70.7	57.6	83.0	105	126	164	200	218	252	285	318	382	429	522	568	732	776
	2.5	10	X	2.5	10	15	25	40	X	65	X	100	X	150	X	250	X	400	X
	Acceptable Quality Levels (tightened inspection)																		

TABLE X-D-2 — SAMPLING PLANS FOR SAMPLE SIZE CODE LETTER: D

Acceptable Quality Levels (normal inspection); each AQL cell gives Ac Re

Type of sampling plan	Cum-lative sample size	Less than 1.5	1.5	2.5	><	4.0	6.5	10	15	25	40	><	65	><	100	><	150	><	250	><
		Ac Re	Ac Re	Ac Re	Ac Re	Ac Re	Ac Re	Ac Re	Ac Re	Ac Re	Ac Re	Ac Re	Ac Re	Ac Re	Ac Re	Ac Re	Ac Re	Ac Re	Ac Re	Ac Re
Single	8	▽	0 1	Use Letter C	Use Letter F	Use Letter E	1 2	2 3	3 4	5 6	7 8	8 9	10 11	12 13	14 15	18 19	21 22	27 28	30 31	41 42
Double	5	▽	*				0 2	0 3	1 4	2 5	3 7	3 7	5 9	6 10	7 11	9 14	11 16	15 20	17 22	23 29
	10						1 2	3 4	4 5	6 7	8 9	11 12	12 13	15 16	18 19	23 24	26 27	34 35	37 38	52 53
Multiple	2	▽	*				# 2	# 2	# 3	# 4	0 4	0 4	0 5	0 6	1 7	1 8	2 9	3 10	4 12	6 15
	4						# 2	0 3	0 3	1 5	1 6	2 7	3 8	3 9	4 10	6 12	7 14	10 17	11 19	16 25
	6						0 2	0 3	1 4	2 6	3 8	4 9	6 10	7 12	8 13	11 17	13 19	17 24	19 27	26 36
	8						0 3	1 4	2 5	3 7	5 10	6 11	8 13	10 15	12 17	16 22	19 25	24 31	27 34	37 46
	10						1 3	2 4	3 6	5 8	7 11	9 12	11 15	14 17	17 20	22 25	25 29	32 37	36 40	49 55
	12						1 3	3 5	4 6	7 9	10 12	12 14	14 17	18 20	21 23	27 29	31 33	40 43	45 47	61 64
	14						2 3	4 5	6 7	9 10	13 14	14 15	18 19	21 22	25 26	32 33	37 38	48 49	53 54	72 73
Acceptable Quality Levels (tightened inspection)		Less than 2.5	2.5	><	4.0	6.5	10	15	25	40	><	65	><	100	><	150	><	250	><	400

D

△ = Use next preceding sample size code for which acceptance and rejection numbers are available.
▽ = Use next subsequent sample size code for which acceptance and rejection numbers are available.
Ac = Acceptance number
Re = Rejection number
* = Use single sampling plan above (or alternatively use letter J)
# = Acceptance not permitted at this sample size.

TABLE X-E — Tables for sample size code letter: E

CHART E — OPERATING CHARACTERISTIC CURVES FOR SINGLE SAMPLING PLANS

(Curves for double and multiple sampling are matched as closely as practicable)

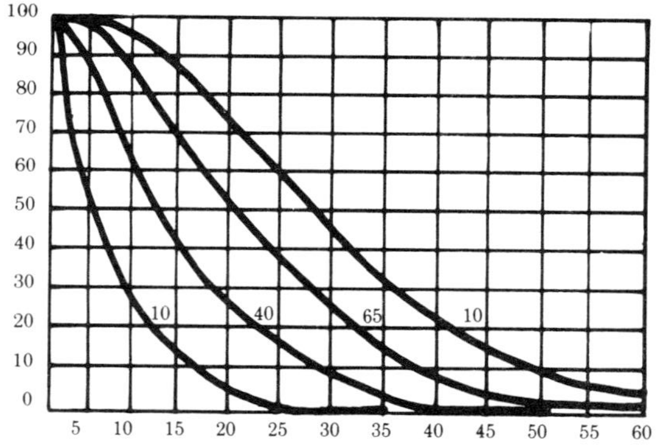

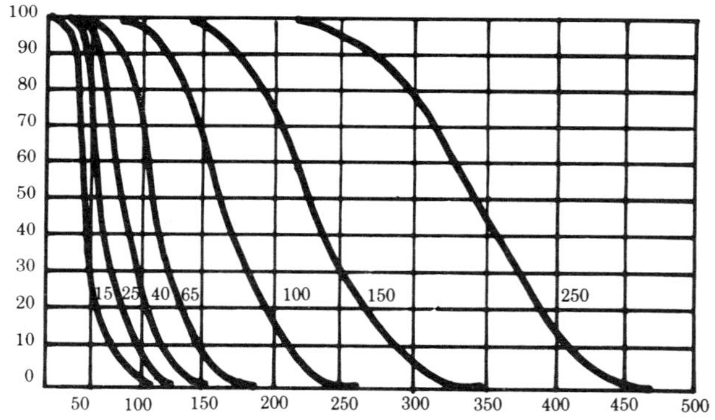

QUALITY OF SUBMITTED LOTS (P_a in percent defective for AQL's ≤ 10; in defects per hundred units for AQL's > 10)

Note: Figures on curves are Acceptable Quality Levels (AQL's) for normal inspection.

TABLE X-E-1 — TABULATED VALUES FOR OPERATING CHARACTERISTIC CURVES FOR SINGLE SAMPLING PLANS

P_a	Acceptable Quality Levels (normal inspection)																			
	1.0	4.0	6.5	10	1.0	4.0	6.5	10	15	25	X	40	X	65	X	100	X	150	X	250
	p (in percent defective)				p (in defects per hundred units)															
99.0	0.077	1.19	3.63	7.00	0.078	1.15	3.35	6.33	13.7	22.4	27.0	36.7	46.9	57.5	79.6	96.7	132	150	219	238
95.0	0.394	2.81	6.63	11.3	0.395	2.73	6.29	10.5	20.1	30.6	36.1	47.5	59.2	71.1	95.7	115	153	173	246	266
90.0	0.807	4.16	8.80	14.2	0.808	4.09	8.48	13.4	24.2	35.8	41.8	54.0	66.5	79.2	105	125	165	185	261	282
75.0	2.19	7.41	13.4	19.9	2.22	7.39	13.3	19.5	32.5	45.8	52.6	66.3	80.2	94.1	122	144	187	208	288	310
50.0	5.19	12.6	20.0	27.5	5.33	12.9	20.6	28.2	43.6	59.0	66.7	82.1	97.5	113	144	168	213	236	321	344
25.0	10.1	19.4	28.0	36.2	10.7	20.7	30.2	39.3	57.1	74.5	83.1	100	117	134	167	192	241	266	355	379
10.0	16.2	26.8	36.0	44.4	17.7	29.9	40.9	51.4	71.3	90.5	100	119	137	155	190	217	269	295	388	414
5.0	20.6	31.6	41.0	49.5	23.0	36.5	48.4	59.6	80.9	101	111	130	150	168	205	233	286	313	409	435
1.0	29.8	41.5	50.6	58.7	35.4	51.1	64.7	77.3	101	123	134	155	176	196	235	264	321	349	450	477
	1.5	6.5	10	X	1.5	6.5	10	15	25	X	40	X	65	X	100	X	150	X	250	X
	Acceptable Quality Levels (tightened inspection)																			

Note: Binomial distribution used for percent defective computations; Poisson for defects per hundred units.

TABLE X-E-2 — SAMPLING PLANS FOR SAMPLE SIZE CODE LETTER: E

Type of sampling plan	Cumulative sample size	Acceptable Quality Levels (normal inspection)																					Cumulative sample size
		Less than 1.0	1.0	1.5	X	2.5	4.0	6.5	10	15	25	X	40	X	65	X	100	X	150	X	250	Higher than 250	
		Ac Re	Ac Re	Ac Re	Ac Re	Ac Re	Ac Re	Ac Re	Ac Re	Ac Re	Ac Re	Ac Re	Ac Re	Ac Re	Ac Re	Ac Re	Ac Re	Ac Re	Ac Re	Ac Re	Ac Re	Ac Re	
Single	13	▽	0 1	Use Letter D	Use Letter G	Use Letter F	1 2	2 3	3 4	5 6	7 8	8 9	10 11	12 13	14 15	18 19	21 22	27 28	30 31	41 42	44 45	△	13
Double	8	▽	*				0 2	0 3	1 4	2 5	3 7	3 7	5 9	6 10	7 11	9 14	11 16	15 20	17 22	23 29	25 31	△	8
	16						1 2	3 4	4 5	6 7	8 9	11 12	12 13	15 16	18 19	23 24	26 27	34 35	37 38	52 53	56 47		16
Multiple	3	▽	*				# 2	# 2	# 3	# 4	0 4	0 4	0 5	0 6	1 7	1 8	2 9	3 10	4 12	6 15	6 16	△	3
	6						# 2	0 3	0 3	1 5	1 6	2 7	3 8	3 9	4 10	6 12	7 14	10 17	11 19	16 25	17 27		6
	9						0 2	0 3	1 4	2 6	3 8	4 9	6 10	7 12	8 13	11 17	13 19	17 24	19 27	26 36	29 39		9
	12						0 3	1 4	2 5	3 7	5 10	6 11	8 13	10 15	12 17	16 22	19 25	24 31	27 34	37 46	40 49		12
	15						1 3	2 4	3 6	5 8	7 11	9 12	11 15	14 17	17 20	22 25	25 29	32 37	36 40	49 55	53 58		15
	18						1 3	3 5	4 6	7 9	10 12	12 14	14 17	18 20	21 23	27 29	31 33	40 43	45 47	61 64	65 68		18
	21						2 3	4 5	6 7	9 10	13 14	14 15	18 19	21 22	25 26	32 33	37 38	48 49	53 54	72 73	77 78		21
		Less than 1.5	1.5	X	2.5	4.0	6.5	10	15	25	X	40	X	65	X	100	X	150	X	250	X	Higher than 250	
		Acceptable Quality Levels (tightened inspection)																					

△ = Use next preceding sample size code letter for which acceptance and rejection numbers are available.
▽ = Use next subsequent sample size code letter for which acceptance and rejection numbers are available.
Ac = Acceptance number.
Re = Rejection number.
* = Use single sampling plan above (alternatively use letter H).
# = Acceptance not permitted at this sample size.

E

TABLE X-F — Tables for sample size code letter: F

CHART F — OPERATING CHARACTERISTIC CURVES FOR SINGLE SAMPLING PLANS

(Curves for double and multiple sampling are matched as closely as practicable)

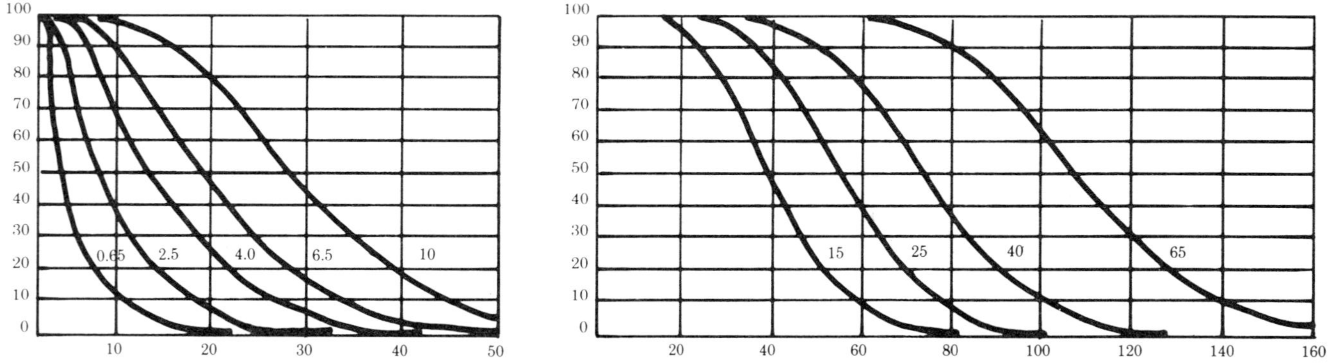

QUALITY OF SUBMITTED LOTS (P_a in percent defective for AQL's ≤10; in defects per hundred units for AQL's >10)
NOTE: Figures on curves are Acceptable Quality Levels (AQL's) for normal inspection.

TABLE X-F-1 — TABULATED VALUES FOR OPERATING CHARACTERISTIC CURVES FOR SINGLE SAMPLING PLANS

P_a	Acceptable Quality Levels (normal inspection)																
	0.65	2.5	4.0	6.5	10	0.65	2.5	4.0	6.5	10	15	×	25	×	40	×	65
	p (in percent defective)					p (in defects per hundred units)											
99.0	0.050	0.75	2.25	4.31	9.75	0.051	0.75	2.18	4.12	8.92	14.5	17.5	23.9	30.5	37.4	51.7	62.9
95.0	0.256	1.80	4.22	7.13	14.0	0.257	1.78	4.09	6.83	13.1	19.9	23.5	30.8	38.5	46.2	62.2	74.5
90.0	0.525	2.69	5.64	9.03	16.6	0.527	2.66	5.51	8.73	15.8	23.3	27.2	35.1	43.2	51.5	68.4	81.2
75.0	1.43	4.81	8.70	12.8	21.6	1.44	4.81	8.68	12.7	21.1	29.8	34.2	43.1	52.1	61.2	79.5	93.4
50.0	3.41	8.25	13.1	18.1	27.9	3.47	8.39	13.4	18.4	28.4	38.3	43.3	53.3	63.3	73.3	93.3	108
25.0	6.70	12.9	18.7	24.2	34.8	6.93	13.5	19.6	25.5	37.1	48.4	54.0	65.1	76.1	87.0	109	125
10.0	10.9	18.1	24.5	30.4	41.5	11.5	19.5	26.6	33.4	46.4	58.9	65.0	77.0	88.9	101	124	141
5.0	13.9	21.6	28.3	34.4	45.6	15.0	23.7	31.5	38.8	52.6	65.7	72.2	84.8	97.2	109	133	151
1.0	20.6	28.9	35.6	42.0	53.4	23.0	33.2	42.0	50.2	65.5	80.0	87.0	101	114	127	153	172
	1.0	4.0	6.5	10	×	1.0	4.0	6.5	10	15	×	25	×	40	×	65	×
	Acceptable Quality Levels (tightened inspection)																

Note: Binomial distribution used for percent defective computations; Poisson for defects per hundred units.

TABLE X-F-2 — SAMPLING PLANS FOR SAMPLE SIZE CODE LETTER: F

Type of sampling plan	Cumulative sample size	Less than 0.65	0.65	1.0	X	1.5	2.5	4.0	6.5	10	15	X	25	X	40	X	65	Higher than 65	Cumulative sample size
		Acceptable Quality Levels (normal inspection)																	
		Ac Re	Ac Re	Ac Re	Ac Re	Ac Re	Ac Re	Ac Re	Ac Re	Ac Re	Ac Re	Ac Re	Ac Re	Ac Re	Ac Re	Ac Re	Ac Re	Ac Re	
Single	20	▽	0 1	Use Letter E	Use Letter H	Use Letter G	1 2	2 3	3 4	5 6	7 8	8 9	10 11	12 13	14 15	18 19	21 22	△	20
Double	13	▽	*				0 2	0 3	1 4	2 5	3 7	3 7	5 9	6 10	7 11	9 14	11 16	△	13
	26						1 2	3 4	4 5	6 7	8 9	11 12	12 13	15 16	18 19	23 24	26 27		26
Multiple	5	▽	*				# 2	# 2	# 3	# 4	0 4	0 4	0 5	0 6	1 7	1 8	2 9	△	5
	10						# 2	0 3	0 3	1 5	1 6	2 7	3 8	3 9	4 10	6 12	7 14		10
	15						0 2	0 3	1 4	2 6	3 8	4 9	6 10	7 12	8 13	11 17	13 19		15
	20						0 3	1 4	2 5	3 7	5 10	6 11	8 13	10 15	12 17	16 22	19 25		20
	25						1 3	2 4	3 6	5 8	7 11	9 12	11 15	14 17	17 20	22 25	25 29		25
	30						1 3	3 5	4 6	7 9	10 12	12 14	14 17	18 20	21 23	27 29	31 33		30
	35						2 3	4 5	6 7	9 10	13 14	14 15	18 19	21 22	25 26	32 33	37 38		35
		Less than 1.0	1.0	X	1.5	2.5	4.0	6.5	10	15	X	25	X	40	X	65	X	Higher than 65	
		Acceptable Quality Levels (tightened inspection)																	

F

△ = Use next preceding sample size code for which acceptance and rejection numbers are available.
▽ = Use next subsequent sample size code for which acceptance and rejection numbers are available.
Ac = Acceptance number
Re = Rejection number
* = Use single sampling plan above (or alternatively use letter J)
# = Acceptance not permitted at this sample size.

TABLE X-G — Tables for sample size code letter: G

CHART G — OPERATING CHARACTERISTIC CURVES FOR SINGLE SAMPLING PLANS

(Curves for double and multiple sampling are matched as closely as practicable)

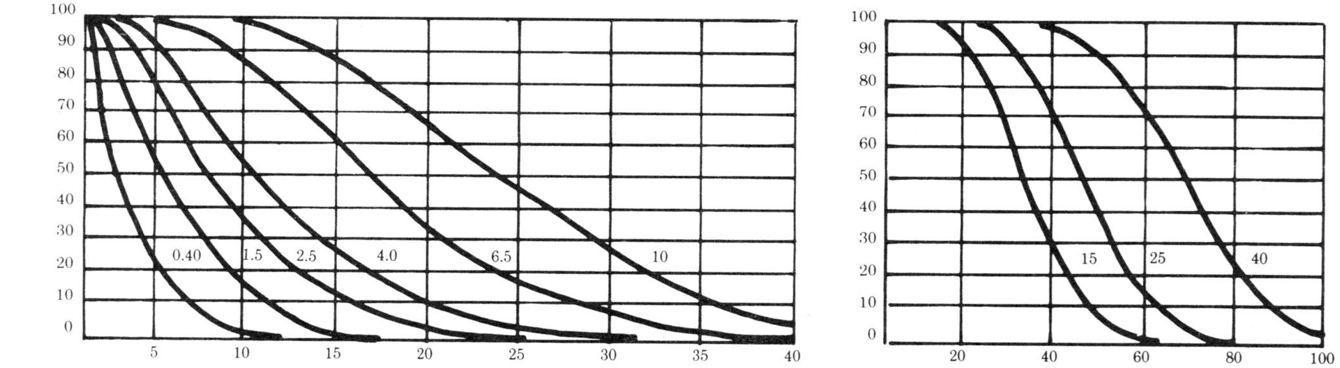

QUALITY OF SUBMITTED LOTS (P_a in percent defective for AQL's ≤10; in defects per hundred units for AQL's >10)

Note: Figures on curves are Acceptable Quality Levels (AQL's) for normal inspection.

TABLE X-G-1 — TABULATED VALUES FOR OPERATING CHARACTERISTIC CURVES FOR SINGLE SAMPLING PLANS

P_a	Acceptable Quality Levels (normal inspection)																	
	0.40	1.5	2.5	4.0	6.5	10	0.40	1.5	2.5	4.0	6.5	10	><	15	><	25	><	40
	p (in percent defective)						p (in defects per hundred units)											
99.0	0.032	0.475	1.38	2.63	5.94	9.75	0.032	0.466	1.36	2.57	5.57	9.08	11.0	14.9	19.1	23.4	32.3	39.3
95.0	0.161	1.13	2.59	4.39	8.50	13.1	0.160	1.10	2.55	4.26	8.16	12.4	14.7	19.3	24.0	28.9	38.9	46.5
90.0	0.329	1.67	3.50	5.56	10.2	15.1	0.328	1.66	3.44	5.45	9.85	14.6	17.0	21.9	27.0	32.2	42.7	50.8
75.0	0.823	3.01	5.42	7.98	13.4	19.0	0.900	3.00	5.39	7.92	13.2	18.6	21.4	26.9	32.6	38.2	49.7	58.4
50.0	2.14	3.19	8.27	11.4	17.5	23.7	2.16	5.24	8.35	11.5	17.7	24.0	27.1	33.3	39.6	45.8	58.3	67.7
25.0	4.23	8.19	11.9	15.4	22.3	29.0	4.33	8.41	12.3	16.0	23.2	30.3	33.8	40.7	47.6	54.4	67.9	78.0
10.0	6.94	11.6	15.8	19.7	27.1	34.1	7.19	12.2	16.6	20.9	29.0	36.8	40.6	48.1	55.6	62.9	77.4	88.1
5.0	8.94	14.0	18.4	22.5	30.1	37.2	9.36	14.8	19.7	24.2	32.9	41.1	45.1	53.0	60.8	68.4	83.4	94.5
1.0	13.5	19.0	23.7	28.0	35.9	43.3	14.4	20.7	26.3	31.4	41.0	50.0	54.4	63.0	71.3	79.5	95.6	107
	0.65	2.5	4.0	6.5	10	><	0.65	2.5	4.0	6.5	10	><	15	><	25	><	40	><
	Acceptable Quality Levels (tightened inspection)																	

Note: Binominal distribution used for percent defective computations; Poisson for defects per hundred units.

TABLE X-G-2 — SAMPLING PLANS FOR SAMPLE SIZE CODE LETTER: G

Type of sampling plan	Cumulative sample size	Acceptable Quality Levels (normal inspection)																Higher than 40	Cumulative sample size
		0.40	0.40	0.65	><	1.0	1.5	2.5	4.0	6.5	10	><	15	><	25	><	40		
		Ac Re	Ac Re	Ac Re	Ac Re	Ac Re	Ac Re	Ac Re	Ac Re	Ac Re	Ac Re	Ac Re	Ac Re	Ac Re	Ac Re	Ac Re	Ac Re	Ac Re	
Single	32	▽	0 1	Use Letter F	Use Letter J	Use Letter H	1 2	2 3	3 4	5 6	7 8	8 9	10 11	12 13	14 15	18 19	21 22	△	32
Double	20	▽	*				0 2	0 3	1 4	2 5	3 7	3 7	5 9	10	7 11	9 14	11 16	△	20
	40						1 2	3 4	4 5	6 7	8 9	11 12	12 13	15 16	18 19	23 24	26 27		40
Multiple	8	▽	*				# 2	# 2	# 3	# 4	0 4	0 4	0 5	0 6	1 7	1 8	2 9	△	8
	16						# 2	0 3	0 3	1 5	1 6	2 7	3 8	3 9	4 10	6 12	7 14		16
	24						0 2	0 3	1 4	2 6	3 8	4 9	6 10	7 12	8 13	11 17	13 19		24
	32						0 3	1 4	2 5	3 7	5 10	6 11	8 13	10 15	12 17	16 22	19 25		32
	40						1 3	2 4	3 6	5 8	7 11	9 12	11 15	14 17	17 20	22 25	25 29		40
	48						1 3	3 5	4 6	7 9	10 12	12 14	14 17	18 20	21 23	27 29	31 33		48
	56						2 3	4 5	6 7	9 10	13 14	14 15	18 19	21 22	25 26	32 33	37 38		56
		Less than 0.65	0.65	><	1.0	1.5	2.5	4.0	6.5	10	><	15	><	25	><	40	><	Higher than 40	
		Acceptable Quality Levels (tightened inspection)																	

G

△ = Use next preceding sample size code for which acceptance and rejection numbers are available.
▽ = Use next subsequent sample size code for which acceptance and rejection numbers are available.
Ac = Acceptance number
Re = Rejection number
* = Use single sampling plan above (or alternatively use letter K)
# = Acceptance not permitted at this sample size.

TABLE X-H — Tables for sample size code letter: H

CHART H — OPERATING CHARACTERISTIC CURVES FOR SINGLE SAMPLING PLANS

(Curves for double and multiple sampling are matched as closely as practicable)

PERCENT OF LOTS EXPECTED TO BE ACCEPTED (P_a)

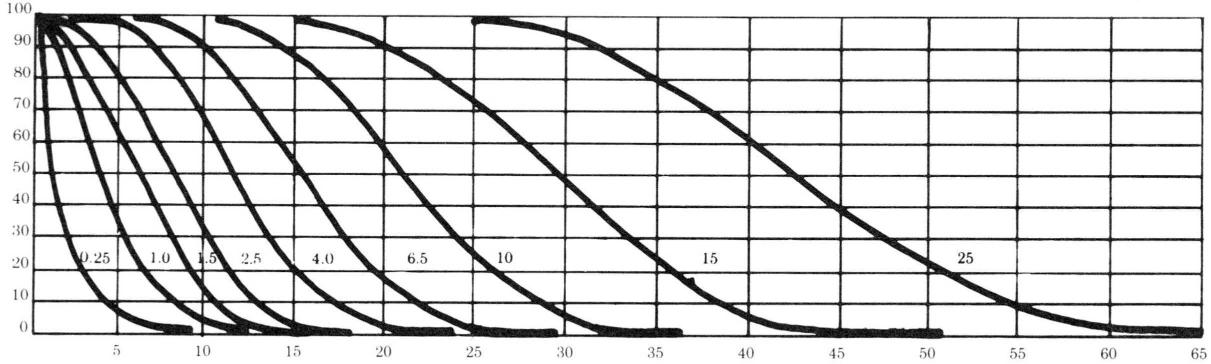

QUALITY OF SUBMITTED LOTS (P_a in percent defective for AQL's ≤ 10; in defects per hundred units for AQL's > 10)

NOTE: Figures on curves are Acceptable Quality Levels (AQL's) for normal inspection.

TABLE X-H-1 — TABULATED VALUES FOR OPERATING CHARACTERISTIC CURVES FOR SINGLE SAMPLING PLANS

P_a	Acceptable Quality Levels (normal inspection)																			
	0.25	1.0	1.5	2.5	4.0	6.5	X	10	0.25	1.0	1.5	2.5	4.0	6.5	X	10	X	15	X	25
	p (in percent defective)								p (in defects per hundred units)											
99.0	0.020	0.306	0.888	1.69	3.66	6.06	7.41	11.1	0.020	0.298	0.872	1.65	3.57	5.81	7.01	9.54	12.2	15.0	20.7	25.1
95.0	0.103	0.712	1.66	2.77	5.34	8.20	9.74	12.9	0.103	0.710	1.64	2.73	5.23	7.96	9.39	12.3	15.4	18.5	24.9	29.8
90.0	0.210	1.07	2.23	3.54	6.42	9.53	11.2	14.5	0.210	1.06	2.20	3.49	6.30	9.31	10.9	14.0	17.3	20.6	27.3	32.5
75.0	0.574	1.92	3.46	5.09	8.51	12.0	13.8	17.5	0.576	1.92	3.45	5.07	8.44	11.9	13.7	17.2	20.8	24.5	31.8	37.4
50.0	1.38	3.33	5.31	7.30	11.3	15.2	17.2	21.2	1.39	3.36	5.35	7.34	11.3	15.3	17.3	21.6	25.3	29.3	37.3	43.3
25.0	2.74	5.30	7.70	10.0	14.5	18.8	21.0	25.2	2.77	5.39	7.84	10.2	14.8	19.4	21.6	26.0	30.4	34.8	43.5	49.9
10.0	4.50	7.56	10.3	12.9	17.8	22.4	24.7	29.1	4.61	7.78	10.6	13.4	18.6	23.5	26.0	30.8	35.6	40.3	49.5	56.4
5.0	5.82	9.13	12.1	14.8	19.9	24.7	27.0	31.6	5.99	9.49	12.6	15.5	21.0	26.3	28.9	33.9	38.9	43.8	53.4	60.5
1.0	8.80	12.5	15.9	18.8	24.3	29.2	31.7	36.3	9.21	13.3	16.8	20.1	26.2	32.0	34.8	40.3	45.6	50.9	61.1	68.7
	0.40	1.5	2.5	4.0	6.5	X	10	X	0.40	1.5	2.5	4.0	6.5	X	10	X	15	X	25	X
	Acceptable Quality Levels (tightened inspection)																			

Note: Binomial distribution used for percent defective computations. Poisson for defects per hundred units.

TABLE X-H-2 — SAMPLING PLANS FOR SAMPLE SIZE CODE LETTER: H

Type of sampling plan	Cumu-lative sample size	Acceptable Quality Levels (normal inspection)																	Cumu-lative sample size
		Less than 0.25	0.25	0.40	><	0.65	1.0	1.5	2.5	4.0	6.5	><	10	><	15	><	25	Higher than 25	
		Ac Re	Ac Re	Ac Re	Ac Re	Ac Re	Ac Re	Ac Re	Ac Re	Ac Re	Ac Re	Ac Re	Ac Re	Ac Re	Ac Re	Ac Re	Ac Re	Ac Re	
Single	50	▽	0 1	Use Letter G	Use Letter K	Use Letter J	1 2	2 3	3 4	5 6	7 8	8 9	10 11	12 13	14 15	18 19	21 22	△	50
Double	32	▽	*				0 2	0 3	1 4	2 5	3 7	3 7	5 9	6 10	7 11	9 14	11 16	△	32
	64		*				1 2	3 4	4 5	6 7	8 9	11 12	12 13	15 16	18 19	23 24	26 27		64
Multiple	13	▽	*				# 2	# 2	# 3	# 4	0 4	0 4	0 5	0 6	1 7	1 8	2 9	△	13
	26						# 2	0 3	0 3	1 5	1 6	2 7	3 8	3 9	4 10	6 12	7 14		26
	39						0 2	0 3	1 4	2 6	3 8	4 9	6 10	7 12	8 13	11 17	13 19		39
	52						0 3	1 4	2 5	3 7	5 10	6 11	8 13	10 15	12 17	16 22	19 25		52
	65						1 3	2 4	3 6	5 8	7 11	9 12	11 15	14 17	17 20	22 25	25 29		65
	78						1 3	3 5	4 6	7 9	10 12	12 14	14 17	18 20	21 23	27 29	31 33		78
	91						2 3	4 5	6 7	9 10	13 14	14 15	18 19	21 22	25 26	32 33	37 38		91
		Less than 0.40	0.40	><	0.65	1.0	1.5	2.5	4.0	6.5	><	10	><	15	><	25	><	Higher than 25	
		Acceptable Quality Levels (tightened inspection)																	

△ = Use next preceding sample size code for which acceptance and rejection numbers are available.

▽ = Use next subsequent sample size code for which acceptance and rejection numbers are available.

Ac = Acceptance number

Re = Rejection number

* = Use single sampling plan above (or alternatively use letter L)

\# = Acceptance not permitted at this sample size.

H

TABLE X-J — Tables for sample size code letter: J

CHART J — OPERATING CHARACTERISTIC CURVES FOR SINGLE SAMPLING PLANS

(Curves for double and multiple sampling are matched as closely as practicable)

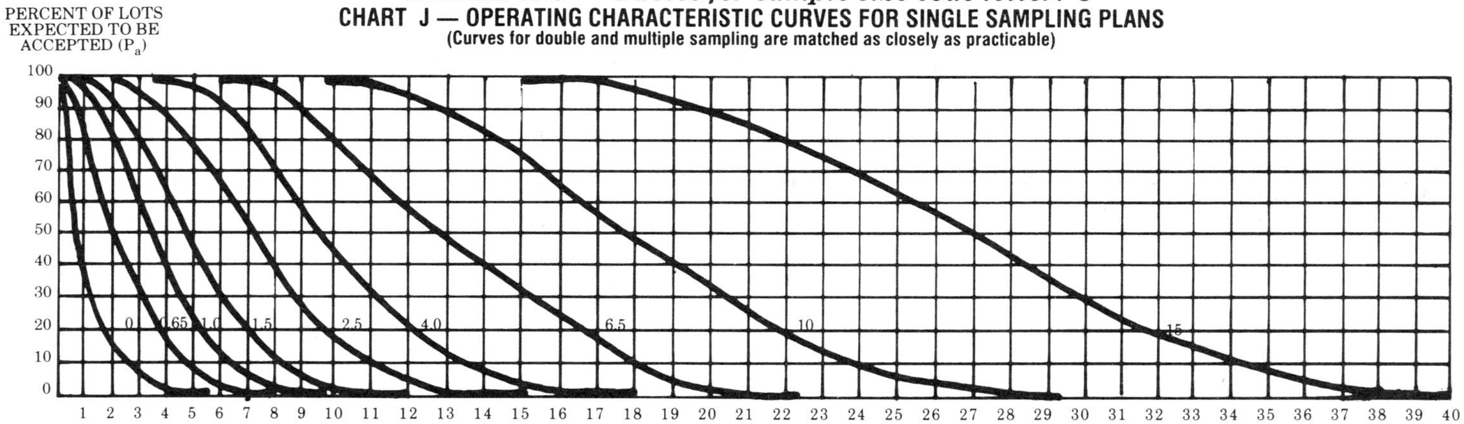

QUALITY OF SUBMITTED LOTS (P_a in percent defective for AQL's ≤10; in defects per hundred units for AQL's >10)
NOTE: Figures on curves are Acceptable Quality Levels (AQL's) for normal inspection.

TABLE X-J-1 — TABULATED VALUES FOR OPERATING CHARACTERISTIC CURVES FOR SINGLE SAMPLING PLANS

P_a	Acceptable Quality Levels (normal inspection)																					
	0.15	0.65	1.0	1.5	2.5	4.0	><	6.5	><	10	0.15	0.65	1.0	1.5	2.5	4.0	><	6.5	><	10	><	15
	p (in percent defective)										p (in defects per hundred units)											
99.0	0.013	0.188	0.550	1.05	2.30	3.72	4.50	6.13	7.88	9.75	0.013	0.186	0.545	1.03	2.23	3.63	4.38	5.96	7.62	9.35	12.9	15.7
95.0	0.064	0.444	1.03	1.73	3.32	5.06	5.98	7.91	9.89	11.9	0.064	0.444	1.02	1.71	3.27	4.98	5.87	7.71	9.61	11.6	15.6	18.6
90.0	0.132	0.666	1.38	2.20	3.98	5.91	6.91	8.95	11.0	13.2	0.131	0.665	1.38	2.18	3.94	5.82	6.79	8.78	10.8	12.9	17.1	20.3
75.0	0.359	1.202	2.16	3.18	5.30	7.50	8.62	1p.9	13.2	15.5	0.360	1.20	2.16	3.17	5.27	7.45	8.55	10.8	13.0	15.3	19.9	23.4
50.0	0.863	2.09	3.33	4.57	7.06	9.55	10.8	13.3	15.8	18.3	0.866	2.10	3.34	4.59	7.09	9.59	10.8	13.3	15.8	18.3	23.3	27.1
25.0	1.72	3.33	4.84	6.31	9.14	11.9	13.3	16.0	18.6	21.3	1.73	3.37	4.90	6.39	9.28	12.1	13.5	16.3	19.0	21.8	27.2	31.2
10.0	2.84	4.78	6.52	8.16	11.3	14.2	15.7	18.6	21.4	24.2	2.88	4.86	6.65	8.35	11.6	14.7	16.2	19.3	22.2	25.2	30.9	35.2
5.0	3.68	5.80	7.66	9.39	12.7	15.8	17.3	20.3	23.2	26.0	3.75	5.93	7.87	9.69	13.1	16.4	18.0	21.2	24.3	27.4	33.4	37.8
1.0	5.59	8.00	10.1	12.0	15.6	18.9	20.5	23.6	26.5	29.5	5.76	8.30	10.5	12.6	16.5	20.0	21.8	25.2	28.5	31.8	38.2	42.9
	0.25	1.0	1.5	2.5	4.0	><	6.5	><	10	><	0.25	1.0	1.5	2.5	4.0	><	6.5	><	10	><	15	><
	Acceptable Quality Levels (tightened inspection)																					

Note: All values given in above table based on Poisson distribution as an approximation to the Binominal

TABLE X-J-2 — SAMPLING PLANS FOR SAMPLE SIZE CODE LETTER: J

Type of sampling plan	Cumulative sample size	Less than 0.15	0.15	0.25	X	0.40	0.65	1.0	1.5	2.5	4.0	X	6.5	X	10	X	15	Higher than 15	Cumulative sample size
		Acceptable Quality Levels (normal inspection)																	
		Ac Re	Ac Re	Ac Re	Ac Re	Ac Re	Ac Re	Ac Re	Ac Re	Ac Re	Ac Re	Ac Re	Ac Re	Ac Re	Ac Re	Ac Re	Ac Re	Ac Re	
Single	80	▽	0 1				1 2	2 3	3 4	5 6	7 8	8 9	10 11	12 13	14 15	18 19	21 22	△	80
Double	50	▽	*	Use Letter H	Use Letter L	Use Letter K	0 2	0 3	1 4	2 5	3 7	3 7	5 9	6 10	7 11	9 14	11 16	△	50
	100						1 2	3 4	4 5	6 7	8 9	11 12	12 13	15 16	18 19	23 24	26 27		100
Multiple	20	▽	*				# 2	# 2	# 3	# 4	0 4	0 4	0 5	0 6	1 7	1 8	2 9	△	20
	40						# 2	0 3	0 3	1 5	1 6	2 7	3 8	3 9	4 10	6 12	7 14		40
	60						0 2	0 3	1 4	2 6	3 8	4 9	6 10	7 12	8 13	11 17	13 19		60
	80						0 3	1 4	2 5	3 7	5 10	6 11	8 13	10 15	12 17	16 22	19 25		80
	100						1 3	2 4	3 6	5 8	7 11	9 12	11 15	14 17	17 20	22 25	25 29		100
	120						1 3	3 5	4 6	7 9	10 12	12 14	14 17	18 20	21 23	27 29	31 33		120
	140						2 3	4 5	6 7	9 10	13 14	14 15	18 19	21 22	25 26	32 33	37 38		140
		Less than 0.25	0.25	X	0.40	0.65	1.0	1.5	2.5	4.0	X	6.5	X	10	X	15	X	Higher than 15	
		Acceptable Quality Levels (tightened inspection)																	

△ = Use next preceding sample size code for which acceptance and rejection numbers are available.

▽ = Use next subsequent sample size code for which acceptance and rejection numbers are available.

Ac = Acceptance number

Re = Rejection number

* = Use single sampling plan above (or alternatively use letter M)

= Acceptance not permitted at this sample size.

J

TABLE X-K — Tables for sample size code letter: K

CHART K — OPERATING CHARACTERISTIC CURVES FOR SINGLE SAMPLING PLANS

(Curves for double and multiple sampling are matched as closely as practicable)

PERCENT OF LOTS EXPECTED TO BE ACCEPTED (P_a)

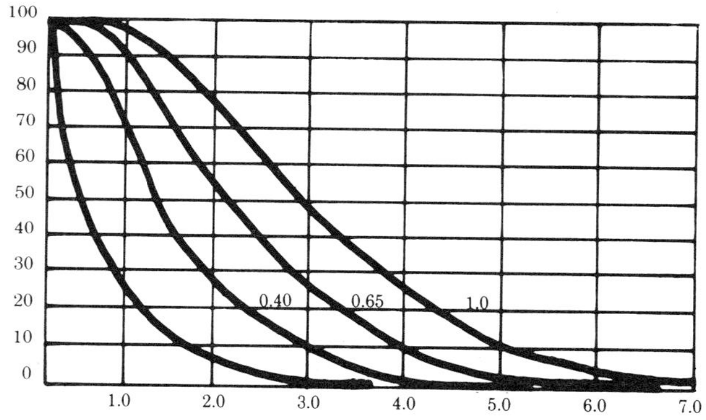

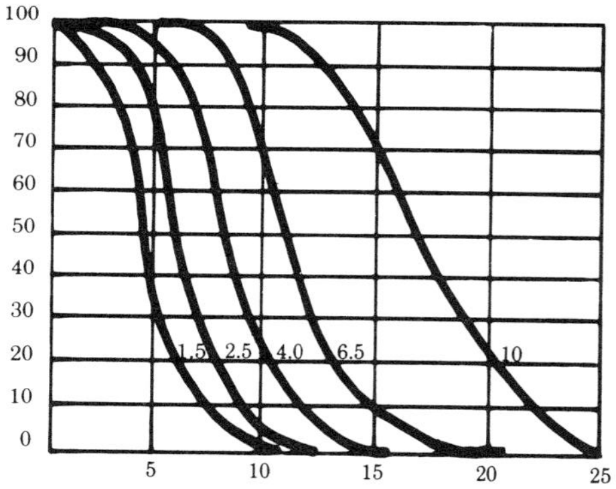

QUALITY OF SUBMITTED LOTS (P_a in percent defective for AQL's ≤ 10; in defects per hundred units for AQL's > 10)
NOTE: Figures on curves are Acceptable Quality Levels (AQL's) for normal inspection)

TABLE X-K-1 — TABULATED VALUES FOR OPERATING CHARACTERISTIC CURVES FOR SINGLE SAMPLING PLANS

P_a	Acceptable Quality Levels (normal inspection)											
	0.10	0.40	0.65	1.0	1.5	2.5	X	4.0	X	6.5	X	10
	p (in percent defective or defects per hundred units											
99.0	0.0081	0.119	0.349	0.658	1.43	2.33	2.81	3.82	4.88	5.98	8.28	10.1
95.0	0.0410	0.284	0.654	1.09	2.09	3.19	3.76	4.94	6.15	7.40	9.95	11.9
90.0	0.0840	0.426	0.882	1.40	2.52	3.73	4.35	5.62	6.92	8.24	10.9	13.0
75.0	0.230	0.769	0.382	2.03	3.38	4.77	5.47	6.90	8.34	9.79	12.7	14.9
50.0	0.554	1.34	2.14	2.94	4.54	6.14	6.94	8.53	10.1	11.7	14.9	17.3
25.0	1.11	2.15	3.14	4.09	5.94	7.75	8.64	10.4	12.2	13.9	17.4	20.0
10.0	1.84	3.11	4.26	5.35	7.42	9.42	10.4	12.3	14.2	16.1	19.8	22.5
5.0	2.40	3.80	5.04	6.20	8.41	10.5	11.5	13.6	15.6	17.5	21.4	24.2
1.0	3.68	5.31	6.73	8.04	10.5	12.8	18.3	16.1	18.3	20.4	24.5	27.5
	0.15	0.65	1.0	1.5	2.5	X	4.0	X	6.5	X	10	X
	Acceptable Quality Levels (tightened inspection)											

Note: All values given in above table based on Poisson distribution as an approximation to the Binominal

TABLE X-K-2 — SAMPLING PLANS FOR SAMPLE SIZE CODE LETTER: K

Type of sampling plan	Cumulative sample size	Acceptable Quality Levels (normal inspection)																	Cumulative sample size
		Less than 0.10	0.10	0.15	×	0.25	0.40	0.65	1.0	1.5	2.5	×	4.0	×	6.5	×	10	Higher than 10	
		Ac Re	Ac Re	Ac Re	Ac Re	Ac Re	Ac Re	Ac Re	Ac Re	Ac Re	Ac Re	Ac Re	Ac Re	Ac Re	Ac Re	Ac Re	Ac Re	Ac Re	
Single	125	▽	0 1	Use Letter J	Use Letter M	Use Letter L	1 2	2 3	3 4	5 6	7 8	8 9	10 11	12 13	14 15	18 19	21 22	△	125
Double	80	▽	*				0 2	0 3	1 4	2 5	3 7	3 7	5 9	6 10	7 11	9 14	11 16	△	80
	160		*				1 2	3 4	4 5	6 7	8 9	11 12	12 13	15 16	18 19	23 24	26 27		160
Multiple	32	▽	*				# 2	# 2	# 3	# 4	0 4	0 4	0 5	0 6	1 7	1 8	2 9	△	32
	64						# 2	0 3	0 3	1 5	1 6	2 7	3 8	3 9	4 10	6 12	7 14		64
	96						0 2	0 3	1 4	2 6	3 8	4 9	6 10	7 12	8 13	11 17	13 19		96
	128						0 3	1 4	2 5	3 7	5 10	6 11	8 13	10 15	12 17	16 22	19 25		128
	160						1 3	2 4	3 6	5 8	7 11	9 12	11 15	14 17	17 20	22 25	25 29		160
	192						1 3	3 5	4 6	7 9	10 12	12 14	14 17	18 20	21 23	27 29	31 33		192
	224						2 3	4 5	6 7	9 10	13 14	14 15	18 19	21 22	25 26	32 33	37 38		224
		Less than 0.15	0.15	×	0.25	0.40	0.65	1.0	1.5	2.5	×	4.0	×	6.5	×	10	×	Higher than 10	
		Acceptable Quality Levels (tightened inspection)																	

K

△ = Use next preceding sample size code for which acceptance and rejection numbers are available.
▽ = Use next subsequent sample size code for which acceptance and rejection numbers are available.
Ac = Acceptance number
Re = Rejection number
* = Use single sampling plan above (or alternatively use letter N)
\# = Acceptance not permitted at this sample size.

TABLE X-L — Tables for sample size code letter: L

CHART L — OPERATING CHARACTERISTIC CURVES FOR SINGLE SAMPLING PLANS

(Curves for double and multiple sampling are matched as closely as practicable)

PERCENT OF LOTS EXPECTED TO BE ACCEPTED (P_a)

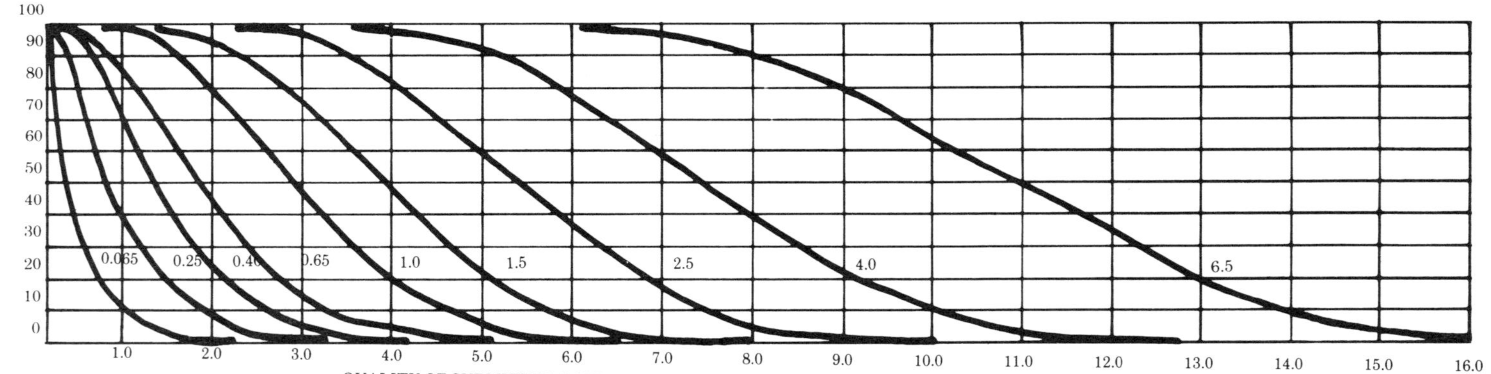

QUALITY OF SUBMITTED LOTS (P_a in percent defective for AQL's $\leq$ 10; in defects per hundred units for AQL's $>$10)

NOTE: Figures on curves are Acceptable Quality Levels (AQL's) for normal inspection)

TABLE X-L-1 — TABULATED VALUES FOR OPERATING CHARACTERISTIC CURVES FOR SINGLE SAMPLING PLANS

P_a	Acceptable Quality Levels (normal inspection)											
	0.065	0.25	0.40	0.65	1.0	1.5	×	2.5	×	4.0	×	6.5
	p (in percent defective or defects per hundred units											
99.0	0.0051	0.075	0.218	0.412	0.893	1.45	1.75	2.39	3.05	3.74	5.17	6.29
95.0	0.0256	0.178	0.409	0.683	1.31	1.99	2.35	3.09	3.85	4.62	6.22	7.45
90.0	0.0525	0.266	0.551	0.873	1.58	2.33	2.72	3.51	4.32	5.15	6.84	8.12
75.0	0.144	0.481	0.864	1.27	2.11	2.98	3.42	4.31	5.21	6.12	7.95	9.34
50.0	0.347	0.839	1.34	1.84	2.84	3.84	4.33	5.33	6.33	7.33	9.33	10.8
25.0	0.693	1.35	1.96	2.56	3.71	4.84	5.40	6.51	7.61	8.70	10.9	12.5
10.0	1.15	1.95	2.66	3.34	4.64	5.89	6.50	7.70	8.89	10.1	12.4	14.1
5.0	1.50	2.37	3.15	3.88	5.26	6.57	7.22	8.48	9.72	10.9	13.3	15.1
1.0	2.30	3.32	4.20	5.02	6.55	8.00	8.70	10.1	11.4	12.7	15.3	17.2
	0.10	0.40	0.65	1.0	1.5	×	2.5	×	4.0	×	6.5	×
	Acceptable Quality Levels (tightened inspection)											

Note: All values given in above table based on Poisson distribution as an approximation to the Binominal

TABLE X-L-2 — SAMPLING PLANS FOR SAMPLE SIZE CODE LETTER: L

Type of sampling plan	Cumulative sample size	Acceptable Quality Levels (normal inspection)																	Cumulative sample size
		Less than 0.065	0.065	0.10	X	0.15	0.25	0.40	0.65	1.0	1.5	X	2.5	X	4.0	X	6.5	Higher than 6.5	
		Ac Re	Ac Re	Ac Re	Ac Re	Ac Re	Ac Re	Ac Re	Ac Re	Ac Re	Ac Re	Ac Re	Ac Re	Ac Re	Ac Re	Ac Re	Ac Re	Ac Re	
Single	200	▽	0 1	Use Letter K	Use Letter N	Use Letter M	1 2	2 3	3 4	5 6	7 8	8 9	10 11	12 13	14 15	18 19	21 22	△	200
Double	125	▽	*				0 2	0 3	1 4	2 5	3 7	3 7	5 9	6 10	7 11	9 14	11 16	△	125
	250						1 2	3 4	4 5	6 7	8 9	11 12	12 13	15 16	18 19	23 24	26 27		250
Multiple	50	▽	*				# 2	# 2	# 3	# 4	0 4	0 4	0 5	0 6	1 7	1 8	2 9	△	50
	100						# 2	0 3	0 3	1 5	1 6	2 7	3 8	3 9	4 10	6 12	7 14		100
	150						0 2	0 3	1 4	2 6	3 8	4 9	6 10	7 12	8 13	11 17	13 19		150
	200						0 3	1 4	2 5	3 7	5 10	6 11	8 13	10 15	12 17	16 22	19 25		200
	250						1 3	2 4	3 6	5 8	7 11	9 12	11 15	14 17	17 20	22 25	25 29		250
	300						1 3	3 5	4 6	7 9	10 12	12 14	14 17	18 20	21 23	27 29	31 33		300
	350						2 3	4 5	6 7	9 10	13 14	14 15	18 19	21 22	25 26	32 33	37 38		350
		Less than 0.10	0.10	X	0.15	0.25	0.40	0.65	1.0	1.5	X	2.5	X	4.0	X	6.5	X	Higher than 6.5	
		Acceptable Quality Levels (tightened inspection)																	

△ = Use next preceding sample size code for which acceptance and rejection numbers are available.
▽ = Use next subsequent sample size code for which acceptance and rejection numbers are available.
Ac = Acceptance number
Re = Rejection number
* = Use single sampling plan above (or alternatively use letter P)
# = Acceptance not permitted at this sample size.

L

TABLE X-M — Tables for sample size code letter: M

CHART M — OPERATING CHARACTERISTIC CURVES FOR SINGLE SAMPLING PLANS

(Curves for double and multiple sampling are matched as closely as practicable)

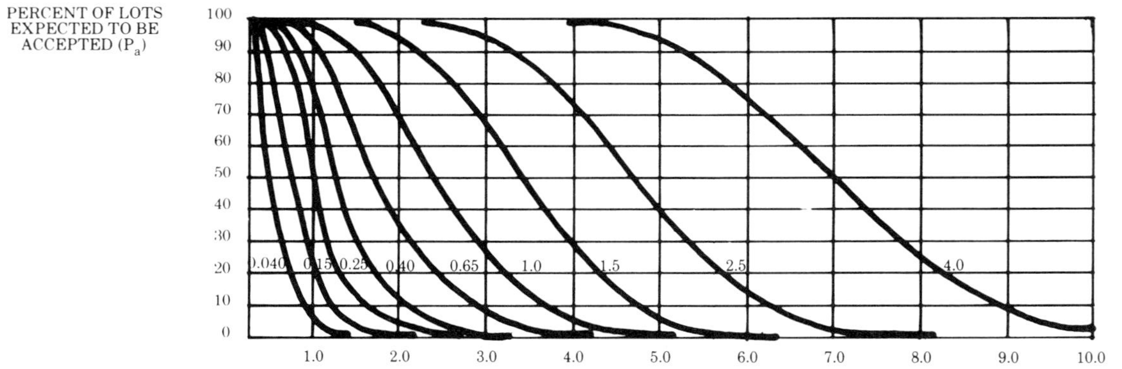

QUALITY OF SUBMITTED LOTS (P_a in percent defective for AQL's ≤ 10; in defects per hundred units for AQL's >10)

NOTE: Figures on curves are Acceptable Quality Levels (AQL's) for normal inspection)

TABLE X-M-1 — TABULATED VALUES FOR OPERATING CHARACTERISTIC CURVES FOR SINGLE SAMPLING PLANS

P_a	Acceptable Quality Levels (normal inspection)											
	0.040	0.15	0.25	0.40	0.65	1.0	X	1.5	X	2.5	X	4.0
	p (in percent defective or defects per hundred units											
99.0	0.0032	0.047	0.138	0.261	0.566	0.922	1.11	1.51	1.94	2.38	3.28	3.99
95.0	0.0163	0.112	0.259	0.433	0.829	1.26	1.49	1.96	2.44	2.94	3.95	4.73
90.0	0.0333	0.168	0.349	0.533	1.00	1.48	1.72	2.23	2.75	3.27	4.34	5.16
75.0	0.0914	0.305	0.580	0.804	1.34	1.89	2.17	2.74	3.31	3.89	5.05	5.93
50.0	0.220	0.532	0.848	1.17	1.80	2.43	2.75	3.39	4.02	4.66	5.93	6.88
25.0	0.440	0.854	1.24	1.62	2.36	3.07	3.43	4.13	4.83	5.52	6.90	7.92
10.0	0.731	1.23	1.69	2.12	2.94	3.74	4.13	4.89	5.65	6.39	7.86	8.95
5.0	0.951	1.51	2.00	2.46	3.34	4.17	4.58	5.38	6.17	6.95	8.47	9.60
1.0	1.46	2.11	2.67	3.19	4.16	5.08	5.53	6.40	7.25	8.08	9.71	10.9
	0.065	0.25	0.40	0.65	1.0	X	1.5	X	2.5	X	4.0	X
	Acceptable Quality Levels (tightened inspection)											

Note: All values given in above table based on Poisson distribution as an approximation to the Binominal

TABLE X-M-2 — SAMPLING PLANS FOR SAMPLE SIZE CODE LETTER: M

Type of sampling plan	Cumulative sample size	Less than 0.040	0.040	0.065	X	0.10	0.15	0.25	0.40	0.65	1.0	X	1.5	X	2.5	X	4.0	Higher than 4.0	Cumulative sample size
		Acceptable Quality Levels (normal inspection)																	
		Ac Re	Ac Re	Ac Re	Ac Re	Ac Re	Ac Re	Ac Re	Ac Re	Ac Re	Ac Re	Ac Re	Ac Re	Ac Re	Ac Re	Ac Re	Ac Re	Ac Re	
Single	315	▽	0 1				1 2	2 3	3 4	5 6	7 8	8 9	10 11	12 13	14 15	18 19	21 22	△	315
Double	200	▽	*	Use Letter L	Use Letter P	Use Letter N	0 2	0 3	1 4	2 5	3 7	3 7	5 9	6 10	7 11	9 14	11 16	△	200
	400						1 2	3 4	4 5	6 7	8 9	11 12	12 13	15 16	18 19	23 24	26 27		400
Multiple	80	▽	*				# 2	# 2	# 3	# 4	0 4	0 4	0 5	0 6	1 7	1 8	2 9	△	80
	160						# 2	0 3	0 3	1 5	1 6	2 7	3 8	3 9	4 10	6 12	7 14		160
	240						0 2	0 3	1 4	2 6	3 8	4 9	6 10	7 12	8 13	11 17	13 19		240
	320						0 3	1 4	2 5	3 7	5 10	6 11	8 13	10 15	12 17	16 22	19 25		320
	400						1 3	2 4	3 6	5 8	7 11	9 12	11 15	14 17	17 20	22 25	25 29		400
	480						1 3	3 5	4 6	7 9	10 12	12 14	14 17	18 20	21 23	27 29	31 33		480
	560						2 3	4 5	6 7	9 10	13 14	14 15	18 19	21 22	25 26	32 33	37 38		560
		Less than 0.065	0.065	X	0.10	0.15	0.25	0.40	0.65	1.0	X	1.5	X	2.5	X	4.0	X	Higher than 4.0	
		Acceptable Quality Levels (tightened inspection)																	

M

△ = Use next preceding sample size code for which acceptance and rejection numbers are available.
▽ = Use next subsequent sample size code for which acceptance and rejection numbers are available.
Ac = Acceptance number
Re = Rejection number
* = Use single sampling plan above (or alternatively use letter Q)
\# = Acceptance not permitted at this sample size.

TABLE X-N — Tables for sample size code letter: N

CHART N — OPERATING CHARACTERISTIC CURVES FOR SINGLE SAMPLING PLANS

(Curves for double and multiple sampling are matched as closely as practicable)

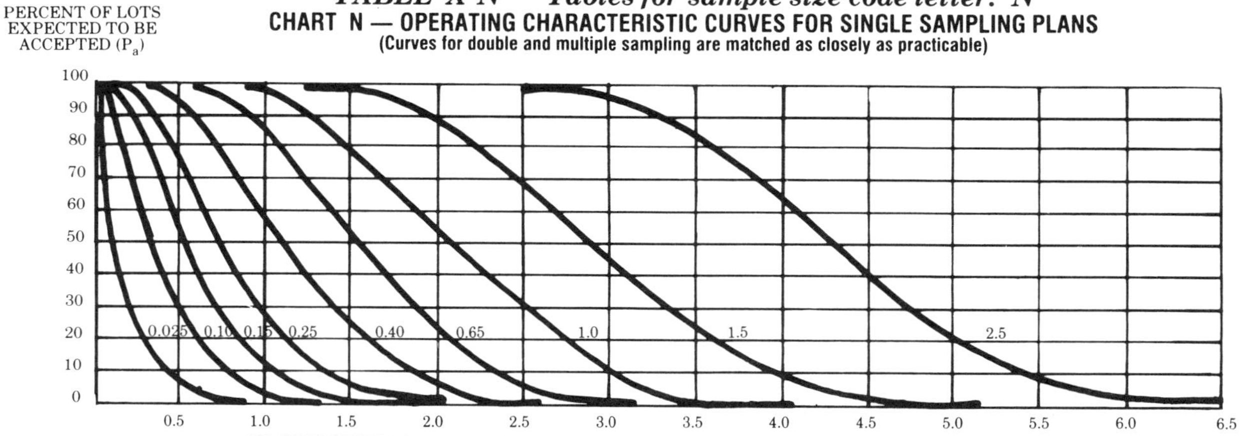

QUALITY OF SUBMITTED LOTS (P_a in percent defective for AQL's ≤ 10; in defects per hundred units for AQL's >10)
NOTE: Figures on curves are Acceptable Quality Levels (AQL's) for normal inspection)

TABLE X-N-1 — TABULATED VALUES FOR OPERATING CHARACTERISTIC CURVES FOR SINGLE SAMPLING PLANS

P_a	Acceptable Quality Levels (normal inspection)											
	0.025	0.10	0.15	0.25	0.40	0.65	X	1.0	X	1.5	X	2.5
	p (in percent defective or defects per hundred units											
99.0	0.0020	0.030	0.087	0.165	0.357	0.581	0.701	0.954	1.22	1.50	2.07	2.51
95.0	0.0103	0.071	0.164	0.273	0.523	0.796	0.939	1.23	1.54	1.85	2.49	2.98
90.0	0.0210	0.106	0.220	0.349	0.630	0.931	1.09	1.40	1.73	2.06	2.73	3.25
75.0	0.0576	0.192	0.345	0.507	0.844	1.19	1.37	1.72	2.08	2.45	3.18	3.74
50.0	0.139	0.336	0.535	0.734	1.13	1.53	1.73	2.13	2.53	2.93	3.73	4.33
25.0	0.277	0.539	0.784	1.02	1.48	1.94	2.16	2.60	3.04	3.48	4.35	4.99
10.0	0.461	0.778	1.06	1.34	1.86	2.35	2.60	3.08	3.56	4.03	4.95	5.64
5.0	0.599	0.949	1.26	1.55	2.10	2.63	2.89	3.39	3.89	4.38	5.34	6.05
1.0	0.921	1.328	1.68	2.01	2.62	3.20	3.48	4.03	4.56	5.09	6.12	6.87
	0.040	0.15	0.25	0.40	0.65	X	1.0	X	1.5	X	2.5	X
	Acceptable Quality Levels (tightened inspection)											

Note: All values given in above table based on Poisson distribution as an approximation to the Binominal

TABLE X-N-2 — SAMPLING PLANS FOR SAMPLE SIZE CODE LETTER: N

Type of sampling plan	Cumulative sample size	Acceptable Quality Levels (normal inspection)																Higher than 2.5	Cumulative sample size
		Less than 0.025	0.025	0.040	><	0.065	0.10	0.15	0.25	0.40	0.65	><	1.0	><	1.5	><	2.5		
		Ac Re	Ac Re	Ac Re	Ac Re	Ac Re	Ac Re	Ac Re	Ac Re	Ac Re	Ac Re	Ac Re	Ac Re	Ac Re	Ac Re	Ac Re	Ac Re	Ac Re	
Single	500	▽	0 1	Use Letter M	Use Letter Q	Use Letter P	1 2	2 3	3 4	5 6	7 8	8 9	10 11	12 13	14 15	18 19	21 22	△	500
Double	315	▽	*				0 2	0 3	1 4	2 5	3 7	3 7	5 9	6 10	7 11	9 14	11 16	△	315
	630						1 2	3 4	4 5	6 7	8 9	11 12	12 13	15 16	18 19	23 24	26 27		630
Multiple	125	▽	*				# 2	# 2	# 3	# 4	0 4	0 4	0 5	0 6	1 7	1 8	2 9	△	125
	250						# 2	0 3	0 3	1 5	1 6	2 7	3 8	3 9	4 10	6 12	7 14		250
	375						0 2	0 3	1 4	2 6	3 8	4 9	6 10	7 12	8 13	11 17	13 19		375
	500						0 3	1 4	2 5	3 7	5 10	6 11	8 13	10 15	12 17	16 22	19 25		500
	625						1 3	2 4	3 6	5 8	7 11	9 12	11 15	14 17	17 20	22 25	25 29		625
	750						1 3	3 5	4 6	7 9	10 12	12 14	14 17	18 20	21 23	27 29	31 33		750
	875						2 3	4 5	6 7	9 10	13 14	14 15	18 19	21 22	25 26	32 33	37 38		875
		Less than 0.040	0.040	><	0.065	0.10	0.15	0.25	0.40	0.65	><	1.0	><	1.5	><	2.5	><	Higher than 2.5	
		Acceptable Quality Levels (tightened inspection)																	

N

△ = Use next preceding sample size code for which acceptance and rejection numbers are available.
▽ = Use next subsequent sample size code for which acceptance and rejection numbers are available.
Ac = Acceptance number
Re = Rejection number
* = Use single sampling plan above (or alternatively use letter R)
\# = Acceptance not permitted at this sample size.

TABLE X-P — Tables for sample size code letter: P

CHART P — OPERATING CHARACTERISTIC CURVES FOR SINGLE SAMPLING PLANS

(Curves for double and multiple sampling are matched as closely as practicable)

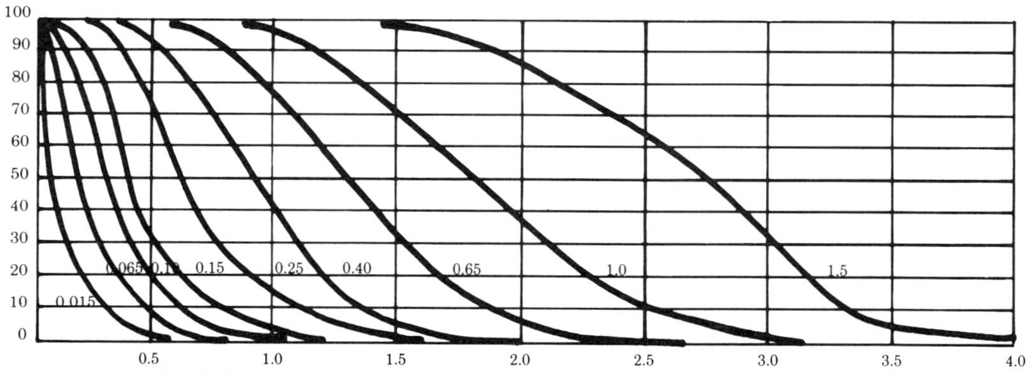

QUALITY OF SUBMITTED LOTS (P_a in percent defective for AQL's $\leqslant$ 10; in defects per hundred units for AQL's $>$10)
NOTE: Figures on curves are Acceptable Quality Levels (AQL's) for normal inspection)

TABLE X-P-1 — TABULATED VALUES FOR OPERATING CHARACTERISTIC CURVES FOR SINGLE SAMPLING PLANS

P_a	Acceptable Quality Levels (normal inspection)											
	0.015	0.065	0.10	0.15	0.25	0.40	><	0.65	><	1.0	><	1.5
	p (in percent defective or defects per hundred units											
99.0	0.0013	0.0186	0.055	0.103	0.223	0.363	0.438	0.596	0.762	0.935	1.29	1.57
95.0	0.0064	0.0444	0.102	0.171	0.327	0.498	0.587	0.771	0.961	1.16	1.56	1.86
90.0	0.0131	0.0665	0.138	0.218	0.394	0.582	0.679	0.878	1.08	1.29	1.71	2.03
75.0	0.0360	0.120	0.216	0.317	0.527	0.745	0.855	1.08	1.30	1.53	1.99	2.34
50.0	0.0866	0.210	0.334	0.459	0.709	0.959	1.08	1.33	1.58	1.83	2.33	2.71
25.0	0.173	0.337	0.490	0.639	0.928	1.21	1.35	1.63	1.90	2.18	2.72	3.12
10.0	0.288	0.486	0.665	0.835	1.16	1.47	1.62	1.93	2.22	2.52	3.09	3.52
5.0	0.375	0.593	0.787	0.969	1.31	1.64	1.80	2.12	2.43	2.74	3.34	3.78
1.0	0.576	0.830	1.05	1.26	1.64	2.00	2.18	2.52	2.85	3.18	3.82	4.29
	0.025	0.10	0.15	0.25	0.40	><	0.65	><	1.0	><	1.5	><
	Acceptable Quality Levels (tightened inspection)											

Note: All values given in above table based on Poisson distribution as an approximation to the Binominal

TABLE X-P-2 — SAMPLING PLANS FOR SAMPLE SIZE CODE LETTER: P

Type of sampling plan	Cumulative sample size	Acceptable Quality Levels (normal inspection)																Higher than 1.5	Cumulative sample size
		0.010	0.015	0.025	×	0.040	0.065	0.10	0.15	0.25	0.40	×	0.65	×	1.0	×	1.5		
		Ac Re	Ac Re	Ac Re	Ac Re	Ac Re	Ac Re	Ac Re	Ac Re	Ac Re	Ac Re	Ac Re	Ac Re	Ac Re	Ac Re	Ac Re	Ac Re	Ac Re	
Single	800	▽	0 1	Use Letter N	Use Letter R	Use Letter Q	1 2	2 3	3 4	5 6	7 8	8 9	10 11	12 13	14 15	18 19	21 22	△	800
Double	500	▽	*				0 2	0 3	1 4	2 5	3 7	3 7	5 9	6 10	7 11	9 14	11 16	△	500
	1000						1 2	3 4	4 5	6 7	8 9	11 12	12 13	15 16	18 19	23 24	26 27		1000
Multiple	200	▽	*				# 2	# 2	# 3	# 4	0 4	0 4	0 5	0 6	1 7	1 8	2 9	△	200
	400						# 2	0 3	0 3	1 5	1 6	2 7	3 8	3 9	4 10	6 12	7 14		400
	600						0 2	0 3	1 4	2 6	3 8	4 9	6 10	7 12	8 13	11 17	13 19		600
	800						0 3	1 4	2 5	3 7	5 10	6 11	8 13	10 15	12 17	16 22	19 25		800
	1000						1 3	2 4	3 6	5 8	7 11	9 12	11 15	14 17	17 20	22 25	25 29		1000
	1200						1 3	3 5	4 6	7 9	10 12	12 14	14 17	18 20	21 23	27 29	31 33		1200
	1400						2 3	4 5	6 7	9 10	13 14	14 15	18 19	21 22	25 26	32 33	37 38		1400
		Less than 0.025	0.025	×	0.040	0.065	0.10	0.15	0.25	0.40	×	0.65	×	1.0	×	1.5	×	Higher than 1.5	
		Acceptable Quality Levels (tightened inspection)																	

△ = Use next preceding sample size code for which acceptance and rejection numbers are available.
▽ = Use next subsequent sample size code for which acceptance and rejection numbers are available.
Ac = Acceptance number
Re = Rejection number
* = Use single sampling plan above.
# = Acceptance not permitted at this sample size.

P

TABLE X-Q — Tables for sample size code letter: Q

CHART Q — OPERATING CHARACTERISTIC CURVES FOR SINGLE SAMPLING PLANS

(Curves for double and multiple sampling are matched as closely as practicable)

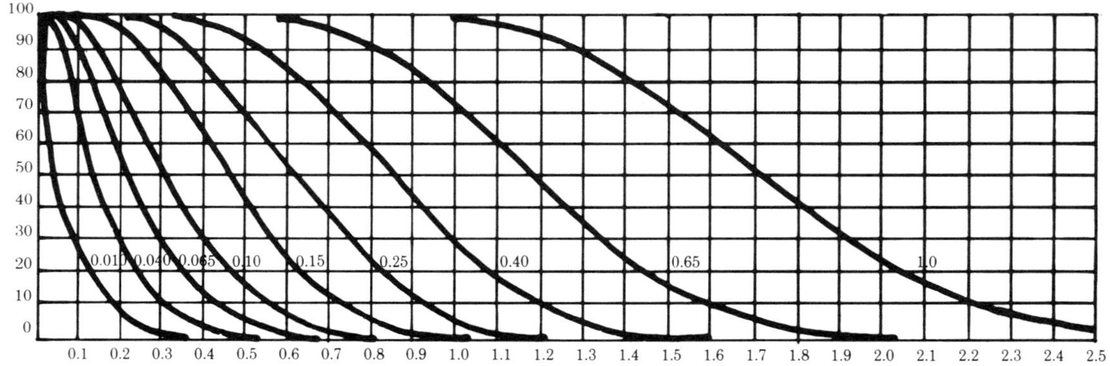

QUALITY OF SUBMITTED LOTS (P_a in percent defective for AQL's ≤ 10; in defects per hundred units for AQL's >10)
Note: Figures on curves are Acceptable Quality Levels (AQL's) for normal inspection)

TABLE X-Q-1 — TABULATED VALUES FOR OPERATING CHARACTERISTIC CURVES FOR SINGLE SAMPLING PLANS

P_a	Acceptable Quality Levels (normal inspection)											
	0.010	0.040	0.065	0.10	0.15	0.25	×	0.40	×	0.65	×	1.0
	p (in percent defective or defects per hundred units											
99.0	0.00081	0.0119	0.0349	0.0656	0.143	0.232	0.281	0.382	0.488	0.598	0.828	1.01
95.0	0.00410	0.0284	0.0654	0.109	0.209	0.318	0.376	0.494	0.615	0.740	0.995	1.19
90.0	0.00840	0.0426	0.0882	0.140	0.252	0.372	0.435	0.562	0.692	0.824	1.09	1.30
75.0	0.0230	0.0769	0.138	0.203	0.338	0.476	0.547	0.690	0.834	0.979	1.27	1.49
50.0	0.0554	0.134	0.214	0.294	0.454	0.614	0.694	0.853	1.01	1.17	1.49	1.73
25.0	0.111	0.215	0.314	0.409	0.594	0.775	0.864	1.04	1.22	1.39	1.74	2.00
10.0	0.184	0.310	0.426	0.534	0.742	0.942	1.04	1.23	1.42	1.61	1.98	2.25
5.0	0.240	0.380	0.504	0.620	0.841	1.05	1.15	1.36	1.56	1.75	2.14	2.42
1.0	0.368	0.531	0.672	0.804	1.05	1.28	1.83	1.61	1.83	2.04	2.45	2.75
	0.015	0.065	0.10	0.15	0.25	×	0.40	×	0.65	×	1.0	×
	Acceptable Quality Levels (tightened inspection)											

Note: All values given in above table based on Poisson distribution as an approximation to the Binominal

TABLE X-Q-2 — SAMPLING PLANS FOR SAMPLE SIZE CODE LETTER: Q

Type of sampling plan	Cumulative sample size	Acceptable Quality Levels (normal inspection)																Higher than 1.0	Cumulative sample size
		><	0.010	0.015	><	0.025	0.040	0.065	0.010	0.015	0.25	><	0.40	><	0.65	><	1.0		
		Ac Re	Ac Re	Ac Re	Ac Re	Ac Re	Ac Re	Ac Re	Ac Re	Ac Re	Ac Re	Ac Re	Ac Re	Ac Re	Ac Re	Ac Re	Ac Re	Ac Re	
Single	1250	Use	0 1	Use	Use	Use	1 2	2 3	3 4	5 6	7 8	8 9	10 11	12 13	14 15	18 19	21 22	△	1250
Double	800	Letter	*	Letter	Letter	Letter	0 2	0 3	1 4	2 5	3 7	3 7	5 9	6 10	7 11	9 14	11 16	△	800
	1600	R		P	S	R	1 2	3 4	4 5	6 7	8 9	11 12	12 13	15 16	18 19	23 24	26 27		1600
Multiple	315		*				# 2	# 2	# 3	# 4	0 4	0 4	0 5	0 6	1 7	1 8	2 9	△	315
	630						# 2	0 3	0 3	1 5	1 6	2 7	3 8	3 9	4 10	6 12	7 14		630
	945						0 2	0 3	1 4	2 6	3 8	4 9	6 10	7 12	8 13	11 17	13 19		945
	1260						0 3	1 4	2 5	3 7	5 10	6 11	8 13	10 15	12 17	16 22	19 25		1260
	1575						1 3	2 4	3 6	5 8	7 11	9 12	11 15	14 17	17 20	22 25	25 29		1575
	1890						1 3	3 5	4 6	7 9	10 12	12 14	14 17	18 20	21 23	27 29	31 33		1890
	2205						2 3	4 5	6 7	9 10	13 14	14 15	18 19	21 22	25 26	32 33	37 38		2205
		0.010	0.015	><	0.025	0.040	0.065	0.10	0.15	0.25	><	0.40	><	0.65	><	1.0	><	Higher than 1.0	
		Acceptable Quality Levels (tightened inspection)																	

Q

△ = Use next preceding sample size code for which acceptance and rejection numbers are available.

Ac = Acceptance number

Re = Rejection number

* = Use single sampling plan above.

= Acceptance not permitted at this sample size.

TABLE X-R — Tables for sample size code letter: R

CHART R — OPERATING CHARACTERISTIC CURVES FOR SINGLE SAMPLING PLANS

(Curves for double and multiple sampling are matched as closely as practicable)

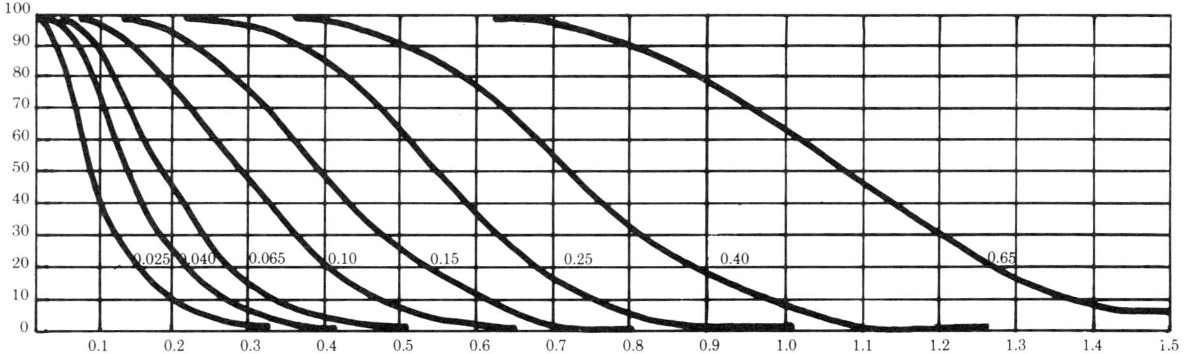

QUALITY OF SUBMITTED LOTS (P_a in percent defective for AQL's ≤10; in defects per hundred units for AQL's >10)
NOTE: Figures on curves are Acceptable Quality Levels (AQL's) for normal inspection.

TABLE X-R-1 — TABULATED VALUES FOR OPERATING CHARACTERISTIC CURVES FOR SINGLE SAMPLING PLANS

P_a	Acceptable Quality Levels (normal inspection)										
	0.025	0.040	0.065	0.10	0.15	X	0.25	X	0.40	X	0.65
	p (in percent defective or defects per hundred units)										
99.0	0.0074	0.0218	0.0412	0.0892	0.145	0.175	0.239	0.305	0.374	0.517	0.629
95.0	0.0178	0.0409	0.0683	0.131	0.199	0.235	0.309	0.385	0.462	0.622	0.745
90.0	0.0266	0.0551	0.0873	0.158	0.233	0.272	0.351	0.432	0.515	0.684	0.812
75.0	0.0481	0.0868	0.127	0.211	0.298	0.312	0.431	0.521	0.612	0.795	0.934
50.0	0.0839	0.134	0.184	0.284	0.384	0.433	0.533	0.633	0.733	0.933	1.08
25.0	0.135	0.196	0.256	0.371	0.484	0.540	0.651	0.761	0.870	1.09	1.25
10.0	0.195	0.266	0.334	0.464	0.589	0.650	0.770	0.889	1.01	1.24	1.41
5.0	0.237	0.315	0.388	0.526	0.657	0.722	0.848	0.972	1.09	1.33	1.51
1.0	0.332	0.420	0.502	0.655	0.800	0.870	1.02	1.14	1.27	1.53	1.72
	0.040	0.065	0.10	0.15	X	0.25	X	0.40	X	0.65	X
	Acceptable Quality Levels (tightened inspection)										

Note: All values given in above table based on Poisson distribution as an approximation to the Binominal

TABLE X-R-2 — SAMPLING PLANS FOR SAMPLE SIZE CODE LETTER: R

Type of sampling plan	Cumulative sample size	Acceptable Quality Levels (normal inspection)															Higher than 0.65	Cumulative sample size
		×	0.010	0.015	×	0.025	0.040	0.065	0.10	0.15	×	0.25	×	0.40	×	0.65		
		Ac Re	Ac Re	Ac Re	Ac Re	Ac Re	Ac Re	Ac Re	Ac Re	Ac Re	Ac Re	Ac Re	Ac Re	Ac Re	Ac Re	Ac Re		
Single	2000	0 1	Use Letter Q	Use Letter P	Use Letter S	1 2	2 3	3 4	5 6	7 8	8 9	10 11	12 13	14 15	18 19	21 22	△	2000
Double	1250	*				0 2	0 3	1 4	2 5	3 7	3 7	5 9	6 10	7 11	9 14	11 16	△	1250
	2500					1 2	3 4	4 5	6 7	8 9	11 12	12 13	15 16	18 19	23 24	26 27		2500
Multiple	500	*				# 2	# 2	# 3	# 4	0 4	0 4	0 5	0 6	1 7	1 8	2 9	△	500
	1000					# 2	0 3	0 3	1 5	1 6	2 7	3 8	3 9	4 10	6 12	7 14	△	1000
	1500					0 2	0 3	1 4	2 6	3 8	4 9	6 10	7 12	8 13	11 17	13 19		1500
	2000					0 3	1 4	2 5	3 7	5 10	6 11	8 13	10 15	12 17	16 22	19 25		2000
	2500					1 3	2 4	3 6	5 8	7 11	9 12	11 15	14 17	17 20	22 25	25 29		2500
	3000					1 3	3 5	4 6	7 9	10 12	12 14	14 17	18 20	21 23	27 29	31 33		3000
	3500					2 3	4 5	6 7	9 10	13 14	14 15	18 19	21 22	25 26	32 33	37 38		3500
		0.010	0.015	×	0.025	0.040	0.065	0.10	0.15	×	0.25	×	0.40	×	0.65	×	Higher than 0.65	
		Acceptable Quality Levels (tightened inspection)																

R

△ = Use next preceding sample size code letter for which acceptance and rejection numbers are available.

Ac = Acceptance number.

Re = Rejection number.

* = Use single sampling plan above.

= Acceptance not permitted at this sample size.

TABLE X-S — Tables for sample size code letter: S

Type of sampling plan	Cumulative sample size	Acceptable Quality Level (normal inspection) × Ac	Re
Single	3150	1	2
Double	2000	0	2
	4000	1	2
Multiple	800	#	2
	1600	#	2
	2400	0	2
	3200	0	3
	4000	1	3
	4800	1	3
	5600	2	3
		0.025 Acceptable Quality Level (tightened inspection)	

Ac = Acceptance number
Re = Rejection number
# = Acceptance not permitted at this sample size.

S

INDEX OF TERMS WITH SPECIAL MEANINGS

Index